T. M. CALLOWAY

RADAR TARGET DETECTION

ELECTRICAL SCIENCE
A Series of Monographs and Texts

Editors: Henry G. Booker and Nicholas DeClaris
A complete list of titles in this series appears at the end of this volume

RADAR TARGET DETECTION

Handbook of Theory and Practice

DANIEL P. MEYER
Sperry Systems Management
Sperry Division
Sperry Rand Corporation
Great Neck, New York

HERBERT A. MAYER
Sperry Gyroscope
Sperry Division
Sperry Rand Corporation
Great Neck, New York

ACADEMIC PRESS, INC.
Harcourt Brace Jovanovich, Publishers
San Diego New York Berkeley Boston
London Sydney Tokyo Toronto

ACADEMIC PRESS, INC.
1250 Sixth Avenue, San Diego, California 92101

United Kingdom Edition published by
ACADEMIC PRESS, INC. (LONDON) LTD.
24/28 Oval Road, London NW1

Library of Congress Cataloging in Publication Data

Meyer, Daniel P
Radar target detection.

(Electrical science)
Includes bibliographies.
1. Radar. I. Mayer, Herbert A., joint author.
II. Title.
TK6580.M46 621.3848'1 72-82642
ISBN 0-12-492850-1

PRINTED IN THE UNITED STATES OF AMERICA

88 89 90 91 10 9 8 7 6 5

To

Marthaann and my parents—H.A.M.

Gunda, Mercedes, and my parents—D.P.M.

CONTENTS

Foreword ix
Preface xi
Acknowledgments xii
List of Symbols xiii

CHAPTER 1 REVIEW OF THE RADAR RANGE EQUATION
1-1 Introduction 1
1-2 The Basic Radar Range Equation 2
1-3 Noise Bandwidth 3
1-4 Noise Factor, Noise Figure, and Effective Noise Temperature 3
1-5 The Radar Range Equation for I.F. Signal-to-Noise Ratio 6
1-6 The Matched (or North) Filter Receiver 6
1-7 Marcum's Use of the Radar Range Equation 10
1-8 Other Useful Forms of the Radar Range Equation 11
1-9 The Energy Radar Range Equation 11
1-10 The Radar Range Equation for a Bistatic System 12
1-11 The Radar Range Equation for a Search Radar 13
1-12 Summary of Search and Tracking Radar System Losses 15
1-13 Illustrative Example Using the Search Radar Equation 16
References 18

CHAPTER 2 STATISTICAL SIGNAL DETECTION
2-1 Detection. 19
2-2 False Alarm Time, Probability of False Alarm, and False Alarm Number 19
2-3 Probability of False Alarm versus Bias Level Setting 21
2-4 Determination of Bias Level Setting 23
2-5 The Advantages and Disadvantages of Pulse Integration 25
2-6 General Theory of Probability of Detection as a Function of Probability of False Alarm and Number of Pulses Integrated 25
2-7 Illustrative Examples for Determining Probability of Detection Using Meyer Plots 27
2-8 Integration Loss 28
2-9 Illustrative Examples for Determining Integration Loss and Integration Efficiency Using Meyer Plots 29
2-10 Collapsing Loss 29
2-11 Illustrative Example for Determining Collapsing Loss Using Meyer Plots 34
References 35

CHAPTER 3 TARGET MODELS AND THE DETERMINATION OF DETECTION PROBABILITY
3-1 Introduction and Historical Background 36
3-2 Scattering Cross Section and the Target Modeling Problem 37
3-3 Single-Hit Probability of Detection 38
3-4 Definition of a Square-Law Detector 58

3-5 The Nonfluctuating Target Model for N Pulse Returns 58
3-6 Fluctuating Targets: Swerling's Cases 1 and 2—The Rayleigh Distribution 59
3-7 Fluctuating Targets: Swerling's Cases 3 and 4—The Chi-Square Distribution with Four Degrees of Freedom 62
3-8 The Relationship between the Five Standard Target Models and Modern Target Models 64
3-9 The Linear-Type Envelope Detector 68
3-10 Best Possible Detector Law 71
3-11 Other Approaches to the Detection Criteria 72
3-12 Application of Pulse Radar Results to CW Radar System 74
3-13 Beam-Shape and Scanning Loss 74
3-14 A Method of Using Meyer Plots for Optimizing the Number of Pulses Integrated When Accounting for Beam-Shape Loss 76
3-15 Some Other Contributing Losses 79
References 81

CHAPTER 4 MEYER PLOTS—DESCRIPTION AND USE
4-1 Basics of Meyer Plots and Target Models 83
4-2 Illustrative Look-Up Problems Using Meyer Plots and Methods of Interpolation 86
4-3 Chi-Square Target Models of Arbitrary Degrees of Freedom 88
4-4 Approximation for Rice Power (One-Dominant-Plus-Rayleigh) Target Model 90
4-5 Approximation for Log-Normal Target Model 90
4-6 Approximation of Other Target Density Functions 95
4-7 Double Threshold and Median Detection 99
4-8 Recirculating Integrators 107
4-9 Meyer Plots 107
References 108
Meyer Plots 109

APPENDIX MEYER PLOTS: TECHNIQUES USED IN THEIR DEVELOPMENT
A-1 Introduction 478
A-2 The Incomplete Gamma Function 478
A-3 The Case-0 Nonfluctuating Target Model 479
A-4 The Case-1 Target Model—Rayleigh-Distributed Signal with Scan-to-Scan Fluctuation 480
A-5 The Case-2 Target Model—Rayleigh-Distributed Signal with Pulse-to-Pulse Fluctuation 480
A-6 The Case-3 Fluctuating Target Model—Four Degrees of Freedom Chi-Square-Distributed Signal with Scan-to-Scan Fluctuation 481
A-7 The Case-4 Fluctuating Target Model—Four Degrees of Freedom Chi-Square-Distributed Signal with Pulse-to-Pulse Fluctuation 484
A-8 Computation of Bias Level Values 486
References 486

INDEX 489

FOREWORD

The original work on radar detection theory published some 18 to 25 years ago has proved surprisingly durable. Nonetheless, the treatment of this subject has been greatly extended and improved since that time, and many of these extensions have not hitherto been collected in a text or handbook under one cover. The present volume takes a sizable step toward that end.

One respect in which the early work has been extended is in the variety of target fluctuation characteristics ("models") for which numerical detection probability results have been obtained. It has been found in the last 20 years that, useful as the original five models (nonfluctuating and Swerling 1–4) may be, they do not well approximate or even bracket the behavior of all physically interesting targets.

Of course, computational techniques exist whereby detection performance can be calculated for any target fluctuation model. Such computations have actually been carried out, either exactly or to very good approximation, for a large variety of cases which can be summarized as follows (the terminology is explained in more detail in the text):

(a) Extended chi-square models with arbitrary correlation.

(b) Rice models with arbitrary correlation (since these are well approximated by a subset of the extended chi-square models).

(c) Log-normal models with pulse-to-pulse independent fluctuations, or fluctuations completely correlated over one scan but scan-to-scan independent, or fluctuations correlated in blocks. [Relatively simple formulas exist for calculating log-normal cases with arbitrary correlation properties—see Eqs. (42) and (26) of my AGARD paper referred to in Chapter 3. However, I do not know of any numerical results which have been computed using these formulas. To be mathematically precise, the correlations referred to are those of the logarithms of amplitudes rather than of the amplitudes themselves.]

Hitherto, these results have appeared in scattered form and often in publications not easily available. One contribution of the present volume is to bring many of these results under one readily available cover.

Another major drawback of the numerical results in the Marcum and Swerling reports was the barbarous method of presentation, i.e., versus a quantity denoted R/R_0 and defined as a normalized range.† Not only did this make the graphs difficult to read accurately (as compared with a presentation versus per-pulse signal-to-noise ratio), but the only viable definition of R/R_0 is in fact not normalized range but "reciprocal fourth root of per-pulse signal-to-noise ratio." There are many physical situations to which the detection calculations apply, but in which signal-to-noise ratio is *not* the inverse fourth power of range. These generally involve external noise sources (e.g., jamming, clutter, or sky noise) or situations involving propagation effects (attenuation, multipath), and there are still other examples. In many of these cases, the same performance calculations apply if signal-to-noise ratio is used as a parameter, and this is vastly preferable to using R/R_0.

Even after the latter improvement is made, however, the task of conveniently presenting graphical results for all parameters of interest over the whole range of interest is very difficult. One has only to think of the thickness of the original Marcum and Swerling reports, and of the fact that even for the five target fluctuation models treated therein, a good deal of cross plotting is necessary to obtain results for intermediate values of several of the parameters. The multiplicity of

† It may also be added that all of Marcum's and Swerling's graphical results were desk-computed using in most cases series approximations. These have since been recomputed many times by digital computer. My only comment to this point is that, considering the methods of desk-computation employed (to which I can personally testify, having done them) the Marcum and Swerling results agree amazingly well with those obtained by digital computer.

cases is increased many times when a larger variety of target fluctuation models is considered. Without a clever method of presentation, many volumes of graphs would be required.

This problem can only be solved by a judicious choice of graphs (particularly of the parameters represented on the ordinate and abscissa versus those represented incrementally), which lend themselves to convenient methods of interpolation and extrapolation for the incrementally represented parameters, supplemented by simple-to-use formulas for calculating the parameters with which to enter the graphs (e.g., relating these parameters to fluctuation models).

It is fair to say that there is no "best" choice among those which satisfy the aforementioned requirements. The Meyer Plot method presented here is one good choice, oriented to maximal convenience of obtaining results for arbitrary false alarm numbers and numbers of postdetection integrated samples.

Finally, I would like to thank the authors and the publisher for this opportunity to write these comments on the current state of computation and presentation of radar detection results.

PETER SWERLING
Technology Service Corporation
Santa Monica, California

PREFACE

The main feature and primary purpose of this text is to present to the designer of radar systems a set of graphical solutions to the detection problem, which we have designated Meyer Plots. These curves represent the most comprehensive and accurate treatment of the detection problem ever to be published and eliminate the need for all tedious work with nonintegrable equations. The curves and their explanations are given in Chapter 4.

The reader will find that graphical solutions to the detection problem have been presented for all five of the standard Marcum and Swerling target models and for all practical values of false alarm numbers, number of pulses integrated, and single-pulse signal-to-noise ratios. In addition, in Chapter 4, it is demonstrated how problems involving the Weinstock, general chi-Square, Rice, and log-normal target models (the so-called modern target models) can be rapidly solved through the use of Meyer Plots. Also, in Chapter 4, a method of solving double threshold detection problems with the aid of Meyer Plots is presented.

In order to aid the reader and facilitate the use of the Meyer Plots a review of basic radar detection theory is presented in the first three chapters. This material is for the most part a survey, reorganization, and condensation of the major published works in the radar detection field. The authors have attempted to present this material in such a manner and in sufficient detail so that most users of this text will find it sufficient for most of their needs. Conversely, reference notes to basic source materials are provided throughout the text for those who require them. Those well versed in basic radar theory will find the following sections in the first three chapters to be of special interest:

- 2-4: Curves for determining bias level.
- 2-7: Examples of use of Meyer Plots.
- 2-9: Accounting for integration loss and integration efficiency when using Meyer Plots.
- 2-11: Accounting for collapsing loss when using Meyer Plots.
- 3-3: Single-hit detection curves.
- 3-8: Discussion of target models.
- 3-14: Analyzing beam-shape loss using Meyer Plots.

ACKNOWLEDGMENTS

The authors would like to thank R. Knox, F. Kurz, and C. LaSpina for the support they gave this program during the period when Meyer Plots and the methods for generating them were first conceived; also, because of their assistance during these early stages of the program our thanks are due to R. Anderson, W. Angle, J. Fare, and W. Trolone for establishing the value of Meyer Plots for determining probability of detection.

The work needed special financial support: The establishment of a nonprofit study program within the Sperry Gyroscope Company was accomplished with the help of R. Colbert, W. Davis, A. Demarkis, and E. Willey. It should also be noted that throughout the program Mr. Colbert assisted and guided in its management. Further final assistance from C. Cook, J. DiFranco, and Dr. W. Rubin helped obtain the necessary funding, with C. Cook and Dr. W. Rubin acting as mentors throughout the rest of this program and through the conceptual stages of this text. Thanks must also be given to the foresight of D. Kahl, H. Harris, Dr. L. Swern, and W. Ziegler for final funding approval. Funding approval was made more difficult on their part because of the fact that, except for the development of a more efficient technique for handling Marcum's and Swerling's theoretical approaches, no profit could be derived from the program.

This only overcame one major obstacle, the financial one. Because of the cumbersomeness of the computer program used to generate the necessary data (see Appendix), the Univac 1107 available to us was not adequate. Therefore, it became necessary to obtain computer time on the largest research computer in the New York City area. Many thanks must be extended to A. Orazio and M. H. Yarnoff for their assistance in cutting through the bureaucratic red tape and for the assistance of Mrs. C. Lieutaud of New York University and J. D. Hart at the New York office of the Atomic Energy Commission, for obtaining the needed time on the CDC 6600 located at the Courant Institute of Mathematics, New York University.

The basic computer program used for this work was originally developed at the Applied Physics Laboratories of The Johns Hopkins University. We are grateful to L. Fehlner for sending a copy of the program he and R. G. Rolls and G. T. Trotter developed. This program formed the basis for our final program. Thanks are also due to R. Belluck for the assistance he gave the authors in developing some of the mathematical relationships needed for the final program.

No words can express the gratitude due to F. Shimamoto, the programmer who worked many times through the night with D. Meyer, for his help in developing the final program and generating the necessary data.

The authors also wish to thank Dr. W. Weinstock of the Radio Corporation of America for furnishing the authors with a copy of his doctoral dissertation, "Target Cross Section Models for Radar System Analysis," University of Pennsylvania, 1964, and for permitting us to use this work in Chapter 3.

Thanks are also due to F. Blatter, D. Corbin, Dr. F. Kreussling, C. Rivera, C. Socci, H. Strauss, and J. Zellers for assisting the authors during the preparation of material for this volume. The authors also thank A. Fine and V. McDonough for permitting the office staff to assist at various times in the preparation of the text.

Mrs. G. Meyer is especially appreciated for the many long thankless hours she contributed to proofreading each of the various stages of the text and for typing the first draft. Thanks are also due to Mrs. E. Dodd, Mrs. J. Glad, Mrs. E. Hamer, Mrs. H. Malarik, and Mrs. C. Webster for typing the final draft.

LIST OF SYMBOLS

α	Median value of x	(3-98)†
γ	Integration efficiency factor	(1-78)
γ	Number of pulses received during T_d	Section 3-12
$\Delta\theta_a$	Separation of pulses in azimuth	(3-137)
$\Delta\theta_e$	Separation of pulses in elevation	(3-137)
Δ_a	Azimuth search interval	Table 2-9
Δe	Elevation search interval	Table 2-9
Δf_{cw}	Overall cw receiver bandwidth	Section 3-12
ΔR	Range search interval	Table 2-9
Δt_{fa}	False alarm pulsewidth (average)	(2-3)
η'	Number of separate velocity channels in the cw receiver	Section 3-12
θ	Angle target is offset from center of beam	(3-135)
θ_a	Azimuth component of θ	(3-136)
θ_e	Elevation component of θ	(3-136)
θ_0	Half the beam width	(3-141)
λ	Wavelength	(1-4)
ν	Number of pulses transmitted in T_d	(3-20)
ν	Number of degrees of freedom of a chi-square probability density function	Fig. 3-18
ν'	Number of separate groups of N pulses in ν	(3-20)
ζ	Transmitter-to-target-to-receiver angle	Section 1-10
ρ	Collapsing ratio	(2-59)
ρ	Ratio of mean to median value of x	(3-99)
ρ_0	Operator-efficiency factor	(3-142)
σ	Back-scattering cross section of target	(1-3)
σ_B	Bistatic scattering cross section of target	(1-67)
σ_F	Forward scattering cross section	(1-68)
σ_k	Scattering cross section of kth element	(3-44)
σ^2	Variance of x or $\bar{x}$	(3-77)
τ_g	Width of range gate	Table 2-8
ϕ_b	Search beam solid angle	(1-71)
ϕ_s	Scanned solid angle	(1-71)
$\phi(y)$	Function of y	(2-30)
χ^2	Variable of chi-square probability density function	(3-72)
ψ_0	Mean square noise voltage	(2-29)
ψ_0	Mean value of Rayleigh component of x	(3-91)
$\psi(v)$	Function of v	(2-31)
ω_a	Azimuth scan speed	Table 2-9
ω_e	Elevation scan speed	Table 2-9
$\omega(\sigma)$	Probability density function of scattering cross section	(3-48)
a	Normalized signal peak amplitude	(2-33)
A	Antenna aperture area	(1-76)
A	Instantaneous amplitude of signal-plus-noise envelope	(2-29)
$\mathscr{A}$	Scaling constant	(3-29)
A_e	Effective antenna aperture area	(1-4)
A_F	Area of target as seen by radar	(1-68)
A_Z	Searched azimuth angle	(1-73)
c	Normalizing factor	(1-27)

† Right-hand column entries indicate where symbol is first used.

B	Bandwidth (video signal or overall receiver)	(2-4)
B	Half-power beam width	(3-134)
B_a	Azimuth component of B (azimuth beam width)	(3-136)
B_e	Elevation component of B (elevation beam width)	(3-136)
B_n	Noise bandwidth	(1-8)
B_v	Video bandwidth	Table 2-9
$C(p)$	Characteristic function	(2-33)
C_B	Bandwidth correction factor	(1-50)
$C_N(p)$	Characteristic function of the sum of N terms	(2-36)
d	Cathode-ray-tube spot diameter	Table 2-9
d	Difference in range	Table 3-1
d_k	Range of kth element	(3-44)
dP	Probability density function	
dP_0	dP for $S/N = 0$	(2-14)
e	2.718, the natural base	
$\bar{E}_i$	Electric field incident on target	(3-42)
$\bar{E}_R$	Received electric field	(3-42)
E	Energy (per pulse)	(1-32)
E_{oc}	Operator coefficient	(3-142)
E_R	Energy received (per pulse)	(1-51)
E_T	Energy transmitted (per pulse)	(1-46)
E_{max}	Maximum elevation angle of search	(1-73)
E_{min}	Minimum elevation angle of search	(1-73)
$f(a, y)$	The generalized probability density function for a single pulse	(2-33)
$f(t)$	Inverse transform of $C(P)$; it is basically $dP(x,Y)$ or $dP(\bar{x},Y)$	Appendix
$f(v)$	Swerling's notation for $dP(y)$ or $dP(Y)$	(3-51)
f_0	Center frequency	(1-8)
${}_1F_1$	Confluent hypergeometric function	(3-66)
F	Number of uncorrelated pulse return groups of N/F correlated pulses	Sections 2-10, 4-3
$F(v)$	Function of v	(2-29)
$g(y)$	Function of y	(3-9)
$G(\ )$	Anticharacteristic function of ( )	(2-37)
$G(\theta)$	Gaussian beam-shape function	(3-141)
G	Gain	
G	Antenna gain	(1-7)
G_{rcvr}	Receiver Gain	(1-10)
G_R	Receiving antenna gain	(1-4)
G_T	Transmitting antenna gain	(1-2)
$h(t)$	Impulse response of filter	(1-29)
$H(f)$ or $H(\omega)$	Frequency response of i.f. amplifier or other filter; $\omega = 2\pi f$	(1-8)
i	Imaginary unit; $\sqrt{-1}$	
$I_0(z)$	Modified Bessel function of the first kind of zero order	(3-6)
$I_{N-1}(z)$	Modified Bessel function of the first kind of $N-1$ order	(3-33)
$I(u,P)$	Incomplete gamma function as defined by Pearson	(2-16)
j	Imaginary unit; $\sqrt{-1}$	
j	Summation index	
k	Summation index	
k	Boltzmann's constant $= 1.38 \times 10^{-23}$	(1-10)
k	Number of duo-degrees of freedom for probability density function of one pulse	(3-72)

k^2	$4 \log_e 2 = 2.77$	(3-134)
K	Number of duo-degrees of freedom for probability density function of N integrated pulses	(3-81)
L	Summation index	
L	Power loss factor (a number that is always $\geqslant 1$)	Section 1-12
L_a	Antenna efficiency loss	Section 1-12
L_b	Beam-shape loss	Section 1-12
L_c	Collapsing loss	(2-58)
L_d	Scan distribution loss	Section 1-12
L_f	Target fluctuation loss	Section 1-12
L_g	Gate and filter spacing loss	Section 1-12
L_i	Integration loss	Section 1-12
L_m	Receiver filter matching loss	Section 1-12
L_p	Plumbing loss	Sections 1-4, 1-12
M	Number of extra noise pulses	(2-59)
M	Number of pulses integrated digitally	Section 4-7
n	Number of scattering elements	(3-44)
n	Marcum's false alarm number	(2-10)
n	Summation index	
n'	Fehlner's false alarm number	(2-12)
n_B	Number of pulses in beam width	(3-139)
n_s	Number of separate beam postions in ϕ_s	(1-71)
$\mathscr{N}$	Noise power spectral density	(1-37)
N	Number of separate pulses transmitted and/or integrated	(1-77)
N	Noise power	(1-23)
N_{fat}	Number of range bins in t_{fa}	(2-5)
N_r	Number of range bins of interest	(2-6)
N_R	Number of range bins per range sweep	(2-10)
N_o	Total noise output power	(1-10)
ΔN_o	Portion of output noise power due to receiver	(1-11)
$\overline{\mathrm{NF}}$	Noise factor or overall receiver noise factor	(1-9)
$\overline{\mathrm{NF}}_o$	Overall system noise factor	(1-11)
p	Transformation variable	(3-30)
P	Cumulative probability	
P	Power	
$P(r)$	Probability of r false alarms in N_r range bins	(2-6)
$P(0) = P_0$	$P(r)$ with $r = 0$	(A-113)
P_{AV}	Average transmitted power	(1-64)
P_D	Probability of detection	
$P_D(N,\nu)$	Probability of detection for ν groups of N integrated pulses	(3-20)
P_{fa}	Probability of false alarm	(2-5)
$P_N = P_D(N)$	Probability of detecting the result of N integrated pulses	(2-39)
P_S	Signal power	(1-44)
P_T	Power transmitted (averaged over pulse)	(1-1)
PRF	Pulse repetition frequency	(1-70)
PRF′	Repetition frequency of groups of N pulses	(2-11)
r	Number of false alarms	(2-6)
$\mathscr{R}_T$	Energy transmitted (per pulse)	(1-57)
$\mathscr{R}_R$	Energy received (per pulse)	(1-57)
R	Range	(1-7)
R_R	Reception leg range	(1-3)
R_T	Transmission leg range	(1-1)
R_0	Range at which S/N = 1; i.e., reference range	(1-45)
s	Cathode-ray tube sweep speed	Table 2-9

s	Ratio of steady to Rayleigh component of x	(3-92)
$s(t)$	Input signal as a function of time	(1-27)
S	Signal power received	(1-5)
S	Peak amplitude of i.f. signal in absence of noise	(3-6)
$S(\omega)$	Frequency spectrum of input signal	(1-27)
$(S/N)_1$ or (S/N)	Signal-to-noise ratio for a single-hit system	(1-77)
$(S/N)_N$	Signal-to-noise ratio after coherent integration of N hits	(1-77)
t	Interval of time	(2-1)
t	Parameter of unit normal probability density function	Fig. 4-12
t_d	Time delay ($> T_P$)	(1-27)
t_d	Time delay (transmit to receive)	(1-59)
t_{fa}	False alarm time (average)	(2-1)
t_s	Search time (duration of search)	(1-80)
t_{vi}	Video integration time	Table 2-9
T	Temperature in degrees Kelvin	(1-10)
T	Threshold for digital detection	Fig. 4-10
T_a	Effective antenna temperature	(1-11)
T_d	Time available for detecting a target	(3-21)
T_e	Effective noise temperature (referred to receiver input)	(1-17)
T_e'	Effective noise temperature (referred to antenna)	(1-21)
T_0	Standard temperature (290°K)	(1-12)
T_P	Pulsewidth	(1-33)
T_p	Plumbing temperature	(1-18)
T_r	Effective receiver noise temperature	(1-18)
T_S	Sample time (observation time)	(1-64)
u	Mean value of binomial and Poisson distribution functions	(2-6)
v	Normalized amplitude of signal-plus-noise envelope	(2-29)
V	Instantaneous video noise voltage	(3-4)
$\mathscr{V}$	Visibility factor (Marcum's)	(1-47)
$\mathscr{V}_{0(50)}$	Visibility factor (Blake's)	Section 1-7
V_{max}	Maximum voltage at receiver output	(1-49)
V_s	Range gate scan rate	Table 2-9
V_{ss}	Steady-state voltage at receiver output	(1-49)
x	Marcum's power signal-to-noise ratio	(1-53)
$\bar{x}$	Average value of x	Table 2-7
y	Signal-plus-noise output of detector of an individual pulse	(2-29)
Y	Normalized noise amplitude at output to square-law detector	(2-14)
Y	Integrated signal-plus-noise voltage after detection	(2-30)
Y_b	Bias level (normalized threshold voltage)	(2-15)

CHAPTER 1

REVIEW OF THE RADAR RANGE EQUATION

1-1 Introduction

Radar is an acronym for "radio detecting and ranging." A radar system's prime purpose is the detection and location of an object by means of a return signal. This signal could be either a reflection or a beacon. Radar enables objects in the atmosphere or in outer space to be detected and located at distances well beyond the limits imposed by the unaided eye. This capability is unimpaired by darkness, fog, cloud, smoke, or most other obstacles of this character.

Because of the frequency range at which radars operate, the resolution capability of a radar is far less than that of the human eye. Radar performs best with gross objects on featureless backgrounds. Thus ships on the open sea, spacecraft and missiles in outer space, aircraft in the atmosphere, islands, and coast lines are ideally suited to radar characteristics.

Another radar characteristic is its ability to present true perspective. A true plan view can be obtained from signal returns regardless of the angular relationship between the target† and radar receiver and is independent of whether the system is bistatic or monostatic. Radar systems can determine a target's relative direction, range, velocity, and acceleration vectors. The superiority of radar for determining these quantities at ranges far in excess of the capability of visual tracking establishes the feasibility of modern day air traffic control, advance warning defense systems, and collision avoidance systems—and these are only a few of its applications.

Radar operates on the basis of transmitting electromagnetic waves from a transmitter powerful enough so that measurable amounts of this electromagnetic energy, when reflected back from a target, are detectable by the receiving device. The received energy is processed in the receiver to determine the characteristics of the object that has caused the echo.‡

Many types of transmission have been proposed, and to some extent most of them have been implemented and used. Despite the multitude of ways in which radar sets can be implemented, one method has been the overwhelming choice of most radar designers: This choice has been pulsed radar, and if not stated otherwise, radar when mentioned in this text will refer to pulsed radar.

A pulsed radar transmitter is modulated in such a way that it sends out high-intensity pulses of duration comparatively short with respect to the time delay between pulses. During the period of time between transmitted pulses the receiver is activated. As time elapses, echoes are received from objects at greater and greater distances. Since this is an electromagnetic wave, the range can be approximated at 164 yd for each microsecond between the transmission of the pulse and the reception of the echo.

Detection of return signals from a distant target would be a simple thing if random fluctuations of electrical energy were not present everywhere in the form of noise. Therefore, despite having the ability to amplify a weak electrical signal by many orders of magnitude, one is limited because the noise is amplified simultaneously. It becomes important to measure the radar's ability to distinguish the signal from the noise that surrounds it. Because of the statistical character of noise, it should be recognized that the minimum detectable signal depends on factors not all of which are determinable or measurable; yet the radar engineer consistently uses the radar range equation for initiating the design of nearly all radar systems.

† From this point throughout the text the "detected object" will be referred to as the "target."

‡ The concept presented in this paragraph is based on general usage. It is noted that beacons can be aboard detected objects and can be used as a source of signal return.

The radar range equation is the basic relationship which permits the calculation of echo signal strength from measurable or known parameters of the radar transmitter, antenna, propagation path, and target. To calculate the performance of a radar system as a detection device, the radar range equation may be extended to express the ratio of signal power to noise power. The radar range equation may also express a radar system's capability in terms of its power aperture product. The reader will recognize that the radar range equation can assume various forms. The steps of its derivation are briefly reviewed in the next section. The presentation generally follows that of Marcum [1, 2] and Barton [3]. The significance of the terms and the assumptions which are implicit in the different forms of the range equation are discussed. Applications of the range equation using Meyer Plots are presented throughout the text.

1-2 The Basic Radar Range Equation

If the energy from a radar transmitter is radiated isotropically and the hypothetical isotropic antenna of the radar is ideal, that is, having a gain of unity, the uniform distribution of radiant energy over the surface of a sphere at range R_T will produce a power density given by

$$\text{power/unit area} = P_T/4\pi R_T^2 \tag{1-1}$$

where

P_T is the power transmitted (usually taken as the average rf power during the pulse), and
R_T the range of transmission.

A directive transmitting antenna of power gain G_T is now supposed. The directive gain G_T of the transmitting antenna referred to an ideal isotropic radiator is the ratio of 4π times the ratio of the power radiated per unit solid angle in the defined direction to the total power delivered to the antenna. A target at range R_T within the transmitted beam width would be illuminated by a wave of radiant energy whose power density is

$$\text{power/unit area} = P_T G_T/4\pi R_T^2 \tag{1-2}$$

The power density of the wave reflected from the target at a distance R_R from the target, assuming that the power is reflected isotropically, is, at the receiving aperture,

$$\frac{\text{reflected power}}{\text{unit area}} = \left(\frac{P_T G_T}{4\pi R_T^2}\right)\left(\frac{\sigma}{4\pi R_R^2}\right) \tag{1-3}$$

where

R_R is the receiving range, and
σ the scattering cross section of the target in units consistent with range.

The amount of radiant energy intercepted by the receiving antenna is the product of the energy power density and the "effective aperture area" of the receiving antenna. Effective aperture is by definition

$$A_e = G_R \lambda^2/4\pi \tag{1-4}$$

(sometimes A_e is denoted as A_{re} for receiving aperture and A_{te} for transmitting aperture) where

λ is the wavelength of the transmitted frequency expressed in units consistent with those used for expressing range, and
G_R the gain of receiving antenna.

Normally the receiving antenna and transmitting antenna are one and the same or, if independent, separated only by a distance sufficient to afford good isolation between the transmitted signal and the received signal (as required in a cw system). The transmitted range R_T, therefore, usually equals the received range R_R. When Eq. (1-3) is multiplied by the effective aperture of the receiving antenna, the actual amount of power delivered to the receiver, assuming no losses, can be expressed as

$$\text{received signal power} = S = \frac{P_T G_T \sigma A_e}{(4\pi)^2 R_R^2 R_T^2} \tag{1-5}$$

By substituting Eq. (1-4) for the effective receiving aperture, the reflected power at the receiver can be expressed as

$$S = P_T G_T G_R \lambda^2 \sigma/(4\pi)^3 R_T^2 R_R^2 \tag{1-6}$$

If similar antennas at the same range are used for both transmitting and receiving, or if these antennas are one and the same, the product $G_T G_R$ can be expressed as G^2 and the product $R_T^2 R_R^2$ can be expressed as R^4, thereby yielding the radar equation in its most common form:

$$S = P_T G^2 \lambda^2 \sigma/(4\pi)^3 R^4 \tag{1-7}$$

1-3 Noise Bandwidth

Noise bandwidth B_n is defined in terms of the bandwidth of an equivalent rectangular filter. If a receiver system has a frequency response $H(f)$, then

$$B_n = \frac{1}{|H(f_0)|^2} \int_{-\infty}^{\infty} |H(f)|^2 \, df \qquad (1\text{-}8)$$

where

f_0 is the frequency at mid-band or frequency of maximum response.

If $H(f)$ is normalized to unity at mid-band frequency, then $H(f_0) = 1$.

The noise bandwidth B_n as defined here is not the half-power bandwidth as used by many electronic engineers. Determination of half-power bandwidth is usually simple; evaluating actual noise bandwidth involves a complete knowledge of the response characteristic $H(f)$. Many receivers have noise bandwidths close enough to the 3-dB bandwidth to make its use a good approximation. A comparison of these two bandwidths for several common filter types was made by Lawson and Uhlenbeck [4] (see Table 1-1).

Table 1-1

COMPARISON OF NOISE BANDWIDTH AND 3-DB BANDWIDTH[a]

Type of receiver coupling circuit	Number of stages	Ratio of noise bandwidth to 3-dB bandwidth
Single-tuned	1	1.57
	2	1.22
	3	1.16
	4	1.14
	5	1.12
Double-tuned[b]	1	1.11
	2	1.04
Staggered triple	1	1.048
Staggered quadruple	1	1.019
Staggered quintuple	1	1.01
Gaussian	1	1.065

[a] From Lawson and Uhlenbeck [4].

[b] Applies to a transitionally coupled double-tuned circuit or to a stagger-tuned circuit with two tuned circuits.

1-4 Noise Factor, Noise Figure, and Effective Noise Temperature

If there were no noise present at the receiver input, and none were generated by the receiver, sufficient amplification would result in the detection of any signal. In reality things are not this simple and noise is always present. Thus amplification of a weak signal results in amplification of the noise as well as the signal. In addition, the amplifier contributes some noise during the amplification process. It can be stated that the minimum detectable signal power in a radar receiver is limited by noise and that noise can usually be attributed to three main sources:

1. thermal or Johnson noise in circuit elements of the radar receiver;
2. shot effect and other noise in the first stage of the receiver (and to some extent succeeding stages);
3. cosmic and atmospheric noise picked up by the antenna.

It should be remembered these three are *not* the only sources; there are also man-made interferences such as radiation from other radars, radio transmitters, and machinery. Weather causes clutter because rain, hail, snow, and clouds cause signal reflections. The use of radars at low elevation angles can result in clutter from sea return and land masses. The effects of clutter and man-made interference are not considered in the radar range equation.

NOISE FACTOR OR NOISE FIGURE† ($\overline{NF}$)

The noise factor of a receiver can be considered as the measure of degradation of the signal-to-noise ratio as the signal passes through the receiver. In general, noise factor is dependent on the design of the first few input stages and the frequency of operation of the receiver. Lower noise factors usually occur at lower frequencies.

The noise factor of a device can be expressed as the ratio of signal-to-noise ratio "in," to signal-to-noise ratio "out" [see Eq. (1-9)]. The equation should be expressed in terms of power or energy to avoid problems caused by circuit nonlinearities.

† Throughout the remainder of the text it will be termed noise factor.

It is common practice to express the noise factor in decibels:

$$\overline{\mathrm{NF}} = 10 \log_{10} \frac{(\mathrm{S/N})_{\mathrm{in}}}{(\mathrm{S/N})_{\mathrm{out}}} \quad \mathrm{dB} \qquad (1\text{-}9)$$

An ideal receiver can be defined as a receiver that adds no noise of its own to the signal being amplified. It should be remembered, however, if the signal being amplified contains noise, the noise will be amplified. The noise factor as defined by Eq. (1-9) would be 0 dB for an ideal receiver.

The noise factor for a nonideal but linear receiver† is

$$\overline{\mathrm{NF}} = \frac{\text{total output noise power}}{\text{output noise power due to source resistor}}$$

$$= \frac{N_o}{kTB_n G_{\mathrm{rcvr}}} \qquad (1\text{-}10)$$

where

N_o is the noise power at the output of the receiver (watts),
k the Boltzmann's constant $= 1.38 \times 10^{-23}$ J/°K,
T the temperature of source resistor (° K), and
G_{rcvr} the receiver power gain.

In the most general application of Eq. (1-10) the source resistor is the Thevenin equivalent resistance of the signal source. This source resistor contributes kTB_n (W) of receiver input noise power. In a real system this resistance cannot be avoided because all real signal sources have a source resistance; for example, an antenna has a radiation resistance. Optimum overall system signal-to-noise ratio is achieved by making the net load resistance on the signal source equal to the source resistance.

If the receiver input is properly terminated (by either the antenna system or a terminating resistor), the internal loads of the receiver input do not contribute noise to the receiver output. This is because noise caused by them is absorbed by the "source resistor." This and other peculiarities of noise power and noise factor calculations are clearly described in a paper by Siegman [5], which the authors recommend.

† A linear relationship exists, in general, when for any input $f(t)$ the output $g(t)$ is $f(t)$ multiplied by C, where C is a constant.

When measuring the noise factor of a receiver, it is standard practice to terminate the input of the receiver in its characteristic impedance. The termination itself is then the source resistor. It is also standard practice to measure and specify noise factor with the termination at temperature $T = 290°\mathrm{K} = T_0$, the standard temperature. This highlights an important weakness in the noise factor concept; that is, noise factor can only be converted to noise power when the input temperature is specified. All noise factors in this text are computed at T_0.

An antenna system usually has an effective temperature other than T_0. Since the actual "source resistor" (antenna system) now has an effective noise temperature that is not the same as that for which the receiver noise factor was computed, overall *system* noise factor ($\overline{\mathrm{NF}}_o$) must be recalculated as follows:

$$\overline{\mathrm{NF}}_o = \frac{N_o}{kT_0 B_n G_{\mathrm{rcvr}}} = \frac{\Delta N_o + kT_a B_n G_{\mathrm{rcvr}}}{kT_0 B_n G_{\mathrm{rcvr}}}$$

$$= (\overline{\mathrm{NF}} - 1) + (T_a/T_0) \qquad (1\text{-}11)$$

where

$$\Delta N_o = (\overline{\mathrm{NF}} - 1) kT_0 B_n G_{\mathrm{rcvr}}$$

is the portion of the output noise power generated within the receiver,
T_a the effective antenna system noise temperature, and
$\overline{\mathrm{NF}}$ the overall receiver noise factor.

NOISE FACTOR OF CASCADE NETWORKS

A receiver usually consists of several stages with various types of networks in cascade. For the simple case of two linear networks in cascade (see Fig. 1-1) having the same noise bandwidth but different gains, G_1 and G_2, and different noise factors, $\overline{\mathrm{NF}}_1$ and $\overline{\mathrm{NF}}_2$, the overall noise power is

$$N_o = \overline{\mathrm{NF}}\, G_1 G_2 kT_0 B_n \qquad (1\text{-}12)$$

where

$\overline{\mathrm{NF}}$ is the overall receiver noise factor.

The overall noise power can also be expressed in terms of the individual noise factors, that is,

$$N_o = \overline{\mathrm{NF}}_1 (G_1 G_2 kT_0 B_n) + \Delta N_{o2} \qquad (1\text{-}13)$$

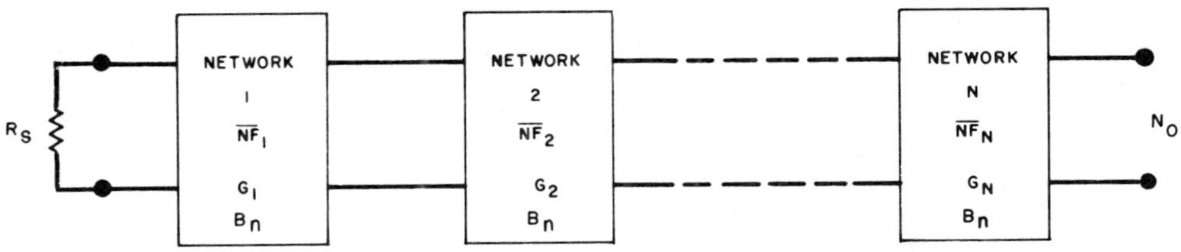

Fig. 1-1 Block diagram for a cascade network.

where

$$\Delta N_{o2} = (\overline{NF}_2 - 1)(kT_0 B_n G_2) \qquad (1\text{-}14)$$

Combining Eqs. (1-12)–(1-14) and dividing by the factor $(G_1 G_2 kT_0 B_n)$ results in a relatively simple expression for the noise factor for a two-stage cascade network in terms of the individual noise factors:

$$\overline{NF} = \overline{NF}_1 + (\overline{NF}_2 - 1)/G_1 \qquad (1\text{-}15)$$

By extension, therefore, the overall noise factor for an nth-order cascade network becomes

$$\overline{NF} = \overline{NF}_1 + \frac{\overline{NF}_2 - 1}{G_1} + \frac{\overline{NF}_3 - 1}{G_1 G_2} + \cdots + \frac{\overline{NF}_N - 1}{G_1 G_2 G_3 \cdots G_{N-1}} \qquad (1\text{-}16)$$

From Eq. (1-16) it is apparent that for gains much greater than unity the noise factor of the first stage is a good approximation to the overall noise factor. Correspondingly, each successive stage contributes less to the overall noise factor by an amount inversely proportional to the product of the gain of the preceding stages. Also from Eq. (1-16) it is apparent that if the first stage does not have any gain, as in the case of a crystal mixer, the overall noise factor would increase. Under such circumstances the gain of the first stage would be less than unity, and the noise contributed by the second stage would have more effect. Additional stages, therefore, would have to be considered when calculating the overall noise factor.

EFFECTIVE NOISE TEMPERATURE

The concept of the noise factor was developed during the early period of radar development when radar receivers were relatively noisy. With the advent of the use of low-noise devices in radar receivers, such as masers and parametric amplifiers, a more convenient measure of noise degradation was required. Effective noise temperature was chosen as this measure, and as with noise factor, effective noise temperature is a means of determining the degradation in the signal-to-noise ratio due to the noise within the receiver. By assigning an effective total receiver input noise temperature T_e, referenced to the input termination of an ideal receiver, the actual receiver output noise power can be related to the sum of component input noise temperatures:

$$N_o = T_e k B_n G_{rcvr} \qquad (1\text{-}17)$$

where

$$T_e = \overline{NF}_o T_0 = \frac{T_a}{L_p} + \frac{T_p(L_p - 1)}{L_p} + T_r \qquad (1\text{-}18)$$

T_a is the effective antenna temperature,
T_p the plumbing temperature,
L_p the plumbing losses (losses between antenna and receiver), and
T_r the effective receiver noise temperature.

The effective noise temperature of a receiver or a network is usually not equal to the physically measured temperature at the input terminals. Returning to Eq. (1-11), we note that the receiver output noise is a result of two main contributing noise sources, that of the antenna or source resistor and that of the receiver itself. The equation may be rewritten as follows:

$$\overline{NF} = 1 + \frac{\Delta N_o}{kB_n T_0 G_{rcvr}} = 1 + \frac{kB_n T_r G_{rcvr}}{kB_n T_0 G_{rcvr}} = 1 + \frac{T_r}{T_0} \qquad (1\text{-}19)$$

Solving Eq. (1-19) for effective receiver noise temperature yields the following useful expression:

$$T_r = (\overline{NF} - 1) T_0 \qquad (1\text{-}20)$$

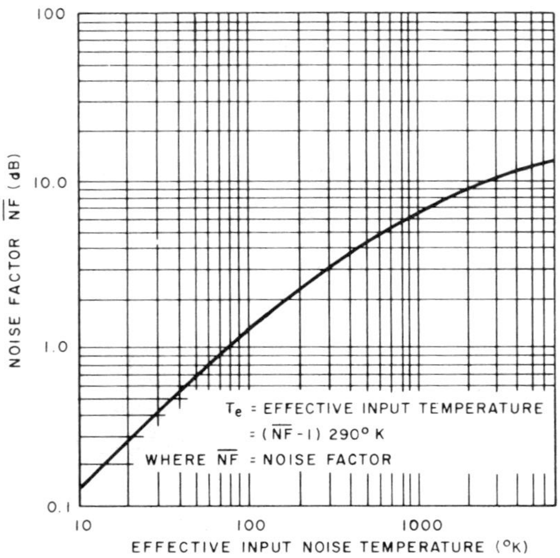

Fig. 1-2 Noise factor versus effective input noise temperature. (From Berkowitz [6].)

It is obvious from these equations that for a receiver with no internal noise the effective receiver noise temperature is 0°K and the receiver's noise factor is unity. Figure 1-2 [6] is a plot of Eq. (1-20).

Effective total noise temperature referenced to the antenna terminal T_e' is sometimes of interest. It follows from Eq. (1-18) that

$$T_e' = T_a + T_p(L_p - 1) + L_p T_r \qquad (1\text{-}21)$$

THE EFFECTIVE NOISE TEMPERATURE OF CASCADE NETWORKS

The effective noise temperature for N networks in cascade can be derived by extending Eq. (1-16) in a manner similar to the derivation of Eq. (1-21):

$$T_e = T_1 + \frac{T_2}{G_1} + \frac{T_3}{G_1 G_2} + \frac{T_4}{G_1 G_2 G_3} + \cdots + \frac{T_N}{G_1 G_2 G_3 \cdots (G_N - 1)} \qquad (1\text{-}22)$$

1-5 The Radar Range Equation for I.F. Signal-to-Noise Ratio

The radar range equation expressed in terms of signal-to-noise ratio is the starting point for most radar performance analyses and initial radar designs. To establish the radar range equation in these terms, a factor representing the "equivalent input noise power" has to be introduced into Eq. (1-7).

Equivalent input noise power is given by

$$N = N_o/G = kT_0 B_n \overline{\mathrm{NF}}_o \qquad (1\text{-}23)$$

and signal-to-noise ratio, therefore, is

$$\mathrm{S/N} = S/kT_0 B_n \overline{\mathrm{NF}}_o \qquad (1\text{-}24)$$

Combining Eqs. (1-7) and (1-24) results in the most commonly used form of the radar range equation:

$$\frac{\mathrm{S}}{\mathrm{N}} = \frac{P_T G^2 \lambda^2 \sigma}{(4\pi)^3 R^4 k T_0 B_n \overline{\mathrm{NF}}_o} \qquad (1\text{-}25)$$

The radar range equation expressed in this manner is usually applied to the case when the input termination is at standard temperature ($T = T_0$). For systems using low-noise receivers, an equation using effective noise temperature is more applicable. In terms of effective noise temperature, Eq. (1-25) becomes

$$\frac{\mathrm{S}}{\mathrm{N}} = \frac{P_T G^2 \lambda^2 \sigma}{(4\pi)^3 R^4 k T_e B_n} \qquad (1\text{-}26)$$

1-6 The Matched (or North) Filter Receiver†, ‡

The matched-filter processor is the optimum linear processor for radar signals, predicated on the assumption that the raw data available to the receiver are corrupted by white Gaussian noise. Its response characteristic can be expressed in terms of a frequency response function or a time response function. It should be remembered that the Fourier transform interrelates these functions. In the frequency domain, the matched-filter transfer function $H(\omega)$ is the complex conjugate function of the spectrum of the signal that is to be optimally processed (see Fig. 1-3a) and can be expressed by

$$H(\omega) = cS^*(\omega) e^{-j\omega t_d} \qquad (1\text{-}27)$$

† D. O. North [7] is generally recognized as being the first to document the properties of the optimum receiver. Van Vleck and Middleton [8] were the first to use the notation "matched filter."

‡ For a rigorous development of matched-filter requirements see Cook and Bernfeld [9].

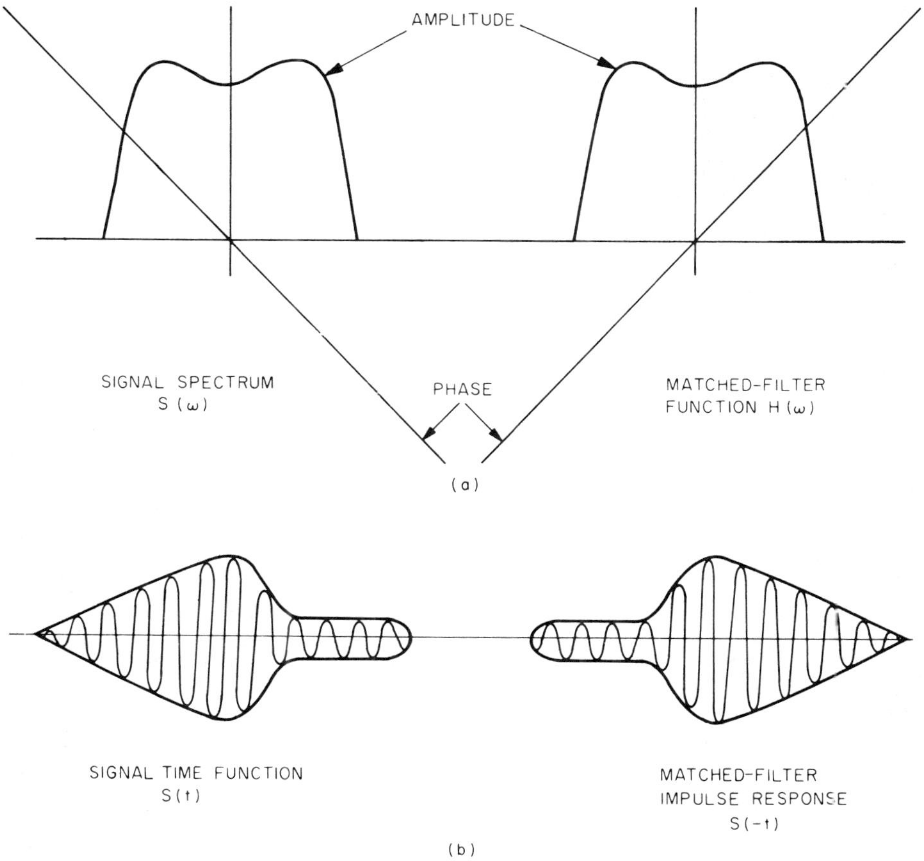

Fig. 1-3 Matched-filter signal relationship. [From M. Bernfeld, C. E. Cook, J. Paolillo, and C. A. Palmieri, Matched Filtering, Pulse Compression, and Waveform Design Part I. *Microwave J.* **7**, No. 10, 57–64 (1965).]

where

c is the normalizing factor,
$S(\omega)$ the frequency spectrum of the input signal $s(t)$,
$S^*(\omega)$ the complex conjugate of $S(\omega)$, and
t_d a time delay constant to establish a feasible filter.

The normalizing factor and the delay constant are generally ignored in formulating the basic significant relationship:

$$H(\omega) = S^*(\omega) \tag{1-28}$$

Correspondingly, by taking the inverse Fourier transform of $H(\omega)$ the time domain relationship between the signal to be operated upon and the matched filter is obtained. From this it can be recognized that the impulse response of a matched filter is a time-reversed replica of the known signal function. Thus if $h(t)$ represents the matched-filter impulse response, the general relationship equivalent to Eq. (1-27) is

$$h(t) = cs(t_d - t) \tag{1-29}$$

As above, the delay constant t_d is usually ignored in order to point out the basic relationship of Eq. (1-30) and Fig. 1-3b:

$$h(t) = cs(-t) \tag{1-30}$$

The criteria used to establish a detection threshold may be either explicit as in an automatic alarm system or implied as when biased by a human observer. The return signal is a train of repeated pulses, each superimposed upon a background of uncorrelated (random) noise waveforms. An optimum detection system, therefore, would maximize the signal return relative to the average value of noise power present.

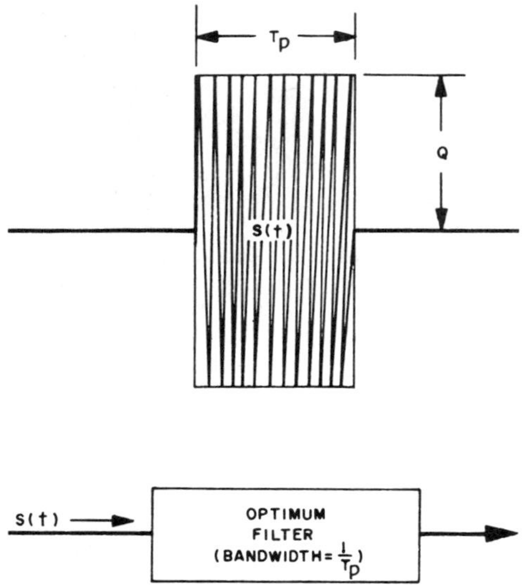

Fig. 1-4 Quasi-matched-filter/signal relationship for a pulsed signal. (From Cook and Bernfeld [9].)

It turns out that, for a matched filter, the maximum (best) signal-to-noise power ratio at the filter output for the case of "white" Gaussian noise is

$$\left(\frac{\mathrm{S}}{\mathrm{N}}\right)_{\text{best}} = \frac{\text{received signal energy}}{\text{noise spectral density (watts per hertz)}} \tag{1-31}$$

The relationship can be derived heuristically for an elementary pulsed radar signal if one defines the energy of the pulse as

$$E = \int_{-\infty}^{\infty} |s(t)|^2\, dt \tag{1-32}$$

and assumes the waveform of the pulse (see Fig. 1-4) to be

$$s(t) = Q\cos\omega_0 t, \qquad \text{when} \quad 0 \leqslant t \leqslant T_P \tag{1-33}$$

and

$$s(t) = 0, \qquad \text{when} \quad 0 > t > T_P \tag{1-34}$$

where T_P is the transmitted pulsewidth. Then

$$E = \int_{-\infty}^{\infty} Q^2 \cos^2 \omega_0 t\, dt \tag{1-35}$$

and the energy of the signal into the matched filter is

$$E = \tfrac{1}{2} Q^2 T_P \tag{1-36}$$

The corresponding noise power is

$$N = \mathcal{N} B_n \tag{1-37}$$

where

$\mathcal{N}$ is the noise power spectral density, and
B_n the noise bandwidth.

Combining the above, the signal-to-noise power ratio during the pulse is

$$\left(\frac{\mathrm{S}}{\mathrm{N}}\right)_{\max} = \frac{Q^2/2}{\mathcal{N} B_n} \tag{1-38}$$

or in terms of energy, it is

$$\left(\frac{\mathrm{S}}{\mathrm{N}}\right)_{\max} = \frac{E}{T_P \mathcal{N} B_n} \tag{1-39}$$

In the early days of radar a classic experiment (classic in the sense that many books written on radar have described it) was performed that demonstrated that an optimum relationship exists between i.f. bandwidth and transmitted pulse width. The experiment indicated that optimum system performance occurred when this product was approximately 1.2 (see Fig. 1-5) [4, 11, 12]. The experiment further indicated that the optimum region was broad and that less than 1.0 dB of additional degradation occurred if the i.f. bandwidth was halved or doubled.

By using practical circuit techniques it is possible to only approximate matched-filter criteria. It is, therefore, of interest to know for various practical transmitted pulse shapes and receiver i.f. bandpass characteristics what the optimum $T_P B$ product is and how much deviation there is from perfect matched-filter performance. Table 1-2 presents such data compiled from the work by Fine [13], and the work of others edited by Lawson and Uhlenbeck [4], and Skolnik [12].

For initial approximations when using the radar range equation, the radar designer often assumes

$$T_P B_n = 1 \tag{1-40}$$

This is done, partially because of the early empirical data, partially because the 6.0-dB reference point values are approximately unity, but mostly for convenience. It follows then that the maximum signal-to-noise ratio at the matched-filter output for the case of white Gaussian noise is

$$(\mathrm{S/N})_{\max} = E/\mathcal{N} \approx P_R/N_o \tag{1-41}$$

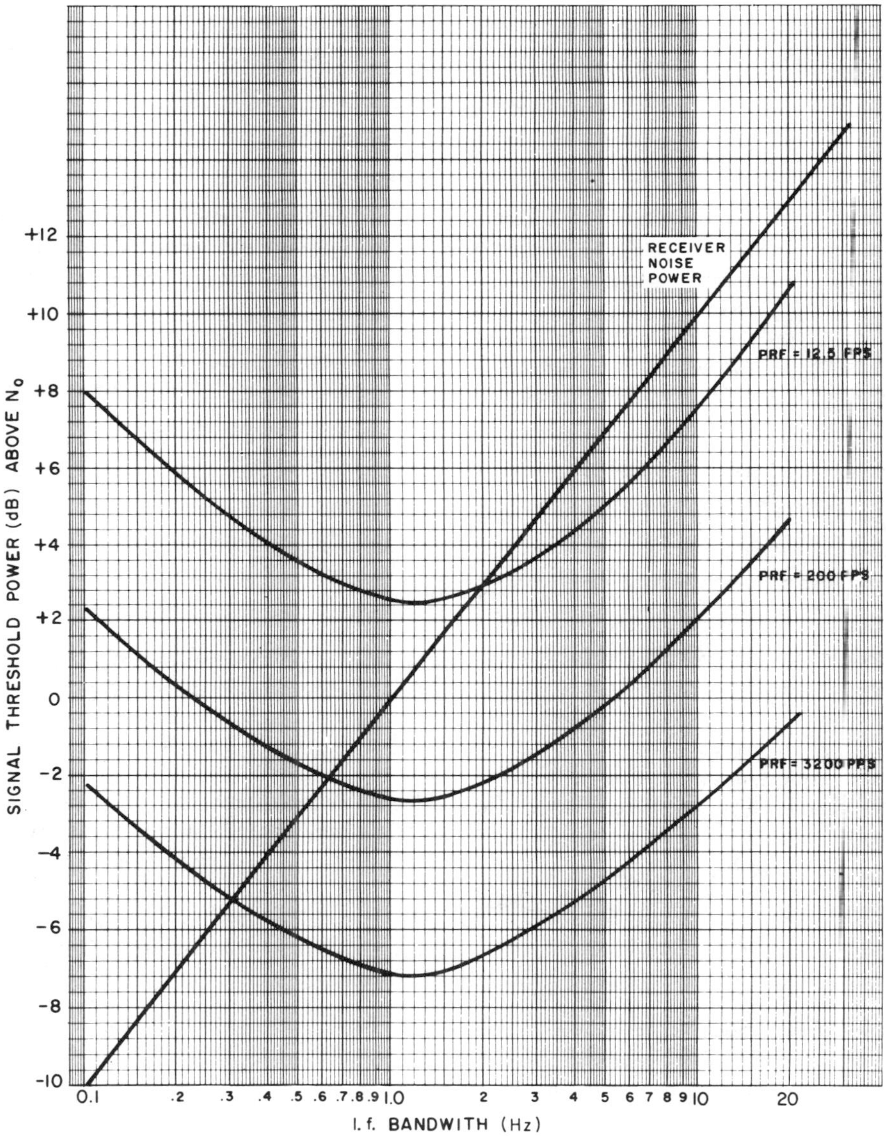

Fig. 1-5 Receiver sensitivity versus receiver i.f. bandwidth. The pulse duration is 1μsec and the signal power is measured in units of the noise power N_o, within a band 1 MHz wide. In these experiments the video bandwidth with 10 MHz, signal presentation time 3 sec, and length of the pulse on the screen of the A-scope 1.7 mm. (From Ridenour [11].)

It is interesting to note from the preceding that, if a matched filter is used in the predetection stages of a receiver, the ability to detect a radar signal is only a function of the energy it contains and in no way depends on the shape of the signal. The proper matched filter will always optimize the signal-to-noise ratio output. Furthermore, the amount of degradation in signal-to-noise ratio between an exact matched-filter solution and a reasonable approximation of the same may not be significant.

From Eq. (1-40), it can be seen that if the receiver has or closely approximates a matched-filter characteristic the radar range equation can be written as follows:

$$\left(\frac{S}{N}\right) = \frac{P_T T_P G^2 \lambda^2 \sigma}{(4\pi)^3 R^4 k T_0 \overline{NF}_o} \qquad (1\text{-}42)$$

Table 1-2

QUASI-MATCHED-FILTER/SIGNAL RELATIONSHIPS

		Optimum $T_P B$			
Pulse shape	Filter (i.f. band-pass characteristic)	Parameters measured at 3-dB points	Parameters measured at 6-dB points	Using effective pulse width and/or noise bandwidth	Mismatch loss L_m (dB)
Gaussian	Gaussian	0.44	0.92	1.0	0.0
Gaussian	Rectangular	0.72	1.04	0.77	0.49
Rectangular	Gaussian	0.72	1.04	0.77	0.49
Rectangular	1-stage single-tuned	0.40	0.70	0.63	0.88
Rectangular	2-stage single-tuned	0.61	0.95	0.74	0.56
Rectangular	3-stage single-tuned	0.65	0.96	0.74	0.53
Rectangular	5-stage single-tuned	0.67	0.97	0.75	0.50
Rectangular	Rectangular	1.37	1.37	1.37	0.85
Rectangular	$x^{-1} \sin x$	0.88			0.0
Rectangular chirp	Gaussian		$(0.86 \times \Delta f)$		0.6

1-7 Marcum's Use of the Radar Range Equation

The basic radar range equation can be solved for R_{max} and expressed in terms of a ratio of transmitted power to a minimum detectable power; for example,

$$R_{max} = \left(\frac{P_T}{P_{min}}\right)^{1/4} \left[\frac{A_e G\sigma}{(4\pi)^2}\right]^{1/4} \tag{1-43}$$

where

P_T is the average transmitted power during the pulse,

P_{min} the minimum detectable signal power,

A_e the effective aperture area of the receiving antenna expressed in units consistent with those used for expressing range, where $A_e = G\lambda^2/(4\pi)$,

G the gain of the transmitting–receiving antenna, and

σ the scattering cross section of the target expressed in units consistent with those used for expressing range.

The nominal range of a radar set R_0 is sometimes defined as that range at which the signal equals the noise. If it is assumed that the only noise source is thermal, then

$$P_S = kT_0 B_n \overline{NF} \tag{1-44}$$

and if Eq. (1-44) is substituted into Eq. (1-43) for P_{min}, then R_0 can be expressed as follows:

$$R_0 = [P_T G A_e \sigma / 16\pi^2 k T_0 B_n \overline{NF_o}]^{1/4} \tag{1-45}$$

(The subscript zero is used on R so as not to leave the impression that the maximum range of a radar can be so simply determined; R_0 is known as the reference range.)

Expressing this equation in terms of energy per pulse results in an alternate form. The energy in a transmitted pulse is

$$E_T = P_T T_P \tag{1-46}$$

By letting

$$\mathscr{V} = 1/T_P B_n \tag{1-47}$$

Eq. (1-45) becomes

$$R_0 = \left[\frac{E_T G A_e \sigma \mathscr{V}}{(4\pi)^2 2kT_0 \overline{NF_o}}\right]^{1/4} \tag{1-48}$$

The term $\mathscr{V}$ is Marcum's visibility factor†:

$$\mathscr{V} = \frac{(V_{max}/V_{ss})^2}{(T_P B_n)} \tag{1-49}$$

† For a complete discussion and development of visibility factor see Van Vleck and Middleton [8] and Marcum [1, 2].

where

V_{max} is the maximum voltage the pulse rises to at the receiver output, and
V_{ss} the steady-state voltage at the receiver output.

It should be noted that the square of the receiver voltage output ratio $(V_{max}/V_{ss})^2$ should be included in Eq. (1-47) but is not. This approximation is justifiable because $(V_{max}/V_{ss})^2 \approx 1$ when $T_P B_n \approx 1$. For the specific case where the band-pass characteristic of the amplifier is the conjugate transform of the pulse, the visibility factor is exactly unity.

In current literature, $\mathscr{V}$ in Eq. (1-48) is replaced by $1/(\mathscr{V}_{0(50)} C_B)$ with the new terms defined by Blake as follows [14]:

$\mathscr{V}_{0(50)}$ is the "visibility factor" for 50% probability of detection and optimum bandwidth; that is, ratio of minimum detectable pulse energy (watt-seconds) at the receiver input terminals to noise power per unit bandwidth (watts per hertz) referred to the same terminals, or ratio of minimum detectable signal power to noise power with $T_P B_n = 1$, and

$$C_B = \frac{T_P B_n}{4.8}\left(1 + \frac{1.2}{T_P B_n}\right)^2 \qquad (1\text{-}50)$$

a bandwidth correction factor empirically derived from Fig. 1-5 or its equivalent. The reciprocal of C_B is directly equivalent to Marcum's $\mathscr{V}$. The use of the $\mathscr{V}_{0(50)}$ term makes it possible to find the actual range of the radar rather than R_0, the reference range.

Marcum [1] expressed his results in a compact form, by working in terms of range and energy ratios. This eliminated the necessity for the appearance of such parameters as E_T, G, A_e, $\overline{\text{NF}}$, σ, and $\mathscr{V}$ in the final results. To accomplish this he first made a variation in Eq. (1-48) that permits calculation of actual range R if the received signal energy E_R is known; that is,

$$R = [E_T G A_e \sigma \mathscr{V}/(4\pi)^2 E_R]^{1/4} \qquad (1\text{-}51)$$

Now, by taking the ratio of Eqs. (1-48) and (1-51) the following simplified relationship is developed:

$$R/R_0 = (kT_0 \overline{\text{NF}}_o/E_R)^{1/4} \qquad (1\text{-}52)$$

Marcum then defined x as the ratio of signal energy to average receiver thermal noise energy†:

$$x = (\text{S/N}) = E_R/kT_0 \overline{\text{NF}}_c \qquad (1\text{-}53)$$

Combining Eqs. (1-52) and (1-53) results in

$$\frac{R}{R_0} = \frac{1}{(\text{S/N})^{1/4}} = \frac{1}{(x)^{1/4}} \qquad (1\text{-}54)$$

This simplified range equation is useful in interpreting the relative improvement offered by different detection schemes.

1-8 Other Useful Forms of the Radar Range Equation [3]

In evaluating the performance of a radar system, specific forms of the radar equation can be advantageously used to highlight specific characteristics of a system. Two variations of Eq. (1-25) follow.

Equation Independent of Operating Wavelength:

$$x = \frac{\text{S}}{\text{N}} = \frac{P_T G A_e \sigma}{(4\pi)^2 R^4 k T_0 B_n \overline{\text{NF}}_o} \qquad (1\text{-}55)$$

Equation with Inverse Dependence on the Operating Wavelength:

$$x = \frac{\text{S}}{\text{N}} = \frac{P_T A_e^2 \sigma}{4\pi R^4 \lambda^2 k T_0 B_n \overline{\text{NF}}_o} \qquad (1\text{-}56)$$

1-9 The Energy Radar Range Equation

The radar range equation can also be expressed in terms of the ratio of received signal energy to thermal noise energy per unit of bandwidth. A form similar to that developed for the basic range equation in terms of transmitted power (Eq. 1-7) can be developed for transmitted energy $\mathscr{R}_T$:

$$\mathscr{R}_R = \mathscr{R}_T G^2 \lambda^2 \sigma/(4\pi)^3 R^4 \qquad (1\text{-}57)$$

where

$$\mathscr{R}_T = \int_0^{T_P} P_T(t)\, dt \qquad (1\text{-}58)$$

and

$$\mathscr{R}_R = \int_{t_d}^{t_d + T_P} P_R(t)\, dt \qquad (1\text{-}59)$$

† Note that x is used extensively to represent (S/N) in this text.

In Eq. (1-59), t_d is the round trip transmit-to-receive time measured from the leading edge of the transmitted pulse and T_P is the signal duration time.

It is apparent from Eq. (1-41) that for a matched-filter system†

$$x = \frac{\mathscr{R}_R}{\mathscr{N}} = \frac{\mathscr{R}_R}{kT_0 \overline{\mathrm{NF}}_o} \tag{1-60}$$

Similarly, for a matched filter, but in terms of transmitted energy,

$$x = \frac{\mathscr{R}_T G^2 \lambda^2 \sigma}{(4\pi)^3 R^4 k T_0 \overline{\mathrm{NF}}_o} \tag{1-61}$$

[see Eq. (1-42)].

The value of $\mathscr{R}_T$, in terms of power P_T and width T_P, is

$$\mathscr{R}_T = P_T T_P \tag{1-62}$$

and, combining Eqs. (1-62) and (1-61) yields

$$x = \frac{P_T T_P G^2 \lambda^2 \sigma}{(4\pi)^3 R^4 k T_0 \overline{\mathrm{NF}}_o} \tag{1-63}$$

Alternately, in terms of the average transmitted power P_{av} and sample or observation time T_S such as for a cw system, x can be expressed as follows:

$$x = \frac{P_{av} T_S G^2 \lambda^2 \sigma}{(4\pi)^3 R^4 k T_0 \overline{\mathrm{NF}}_o} \tag{1-64}$$

In terms of effective noise temperature,

$$x = \frac{P_{av} T_S G^2 \lambda^2 \sigma}{(4\pi)^3 R^4 k T_e} \tag{1-65}$$

In both Eqs. (1-64) and (1-65)

$$\mathscr{R}_T = P_{av} T_S = P_{av}/B_n \tag{1-66}$$

1-10 The Radar Range Equations for a Bistatic System

If the radar transmitter and receiver use separate antennas and they are widely separated, the radar system is termed bistatic (see Fig. 1-6). The radar range equation for this case is

$$x = \frac{P_T G_T G_R \lambda^2 \sigma_B}{(4\pi)^3 R_T^2 R_R^2 k T_0 B_n \overline{\mathrm{NF}}_o} \tag{1-67}$$

† For a detailed review of matched-filter receivers see Cook and Bernfeld [9]; see also Woodward [15].

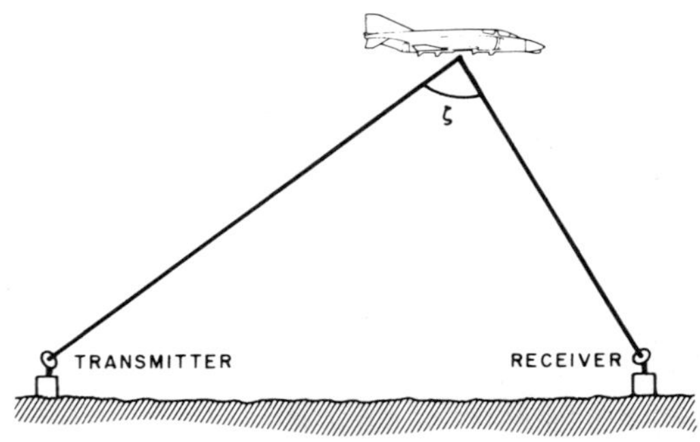

Fig. 1-6 A bistatic radar configuration.

The subscripts T and R are used to identify the transmit and receive portions of the system. The scattering cross section σ_B in Eq. (1-67) accounts for the reflection of radiant energy in the direction of the receiver and is the "bistatic scattering cross section." The bistatic scattering cross section for a particular target is usually comparable in magnitude to the monostatic scattering cross section, except where the angle ζ the target makes between the transmitting and receiving stations approaches 180°. This does not mean that a monostatic and a bistatic radar viewing the same target will see exactly the same value of σ, as in some instances the monostatic cross section will be greater and in others the bistatic cross section will be greater [16, 17].

When the angle ζ approaches 180°, the bistatic forward scattering cross section σ_F can be many times larger than the monostatic cross section σ or the bistatic cross section σ_B at angles less than 180°. It has been shown that the forward scattering cross section of a target is [18]

$$\sigma_F = 4\pi A_F{}^2/\lambda^2 \tag{1-68}$$

where

A_F is the forward cross-sectional area of the target, and

λ the wavelength of the radiation. For Eq. (1-68) to be valid λ must be small in comparison to the target dimensions.

Equation (1-68) is related to the equation for the scattering cross section of a round flat plate measured normal to the plate and is in terms of a projected area of the target as viewed from the receiving station.

For a sphere, the ratio of forward scattering to back scattering cross section is readily determined and serves to illustrate the significant relationships:

$$\sigma_F/\sigma = 2\pi D^2/\lambda^2 \tag{1-69}$$

where D is the diameter of the sphere.

The forward scattering from an absorbing body is comparable to the forward scattering from a reflecting body of identical shape. Conversely, the back scattering from an absorbing body will be much less than that for a reflecting body of identical shape.

With the use of Meyer Plots the bistatic system is evaluated in the same way as a normal monostatic system. As the transmitter-to-target-to-receiver angle approaches 180°, the value of x obtained using a Meyer Plot, as described in Chapter 4, will become conservative and must be corrected by the ratio of σ to σ_F, if known.

1-11 The Radar Range Equation for a Search Radar

A search radar is designed to detect and locate a target within a defined volume of space during a specified time interval. Discrimination between target returns and random noise is accomplished through the use of display equipment and a human observer, through automatic detection equipment, or some combination thereof. Obviously these techniques produce results that have statistical characteristics.

The basic radar equation for a search radar will be determined in this section in terms of certain characteristics that will be idealized, namely, receiver beam shape, search pattern, antenna efficiency, and gain from coherent integration. The effects of practical loss factors, however, will be imposed so that estimates of the performance of feasible systems can be made. Thus the design engineer will be permitted to compare the performance of radically different search radar systems to that of a defined ideal system. Also, the designer is permitted to evaluate what factors contribute to superior performance for specific requirements.

The detection of a target will be the only criterion of a search radar's performance considered in this text; accuracy and resolution are determined by factors that are not considered herein. Because of this, detection performance of a search radar will be expressed in terms independent of the frequency and type of modulation imposed on the transmitted signal.

For the purpose of this text, therefore, an ideal search radar system will consist of the following.

Matched-Filter Receiver It will be assumed that the search radar receiver will be matched to the signal spectrum, so that the pulse width when multiplied by the bandwidth is unity if a rectangular pulse is actually used. The peak and average powers can now be related for a pulsed transmission at a specified pulse repetition frequency (PRF) by

$$P_{av} = P_T T_P(\mathrm{PRF}) = P_T(\mathrm{PRF})/B_n \tag{1-70}$$

Beam Shape The search radar beam shape will be assumed to be rectangular and uniform; adjacent beams will not overlap or establish gaps. The same antenna will be used for both transmitting and receiving.

Uniform Search Pattern In order to calculate the number of pulses transmitted to and received from the target during the search time t_s, it is assumed that a solid angle ϕ_s is scanned in a uniform fashion by the radar beam. The search radar beam has a solid angle ϕ_b. Therefore, the number of beam positions to be searched is

$$n_s = \phi_s/\phi_b \tag{1-71}$$

If A_e represents the effective antenna aperture, and λ is the wavelength, the included solid beam angle is commonly assumed to be

$$\phi_b = \lambda^2/A_e \tag{1-72}$$

If A_Z defines the azimuth angle, E_{max} the maximum elevation angle, and E_{min} the minimum elevation angle within the solid angle ϕ_s (see Fig. 1-7), then

$$\phi_s = A_Z(\sin E_{max} - \sin E_{min}) \tag{1-73}$$

It should be noted that for low elevation, Eq. (1-73) simplifies to

$$\phi_s = A_Z(E_{max} - E_{min}) \tag{1-74}$$

Now, by combining Eqs. (1-71)–(1-73), one obtains

$$n_s = \frac{A_Z(\sin E_{max} - \sin E_{min}) A_e}{\lambda^2} \tag{1-75}$$

Antenna Efficiency The antenna will be assumed to have a 100% aperture efficiency. Then the beam angle ϕ_b, antenna aperture area A, and wavelength λ are, by definition, related to the antenna gain G as follows (a common trans-

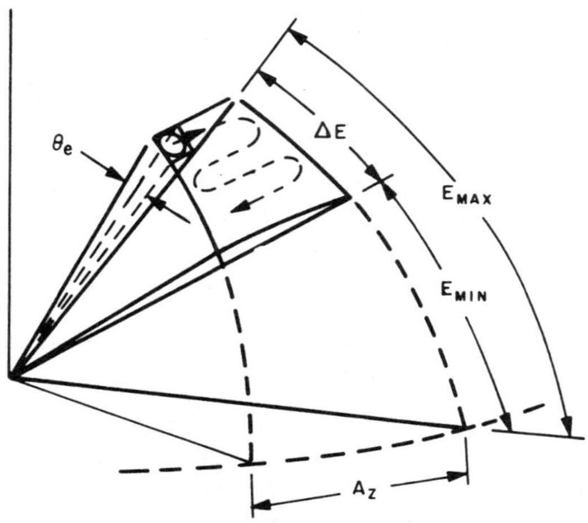

Fig. 1-7 Search pattern parameters.

mitting–receiving antenna is assumed):

$$G = 4\pi/\phi_b = 4\pi A/\lambda^2 \qquad (1\text{-}76)$$

Integration Signal-to-noise ratio is usually defined for single-hit detection. The effective signal-to-noise ratio for an ideal coherent multihit system $(S/N)_N$, in terms of the signal-to-noise ratio for a single-hit detection $(S/N)_1$, is

$$(S/N)_N = N(S/N)_1 \qquad (1\text{-}77)$$

where N is the number of hits coherently integrated.

The effective signal-to-noise ratio for noncoherent integration can also be determined if an integration efficiency γ is defined such that N^γ is the equivalent number of coherently integrated hits. If γ is known, the probability of detection for any noncoherently integrating system may be determined (see Sections 2-8 and 2-9).

Based on the aforementioned assumptions, the basic range equation for search radar i.f. signal-to-noise ratio in terms of noise factor is

$$\frac{S}{N} = \frac{P_T A^2 \sigma N^\gamma}{(4\pi) R^4 \lambda^2 k T_0 B_n \overline{NF}_o} \qquad (1\text{-}78)$$

and in terms of effective noise temperature

$$\frac{S}{N} = \frac{P_T A^2 \sigma N^\gamma}{4\pi R^4 \lambda^2 k T_e B_n} \qquad (1\text{-}79)$$

As previously stated, the number of pulses transmitted to and received from the target is termed N. The product of the search time t_s and the pulse repetition frequency (PRF) is the total number of pulses transmitted while searching a defined volume. The term n_s defined by Eq. (1-75) represents the total number of separate beam positions in the searched volume. The number of pulses integrated, therefore, for any beam position is

$$N = t_s(\text{PRF})/n_s \qquad (1\text{-}80)$$

By substituting Eq. (1-75) into Eq. (1-80) one obtains, noting that A_e is now A,

$$N = \frac{t_s(\text{PRF})\,\lambda^2}{A A_Z(\sin E_{max} - \sin E_{min})} \qquad (1\text{-}81)$$

Finally, by combining Eqs. (1-81) and (1-78), one obtains the complete radar range equation for a coherently integrating search system ($\gamma = 1$):

$$\left(\frac{S}{N}\right)_N = \frac{P_T(\text{PRF})\,A t_s \sigma}{4\pi R^4 A_Z(\sin E_{max} - \sin E_{min}) k T_0 B_n \overline{NF}_o} \qquad (1\text{-}82)$$

For the case of coherent integration probability of detection P_D can now be obtained directly using the single-hit detection curves of Chapter 3. To determine P_D for a noncoherent integrator, see Sections 1-13, 2-8, and 2-9.

The search radar equation can also be expressed in terms of energy ratio, as was done earlier for a matched-filter receiver:

$$\left(\frac{S}{N}\right)_N = \frac{\mathscr{R}_T(\text{PRF})\,A t_s \sigma}{4\pi R^4 A_Z(\sin E_{max} - \sin E_{min}) k T_0 \overline{NF}_o} \qquad (1\text{-}83)$$

It is worthwhile considering the search radar range equation from a physical point of view. The total energy transmitted is the product of average transmitted power $[\mathscr{R}_T(\text{PRF})]$ and the search time (t_s). In the denominator of Eqs. (1-82) and (1-83) the factor

$$R^2 A_Z(\sin E_{max} - \sin E_{min})$$

gives the total area to be searched and over which the power is distributed. The ratio $A/4\pi R^2$ in the search radar equation represents that portion of the power intercepted by the target and returned to the radar receiver.

According to the aforementioned interpretation, the search range equation is independent of the type or size of the transmitting antenna used but does depend upon uniform distribution of the transmitted energy over the designated search area. In the search radar range equation it is also assumed that the reflected transmitted energy is collected efficiently. This in turn necessitates that the receiving beam or beams should be pointing to the direction from which target energy arrives. The simplest manner of achieving this is to establish a common antenna for transmission and reception and a scan rate slow enough to permit each pulse to be received before the antenna changes angle appreciably. As the scanning rate is increased to a point where the antenna gain changes between transmission and reception, a scanning loss is encountered.

If the antenna is scanned sufficiently slowly, more than one pulse may be transmitted and received while the antenna beam sweeps across a given reflecting point. These pulses are integratable on a pulse-to-pulse basis. It is also possible to integrate on a scan-to-scan basis. If the target is fluctuating, these two types of integration lead to different probabilities of detection (see Chapter 3).

1-12 Summary of Search and Tracking Radar System Losses

System losses of varied types must be accounted for in a real radar. Search radar systems experience most of the losses common to tracking radars plus several losses common to only search systems. A brief list of various losses is presented below and a short description of each loss will follow. It is common practice to state losses in numbers greater than unity and include them in the denominator of the radar range equation. By doing this the radar engineer can always work in positive decibels.

Integration loss (L_i).
Receiver filter matching loss (L_m).
Antenna efficiency loss (L_a).
Beam-shape loss (L_b).
Scanning loss (L_s).
Target fluctuation loss (L_f).
Scan distribution loss (L_d).
Gate and filter spacing loss (L_g).
Plumbing loss (L_p).

Integration Loss (L_i) Integration loss occurs in noncoherent integrators because integration of the detected signal-plus-noise envelope results in less self-noise cancellation (see Sections 2-8 and 2-9). The Meyer Plots presented in Chapter 4 incorporate the effects of integration loss due to noncoherent integration.

Receiver Filter Matching Loss (L_m) Receiver filter matching loss occurs when the predetector filter response (normally the i.f. amplifier response) is not matched to the spectrum of the received signal. The filter's frequency response should be the complex conjugate of the spectrum of the received signal. The response of a common stagger tuned i.f. amplifier has a flat top and relatively steep leading and trailing edges. It was found that for this type of "rectangular" filter, the optimum bandwidth pulse-width product is 1.2 (see Section 1-6).

Antenna Efficiency Loss (L_a) For the previous search radar range equations the assumption was made that the antenna aperture is 100% effective. Actually, the effective aperture of a parabola or metal lens is 0.5–0.7A, where A is the actual aperture.

Beam-Shape Loss (L_b) Beam-shape loss must be accounted for because the shape of the antenna pattern is not accounted for in the radar range equation. It is usually assumed that the field strength is a constant over the width of the beam. In actuality, as a target passes through a beam, the signal return is modulated. This causes a loss which is termed beam-shape loss (see Sections 1-11 and 3-13).

Scanning Loss (L_s) Scanning loss is caused by motion of the beam between pulse transmission and pulse reception (see Sections 1-11 and 3-13).

Target Fluctuation Loss (L_f) The radar range equation is normally written for fixed target size. In reality this is seldom the case. In general, it is necessary to work with the average radar cross section $\bar{\sigma}$. The thus calculated S/N ratio can then be used with the Meyer Plot appropriate for one of the five classical target models (see Chapter 4) to determine P_D and P_{fa}.

Scan Distribution Loss (L_d) Scan distribution loss is caused by the failure to integrate as a single group all the pulses received from a given target before detection. With some search radars the beam will scan at a rapid rate and will pass the target location a few times during the search time t_s. The actual number of pulses integrated on each scan is some fraction of the number of pulses that can be integrated from a beam position during the search time, that is,

$$\left[\begin{array}{c}\text{actual number of pulses}\\ \text{integrated per scan}\end{array}\right] \ll N = \frac{(\mathrm{PRF})\,t_s}{n_s} \tag{1-84}$$

On the other hand, the increased number of scans does establish more than one opportunity to detect a target. Therefore, the increased number of scans compensates in part for the reduction in integration gain. Since this results in partial compensation only, the net loss is termed the scan distribution loss L_d.

Gate and Filter Spacing Loss (L_g) The range gates and Doppler filters introduce losses due to their required separation in the time–frequency domain. These losses depend upon the degree of overlap between adjacent channels and the rate of scan through the time–frequency domain. Because of present day filter design capabilities, no real minimum value of loss exists except possibly that dictated by complexity, size, weight, and cost considerations.

Plumbing Loss (L_p) Common plumbing losses are loss per unit length of waveguide, loss per waveguide bend, rotary joint loss, duplexer loss, and losses caused by poor connections. Based on a system operating at 3 GHz, typical values for these losses are as follows:

duplexer loss:	1.2–1.7 dB
rotary joint loss:	0.3–0.5 dB
loss per 10 linear ft of waveguide:	0.1 dB
loss per bend in waveguide:	0.1–0.2 dB

1-13 Illustrative Example Using the Search Radar Equation

The following example of the search radar range equation applied to a problem is presented to unify the material presented in this chapter.

It is required to maintain a final 90% probability of detection P_D and false alarm number† n' of 10^8. This requirement coincides with what is considered to be the minimum acceptable P_D and n' for a search radar system for both Marcum's fixed target model and Swerling's fluctuating target models.‡ The probability of detection and false alarm number will be computed for 10 pulses being noncoherently integrated.

From the Meyer Plots presented in Chapter 4, the following values of minimum required single pulse signal-to-noise ratio $(S/N)_1$ can be established (Table 1-3). To meet the stated requirement, therefore, Target Model Case 1 is the governing factor and the minimum signal-to-noise ratio acceptable is 14.7 dB. Table 1-4 presents the probabilities of detection one would now obtain based on a 14.7-dB minimum S/N ratio.

† False alarm number is defined in Section 2-2.
‡ Target models are described in Chapter 3.

Table 1-3

MINIMUM REQUIRED $(S/N)_1$ VERSUS TARGET MODEL

Target model case	P_D (%)	$(S/N)_1$ minimum required (dB)
0	90	6.2
1	90	14.7
2	90	8.2
3	90	10.6
4	90	6.7

Table 1-4

P_D VERSUS TARGET MODEL

Target model case	$(S/N)_1$ minimum (dB)	P_D (%)
0	14.7	99.99+
1	14.7	90
2	14.7	99.99+
3	14.7	99.99+
4	14.7	99.99+

Equation (1-82) was derived for a search radar that used coherent integration, but can be modified for a noncoherent integrator by the inclusion of an integration loss term L_i, as follows:

$$\left(\frac{S}{N}\right)_N = N\left(\frac{S}{N}\right)_1$$

$$= \frac{P_T(\mathrm{PRF})\, t_s A\sigma}{4\pi R^4 A_Z(\sin E_{max} - \sin E_{min})\, kT_0 B_n \overline{\mathrm{NF}}_o L_i} \tag{1-85}$$

Although the exact value of L_i, which is a function of $(S/N)_N$ and N, can be determined by the appropriate use of Meyer Plots, for the purpose of this problem an integration efficiency γ of 80%

Table 1-5

ASSUMED PERFORMANCE CRITERIA

Parameter	Symbol	Value
Azimuth to be searched	A_Z	360°
Maximum elevation to be searched	E_{max}	45°
Minimum elevation to be searched	E_{min}	0°
Duration of search	t_s	15 sec
Range	R	100 nmi
Target cross section	σ	1 m²
Pulse width	T_P	1 μsec
Pulse repetition rate	PRF	300/sec
Effective noise temperature of receiver	T_e	290° K
Aperture efficiency loss	L_a	2.0 dB
Scanning loss	L_s	1.0 dB
Beam-shape loss	L_b	1.5 dB
Scan distribution loss	L_d	1.0 dB
Receiver filter matching loss	L_m	0.5 dB
Plumbing loss	L_p	2.5 dB
Gate and filter spacing loss	L_g	0.5 dB
Integration loss	L_i	2.0 dB
Total of above losses	L	11.0 dB
Number of pulses integrated	N	10

is assumed. Thus (see Sections 2-7 and 2-8)

$$L_i = N^{(1-\gamma)} = 10^{(1-0.8)} = 10^{0.2} = 1.59$$

or

$$L_i = 2.0 \quad \text{dB} \tag{1-86}$$

Let us assume that we wish to design a radar to meet the additional performance criteria listed in Table 1-5 and that we further wish to calculate the power and antenna aperture that will be required.

In order to calculate the desired quantities, it will be necessary to solve Eq. (1-85) for the power aperture product, replace $T_0\overline{\mathrm{NF}}_o$ with T_e, and include all the desired loss terms:

$$P_T A = \frac{[N(S/N)_1]\, 4\pi R^4 A_Z (\sin E_{max} - \sin E_{min})\, k T_e B_n L_a L_s L_b L_d L_m L_p L_g L_i}{(\mathrm{PRF})\, t_s \sigma} \tag{1-87}$$

It is common practice when working with the radar range equation to convert all parameters to decibels (dB), a logarithmic quantity, in order to reduce the amount of computational labor. It is also necessary to put all terms in a consistent set of units. Both the former and latter are done in Table 1-6.

Table 1-6

CONVERSION OF ESSENTIAL PARAMETERS TO DECIBELS

Symbol	Value	Decibels
σ	1 m²	0.00
t_s	15 sec	11.76
$R_m{}^4$	(100 nm)⁴ = (1.853 × 10⁵)⁴ m	210.60
A_Z	360° = 2π rad	7.96
($\sin E_{max} - \sin E_{min}$)	(sin 45° − sin 0°) = (0.707)	−1.50
($T_e k$)	290 (1.38 × 10⁻²³)	−204.00
4π	12.56	10.96
(PRF)	300	24.76
$B_n = 1/T_P$	1 MHz	60.00
N	10	10.00
(S/N)₁	—	14.70
L	Total losses	11.00

Using Eq. (1-87), we find for $P_T A$ (in decibels relative to 1 Wm²)

$$\begin{aligned} P_T A = {} & 10.00 + 14.70 + 10.96 \\ & + 210.60 + 7.96 - 1.50 - 204.00 \\ & + 60.00 + 11.00 - 24.76 \\ & - 11.76 - 0.00 = 83.20 \end{aligned} \tag{1-88}$$

Now, if we decide to let the transmitted power P_T be 5 MW (67.0 dBW) during a pulse, the actual aperture is 83.20 dB − 67.0 dBW = 16.2 dBm² (41.8 m²). The effective common transmitting–receiving aperture is 16.2 dB − 2.0 dB = 14.2 dBm² (26.4 m²).

If the reader reviews the derivation of the preceding equations, he will note that at this point the wavelength of the radar has also been determined for the case where a single search beam is generated by a common transmitting–receiving aperture. For such an aperture, λ is 0.51 m (L-band), a rather long wavelength, and the resulting antenna gain is 31 dB.

This problem is typical of calculations made to find the approximate magnitudes of a radar system's major parameters. By working with decibels, it is easy to trade off different parameters in order to fulfill a system requirement. Savings in cost, size, power, weight, and other parameters can be introduced heuristically in the problem to achieve the optimum design.

References

1. J. I. Marcum, Studies of Target Detection by a Pulsed Radar. Rand Res. Memo, RM-754 (Dec. 1947); reprinted in *IRE Trans. Inform. Theory* **IT-6** (April 1960); also AD-101287.
2. J. I. Marcum, Studies of Target Detection by a Pulsed Radar—Mathematical Appendix. Rand Res. Memo RM-753 (July 1948); reprinted in *IRE Trans. Inform. Theory* **IT-6** (April 1960); also AD-101882.
3. D. K. Barton, "Radar System Analysis." Prentice-Hall, Englewood Cliffs, New Jersey, 1964.
4. J. L. Lawson and G. E. Uhlenbeck (eds.), "Threshold Signals," M.I.T. Radiat. Lab. Ser., Vol. 24, p. 177. McGraw–Hill, New York, 1950.
5. A. E. Siegman, Thermal Noise in Microwave Systems. *Microwave J.* March (1961).
6. R. S. Berkowitz, "Modern Radar-Analysis, Evaluation, and System Design." Wiley, New York, 1965.
7. D. O. North, Analysis of Factors which Determine Signal to Noise Discrimination in Radar. R.C.A. Lab., Princeton, New Jersey, Rep. PTR-6C (June 1963); reprinted in *Proc. IEEE* July (1963).
8. J. H. Van Vleck and D. Middleton, A Theoretical Comparison of Visual, Aural, and Meter Reception of Pulsed Signals in the Presence of Noise. *J. Appl. Phys.* **17**, 940–971 (1946).
9. C. E. Cook and M. Bernfeld, "Radar Signals—An Introduction to Theory and Application." Academic Press, New York, 1967.
10. M. Bernfeld, C. E. Cook, J. Paolillo, and C. A. Palmieri, Matched Filtering, Pulse Compression, and Waveform Design Parts I–IV. *Microwave J.* **7**, No. 10, 57–64; No. 11, 81–90; No. 12, 70–76; **8**, No. 1, 76–81 (1965).
11. L. N. Ridenour, "Radar Systems Engineering," M.I.T. Radiat. Lab. Ser., Vol. 1, p. 34, McGraw–Hill, New York, 1947.
12. M. I. Skolnik, "Radar Handbook," Chapters 2 and 5. McGraw–Hill, New York, 1970.
13. A. M. Fine, Optimum Filters for Pulsed Signals in Noise. Thesis . . . (for) . . . M.E.E. . . . (from) . . . Polytechnic Inst. of Brooklyn (June 1957).
14. L. V. Blake, A Guide to Basic Pulse Radar Maximum-Range Calculation—Part 1. U.S. Naval Res. Lab. Rep. 5868 (December 10, 1963).
15. P. M. Woodward, "Probability and Information Theory with Applications to Radar." Pergamon, Oxford, 1953.
16. J. W. Crispin, Jr., R. F. Goodrich, and K. M. Siegel, A Theoretical Method for the Calculation of the Radar Cross Sections of Aircraft and Missiles. Univ. of Michigan Radiat. Lab., Ann Arbor, Michigan. Rep. No. 2591-1-H on contract AF-19(604)-1949 (July 1959).
17. M. I. Skolnik, An Analysis of Bistatic Radar. *IRE Trans. Aeronaut. Navig. Electron.* **ANE-8**, No. 1 (1961).
18. K. M. Siegel, H. A. Alperin, R. R. Bronski, J. W. Crispin, A. L. Moffett, C. E. Schensted, and I. V. Schensted, Bistatic Radar Cross-Sections of Surfaces of Revolution. *J. Appl. Phys.* **26** (1955).

Recommended reading for further study in addition to the references cited above:

M. I. Skolnik, "Introduction to Radar Systems," Chapter 2. McGraw–Hill, New York, 1962.

F. E. Nathanson, "Radar Design Principles—Signal Processing and the Environment." McGraw–Hill, New York, 1969.

CHAPTER 2

STATISTICAL SIGNAL DETECTION

2-1 Detection

All peak detection decisions are based on comparing the output of a receiver with some threshold or bias level. In the case of the human operator, the bias level is established by means of the sensitivity of both the display and the human operator's visual perception. To be sure, human visual perception is a variable between different operators and even within the same operator, as it is a function of an individual's alertness, incentive, and other nondimensional quantities.

For the purpose of this text, detection of a signal is said to occur whenever the output of the receiver exceeds a certain predetermined value called the bias level. This level is sometimes termed the threshold level or detection threshold.

2-2 False Alarm Time, Probability of False Alarm, and False Alarm Number

In detection systems the concept of false alarm time is often found useful because it measures the rate at which noise pulses are mistaken for target echoes. There are at least two major definitions of false alarm time:

1. *Marcum's Definition* The false alarm time is the time during which the probability is $\frac{1}{2}$ that the noise will not exceed the bias level [1–3].

2. *Kaplan and McFall's Definition* The more common definition is the one given by Kaplan and McFall—the false alarm time is the average time interval between false target indications [4].

Application of the second (Kaplan and McFall's) definition to Fig. 2-1 results in

$$t_{fa} = \frac{t_1 + t_2 + t_3 + \cdots + t_n + t_{n+1} + \cdots + t_k}{k} \qquad \text{(2-1)}$$

which in the limit is

$$t_{fa} = \lim_{k \to \infty} \left(\frac{1}{k} \sum_{n=1}^{k} t_n \right) \qquad \text{(2-2)}$$

Also it can be seen by referring to Fig. 2-1 that the mean or average time a noise pulse is greater than the bias level is

$$\Delta t_{fa} = \lim_{k \to \infty} \left(\frac{1}{k} \sum_{n=1}^{k} \Delta t_n \right) \qquad \text{(2-3)}$$

This quantity is usually approximated to be the reciprocal of the bandwidth B of the video signal with which it is associated; that is,

$$\Delta t_{fa} \approx 1/B \qquad \text{(2-4)}$$

The term $1/B$ is also a measure of the range resolution of the radar and is also often used to define a quantity called a "range bin." One may say, therefore, that according to Kaplan and McFall, on the average there will be one range bin that contains noise of a magnitude greater than the bias level during each false alarm time. The noise energy in adjacent range bins is assumed to be statistically independent; that is, the bandwidth of the radar receiver approximately determines the noise correlation interval.

It follows then that Kaplan and McFall's probability of false alarm (P_{fa}) is the period of time that the envelope is above the bias level as compared with the average period of time between such occurrences. In other words, the probability of false alarm is the reciprocal of the average number of range bins (N_{fat}) in the false alarm time. Therefore,

$$P_{fa} = 1/N_{fat} \cong 1/Bt_{fa} \qquad \text{(2-5)}$$

The reason P_{fa} is only approximately equal to the right-hand term of Eq. (2-5) is explained later in this section [see Eq. (2-11)].

The assumption of stationary noise statistics also presupposes that the probability that a false alarm will occur at any given instant within the false alarm time is a constant. The cumulative probability of false alarm over a number of range bins, therefore, is governed by a binomial distribution. The probability P_{fa} is usually so small and the number of possible range bins in the search

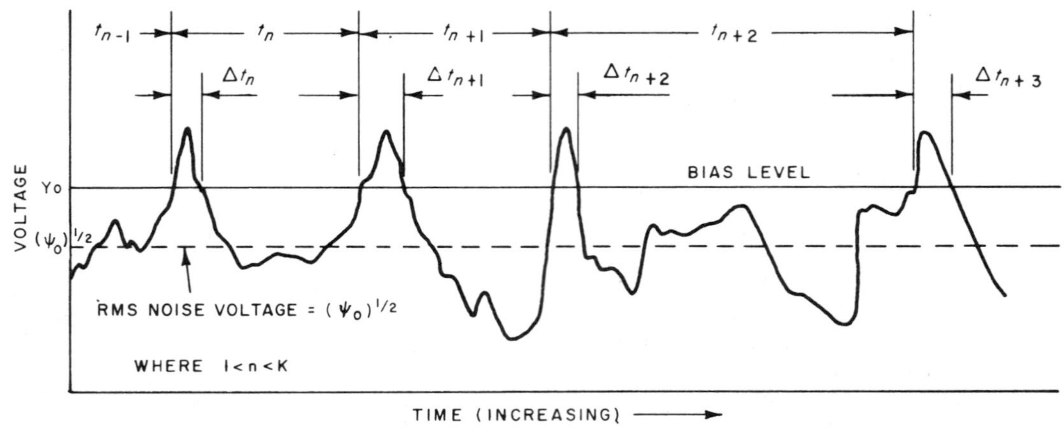

Fig. 2-1 Receiver output noise voltage envelope versus time. The various parameters that define Kaplan and McFall's false alarm time are illustrated.

time usually so great, however, that the cumulative probability approaches the limiting case, a Poisson distribution; that is,

$$P(r) = \frac{N_r!}{r!(N_r - r)!}(1 - P_{fa})^{N_r - r}(P_{fa})^r \cong \frac{e^{-u}u^r}{r!} \qquad (2\text{-}6)$$

where

$P(r)$ is the probability of r false alarms,
$u = N_r P_{fa}$ the mean value of binomial and Poisson distribution functions, and
N_r the number of range bins over which the cumulative probability is being calculated.

A better understanding of the difference between the two definitions for false alarm time is obtained by applying Eq. (2-6) to them. By Kaplan and McFall's definition $N_r = N_{fat}$; therefore, $u = 1$, and the probability of exactly one false alarm within the false alarm time interval is

$$P(1) = e^{-1}(1)^1/1! = 0.368 \qquad (2\text{-}7)$$

According to Marcum's definition of false alarm time, however, the false alarm time is the time when the probability is 50% that noise alone will *not* exceed the bias level ($r = 0$ case). Applying the Poisson distribution to this definition results in $\frac{1}{2} = e^{-u}$ and, therefore, $u = 0.693$. The probability of one false alarm ($r = 1$ case) in this interval is

$$P(1) = 0.693\,e^{-0.693} = 0.346 \qquad (2\text{-}8)$$

Table 2-1[a]

Number of false alarms	Probability for t_{fa} (Kaplan and McFall case)	Probability for 0.693 t_{fa} (Marcum case)
0	0.368	0.500
1	0.368	0.346
2	0.184	0.120
3	0.061	0.028
4	0.015	0.005

[a] From Hollis [5].

It is seen that since $u = P_{fa} Bt_{fa} = \text{constant} \times t_{fa}$ (the terms B and P_{fa} are fixed for any real situation), the false alarm time as defined by Kaplan and McFall ($u = 1$ case) is about 44% longer than Marcum's false alarm time ($u = 0.693$ case), although the probabilities of one false alarm within their respective alarm times differ by only about 6%. This difference between the definitions was first noted by Hollis [5]. He also tabulated the probabilities (see Table 2-1) for different numbers of false alarms during each time interval t_{fa}.

It should be noted that Eq. (2-5), which defines Kaplan and McFall's probability of false alarm, must be modified in order to account for Marcum's definition of false alarm time. The relationship becomes

$$P_{fa} = \frac{\log_e 2}{N_{fat}} \cong \frac{\log_e 2}{Bt_{fa}} \qquad (2\text{-}9)$$

The false alarm number is also defined in two ways. The first and most common definition of false alarm number is the one used by Marcum in plotting his P_D versus S/N curves. According to this definition, false alarm number is the number of opportunities for error that are possible in the false alarm time if each and every pulse is separately thresholded. It is given by

$$n = (\mathrm{PRF})\, N_R\, t_{fa} \qquad (2\text{-}10)$$

where

PRF is the pulse repetition frequency (average PRF if it varies or pulses are transmitted in bursts),†
N_R the number of range bins per range sweep, and
t_{fa} the false alarm time.

It should be noted that Marcum's definition of n is equivalent to Kaplan and McFall's definition for N_{fat}, since for most radars the range sweep occupies nearly the entire time between pulses (or pulse bursts), and therefore

$$B \cong (\mathrm{PRF}')\, N_R = (\mathrm{PRF})\, N_R/N \qquad (2\text{-}11)$$

where (PRF′) is the repetition frequency of groups of N pulses.

An alternative definition for the false alarm number is

$$n' = n/N = (\mathrm{PRF})\, N_R\, t_{fa}/N \qquad (2\text{-}12)$$

where

N is the number of pulses integrated.

This false alarm number is the number of independent samples delivered to the data processor for threshold detection during the false alarm time. Although originally defined by Marcum, n' is commonly known as Fehlner's false alarm number because it is the false alarm number he used in plotting his P_D versus S/N curves. In this text, the custom of calling n' Fehlner's false alarm number will be continued.

When using Meyer Plots (Chapter 4), the false alarm number in the upper left-hand corner of the graph is always Fehlner's false alarm number (n').

As an illustrative example of the difference between Marcum's definition and that of Fehlner, we will assume a false alarm number $n = 10^{10}$

Table 2-2

Number of pulses integrated	False alarm number n	False alarm number n'
1	10^{10}	10^{10}
10	10^{10}	10^{9}
100	10^{10}	10^{8}
10,000	10^{10}	10^{6}

and then determine the false alarm number n' for 1, 10, 100, and 10,000 pulses being integrated using Eq. (2-12). The results are presented in Table 2-2.

The false alarm number as defined by Fehlner can be converted to *Kaplan and McFall's* [not Marcum's—see Eq. (2-24)] probability of false alarm by

$$P_{fa} = 1/N_{fat} = 1/n' = N/n \qquad (2\text{-}13)$$

2-3 Probability of False Alarm versus Bias Level Setting

The purpose of the bias level is to divide the output into two energy zones: one zone greater than the bias level known as the detection zone, and one zone smaller than the bias level known as the nondetection zone. The thresholding device will on occasion be subjected to noise pulses of magnitudes greater than the bias level setting. By raising or lowering the bias level, the number of times a noise pulse surpasses the bias level either decreases (as the bias level is raised) or increases (as the bias level is lowered). How the bias level will be set depends on the number of times noise will be permitted to exceed the bias level during a given period of time. To mistake a noise pulse for a signal return is called a false alarm, which in statistical decision theory is termed a type-1 error. To mistake a signal return for a noise pulse is called miss detection, which in statistical decision theory is termed a type-2 error. Miss detection increases as the bias level increases. Miss detection and false alarm, therefore, are subject to tradeoff.

The probability density function for noise alone, $dP_0(N)$ (power signal-to-noise ratio is zero), at the output of a square-law detector was determined by Marcum [3, Eqs. 38–48] to be

$$dP_0(N) = \frac{Y^{N-1}e^{-Y}}{(N-1)!}\,dY \qquad (2\text{-}14)$$

† Some radars transmit a group or burst of pulses for each range sweep when the sweep is of considerable length. This is to keep the average power high without sacrificing range resolution. The individual pulses within a group are then integrated.

where

N is the number of pulses integrated, and
Y the instantaneous noise amplitude at the output of a square-law detector normalized to the rms video noise voltage.

The probability of false alarm is determined by integrating Eq. (2-14) over appropriate limits. Thus

$$P_{fa} = \int_{Y_b}^{\infty} \frac{Y^{N-1}e^{-Y}}{(N-1)!}\,dY \tag{2-15}$$

where

Y_b is the bias level.

The incomplete gamma function as defined by Pearson [6] is

$$I(u,P) = \int_0^{u(P+1)^{1/2}} \frac{e^{-y}y^P}{P!}\,dy \tag{2-16}$$

and in terms of the incomplete gamma function, Eq. (2-15) is

$$P_{fa} = 1 - I[Y_b/N^{1/2}, (N-1)] \tag{2-17}$$

In Pearson's "Table of Incomplete Gamma Functions," the tables for $I(u,P)$ extend to $P = 50$ and values are given to seven significant figures. For $N \leqslant 51$ and $n' \leqslant 10^6$, therefore, the bias level can be obtained directly from Pearson's table, in which case, as indicated above,

$$u = Y_b/N^{1/2} \tag{2-18}$$

$$P = N - 1 \tag{2-19}$$

Marcum succeeded in developing a couple of expressions relating to Y_b that are somewhat useful when $N > 51$ [3]. By successively applying integration by parts to Eq. (2-15) and by treating the resultant finite series as an infinite series, the following approximation for P_{fa} can be obtained:

$$P_{fa} \approx \frac{N Y_b{}^N \exp(-Y_b)}{N!(Y_b - N + 1)} \tag{2-20}$$

Further simplication was achieved using Stirling's approximation for $N!$ [7], that is,

$$N! \approx (2\pi)^{1/2} N^{(N+1/2)} e^{-N} \tag{2-21}$$

Stirling's approximation contributes errors of the magnitude given in Table 2-3. Applying Stirling's approximation to Eq. (2.20) results in

$$P_{fa} \cong \left(\frac{N}{2\pi}\right)^{1/2} \left\{\frac{\exp[-Y_b + N(1 + \log_e Y_b/N)]}{(Y_b - N + 1)}\right\} \tag{2-22}$$

Table 2-3

$N!$	Error using Stirling's approximation (%)
1!	8.0
2!	4.0
5!	2.0
10!	0.8
100!	0.08

Equation (2-22) can be related to Fehlner's false alarm number by use of Marcum's definition of false alarm time. This definition states that the probability is 50% $[P(r = 0) = 0.5]$ that noise will not exceed the bias level in any given false alarm time. Solving Eq. (2-6) for P_{fa} and letting $r = 0$ and $N_r = n'$ (since n' is the number of range bins investigated during the false alarm time) yields

$$P_{fa} = 1 - P(0)^{1/n'} \tag{2-23}$$

For $n/N = n' \gg 1$, a condition that usually exists in practice, and $P(0) = 0.5$, this reduces to

$$P_{fa} \cong \frac{\log_e[1/P(0)]}{n'} = \frac{\log_e 2}{n'} = \frac{\log_e 2}{n/N} \tag{2-24}$$

Equating Eq. (2-24) to Eq. (2-22) results in a closed form relationship between Y_b, N, and n†:

$$\log_{10} n = 0.24 + 0.5 \log_{10} N + \log_{10}(Y_b - N + 1) + 0.434(Y_b - N) - N \log_{10} Y_b/N \tag{2-25}$$

Equation (2-25) for $N = 1$ reduces to

$$Y_b = 2.302(\log_{10} n) + 0.45 \tag{2-26}$$

whereas the exact expression for Y_b with $N = 1$ derived from Eq. (2-15) or (2-17) is

$$Y_b = (\log_e n) - \log_e(\log_e 2) = 2.302(\log_{10} n) + 0.367 \tag{2-27}$$

demonstrating that the effect of the approximation for $N!$ is small.

† Figures 8 and 9 of Marcum [3] are graphical representations of this equation in terms of n and n', respectively.

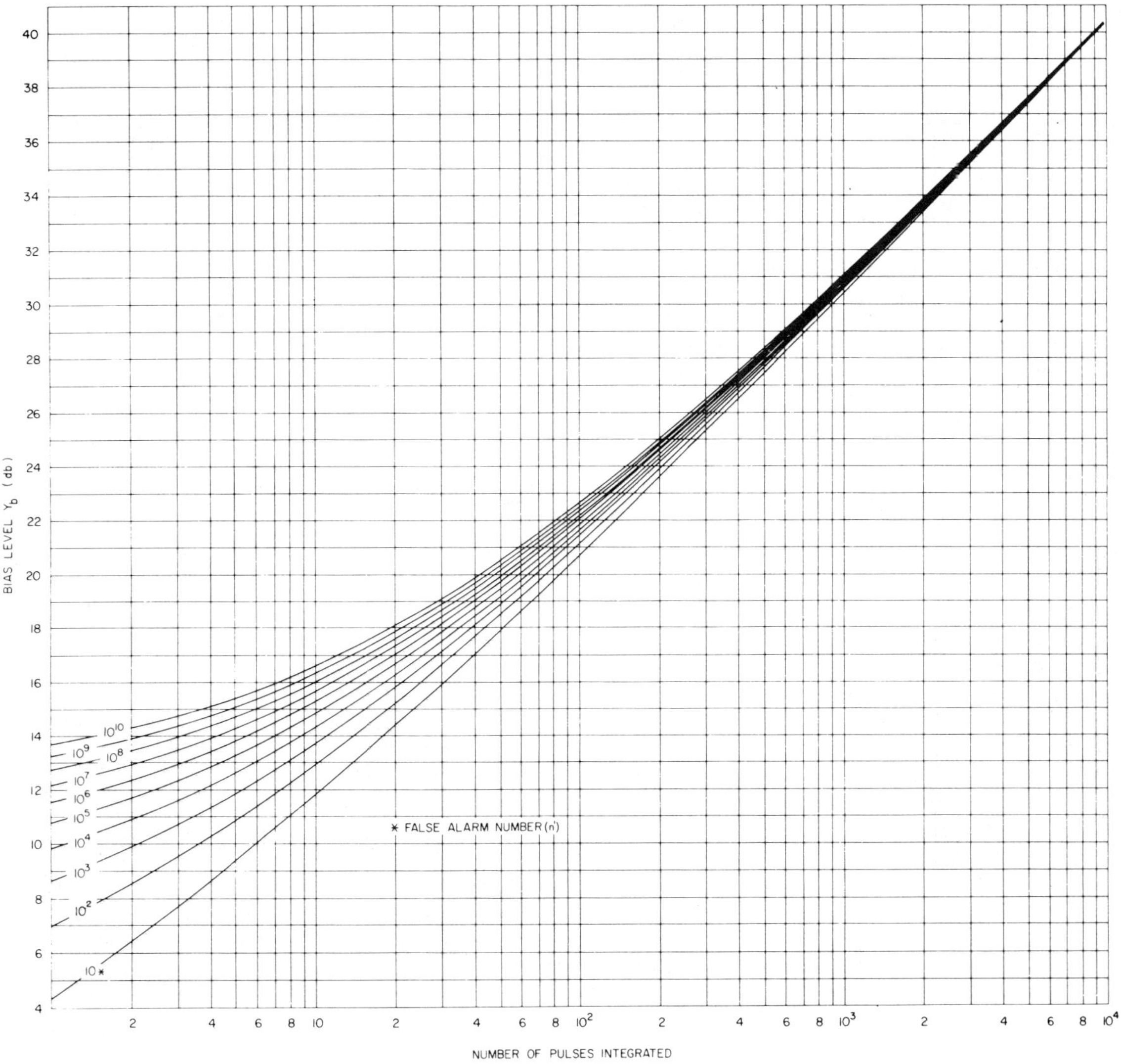

Fig. 2-2 Bias level Y_b versus number of pulses integrated for a square-law detector.

2-4 Determination of Bias Level Setting

None of the equations developed in the previous section lend themselves to ready determination of the bias level setting Y_b as a function of P_{fa} or n' if $N > 1$. The only useful one is Eq. (2-27), which applies only to the case where $N = 1$. It is not possible to determine P_D accurately as a function of S/N, n', and N, especially for large values of N, if Y_b is not accurately known. The authors, by use of a convergent series solution (see Appendix, Section A-8) and a digital computer, generated a table of exact values for Y_b for the case of a square-law detector.† The table has been reduced to graphical form (see Fig. 2-2), from which Y_b can be determined with sufficient accuracy for radar design work. The figure presents Y_b (in decibels relative to the rms video noise voltage) as a function of N and n'. The constant of the detector law is $\mathscr{A} = 0.5$, that is, $y = 0.5v^2$ (see Section 3-4).

† It was not possible to use the tables for Y_b generated by Pachares [8], as they were not sufficiently accurate for large false alarm numbers.

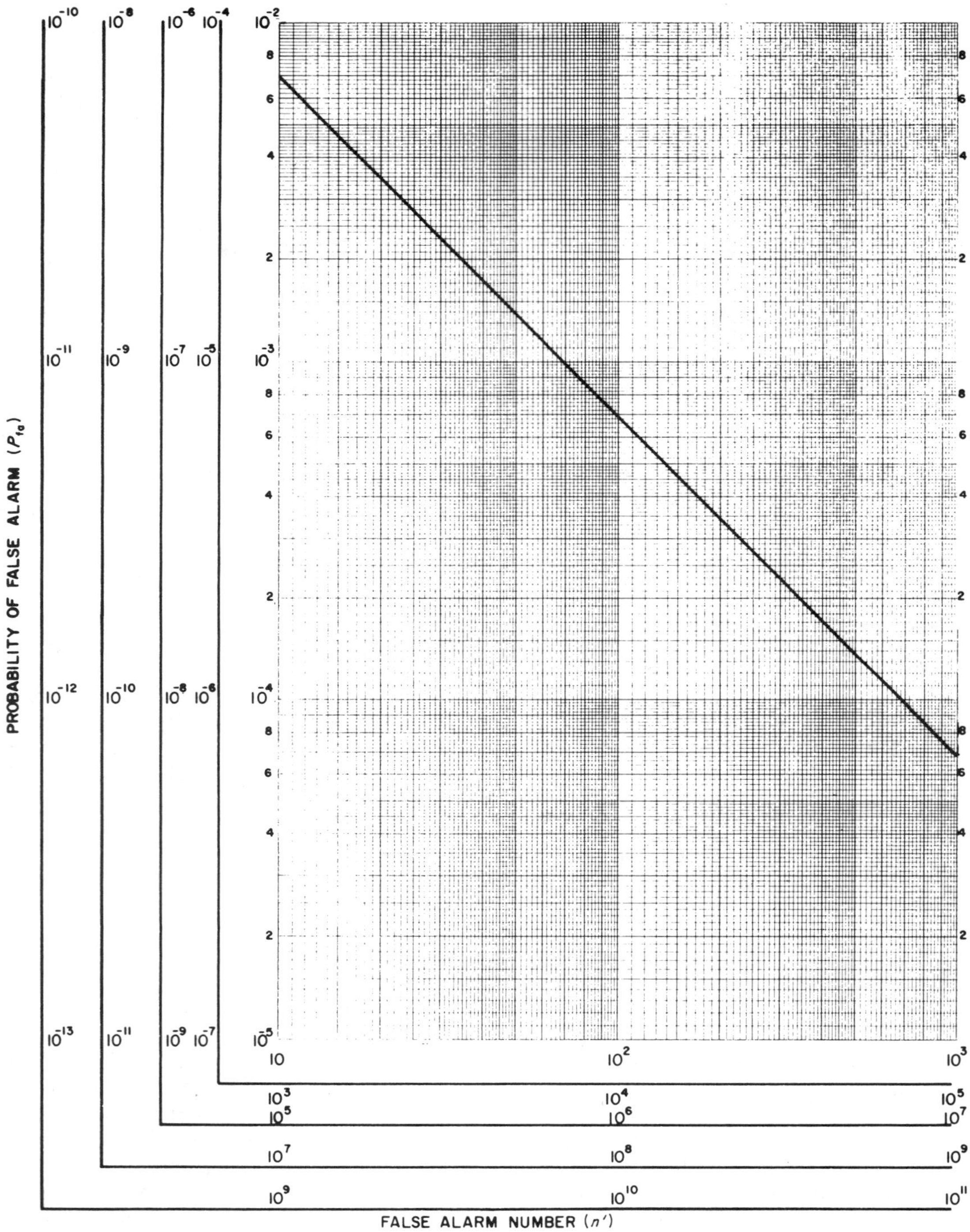

Fig. 2-3 Probability of false alarm as a function of false alarm number.

Table 2-4

Multiplication factor of false alarm number (a)	If $n' = a \times 10^b$, then $P_{fa} = (0.693/a) \times 10^{-(b+1)}$ (P_{fa})
1	$6.93 \times 10^{-(b+1)}$
2	$3.475 \times 10^{-(b+1)}$
4	$1.733 \times 10^{-(b+1)}$
6	$1.155 \times 10^{-(b+1)}$
8	$8.663 \times 10^{-(b+2)}$

The Fehlner false alarm numbers given can be readily converted to values of probability of false alarm (Marcum's) by use of Table 2-4, which is set up for converting the specific false alarm numbers used in Meyer Plots, or by use of Fig. 2-3.

To illustrate the use of Table 2-4, assume we are using the Meyer Plot where the false alarm number is $n' = 8 \times 10^7$. Therefore $a = 8$ and $b = 7$. The equivalent probability of false alarm is then $P_{fa} = 8.663 \times 10^{-9}$. Alternatively, the Kaplan and McFall probability of false alarm is simply the inverse of n'.

2-5 The Advantages and Disadvantages of Pulse Integration

Integration involves adding a number of successive signal pulses and attempting to detect the sum signal rather than the signal from an individual pulse. An integrator cannot distinguish between signal and noise; therefore, in the absence of a signal it will add up to N successive noise pulses. The comparison the detection device must make is between N signal-plus-noise pulses and N noise pulses alone as contrasted with one signal-plus-noise pulse and one noise pulse alone for a nonintegrating detector. Fortunately, the fluctuation due to noise of the sum voltage as a result of N integrations is not equal to N times the fluctuation voltage of one pulse. If it were, integration would have the same effect as a common gain and would be a useless process. Noise voltage fluctuates about some mean value, which can be biased out, and in actuality it is the ratio of signal to standard deviation† of noise and not the signal-to-average-noise-value ratio that is of paramount importance. The sum fluctuation voltage for the linear addition of N random noise pulses is on the average only $(N)^{1/2}$ times the fluctuation voltage of a single pulse.

† The statistical term that measures fluctuation about a mean is termed standard deviation and is denoted by σ.

As the number of pulses integrated is increased, the greater the ratio of signal to standard deviation of noise and the more likely it is that a submerged-in-noise signal will be detected. At the same time, however, it should be noted that as the number of pulses integrated is increased, the more sensitive the probability of false alarm P_{fa} or false alarm number n' becomes in terms of the bias setting Y_b (see Fig. 2-2), and the longer the period of time required for determining if a target is detected.

2-6 General Theory of Probability of Detection as a Function of Probability of False Alarm and Number of Pulses Integrated

Pulse integration improves signal-to-noise ratio and correspondingly the probability of detection, but the amount of improvement depends upon the method of integration. Primarily, integration may be accomplished in either of two sections of the radar receiver: in the i.f. section prior to the second detector, or in the video section after the second detector. There is a considerable difference between these two types of integration. Integration *before the second detector* is defined as coherent or predetection integration. Integration *after the second detector* is defined as noncoherent or postdetection integration. Furthermore it must be noted that postdetection integration may in some systems be accomplished by a human observer.

COHERENT INTEGRATION

Coherent integration requires that the phase of the echo signal be preserved during the summing process; that is, in a coherent integration processor the phase of the received signal must remain constant with respect to the phase of a known sinusoid. Thus, all Doppler information about the target and Doppler information introduced by relative target and radar motion into the signal must also be preserved.

The i.f. predetection filter is usually narrow band (high Q); therefore, integration of a moving target's signal return could produce a Doppler-shifted signal such that the resultant signal frequency would lie outside the passband of the predetection pulse integrator. To overcome this limitation, a number of similar integrators, each tuned to a slightly different frequency, can be used to cover the range of frequencies in which the signal return is expected. A speed that would cause a Doppler shift beyond the upper or lower limits of the filter bank would, however, result in no detection of the target.

The sum signal voltage due to coherent pulse integration is the algebraic sum of the individual pulse voltages being integrated. If N pulses are integrated coherently, therefore, the output signal would have N times the signal-to-noise power ratio of one pulse. The Meyer Plots relating to multiple-hit detection described in Chapter 4 *are not for this type of integration.* The resultant sum of the pulses being coherently integrated could be represented by a new single pulse having a new signal-to-noise ratio equal to $N(\mathrm{S/N})$ or in the case of varying pulse amplitudes by

$$\left(\frac{\mathrm{S}}{\mathrm{N}}\right)_N = \sum_{k=1}^{N} \left(\frac{\mathrm{S}}{\mathrm{N}}\right)_k \tag{2-28}$$

Thus for the case of coherent integration the curves for single-hit detection (see Figs. 3-2–3-16) are used in conjunction with Eq. (2-28) to determine probability of detection.

NONCOHERENT INTEGRATION

The primary concern of this text is with the characteristics of noncoherent (postdetection) integration. This is the most common type of integration used in radar systems. If the same N pulses described in the previous section on coherent integration were now integrated noncoherently, the final signal-to-noise ratio would be less than that for coherent integration, that is, less than N times that of a single pulse. The relative increase in noise energy is caused by the rectifying characteristic of the second detector, which prevents some of the noise self-cancellation that would otherwise occur. The integration efficiency of a postdetection integrator is thus always less than that of a predetection integrator. Furthermore, noncoherent integration cannot preserve information, such as Doppler data, that is already lost.

The ease of implementing a postdetection pulse integrator usually outweighs any advantages achieved from the improvement in integration efficiency that would be obtained by the use of a predetection pulse integrator. Postdetection pulse integration, therefore, is usually implemented, although not ideally preferred. In either case, however, if the output of the receiver can be observed over a period of several pulse returns, it is always advantageous to integrate the output.

The i.f. filter that precedes the video postdetection integrator is ideally the filter matched for a single pulse. This filter cannot be exactly realized in practice, but it is feasible to approach it with standard circuit techniques. A feasible filter matched to a radar transmitting a rectangularly shaped pulse can be characterized as having a bandwidth B, approximately the reciprocal of the pulse width T_P. Consequently $BT_\mathrm{P} \cong 1$, or to be more exact, $BT_\mathrm{P} = 1.2$ (see Section 1-6 and Fig. 1-5). The output of a matched-filter receiver is mathematically the cross correlation of the received waveform and the impulse response of the filter.

The standard integrator is a device which linearly adds N samples of the voltage output from the video detector. When the sum of N such samples of signal-plus-noise exceeds a bias level, then a signal is said to be detected. The bias level is calculated from the probability density function for the sum of N samples of noise alone so as to give a specific false alarm rate. The weighting function which the integrator applies to a set of pulses is called the law of the integrator. If the weighting function applied to each pulse in a processing period is constant, the integrator is said to have a linear law. This is the usual case. If, however, the received pulse train has been modulated, as for instance, by the antenna beam pattern while the antenna is scanning (see Sections 3-13 and 3-14), then the optimum integrator is one that is "matched" to the envelope of the received pulse train. In other words, if possible, the "matched-filter" concept should be applied to the received pulse train as a whole. The most common example of nonoptimum integration is that provided by a cathode ray tube when traces are superimposed and the resultant image is intensity modulated.

Marcum [3] defines a detector as any device whose instantaneous output is a function of only the envelope of the input signal; that is,

$$y = F(v) \tag{2-29}$$

where

y is the signal-plus-noise output of the detector,

$v = A/(\psi_0)^{1/2}$ the normalized amplitude of the signal-plus-noise envelope,

A the instantaneous amplitude of the signal-plus-noise envelope, and

$(\psi_0)^{1/2}$ the rms noise voltage to which A is normalized.

The law of the linear integrator follows the law of the detector. If the detector output can be expressed as Eq. (2-29), then the integrator output is

$$Y = \sum_{1}^{N} \phi(y) = \sum_{1}^{N} \phi[F(v)] \qquad (2\text{-}30)$$

Mathematically, the only function of importance is

$$\psi(v) = \phi[F(v)] \qquad (2\text{-}31)$$

There is an infinite number of combinations of ϕ and F functions which can produce the same function ψ and hence the same theoretical results. Marcum [3] lets the output of the combination of the detector and integrator for one independent pulse be called $y = \psi(v)$, and the sum of N variates

$$Y = \sum_{1}^{N} y \qquad (2\text{-}32)$$

Then he employs the characteristic function $C(p)$ as the first step in an attempt to obtain a general solution for P_{D} and P_{fa}:

$$C_1(p) = \int_{-\infty}^{\infty} f(a, y) e^{-py}\, dy \qquad (2\text{-}33)$$

where $f(a, y)$ is the generalized probability density function for a single pulse, and a the peak amplitude of the desired signal normalized to $(\psi_0)^{1/2}$. In $f(a, y)$, y is considered to be the variable. It is the normalized instantaneous detector output voltage for a signal-plus-noise pulse. The term a is considered a constant and only modifies the shape of the probability density function for y. The term p is a transformation variable†:

$$p = \sqrt{-1}\, 2\pi f \qquad (2\text{-}34)$$

where f is the frequency variable. The characteristic function for the sum of N identical pulses is then

$$C_N = (C_1)^N \qquad (2\text{-}35)$$

and its probability density function is

$$dP_N = dY\left(\int_{-i\infty}^{+i\infty} C_N(p)\, e^{+pY} \frac{dp}{2\pi i}\right) \qquad (2\text{-}36)$$

† In most references the characteristic function is defined such that $p = -i2\pi f$, where $i = \sqrt{-1}$ and $2\pi f$ is sometimes equal to ω or t. In Campbell and Foster [9] $p = +i2\pi f$. These differences do not matter as long as one is consistent. Campbell and Foster's transforms are followed throughout this text.

The same probability density function in terms of Campbell and Foster's anticharacteristic function [9] is

$$dP_N = G(a, N, Y)\, dY \qquad (2\text{-}37)$$

Similarly, the probability of false alarm as determined from Eq. (2-36) is

$$P_{\mathrm{fa}} = \int_{Y_{\mathrm{b}}}^{\infty} G(0, N, Y)\, dY \qquad (2\text{-}38)$$

and the probability of detection is

$$P_{\mathrm{D}}(N) = P_N = \int_{Y_{\mathrm{b}}}^{\infty} G(a, N, Y)\, dY \qquad (2\text{-}39)$$

If Y_{b} is eliminated from Eqs. (2-38) and (2-39), the solution for the probability of detection P_{D} is a function in terms of the probability of false alarm, the number of pulses integrated N, and the amplitude of the desired signal normalized to the rms noise level a. In most cases the integrations required for Eqs. (2-32)–(2-39) are not feasible in terms of known functions. This is the reason Meyer Plots are useful.

2-7 Illustrative Examples for Determining Probability of Detection Using Meyer Plots

PROBLEM To determine the signal-to-noise power ratio (S/N) required for a radar receiver using a square-law detector to achieve a 90% probability of detection at a Marcum false alarm number of $n = 10^6$. A propeller aircraft target, where the propellers contribute a significant portion of the echo area, is assumed, and therefore the Swerling Case-2 target model (Rayleigh distribution with pulse-to-pulse fluctuation) is chosen. (The use of target models is described in Chapter 3.) The 90% probability of detection is to be obtained for both the cases of 10 pulses integrated and 1000 pulses integrated.

From the statement of the problem, n is fixed, and therefore the false alarm time is fixed. Fehlner's false alarm number n', however, is a function of the number of pulses integrated; that is,

$$n' = n/N \qquad (2\text{-}40)$$

Therefore, for $N = 10$

$$n' = 10^5 \qquad (2\text{-}41)$$

and for $N = 1000$

$$n' = 10^3 \qquad (2\text{-}42)$$

Table 2-5

REQUIRED (S/N) FOR TARGET MODEL CASE 2 AND $P_D(N) = P_N = 0.90$ IF n IS CONSTANT

Marcum's false alarm number (n)	Fehlner's false alarm number (n')	Number of pulses integrated (N)	Signal-to noise ratio (S/N) (dB)
10^6	10^5	10	5.8
10^6	10^3	1000	−8.2

Table 2-6

REQUIRED (S/N) FOR TARGET MODEL CASE 2 AND $P_N = 0.90$ IF n' IS CONSTANT

Marcum's false alarm number (n)	Fehlner's false alarm number (n')	Number of pulses integrated (N)	Signal-to-noise ratio (S/N) (dB)
10^7	10^6	10	6.3
10^9	10^6	1000	−6.9

Solution Find the Meyer Plots corresponding to Target Model Case 2 and false alarm numbers n' equal to 10^5 and 10^3. The false alarm number is shown in the upper left corner of the plot and the target model case in the upper right. The values of (S/N) obtained for $P_N = 0.90$ are tabulated in Table 2-5.

Alternatively, if Fehlner's false alarm number is maintained constant at $n' = 10^6$, then from the Meyer Plot for Target Model Case 2 and $n' = 10^6$, the values of (S/N) for $P_N = 0.90$ are as shown in Table 2-6.

2-8 Integration Loss

It has already been noted that noncoherent integration is less efficient than coherent integration. Marcum treats this property of noncoherent integration as a loss in (S/N) relative to what would have been obtained using coherent integration [3, Eq. 100a]. This was done by defining integration loss, in decibels, for the case when all pulses are of equal amplitude:

$$L_i = 10 \log_{10} \frac{N(S/N)_2}{(S/N)_1} \tag{2-43}$$

where

N is the number of pulses noncoherently integrated,

$(S/N)_1$ the required value of signal-to-noise power ratio to produce a given probability of detection for $N = 1$, and

$(S/N)_2$ the required value of signal-to-noise power ratio to produce the same probability of detection for $N = N$.

The integration loss L_i is a function of probability of detection and false alarm number; however, this dependency is small. In the case of coherent integration, the signal-to-noise ratio of the result of N pulse integrations for a specific probability of detection is theoretically always equal to $1/N$ times the signal-to-noise ratio of a single pulse that produces the same probability of detection. The integration loss for this case, therefore, is considered to be always zero; that is, when

$$(S/N)_2 = (S/N)_1/N \tag{2-44}$$

then

$$L_i = 0 \tag{2-45}$$

For noncoherent integration (postdetection integration), however,

$$(S/N)_2 > (S/N)_1/N \tag{2-46}$$

The integration efficiency variable γ is often used instead of integration loss. These two measures of integration loss are related as follows:

$$N^\gamma = N/L_i \tag{2-47}$$

or, in decibels,

$$L_i = 10(1-\gamma) \log_{10} N \tag{2-48}$$

The parameter γ will always be less than 1; when only a small number of pulses are integrated, the integration efficiency will be between 75–90% depending upon the probability of detection and the probability of false alarm. [This can be verified by analysis of the Meyer Plots presented in Chapter 4 using Eq. (2-48)—see Section 2-9.] When a large number of pulses are integrated, the difference between postdetection and predetection pulse integration is larger; postdetection integration efficiency decreases to approximately 50–60%, again depending on the probability of detection and the probability of false alarm one is working with. The combination of a high probability of false alarm (e.g., 10^{-2}) and a low probability of detection (e.g., 50%) results in a lower efficiency of integration. Alternatively, a high probability of detection (e.g., 90–99%) and a low probability of false alarm (e.g., 10^{-8}–10^{-10}) results in a higher integration efficiency.

2-9 Illustrative Examples for Determining Integration Loss and Integration Efficiency Using Meyer Plots

A radar system is specified to have a 90% probability of detection for a nonfluctuating target model and a probability of false alarm of 6.93×10^{-9}. It noncoherently integrates 10, 100, or 10,000 pulse returns.

PROBLEM To determine the integration loss for both coherent and noncoherent integration for each of 10, 100, or 1000 pulse integrations.

Solution Based on Eq. (2-43), the integration loss is

$$L_i = 10 \log N + 10 \log_{10}(S/N)_2 - 10 \log_{10}(S/N)_1 \quad (2\text{-}49)$$

From Meyer Plot Case 0, false alarm number 10^8 (see Table 2-4); from Eq. (2-49); and for 10 pulses being integrated, one obtains the following integration loss:

$$L_i = 10 + 6.2 - 14.25 = 1.95 \quad \text{dB} \quad (2\text{-}50)$$

For 100 pulses integrated,

$$L_i = 20 - 0.4 - 14.25 = 5.35 \quad \text{dB} \quad (2\text{-}51)$$

and for 10,000 pulses integrated,

$$L_i = 40 - 11.4 - 14.25 = 14.35 \quad \text{dB} \quad (2\text{-}52)$$

By definition the integration loss for coherent integration is 0 dB for all three cases.

PROBLEM To determine the integration efficiencies for the preceding problem.

Solution Equation (2-48) gives integration loss in terms of integration efficiency and the number of pulses being integrated; thus

$$L_i = 10(1-\gamma) \log_{10} N \quad (2\text{-}53)$$

Therefore,

$$\gamma = \frac{(10 \log_{10} N) - L_i}{10 \log_{10} N}(100) \quad (2\text{-}54)$$

For 10 pulses integrated one obtains the following integration efficiency:

$$\gamma = \frac{10 - 1.95}{10}(100) = 80.5\% \quad (2\text{-}55)$$

For 100 pulses integrated,

$$\gamma = \frac{20 - 5.35}{20}(100) = 73.25\% \quad (2\text{-}56)$$

and for 10,000 pulses integrated,

$$\gamma = \frac{40 - 14.35}{40}(100) = 64.13\% \quad (2\text{-}57)$$

2-10 Collapsing Loss

Collapsing loss L_c can be defined as the ratio of the signal-to-noise ratio required with M extra noise pulses, when noncoherently integrating N signal-plus-noise pulses, to the signal-to-noise ratio required with no extra noise pulses, such that the probability of detection is the same in both cases. The definition expressed mathematically in power decibels (dB) is

$$L_c = 10 \log_{10}\left(\frac{S}{N}\right)_{N+M} - 10 \log_{10}\left(\frac{S}{N}\right)_N \quad (2\text{-}58)$$

where

$(S/N)_{N+M}$ is the signal-to-noise ratio required with N signal-plus-noise pulses plus M extra noise pulses, and

$(S/N)_N$ the signal-to-noise ratio required with N signal-plus-noise pulses but no extra noise pulses.

The majority of operational radar detection devices have some collapsing loss. The loss can be caused by improper receiver bandwidth, improper range gate width, compression of three-dimensional data into two-dimensional data, excessive cathode-ray-tube spot size, or any other phenomenon that causes loss of resolution. (Table 2-9 gives examples of these.)

Collapsing loss occurs when noise originating outside the radar detection system resolution element that contains the signal is mixed, after envelope detection, with the signal and its asso-

Table 2-7

CHARACTERISTIC FUNCTIONS FOR MARCUM AND SWERLING TARGET MODELS SHOWING EFFECT OF COLLAPSING LOSS TERMS[a]

Case	Characteristic function for density function of N signal-plus-noise pulses[b]	Characteristic function for density function of M noise pulses plus N signal-plus-noise pulses (collapsing loss case)[c]	Comments
0	$C_N(p) = \dfrac{\exp\left[-Nx\left(\dfrac{p}{p+1}\right)\right]}{(p+1)^N}$	$C_{N+M}(p) = \dfrac{\exp\left[-N\rho\left(\dfrac{x}{\rho}\right)\left(\dfrac{p}{p+1}\right)\right]}{(p+1)^{N\rho}}$	Collapsing loss can be determined from Meyer Plots by determining $N\rho$ and (x/ρ). See Section 2-11 for illustration of technique
1	$\bar{C}_N(p) = \dfrac{1}{(1+p)^{N-1}(1+p(1+N\bar{x}))}$	$\bar{C}_{N+M}(p) = \dfrac{1}{(1+p)^{N\rho-1}\{1+p[1+N\rho(\bar{x}/\rho)]\}}$	Collapsing loss can be determined from Meyer Plots by determining $N\rho$ and $(\bar{x}/\rho)$. See Section 2-11 for illustration of technique
2	$\bar{C}_N(p) = \dfrac{1}{[1+p(1+\bar{x})]^N}$	$\bar{C}_{N+M}(p) = \dfrac{1}{(1+p)^{N(\rho-1)}[1+p(1+\bar{x})]^N}$	Collapsing loss can be determined if the procedure described in Section 4-3 is used to keep the degrees of freedom of the target model equal to N
3	$\bar{C}_N(p) = \dfrac{1}{(1+p)^{N-2}\{1+p[1+(N\bar{x}/2)]\}^2}$	$\bar{C}_{N+M}(p) = \dfrac{1}{(1+p)^{N\rho-2}\left\{1+p\left[1+\dfrac{N\rho(\bar{x}/\rho)}{2}\right]\right\}^2}$	Collapsing loss can be determined from Meyer Plots by determining $N\rho$ and $(\bar{x}/\rho)$. See Section 2-11 for illustration of technique
4	$\bar{C}_N(p) = \dfrac{(1+p)^N}{\{1+p[1+(\bar{x}/2)]\}^{2N}}$	$\bar{C}_{N+M}(p) = \dfrac{(1+p)^{N(2-\rho)}}{\{1+p[1+(\bar{x}/2)]\}^{2N}}$	Collapsing loss can be determined if the procedure described in Section 4-3 is used to keep the degrees of freedom of the target model equal to $2N$

[a] It should be noted that in this table x represents the single-pulse signal-to-noise power ratio often represented as (S/N) and that $\bar{x}$ is its average value for the fluctuating cases.

[b] The characteristic functions presented in column 2 are from Swerling [10].

[c] In determining the column 3 equations, the column 2 functions were multiplied by the characteristic function for M noise pulses. The characteristic function for a single-noise variate is $1/(1+p)$ and for M noise variates $1/(1+p)^M$.

ciated noise. Marcum defined the following collapsing ratio to handle this effect [3, Eq. 226]:

$$\rho = (M+N)/N \quad (2\text{-}59)$$

where

N is the number of signal-plus-noise pulses integrated, and

M the effective number of additional noise pulses integrated.

Mathematically, to determine the effect of M extra noise pulses, it is only necessary to multiply the characteristic function for N signal-plus-noise pulses by the characteristic function for M noise pulses alone. These characteristic functions are of the form shown in Table 2-7 if a square-law detector is used. The table gives for the five standard target models the characteristic function for N signal-plus-noise pulses and the characteristic function for N signal-plus-noise pulses combined with M additional noise pulses.

Examination of Table 2-7 reveals that as $M \to 0$ or $\rho \to 1$, the equations of the characteristic functions that include a collapsing factor become the same as the equations of the characteristic function having no collapsing loss factor. The table also reveals that only for Target Model Cases 0, 1, and 3 can Meyer Plots be used directly to determine collapsing loss. For Target Model Cases 2 and 4 (the pulse-to-pulse fluctuation cases), in addition to making the parameter substitutions described below, it is necessary to modify the degrees of freedom of the target model using the technique described in Chapter 4, Section 4-3.

Marcum's false alarm number n is related to Fehlner's false alarm number n' for the case of no collapsing loss by

$$n = n'N \quad \text{(no collapsing loss case)} \quad (2\text{-}60)$$

In the presence of collapsing loss, however, this relationship is usually not valid. For certain types of collapsing loss n remains constant, while n' is reduced as a function of increasing ρ. Conversely, for other types of collapsing loss the reverse condition exists. Table 2-6 notes for each type of collapsing loss listed whether n or n' remains constant as ρ increases. Marcum, in order to account for the effects of collapsing loss, found it necessary to revise the relationship between n and n', as follows [3]:

$$n = n'N\rho \quad \text{(collapsing loss case)} \quad (2\text{-}61)$$

For the case where n remains constant, this result is solved for n':

$$n' = n/\rho N \quad (2\text{-}62)$$

In order to use Meyer Plots when collapsing loss exists, the parameter substitution indicated in Table 2-7 must be made. In other words, the procedure is as follows:

(1) Calculate n' using Eq. (2-62). For the case where one wishes to keep Fehlner's false alarm number n' constant, this step is omitted.

(2) Use ρN instead of N as the number of pulses integrated.

(3) Multiply the signal-to-noise ratio thus obtained from the Meyer Plot by ρ.

(4) For Target Model Cases 2 and 4 the number of pulses integrated is chosen to be ρN as stated in (2), but the number of degrees of freedom of the chi-square target model must remain $K = kN$, where $k = 1$ for Case 2 and $k = 2$ for Case 4. In addition, step (3) is skipped and $\bar{x}$ is read directly off the Meyer Plots. The reader will not appreciate the significance of this until he has reviewed the sections (Sections 3-8 and 4-3) of this text relating to the general chi-square target model. In general, whenever collapsing loss must be determined for a target that fluctuates on a pulse-to-pulse basis (completely uncorrelated pulses), the procedure described in Section 4-3 must be employed with ρN as the number of pulses integrated and $k = kN$ as the number of degrees of freedom. For block-correlated pulse returns (see Section 4-3) $K = kF$, where F is the number of uncorrelated blocks of N/F correlated pulses.

Table 2-8 summarizes which of the procedural steps must be invoked in order to calculate collapsing loss using Meyer Plots.

Table 2-8

Target Model	Constant n problem	Constant n' problem
Cases 0, 1, and 3 (constant or completely correlated pulse returns)	Steps 1, 2, and 3	Steps 2 and 3
Cases 2 and 4 (completely uncorrelated pulse returns)	Steps 1, 2, and 4	Steps 2 and 4
Block-correlated pulse returns	As above but $K = kF$	As above but $K = kF$

Table 2-9

COLLAPSING RATIO FOR SPECIFIC CAUSES

Cause (description and use)	Equation[a]	Definition of terms
General form of collapsing ratio ρ	$\rho_1 = \dfrac{M+N}{N}$	M number of effective additional noise pulses integrated N number of signal-plus-noise pulses integrated
Video		
Collapsing ratio when collapsing loss is caused by insufficient video bandwidth. For this type of collapsing loss, n remains constant as ρ increases[b]	$\rho_2 = \dfrac{2B_v + B}{2B_v}$	B_v video bandwidth (narrower than one-half the matched i.f. bandwidth) B $1/T_P$ where T_P is the pulse width
Collapsing ratio when collapsing loss is caused by excessive bandwidth. Use when radar range equation is based on pulse energy or average power and the i.f. bandwidth does not appear. The loss caused by excessive bandwidth can be accounted for by introducing a filter matching loss when threshold detection is dependent upon peak amplitude. A collapsing loss replaces the matching loss when the video bandwidth is made to match the spectrum of the detected signal. For this type of collapsing loss, n remains constant as ρ increases[b]	$\rho_3 = \dfrac{2B_v + B_n}{2B_v}$	B_n noise bandwidth of receiver (wider than $1/T_P$, the matched i.f. bandwidth. For this case $B_v = 1/T_P$)
Collapsing ratio when collapsing loss is caused by the excessive width of a fixed range gate used to detect target returns from a given range interval. For this type of collapsing loss, n remains constant as ρ increases. It should be noted that relatively long gates in some instances permit integration of a sufficient number of pulses from target returns from a fast moving target as it passes through the gate to partially or more than compensate for the collapsing loss	$\rho_4 = \tau_g/T_P$	τ_g width of gate T_P pulse width
Collapsing ratio when collapsing loss is caused by the slow sweep speed of a cathode ray tube. For this type of collapsing loss, n remains constant as ρ increases	$\rho_5 = \dfrac{d + sT_P}{sT_P}$	d spot diameter s sweep speed
C-scope (display elevation angle vs. azimuth angle)		
Collapsing loss due to the use of two-dimensional displays where the integration time exceeds the time on target. For this type of collapsing loss, n remains constant as ρ increases	$\rho_6 = \Delta R/T_P$	ΔR range search interval expressed in units of range delay time

Table 2-9—continued

Cause (description and use)	Equation[a]	Definition of terms
Collapsing loss when a narrow range gate scans the range search interval. For this type of collapsing loss, n remains constant as ρ increases	$\rho_7 = \dfrac{V_s t_{vi}}{T_P}$	V_s rate at which gate is scanned t_{vi} video integration time = N/PRF N number of pulses integrated PRF pulse repetition frequency
B-scope or PPI (range vs. azimuth)		
Collapsing loss due to the use of a two-dimensional display when the integration time exceeds the time on target. For this type of collapsing loss, n remains constant as ρ increases	$\rho_8 = \dfrac{\omega_e t_{vi}}{B_e}$	B_e elevation beam width ω_e elevation scan speed
Collapsing loss due to the use of two-dimensional display when the search interval is covered in less than the integration time. For this type of collapsing loss, n remains constant as ρ increases	$\rho_9 = \dfrac{\Delta e}{B_e}$	Δe elevation search interval = $E_{max} - E_{min}$
E-scope or RHI (display elevation angle vs. range)		
Collapsing loss due to the use of a two-dimensional display when the integration time exceeds the time on target. For this type of collapsing loss, n remains constant as ρ increases	$\rho_{10} = \dfrac{\omega_a t_{vi}}{B_a}$	B_a azimuth beam width ω_a azimuth scan speed
Collapsing loss due to the use of a two-dimensional display when the search interval Δa is covered in less than the integration time. For this type of collapsing loss, n remains constant as ρ increases	$\rho_{11} = \Delta a/B_a$	Δa azimuth search interval
Video mixing		
Mixing and summing receiver outputs, etc. For this type of collapsing loss, n' remains constant as ρ increases		If the signal is received in only one receiver channel of a multichannel system and two or more of these channels are mixed or summed and applied to a single threshold detector a collapsing loss occurs. The collapsing loss ratio is proportioned to the number of receivers whose outputs are mixed. This collapsing loss also exists for receivers connected to separate antenna feeds as well as receivers having parallel Doppler filter channels as inputs to a single detection device

[a] ρ_2 thru ρ_{11} are from Barton [11].

[b] Barton has replaced Marcum's [3] B_v with $2B_v$. Skolnik has taken exception to ρ_2 and ρ_3 as given herein because ρ does not approach unity when, respectively, B or B_n is approximately $2B_v$ [12, pp. 2–60]. Actually, some collapsing loss should occur for this condition ($\rho = 2$), as overall bandwidth is reduced when networks of the same bandwidth are cascaded. The condition $\rho = 2$ typically leads to collapsing losses less than 1 dB (see Section 2-11).

In any system, there may be more than one source of collapsing loss. Where the collapsing losses are all of the same type, constant n or constant n', the overall collapsing loss will be somewhat less than the sum of the separate losses. If losses of both constant n and n' types exist in a given system, then the collapsing loss for each type of loss should be determined separately. The overall collapsing loss will then be somewhat less than the sum of the constant n and constant n' collapsing loss subtotals.

Table 2-9 lists 12 of the most common sources of collapsing loss. For conditions not specified in this table, the existence of a collapsing loss can be established by answering the question: Is the signal sample forced to compete with more than one noise sample? If it is, then a collapsing loss exists.

Examples of the use of Meyer Plots for solving typical collapsing loss problems are presented in Section 2-11.

2-11 Illustrative Example for Determining Collapsing Loss Using Meyer Plots

For convenience these examples will be confined to the use of the general form of the collapsing ratio, that is,

$$\rho = (M+N)/N \tag{2-63}$$

Both types of collapsing loss will be considered: constant n' and constant n. All examples will be for Target Model Case 0; however, collapsing loss can also be obtained from Meyer Plots for Cases 1 and 3 by means of the same procedure.

PROBLEM A detection system is required to have a 90% probability of detection for a non-fluctuating target (Case 0) and a Marcum false alarm number of 10^7. Determine the required signal-to-noise ratio for 10 pulses integrated ($N = 10$).

Solution From Eq. (2-12), it is found that

$$n' = n/N = 10^7/10 = 10^6 \tag{2-64}$$

In Chapter 4, from the Meyer Plot that corresponds to Case 0, false alarm number $n' = 10^6$ and $N = 10$, the required signal-to-noise ratio is found to be 5.3 dB for a probability of detection of 90%.

PROBLEM Determine the collapsing ratio ρ for 10 additional noise pulses ($M = 10$).

Solution From Eq. (2-63) the collapsing ratio (ρ) is 2.

PROBLEM Determine the resulting collapsing loss for a collapsing ratio of 2.

Solution Basically, there are two types of collapsing loss and both will be considered here. For either type of collapsing loss the number of pulses integrated must be modified by the factor ρ; accordingly $\rho N = 2(10) = 20$.

(a) For the type of collapsing loss where n' remains constant, the amount of collapsing loss that occurs is determined as follows:

In Chapter 4, from the Meyer Plot for Case 0 where the false alarm number $n' = 10^6$ and $N = 20$, the indicated signal-to-noise ratio for a probability of detection of 90% is 3.2 dB. Since this value is $(S/N)/\rho$, and since $\rho = 2$, the actual signal-to-noise required ratio is $3.2+3 = 6.2$ dB. Using Eq. (2-58), we find the following collapsing loss:

$$L_c = 6.2 - 5.3 = 0.9 \quad \text{dB} \tag{2-65}$$

(b) Now, the second type of collapsing loss where the false alarm number n is constant is determined. By definition, if the false alarm number n is constant, a variation in the probability of false alarm and n' exists.

The false alarm number n', adjusted for collapsing loss while maintaining a constant false alarm number n, is found from Eq. (2-62):

$$n' = \frac{n}{\rho(N)} = \frac{10^7}{2(10)} = 5 \times 10^5 \tag{2-66}$$

For $n' = 5 \times 10^5$, interpolating between Case 0 Meyer Plots $n' = 4 \times 10^5$ and $n' = 6 \times 10^5$, the signal-to-noise ratio indicated for $N = 20$ and a probability of detection of 90% is 3.1 dB (see Section 4-2 relating to methods of interpolation). Again, since this value is for $(S/N)/\rho$, and since $\rho = 2$, the actual signal-to-noise ratio required is $3.1+3 = 6.1$ dB. Again, by using Eq. (2-58) the collapsing loss is as follows:

$$L_c = 6.1 - 5.3 = 0.8 \quad \text{dB} \tag{2-67}$$

References

1. J. I. Marcum, Douglas Aircraft Co. Rep. No. RA-15061.
2. J. I. Marcum, Studies of Target Detection by a Pulsed Radar. Rand. Res. Memo, RM-754 (Dec. 1947); reprinted in *IRE Trans. Inform. Theory* **IT-6** (April 1960); also AD-101287.
3. J. I. Marcum, Studies of Target Detection by a Pulsed Radar—Mathematical Appendix. Rand Res. Memo RM-753 (July 1948); reprinted in *IRE Trans. Inform. Theory* **IT-6** (April 1960); also AD-101882.
4. S. M. Kaplan and R. W. McFall, The Statistical Properties of Noise Applied to Radar Performance. *Proc. IRE* **39**, 56–60 (1951).
5. R. Hollis, False Alarm Time Pulse Radar. *Proc. IRE* (Correspondence) **42**, 1189 (1954).
6. K. Pearson, "Table of Incomplete Gamma Functions." Cambridge Univ. Press, London and New York, 1946.
7. J. Stirling, "Methods Differentialis." 1730.
8. J. Pachares, A Table of Bias Levels Useful in Radar Detection Problems. *IRE Trans. Inform. Theory* **IT-4**, No. 1, 38–45 (1958).
9. G. A. Campbell and R. M. Foster, "Fourier Integrals for Practical Applications." Van Nostrand Reinhold, Princeton, New Jersey, 1947; Bell Telephone Syst. Monograph No. B-584.
10. P. Swerling, Probability of Detection for a Fluctuating Target. Rand Res. Memo, RM-1217 (March 17, 1954); reprinted in *IRE Trans. Inform. Theory* **IT-6** (1960).
11. D. K. Barton, "Radar System Analysis." Prentice-Hall, Englewood Cliffs, New Jersey, 1964.
12. M. I. Skolnik (ed.), "Introduction to Radar Systems," McGraw-Hill, New York, 1962.

Recommended reading for further study in addition to the references cited above:

J. V. DiFranco and W. L. Rubin, "Radar Detection." Prentice-Hall, Englewood Cliffs, New Jersey, 1968.

CHAPTER 3

TARGET MODELS AND THE DETERMINATION OF DETECTION PROBABILITY

3-1 Introduction and Historical Background

Historically, the earliest descriptions of a target were in terms of a single cross-sectional area value.† This quantity was usually some type of an average cross section over the aspect angles which the system designer considered most probable. The approach led system designers to associate a single cross-sectional area of unvarying value, with a target and thus led to the model of a steady-state target. Following this concept systems were designed to achieve a somewhat arbitrarily specified signal-to-noise ratio at some specified maximum operating range. The inherent basic assumption was that the required detection capability and parameter estimation accuracy could be achieved if this goal were met. The dominant role of the human observer in early detection systems tended to make this approach acceptable. As the requirements on radar systems became more demanding, and as automatic processing of radar returns was developed, the situation underwent a change. The desire to optimize radar designs established the need for more precise methods of analysis.

The initial work of J. I. Marcum of the RAND Corporation in the late 1940s applied the work of Rice [2] of the Bell Telephone Laboratories, relating to steady-state signals immersed in noise, to the problem of radar signal processing. Marcum effectively gave a complete treatment to the statistical problem of a group of *constant amplitude* signal pulses in the presence of noise [3, 4]. His work resulted in an evaluation of previously untabulated functions and a direct application of the results to a gamut of problems in automatic detection. The statistical approach used in Marcum's studies is the basis for much of the work that followed.

Radar system behavior with respect to the detection of signals resulting from a fluctuating target is basically different than for signals resulting from a nonfluctuating target. A requirement remained, therefore, to produce the same type of analysis and supporting tabulation of basic functions for a fluctuating target as had been done by Marcum for the nonfluctuating target. This was done by Swerling [5], in the early 1950s. He extended Marcum's approach by employing for target cross-sectional area two different density functions in conjunction with two extremes of correlation.

Swerling's Cases 1 and 2 are the target models which describe large complex targets, such as aircraft. They represent two extremes of correlation, and the statistical model is derivable from the physical scattering characteristics of such bodies. It can also be stated that empirically derived data on such targets are in very good agreement with those obtained from the mathematical model (see Section 3-6).

At the time Swerling was doing his work, it was realized that the model suitable for large complex targets would not give an adequate description of large simple structures such as terminal tankage from a missile. The problem of selecting a model to describe this type or class of target is a difficult one to handle directly. There is no obvious parallel development for it starting from the scattering characteristics of a specific body type. The approach employed was to establish qualitatively the basic statistical behavior of the target cross sections of interest. Having done this, a convenient form of distribution was selected.

Swerling picked a class of distributions known as the chi-square class, which has the aircraft density function distribution as one member. By an appropriate selection of a class parameter,

† Historical comments are based on Weinstock [1].

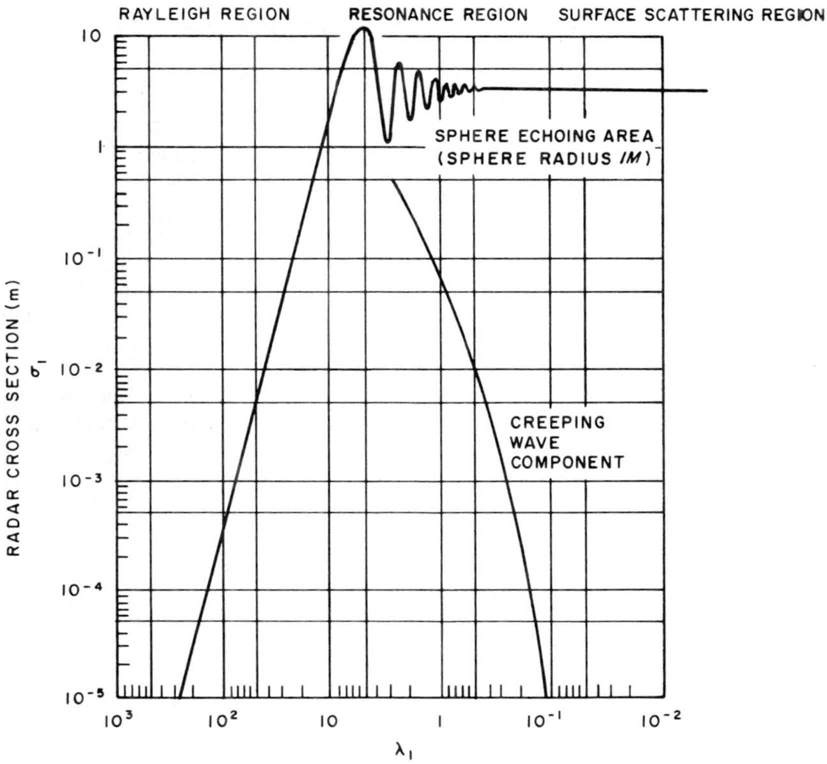

Fig. 3-1 Radar cross section of a spherical target versus wavelength. [Reprinted with permission from J. S. Hey, H. Gent, and P. G. Smith, Echoing Area Characteristics, from "Avionics Research: Satellites and Problems of Long Range Detection and Tracking," published for the Advisory Group for Aerospace Research and Development, NATO, by Pergamon Press, Oxford, 1960. Editors: E. V. D. Glazier, E. Richtin, and J. Boge.]

namely, the number of degrees of freedom, the qualitative properties desired for the distribution associated with large simple structures was obtained (see Section 3-7). Swerling's Cases 3 and 4 represent targets that behave as if they have four degrees of freedom and are valid for a stabilized cylinder.

Swerling's two classes of fluctuating targets evaluated at two extremes of correlation, together with Marcum's work, have been until recently the bases of virtually all radar systems analyses. The next few sections of this chapter present a review of this work. Later, in Section 3-8 several important new target models are described. Chapter 4 describes how Meyer Plots can be used to solve problems involving most any target model. In conjunction with this, a concise review of the properties of all target models of major interest is presented in Chapter 4, Table 4-1B.

3-2 Scattering Cross Section and the Target Modeling Problem

According to Barton there are three common terms used to describe a target's reflection characteristic, namely, the "effective echoing area," the "back-scattering coefficient," and, the most common in use, the "radar cross-section."† The definition of the equivalent radar cross section (RCS) of a target is based on the assumption that the target is in free space at a range so distant that the incident radiation is effectively a plane wave in the area of the target. This equivalent echoing area or radar cross-sectional area is usually denoted as

† Barton [6, p. 66] notes that the IRE Dictionary gives "cross section" as a depreciated term referring to the equivalent echoing area of the target, but does not define "equivalent echoing area". Under "back scattering coefficient (echoing area)," he further notes that an appropriate definition is given, as paraphrased above. In this text "radar cross section" will be used exclusively except in the case of the discussion of bistatic radar systems.

Table 3-1

RADAR CROSS SECTION OF A SPHERICAL TARGET[a]

Type scattering	Target wavelength	Characteristic
Rayleigh scattering	Wavelength large compared with the target dimensions	In this region fine details of the target do not affect the echoing area
Resonance region	Wavelength of the same order as the target dimensions	Characteristics correspond to an oscillatory circuit consisting of a capacitance and inductance. In this region no detail or even target size may characterize the true target shape except in its relationship wavelength
Surface and edge scattering	Wavelength small compared with the target dimensions	Scattering can follow a target's shape in a very broad sense. For two scatters in this region σ is given by $\sigma = \sigma_1 + \sigma_2 + 2(\sigma_1 \sigma_2)^{1/2} \cos\left(\frac{4\pi d}{\lambda} + \phi\right)$ (3-2) where d is the difference in range between scatters, λ the wavelength, ϕ the phase of scattering field, and σ_1 and σ_2 the radar cross section of separate scatters

[a] Reprinted with permission from J. S. Hey, H. Gent, and P. G. Smith, Echoing Area Characteristics, from "Avionics Research: Satellites and Problems of Long Range Detection and Tracking," published for the Advisory Group for Aerospace Research and Development, NATO, by Pergamon Press, Oxford, 1960. Editors: E. V. D. Glazier, E. Richtin, and J. Boge.

σ, but sometimes denoted as RCS. The definition is as follows:

$$\sigma = 4\pi \left(\frac{\text{power per unit of solid angle scattered back to the receiver}}{\text{power per unit area incident on the target}} \right) \tag{3-1}$$

The above expression oversimplifies the scattering characteristic of a target. In reality σ depends on a large number of parameters. The polarization characteristic of any target's reflection characteristic for a monostatic system is a six-valued function of three independent variables; for a bistatic system it is an eight-valued function of five independent variables. The difficulty in obtaining a generalized solution for σ is that a detailed electromagnetic scattering problem has to be solved for each set of variables in order to determine a single value of radar cross section.

One of the simpler situations to evaluate is the case of a spherical target of fixed size with only the wavelength (frequency) of the radar system varying. It has been shown, for this case, that the scattering characteristic can be divided into three principal regions according to the relation of wavelength to target size (see Fig. 3-1 and Table 3-1) [7]. These are the Rayleigh region, the resonance region, and the surface scattering region. The major portion of this chapter concerns the third region, the one relating to surface or edge scattering. This is the region that prevails for a normal aircraft at normal radar frequencies.

3-3 Single-Hit Probability of Detection

The most commonly used criterion of system sensitivity is the single-hit detection probability

P_D, usually termed the "blip-scan ratio."† The calculations required to evaluate this probability of detection, when a signal from only one correlation interval is of interest, are simple. A single correlation interval for a pulsed system is simply the duration of a single pulse; for a continuous-wave (cw) system it is equivalent to observing the output for a period of time defined by

$$t = 1/B \tag{3-3}$$

where

B is the overall effective bandwidth of the radar receiver.

In either case, pulse or cw, if the output resulting from one correlation interval exceeds the bias level, a signal is assumed to be detected. It should be noted that Marcum has shown that for single-hit detection, the probability of detecting a given signal-plus-noise radar return is independent of the type of detector, everything else being held constant.

As a basic assumption preliminary to developing the equations for probability of detection, Marcum [4, Eq. 1] let a normal distribution characterize the thermal noise voltage in resistive elements, the noise voltage due to cosmic sources, and the noise voltage due to the shot effect in vacuum tubes. The probability density function has the form

$$dP = \frac{\exp(-V^2/2\psi_0)\, dV}{(2\pi\psi_0)^{1/2}} \tag{3-4}$$

where

V is the instantaneous video noise voltage, and
ψ_0 the mean-square value of the noise voltage.

The above is valid only if all the elements involved in the generation and processing of the total noise voltage are linear.

If the noise energy is passed through a linear filter having a passband that is narrow as compared to its center frequency, the output *envelope* will have a probability density function of the form

$$dP = \frac{A \exp(-A^2/2\psi_0)\, dA}{\psi_0} \tag{3-5}$$

where A is the instantaneous amplitude of the envelope. If a sine wave having the same center frequency as the filter is now added to the above noise energy, the output envelope's probability density function is modified by S and has the form

$$dP = \frac{A \exp[-(A^2+S^2)/2\psi_0]}{\psi_0} I_0\left(\frac{AS}{\psi_0}\right) dA \tag{3-6}$$

for $A > 0$, and

$$dP = 0 \tag{3-7}$$

for $A < 0$, where

A is the amplitude of envelope as in Eq. (3-5),
S the peak amplitude of the sine wave at the output of the filter in the absence of noise, and
I_0 the modified Bessel function of the first kind of zero order [8].

The output envelope has a correlation time that is approximately equal to the reciprocal of the filter bandwidth. This means that the envelope will not vary by an appreciable percentage in a period of time that is a small fraction of the correlation time. It also means that for a period of time considerably greater than the correlation time, the change could be a significant percentage of the envelope amplitude. The fact of importance, however, is that for the approximate time period equal to the reciprocal of the filter bandwidth, samples of the output can be assumed to be statistically independent. While this approximation is not exact, it is sufficient to obtain results for most practical problems.

For the case of a pulsed radar the preceding assumption can be easily verified. The reciprocal of the filter bandwidth, $1/B$, can be written as k/B. Now, the implied change in correlation time has, for the most part, only the effect of changing the false alarm time t_{fa} by some constant factor k. Furthermore, examination of various single-hit detection curves will reveal that the probability of detection is an insensitive function of false alarm time, which is proportional to false alarm number. If k is a factor on the order of unity, therefore, the end result will be negligibly affected by assuming the correlation time to be $1/B$.

† An alternate definition of the "blip-scan ratio" is the ratio of the number of times that a particular target is observed as a blip on a scope at a particular range relative to the number of times it could have been observed at that range, based on the actual number of scans made.

As previously noted in Chapter 2 [Eq. (2-29)], a detector can be defined as any device whose instantaneous output is a function of the envelope of the input wave only, such that [4, Eqs. (5, 7, and 8)]

$$y = F(v) \tag{3-8}$$

and conversely

$$v = A/(\psi_0)^{1/2} = g(y) \tag{3-9}$$

where

y is the output of the detector, and
v the normalized amplitude of the envelope.

If in Eqs. (3-6) and (3-7) $S/(\psi_0)^{1/2}$ is replaced by a and $A/(\psi_0)^{1/2}$ is replaced by $g(y)$, then dP can be written in the form

$$dP = f(a, y)\, dy \tag{3-10}$$

from which generalized solutions for P_D and P_{fa} independent of the detector function $y = F(v)$ can be developed.

It has been experimentally shown that the thresholding of a signal at the output of a radar receiver video amplifier is not significantly affected by the video bandwidth until B_v becomes less than half the i.f. bandwidth. This effect of improper video bandwidth can be treated in an approximate manner by establishing a collapsing loss correction (see Sections 2-10 and 2-11; also see Tables 2-8 and 2-9). Usually, however, the video bandwidth is greater than half the i.f. bandwidth and, therefore, for simplicity, the effect of improper video bandwidth will not be considered in the development of equations for P_D and P_{fa}.

In Chapter 2 it was established that Marcum's false alarm time t_{fa} is the time in which the probability is $P(0) = 0.5$ that noise alone will not exceed the bias level, and that the probability of false alarm for a single pulse is

$$P_{fa}(1) = 1 - P(0)^{1/n} \tag{3-11}$$

and for N integrated pulses

$$P_{fa}(N) = 1 - P(0)^{N/n} \tag{3-12}$$

where

$$n = t_{fa} B$$

The probability of false alarm can also be expressed in terms of Eq. (3-10) if we let $a = 0$ and $y_b = Y_b$, the bias level:

$$P_{fa}(1) = \int_{Y_b}^{F(\infty)} f(0, y)\, dy \tag{3-13}$$

Returning to Eq. (3-5) and letting $v = A/(\psi_0)^{1/2}$, we note that Eq. (3-13) becomes

$$P_{fa}(1) = \int_{g(Y_b)}^{\infty} v \exp\left(-\frac{v^2}{2}\right) dv = \exp\left[-\frac{g^2(Y_b)}{2}\right] \tag{3-14}$$

and that solving this result for $g(Y_b)$ yields

$$g(Y_b) = \{-2 \log_e[P_{fa}(1)]\}^{1/2} \tag{3-15}$$

The probability of detection can be similarly determined from Eq. (3-10) by solving instead

$$P_D(1) = \int_{Y_b}^{F(\infty)} f(a, y)\, dy \tag{3-16}$$

and by using the solution for $g(Y_b)$ obtained above in conjunction with Eqs. (3-6) and (3-7). Thus

$$P_D(1) = \int_{\{-2 \log_e[P_{fa}(1)]\}^{1/2}}^{F(\infty)} v \exp\left[-\left(\frac{v^2 + a^2}{2}\right)\right] \times I_0(av)\, dv \tag{3-17}$$

It should be noted that this solution is independent of the detector function.

Equation (3-17) is in the form of a Q function; that is [8]

$$Q(\alpha, \beta) = \int_{\beta}^{\infty} \alpha \exp\left[-\left(\frac{\alpha^2 + \beta^2}{2}\right)\right] I_0(\alpha\beta)\, d\alpha \tag{3-18}$$

and thus the probability of detection for the single-hit case may be simply written

$$P_D(1) = Q(a, \{-2 \log_e[P_{fa}(1)]\}^{1/2}) \tag{3-19}$$

These solutions for P_D apply only to the case where the time available for detecting a target T_d permits only the observation of one signal return. It is common, however, for $T_d \gg 1/(\mathrm{PRF})$ and for groups of pulses to be integrated prior to thresholding. In general, therefore, the final or overall probability of detection is

$$P_D(N, \nu) = 1 - [1 - P_D(N)]^{\nu'} \quad (3\text{-}20)$$

where

$$\nu' = \nu/N = (\text{PRF})\,T_d/N \quad (3\text{-}21)$$

N is the number of pulses integrated prior to thresholding, and

T_d the time available for detection.

Only whole groups of N pulses count; therefore ν' is required to be an integer. Also, it will later become apparent that optimum radar performance is obtained when $\nu' = 1$, the condition generally assumed in this text, for which case $P_D(N, N)$ is sometimes written simply as $P_D(N)$ or P_N, and $P_D(1, 1)$ is written simply as $P_D(1)$ or P_1. The method of determining $P_D(N)$ is shown in Section 3-5. $P_D(1, \nu)$ is determined by use of Eqs. (3-19) and (3-20), the tables of Q functions available [8], and the following relationship derived from Eq. (1-54):

$$\frac{R}{R_0} = \frac{1}{x^{1/4}} = \frac{1}{(S^2/2\psi_0)^{1/4}} = \frac{2^{1/4}}{a^{1/2}} \quad (3\text{-}22)$$

where $x = a^2/2$, the power signal-to-noise ratio.

In Chapter 2 an excellent approximation for the probability of false alarm is given in Eq. (2-24), namely,

$$P_{fa} \cong -\frac{1}{n'} \log_e[P(0)] \quad (3\text{-}23)$$

which is valid for $n' \gg 1$, a condition always true in practice. For Marcum's definition of P_{fa}, $P(0) = 0.5$ and Eq. (3-23) becomes

$$P_{fa}(N) = 0.693/n' \quad (3\text{-}24)$$

For the case where $N = 1$, $n' = n$; therefore

$$P_{fa}(1) = 0.693/n \quad (3\text{-}25)$$

Now, by solving Eqs. (3-22) and (3-25) for a and $P_{fa}(1)$, respectively, Eq. (3-19) can be expressed entirely in terms of design parameters:

$$P_D(1) = Q\left\{\sqrt{2}\left(\frac{R_0}{R}\right)^2, [4.60 \log_{10}(n) + 0.733]^{1/2}\right\} \quad (3\text{-}26)$$

Thus far only the development of the single-hit probability of detection equations for Marcum's nonfluctuating target model (Case 0) has been presented. The development of similar equations for Swerling's fluctuating target models (Cases 1–4) is presented in Sections 3-6 and 3-7, but for the more general case where pulse integration prior to thresholding is employed. The single-hit versions of these equations will be just stated here for reference. Thus for Target Model Cases 1 and 2, the probability of detection for a single hit was found by Swerling [5, Eq. II.1] to be

$$P_D(1) = \exp\left[-\left(\frac{Y_b}{1 + \bar{x}}\right)\right] \quad (3\text{-}27)$$

where $\bar{x}$ is the *average* signal-to-noise ratio. Similarly, for Swerling's Cases 3 and 4, the probability of detection for a single hit is

$$P_D(1) = \left[1 + \frac{2Y_b\bar{x}}{(2 + \bar{x})^2}\right] \exp\left[-\frac{Y_b}{1 + (\bar{x}/2)}\right] \quad (3\text{-}28)$$

Table 3-2 lists and restates the equations needed to find the single-hit probability of detection for each of the five important target models. Also listed are the characteristic equations that describe the statistical structure of signal returns for each target model. Although these equations and the background to them presented in this section are pertinent to the understanding of the single-hit detection problem, they do not lead to easy solutions of real problems. Figures 3-2 through 3-16, however, give graphical solutions for the five target models.† These figures (single-hit detection curves) cover a range of probability of false alarm‡ from 6.93×10^{-2} to 6.93×10^{-12}. The running parameter is power signal-to-noise ratio, which for values of interest can be determined to an accuracy of better than 0.1 dB.

† These graphical solutions were first published in Meyer [9].

‡ It should be noted that these curves are actually expressed in terms of the false alarm number n', but this requires merely a simple conversion:

$$n' = 0.693/P_{fa} = n, \qquad N = 1$$

Table 3-2

CHARACTERISTIC EQUATION AND PROBABILITY OF DETECTION FOR SINGLE-HIT DETECTION

Case	Characteristic equation for the probability density function of a single pulse	Probability of detection for single hit	Comments
0	$C_1(p) = \dfrac{\exp\{-x[p/(p+1)]\}}{(p+1)}$	$P_D(1) = Q\{(2x)^{1/2}, [4.60 \log_{10}(n) + 0.733]^{1/2}\}$	$Q(\alpha,\beta)$ is the Q function of α and β [8] x the power signal-to-noise ratio, and n the false alarm number
1 and 2	$\bar{C}_1(p) = \dfrac{1}{1+p(1+\bar{x})}$	$P_D(1) = \exp\left(-\dfrac{Y_b}{1+\bar{x}}\right)$	It should be noted that the characteristic equation for single-pulse detection is the same for Cases 1 and 2. Refer to Table 2-7: If we let $N=1$ in the characteristic equation for N signal-plus-noise pulses for both Case 1 and Case 2, the resulting equation for $\bar{C}_1(p)$ is the same
3 and 4	$\bar{C}_1(p) = \dfrac{(1+p)}{\{1+p[1+(\bar{x}/2)]\}^2}$	$P_D(1) = \left[1 + \dfrac{2Y_b\bar{x}}{(2+\bar{x})^2}\right] \exp\left[-\dfrac{Y_b}{1+(\bar{x}/2)}\right]$	It should be noted that the characteristic equation for single-pulse detection is the same for Cases 3 and 4. Refer to Table 2-7: If we let $N=1$ in the characteristic equation for N signal-plus-noise pulses for both Case 3 and Case 4, the resulting equation for $\bar{C}_1(p)$ is the same

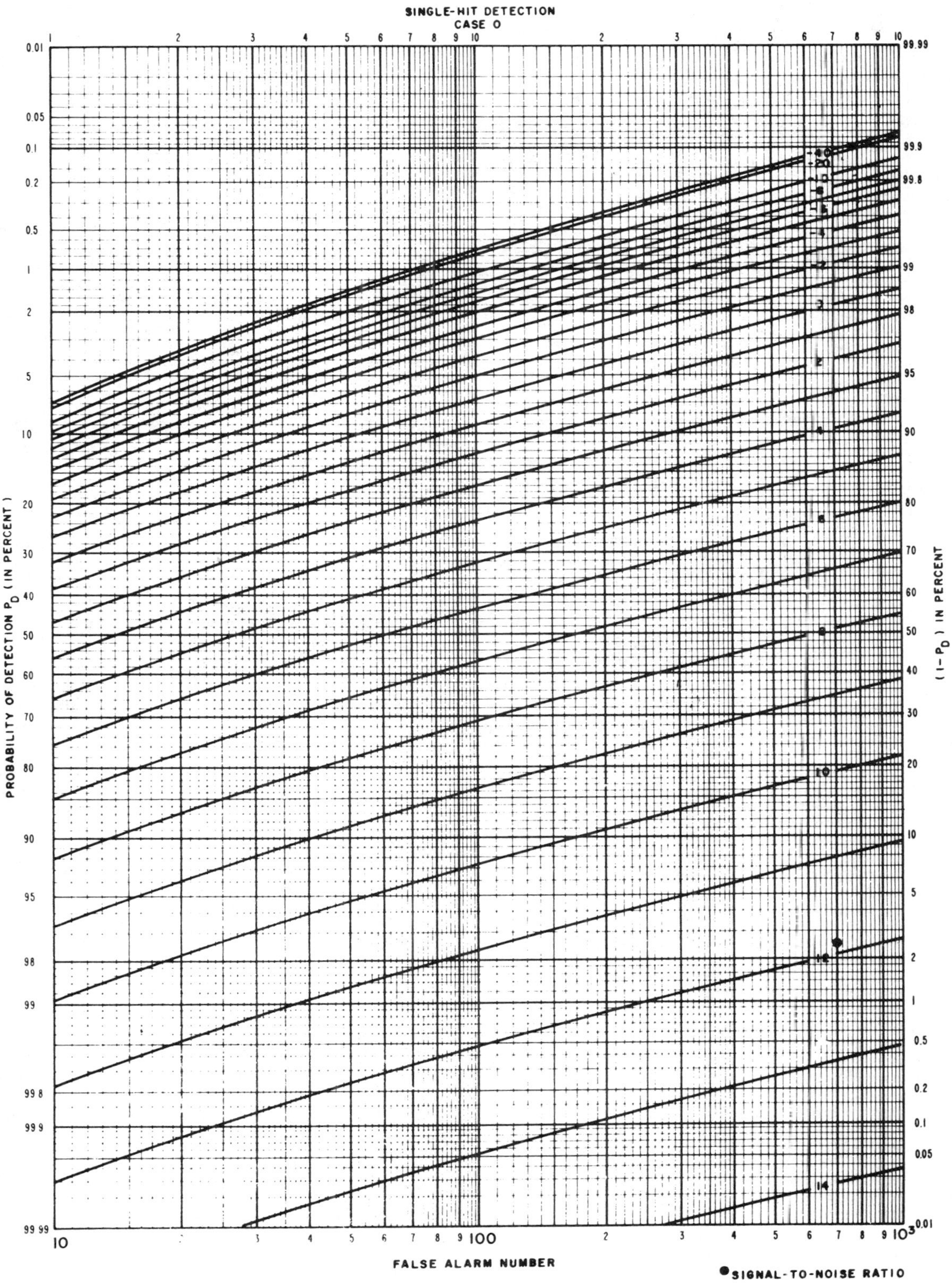

Fig. 3-2

(Figures 3-2–3-6: Single-hit detection curves for Case 0.) (From Meyer [9].)

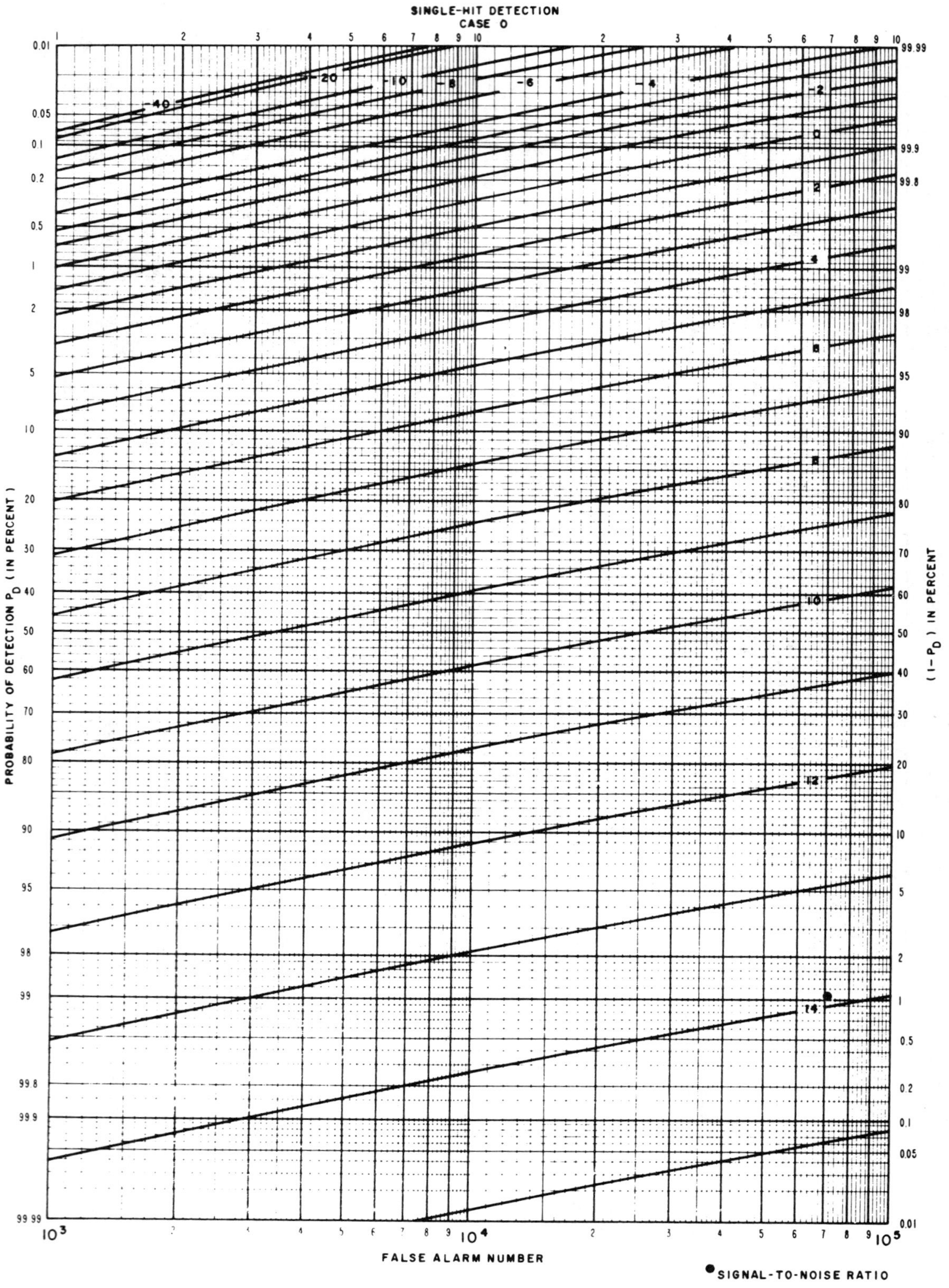

SINGLE-HIT DETECTION
CASE 0
PROBABILITY OF DETECTION P_D (IN PERCENT)
$(1-P_D)$ IN PERCENT
FALSE ALARM NUMBER
●SIGNAL-TO-NOISE RATIO
-40
-20
-10
-8
-6
-4
-2
0
2
4
6
8
10
12
14

Fig. 3-3

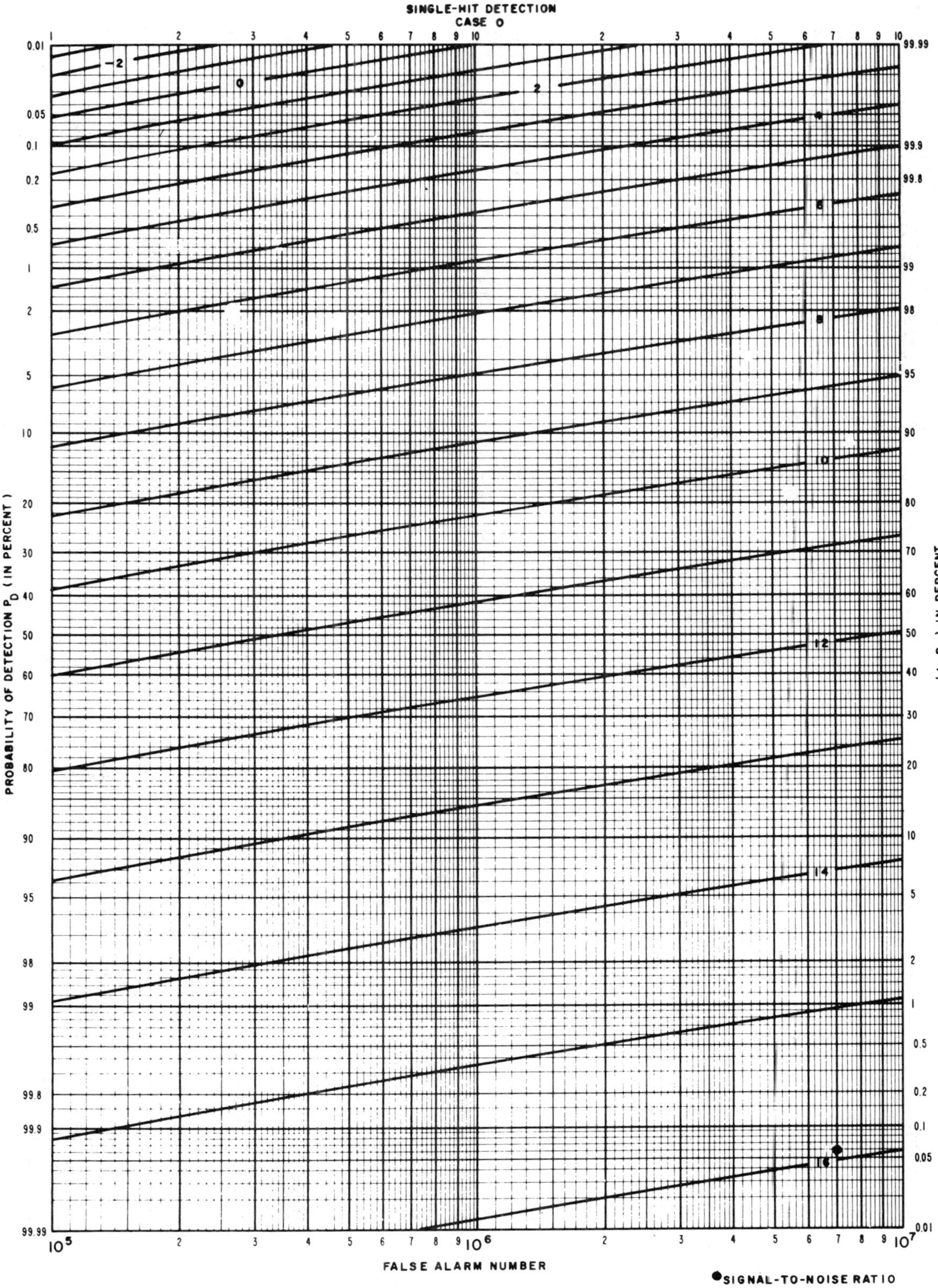
SINGLE-HIT DETECTION
CASE 0
PROBABILITY OF DETECTION P_D (IN PERCENT)
$(1-P_D)$ IN PERCENT
FALSE ALARM NUMBER
●SIGNAL-TO-NOISE RATIO
-2
0
2
4
6
8
10
12
14
16

Fig. 3-4

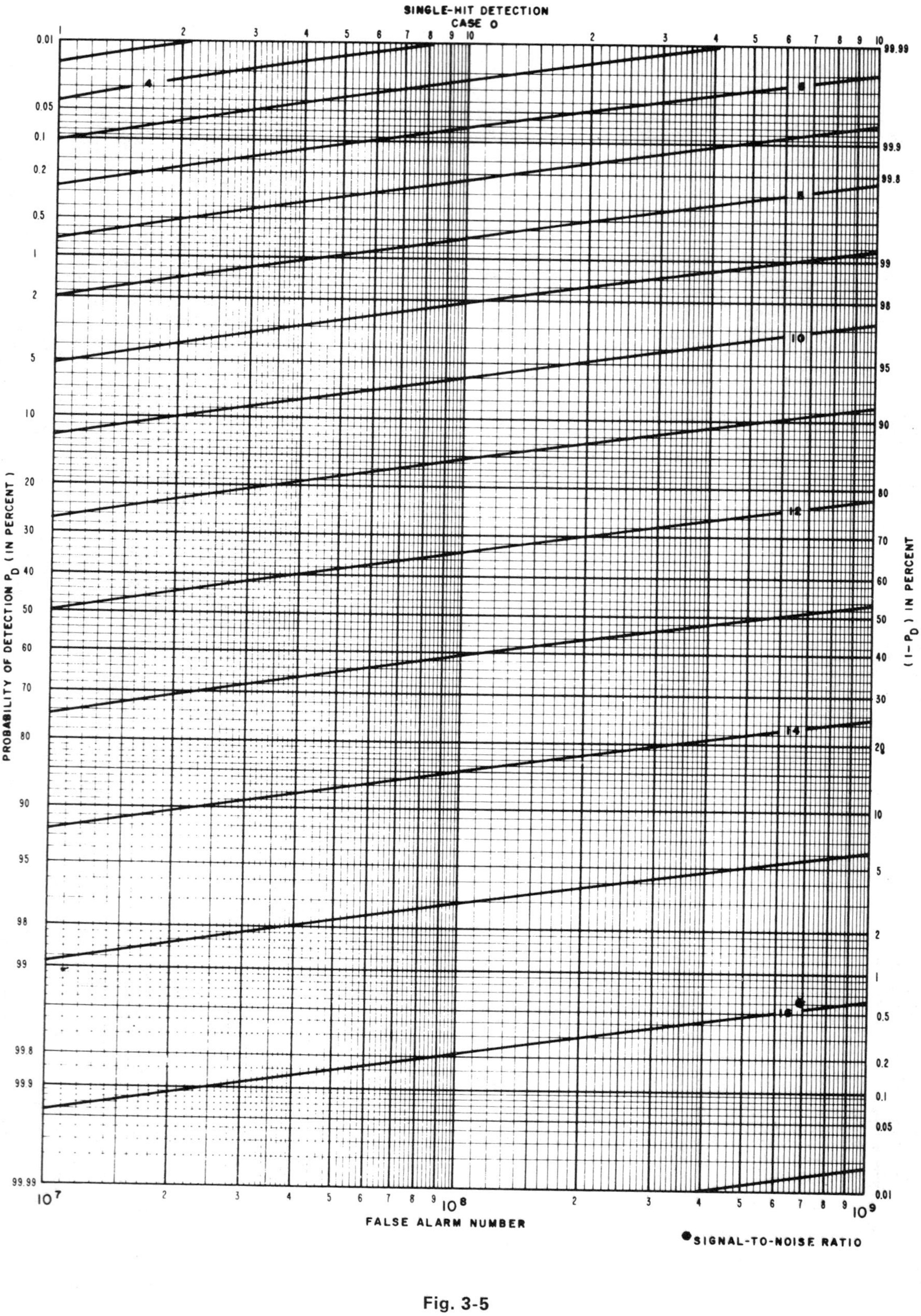
SINGLE-HIT DETECTION
CASE 0
PROBABILITY OF DETECTION P_D (IN PERCENT)
$(1-P_D)$ IN PERCENT
FALSE ALARM NUMBER
10^7
10^8
10^9
4
6
8
10
12
14
16
●SIGNAL-TO-NOISE RATIO

Fig. 3-5

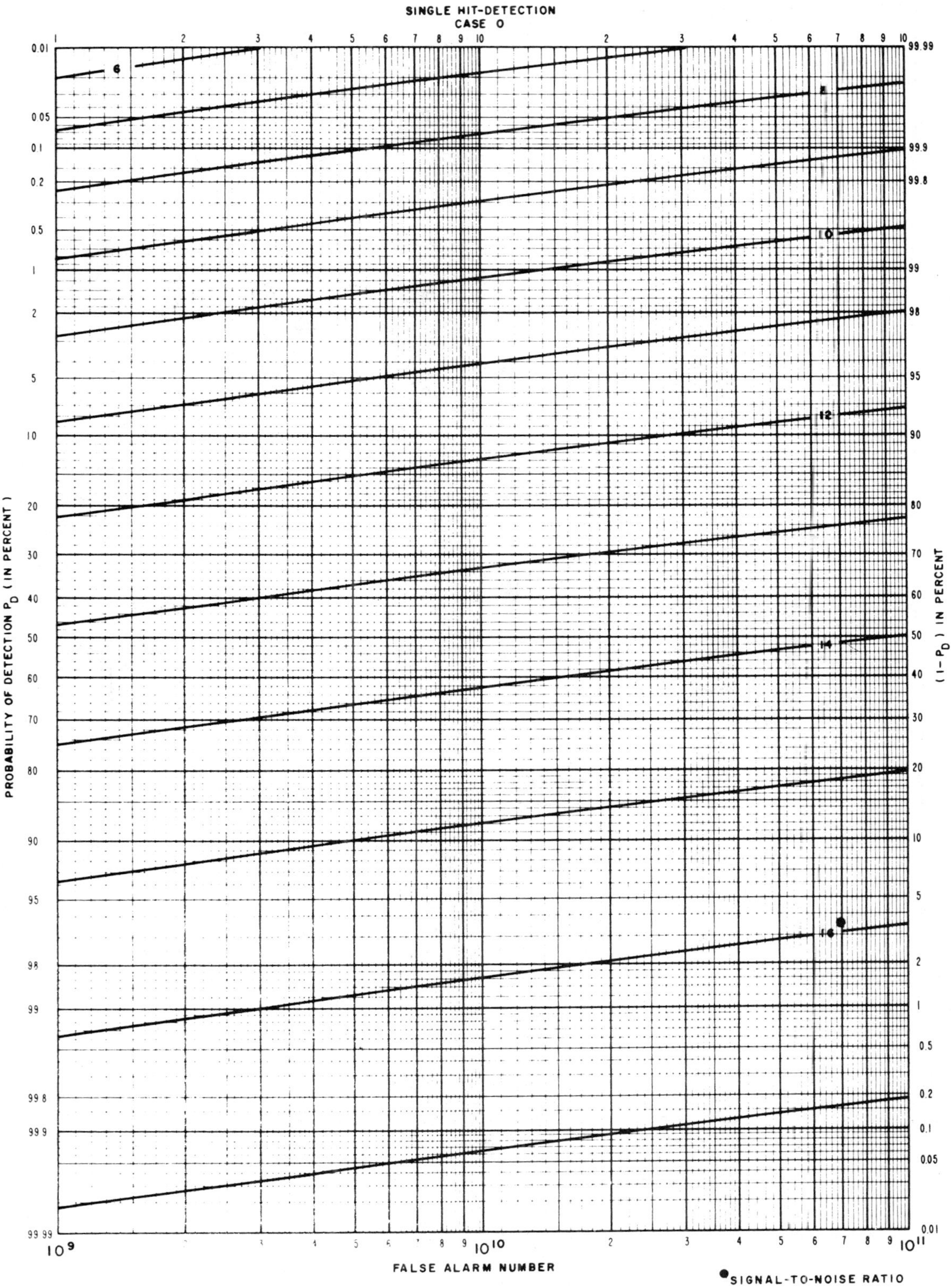
SINGLE HIT-DETECTION
CASE O
PROBABILITY OF DETECTION P_D (IN PERCENT)
$(1-P_D)$ IN PERCENT
FALSE ALARM NUMBER
●SIGNAL-TO-NOISE RATIO
6
8
10
12
14
16

Fig. 3-6

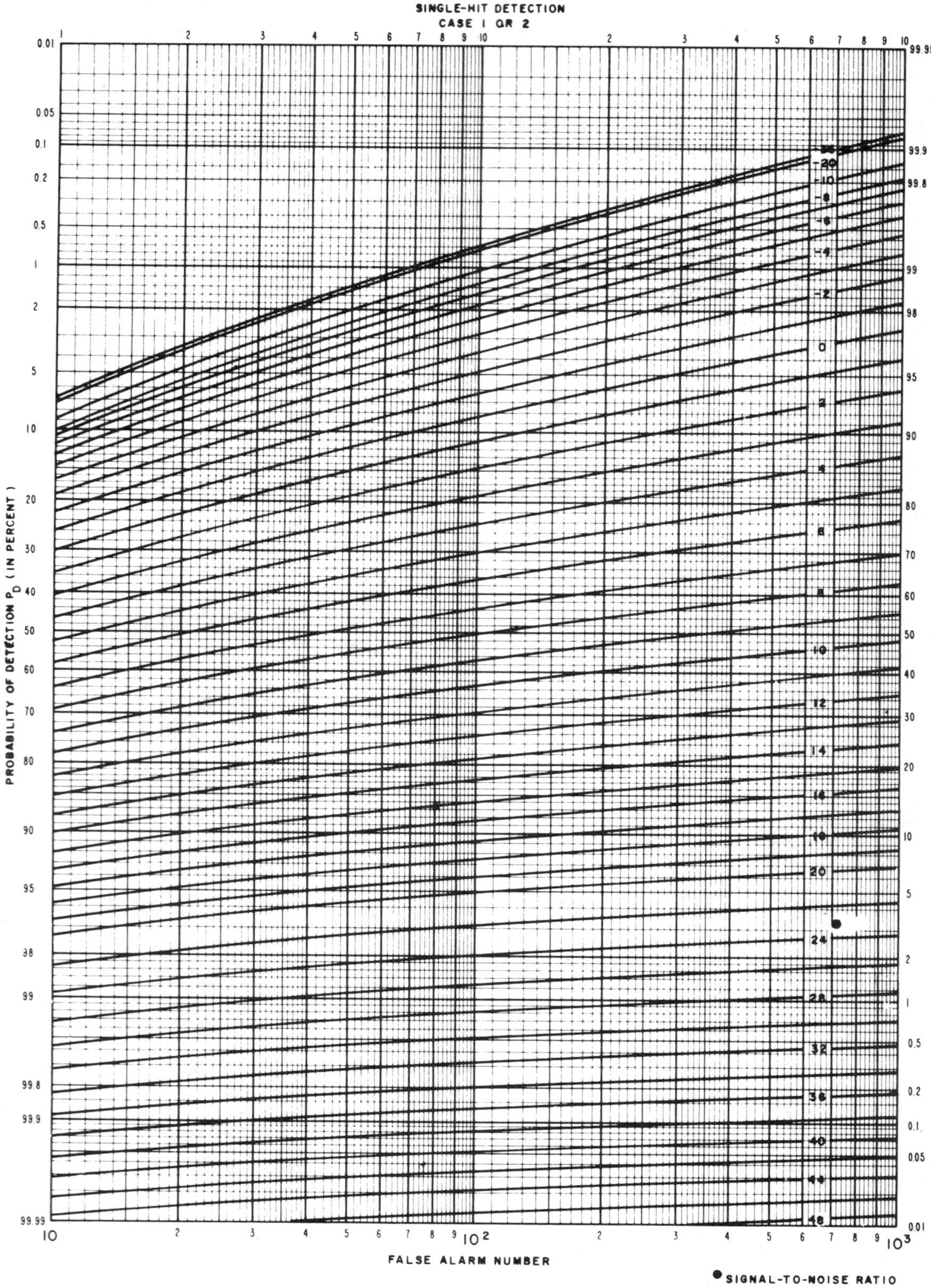

Fig. 3-7

(Figures 3-7–3-11: Single-hit detection curves for Cases 1 or 2.) (From Meyer [9].)

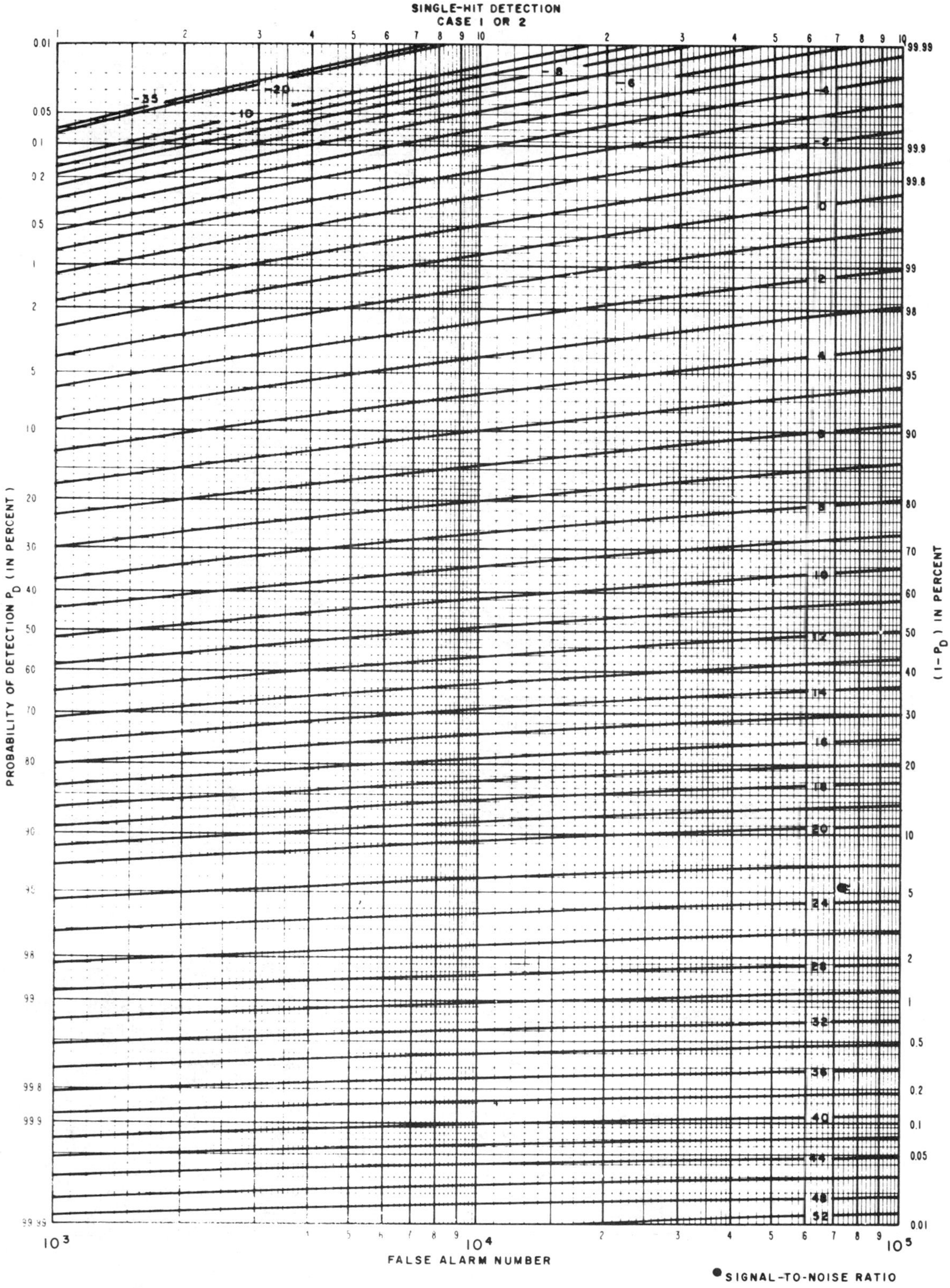
SINGLE-HIT DETECTION
CASE 1 OR 2
PROBABILITY OF DETECTION P_D (IN PERCENT)
$(1-P_D)$ IN PERCENT
FALSE ALARM NUMBER
10^3
10^4
10^5
● SIGNAL-TO-NOISE RATIO

Fig. 3-8

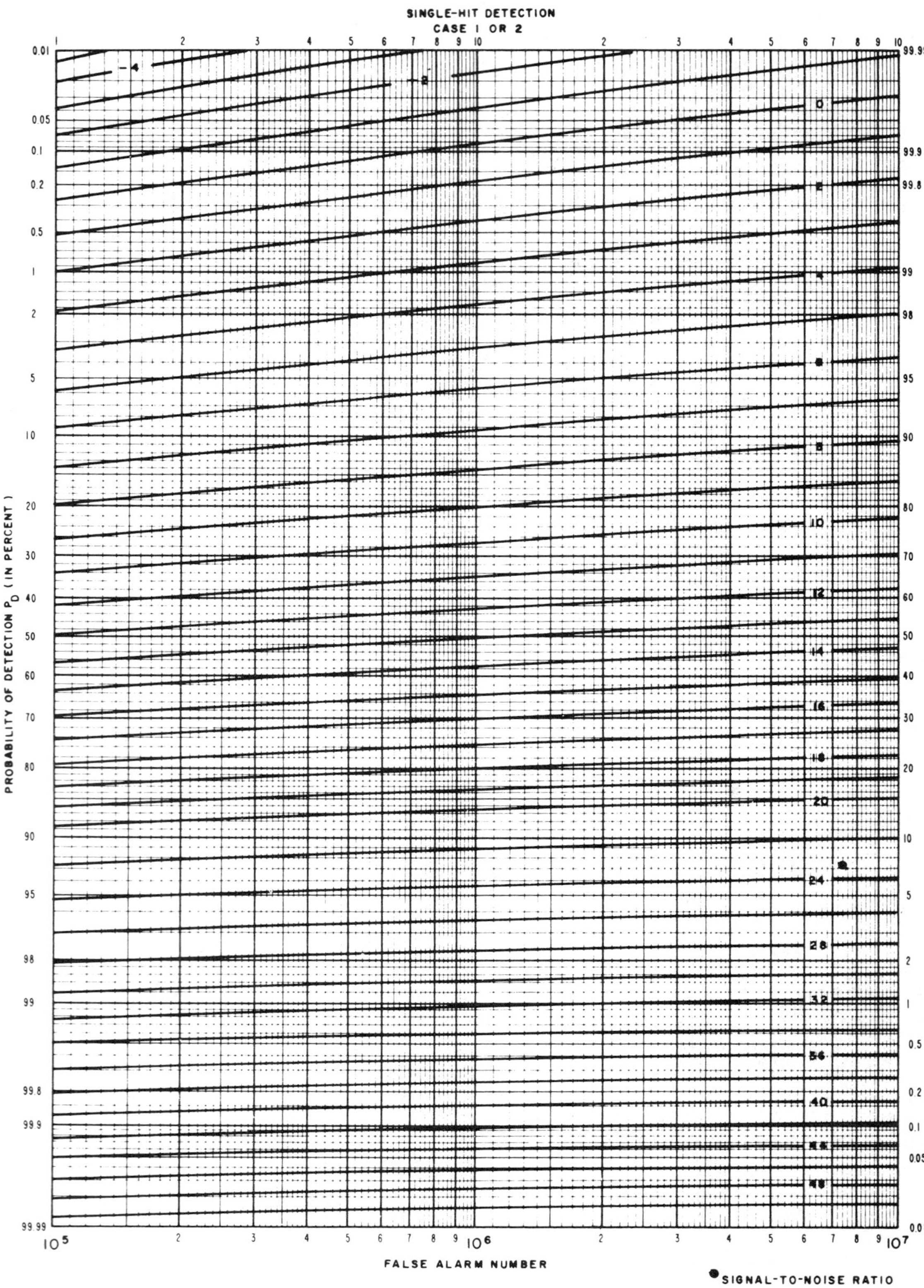
SINGLE-HIT DETECTION
CASE I OR 2
PROBABILITY OF DETECTION P_D (IN PERCENT)
$(1-P_D)$ IN PERCENT
FALSE ALARM NUMBER
●SIGNAL-TO-NOISE RATIO

Fig. 3-9

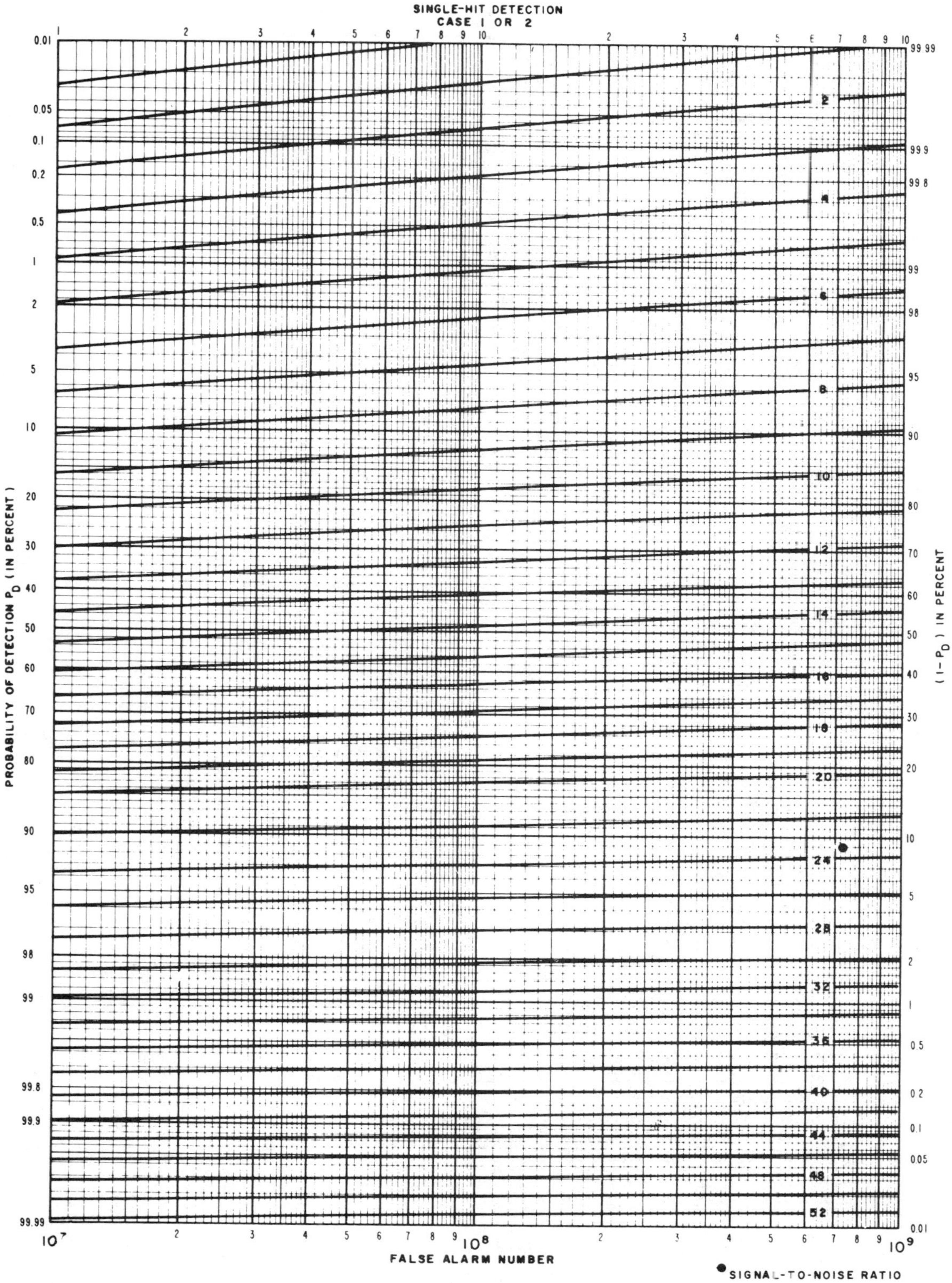
SINGLE-HIT DETECTION
CASE I OR 2
PROBABILITY OF DETECTION P_D (IN PERCENT)
$(1-P_D)$ IN PERCENT
FALSE ALARM NUMBER
SIGNAL-TO-NOISE RATIO

Fig. 3-10

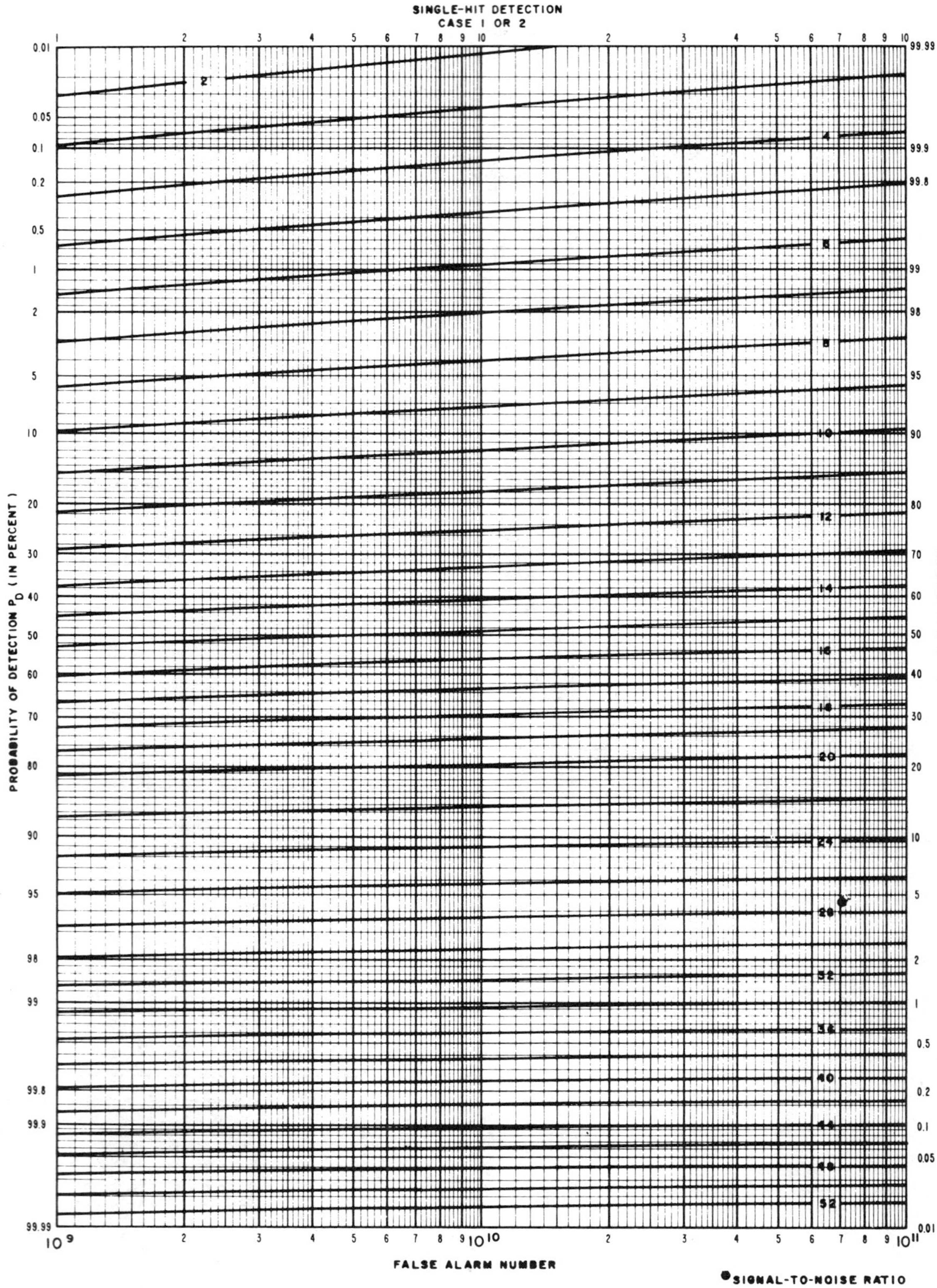
SINGLE-HIT DETECTION
CASE I OR 2
PROBABILITY OF DETECTION P_D (IN PERCENT)
$(1-P_D)$ IN PERCENT
FALSE ALARM NUMBER
●SIGNAL-TO-NOISE RATIO

Fig. 3-11

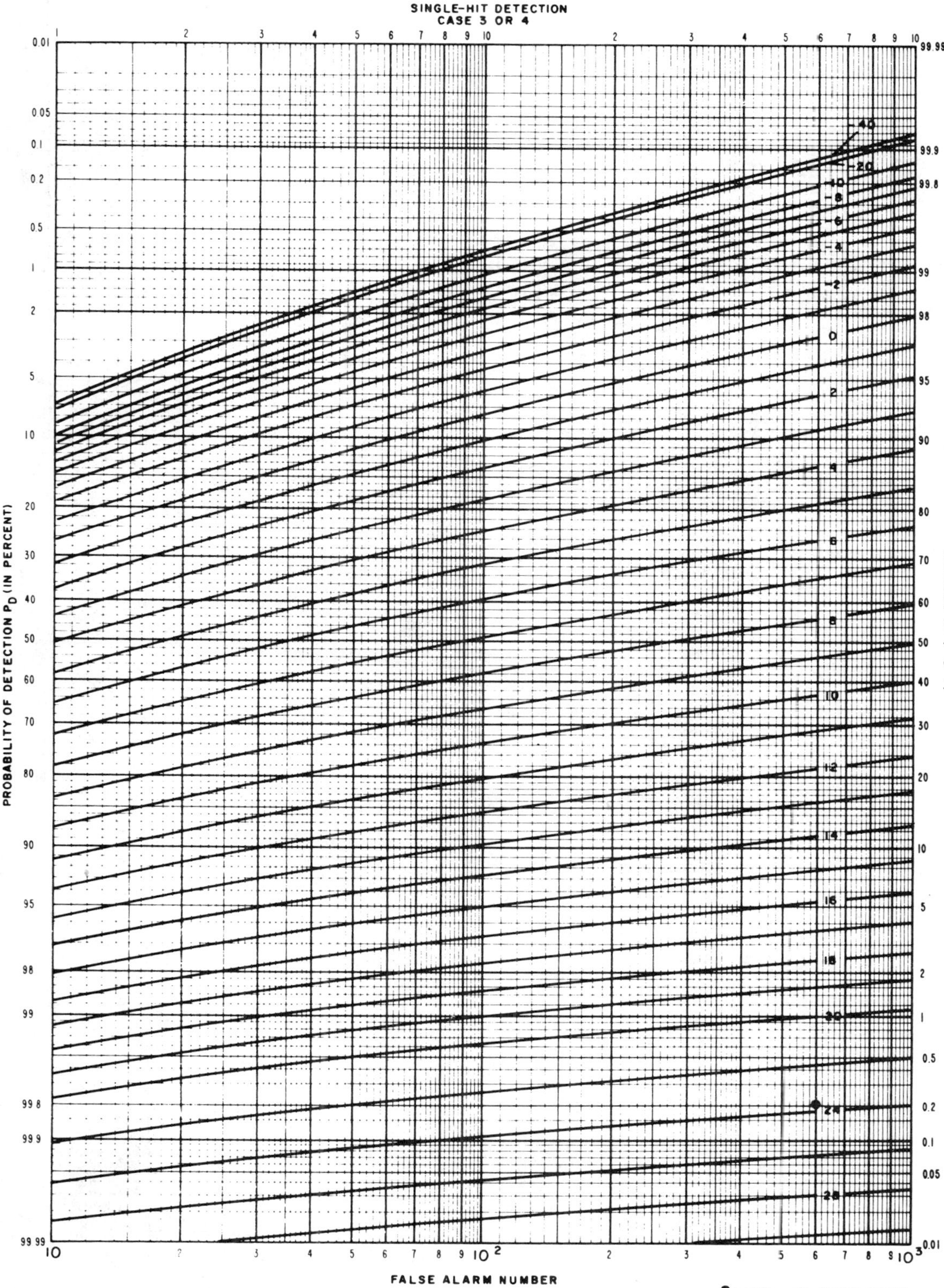

Fig. 3-12

(Figures 3-12–3-16: Single-hit detection curves for Cases 3 or 4.) (From Meyer [9].)

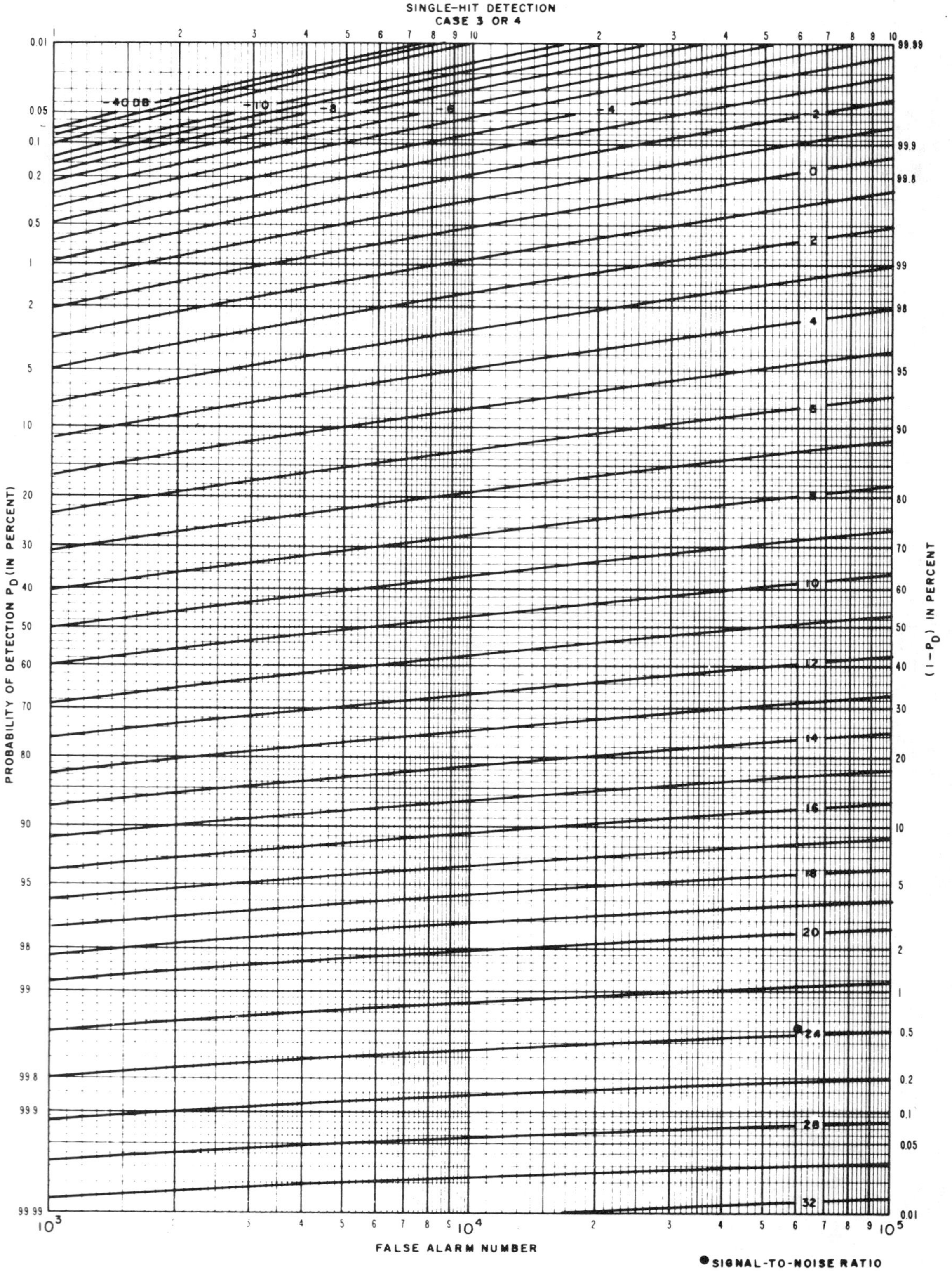

SINGLE-HIT DETECTION
CASE 3 OR 4
PROBABILITY OF DETECTION P_D (IN PERCENT)
$(1-P_D)$ IN PERCENT
FALSE ALARM NUMBER
SIGNAL-TO-NOISE RATIO

Fig. 3-13

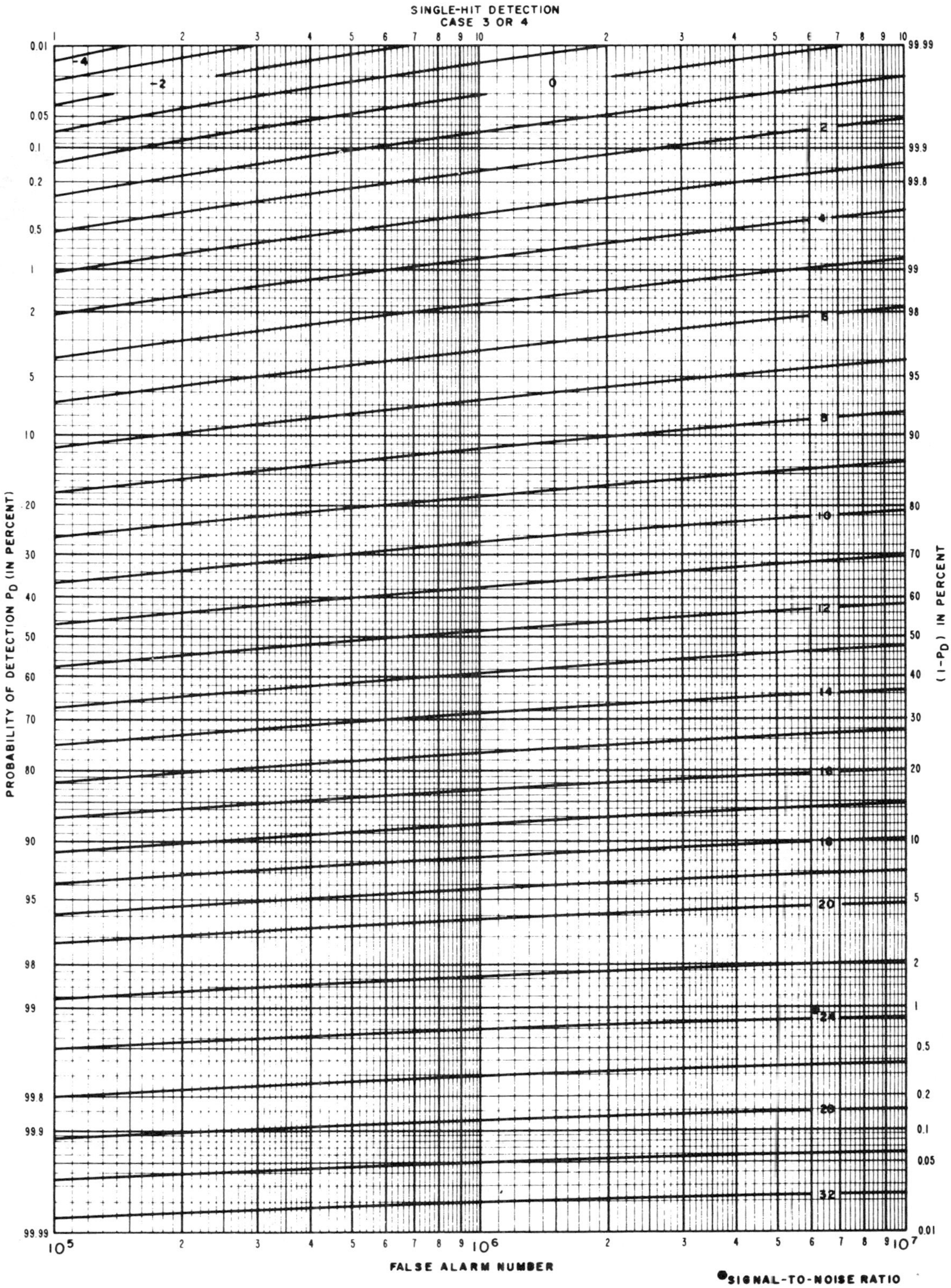
SINGLE-HIT DETECTION
CASE 3 OR 4
PROBABILITY OF DETECTION P_D (IN PERCENT)
$(1-P_D)$ IN PERCENT
FALSE ALARM NUMBER
●SIGNAL-TO-NOISE RATIO
-4
-2
0
2
4
6
8
10
12
14
16
18
20
24
28
32

Fig. 3-14

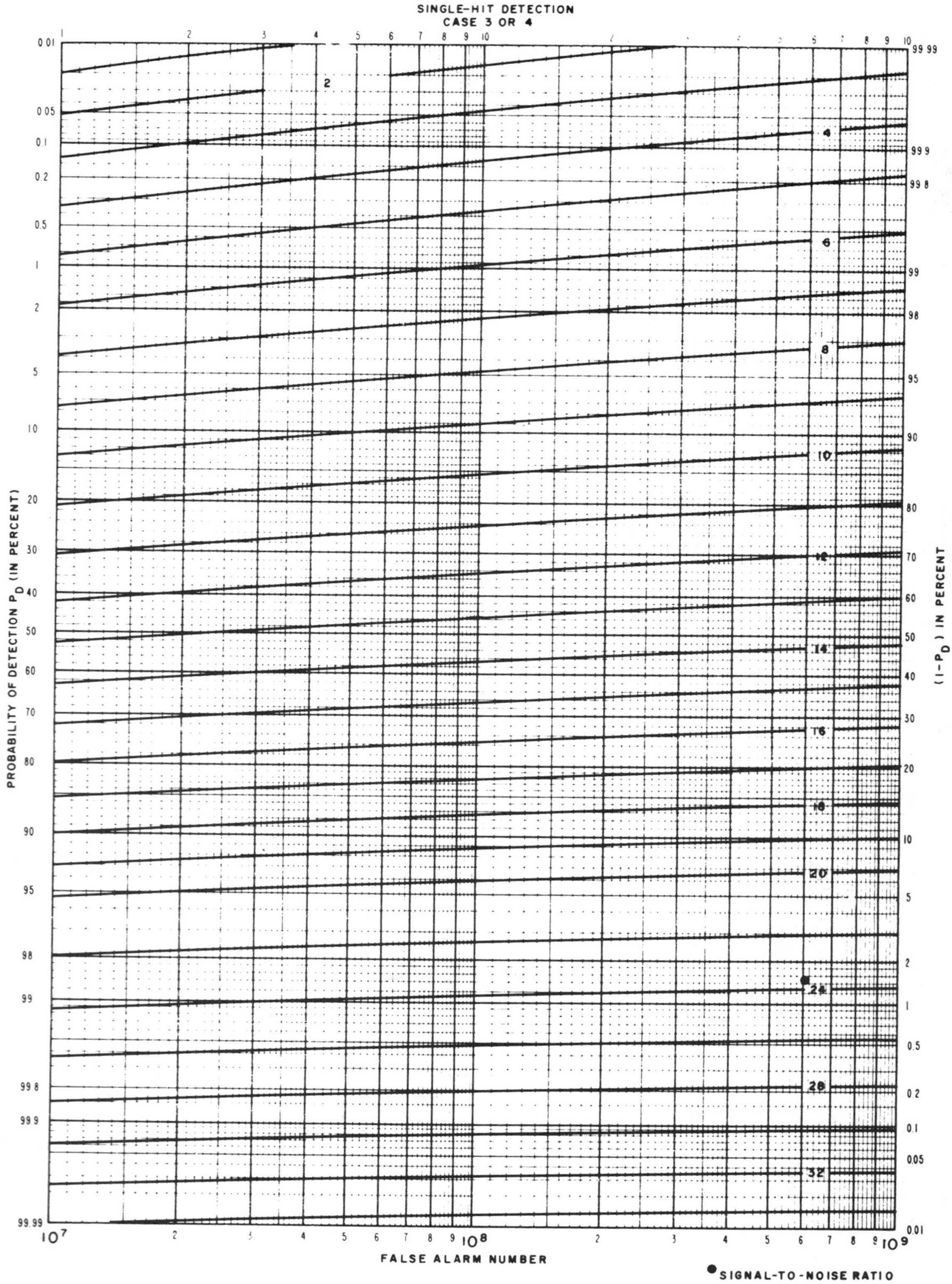
SINGLE-HIT DETECTION
CASE 3 OR 4
PROBABILITY OF DETECTION P_D (IN PERCENT)
$(1-P_D)$ IN PERCENT
FALSE ALARM NUMBER
10^7
10^8
10^9
2
4
6
8
10
12
14
16
18
20
24
28
32
● SIGNAL-TO-NOISE RATIO

Fig. 3-15

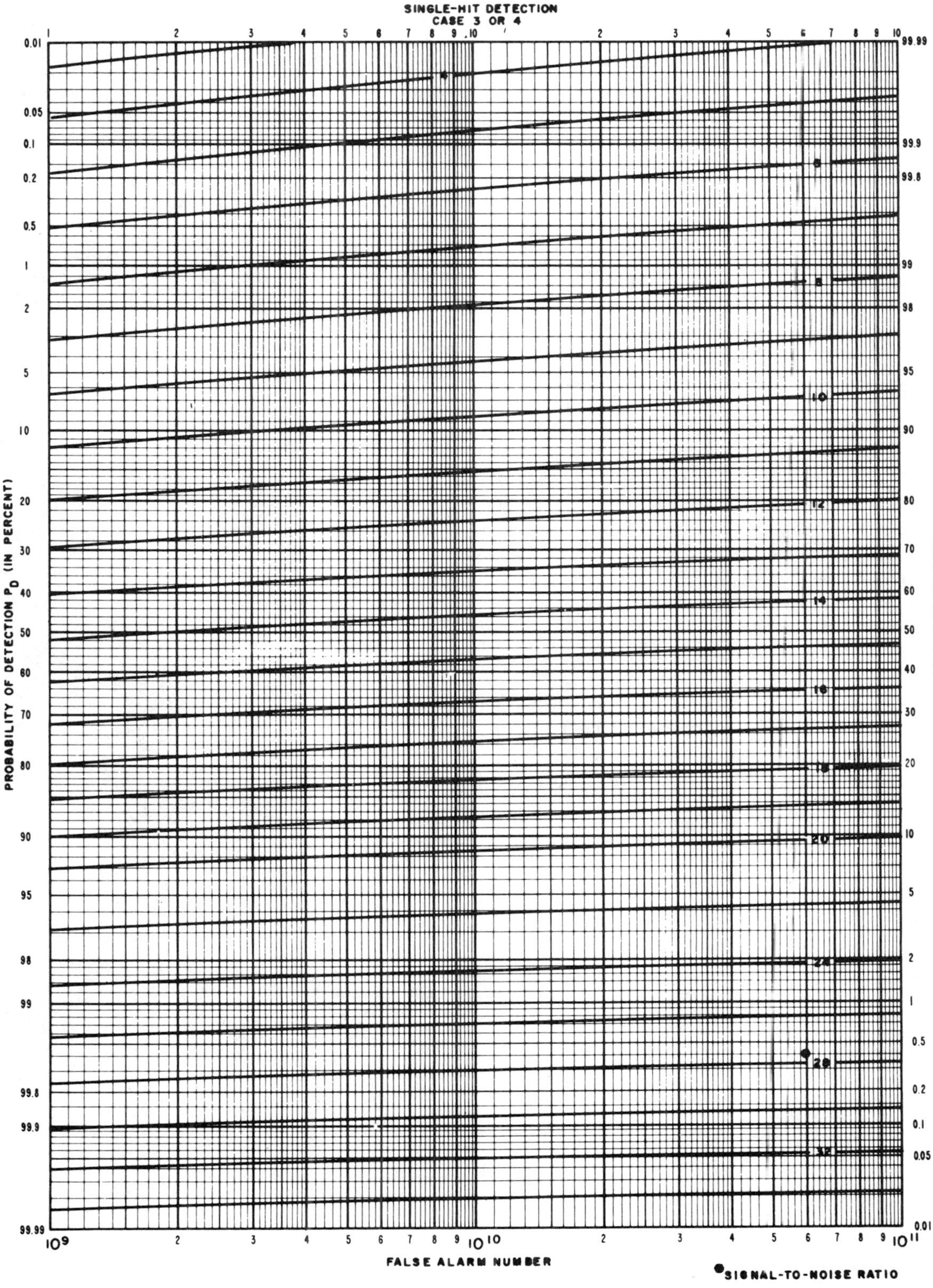
SINGLE-HIT DETECTION
CASE 3 OR 4
PROBABILITY OF DETECTION P_D (IN PERCENT)
$(1-P_D)$ IN PERCENT
FALSE ALARM NUMBER
10^9
10^{10}
10^{11}
●SIGNAL-TO-NOISE RATIO
10
12
14
16
18
20
24
28
32

Fig. 3-16

3-4 Definition of a Square-Law Detector

A quadratic or square-law detector can be defined as a device from which the output is proportional to the square of the input envelope. A full-wave square-law detector has a transfer characteristic given by

$$\psi(v) = \mathscr{A}v^2 = y \qquad (3\text{-}29)$$

By trial and error, Marcum [4] found that Eq. (3-29) represents an optimum function for $\psi(v)$ in Chapter 2, Eq. (2-31). This was in terms of producing integrable functions for Eqs. (2-33)–(2-39). Mathematically, this function results in a square-law for the combined detector and integrator law. Physically, however, it usually is a square-law detector coupled with a linear-law integrator (see Fig. 3-17).

The term $\mathscr{A}$ in Eq. (3-29) is a scaling constant. Its only effect is to multiply the bias level Y_b by $\mathscr{A}$. Probability of detection P_D, as expressed in Eq. (2-39), is independent of $\mathscr{A}$. For the bias level curves presented in Chapter 2, Fig. 2-2, $\mathscr{A} = \frac{1}{2}$. This is true also for all the work to follow in this text.

The square-law detector is one of the simpler of the nonlinear devices that can be analyzed and, therefore, is the one generally assumed when determining probability of detection, etc. Fortunately, most radar detectors behave in a square-law manner for weak signals.

3-5 The Nonfluctuating Target Model for *N* Pulse Returns

The nonfluctuating target model applies to spherical or nearly spherical objects, such as balloons, many wavelengths in diameter. The target model, therefore, can be represented by a constant-valued radar cross section σ.

Marcum found that for a square-law detector, the characteristic function for the sum of N equal amplitude pulses submerged in noise is [4]

$$C_N = C_1{}^N = \frac{e^{-Nx}e^{Nx/(p+1)}}{(p+1)^N} \qquad (3\text{-}30)$$

To obtain this result, Eqs. (3-6) and (3-7) were rewritten with

$$x = S^2/2\psi_0 = a^2/2 \qquad (3\text{-}31)$$

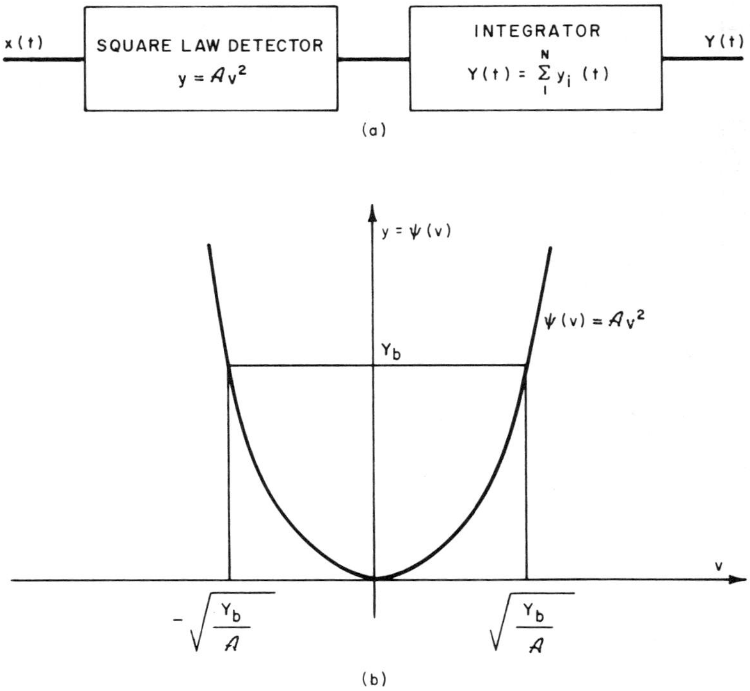

Fig. 3-17 (a) Detector–integrator block diagram; (b) square-law detector function.

and

$$y = A^2/2\psi_0 = v^2/2 \qquad (3\text{-}32)$$

Then $C_1(p)$ was determined using pair 655.1 from Campbell and Foster [10]. In writing Eq. (3-32) it has been assumed that $\mathscr{A} = 0.5$ in Eq. (3-29). The characteristic function of $dP(x, y) = f(a, y)$ was defined in Chapter 2 by Eq. (2-33).

The corresponding probability density function can be determined by using pair 650.0 from Campbell and Foster [10]:

$$dP = \left[\frac{Y}{Nx}\right]^{(N-1)/2} e^{-(Y+Nx)} I_{N-1}[2(NxY)^{1/2}]\, dY \qquad (3\text{-}33)$$

for $Y \geqslant 0$, and $dP = 0$ for $Y < 0$, where I_{N-1} is the modified Bessel function of the first kind of order $N-1$.

Once the required bias level for a given false alarm interval is determined (see Section 2-3), it is possible to determine the probability of detection. The direct approach is to integrate Eq. (3-33) over appropriate limits, as follows:

$$P_D = \int_{Y_b}^{\infty} \left[\frac{Y}{Nx}\right]^{(N-1)/2} \times e^{-(Y+Nx)} I_{N-1}[2(NxY)^{1/2}]\, dY \qquad (3\text{-}34)$$

Equation (3-34), however, is not directly integrable. It is possible, but not practical, to reduce the order of the Bessel function in single steps by successive integrations by parts so that the final integral is of a form that can be expressed in terms of the Q function [see Eq. (3-18)] [8]. The same result is obtainable through the use of the characteristic function. The cumulative distribution from $-\infty$ to Y_b of any density function is obtained from the anticharacteristic function of C/p, where C represents the characteristic function of the original density function. Accordingly, therefore, from Eq. (3-30) it is seen that

$$P_D = 1 - \int_{-i\infty}^{i\infty} \frac{C(p)}{p} e^{pY_b} \frac{dp}{2\pi i} = 1 - e^{-Nx} \int_{-i\infty}^{i\infty} \frac{\exp\{[Nx/(p+1)] + pY_b\}\, dp}{p(p+1)^N 2\pi i} \qquad (3\text{-}35)$$

If the integrand is expanded by use of the identity

$$\frac{1}{p(p+1)^N} = \frac{1}{p(p+1)} - \sum_{r=2}^{N} \frac{1}{(p+1)^r} \qquad (3\text{-}36)$$

then with the aid of pairs 210, 650.0, and 655.1 from Campbell and Foster [10] and the use of Eq. (3-18), it can be shown that

$$P_D(N) = Q[(2Nx)^{1/2}, (2Y_b)^{1/2}] + e^{-Y_b - Nx} \times \sum_{r=2}^{N} \left(\frac{Y_b}{Nx}\right)^{(r-1)/2} I_{r-1}[2(NxY_b)^{1/2}] \qquad (3\text{-}37)$$

For all practical purposes, however, this equation is only useful for numerical calculations when N is small.

The special condition where $Y_b = Nx$ permits writing the following simplified expression for the preceding:

$$P_D = \tfrac{1}{2} + e^{-2Y_b}[\tfrac{1}{2}I_0(2Y_b) + I_1(2Y_b) + I_3(2Y_b) + \cdots + I_{N-1}(2Y_b)] \qquad (3\text{-}38)$$

This formula is useful for cross-checking specific values of P_D for low values of N when $Y_b = Nx$.

An alternate solution for P_D of complementary form can be obtained by use of a different expansion of Eq. (3-36), namely,

$$\frac{1}{p(p+1)^N} = \sum_{r=(N+1)}^{r=\infty} \frac{1}{(p+1)^r} \qquad (3\text{-}39)$$

Employing integration by parts and using Campbell and Foster's tables as before, we obtain

$$P_D = 1 - e^{-(Y_b + Nx)} \times \sum_{r=N+1}^{r=\infty} \left(\frac{Y_b}{Nx}\right)^{(r-1)/2} I_{r-1}[2(NxY_b)^{1/2}] \qquad (3\text{-}40)$$

None of the expressions developed in this section are suitable for numerical determination of P_D if N is large. The general method of the Gram–Charlier Series was used by Marcum to somewhat overcome this limitation [4]. The approach is useful for a number of problems concerning distribution functions where there is a large range of variation of N. The equations used by the authors to generate Meyer Plots are given in the Appendix [see Eqs. (A-27) and (A-29)].

3-6 Fluctuating Targets: Swerling's Cases 1 and 2—The Rayleigh Distribution

These distributions characterize the return from a group of independent random scatterers and are used to describe the type of returns from aircraft and other large complex targets [1, 5]. The mathematically formulated statistical models of Cases 1 and 2 have been adequately corroborated empirically to gain wide acceptance. Most actual targets fall between Cases 1 and 2, and therefore these cases determine the limiting conditions.

CASE 1 Scan-to-Scan Fluctuations of a Rayleigh Nature The scan-to-scan fluctuations of a Rayleigh nature apply to targets such as jet aircraft, when radars having a fairly high pulse repetition rate and scan rate are employed.

CASE 2 Pulse-to-Pulse Fluctuations of a Rayleigh Nature Pulse-to-pulse fluctuations of a Rayleigh nature apply to targets such as propeller-driven aircraft (if the propellers contribute a significant portion of the echo area), to targets where small changes in orientation would establish significant changes in echoing area (such as long thin objects subjected to a high-frequency signal), or to targets viewed by radars with sufficiently low repetition rates.

The statistical cross-section model which leads to a Rayleigh distribution is based on a complex of targets with radar cross sections $\sigma_1, \sigma_2, \ldots, \sigma_n$ at distances from the radar $d_1, d_2, \ldots, d_n$ [1]. The radar cross section of a target (σ) is the area of a hypothetical reflector that scatters all radar beam energy it intercepts omnidirectionally, such that it produces an echo at the radar antenna equal to that energy actually received from the target; that is,

$$\sigma = \frac{\text{power reflected toward antenna per unit solid angle}}{\text{incident power density per } 4\pi \text{ rad}} \tag{3-41}$$

or†

$$\sigma = \lim_{R\to\infty} 4\pi R^2 \frac{|\bar{E}_R|^2}{|\bar{E}_i|^2} \tag{3-42}$$

where

$\bar{E}_i$ is the electric field incident on the target,
$\bar{E}_R$ the reflected electric field as measured at the radar receiver antenna, and
R the distance from the target to the radar receiver antenna.

† The far-field assumption is inherent in this definition ($R \gg \lambda$).

The total electric field received from a complex target can be expressed as a summation, namely,

$$E_R = \left| \sum_{k=1}^{n} |\bar{E}_{R_k}| \exp\left(\frac{j4\pi d_k}{\lambda}\right) \right| \tag{3-43}$$

and therefore the radar cross section is

$$\sigma = \left| \sum_{k=1}^{k=n} (\sigma_k)^{1/2} \exp\left(\frac{j4\pi d_k}{\lambda}\right) \right|^2 \tag{3-44}$$

where

σ_k is the cross section of individual scattering elements,
n the number of scattering elements,
d_k the range of the kth element, and
λ the wavelength.

In order to readily relate the preceding to a real target, two assumptions must be made:

1. On a short term basis the $4\pi d_k/\lambda$ is a random variable which can take on any value from 0 to 2π with equal probability.
2. The individual scattering elements are of identical area, that is, $\sigma_k = \sigma_0$.

The problem of determining the probability density function of σ is identical to the problem of determining the distance moved from an origin in a two-dimensional walk problem of n steps, where the length of each step is $(\sigma_0)^{1/2}$ and the direction of each step is perfectly random. On this basis, the x component of the kth step can be expressed as $(\sigma_0)^{1/2} \cos(4\pi d_k/\lambda)$ and the y component of the kth step can be expressed as $(\sigma_0)^{1/2} \sin(4\pi d_k/\lambda)$. The probability of having a component (x, y) after n steps where n is a comparatively large number is given by [11, p. 50]

$$\omega(x,y)\,dx\,dy = \frac{\exp[-(x^2+y^2)/n\sigma_0]}{\pi n \sigma_0}\,dx\,dy \tag{3-45}$$

Converting from rectangular coordinates (x, y) to polar coordinates (R, θ) yields

$$\omega(R,\theta)\,dR\,d\theta = \frac{R}{\pi\sigma_0 n} \exp\left(-\frac{R^2}{n\sigma_0}\right) dR\,d\theta \tag{3-46}$$

The marginal distribution of R, obtained by integrating the preceding with respect to θ, is

$$\omega(P)\,dR = \frac{2R}{n\sigma_0} \exp(-R^2/n\sigma_0)\,dR \tag{3-47}$$

Now, since $\sigma = R^2 = x^2 + y^2$ and $d\sigma = 2R\,dR$, Eq. (3-47) becomes

$$\omega(\sigma)\,d\sigma = \frac{\exp[-(\sigma/n\sigma_0)]\,d\sigma}{n\sigma_0} \tag{3-48}$$

for $\sigma > 0$, and $\omega(\sigma) = 0$ for $\sigma \leqslant 0$. This demonstrates that the probability density function of the cross section is exponentially distributed with an average cross section $\bar{\sigma} = n\sigma_0$. Thus, the average total echoing area is the sum of the individual echoing areas of the individual elements. It is interesting to note that it has been empirically demonstrated that $\bar{\sigma} = n\sigma_0$ is an excellent approximation even when n is as small as 4 or 5 [12].

The density function derived is sometimes called the Rayleigh fluctuation model; that is, the received voltage as expressed by Eq. (3-43) is Rayleigh distributed according to normally used definitions for this distribution. The echoing area or scattered power, however, is exponentially distributed.

Normally, for aircraft-type targets, larger cross sections occur to a greater degree than would be predicted from the derived distribution. This implies that the returns could be associated with a smaller number of reflecting elements thus establishing a condition that results in a perturbation of the derived distribution in the proper manner. Delano found there to be a 20° region out of the possible 360° of azimuth where the return is associated with only a few strong reflectors [13–15]. In this region the echo is not Rayleigh distributed. Over the remaining 340° of azimuth, however, the returns are remarkably Rayleigh. His conclusions were based on analysis of scale model data.†

In most system applications, one is concerned with a target ensemble. Furthermore, one is interested in the radar parameters required to achieve a given level of probability of detection. In general, if a system can meet its basic objectives, one is not necessarily interested in the possibility that over a small part of the target ensemble even better performance can be achieved. If an aircraft ensemble has its echoes approximately Rayleigh distributed, it is not significant that a small percentage of the time very large cross sections that are unpredictable from the Rayleigh model are evidenced. All this means is that the Rayleigh model establishes a conservative estimate.

For Case 1, Swerling derives the probability of detection by initially defining a target model where the received voltage return is Rayleigh distributed, namely,

$$\omega(x, \bar{x}) = \frac{\exp(-x/\bar{x})}{\bar{x}} \tag{3-49}$$

for $x > 0$, and $\omega(x, \bar{x}) = 0$ for $x \leqslant 0$. The characteristic equation for the probability density function resulting from the integration of N pulse returns from a Rayleigh fluctuating target, where there is complete correlation between pulses, is derived from the nonfluctuating case (Marcum's Case 0) characteristic function as follows:

$$\bar{C}(p) = \int_0^\infty \omega(x, \bar{x}) \frac{\exp\{-Nx[p/(p+1)]\}}{(p+1)^N}\,dx$$

$$= \frac{1}{(p+1)^{N-1}[1 + p(1 + N\bar{x})]} \tag{3-50}$$

For $N = 1$ the probability density function is given by pair 438 from Campbell and Foster [10]:

$$f(v) = \frac{\exp[-v/(1+\bar{x})]}{1+\bar{x}} \qquad (v \geqslant 0) \tag{3-51}$$

[Note that Swerling's notation is $f(v)$, where Marcum's was $dP(y)$ or $dP(Y)$.] The corresponding probability of detection expression is

$$P_D = \exp[-Y_b/(1+\bar{x})] \tag{3-52}$$

The probability density function for $N > 1$ is determined by using pair 581.7 from Campbell and Foster [10] and by noting that

$$\gamma(\nu, z) = \Gamma(\nu) I[z/(\nu)^{1/2}, \nu - 1] \tag{3-53}$$

The result is

$$f(v) = \left(\frac{1}{N\bar{x}}\right)\left(1 + \frac{1}{N\bar{x}}\right)^{N-2} \times I\left\{\frac{v}{[1 + (1/N\bar{x})](N-1)^{1/2}}, N-2\right\} \times \exp\left(-\frac{v}{1+N\bar{x}}\right) \tag{3-54}$$

where $I(u, P)$ is Pearson's form of the incomplete gamma function [17].

By rearranging Eq. (3-50) and by use of pairs 101, 208, and 431 from Campbell and Foster and

† An excellent review of other published works on this subject is presented in Nathanson [16, Chapter 5].

a subsequent integration, Swerling was able to write [5]

$$1 - P_D(N) = \int_0^{Y_b} f(v)\, dv = I\left[\frac{Y_b}{(N-1)^{1/2}}, N-2\right] - (1+N\bar{x})f(Y_b) \tag{3-55}$$

which leads to

$$P_D(N) = 1 - I\left[\frac{Y_b}{(N-1)^{1/2}}, N-2\right] + \left(1+\frac{1}{N\bar{x}}\right)^{N-1} \times I\left\{\frac{Y_b}{[1+(1/N\bar{x})](N-1)^{1/2}}, N-2\right\} \times \exp\left(\frac{-Y_b}{1+N\bar{x}}\right) \tag{3-56}$$

The Meyer Plots for Case 1 (see Chapter 4) accurately reduce Eq. (3-56) to graphical form. The plots cover values of N (number of pulses noncoherently integrated) from 1 through 10,000.

For Case 2, Swerling uses a target model where the amplitude of the received signal returns are also Rayleigh distributed, but for this case the signal is completely decorrelated (pulse-to-pulse fluctuation exists). Now, by letting $N = 1$ in Eq. (3-50), a characteristic equation for a single pulse having a Rayleigh distribution is produced, namely,

$$\bar{C}_1(p) = 1/[1 + p(1+\bar{x})] \tag{3-57}$$

The characteristic equation for the sum of N pulses that are completely decorrelated (fluctuating independently from pulse-to-pulse) is the Nth power of Eq. (3-57):

$$\bar{C}_N(p) = [\bar{C}_1(p)]^N = \frac{1}{[1+p(1+\bar{x})]^N} \tag{3-58}$$

The corresponding probability density function using pair 431 from Campbell and Foster [10] is

$$f(v) = \frac{v^{N-1}\exp[-v/(1+\bar{x})]}{(1+\bar{x})^N (N-1)!} \tag{3-59}$$

and the probability of detection expression is

$$P_D(N) = 1 - I\left[\frac{Y_b}{(1+\bar{x})(N)^{1/2}}, N-1\right] \tag{3-60}$$

where $I(u, P)$ is Pearson's form of the incomplete gamma function.

For single-hit detection the preceding equation for P_D states that system performance is independent of the manner in which the target returns correlate. That is mathematically obvious from the fact that the charcateristic equation for Case 1 was used to obtain the initial single-hit characteristic equation used in the development of Case 2 equations. Graphical solutions for single-hit detection problems have been presented in Section 3-3.

The Meyer Plots for Case 2 (see Chapter 4) accurately reduce Eq. (3-60) to graphical form. The plots cover values of N (number of pulses noncoherently integrated) from 1 through 10,000.

3-7 Fluctuating Targets: Swerling's Cases 3 and 4—The Chi-Square Distribution with Four Degrees of Freedom

The target models defined by Swerling's Cases 3 and 4 are sometimes termed one dominant plus Rayleigh, but in reality they are only an approximation of one-specific one-dominant-plus-Rayleigh case. Swerling's Cases 3 and 4 are based on chi-square distributions with four degrees of freedom. The density function is commonly associated with stabilized missile tankage (see Section 3-8).

Swerling's Case 3 and 4 target models do not have as large an amount of inherent randomness as the group of random scatters discussed in Section 3-6. The chi-square density function given in Eq. (3-61) effectively represents a large reflector subject to small changes in orientation.

$$\omega(\sigma, \bar{\sigma}) = \frac{4\sigma}{\bar{\sigma}^2}\exp\left(-\frac{2\sigma}{\bar{\sigma}}\right) \tag{3-61}$$

for $\sigma > 0$, and $\omega(\sigma, \bar{\sigma}) = 0$ for $\sigma \leqslant 0$.

For Cases 3 and 4 the received voltage is expressed in the format of Eq. (3-61) but in terms of power signal-to-noise ratio, namely,

$$\omega(x, \bar{x}) = \frac{4x}{\bar{x}^2}\exp\left(-\frac{2x}{\bar{x}}\right) \tag{3-62}$$

The characteristic equation for Case 3 which represents the condition of complete correlation (no pulse-to-pulse fluctuation) is given by

$$\bar{C}(p) = \frac{1}{(1+p)^{N-2}\{1+p[1+(N\bar{x}/2)]\}^2} \tag{3-63}$$

As obtained from the Campbell and Foster [10] pairs 201, 442, and 449.5, the corresponding probability density functions for $N = 1$ and $N = 2$ are, respectively,

$$f(v) = \frac{1}{[1+(\bar{x}/2)]^2}\left[1 + \frac{v}{1+(2/\bar{x})}\right]\exp\left[\frac{-v}{1+(\bar{x}/2)}\right] \tag{3-64}$$

and

$$f(v) = \frac{v}{(1+\bar{x})^2}\exp\left[\frac{-v}{1+\bar{x}}\right] \tag{3-65}$$

For $N > 2$, the probability density function can be determined from Campbell and Foster [10] pair 581.1 after appropriate transformations are made. Doing so, it is found that

$$f(v) = \left[\frac{v^{N-1}e^{-v}}{(N-1)!\,[1+(N\bar{x}/2)]^2}\right] \times {}_1F_1\left[2, N, \left(\frac{v}{1+(2/N\bar{x})}\right)\right] \tag{3-66}$$

where ${}_1F_1(a, b, c)$ is a confluent hypergeometric function.

Swerling uses two identities relating to confluent hypergeometric functions in order to expand the preceding into the more familiar Pearson's form of the incomplete gamma function, $I(u, P)$ [5, 17]:

$$f(v) = \frac{[1+(2/N\bar{x})]^{N-2}v}{[1+(N\bar{x}/2)]^2} I\left[\frac{v}{[1+(2/N\bar{x})](N-1)^{1/2}}, N-2\right]\exp\left[\frac{-v}{1+(N\bar{x}/2)}\right] - \frac{(N-2)[1+(2/N\bar{x})]^{N-1}}{[1+(N\bar{x}/2)]^2} \times I\left[\frac{v}{[1+(2/N\bar{x})](N-1)^{1/2}}, N-2\right]\exp\left(\frac{-v}{1+N\bar{x}}\right) + \frac{v^{N-1}e^{-v}}{[1+(N\bar{x}/2)](N-2)!} \tag{3-67}$$

If one assumes the false alarm probability to be negligible and $N\bar{x}/2 > 1$, then for $v > Y_b$, the portion of Eq. (3-67) represented by the incomplete gamma function is found to be practically equal to 1. Also, the integral from Y_b to ∞ of the third term is found to be negligible for $N > 2$. A valid approximation for P_D, therefore, is

$$P_D(N) \approx \int_{Y_b}^{\infty} \frac{[1+(2/N\bar{x})]^{N-1}}{[1+(N\bar{x}/2)]^2} \times \left[\frac{v}{[1+(2/N\bar{x})]} - (N-2)\right] \times \exp\left[\frac{-v}{1+(N\bar{x}/2)}\right] dv \tag{3-68}$$

or

$$P_D(N) \approx \left(1 + \frac{2}{N\bar{x}}\right)^{N-2} \times \left[1 + \frac{Y_b}{1+(N\bar{x}/2)} - \frac{2(N-2)}{N\bar{x}}\right] \times \exp\left[\frac{-Y_b}{1+(N\bar{x}/2)}\right] \tag{3-69}$$

The above is exact for $N = 1$ and $N = 2$.

An exact expression for the Case 3 probability of detection for $N > 2$ in terms of a power series is given by Eq. (A-85) in the Appendix.

For Case 4 (pulse-to-pulse fluctuation with no correlation), the characteristic equation for a single hit obtained by letting $N = 1$ in Eq. (3-63) is

$$\bar{C}(p) = \frac{1+p}{\{1 + p[1+(\bar{x}/2)]\}^2} \tag{3-70}$$

Then for the sum of N pulses fluctuating independently from pulse-to-pulse the characteristic equation is the Nth power of the preceding, that is,

$$\bar{C}_N(p) = [\bar{C}_1(p)]^N = \frac{(1+p)^N}{\{1 + p[1+(\bar{x}/2)]\}^{2N}} \tag{3-71}$$

Swerling's exact expression for the density function proved to be a polynomial multiplied by $\exp\{-v/[1+(\bar{x}/2)]\}$ [5]. Except for small values of N (number of pulses integrated), this polynomial is too long to be useful for computational purposes. For large values of N he found that one can use the Edgeworth series. For $N = 1$, the exact formula is shown in Table 3-2.

In the Appendix, solutions in series form for Case 4 are determined. The probability density function is given by Eq. (A-99) and probability of detection by Eq. (A-107) and (A-111). Neither of these equations is suitable for hand calculation and, indeed, their evaluation took much time even with the aid of a digital computer. Nonetheless, they were successfully used to develop Meyer Plots for Case 4.

3-8 The Relationship between the Five Standard Target Models and Modern Target Models

The advantage of using Marcum's and Swerling's target models is that they are reasonable approximations to a variety of actual targets and that through the use of Meyer Plots extensive data are readily available for problem solving. It has been demonstrated, however, by various current workers in modern detection theory that the five standard target models are inadequate for many problems of current interest [18]. Various "second generation" target models have been developed that more accurately describe certain target types. These are examined in this section. (In Chapter 4 it is shown how the Marcum and four Swerling target models can be used to obtain solutions for actual target returns using almost any target model.)

WEINSTOCK TARGET MODELS

W. Weinstock investigated the possibility of improving upon the five standard target models covered by this handbook [1]. His work was directed toward obtaining a better fit to empirically obtained scattering cross-section density data, especially for the randomly oriented cylinder class of targets. He found that in order to obtain a good fit to his density data, in particular the class of targets supposedly covered by Swerling's Cases 3 and 4, it was necessary to use a Laguerre polynomial. The leading term of a Laguerre polynomial expansion is a chi-square density function. The number of degrees of freedom is twice the square of the mean cross section divided by the variance of the cross section. The use of the chi-square approximation of a Laguerre polynomial, however, tends to result in an underestimate of the values in the tails of the density functions obtained from measurements on real targets and scale models of real targets. For this reason it is not possible to obtain reliable *high* values of $P_D(1)$ (probability of detection for a single hit) using a chi-square target model.

Although Weinstock would have preferred to represent a target model using the first two terms of a Laguerre polynomial, he conceded that it is more practical to use the chi-square function. Furthermore, he concluded that when the chi-square approximation is used, the degree of freedom should be chosen as shown in Table 3-3.

Table 3-3

Target class	Chi-square degrees of freedom
Randomly oriented cylinder	0.6–1.4
Stabilized cylinder	4.0
Randomly oriented smooth bodies, such as a prolate spheroid	4.0

It should be noted at this point that the Rayleigh function, used for Target Model Cases 1 and 2, is a chi-square function with two degrees of freedom. Weinstock, therefore, if limited to the use of published probability of detection data, preferred the Case 1 and Case 2 target models for the randomly oriented cylinder class of targets. Furthermore, he recommended that an average signal-to-noise ratio $\bar{x}$ based on the median rather than the mean target cross section be used in order to minimize the potential spread in system requirements that would otherwise occur.

GENERAL CHI-SQUARE TARGET MODELS

At this point it is apropos to examine the chi-square family of density functions and to introduce the general chi-square target model, which is an outgrowth of Weinstock's work [18]. The basic chi-square probability density function is

$$f(\chi^2) = \frac{(\chi^2)^{k-1}\exp(-\chi^2/2)}{2^k(k-1)!} \tag{3-72}$$

where $k = \nu/2$ is the number of duo-degrees of freedom. It is the probability density function for the "radial" mean-square error in ν-dimensional space when the error in each dimension is normally distributed. Figure 3-18a presents plots of several chi-square density functions [19]. When related to a single signal-plus-noise pulse return, the density function becomes

$$\omega(x,\bar{x}) = \frac{k}{(k-1)!\,\bar{x}}\left(\frac{kx}{\bar{x}}\right)^{k-1}\exp\left(\frac{-kx}{\bar{x}}\right) \tag{3-73}$$

for $x > 0$, and $\omega(x,\bar{x}) = 0$ for $x \leqslant 0$, where

x is the instantaneous single-pulse signal-to-noise ratio, and
$\bar{x}$ the average single-pulse signal-to-noise ratio.

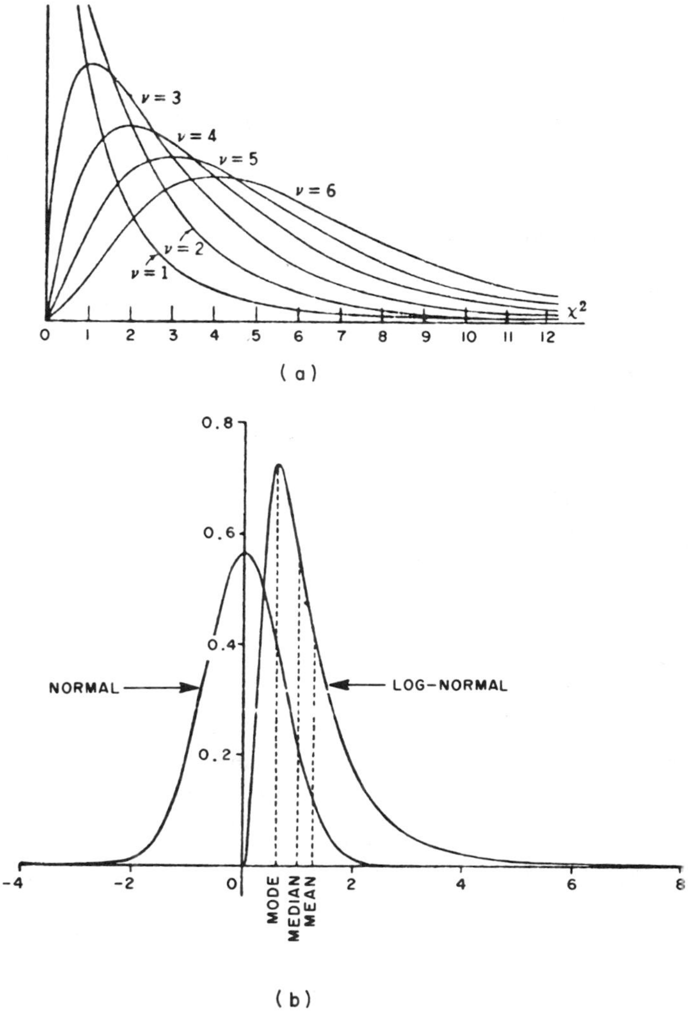

Fig. 3-18 (a) Chi-square probability density function for various degrees of freedom. (From Hoel [19].) (b) Typical normal and log-normal probability density functions. (From Aitchison and Brown [20].)

In this general chi-square target model

$$\chi^2 = 2kx/\bar{x} \qquad (3\text{-}74)$$

Properly speaking, therefore, this probability density function is a normalized mean-square distribution with $2k$ degrees of freedom. We will stay with common practice, however, and use the chi-square nomenclature.

The properties of this probability density function are as follows [21]:

$$\text{mean} = \bar{x} = \text{arithmetic average value} \qquad (3\text{-}76)$$

$$\text{variance} = \sigma^2 = \bar{x}^2/2 = \text{second moment about mean} \qquad (3\text{-}77)$$

$$\text{skewness}^{\dagger} = (4/k)^{1/2} = \text{third moment about mean}/\sigma^3 \qquad (3\text{-}78)$$

$$\text{kurtosis}^{\ddagger} = 3 + (6/k) = \text{fourth moment about mean}/\sigma^4 \qquad (3\text{-}79)$$

$$\text{mode} = [(k-1)/k]\,\bar{x} = \text{most frequent value} \qquad (3\text{-}80)$$

and

$$d\chi^2/dx = 2k/\bar{x} \qquad (3\text{-}75)$$

† Skewness is a measure of asymmetry.

‡ Kurtosis is a measure of concentration about the mean.

When N *independent* chi-square distributed variates are summed, the density function of the result is also chi-square with

$$K = \sum_{i=1}^{N} k_i \tag{3-81}$$

duo-degrees of freedom, and

$$\bar{x}_N = \sum_{i=1}^{N} \bar{x}_i \tag{3-82}$$

If the N summed variates are completely correlated, however, then

$$K = k \tag{3-83}$$

The characteristic function of the probability density function of $\bar{x}$ for a single pulse return is

$$\bar{C}(p) = \int_0^\infty \omega(x,\bar{x})\, C(p)\Big|_{\text{Case 0}} dx$$

$$= \frac{1}{(1+p)\left[1 + \left(\frac{p}{1+p}\right)\left(\frac{\bar{x}}{k}\right)\right]^k} \tag{3-84}$$

and for N noncoherently integrated pulse returns

$$\bar{C}(p) = \frac{1}{(1+p)^N\left[1 + \left(\frac{p}{1+p}\right)\left(\frac{N\bar{x}}{K}\right)\right]^K} \tag{3-85}$$

(it is assumed that all $\bar{x}_i$ are equal.)

It is now readily demonstrated that the five standard target models are related to the general chi-square target model as follows [18]:

Marcum—Case 0: $K = \infty$

Swerling—Case 1: $K = 1$

Swerling—Case 2: $K = N$

Swerling—Case 3: $K = 2$

Swerling—Case 4: $K = 2N$

(The authors suggest that the reader attempt the proof of the above as an educational exercise. Hint: For Marcum's Case 0 it is necessary to show that

$$\lim_{K\to\infty}\left[1 + \left(\frac{p}{1+p}\right)\left(\frac{N\bar{x}}{K}\right)\right]^K = \exp\left(\frac{p}{1+p}\right)(N\bar{x}) \tag{3-86}$$

by use of series expansions of both sides of the equation and term-by-term comparisons.)

In general, when the target returns are completely correlated,

$$K = k \tag{3-87}$$

and when the returns are completely uncorrelated,

$$K = kN \tag{3-88}$$

The probability density function corresponding to Eq. (3-85) as determined through the use of pair 581.1 from Campbell and Foster [10] is

$$f(v) = \frac{1}{(N-1)!}\left(1 + \frac{N\bar{x}}{K}\right)^{-K} v^{N-1} e^{-v} \times {}_1F_1\left\{K, N, \left[\frac{v}{1 + (K/N\bar{x})}\right]\right\} \tag{3-89}$$

for $v \geqslant 0$, and $f(v) = 0$ for $v < 0$.

The probability of detection is

$$P_{\mathrm{D}} = 1 - \int_0^{Y_b} f(v)\, dv \tag{3-90}$$

The general chi-square target model is used extensively for problem solving in Chapter 4. The method of determining probability of detection for the general chi-square target model by use of Meyer Plots is presented in Section 4-3.

RICE TARGET MODEL

The Rice target model is a model that combines the return of a steady reflector with that of a field of many random scatterers. The random scatterers by themselves produce a Rayleigh (chi-square with $k = 1$) distributed signal return. The Rice is the exact one-dominant-plus-Rayleigh target model [22]. Its probability density function for a single signal-plus-noise pulse return is

$$\omega(x,\bar{x}) = \frac{1}{\psi_0}\exp\left(-s - \frac{x}{\psi_0}\right) I_0\left[2\left(\frac{sx}{\psi_0}\right)^{1/2}\right] \tag{3-91}$$

for $x > 0$, and $\omega(x,\bar{x}) = 0$ for $x \leqslant 0$, where

$$s = \frac{\text{radar cross section of steady reflector}}{\text{combined average cross section of Rayleigh scatterers}} \tag{3-92}$$

ψ_0 is the mean value of Rayleigh component of x, and

$I_0(z)$ the modified Bessel function of the first kind of zero order.

The properties of this probability density function are as follows:

$$\text{mean} = \bar{x} = \psi_0(1+s) \quad (3\text{-}93)$$

$$\text{variance} = \sigma^2 = \psi_0^2(1+2s) \quad (3\text{-}94)$$

The characteristic function for the probability density function of $\bar{x}$ for N noncoherently integrated pulse returns is

$$\bar{C}(p) = \frac{\exp\left\{\dfrac{-sN\psi_0 p}{[(N\psi_0/K)+1]\,p+1}\right\}}{(1+p)^N\{1+[p/(1+p)](N\psi_0/K)\}^K} \quad (3\text{-}95)$$

where for when the target returns are completely correlated

$$K = 1 \quad (3\text{-}96)$$

and for when the returns are completely uncorrelated

$$K = N \quad (3\text{-}97)$$

This characteristic function reduces to that of a standard target model under the following conditions:

Case 0: $\psi_0 \to 0, \quad s \to \infty, \quad s\psi_0 = \bar{x}$

Case 1: $s = 0, \quad K = 1$

Case 2: $s = 0, \quad K = N$

It is apparent that it would not be difficult to extend the Rice target model to include any member of the general chi-square family for the random component of the target return by letting $K = k$ or $K = kN$, as required.

A method of determining the probability of detecting a target that exhibits Rice characteristics by use of Meyer Plots is presented in Section 4-4.

LOG-NORMAL TARGET MODEL

The log-normal distribution, although not as widely known, is as fundamental a probability distribution in statistics as the normal (Gaussian) distribution [20, 23]. Whereas the normal distribution arises from a theory of elementary errors combined by an additive process, the log-normal distribution arises from a theory of elementary errors combined by a multiplicative process. The appropriateness of the use of either distribution to describe a particular situation is related to use of either the arithmetic or geometric mean to find the expected value for that situation. The significant feature of this distribution is that it can produce a large mean to median ratio (ρ), which has the effect of emphasizing the large cross-sectional elements of a target. The log-normal distribution, therefore, is suited for situations where a target produces large specular returns at only a relatively few aspect angles or for when a few large surfaces predominate in a target.

The log-normal distribution can be considered a result of the theory of proportionate effect, which makes the change in a variate at any step of a process to be a random proportion of the previous value of the variate. The theory is found to apply in the analysis of a number of biological processes, in connection with partitioning processes (placing objects in logically derived categories), in the prediction of maintenance downtimes for electronic systems, in economics in the statistical analysis of income or consumer demand, and in the theory of breakage [20, 23].

It has become popular to use the log-normal target model to describe large metal objects of irregular shape, especially ocean vessels with complex superstructures [18, 24]. Apparently, a large group of randomly sized and placed corner reflectors emulates the multiplicative effect that leads to the log-normal probability distribution [25]. The superstructure of a ship is made up of many such corner reflectors. Some of the larger, more irregularly shaped space vehicles and satellites also fall into this category.

The probability density function for a single signal-plus-noise log-normally distributed pulse return is

$$\omega(x,\bar{x}) = \frac{\exp\{-\log_e^2(x/\alpha)/4\log_e(\rho)\}}{x[4\pi\log_e(\rho)]^{1/2}} \quad (3\text{-}98)$$

for $x > 0$, and $\omega(x,\bar{x}) = 0$ for $x \leqslant 0$, where

$$\alpha = \text{median value of } x$$

and

$$\rho = \frac{\text{mean of } x}{\text{median of } x} = \frac{\bar{x}}{\alpha} \quad (3\text{-}99)$$

The properties of this density function are as follows [24]:

$$\text{mean} = \bar{x} = \alpha\rho \quad (3\text{-}100)$$

$$\text{variance} = \sigma^2 = (\bar{x})^2(\rho^2-1) \quad (3\text{-}101)$$

$$\text{skewness} = (\rho^2-1)^{1/2}(\rho^2+2) \quad (3\text{-}102)$$

$$\text{kurtosis} = \rho^8 + 2\rho^6 + 3\rho^4 - 6 \quad (3\text{-}103)$$

$$\text{mode} = \alpha/\rho^2 \quad (3\text{-}104)$$

A sketch of a typical normal and a typical log-normal probability density function is presented in Fig. 3-18b [20].

The characteristic function of the probability density function of $\bar{x}$ for N noncoherently integrated pulse returns from a slowly fluctuating target is

$$\bar{C}_N(p) = \frac{1}{(p+1)^N} \sum_{k=0}^{\infty} \left(\frac{Np\alpha}{p+1}\right)^k \frac{\exp[k^2 \ln(\rho)]}{k!} \tag{3-105}$$

and from a rapidly fluctuating target

$$\bar{C}_N(p) = \frac{1}{(p+1)^N} \left\{ \sum_{k=0}^{\infty} \left(\frac{p\alpha}{p+1}\right)^k \frac{\exp[k^2 \ln(\rho)]}{k!} \right\}^N \tag{3-106}$$

Unfortunately, these characteristic functions are not absolutely convergent or useful for determining $f(v)$.

A method of determining the probability of detecting a target that exhibits log-normal characteristics by use of Meyer Plots is presented in Chapter 4, Section 4-5.

Table 4-1B in Chapter 4 presents a concise review of the major characteristics and applications of the standard and modern target models described in this chapter. Chapter 4 also shows how Meyer Plots can be efficiently used to solve problems involving these target models.

3-9 The Linear-Type Envelope Detector

An envelope detector is assumed to be of the linear type if the output is directly proportional to the envelope of the input. Marcum's calculations for a linear-type envelope detector reveal the difference in the resulting probability of detection as compared with that of a square-law detector to be so small that extreme accuracy is required in the calculations to show the relationship in true form. Figure 3-19 presents a graph by Marcum [4, Fig. 42] relating to probability of detection as a function of signal-to-noise power ratio (not in decibels). It compares the linear-law detector to the square-law detector for different values of N (number of pulses integrated) at a constant Marcum false alarm n of 10^6. Each curve of Fig. 3-19, therefore, corresponds to a different probability of false alarm. It can be seen that the variation in performance caused by the two types of detectors is small. Figure 3-20 presents a graph by Marcum [4, Fig. 43] relating to the difference in signal-to-noise requirements between a square-law and a linear-law detector, as a function of N, for a probability of detection P_D of 50% and a Marcum false alarm number n of 10^6. It is seen that both detectors perform identically at $N = 1$; the linear-law detector becomes better than the square-law detector as the number of pulses integrated N is increased until a maximum occurs at $N = 10$, where the difference signal-to-noise ratio is 0.11 dB; the improvement in the linear over the square-law detector decreases as the number of pulses increases above 10 until at $N = 70$ the linear and square-law detectors perform identically; above $N = 70$ the square-law becomes better and asymptotically exceeds the linear detector by 0.19 dB as N becomes infinite, having reached 0.16 dB at $N = 1000$.

Basically, there is little to choose from between a linear-law and a square-law detector as far as theoretical signal threshold is concerned.† It is apparent, therefore, that Meyer Plots are applicable to linear detectors. From the plots it is seen that an error of 0.2 dB in signal-to-noise ratio results in a maximum error of plus or minus a few percent in the probability of detection. When evaluating a linear-type envelope detector, the signal-to-noise ratio determined by use of Meyer Plots should be considered conservative for up to 70 pulses being integrated and optimistic for more than 70 pulses being integrated.

Marcum also developed an expression to determine the bias level Y_b for a linear-type envelope detector, which is useful when the number of pulses integrated N is greater than 2. To obtain it, he first derived an expression for the cumulative distribution function for noise alone ($x = 0$) in the form of a Gram–Charlier series, namely [4, Eq. 116],

$$P_{fa} = \frac{1}{2}[1 - \phi^{-1}(T)] + \frac{0.1053}{N^{1/2}}\phi^2(T) - \frac{0.0108}{N}\phi^3(T) - \frac{0.00555}{N}\phi^5(T) - \cdots \tag{3-107}$$

where the parameter

$$T = \frac{Y_b - N(\pi/2)^{1/2}}{\{N[2-(\pi/2)]\}^{1/2}} \tag{3-108}$$

† According to Hall [26] collapsing loss is more significant if a linear-law detector is used.

Fig. 3-19

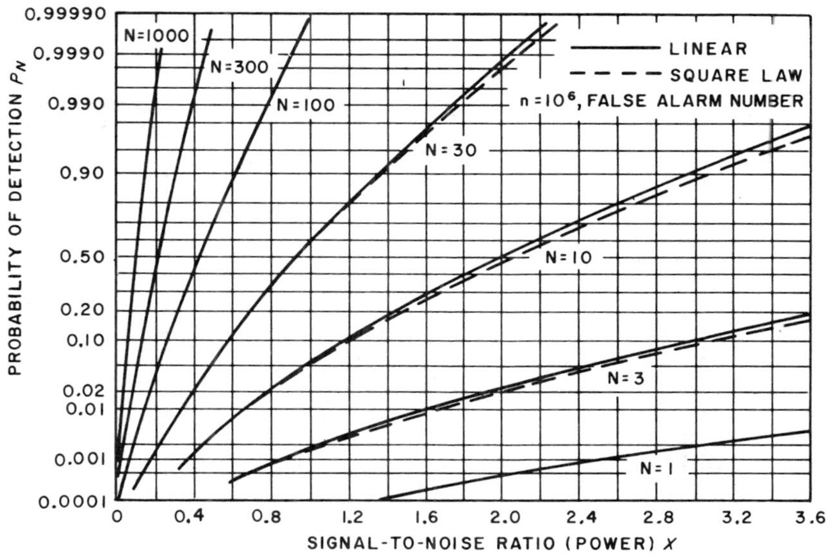

Fig. 3-20

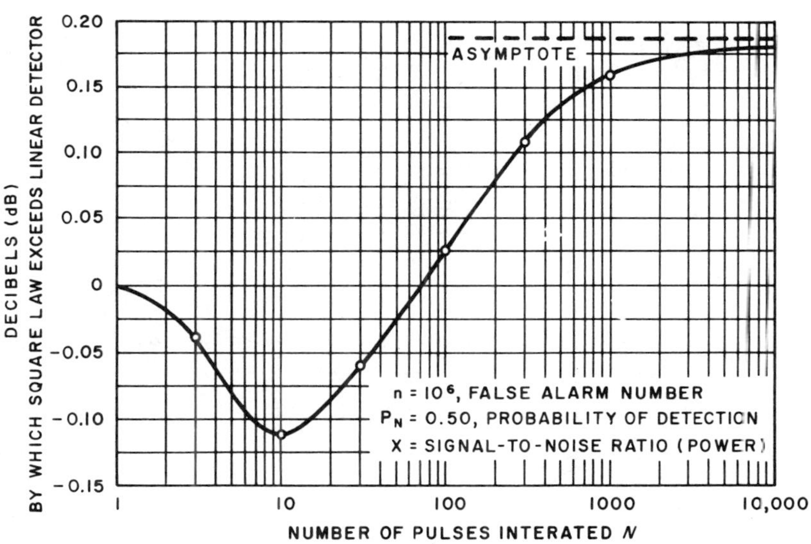

Fig. 3-19 Comparison of linear and square-law detector performance (abscissa expressed as power ratio, not in terms of decibels). (From Marcum [4].)

Fig. 3-20 Ratio of square-law to linear-law performance. (From Marcum [4].)

and

$$\phi(T) = \frac{\exp(-T^2/2)}{\sqrt{2\pi}} \tag{3-109}$$

For any given value of false alarm number n and number of pulses integrated N, the bias level Y_b can be determined by using Eqs. (2-24), (3-107)–(3-109), and the technique of trial and error.

An approximate value of T is found by use of Eq. (2-24) and by neglecting all but the first term of Eq. (3-107), namely,

$$0.693N/n = \tfrac{1}{2}[1-\phi^{-1}(T)] \tag{3-110}$$

A more accurate value T obtained through the use of Newton's method is

$$T_2 = T_1 - [f(T_1)/f'(T_1)] \tag{3-111}$$

Marcum found it better to plot Eq. (3-107) such that n is a function of T and N. From this the bias level graph (see Fig. 3-21) showing Y_b as a function of n and N for the linear detector was obtained.

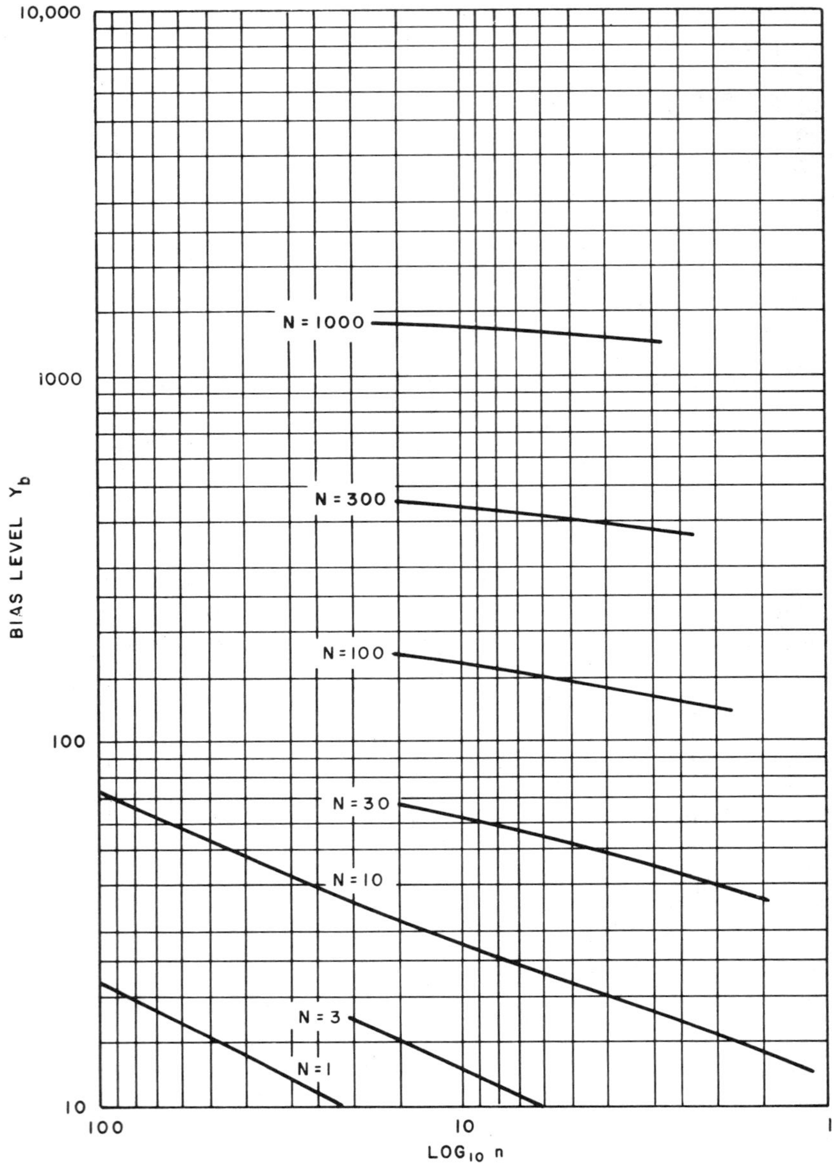

Fig. 3-21 Bias level for a linear-law detector versus Marcum false alarm number. (From Marcum [4].)

To calculate the probability of detection (for a nonfluctuating target) when noncoherent integration is employed following a linear detector, Marcum, as before to calculate P_{fa}, was forced to resort to the Gram–Charlier series technique. This time, however, it was necessary to regenerate the series for each value of x. It is feasible to write a general solution for P_D as one equation, but it would be several pages long and impractical to work with.

Marcum's technique for generating a Gram-Charlier series is tedious but explained in detail in his mathematical appendix [4]. The series has the form

$$f(y) = g(T) = \sum_{i=0}^{\infty} c_i \phi^i(T) \qquad (3\text{-}112)$$

where

$$T = (y-\bar{y})/\sigma \qquad (3\text{-}113)$$

and

$$\sigma = \text{standard deviation of } y \quad (3\text{-}114)$$

The coefficients c_i can be related to the standard central moments of the distribution function of interest. Basically, the technique is as follows:

(a) Find the standard central moments of the cumulative distribution function of interest, for example, of $P_{fa}(1)$ or $P_D(1)$.

(b) Use the relationships given by Cramer [27] to find the corresponding standard central moments for the noncoherently integrated sum of N pulses.

(c) Use the identities given by Marcum to calculate the c_i's from the moments determined in (b).

Prior to resorting to the Gram-Charlier series, Marcum explored the usual technique of working with the characteristic function of the probability density function of the output of a linear detector for a single pulse. For noise alone

$$\begin{aligned} dP_1 &= y \exp(-y^2/2)\, dy, \quad && y \geqslant 0 \\ &= 0, && y < 0 \end{aligned} \quad (3\text{-}115)$$

and

$$C_1(p) = -(2\pi)^{1/2} p \exp(p^2/2) \quad (3\text{-}116)$$

For signal-plus-noise

$$\begin{aligned} dP_1 &= y \exp\left[-\left(\frac{y^2+a^2}{2}\right)\right] I_0(ay)\, dy, \quad && y \geqslant 0 \\ &= 0, && y < 0 \end{aligned} \quad (3\text{-}117)$$

and

$$C_1(p) = 1 - (\pi/2)^{1/2} p \exp(p^2/2) \operatorname{erfc}(P/\sqrt{2}) \quad (3\text{-}118)$$

The usual method of finding $P_D(N)$ is first to let $C_N(p) = [C_1(p)]^N$, then find the anticharacteristic function of $C_N(p)$, and finally to integrate this result from Y_b to ∞. Marcum found this to be hopeless for the signal-plus-noise case, but for the noise-alone case he obtained the following approximate relationship:

$$P_{fa} = \frac{0.693N}{n} \approx \left(\frac{2\pi}{N}\right)^{N/2} \phi^{N-1}\left(\frac{Y_b}{N}\right) \quad (3\text{-}119)$$

The equation is impractical for values of N above 20 because of the limits imposed by the tables available for ϕ^i. It is of interest that no such approximation is derivable for the square-law case.

3-10 Best Possible Detector Law†

It is of considerable importance to know whether there may be some detector law which will give results that are appreciably better than those obtained from linear or square-law detectors.

The problem may be stated as follows. There are N samples available:

$$v_1, v_2, \ldots, v_N \quad (3\text{-}120)$$

which, it is assumed, are known to have come from either the distribution

$$dP_1 = v \exp(-v^2/2)\, dv \quad (3\text{-}121)$$

or the distribution

$$dP_2 = v \exp\left[-\left(\frac{v^2+a^2}{2}\right)\right] I_0(av)\, dv \quad (3\text{-}122)$$

the former being the distribution of the envelope of noise alone, and the latter the distribution of signal plus noise.

The probability that all of the variates $v_1, \ldots, v_N$ came from the second distribution is simply

$$dP_{N2} = dP_2(v_1)\, dP_2(v_2) \cdots dP_2(v_N) \quad (3\text{-}123)$$

whereas the probability that they all came from the first distribution is

$$dP_{N1} = dP_1(v_1)\, dP_1(v_2) \cdots dP_1(v_N) \quad (3\text{-}124)$$

The ratio of dP_{N2} to dP_{N1} is the best measure of the likelihood that all the variants came from the signal-plus-noise distribution. It can be shown that any monotonic function of this ratio gives an equally good significance test. One arbitrarily picks a constant which the ratio must exceed to say that it shows that the variants came from the signal-plus-noise distribution. This constant determines the false alarm time.

Taking the ratio of Eq. (3-124) to Eq. (3-123) and substituting values from Eqs. (3-121) and (3-122) yields

$$\frac{dP_{N2}}{dP_{N1}} = \frac{\prod_{i=1}^{N} v_i \exp[(v_i^2+a^2)/2]\, I_0(av_i)}{\prod_{i=1}^{N} v_i \exp(v_i^2/2)} \quad (3\text{-}125)$$

† This section is taken directly from Marcum [4].

or

$$\exp\left(-\frac{a^2}{2}\right)\prod_{i=1}^{N} I_0(av_i) \geqslant \lambda \qquad (3\text{-}126)$$

where λ is the constant which determines the false alarm time.

Taking the log of both sides of Eq. (3-126) yields

$$\sum_{i=0}^{N} \log_e I_0(av_i) \geqslant \log_e \lambda + (a^2/2) \qquad (3\text{-}127)$$

Note that nothing has been said in the foregoing discussion about integration. Now, however, Eq. (3-127) indicates that the best thing to do is take the log of I_0 of each variate, add these functions for each variate, and require the sum to exceed a certain value. Clearly this calls for a detector and integrator which has the combined law

$$y = \log_e I_0(av) \qquad (3\text{-}128)$$

Marcum found this result to be quite remarkable. It says, in effect, that by having the sum only of N variates which have been subjected to the law $y = \log I_0(av)$, one has as much useful information as if the individual values of each of the variates were known (as far as determining to which distribution the variates belong).†

Suppose that the signal strength is very small (which would make N large for any reasonable probability of detection). Then

$$I_0(av) \approx 1 + a^2v^2/4 \qquad (3\text{-}129)$$

and

$$y = \log_e I_0(av) \approx \log_e\left(1 + \frac{a^2v^2}{4}\right) \approx \frac{a^2v^2}{4} \qquad (3\text{-}130)$$

In this case, the square law is seen to be the best possible choice. If, on the other hand the signal strength is large,

$$I_0(av) \approx e^{av}/(2\pi av)^{1/2} \qquad (3\text{-}131)$$

† If the two distribution functions to be distinguished are normal, then the simple sum of the N variates, or the mean, is the best criterion. In other words, a linear law would be the best if the envelopes of noise and signal plus noise were normally distributed.

and

$$y = \log_e I_0(av) \approx \log_e[e^{av}/(2\pi av)^{1/2}] \approx av - \tfrac{1}{2}\log_e 2\pi av \approx av \qquad (3\text{-}132)$$

Thus, for large signals (N usually is small) the linear law is best.

It should be pointed out again that the results for the two extreme cases, square and linear law, are not very different (see Figs. 3-19 and 3-20), and in practice a linear detector would usually be preferred on account of its relative immunity to saturation by large signals.

3-11 Other Approaches to the Detection Criteria

SIGNAL-PLUS-NOISE MINUS NOISE

Sometimes the gain of an amplifier and/or the bias level of a detector is controlled by the average value of noise. This results in a signal-plus-noise minus noise-type detector. When a noise pulse is subtracted from a signal-plus-noise pulse, the result is defined as a composite pulse. The theoretical distribution functions for this composite pulse are entirely different from the distribution functions obtained previously for signal-plus-noise pulses and can be treated in any reasonable fashion only for the square-law detector case.

Marcum succeeded in obtaining theoretical results for the case of a system which integrated composite pulses using a square-law detector.† His analysis indicates there is no significant difference in the probabilities of detection for values of N ranging between 1 to 10 composite pulses integrated. For between 100 and 1000 composite pulses integrated, an effective signal-to-noise ratio approximately 1.0 dB lower than that obtained for the signal-plus-noise square-law detector case results. As can be seen by referring to various Meyer Plots, a 1-dB reduction in signal-to-noise ratio results in approximately a 30% reduction in probability of detection (typical for N between 100 to 1000 and $P_D = 90\%$).

An advantage of the signal-plus-noise minus noise scheme is that substantially lower bias levels result when N is large.

† Marcum [4] has a detailed review of this case.

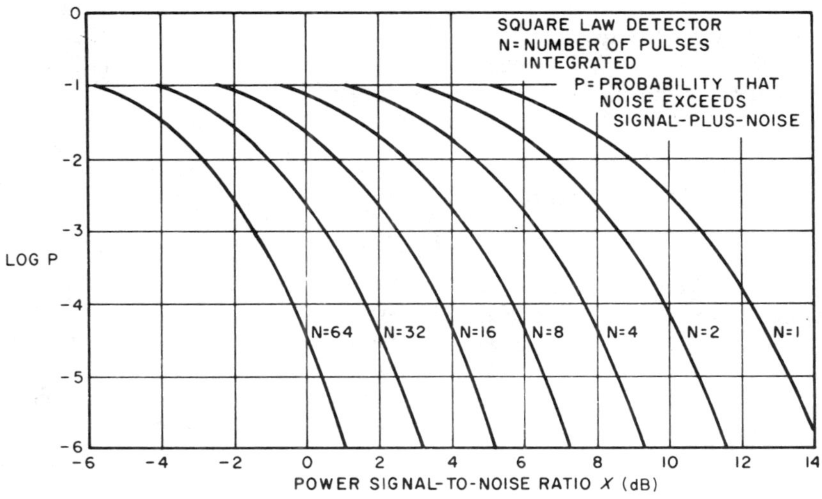

Fig. 3-22 Probability that noise exceeds signal-plus-noise. (From Marcum [4].)

SIGNAL-PLUS-NOISE EXCEEDING NOISE ALONE OVER A DEFINED PERIOD OF TIME

The common method of designating a bias or threshold level and calling any signal-plus-noise or noise-alone voltage that exceeds this level a detected signal is not the only way of defining detection. Another method of determining if a target is detected is by asking what the probability is that any given signal-plus-noise voltage will be larger than a noise-alone voltage during a given period of time. This period of time is usually taken as the false alarm time. The criterion is not useful for search modes because of the difficulty of picking out the largest signal over a period of time and because of the difficulty of deciding what to do when many signals are present. The criterion is useful, however, for determining the probability that an automatic tracking beam will lose its target due to noise. The calculation of these probabilities has been accomplished by Marcum and his results are presented in Fig. 3-22 [4, Fig. 48].

THE DEFLECTION CRITERION

Lawson and Uhlenbeck [11] have made use of a quantity which is the shift in average value of a signal-plus-noise distribution from that of a noise-alone distribution divided by the standard deviation of the noise-alone distribution. This quantity, which they called the detection criterion, is also called the deflection criterion. Marcum shows that the deflection criterion makes detection a function of $N^{1/2}$, whereas the bias level criterion makes detection a function of N raised to powers from about 0.75 to 1, depending on the value of N [4]. The deflection criterion, therefore, is not generally used. Marcum, however, used the deflection criterion to develop a proof that a square law for the detector–integrator is the best possible law for comparatively small signal-to-noise ratios (see Section 3-10).

MEDIAN, TRIMMED-MEAN, AND OTHER ORDERED CRITERIA

An ordered detection criterion is one where a group of pulse returns from a target are ordered according to their relative signal strengths. A target is declared to be present if, for instance, the Mth strongest signal return out of the group of N signal returns exceeds some predetermined threshold (bias) level. If $M = (N+1)/2$, then the detector is a median detector. If, for instance, all but the M_xth through M_yth strongest signal returns are deleted and those kept are integrated, then the detector is a trimmed-mean detector. Trunk and George have demonstrated that the median and trimmed-mean detectors will achieve a greater probability of detection for a given S/N when the probability density function of the contaminating signal has long tails [28, 29]. Such a condition is typical of clutter caused by high-sea states. A method of solving detection problems involving median detectors by use of Meyer Plots is presented in Chapter 4, Section 4-7.

3-12 Application of Pulse Radar Results to CW Radar System

The results arrived at for a pulsed radar system are directly applicable to a continuous-wave (cw) system. The relationships required for this conversion are presented here as established by Marcum [3].

The following symbology is introduced for a cw system:

P_{av} is the average cw transmitter power,

Δf_{cw} the effective combined rf and i.f. bandwidth of the cw receiver, and

η' the number of separate velocity channels incorporated into the receiver. (For both the pulse and cw analysis it has been assumed that the range and/or velocity gates are fixed in position. In the case where such gates sweep as a function of time, the analysis is not strictly valid. A rule of thumb is that the gate should move through the amplifier passband in a time equal to the reciprocal of the amplifier passband, which gives a visibility factor of approximately 0.8 [4]).

The conversion from a pulse system to a cw system requires the parameter substitutions shown in Table 3-4.

The parameter N must be less than or equal to ν for the conversion to be valid. If N is greater than ν, a form of collapsing loss occurs (see Sections 2-10 and 2-11).

An optimum cw system is one in which

$$\nu = 1, \quad N = 1, \quad \text{and} \quad T_d = 1/\Delta f_{cw} \tag{3-133}$$

This gives the greatest range for a given energy expended during the detection time T_d, as it corresponds to the case $N = 1$ and $\nu = 1$ in a pulsed system. It should be noted that if the number of range channels η in the pulsed system is equal to the number of velocity channels η' in the cw system, then the two systems, with $N = 1$ and $\nu = 1$, will have identical ranges for the same average power.

Reducing N reduces the system's capability of tracking a target that does not exhibit a reasonably constant velocity. If the velocity does vary considerably, the received signal will wander in and out of the passband of the receiver.

3-13 Beam-Shape and Scanning Loss

BEAM-SHAPE LOSS

Marcum assumed that the antenna pattern is flat over the half-power points and zero elsewhere [4]. Normally, the variation in the antenna pattern is not accounted for in the radar range equation. The usual assumption that the antenna pattern field strength is a constant can be inter-

Table 3-4

CONVERSIONS FROM PULSE TO CW SYSTEM

Pulse system parameter	Symbol	CW system equivalent	Definition of cw parameter(s)
Transmitted energy per pulse	E	$P_{av}/\Delta f_{cw}$	See beginning of this section
Number of pulses received during detection time	ν	$T_d \cdot \Delta f_{cw}$	T_d is the detection time
False alarm number	n	$t_{fa} \cdot \Delta f_{cw} \cdot \eta'$	t_{fa} is the false alarm interval
Number of pulses integrated	N	N	Number of variates (of duration $1/\Delta f_{cw}$) which are integrated after detection[a]

[a] Integration before detection is effectively the same as narrowing Δf_{cw} by the factor $1/N$.

preted to mean that as a constant target passes through a radar beam, the beam strength of the signal return will be rectangular in form, going from zero to some constant value and back to zero. In reality, as a constant target passes through a radar beam or a radar beam scans a constant target, or any combination of events thereof, the signal return is modulated by the shape of the antenna beam pattern.

The loss in the amount of received energy may be taken into account in the radar range equation by either

(1) taking an average value of antenna gain, or

(2) assuming the antenna's maximum power gain and modifying the radar range equation by introducing a beam-shape loss L_b.

Either approach is acceptable, but the latter is normally used in practice.

Beam shape (relative power gain) is often approximated by a Gaussian function, such that

$$\text{beam shape} \approx \exp(-k^2\theta^2/B^2) \quad (3\text{-}134)$$

where

$$\theta = \text{angular offset from the center of the beam} \quad (3\text{-}135)$$

$$k^2 = 4 \log_e 2 = 2.77$$

$$B = \text{the half-power beam width}$$

This approximation holds reasonably well out to plus or minus half the beam width from the point of maximum gain; the approximation deviates considerably from the actual antenna pattern at angles in excess of the main beam width because a Gaussian function does not reproduce the sidelobe radiation pattern of an antenna.

In the definition of beam-shape loss, it is assumed that when the beam is searchlighting a target, the transmitted pulse and its corresponding echo pulse will result from the same point in the beam pattern. If not, a scanning loss also exists.† For a single pulse then, the total two-way (transmit and receive) beam-shape loss when the target is off axis both in azimuth and elevation is

$$L_b = \exp\left[2k^2\left(\frac{\theta_a^2}{B_a^2} + \frac{\theta_e^2}{B_e^2}\right)\right] \quad (3\text{-}136)$$

† See latter part of this section.

where

L_b is the beam-shape power loss,

θ_a the azimuth angle between target and antenna axis,

θ_e the elevation angle between target and antenna axis,

B_a the azimuth half-power beam width, and

B_e the elevation half-power beam width.

The loss incurred when integrating a series of N pulses due to beam shape can be readily determined. For simplicity, assume that, in the series of N pulses, one pulse is coincident with the center of a Gaussian antenna beam pattern. The assumption will also be made that, for the purpose of this analysis, no scanning loss exists. The total signal power of N pulses received and integrated, considering no other losses, under these circumstances, is

$$S_N = S_1\left(1 + 2\sum_{i=1}^{(N-2)/2} \exp\left\{-2k^2 i^2 \times \left[\left(\frac{\Delta\theta_a}{B_a}\right)^2 + \left(\frac{\Delta\theta_e}{B_e}\right)^2\right]\right\}\right) \quad (3\text{-}137)$$

where

S_N is the total signal power representing N pulses,

S_1 the signal power received by the radar from a pulse that is both transmitted and received from the beam center,

$\Delta\theta_a$ the angular separation of pulses in azimuth, and

$\Delta\theta_e$ the angular separation of pulses in elevation.

For a radar integrating N pulses, the beam-shape loss

$$L_b = NS_1/S_N \quad (3\text{-}138)$$

Combining the preceding equations and letting

$$n_B^2 = \left[\left(\frac{\Delta\theta_a}{B_a}\right)^2 + \left(\frac{\Delta\theta_e}{B_e}\right)^2\right]^{-1} \quad (3\text{-}139)$$

permits relating the beam-shape loss to both the number of pulses integrated by the receiver N and the number of pulses that fall between the half-power points of the antenna pattern n_B. Thus

$$L_b = \left[\frac{N}{1 + 2\sum_{i=1}^{(N-2)/2} \exp\left(-\frac{2k^2 i^2}{n_B^2}\right)}\right] \quad (3\text{-}140)$$

Note that N should be less than n_B for this equation to give totally satisfactory results. As explained previously, a Gaussian function does not adequately approximate an antenna pattern beyond its half-power points. In addition, the equation as written is valid only for the case where one pulse is from the center of the beam.

If the target passes through other than the center of the beam, the maximum signal return will be somewhat less than indicated. The actual increase in beam-shape loss is the ratio of the square of the maximum antenna gain experienced, divided by the square of the antenna gain at beam center.

Antenna scanning speed was not considered in developing the equations for beam-shape loss, as it was assumed that the antenna scanning rate was slow enough so that the transmission gain and reception gain were the same. Scanning loss becomes significant for rapid scan antennas or for very long-range radars.

SCANNING LOSS

Because of its close relationship with beam-shape loss, scanning loss is often confused with it. Scanning loss is caused by beam motion and is due to the change in antenna position during the interval in which the signal travels from the antenna to the target and back. It increases sharply as the beam moves more than its beam width during the range delay time.† In the extreme case the pulse return will be too late to fall into the main lobe of the antenna pattern.

3-14 A Method of Using Meyer Plots for Optimizing the Number of Pulses Integrated When Accounting for Beam-Shape Loss [30]

As limited by beam-shape effects, the optimum number of pulses that should be integrated in order to obtain the highest probability of detection can be accurately determined using Meyer Plots. The approach described assumes that no scanning losses exist, that the beam is symmetrical about its center and does not have any beam-shape anomalies, and that the pulses of interest are symmetrically arranged about the center of the beam. It is also assumed that for any desired number of return signal pulses integrated, a mean antenna gain can be applied to all samples. The latter assumption is an approach that has been shown by Blake to result in a conservative estimate of performance [31].

As one applies this technique, it will become evident that integration of pulses that fall between only the half-power points of the beam is required, as this approximates the maximum number of pulses that can be advantageously integrated to improve the probability of detection. The end result is a good approximation to the optimum solution.

The assignment of target model and false alarm number will dictate the Meyer Plot to be used. An initial point on the Meyer Plot corresponding to the single-pulse signal-to-noise ratio that is approximately correct is selected. The calculated loss due to beam shape [using Eq. (3-140)] is subtracted from this parametric signal-to-noise value for each value of N selected and then marked on the Meyer Plot. All points are plotted relative to the same S/N curve. As the number of pulses integrated is increased, a greater portion of the beam is employed. The probability of detection will improve until the loss due to the beam shape becomes excessive. At that point the curve connecting these points is at its minimum; thus there will be a deterioration in the probability of detection for any further increase in the number of pulses integrated. For the examples tried by the authors the minimum point occurred consistently for a value of N associated with points on the beam approximately at the beam's half-power points (see illustrative problem below).

ILLUSTRATIVE PROBLEM RELATIVE TO THE METHOD OF OPTIMIZING THE NUMBER OF PULSES INTEGRATED WHEN ACCOUNTING FOR BEAM-SHAPE LOSS WITH THE AID OF MEYER PLOTS

Let us assume a search radar with a Gaussian beam shape. It is desired to determine the maximum number of pulses that can be integrated advantageously, as limited by beam-shape effects, to obtain the maximum probability of detection.

† Range delay time is the time the signal takes to travel from the antenna to the target and back to the antenna.

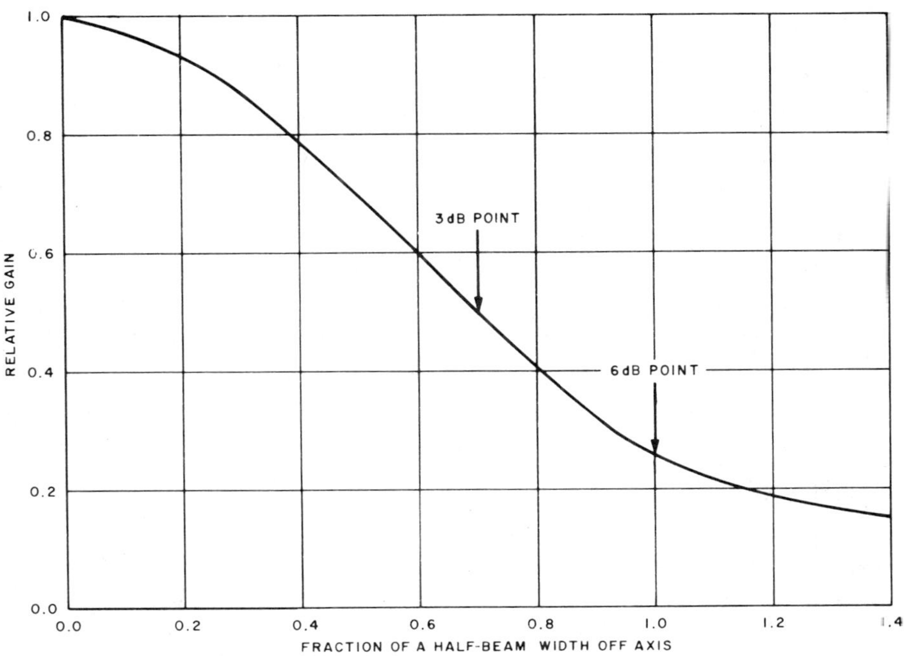

Fig. 3-23 Gaussian-shaped beam-power gain squared versus angular offset from beam axis.

Based on the given Gaussian beam shape, the expression for the ratio of the beam center power to the power at any given point within the main beam is

$$\frac{G(\theta)}{G_0} = \exp\left[-0.694\left(\frac{\theta}{\theta_0}\right)^2\right] \quad (3\text{-}141)$$

where

θ_0 is one-half the beam width based on a beam width as defined by the half-power points, and

θ the angle between target and the center of the beam. It represents the degree the target is off beam center.

Figure 3-23 presents a plot of the beam-shape power gain pattern squared $[G(\theta)/G_0]^2$. Since it is G^2 that appears in the radar range equation, it is this ratio that is of importance.

To simplify the example, it is assumed that the integrated pulses are uniformly and symmetrically distributed about the center of the beam and that exactly 21 pulses are available for integration between the half-power points. Now, by using Eq. (3-141) or Fig. 3-23 it is possible to determine the relative squared gain for each pulse by noting that the corresponding angular beam ratio (θ/θ_0)

Table 3-5

BEAM-SHAPE GAIN REDUCTION VERSUS N FOR GAUSSIAN GAIN PATTERN AND A MAXIMUM OF 21 PULSES BETWEEN 3-DB POINTS

Number of pulses integrated (N)	Angular position of extreme pulses (θ/θ_c)	Mean (gain)^2 loss (dB)
1	0	0.00
3	0.1	0.07
5	0.2	0.16
7	0.3	0.28
9	0.4	0.44
11	0.5	0.63
13	0.6	0.85
15	0.7	1.08
17	0.8	1.34
19	0.9	1.60
21	1.0	1.87
23	1.1	2.08
25	1.2	2.44
27	1.3	2.70
29	1.4	2.95
31	1.5	3.24

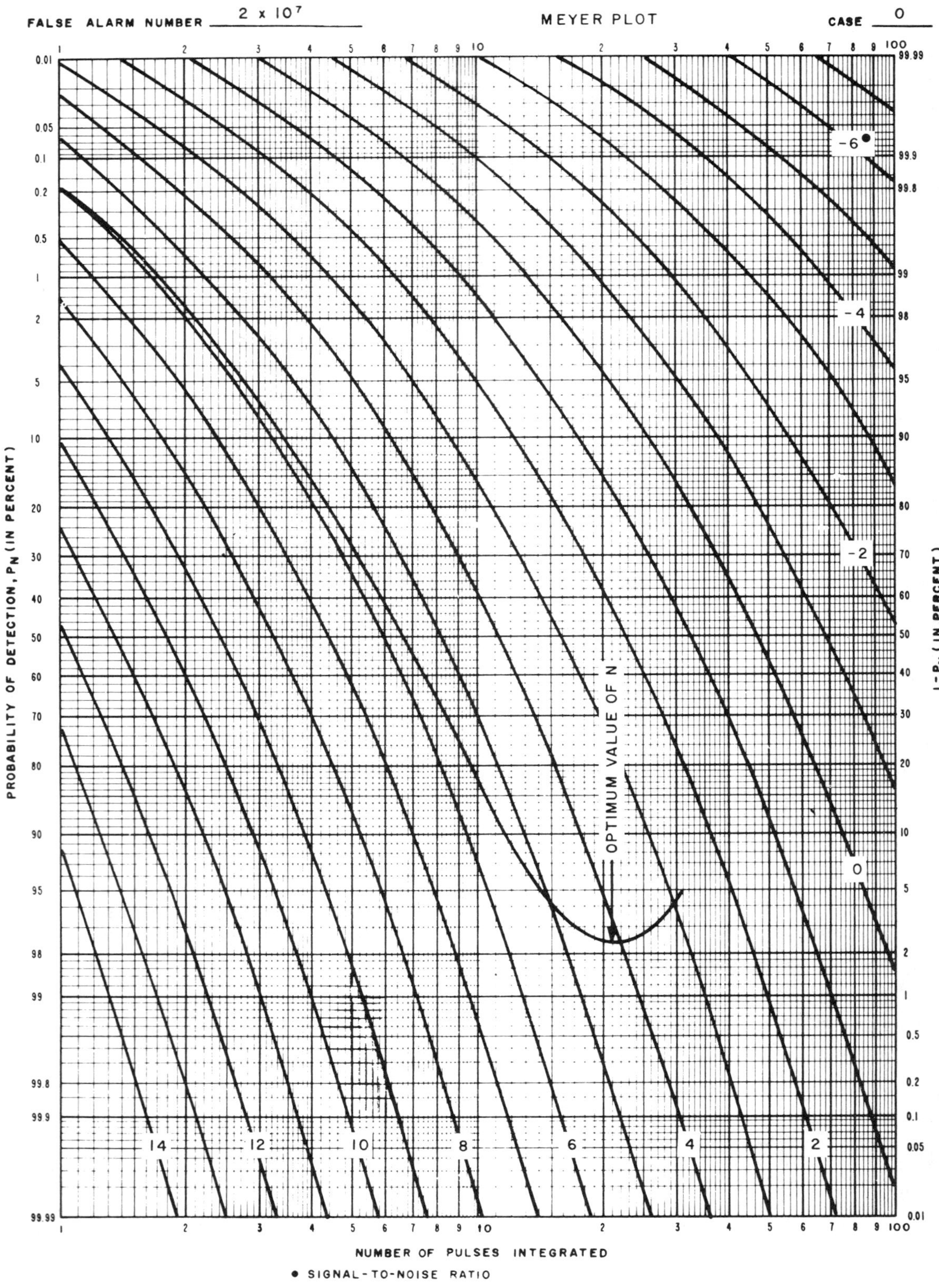

Fig. 3-24 Graphical method of optimizing the number of pulses integrated when accounting for beam-shape loss.

for successive pulse pairs is 0.1, 0.2, 0.3, etc. Next, the square of the individual gain terms is numerically integrated and then divided by N, as per Eq. (3-140), to determine the mean of the square of the gain. This is done for each value of N (see Table 3-5). If the target model of interest is the Case-0 nonfluctuating target, the false alarm number n' is 2×10^7, and the signal-to-noise power ratio with no beam-shape loss and $N = 1$ is 6.0 dB, the curve shown in Fig. 3-24 may be plotted on the Meyer Plot indicated. Note that the results verify the thesis that it is only profitable to integrate pulses that fall between the half-power points of the beam.

The technique just described works equally well with any antenna beam shape. Normal antenna beam shapes are a $(1/x)\sin(x)$-type function, not Gaussian as used in the preceding example. The Gaussian function, however, is easy to evaluate and yields very typical results, that is, that the optimum procedure is to integrate only pulses that fall between the half-power points of the antenna pattern and that the resulting beam-shape loss is approximately 2.0 dB.

3-15 Some Other Contributing Losses

LIMITING LOSS

When limiting occurs anywhere within a receiver, probability of detection is lowered, everything else being constant. Limiting usually occurs first in the video amplifiers. Marcum suggests handling this by defining a limiting ratio [4]. It is the ratio of the limit level to the rms noise level (both are voltages). He concludes that if the number of pulses integrated is large, the limiting loss is only a fraction of a decibel if the limiting

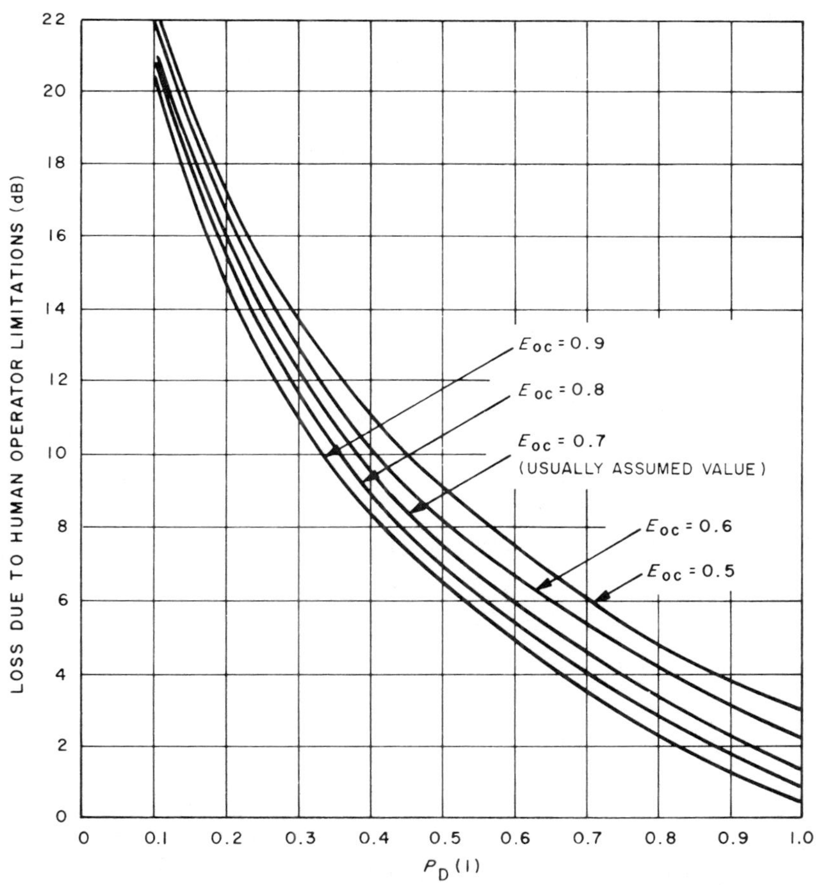

Fig. 3-25 Operator loss in decibels versus probability of detection.

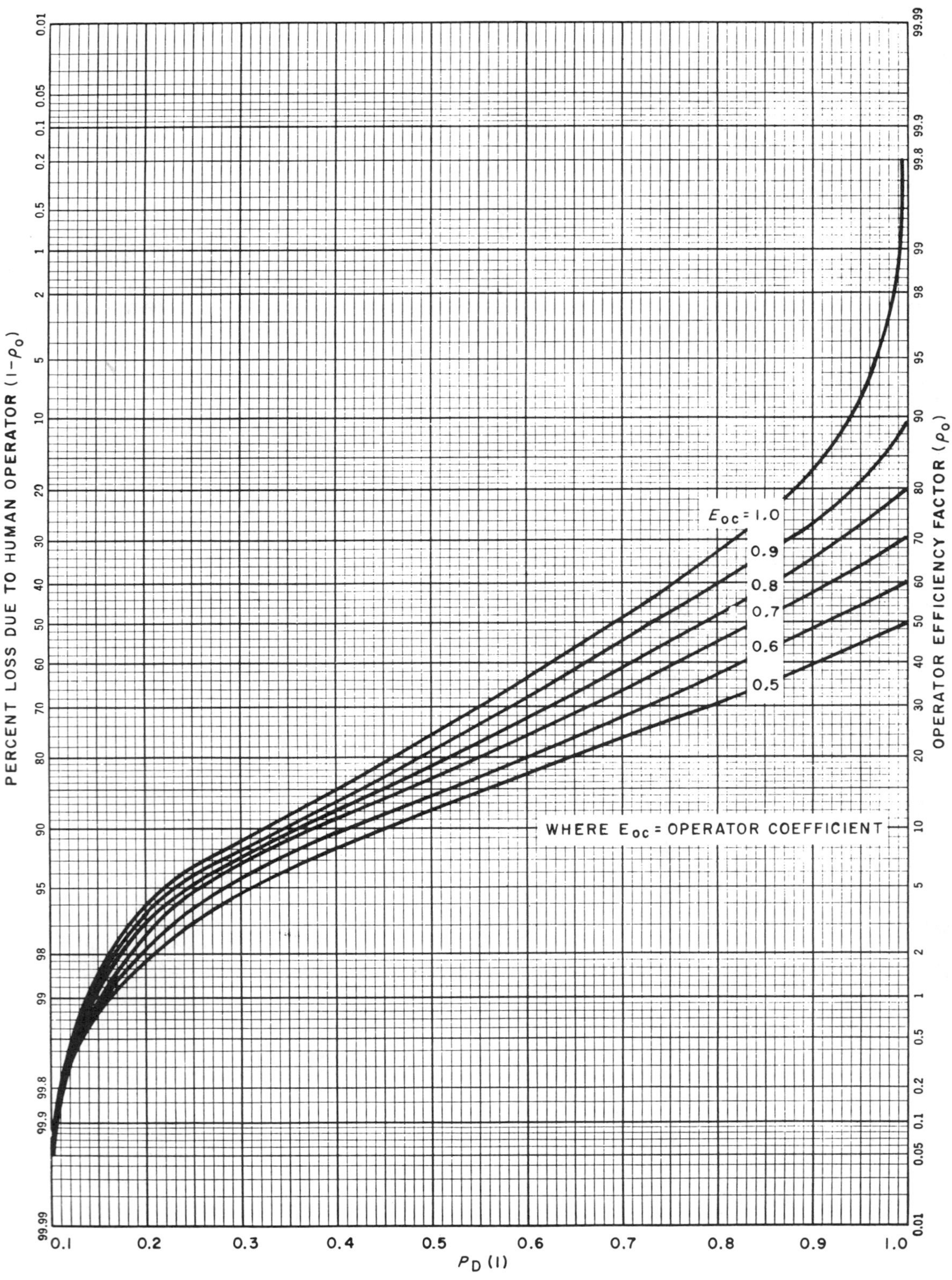

Fig. 3-26 Operator loss in percent versus probability of detection.

ratio is greater than 2 or 3. If no integration is performed, or just two or three pulses are integrated, the limiting ratio must be on the order of 10 to prevent a serious loss.

Limiting in the integrator can also cause a loss. This loss, however, in most practical cases is small compared to the loss caused by limiting of the individual pulses.

OPERATOR LOSS

An operator's ability to scan a CRT display visually in order to determine the presence of targets is limited to an information bandwidth of approximately 10 Hz (20 samples/sec). If an operator is fatigued, his abilities are degraded and correspondingly his information bandwidth may be significantly less.

Using empirical and experimental data, Skolnik defined an operator-efficiency factor (ρ_O), as follows [32]:

$$\rho_O = E_{oc}[P_D(1)]^2 \qquad (3\text{-}142)$$

where

E_{oc} is the operator coefficient (usually about 0.7), and

$P_D(1)$ the single-hit probability of detection or blip-scan ratio (see Figs. 3-2 through 3-16).

The above is based on the assumption that a good operator is observing a plan position indicator (PPI) display under good operating conditions. Under less favorable conditions the operator efficiency will be somewhat less (see Figs. 3-25 and 3-26). In many instances in the past, when operator efficiency has been analyzed, other sources of loss have been attributed to the operator and included as part of the operator loss. The most common of these are unaccounted-for propagation effects and loss due to field degradation. If possible, these losses should be separated from operator loss.

Figure 3-25 gives operator efficiency factor in terms of a loss expressed in decibels; that is

$$\text{operator loss} = 10\log_{10}(1/\rho_O) \qquad (3\text{-}143)$$

ATMOSPHERIC-ABSORPTION LOSS

Radar signals are to some degree attenuated by absorption during their round trip through the troposphere. For example, for a radar operating at 3 GHz with its antenna at zero elevation angle, the atmospheric-absorption loss will be approximately 1 dB if the target is 37 n.mi. distant. In general, the loss increases with increasing frequency and target range, and decreases with increasing elevation angle. Blake has generated a series of useful curves for determining this loss using the Van Vleck theory of oxygen and water-vapor absorption [33]. These curves are reproduced in the *Radar Handbook* [34]. Additional information can be found in the *Handbook of Geophysics* [35].

MISCELLANEOUS LOSSES

In addition to the radar system losses described in this handbook, there are numerous others that may occur particularly in specialized circumstances. A few of these are:

Polarization Loss The polarization of the received signal sometimes does not agree with the polarization of the receiving antenna because of unusual propagation effects or target characteristics.

Pulse Length Loss With some feed structures the rf pulse does not arrive at all elements of a phased-array antenna at the same time.

Squint Loss The effective antenna aperture of a phased-array antenna is reduced when the antenna is not broadside to the target.

References

1. W. Weinstock, Target Cross Section Models for Radar Systems Analysis. Ph.D. Dissertation in Electrical Engineering, Univ. of Pennsylvania (1964).
2. S. O. Rice, Mathematical Analysis of Random Noise. *Bell Syst. Tech. J.* **23**, 282 (1944); **24**, 44 (1945).
3. J. I. Marcum, Studies of Target Detection by a Pulsed Radar. Rand Res. Memo RM-754 (Dec. 1947); reprinted in *IRE Trans. Inform. Theory* **IT-6** (April 1960); also AD-101287.
4. J. I. Marcum, Studies of Target Detection by a Pulsed Radar—Mathematical Appendix. Rand Res. Memo RM-753 (July 1948); reprinted in *IRE Trans. Inform. Theory* **IT-6** (April 1960); also AD-101882.
5. P. Swerling, Probability of Detection for a Fluctuating Target. Rand. Res. Memo, RM-1217

(March 17, 1954); reprinted in *IRE Trans.* **IT-6** (April 1960).
6. D. K. Barton, "Radar System Analysis." Prentice-Hall, Englewood Cliffs, New Jersey, 1964.
7. J. S. Hey, H. Gent, and P. G. Smith, Echoing Area Characteristics, in "Avionics Research: Satellites and Problems of Long Range Detection and Tracking" (E. V. D. Glazier, E. Richtin, and J. Voge, eds.). Pergamon, Oxford, 1960.
8. G. N. Watson, "A Treatise on the Theory of Bessel Functions." Macmillan, New York, 1944.
9. D. P. Meyer, Further Studies of Target Detection by a Pulsed Radar, Part II. Final Rep. PN-35211-10, Sperry Gyroscope Div., Div. of Sperry Rand Corp., Great Neck, New York, 11020.
10. G. A. Campbell and R. M. Foster, "Fourier Integrals for Practical Applications." Van Nostrand Reinhold, Princeton, New Jersey, 1947; Bell Telephone Syst. Monograph No. B-584.
11. J. L. Lawson and G. E. Uhlenbeck, "Threshold Signals" (Radiation Lab. Ser., Vol. 24). McGraw–Hill, New York, 1950.
12. D. Kerr, "Propagation of Short Radio Waves." McGraw–Hill, New York, 1951.
13. R. H. Delano, "Further reply to comments by L. Peters, Jr., and F. C. Weiner" [14, 15] *IRE Trans. Antennas Propagations* **AP-9**, 110 (1961).
14. L. Peters, Jr. and F. C. Weiner, "Concerning the Assumption of Random Scatterers as a Model of an Aircraft for Tracking Radars." *IRE Trans. Antennas Propagation* **AP-9**, 110 (1961).
15. R. B. Muchmore, Reply to comments by L. Peters, Jr., and F. C. Weiner. *IRE Trans. Antennas Propagation* **AP-9**, 112 (1961).
16. F. E. Nathanson, "Radar Design Principles—Signal Processing and the Environment." McGraw–Hill, New York, 1969.
17. K. Pearson, "Tables of the Incomplete Gamma Function." Cambridge Univ. Press, London and New York, 1946.
18. P. Swerling, Recent Developments in Target Models for Radar Detection Analysis. AGARD Avionics Tech. Symp. Proc. (Istanbul, Turkey, May 25–29, 1970); Lecture Notes on Radar Target Signatures: Measurements, Statistical Models, and Systems Analysis. Technol. Serv. Corp., Santa Monica, Calif. (August 1968). The former is a condensation of the latter.
19. P. G. Hoel, "Introduction to Mathematical Statistics," 3rd ed. Wiley, New York, 1962.
20. J. Aitchison and J. A. C. Brown, "The Lognormal Distribution." Cambridge Univ. Press, London and New York, 1957.
21. R. S. Burington and D. C. May, Jr., "Handbook of Probability and Statistics with Tables." Handbook Publ., Sandusky, Ohio, 1953.
22. P. H. R. Scholefield, Statistical Aspects of Ideal Radar Targets. *Proc. IEEE* **55**, No. 4 (1967).
23. A. S. Goldman and T. B. Slattery, "Maintainability." Wiley, New York, 1964.
24. G. R. Heidbreder and R. L. Mitchell, Detection Probabilities for Log-Normally Distributed Signals *IEEE Trans. Aerospace Electron. Syst.* **AES-3**, No. 1 (1967); also Aerospace Corp. Report No. TR-669(9990)-6 (April 1966).
25. B. J. DuWaldt, private communication with Rice [2] Aerospace Corp., El Segundo, Calif.
26. W. M. Hall, Prediction of Pulse Radar Performance. *IEEE Trans. Aerospace Electron. Syst.* **AES-42**, 224–231 (1956).
27. H. Cramer, "Mathematical Methods of Statistics." Princeton Univ. Press, Princeton, New Jersey, 1946.
28. G. V. Trunk and S. F. George, Detection of Targets in Non-Gaussian Sea Clutter. *IEEE Trans. Aerospace Electron. Syst.* **AES-6**, 620–628 (1970).
29. G. V. Trunk (correspondence), Further Results on the Detection of Targets in Non-Gaussian Sea Clutter. *IEEE Trans. Aerospace Electron. Syst.* **AES-7**, 553–556 (1971).
30. Based on work summarized in Sperry Systems Management Div., internal memo by A. O. Sperber (Sept. 10, 1968).
31. L. V. Blake, The Effective Number of Pulses per Beamwidth for a Scanning Radar. *Proc. IRE* **41**, No. 6 (1953).
32. M. I. Skolnik, "Introduction to Radar Systems." McGraw–Hill, New York, 1962.
33. L. V. Blake, A Guide to Basic Pulse Radar Maximum-Range Calculation, U.S. Naval Res. Lab. Rep. 5868 (December, 1962, reprinted with revisions, December, 1963); 2nd ed., pt. 1, U.S. Naval Res. Lab. Rep. 6930 (1969).
34. M. I. Skolnik (ed.), "Radar Handbook," Chapter 2. McGraw–Hill, New York, 1970.
35. U.S.A.F. Air Research and Development Command, Air Force Research Division, Geophysics Research Directorate, "Handbook of Geophysics," Macmillan, New York, 1960.

Recommended reading for further study in addition to the references cited above:

J. V. DiFranco and W. L. Rubin, "Radar Detection." Prentice-Hall, Englewood Cliffs, New Jersey, 1968.

CHAPTER 4

MEYER PLOTS—DESCRIPTION AND USE

4-1 Basics of Meyer Plots and Target Models

Meyer Plots are no more than a graphical presentation of probability of detection P_D as a function of the number of pulses noncoherently integrated N with the average single-pulse signal-to-noise ratio, $(S/N)_1 \equiv \bar{x}$, as a running parameter. The plots are made on probability-versus-logarithmic paper.

The series of plots is subdivided into five groups, each group representing one of the five basic target models defined by Marcum and Swerling. Their defining characteristics are summarized in Table 4-1A. In the upper right-hand corner of each Meyer Plot, the applicable target model case number is indicated. In the upper left-hand corner of each Meyer Plot the applicable Fehlner false alarm number is indicated. Fehlner's false alarm number n' is related to Marcum's false alarm number n and the probability of false alarm P_{fa} by

$$n' = \frac{n}{N} = \frac{1}{P_{fa}} \log_e \left[\frac{1}{P(0)}\right] = \frac{1}{P_{fa}} \log_e(2) = \frac{0.693}{P_{fa}} \qquad (4\text{-}1)$$

where

$P(0) = 0.5$, the probability that noise will not exceed the bias level during Marcum's false alarm time (see Section 2-2).

Each plot is divided into two parts, one part on each of two pages. The first page covers values of N in the range 1–100 pulses being integrated. The second page covers values of N in the range of 100–10,000 pulses being integrated.

The constant average single-pulse signal-to-noise ratio curves (the running parameter) are plotted in increments of 1 dB (decibels of per-pulse average signal power divided by average noise power) over the range of -20 to $+20$ dB. Below -20 dB the increments are in steps of 5.0 dB, and above $+20$ dB, the increments are in steps of 2.0 dB. Along the left-hand margin of each plot is printed the probability of detection P_D and, correspondingly, along the right-hand margin is printed the probability of miss detection, $1-P_D$.

Table 4-1A

TARGET MODELS USED FOR MEYER PLOTS

Target Model Case	Density function used for target cross-section model	Fluctuation of signal return
0 (Marcum)	Nonfluctuating	No fluctuation
1 (Swerling 1)	Rayleigh	Slow (scan-to-scan) fluctuation
2 (Swerling 2)	Rayleigh	Rapid (pulse-to-pulse) fluctuation
3 (Swerling 3)	Chi-square with 4 degrees of freedom	Slow (scan-to-scan) fluctuation
4 (Swerling 4)	Chi-square with 4 degrees of freedom	Rapid (pulse-to-pulse) fluctuation

The target models of primary interest in modern radar work are the general chi-square (Weinstock), Rayleigh (Swerling 1 and 2), Swerling 3 and 4, Rice, log-normal, and nonfluctuating (Marcum). This chapter will describe procedures that permit easy solution of probability-of-detection problems involving these as well as other target models [1]. Table 4-1B is a detailed summary of particulars relating to these target models giving all the essential information needed to apply successfully the calculation procedures presented in the next seven sections of this handbook.

Table 4-1B

Target model (all target models listed below except the log-normal can be represented by the steady-state (Case 0) model if $K \geqslant N \geqslant 100$)	Case designation	Density function for one sample: $\omega(x, \bar{x})$ where x = single-pulse signal-to-noise ratio, $\bar{x}$ = mean of x, and σ^2 = variance of x	Type of target fluctuation	Number of statistically independent (groups of) pulses integrated
General chi-square with $2k$ degrees of freedom	Weinstock cases (especially if $0.3 \leqslant k \leqslant 0.7$)	$\frac{k}{(k-1)!\,\bar{x}}\left(\frac{kx}{\bar{x}}\right)^{k-1}\exp\left(\frac{-kx}{\bar{x}}\right)$ for $x > 0$, and $\omega = 0$ for $x \leqslant 0$; $\sigma^2 = \bar{x}^2/k$	Slow (scan-to-scan)	1
			Block-correlated (group-to-group)	$1 \leqslant F \leqslant N$
			Rapid (pulse-to-pulse)	N
Rayleigh power (chi-square with two degrees of freedom, i.e., $k = 1$)	Swerling 1 (Case 1)	$\frac{1}{\bar{x}}\exp\left(\frac{-x}{\bar{x}}\right)$ for $x > 0$, and $\omega = 0$ for $x \leqslant 0$; $\sigma^2 = \bar{x}^2$	Slow (scan-to-scan)	1
	None		Block-correlated (group-to-group)	$1 \leqslant F \leqslant N$
	Swerling 2 (Case 2)		Rapid (pulse-to-pulse)	N
Chi-square with four degrees of freedom, i.e., $k = 2$	Swerling 3 (Case 3)	$\frac{4x}{\bar{x}^2}\exp\left(\frac{-2x}{\bar{x}}\right)$ for $x > 0$, and $\omega = 0$ for $x \leqslant 0$; $\sigma^2 = \frac{\bar{x}^2}{2}$	Slow (scan-to-scan)	1
	None		Block-correlated (group-to-group)	$1 \leqslant F \leqslant N$
	Swerling 4 (Case 4)		Rapid (pulse-to-pulse)	N
Steady-state	Marcum (Case 0)	1	no fluctuation	1
Rice Power	None	$\frac{1}{\psi_0}\exp\left(-s-\frac{x}{\psi_0}\right)I_0\left[2\left(\frac{sx}{\psi_0}\right)^{1/2}\right]$ for $x > 0$, and $\omega = 0$ for $x \leqslant 0$; $\bar{x} = \psi_0(1+s)$; $\sigma^2 = \psi_0^2(1+2s)$	Slow (scan-to-scan)	1
			Block correlated (group-to-group)	$1 \leqslant F \leqslant N$
			Rapid (pulse-to-pulse)	N
Log-normal	None	$\exp\left[\frac{-\ln^2\left(\frac{x}{\alpha}\right)}{4\ln(\rho)}\right]\Big/ x[4\pi\ln(\rho)]^{1/2}$ for $x > 0$, and $\omega = 0$ for $x \leqslant 0$; $\bar{x} = \alpha\rho$; $\sigma^2 = (\bar{x})^2(\rho^2-1)$	Slow (scan-to-scan)	1
			Block correlated (group-to-group)	$1 \leqslant F \leqslant N$
			Rapid (pulse-to-pulse)	N
Other	None	Usually a density function obtained by empirical means, e.g., field test measurements on a specific real target	Slow (scan-to-scan)	1
			Block-correlated (group-to-group)	$1 \leqslant F \leqslant N$
			Rapid (pulse-to-pulse)	N

SUMMARY OF TARGET MODEL PARTICULARS[a]

Number of duo-degrees of freedom for an (exact/equivalent) chi-square model of $X=\sum_{i=1}^{N} x_i$ (K)	Ratio of mean to median of density function of X (ρ)	Physical interpretation of target model (see Sections 3-1, 3,2, 3-6, 3-7, and especially 3-8 for additional information)	Application procedure and instructions to determine required $\bar{x}$, the average single-pulse signal-to-noise ratio as a function of N, n', P_D, and target model
k (exact)	$\rho = 2.20$ for $K = 0.5$	Except for the specific cases described below there is no known theoretical scattering mechanism that results in chi-square density functions. The chi-square, however, is the leading term of Weinstock's Laguerre polynomial general density function	This density function, properly used, will be an adequate approximation to almost all other density functions of interest. Use appropriate Meyer Plots and the procedure described in Section 4-3 to obtain $\bar{x}$ vs. K for specific values to N, n', and P_D. Use Case-0 curves when $K \geqslant N > 100$
Fk (exact)			
Nk (exact)	$\rho \to 1$ rapidly as $K \to \infty$		
1 (exact)	$\rho = 1.44$	Corresponds to an assembly of random scatterers where no single one is dominant. Often, but not necessarily, identified with aircraft targets (Case 1 for jet and Case 2 for propeller)	Use applicable Meyer Plot to obtain exact solution for Cases 1 and 2; follow procedure for general chi-square to obtain solution for block-correlated case (see Section 4-3). Use Case-0 curves when $K \geqslant N > 100$
F (exact)	$1 \leqslant \rho \leqslant 1.44$		
N (exact)	$\rho \to 1$ rapidly as $K \to \infty$		
2 (exact)	$\rho = 1.18$	This chi-square function closely approximates the Rice case with $s = 2.4$. Weinstock found that these cases can be used to represent a stabilized cylinder or a randomly oriented smooth body if $\bar{x}$ is replaced by $\bar{x}/\rho$	Use applicable Meyer Plot to obtain exact solution for Cases 3 and 4; follow procedure for general chi-square to obtain solution for block-correlated case (see Section 4-3). Use Case 0 curves when $K \geqslant N > 100$
$2F$ (exact)	$1 \leqslant \rho \leqslant 1.18$		
$2N$ (exact)	$\rho \to 1$ rapidly as $K \to \infty$		
∞ (exact)	$\rho = 1$	Corresponds to a steady reflector such as a large spherical object, e.g., a balloon many wavelengths in diameter	Use Meyer Plot for Case 0 to obtain exact solution for a nonfluctuating target
k_{eq}[b]	$\rho < 1.44$	Also known as one-dominant-plus-Rayleigh target model. Corresponds to a steady reflector plus Rayleigh scatterers where s is the ratio of the radar cross sections of the former to the latter. Reduces to a Rayleigh when $s = 0$	Excellent approximate solutions for this case can be obtained by using the appropriate Meyer Plot and the procedure described in Section 4-4
Fk_{eq}	$1 \leqslant \rho < 1.44$		
Nk_{eq}	$\rho \to 1$ rapidly as $K \to \infty$		
k_{eq}	$\rho = \bar{x}/\alpha$, where α is median of x	Corresponds to a directive scatterer viewed from random spherical aspect angles. Often, but not necessarily, identified with ships and space vehicles. Typically $\sqrt{2} < \rho < 2$. Reduces to Case 0 when $\rho = 1$	Fair to excellent approximate solutions for this case can be obtained by using the appropriate Meyer Plot and the procedure described in Sections 4-5. Solutions for $\bar{x}$ are approximately within: 0.0–1.0 dB if $\rho \leqslant 2$; 1.0–2.5 dB if $2 < \rho \leqslant 4$
Fk_{eq}	$1 \leqslant \rho \leqslant \bar{x}/\alpha$		
Nk_{eq}	$\rho \to 1$ slowly as $K \to \infty$		
k_{eq}	—	Usually a model of a specific real target	The general chi-square density function will be an adequate approximation to almost all other well-behaved density functions except if ρ is large *and* K is small (see Section 4-6 for fitting procedure). If K is large, the Meyer Plots for Case 0 can be used
Fk_{eq}	—		
Nk_{eq}	$\rho \to 1$ as $K \to \infty$		

[a] Source: See [1].
[b] Equivalent value of k.

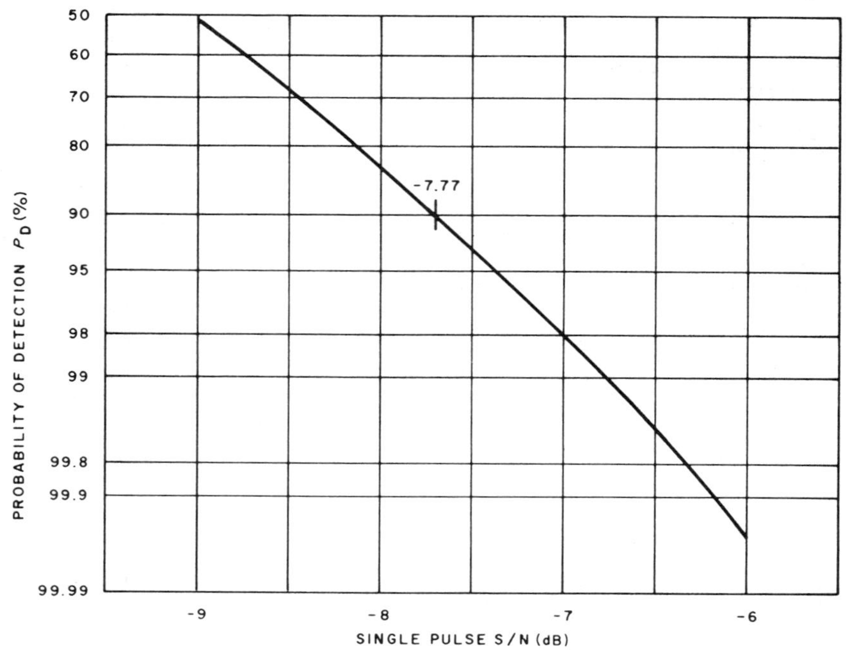

Fig. 4-1 Example of method of interpolation within the same Meyer Plot for Case 0, $N = 1000$, $n' = 10^4$.

The reader will note that in Table 4-1B the concept of block correlation has been introduced for the first time. This is the case where individual pulses fluctuate on a group-to-group basis rather than on a rapid (pulse-to-pulse) basis, where there is no correlation, or on a slow (scan-to-scan) basis, where all N pulses are correlated. Procedures for handling block correlation will also be described in this chapter [1].

Meyer Plots are also useful in solving probability-of-detection problems relating to radar signal processors employing double threshold and median detectors and/or recirculating integrators. The procedures for handling these situations are presented, respectively, in Sections 4-7 and 4-8.

4-2 Illustrative Look-Up Problems Using Meyer Plots and Methods of Interpolation

Interpolation between plotted values of signal-to-noise ratio and false alarm number (probability of false alarm) is often necessary. Graphical interpolation permits accurate coupling of any value of required single-pulse signal-to-noise ratio with any false alarm number n' and number of pulses integrated N within the range covered by the plots.

Table 4-2

MEYER PLOT DATA FOR EXAMPLE 1

$\bar{x}$ (dB)	P_D when $n' = 10^4$, $N = 1000$ (%)
-6	99.96
-7	98.05
-7.77[a]	90.0[a]
-8	83.5
-9	51.5

[a] Value determined from graph.

It should be noted that the spacing of false alarm numbers used for the plots is quasi-logarithmic in character, that is,

$$10,\ 2 \times 10,\ 4 \times 10,\ 6 \times 10,\ 8 \times 10,$$

$$10^2,\ 2 \times 10^2,\ 4 \times 10^2,\ \ldots,$$

$$2 \times 10^9,\ 4 \times 10^9,\ 6 \times 10^9,\ 8 \times 10^9,\ 10^{10}$$

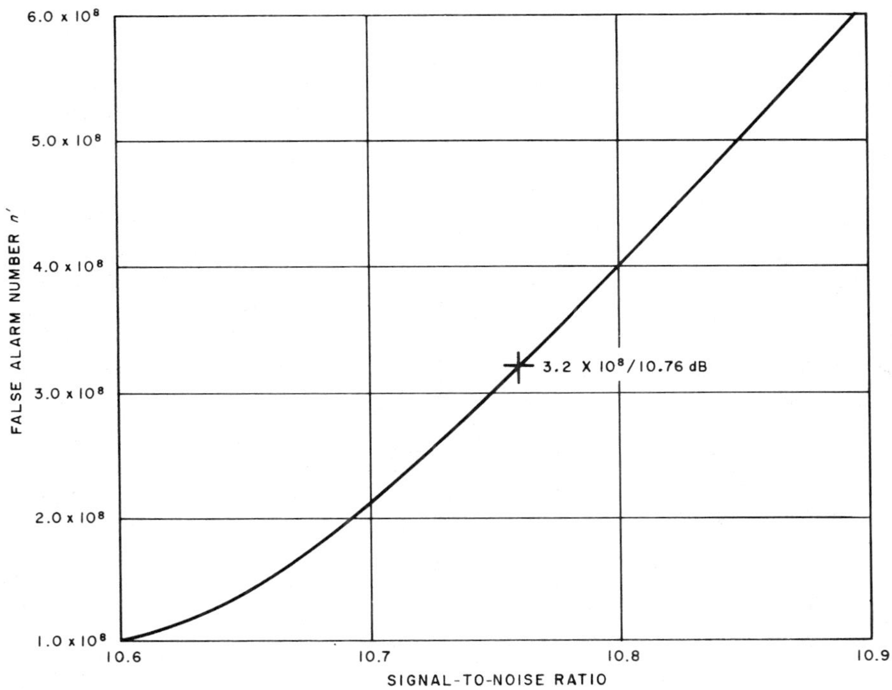

Fig. 4-2 Example of method of interpolation between adjoining Meyer Plots for Case 3, $N = 10$, $P_D = 90\%$.

In order to work with both a nonstandard signal-to-noise ratio and a nonstandard false alarm number, a two-step interpolation is required. This is demonstrated in the following examples:

EXAMPLE 1 Interpolation of S/N to Obtain a Given P_D

PROBLEM Determine the signal-to-noise ratio required for a 90% probability of detection when 1000 pulses are noncoherently integrated and the false alarm number n' is 10^4. Assume Target Model Case 0.

Solution The desired value of S/N could be closely estimated by a visual inspection and the result would be reasonably accurate. If the system designer desires a more accurately determined value of signal-to-noise ratio, he should select the two values of probability of detection above and the two below the 90% probability-of-detection line along the line corresponding to 1000 pulses being integrated. Examples of such values are shown in Table 4-2, and they should be plotted as in Fig. 4-1 on probability-versus-linear paper. The desired value of S/N is then easily determined from this plot. For the stated example the signal-to-noise ratio required for a 90% probability of detection is 7.77 dB.

Table 4-3

MEYER PLOT DATA FOR EXAMPLE 2

False alarm number n'	$\bar{x}$ for $P_D = 90\%$, $N = 10$ (dB)
10^8	10.6
2×10^8	10.7
3.2×10^8	10.76[a]
4×10^8	10.8
6×10^8	10.9

[a] Value determined from graph.

EXAMPLE 2 Interpolation of False Alarm Number n' to Obtain a Specific Probability of Detection P_D

PROBLEM For Target Model Case 3 determine the signal-to-noise ratio required for a 90% probability of detection with a false alarm number n' of 3.2×10^8 and with 10 pulses being noncoherently integrated.

Solution This requires intergraphical interpolation of values of S/N ratio relative to a specific value of P_D and N. Proceed by establishing a table similar to Table 4-3. Obtain the values of S/N from the four separate Meyer Plots with n' nearest to 3.2×10^8. Note that an interpolation of the type described in Example 1 above must be performed to obtain each of the four values of S/N. Again a simple estimate could be made and reasonably accurate results obtained as $\bar{x}$ is an insensitive function of n'. Best results are obtained, however, by use of a graphical solution. Construct a plot on linear-versus-linear paper similar to that shown in Fig. 4-2. The signal-to-noise ratio thus obtained for the stated example at a false alarm number $n' = 3.2 \times 10^8$ is 10.76 dB.

4-3 Chi-Square Target Models of Arbitrary Degrees of Freedom

In this section we describe a procedure by which Meyer Plots can be used to determine the required average single-pulse signal-to-noise ratio $\bar{x}$ when the degrees of freedom of the chi-square target model do not correspond to that of one of the five standard target models. This will make it possible to obtain exact solutions for the block-correlated and Weinstock cases, nearly exact solutions for all the Rice cases, and approximate solutions for most of the log-normal and other cases. Procedures for the Rice, log-normal, and other cases are presented in Sections 4-4, 4-5, and 4-6, respectively. The procedure for the general chi-square case is presented by means of an illustrative problem.

PROBLEM Determine the required average single-pulse signal-to-noise ratio $\bar{x}$ for a slowly fluctuating target that exhibits chi-square characteristics with $k = 0.5$ (typical Weinstock case). The following system parameters are specified:

$$P_D = 90\%, \qquad N = 20, \quad \text{and} \quad n' = 10^8$$

Furthermore, under varying operating conditions the 20 pulses will be evenly divided between either 1, 5, 10, or 20 different widely separated transmitter frequencies. These varying degrees of frequency diversity will cover the full range of block correlation.

Table 4-4

Meyer plot source[a]	$\bar{x}$ (dB)	Corresponding value of K
Case 1	12.5	1
Case 3	8.5	2
Case 2	4.6	$N = 20$
Case 4	4.4	$2N = 40$
Case 0	4.0	Asymptotic value as $K \to \infty$

[a] $n' = 10^8$; $N = 20$; $P_D = 90\%$.

Table 4-5

Number of different transmitter frequencies (F)	Corresponding value of K ($K = kF$)	Required $\bar{x}$ for $P_D = 90\%$, $n' = 10^8$ (dB)
1	$(0.5)(1) = 0.5$	~18.0
5	$(0.5)(5) = 2.5$	7.7
10	$(0.5)(10) = 5$	6.2
$N = 20$	$(0.5)(20) = 10$	5.2

Solution The first step in the solution is to obtain five values of single-pulse signal-to-noise ratios for the specified parameters, one each from the appropriate Meyer Plot of each standard target model. These $\bar{x}$'s correspond to values of K as shown in Table 4-4. It is next necessary to plot these data (S/N versus K) on semilog paper (preferably 4 or 5 cycle) as illustrated in Fig. 4-3. The required average single-pulse signal-to-noise ratios can now be obtained from this curve for the different aforementioned frequency diversity conditions by noting that $K = kF$. The parameter F is the number of statistically independent groups of pulses integrated and is equal to the number of widely separated transmitter frequencies used (see Table 4-5).

In order to obtain the desired solution for the $F = 1$ case, it is necessary to extrapolate further than one would normally desire; nevertheless, since the plotted function is "well behaved" in

Fig. 4-3 Chi-square target model with nonstandard K. P_D=90%, n'=10^8, N=20, k=0.5 (see [1]).

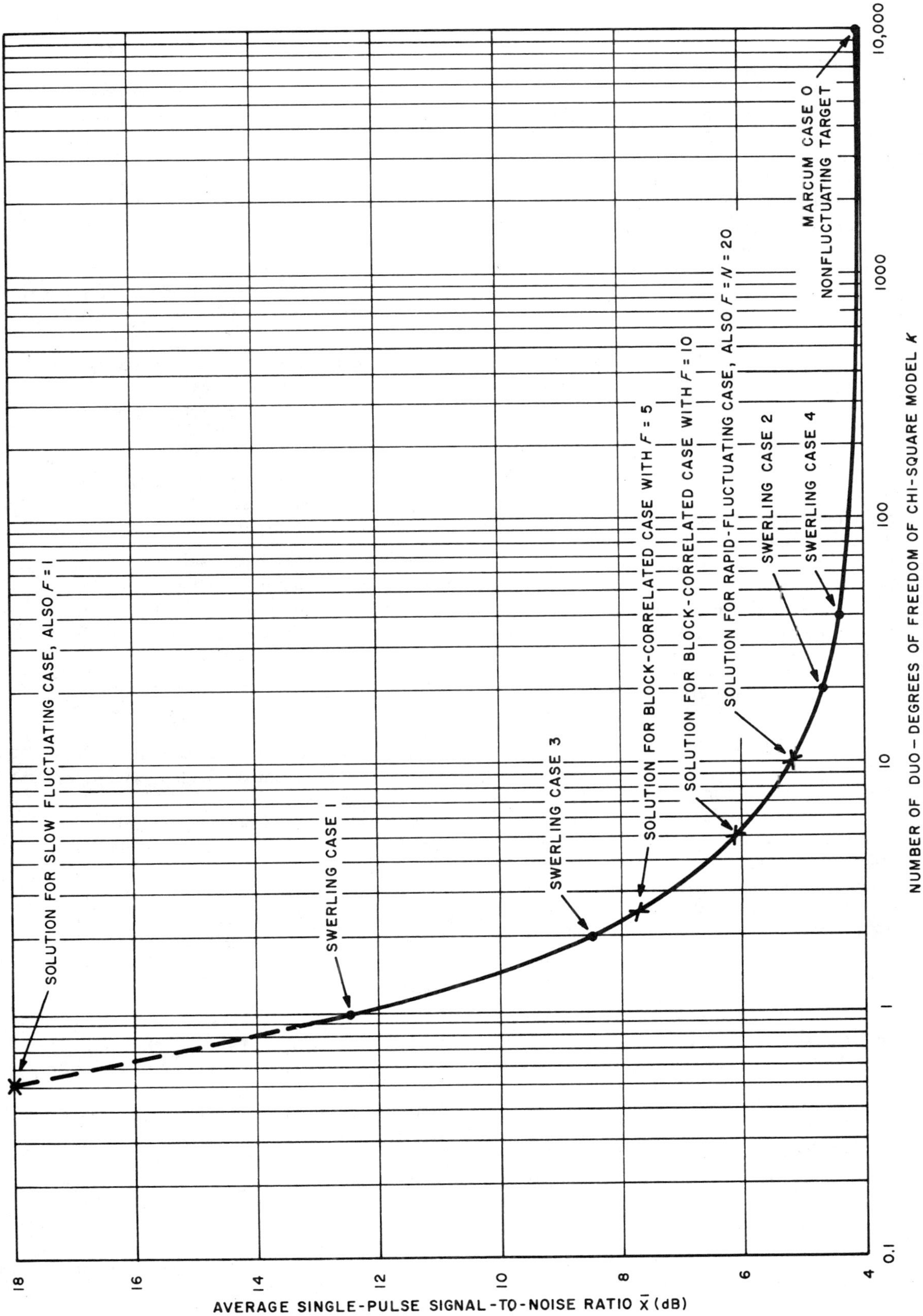

SOLUTION FOR SLOW FLUCTUATING CASE, ALSO F = 1
SWERLING CASE 1
SWERLING CASE 3
SOLUTION FOR BLOCK-CORRELATED CASE WITH F = 5
SOLUTION FOR BLOCK-CORRELATED CASE WITH F = 10
SOLUTION FOR RAPID-FLUCTUATING CASE, ALSO F = N = 20
SWERLING CASE 2
SWERLING CASE 4
MARCUM CASE 0
NONFLUCTUATING TARGET
0.1
1
10
100
1000
10,000
NUMBER OF DUO-DEGREES OF FREEDOM OF CHI-SQUARE MODEL K
18
16
14
12
10
8
6
4
AVERAGE SINGLE-PULSE SIGNAL-TO-NOISE RATIO X̄ (dB)

the region of interest, with care, a useful result is obtained, i.e., $\bar{x} \cong 18$ dB.

From the preceding example it would appear that frequency diversity improves radar performance. This is generally true for $P_D > 70\%$; for $P_D < 50\%$ the opposite is true. The demonstration of this is left as an exercise for the reader. Try the following values for P_D: 1%, 10%, 50%, 90%, and 99%. Plot all five as in Fig. 4-3 on a sheet of five-cycle semilog graph paper. The reader can anticipate the results by inspecting Fig. 4-9.

4-4 Approximation for Rice Power (One-Dominant-Plus-Rayleigh) Target Model

The Rice Power Target Model is the exact solution for the case of one steady-state reflector immersed in a field of random scatterers. The ratio of the radar cross sections of the one-dominant component to the Rayleigh component is denoted by s. Details of the target model are presented in Section 3-8.

An excellent approximation to the Rice Power result can be obtained using the chi-square target model with parameters defined as follows:

$$\bar{x}\text{ (chi-square)} = \bar{x}\text{ (Rice)} \qquad (4\text{-}2)$$

$$k\text{ (chi-square)} = 1 + \frac{s^2}{1+2s} \qquad (4\text{-}3)$$

According to Swerling [2] the value of $\bar{x}$ thus obtained will be within 1.0 dB of the exact Rice Power result over the range of P_D from 1% to 99%. The above relationships were obtained by "moment matching," a technique that will be described in Section 4-6.

The following illustrative problem will serve to demonstrate the details of the approximation procedure and the accuracy of the end result:

PROBLEM Determine the required average single-pulse signal-to-noise ratio $\bar{x}$ for a target that exhibits Rice Power characteristics with $s = 1$. The following system parameters are specified:

$$P_D = 90\%, \qquad N = 10, \quad \text{and} \quad n' = 10^5$$

Furthermore, the pulse and scan rates are yet to be fixed; so it is desired to explore three possible conditions of correlation:

(1) complete correlation for when all pulses are from one scan;

(2) block correlation for when the pulses are divided between two scans;

(3) no correlation for when each pulse is from a different scan.

The assumption is that correlation will exist pulse-to-pulse but not scan-to-scan. This is the usual case, but it is not always true.

Solution It is first necessary to plot a chi-square $\bar{x}$-versus-K curve following the procedure presented in the previous section. The result is shown in Fig. 4-4. The value of k for the best fitting chi-square approximation is

$$k = 1 + \frac{s^2}{1+2s} = 1 + \frac{1}{1+2} = \frac{4}{3} \qquad (4\text{-}4)$$

The approximate required average single-pulse signal-to-noise ratios are read off the curve as shown in Table 4-6.

Table 4-6

Number of scans for N pulses (F)	Corresponding value of K ($K = kF$)	Required $\bar{x}$ for $P_D = 90\%$, $n' = 10^5$ (dB)
1	$(4/3)(1) = 1.33$	11.0
2	$(4/3)(2) = 2.67$	8.2
10	$(4/3)(10) = 13.3$	5.5

According to Scholefield [3], for $F = 1$ the value of $\bar{x}$ should be 12.0 dB instead of 11.0 dB. Thus, at worst, the approximation results in an error of 1 dB. The error is less for larger values of F; that is, as $F \to \infty$, the target model approaches Case 0. It should also be noted that for $s = 0$ the Rice Power and the Rayleigh are one and the same.

4-5 Approximation for Log-Normal Target Model

The log-normal function permits the modeling of fluctuating-target radar cross sections that exhibit a pronounced tail for high values of cross section. It is characterized by mean-to-median ratios (ρ) usually larger than those for the five

Fig. 4-4 Rice Power approximation using chi-square data. $P_D = 90\%$, $n' = 10^5$, $N = 10$, and $s = 1$ (see [1]).

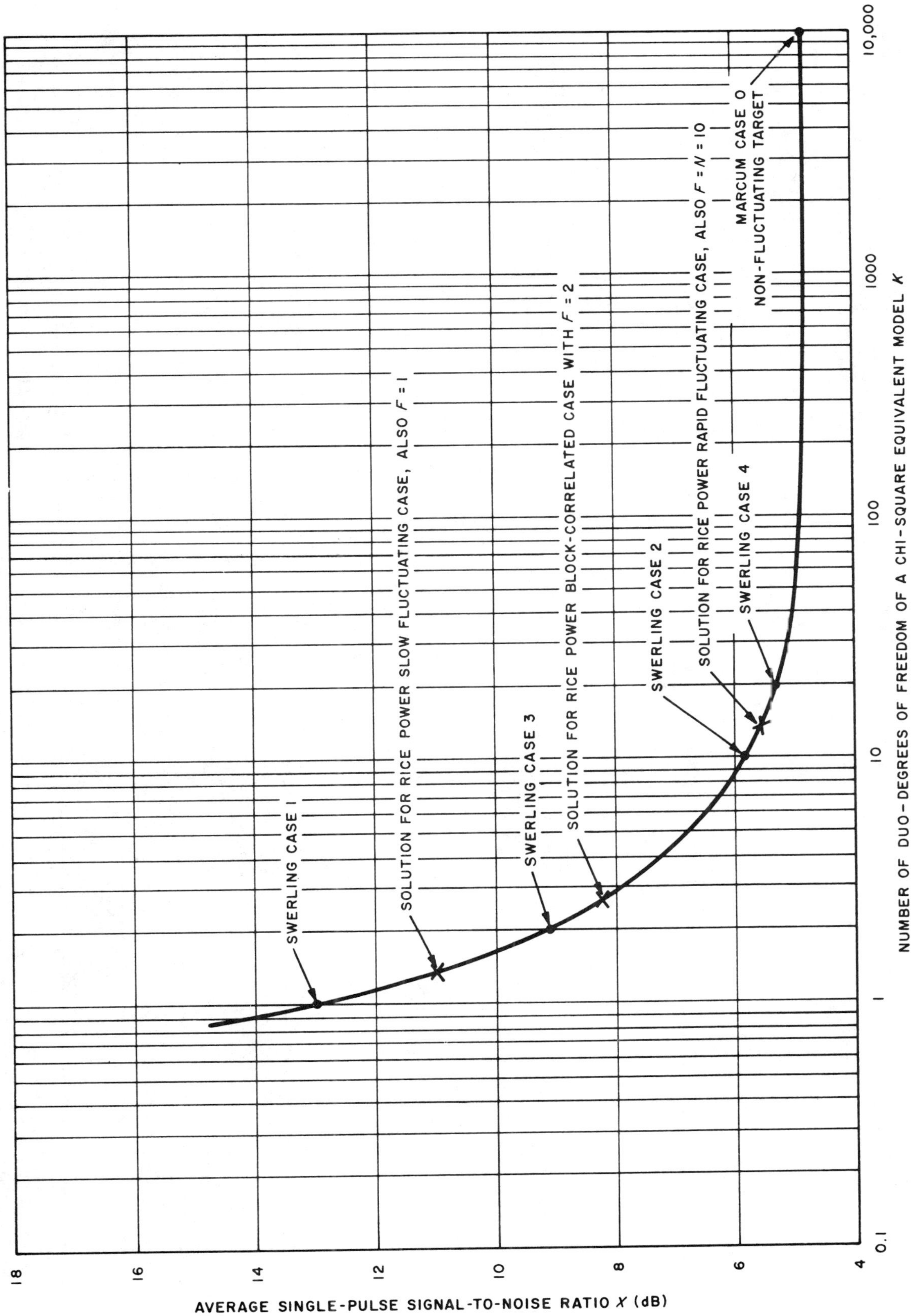
SWERLING CASE 1
SOLUTION FOR RICE POWER SLOW FLUCTUATING CASE, ALSO $F = 1$
SWERLING CASE 3
SOLUTION FOR RICE POWER BLOCK-CORRELATED CASE WITH $F = 2$
SWERLING CASE 2
SOLUTION FOR RICE POWER RAPID FLUCTUATING CASE, ALSO $F = N = 10$
SWERLING CASE 4
MARCUM CASE 0
NON-FLUCTUATING TARGET
0.1
1
10
100
1000
10,000
NUMBER OF DUO-DEGREES OF FREEDOM OF A CHI-SQUARE EQUIVALENT MODEL K
18
16
14
12
10
8
6
4
AVERAGE SINGLE-PULSE SIGNAL-TO-NOISE RATIO $\bar{X}$ (dB)

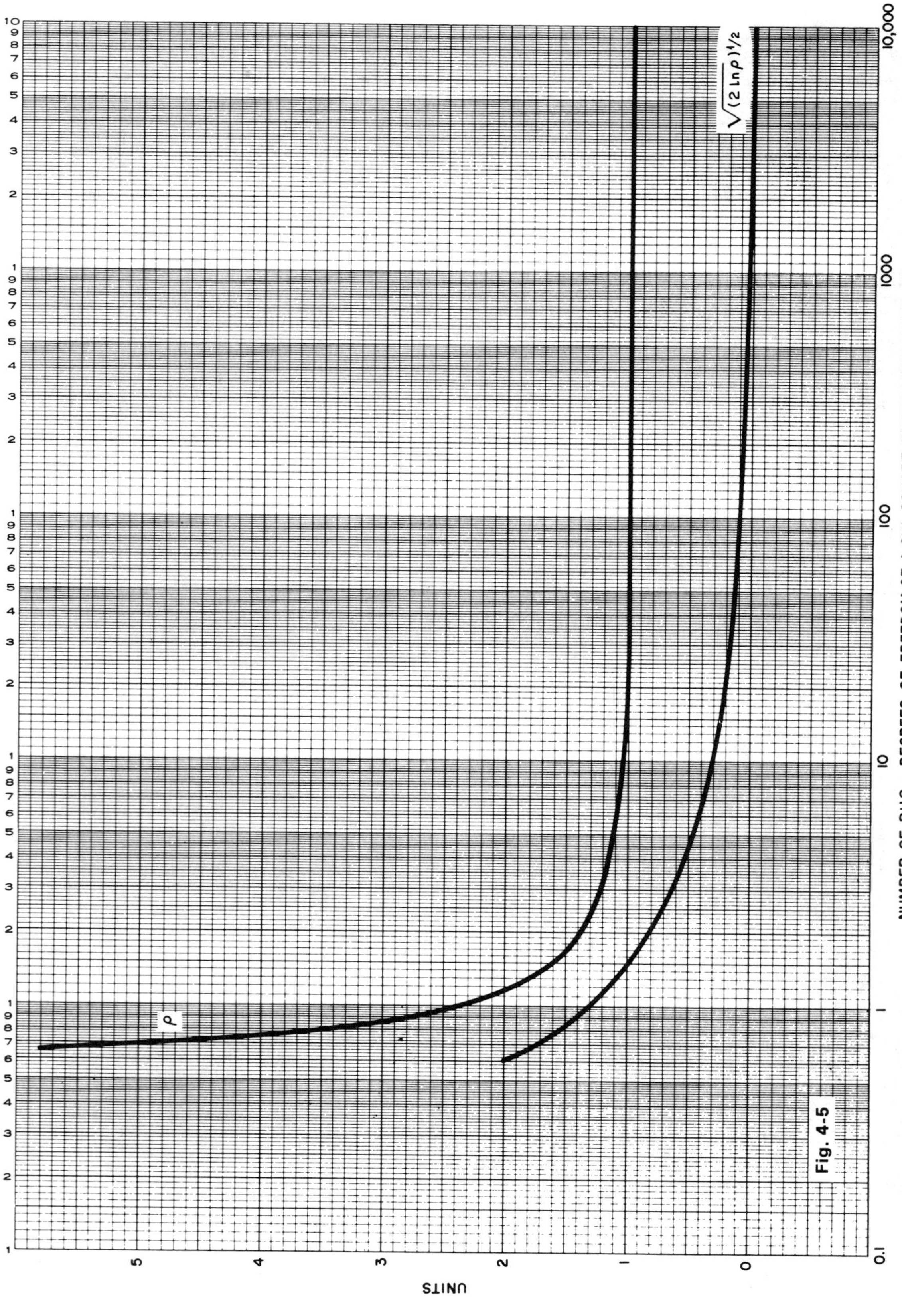
NUMBER OF DUO - DEGREES OF FREEDOM OF A CHI-SQUARE EQUIVALENT MODEL K
UNITS
ρ
√(2 ln ρ)^{1/2}
0.1
1
10
100
1000
10,000
0
1
2
3
4
5

Fig. 4-5

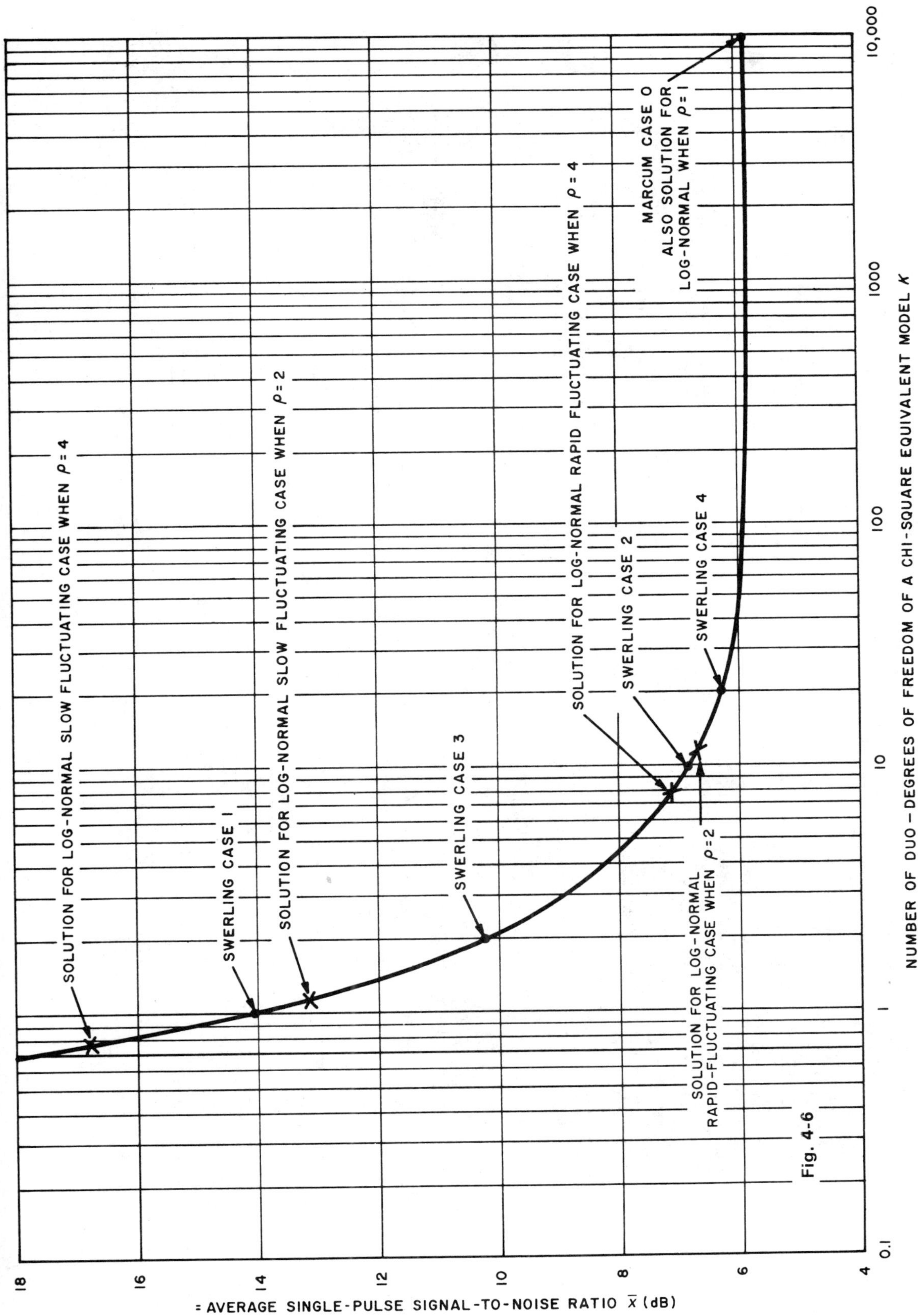
SOLUTION FOR LOG-NORMAL SLOW FLUCTUATING CASE WHEN ρ = 4
SWERLING CASE 1
SOLUTION FOR LOG-NORMAL SLOW FLUCTUATING CASE WHEN ρ = 2
SWERLING CASE 3
SOLUTION FOR LOG-NORMAL RAPID FLUCTUATING CASE WHEN ρ = 4
SWERLING CASE 2
SWERLING CASE 4
MARCUM CASE 0
ALSO SOLUTION FOR
LOG-NORMAL WHEN ρ = 1
SOLUTION FOR LOG-NORMAL
RAPID-FLUCTUATING CASE WHEN ρ = 2
0.1
1
10
100
1000
10,000
NUMBER OF DUO-DEGREES OF FREEDOM OF A CHI-SQUARE EQUIVALENT MODEL K
18
16
14
12
10
8
6
4
= AVERAGE SINGLE-PULSE SIGNAL-TO-NOISE RATIO x̄ (dB)
Fig. 4-6

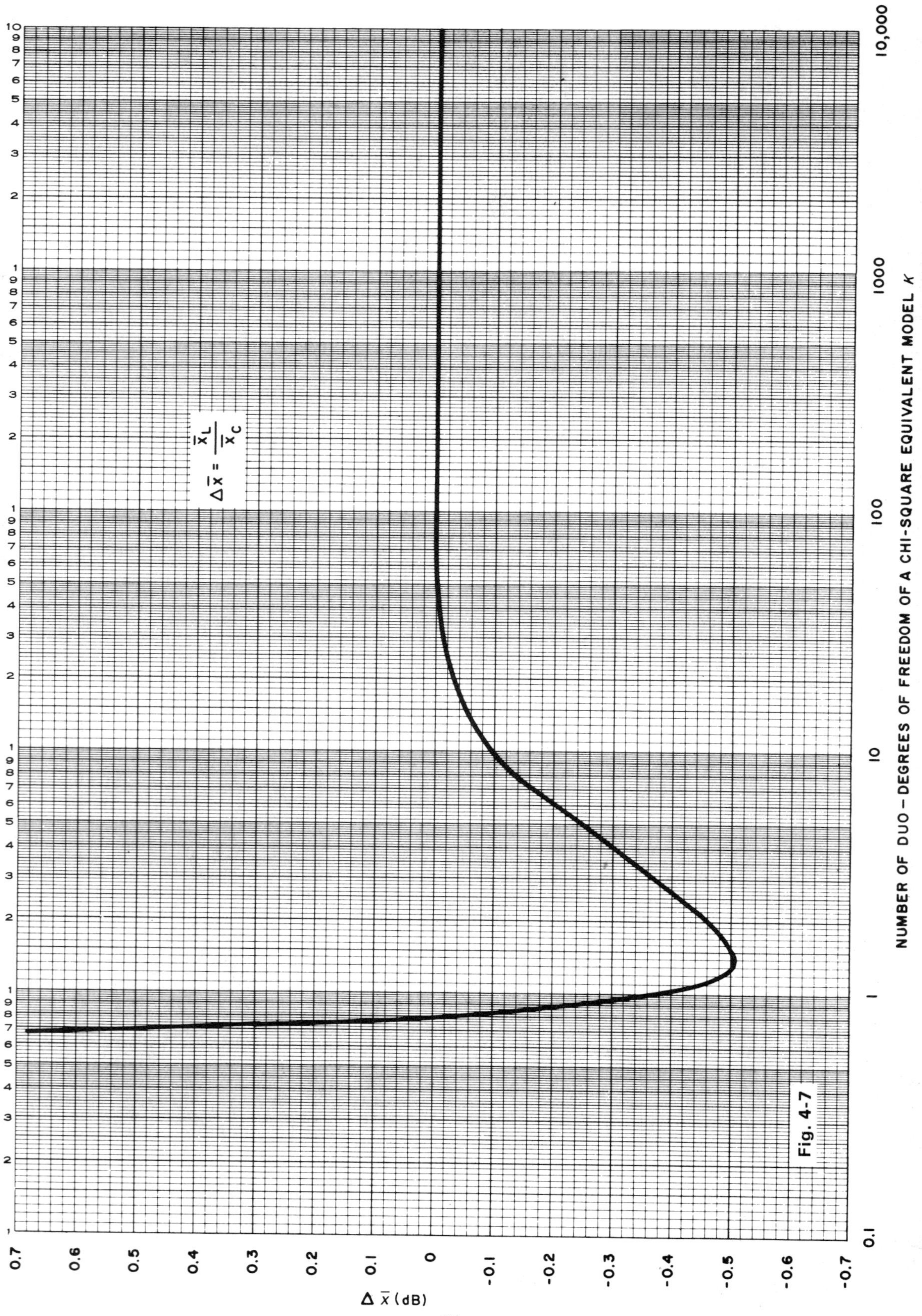

$\Delta \bar{x} = \frac{\bar{x}_L}{\bar{x}_C}$
NUMBER OF DUO-DEGREES OF FREEDOM OF A CHI-SQUARE EQUIVALENT MODEL κ
0.1
1
10
100
1000
10,000
$\Delta \bar{x}$ (dB)
0.7
0.6
0.5
0.4
0.3
0.2
0.1
0
-0.1
-0.2
-0.3
-0.4
-0.5
-0.6
-0.7

Fig. 4-7

Table 4-7

Fluctuating model	Corresponding value of K ($K = kF$)	Value of $\bar{x}$ obtained from curve (dB)	Correction for difference in means (dB)	Required $\bar{x}$ (dB)	Correct $\bar{x}$ obtained from [4] (dB)
$\rho = 1$					
Slow	$(\infty)(1) = \infty$	5.8	0	5.8	5.8
Rapid	$(\infty)(10) = \infty$	5.8	0	5.8	5.8
$\rho = 2$					
Slow	$(1.17)(1) = 1.17$	13.1	+0.47	13.57	14
Rapid	$(1.17)(10) = 11.7$	6.65	+0.47	7.12	7.5
$\rho = 4$					
Slow	$(0.76)(1) = 0.76$	16.8	−0.14	16.66	20
Rapid	$(0.76)(10) = 7.6$	7.14	−0.14	7.00	9.7

standard target models. Details of the target model are presented in Section 3-8.

An excellent approximation of the log-normal target model for $1 \leqslant \rho \leqslant 2$ can be achieved by use of the appropriate chi-square target model. The approximation deteriorates for larger values of ρ. The approximation procedure is demonstrated by means of the following illustrative problem.

PROBLEM Determine the required single-pulse signal-to-noise ratio $\bar{x}$ for a class of targets that exhibit log-normal characteristics with $1 \leqslant \rho \leqslant 4$. The following system parameters are specified:

$$P_D = 90\%, \quad N = 10, \quad \text{and} \quad n' = 10^7$$

The problem will be solved for $\rho = 1$, 2, and 4 and for both slowly and rapidly fluctuating targets.

Solution It is first necessary to find the value of k for the best fitting chi-square target model. This is done with the aid of Fig 4-5. Thus, for the stated problem:

$$\rho = 1, \quad 2, \quad 4$$

$$k = \infty, \quad 1.17, \quad 0.76$$

It is also necessary to plot a chi-square $\bar{x}$-versus-K curve, as shown in Fig. 4-6, following the procedure described in Section 4-3. Since the mean value of the log-normal is somewhat less than that of the best fitting chi-square, a correction factor $\Delta\bar{x}$ will have to be applied to the $\bar{x}$'s obtained from this curve. The amount of correction is plotted in Fig. 4-7 as a function of k. The approximate required average single-pulse signal-to-noise ratios can now be determined from

$$\bar{x}_1 = (\bar{x}_c)(\Delta\bar{x}) \tag{4-5}$$

and are given in Table 4-7.

The last column presents solutions for the stated problem obtained from Heidbreder and Mitchell [4]. It is apparent that for $\rho = 1$ the chi-square approximation is exact; for $\rho = 2$ the approximation is excellent; and for $\rho = 4$ the error has become significant.

Fig. 4-5 Value of k for best fitting chi-square target model. ρ is the corresponding log-normal parameter.

Fig. 4-6 Log-normal approximation using chi-square data ($\bar{x}$'s obtained from curve must be corrected for difference in means between log-normal and chi-square). $P_D = 90\%$. $n' = 10^7$, $N = 10$ (see [1]).

Fig. 4-7 Value of $\Delta\bar{x}$ for best fitting chi-square target model.

4-6 Approximation of Other Target Density Functions

If the density function that describes a fluctuating target is other than chi-square, Rayleigh, Rice, or log-normal, or if the density function is based on empirical data, it is still usually possible to find a chi-square function that is a good fit. If this can be done, Meyer Plots can be used to calculate radar performance.

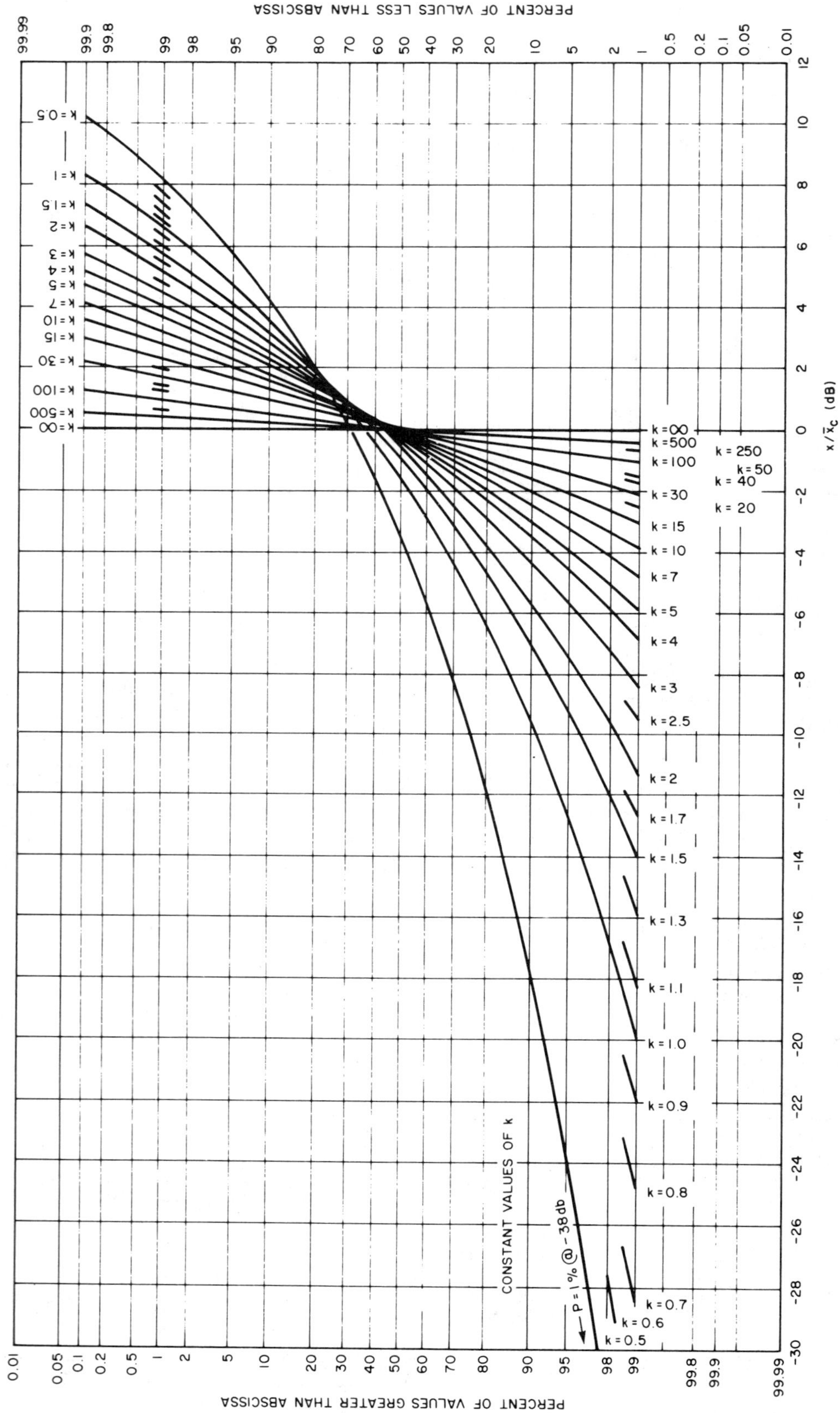

Fig. 4-8 Chi-square distributions with 2*k* degrees of freedom—constant values of *k*.

Table 4-8

Parameter	Chi square	Rice Power
First moment (mean) $\equiv \bar{x}$	$\bar{x}$	$\psi_0(1+s)$
Second moment $\equiv \bar{x^2}$	$\left(1+\frac{1}{k}\right)\bar{x}^2$	$\psi_0^2(2+4s+s^2)$
Variance $\equiv \sigma^2 \equiv \bar{x^2} - \bar{x}^2$	$\left(\frac{1}{k}\right)\bar{x}^2$	$\psi_0^2(1+2s)$

There are two methods of finding the best fitting chi-square function: moment matching and curve fitting. The former is usually easier and the latter is usually more reliable.

MOMENT MATCHING

Two parameters are required to define a chi-square density function: the mean $\bar{x}$ and the degrees of freedom k. With these two parameters fixed the second moment and variance are also fixed. In practice, it is usually easiest to determine first a function's first moment $\bar{x}$ and second moment $\bar{x^2}$.

An example of the moment-matching technique is the approximation made for the Rice Power Target Model in Section 4-4. The moments to be matched are presented in Table 4-8. Thus

$$\bar{x} = \psi_0(1+s) \tag{4-6}$$

and

$$\frac{\bar{x}^2}{k} = \left(\frac{1}{k}\right)\psi_0^2(1+2s+s^2) = \psi_0^2(1+2s) \tag{4-7}$$

resulting in

$$k = 1 + [s^2/(1+2s)] \tag{4-8}$$

which is the relationship used in Section 4-4 to define k for the best fitting chi-square function.

CURVE FITTING

Curve fitting is a graphical method of finding a chi-square distribution function that approximates the arbitrary distribution function of interest. It is a trial and error procedure whereby a chi-square distribution curve selected from Fig. 4-8 is fitted to the arbitrary distribution function. To do this, both values of k and $\bar{x}$ are chosen, such that optimum slope and lateral translation for the chi-square distribution curve is achieved.

Table 4-9

$P(x/\alpha < x_a/\alpha)$ (%)	(x_a/α) (dB)
1	$-14.3(\ln \rho)^{1/2} = -9.9$[a]
50	0.0
99	$+14.3(\ln \rho)^{1/2} = +9.9$[a]

[a] For $\rho = 2$.

For convenience, the arbitrary distribution function should be plotted on the same type of log-probability graph paper as that used for the chi-square distribution curves. An excellent technique would be to make several XEROX copies of Fig. 4-8 on vellum paper. The arbitrary distribution function could then be plotted on one copy and set on top of another copy of the figure laterally translated in such a manner that a curve of specific k value will give a satisfactory fit. The amount of lateral translation and the k value are then noted. According to Swerling [2] the areas of misfit should be evenly divided between both sides of the curve, and particular emphasis should be placed on achieving a good fit between the 1% and 99% probability range. Furthermore, for well-behaved functions the end error in finding required $\bar{x}$ versus P_D will be at worst on the same order as the misfit error.

An example of the curve-fitting technique is the approximations made for the log-normal target model in Section 4-5. Consider the log-normal distribution function with $\rho = 2$; plotted on log-probability paper, it is a straight line as shown in Table 4-9. Following the procedure just described, it is found that the best fitting chi-square distribution function has $k = 1.17$. Furthermore, the lateral translation is such that $x/\bar{x}_c = -3.48$ dB on the chi-square plot corresponds to $x/\alpha = 1$ on the log-normal plot, where α is the median value of the log-normal distribution function. Noting that $\alpha = \bar{x}_1/\rho = 0.5\bar{x}_1$ for the log-normal, it is apparent that

$$\frac{\bar{x}_1}{\bar{x}_c} = \frac{\text{mean of log-normal}}{\text{mean of chi-square}}$$

$$= -3.48 + 3.01 = -0.47 \quad \text{dB} \tag{4-9}$$

The effect of this is that 0.47 dB must be added to the values of required average single-pulse signal-to-noise ratio $\bar{x}_c$ determined by use of the chi-square approximation for the log-normal target model. These results, $k = 1.17$ and $\Delta\bar{x} = -0.47$ dB, agree with the values given in Section 4-5 and Figs. 4-6 and 4-7 for the best fitting chi-square

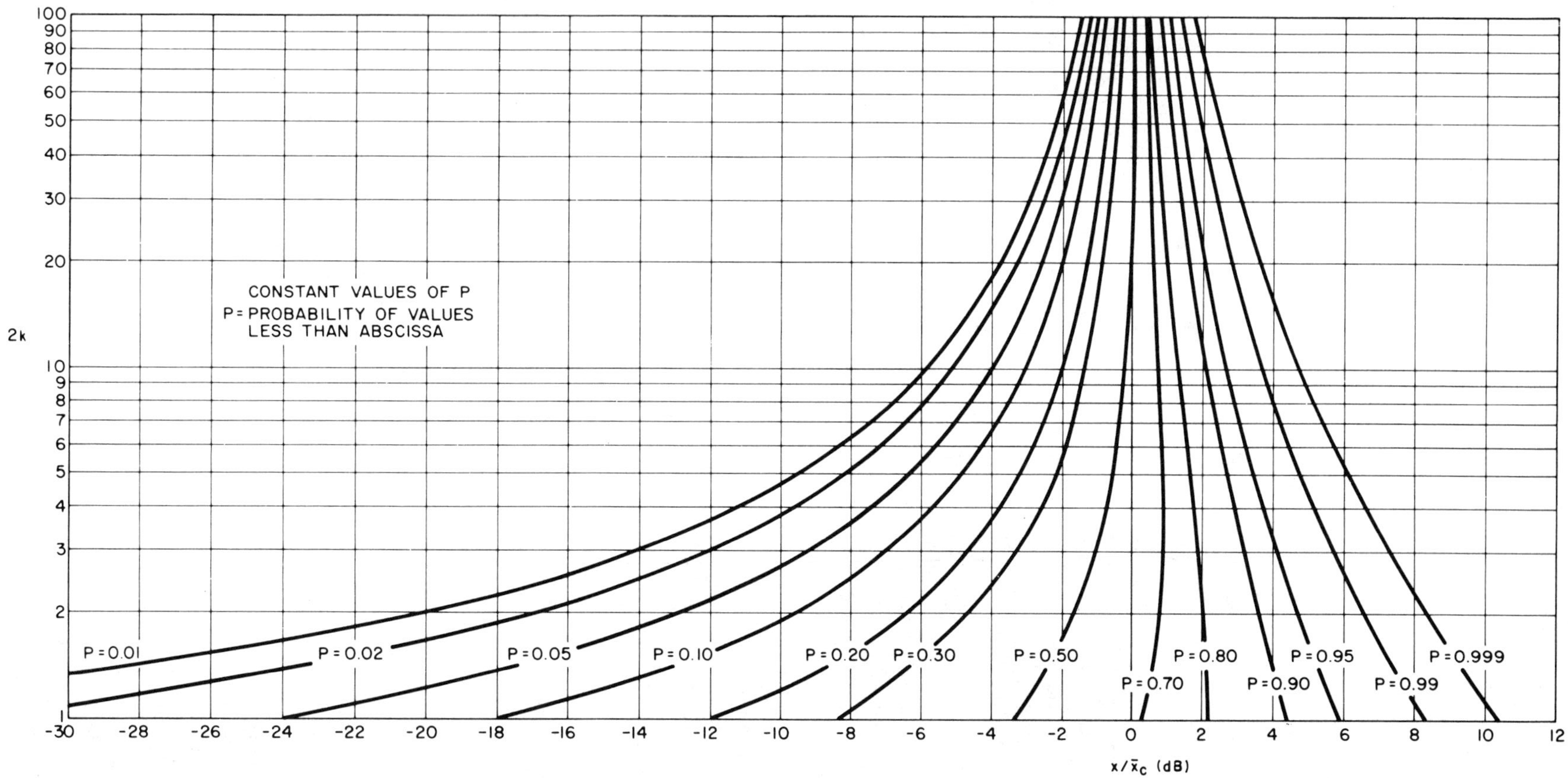

Fig. 4-9 Chi-square distributions with $2k$ degrees of freedom—constant values of P.

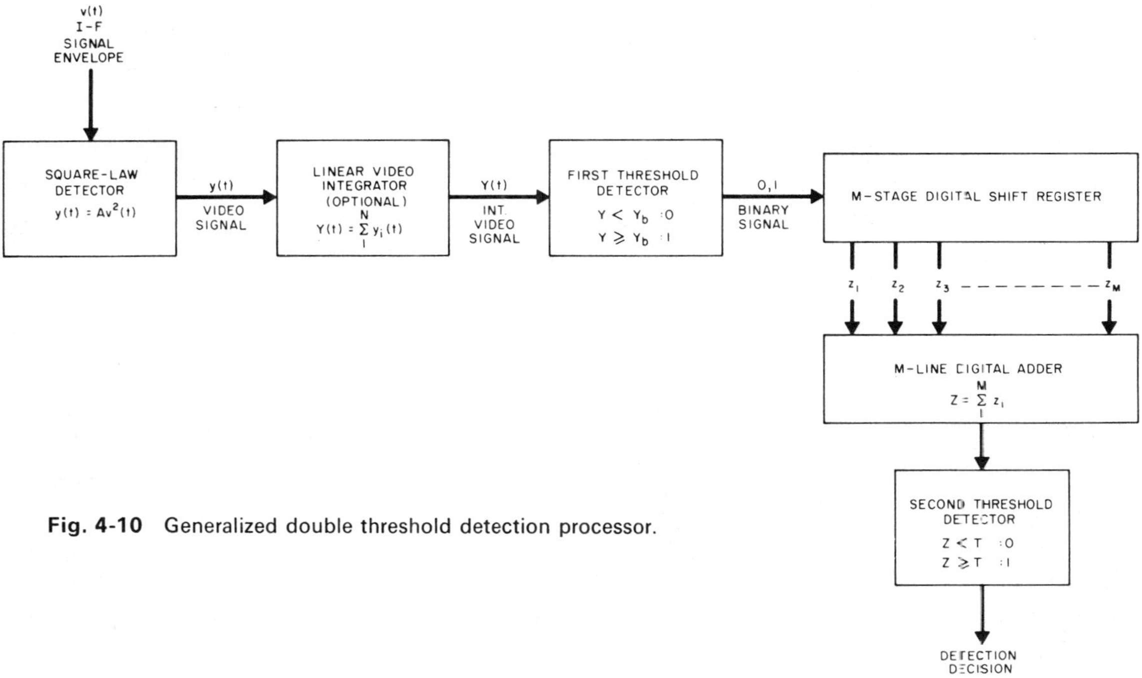

Fig. 4-10 Generalized double threshold detection processor.

approximation. It is also interesting to note that the greatest misfit between the log-normal and its chi-square approximation occurs at $P = 0.01$, 0.99, and around 0.40 and is approximately ± 2.1 dB. The corresponding error in determining $\bar{x}_c$ for a particular set of detection criteria (see Table 4-7) is $14.0 - 13.57 = 0.43$ dB. This indicates that Swerling's contentions (stated above) with regard to misfit errors [2] is conservative.

Figure 4-9 shows how cumulative probability for a chi-square distribution varies as a function of k. It is useful when interpolating between values of k in Fig. 4-8, as it clearly shows how P varies as a function of k. The curves converge at $x/x_c = 1$ (0 dB) as $k \to \infty$. It is clear that if the final system P_D is to be greater than approximately 70%, it is advantageous to decorrelate the target returns, as by use of frequency diversity. The opposite is true, however, for very weak target returns as frequency diversity will further reduce their probability of being detected.

4-7 Double Threshold and Median Detection [1]

Some radars impose detection criteria whereby the threshold of the first detector, the conventional postintegration video threshold detector, must be exceeded several times in a defined period of time before a target is certified. It is a binomial process that in effect provides additional integration in a form that is less efficient than conventional video integration. A generalized block diagram of such a radar processor is given in Fig. 4-10. The first three blocks are the same as in a conventional radar video processor; however, the video integrator is often not present when double threshold detection is employed. The next two blocks determine the number of true decisions in the previous M decisions made by the first threshold detector. A shift register or other memory device is used to retain the input data for the digital adder. Alternately, a digital counter could be used, but then the final detection decisions are made once for every Mth pulse instead of once for each new pulse, as the counter must be reset to zero after every Mth pulse. The last block tests that there were at least T true decisions in the M previous decisions. In actuality, the last three blocks are an instrumentation of the binomial law of probability.

Referring to the output of the first threshold detector, probability of detection P_{D1}, and probability of false alarm P_{fa1} are determined through the use of Meyer Plots using the procedures described in the preceding sections of this chapter. At the output of the second threshold detector, probability of detection P_{D2} and probability of false alarm P_{fa2} for a nonfluctuating target (Case 0) are related to the former simply by cumulative binomial relationships as follows:

$$P_{D2}(x, N, n_2', K) = \sum_{i=T}^{M} C_i^M [P_{D1}(x, N, n_1', K)]^i \times [1 - P_{D1}(x, N, n_1', K)]^{M-i} \tag{4-10}$$

$$P_{fa2}(N, n_2') = \sum_{i=T}^{M} C_i^M [P_{fa1}(N, n_1')]^i \times [1 - P_{fa1}(N, n_1')]^{M-i} \tag{4-11}$$

where

$$K = \infty = \text{defined value for a nonfluctuating target (Case 0)} \tag{4-12}$$

$$C_i^M = \frac{M!}{i!(M-i)!}, \quad \text{the binomial coefficient} \tag{4-13}$$

$$P_{fa1} = \frac{0.693}{n_2'} = \frac{0.693N}{n_1} \tag{4-14}$$

$$P_{fa2} = \frac{0.693}{n_2'} = \frac{0.693NM}{n_2} \tag{4-15}$$

Since for many problems of normal interest $P_{fa1} \ll 1$, Eq. (4-11) can be written as

$$P_{fa2} \cong C_T^M (P_{fa1})^T \tag{4-16}$$

which permits writing the following most useful relationship:

$$n_1' \cong (0.693)^{(T-1)/T} (C_T^M n_2')^{1/T} \tag{4-17}$$

Since false alarms are not caused by the target, these last equations for probability of false alarm and false alarm number are valid for all manner of target fluctuation. The situation for probability of detection, however, is not so simple; the effect on P_{D2} of target fluctuation is complicated. For slow chi-square target fluctuations, P_{D2} of Eq. (4-10) is modified as follows:

$$P_{D2}(\bar{x}, N, n_2', K) = \int_0^\infty dP(\chi^2) \times P_{D2}(x, N, n_2', \infty)\, d(\chi^2) \tag{4-18}$$

where

$$dP(\chi^2) = \text{chi-square density function} \tag{4-19}$$

$$\chi^2 = 2Kx/\bar{x} \tag{4-20}$$

$$K = k \quad \text{(the defined relationship for a slowly fluctuating target, e.g., Cases 1 or 3)} \tag{4-21}$$

Alternatively, for rapid chi-square target fluctuations, P_{D2} is

$$P_{D2}(\bar{x}, N, n_2', K) = \sum_{i=T}^{M} C_i^M [P_{D1}(\bar{x}, N, n_1', K)]^i \times [1 - P_{D1}(\bar{x}, N, n_1', K)]^{M-i} \tag{4-22}$$

where

$$P_{D1}(\bar{x}, N, n_1', K) = \int_0^\infty dP(\chi^2) \times P_{D1}(x, N, n_1', \infty)\, d(\chi^2) \tag{4-23}$$

$$dP(\chi^2) = \text{chi-square density function} \tag{4-24}$$

$$\chi^2 = 2Kx/\bar{x}$$

$$K = Nk \quad \text{(the defined relationship for a rapidly fluctuating target, e.g., Cases 2 or 4)} \tag{4-25}$$

Note that for slow fluctuation the averaging for target fluctuation is performed after the binomial summation, while for rapid fluctuation the averaging is performed before the summation. This will have a major impact in the procedures that will be followed to solve these problems.

For the special case when the video integrator is deleted and $T = (M+1)/2$, the double threshold detector is effectively equivalent to a median detector. This is because for a second threshold detection to occur, the majority of pulses must be greater than the first threshold. Obviously, the equivalence is exact only in the limit as the sample size becomes infinite. For all practical purposes, however, such a configuration can be considered to be a median detector.

The method of solving double threshold detection problems using Meyer Plots will be demonstrated by means of illustrative problems where the following are held constant for all examples:

Overall probability of detection $P_{D2} = 0.90$.
Overall Fehlner false alarm number $n_2' = 10^8$.
Number of pulses processed $N \cdot M = 9$.

The calculation procedures for the nonfluctuating and rapid-fluctuating target cases as suggested by Eqs. (4-10) and (4-22) are similar; the step-by-step calculations are presented in Table 4-10A. Three types of detectors are compared: the

Table 4-10A

COMPARISON OF THREE SIGNAL PROCESSORS FOR NINE PULSE RETURNS[a]

	System		
	Configuration A (median detector)	Configuration B (hybrid detector)	Configuratoin C (mean detector)
System design parameters/statement of problem			
Parameter			
P_{D2}	0.90	0.90	0.90
n_2'	10^8	10^8	10^8
N	1	3	9
M	9	3	1
$T = (M+1)/2$	5	2	1
Intermediate calculation			
C_T^M	126	3	1
Calculated Meyer Plot parameters			
n_1'	78	1.45×10^4	10^8
P_{D1}	0.6990	0.8042	0.9000
Resulting system performance in terms of required $\bar{x}$ (dB)			
Case 0	7.7	7.5	6.5
Case 4	9.2	8.5	7.1
Case 2	10.9	9.4	7.6

[a] See text for details of method of calculation.

median, which employs no video integration; the mean, which employs only video integration; the hybrid, which as the name implies is a combination of the mean and the median. The parameters M, N, and T define processor configuration. The parameter C_T^M may be either calculated or obtained from tables of the binomial coefficient [5]. For the chosen illustration, n_1' can be calculated with sufficient accuracy using Eq. (4-17). For other cases it may be necessary to refine this value working with several terms of Eq. (4-11). The calculation of P_{D1} involves working the binomial relationship [Eq. (4-10)] in reverse, which is best accomplished by using tables of the incomplete beta function [6]. Tables of binomial function [7] are available for working the binomial relationship in the forward direction. These tables are too extensive to be reproduced here; however, Fig. 4-11 presents graphically some useful inverse binomial values for the median processor [$T = (M+1)/2$]. Now, having determined n_1' and P_{D1} and knowing N, the desired value of average single-pulse signal-to-noise ratio $\bar{x}$ is simply read off the Meyer Plot of the appropriate case number. For nonstandard values of k ($k \neq 1, 2$), $\bar{x}$ can be determined as described in Section 4-3.

The calculation procedures for the slowly fluctuating target cases, as suggested by Eq. (4-18), are somewhat more involved than those described in the preceding paragraph for the nonfluctuating and rapidly fluctuating target cases. The step-by-step procedures are outlined in Table 4-10B. The same three types of detectors are compared as were compared in Table 4-10A. Since the calculation procedure is complex, for added clarity a brief explanatory note is given in the right-hand column of Table 4-10B alongside each major parameter calculated.

Equation (4-18) indicates that the first general step in the procedure is to find a Case-0 P_{D2} that corresponds to the value of $\bar{x}$ for Case $K = k$ and $N = 1$. It develops, however, that such values of P_{D2} are usually almost equal to unity and well above $P_D = 99.99\%$ (the maximum value of P_D covered by the Meyer Plots or single-hit detection curves for Case 0 contained in this handbook). Since the values of $\bar{x}$ that are being worked with

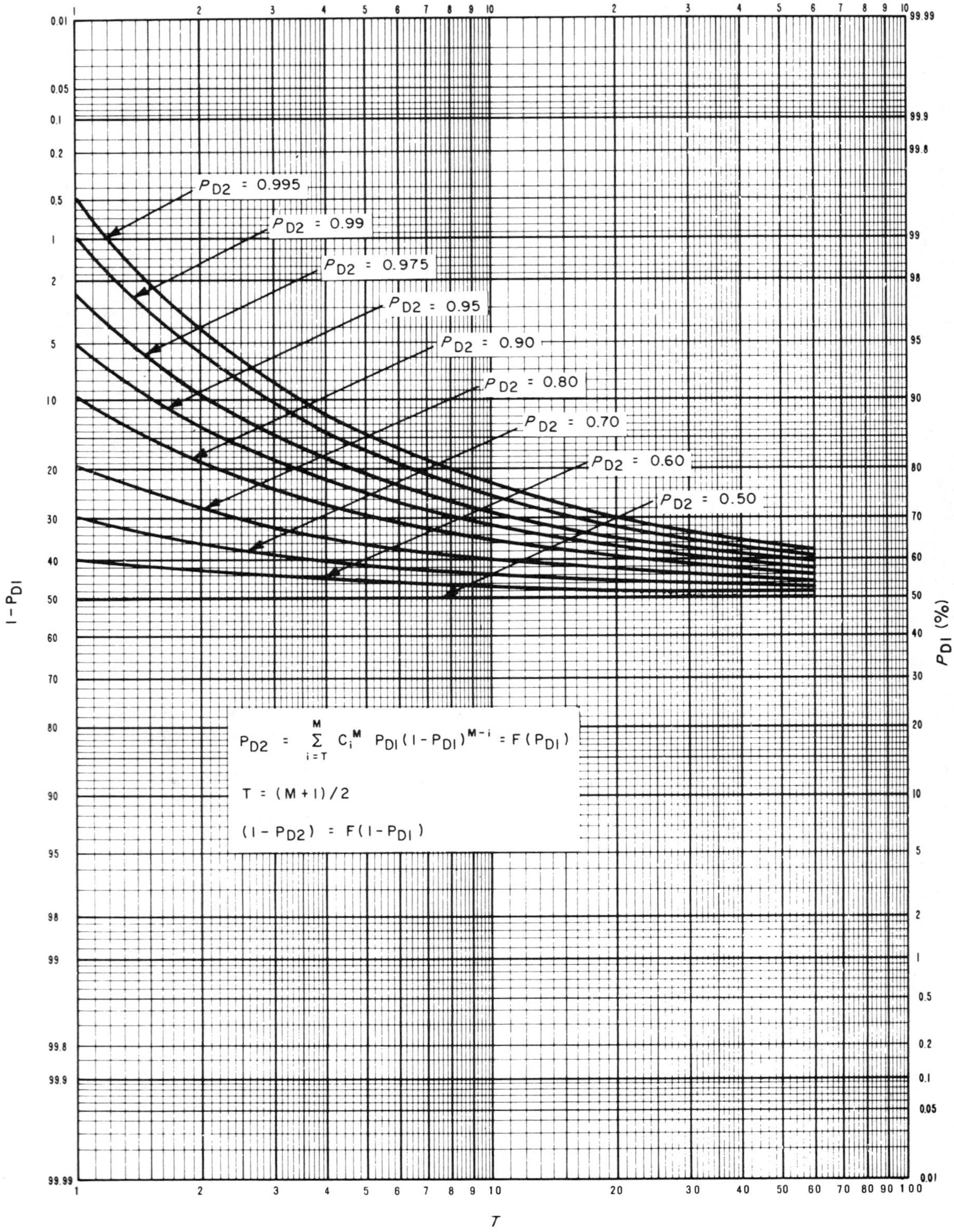

Fig. 4-11 First-threshold probability of detection for a median detector.

Table 4-10B

COMPARISON OF THREE SIGNAL PROCESSORS FOR NINE PULSE RETURNS

	System						
	Configuration A (median detector)		Configuration B (hybrid detector)		Configuration C (mean detector)		
Parameter	Case 1	Case 3	Case 1	Case 3	Case 1	Case 3	How obtained
System design parameters/statement of problem							
P_{D2}	0.90	0.90	0.90	0.90	0.90	0.90	
n_2'	10^8	10^8	10^8	10^8	10^8	10^8	
N	1	1	3	3	9	9	$N \cdot M = 9$
M	9	9	3	3	1	1	$N \cdot M = 9$
T	5	5	2	2	1	1	$T = (M+1)/2$
Calculations to find equivalent Case 0 P_{D2}							
$\bar{x}_2$ (dB)	22.5	18.6	22.5	18.6	22.5	18.6	Meyer Plots: $K = k$, $N = 1$ P_{D2}, n_2'
Y_{b2} (dB)	12.75	12.75	12.75	12.75	12.75	12.75	Fig. 2-2: $N = 1$, n_2'
t_2	12.68	5.88	12.68	5.88	12.68	5.88	Eq. (4-26)
$-\log_{10}(1-P_{D2})$	36.4	8.7	36.4	8.7	36.4	8.7	Fig. 4-12
Calculations to find P_{D1} and t_1							
T^1	5	5	2	2	1	1	Eq. (4-30)
$\log_{10}(C^M_{T^1})$	2.1	2.1	0.477	0.477	0	0	$\log_{10}$ (Eq. 4-13)
$-\log_{10}(1-P_{D1})$	7.7	2.16	18.44	4.59	36.4	8.7	Eq. (4-31)
t_1	5.5	2.45	8.9	4.05	12.68	5.88	Fig. 4-12
Calculations to find Required $\bar{x}$							
n_1'	78	78	1.45×10^4	1.45×10^4	10^8	10^8	Eq. (4-17)
Y_{b1}	6.75	6.75	11.7	11.7	15.75	15.75	Fig. 2-2: $N = N$, n_1'
$\bar{x}_1$ (dB)	15.6	11.7	14.9	11.1	14.1	10.7	Eq. (4-32)
$\bar{x}$ (dB)	—	—	—	—	15.0	10.9	Meyer Plots: $K = k$, $N = 9$, P_{D2}, n_2'

are usually large, however, a normal approximation of dP_{D2} can be used to find the desired value of P_{D2}. Figure 4-12 presents graphically the extreme values of interest for the normal distribution. It is necessary to plot the $\log_{10}(1-P_D)$ as a function of the index t instead of P_D or $1-P_D$ in order to handle the large range of values that are of interest. To find $P_{D2}(x, N, n_2', \infty)$ let

$$t_2 = (2\bar{x}_2)^{1/2} - (2Y_{b2})^{1/2} \qquad (4\text{-}26)$$

where

$$\bar{x}_2 = \bar{x} \quad \text{for} \quad P_{D2} = 0.90, \quad N = 1, \quad n_2' = 10^8, \quad \text{and} \quad K = k \qquad (4\text{-}27)$$

and

$$Y_{b2} = Y_b \quad \text{for} \quad N = 1 \quad \text{and} \quad n_2' = 10^8 \quad \text{from Fig. 2-2} \qquad (4\text{-}28)$$

The next general step in the procedure is to find P_{D1}, which is related to P_{D2} by the binomial process. Since near unity values of P_{D1} and P_{D2} are of interest here, the relationship may be simplified as follows:

$$(1-P_{D2}) = C^M_{T^1}(1-P_{D1})^{T^1} \qquad (4\text{-}29)$$

where

$$T^1 = M - T + 1 \qquad (4\text{-}30)$$

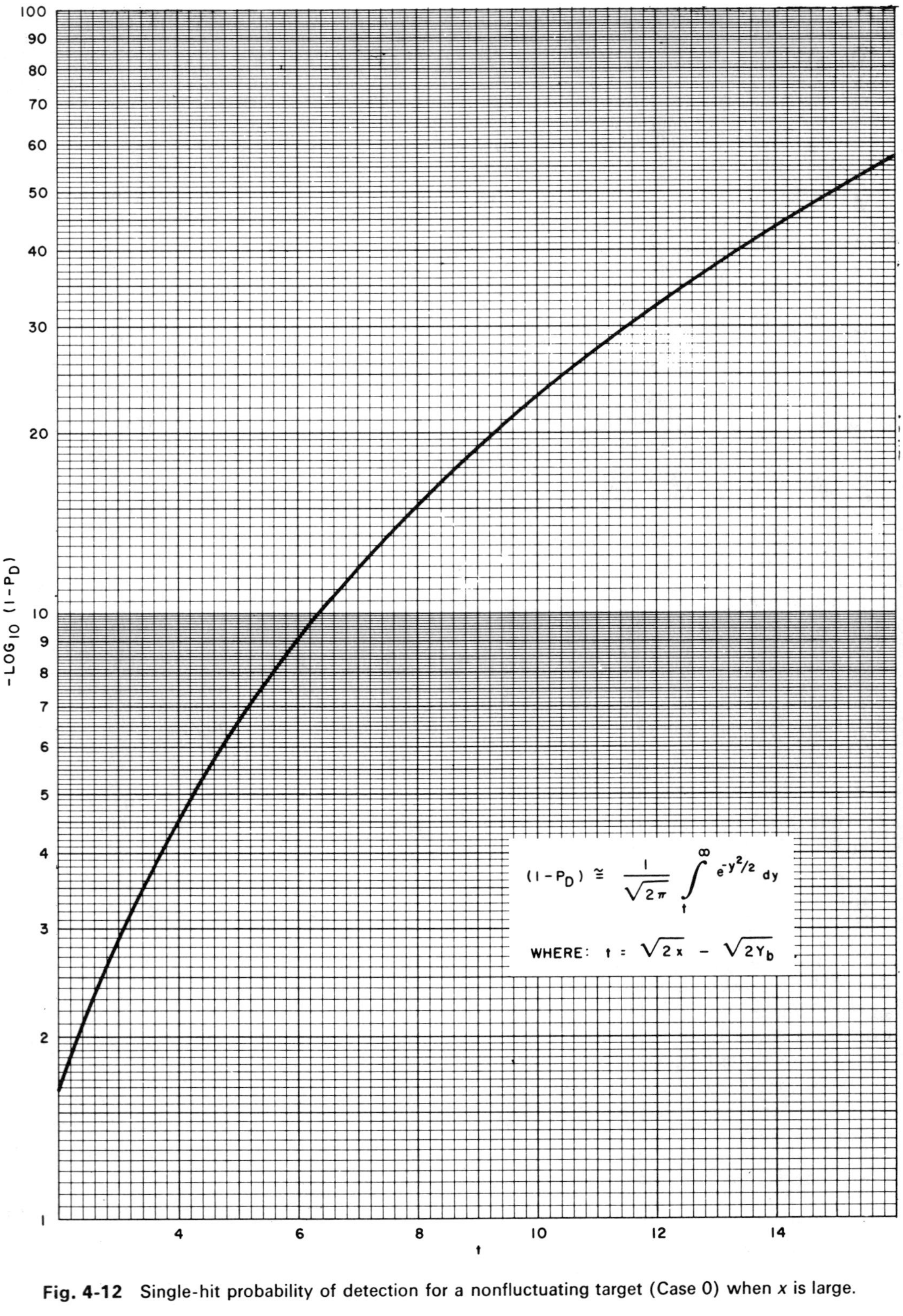

Fig. 4-12 Single-hit probability of detection for a nonfluctuating target (Case 0) when x is large.

Table 4-10C

SUMMARY: COMPARISON OF THREE SIGNAL PROCESSORS FOR NINE PULSE RETURNS

	System		
	Configuration A (median detector) (dB)	Configuration B (hybrid detector) (dB)	Configuration C (mean detector) (dB)
Required $\bar{x}$			
Case 0	7.7	7.5	6.5 ($K=\infty$)
Case 4	9.2	8.5	7.1 ($K=18$)
Case 2	10.9	9.4	7.6 ($K=9$)
Block ($k=2$)	—	10.25	8.2 ($K=6$)
Block ($k=1$)	—	12.9	9.6 ($K=3$)
Case 3	11.7	11.1	10.7 ($K=2$) (10.9)[a]
Case 1	15.6	14.9	14.1 ($K=1$) (15.0)[a]

[a] Correct value from appropriate Meyer Plot.

Furthermore, since at this point in the procedure it is more convenient to work with logarithms, a more useful form of Eq. (4-29) is

$$-\log_{10}(1-P_{D1}) = \frac{1}{T^1}[\log_{10}(C_{T^1}^M) - \log_{10}(1-P_{D2})] \tag{4-31}$$

The last general step in the procedure is to find the required value of $\bar{x}_1$. Determining

$$-\log_{10}(1-P_{D1})$$

permits determining t_1 from Fig. 4-12, from which $\bar{x}_1$ can be determined by

$$\bar{x}_1 = \frac{1}{2N}[t_1+(2Y_{b1})^{1/2}]^2 \tag{4-32}$$

where

Y_{b1} is determined from Fig. 2-2

$$\text{for} \quad N = N \quad \text{and} \quad n' = n_1' \tag{4-33}$$

This last step also involved using the normal approximation for dP_D.

The procedure for determining $\bar{x}$ for the slowly fluctuating target case described in the last few paragraphs is susceptible to error in several places. The last line in Table 4-10B gives two values of $\bar{x}$ for the video integration only processor determined from the appropriate Meyer Plots. As can be seen, for the two values of $\bar{x}$ that can be compared the procedure leads to errors of 0.2 and 0.9 dB, respectively.

The procedures for handling block correlation are even more complex than those described in the preceding paragraphs. There is one exception—when the slow fluctuation can be attributed to those pulses which are video integrated by a hybrid processor, a situation which is physically reasonable. The procedure for this case is the same as that followed in Table 4-10A except that the required value of $\bar{x}$ is found on a Case $K=k$ Meyer Plot. The results of following this through are presented in Table 4-10C.

Table 4-10C and Fig. 4-13 which summarize the work of this section, verifies that the performance of a double threshold detector is always somewhat worse than that of a mean detector. Note that in Fig. 4-13 for both the median and hybrid detector, the portion of curve relating to completely correlated pulse returns (left-hand portion) does not, as in all previous such curves presented in this text (Figs. 4-3, 4-4, and 4-5), align with the portion of the curve relating to completely uncorrelated pulse returns (right-hand portion). Also note the error introduced into the mean detector curve by the normal approximation procedure outlined in Table 4-10B. At this point it must be noted that the median detector is not the optimum double threshold detector with

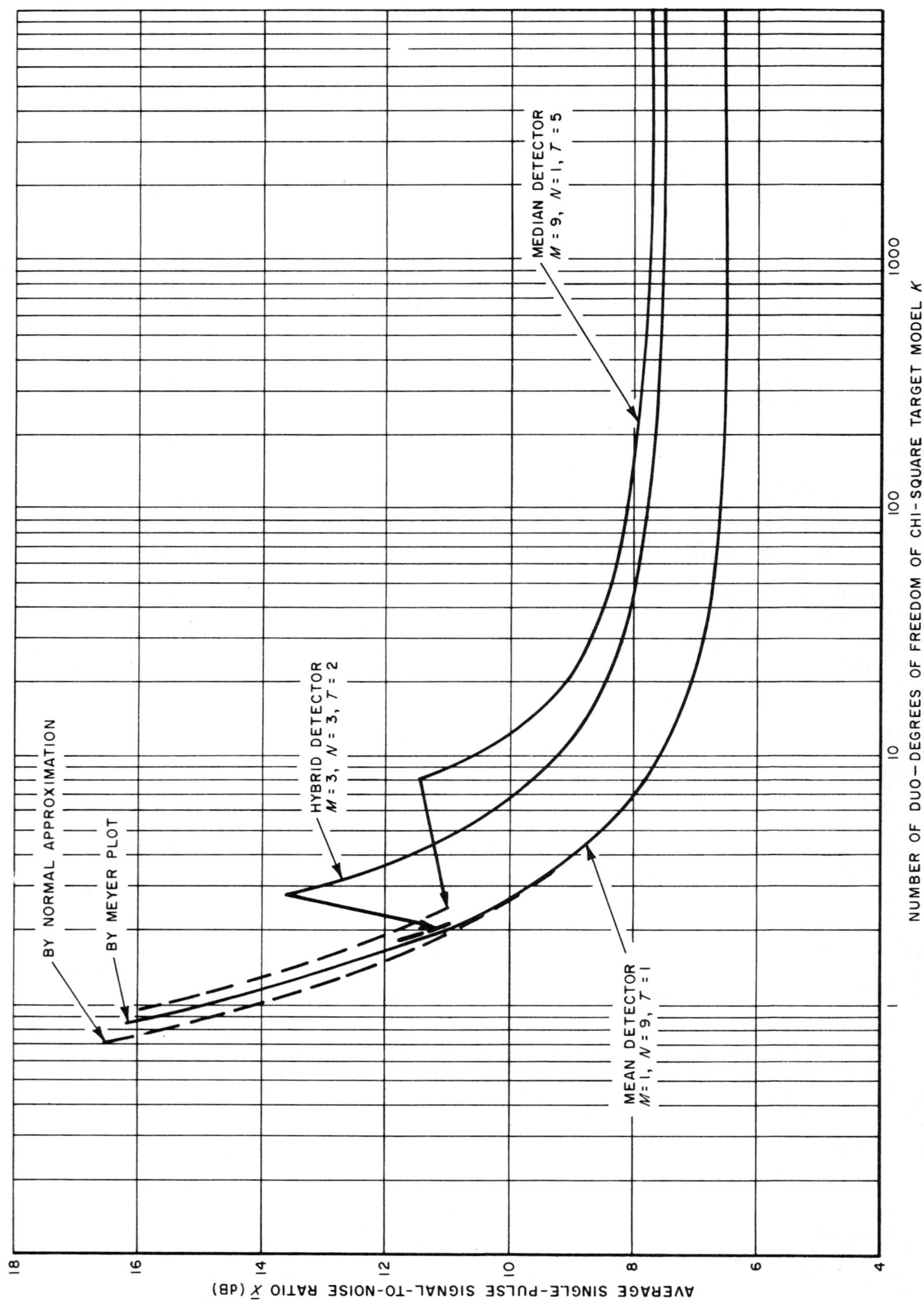

NUMBER OF DUO-DEGREES OF FREEDOM OF CHI-SQUARE TARGET MODEL K
AVERAGE SINGLE-PULSE SIGNAL-TO-NOISE RATIO X̄ (dB)
1
10
100
1000
18
16
14
12
10
8
6
4
MEDIAN DETECTOR
M = 9, N = 1, T = 5
HYBRID DETECTOR
M = 3, N = 3, T = 2
MEAN DETECTOR
M = 1, N = 9, T = 1
BY NORMAL APPROXIMATION
BY MEYER PLOT

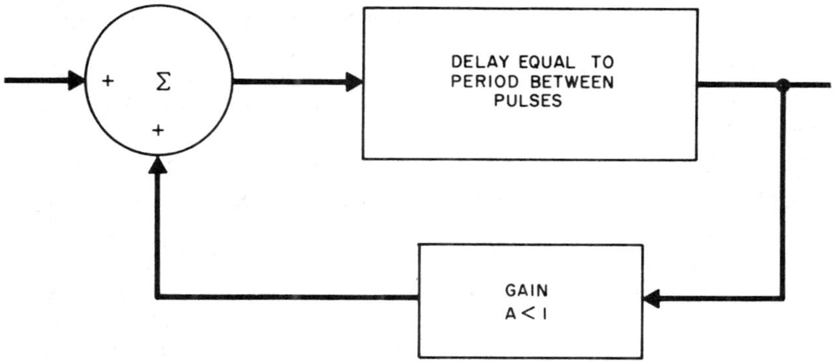

Fig. 4-14 Video integration by signal recirculation. $N_{eq} = 1/(1-A)$: equivalent number of pulses integrated.

respect to minimizing required $\bar{x}$. Walker has determined that

$$0.2M < T_{opt} < 0.5M \tag{4-34}$$

with the exact value depending on the value of K and M, but not being at all critical [8, 9]. It is also interesting to note that Capon has demonstrated that in the limit for large M, the required $\bar{x}$ for the double threshold detector is always 0.94 dB greater than that required for the optimum noncoherent video integration (mean detector) [10].

4-8 Recirculating Integrators

Some radar video processors employ a simplified and low-cost form of video integration that involves the constant recirculation of the video signal with a loop gain less than unity. A generalized block diagram of such an integrator is presented in Fig. 4-14. In order to use Meyer Plots and other analytical tools presented in this handbook to solve problems related to radars employing such integrators, it is necessary to determine the equivalent number of pulses integrated, N. It develops that this type of integrator is an instrumentation of a summed geometric series. For the case in which the integration is not interrupted, the series is infinite and

$$N = N_{eq} = 1/(1-A) \tag{4-35}$$

where

$$1 > A = \text{loop gain}$$

For the case in which the integration is interrupted after r recirculations,

$$N = N_{eq} = (A^r - 1)/(A - 1) \tag{4-36}$$

4-9 Meyer Plots

Following this section 368 Meyer Plots are presented.† They are arranged as follows: The first level of order is by target model case number, i.e., Case 0, Case 1, Case 2, Case 3, and Case 4 (see Table 4-1 for details); the second level of order is by Fehlner's false alarm number, i.e., $n' = 10, 20, 40, 60, 80, 100, 200, 400, \ldots, 10^{10}$; the third level of order is, with two exceptions, by the two ranges of number of pulses integrated, i.e., $1 \leqslant N \leqslant 100$ and $100 \leqslant N \leqslant 10^4$. The two exceptions are that for both Target Model Cases 2 and 4, the Meyer Plots for $100 \leqslant N \leqslant 10^4$ have been deleted because the Case-0 Meyer Plots for $100 \leqslant N \leqslant 10^4$ can be used instead. Examination of Figs. 4-3–4-5 will reveal that for $K \geqslant N \geqslant 100$ the error introduced by using the Case-0 Meyer Plots in place of the Case-2 or -4 Meyer Plots for $100 \leqslant N \leqslant 10^4$ is less than 0.1 dB.

Before using the Meyer Plots it would be expedient to remember the following:

$$n' = n/N = \log_e(2)/P_{fa} \tag{4-37}$$

Fig. 4-13 Double-threshold detector performance. $P_{D2} = 0.90$, $M \cdot N = 9$, $n_2' = 10^8$.

† These were first published in Meyer [11].

where

n' is the Fehlner false alarm number,
n the Marcum false alarm number, and
P_{fa} the probability of false alarm.

Ideal square-law detection and linear noncoherent video integration are assumed; however, the results that would be obtained using a linear-law detector would not be significantly different (see Figs. 3-18 and 3-19).

Gaussian noise is the assumed contaminating signal.

The data presented are "required average single-pulse signal-to-noise ratios" (required average single-pulse signal power divided by average noise power).

N is the number of pulses noncoherently integrated.

The false alarm number indicated at the upper left-hand corner of each plot is n', the Fehlner false alarm number.

To determine the bias level (threshold voltage) required to achieve a particular value of n', see Section 2-4.

For problems where $N = 1$ (no video integration is employed), the single-hit detection curves presented in Figs. 3-2–3-16 may be more convenient to use than the Meyer Plots that follow.

In addition to the several examples given in this chapter, other examples of the use of Meyer Plots are presented in Chapters 1–3. They describe the use of Meyer Plots in conjunction with solving the search radar equation (Section 1-13), determining probability of detection (Section 2-7), determining integration loss and integration efficiency (Section 2-9), determining collapsing loss (Section 2-11), and determining optimum N as a function of beam shape (Section 3-14).

References

1. Based on work summarized in Sperry Gyroscope Div. internal memo by H. A. Mayer (Sept. 15, 1971).
2. P. Swerling, Recent Developments in Target Models for Radar Detection Analysis. *AGARD Avionics Tech. Symp. Proc., Istanbul, Turkey* (May 25–29, 1970); Also, Lecture Notes on Radar Target Signatures: Measurements, Statistical Models, and Systems Analysis. Technol. Service Corp., Santa Monica, Calif. (August 1968). The former is a condensation of the latter.
3. P. H. R. Scholefield, Statistical Aspects of Ideal Radar Targets. *Proc. IEEE* **55**, No. 4 (1967).
4. G. R. Heidbreder and R. L. Mitchell, Detection Probabilities for Log-Normally Distributed Signals. *IEEE Trans. Aerospace Electron. Syst.* **AES-3**, No. 1 (1967); also Aerospace Corp. Rep. No. TR-669(9990)-6 (April 1966).
5. M. Abramowitz and I. A. Stegun (eds.), "Handbook of Mathematical Functions with Formulas, Graphs, and Mathematical Tables." Nat. Bur. of Std. Appl. Math. Ser. 55. U.S, Government Printing Office, Washington, D.C., 1964.
6. K. Pearson (ed.), "Tables of the Incomplete Beta Function." Biometrika Office, Univ. College, Cambridge Univ. Press, London and New York, 1948.
7. "Tables of the Bionomial Probability Distribution." Nat. Bur. of Std. Appl. Math. Ser. 6. U.S. Government Printing Office, Washington D.C., 1950.
8. J. F. Walker, Performance Data for a Double-Threshold Detection Radar. *IEEE Trans. Aerospace Electron. Syst.* **AES-7**, No. 1 (1971).
9. V. G. Hansen, Comments on "Performance Data for a Double-Threshold Detection Radar. *IEEE Trans. Aerospace Electron. Syst.* **AES-7**, No. 3 (1971).
10. J. Capon, Optimum Coincidence Procedures for Detecting Weak Signals in Noise. *IRE Int. Conv. Rec.* pt. 4, pp. 154–166 (1960).
11. D. P. Meyer, Further Studies of Target Dection by a Pulsed Radar. Final Rep. PN-51601-21, Sperry Gyroscope Co., Sperry Rand Corp., Great Neck, N.Y. (July, 1967).

Meyer Plots

Page 110	Case 0:	$N = 1\text{–}10{,}000$ $n' = 10, 20, 40, 60, 80, 100, 200, 400, \ldots, 10^{10}$
Page 202	Case 1:	$N = 1\text{–}10{,}000$ $n' = 10, 20, 40, 60, 80, 100, 200, 400, \ldots, 10^{10}$
Page 294	Case 2:	$N = 1\text{–}100$ $n' = 10, 20, 40, 60, 80, 100, 200, 400, \ldots, 10^{10}$
Page 340	Case 3:	$N = 1\text{–}10{,}000$ $n' = 10, 20, 40, 60, 80, 100, 200, 400, \ldots, 10^{10}$
Page 432	Case 4:	$N = 1\text{–}100$ $n' = 10, 20, 40, 60, 80, 100, 200, 400, \ldots, 10^{10}$

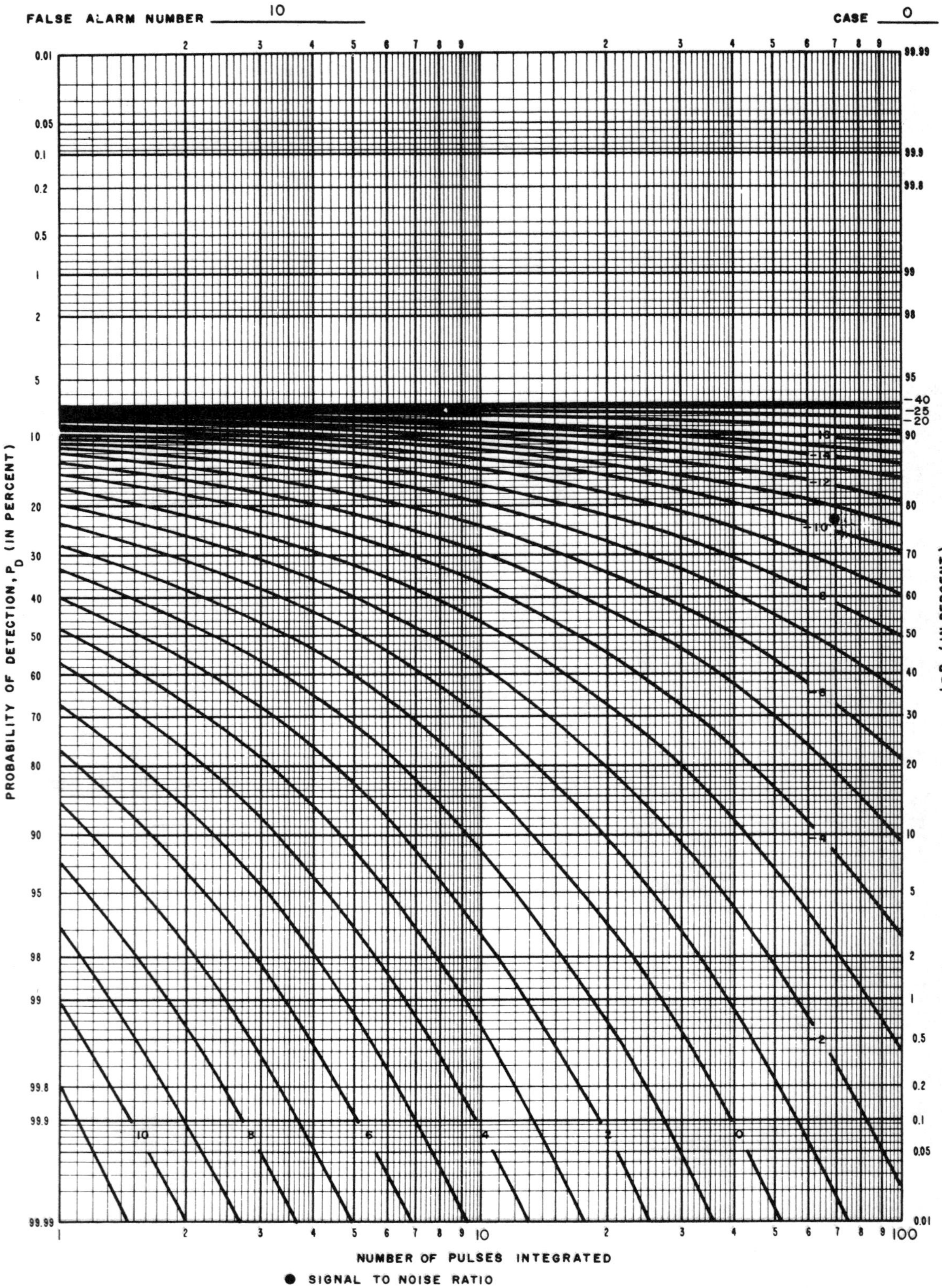
FALSE ALARM NUMBER 10
CASE 0
PROBABILITY OF DETECTION, P_D (IN PERCENT)
$1-P_D$ (IN PERCENT)
NUMBER OF PULSES INTEGRATED
● SIGNAL TO NOISE RATIO

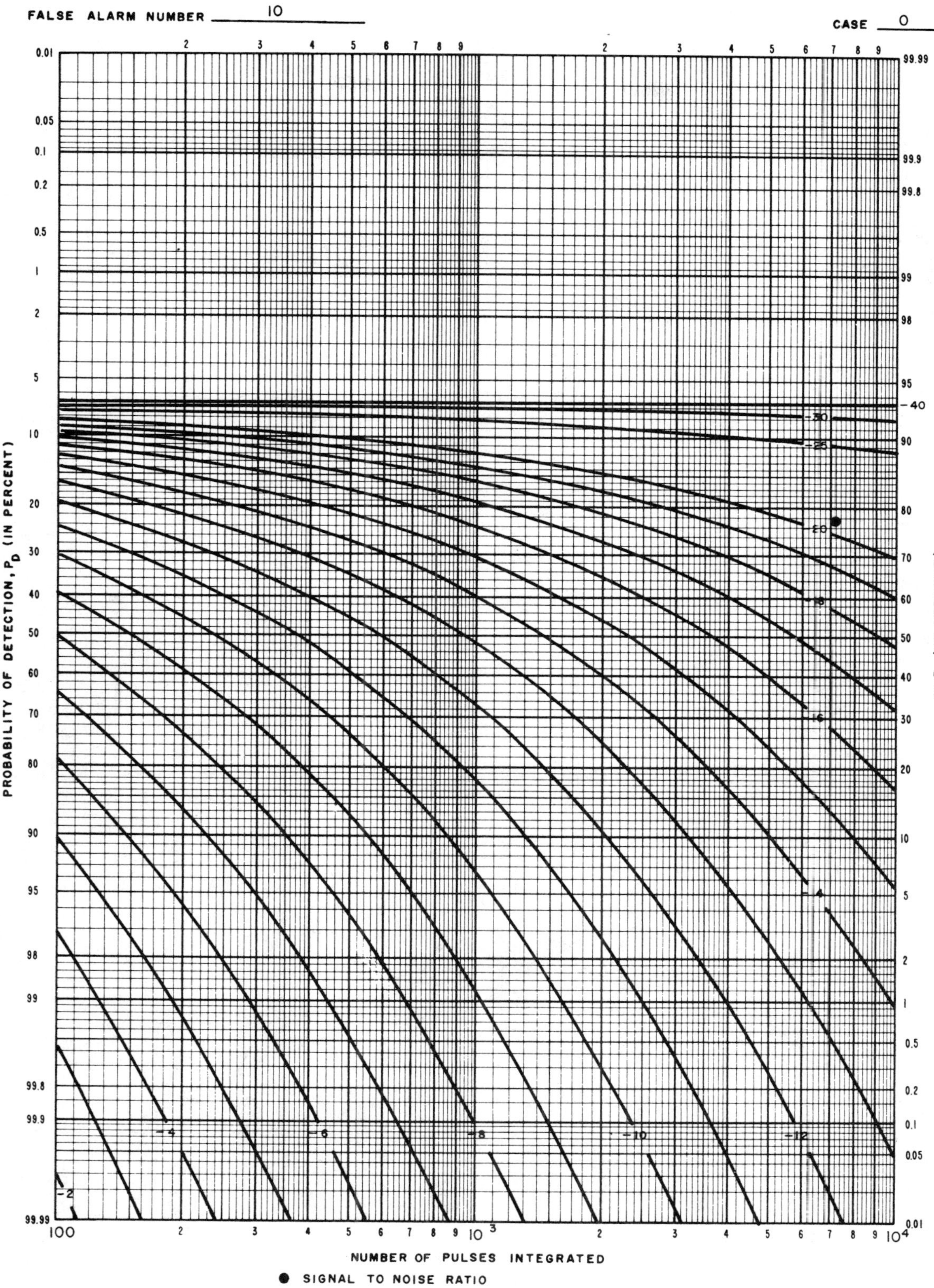
FALSE ALARM NUMBER 10
CASE 0
PROBABILITY OF DETECTION, P_D (IN PERCENT)
$1-P_D$ (IN PERCENT)
NUMBER OF PULSES INTEGRATED
● SIGNAL TO NOISE RATIO
-40
-30
-25
-20
-18
-16
-14
-12
-10
-8
-6
-4
-2
100
10^3
10^4

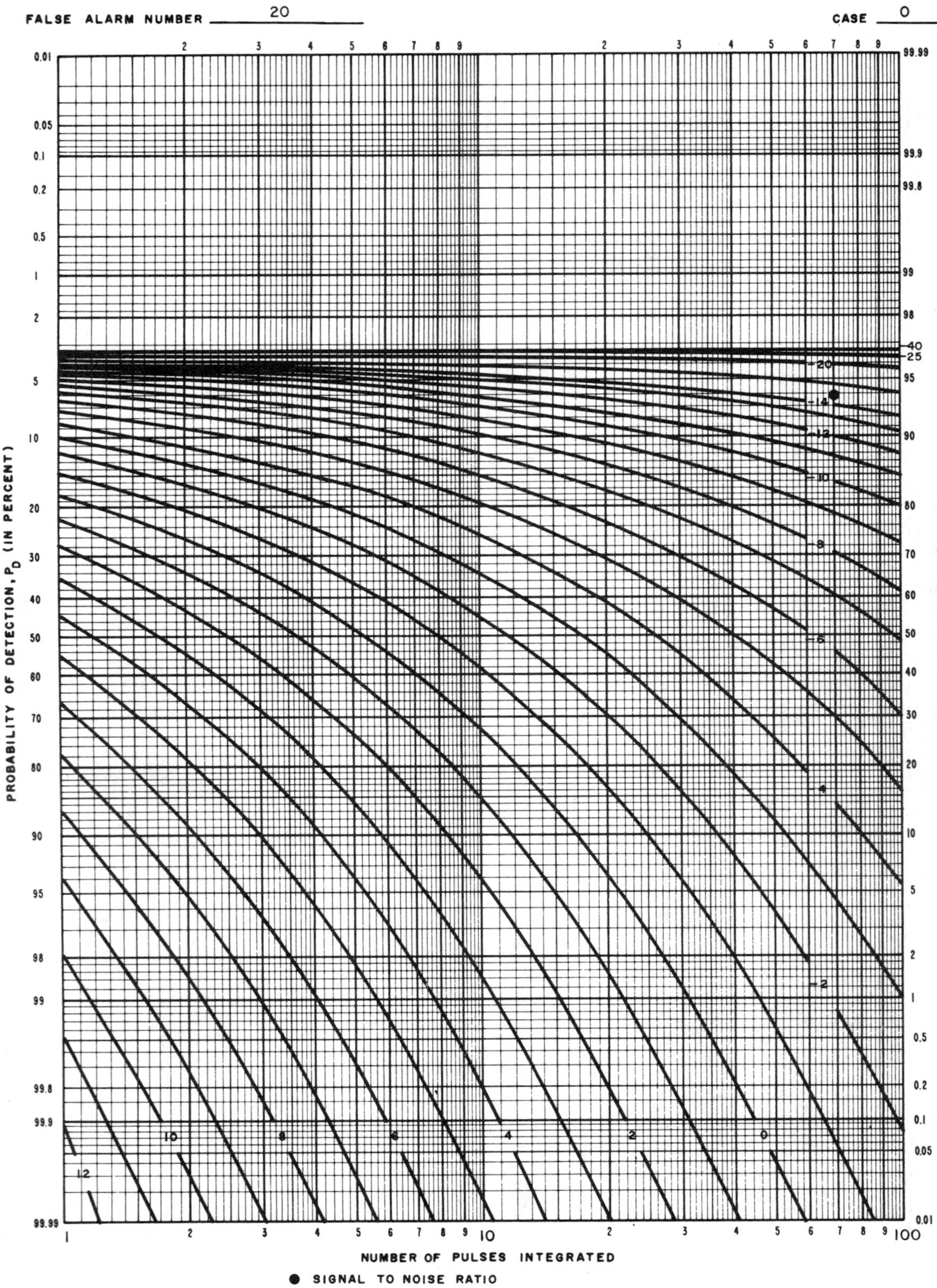
FALSE ALARM NUMBER 20
CASE 0
PROBABILITY OF DETECTION, P_D (IN PERCENT)
1 - P_D (IN PERCENT)
NUMBER OF PULSES INTEGRATED
● SIGNAL TO NOISE RATIO
-40
-25
-20
-14
-12
-10
8
6
4
-2
12
10
8
6
4
2
0

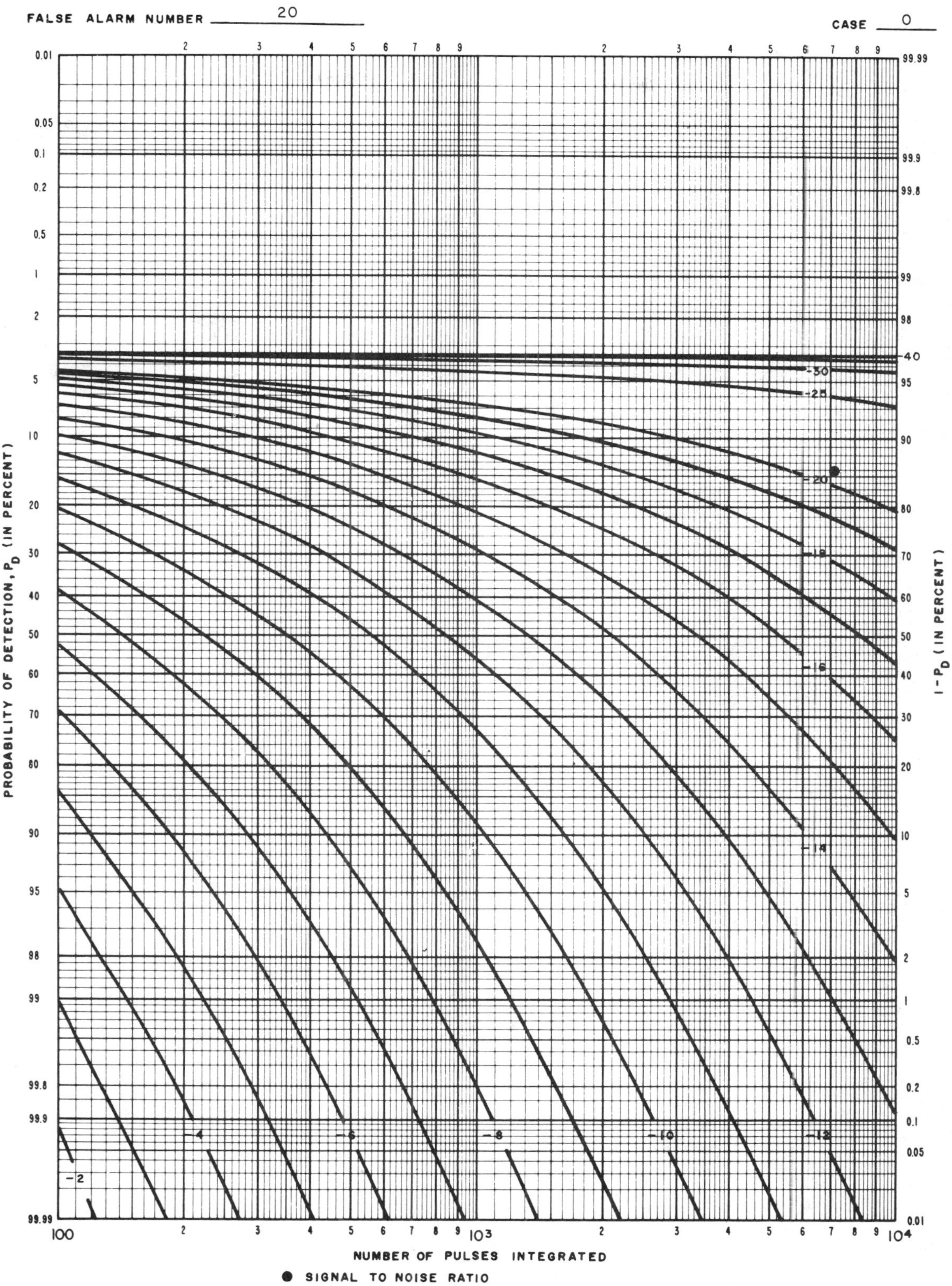
FALSE ALARM NUMBER 20
CASE 0
PROBABILITY OF DETECTION, P_D (IN PERCENT)
1-P_D (IN PERCENT)
NUMBER OF PULSES INTEGRATED
● SIGNAL TO NOISE RATIO
-2
-4
-6
-8
-10
-12
-14
-16
-18
-20
-25
-30
-40

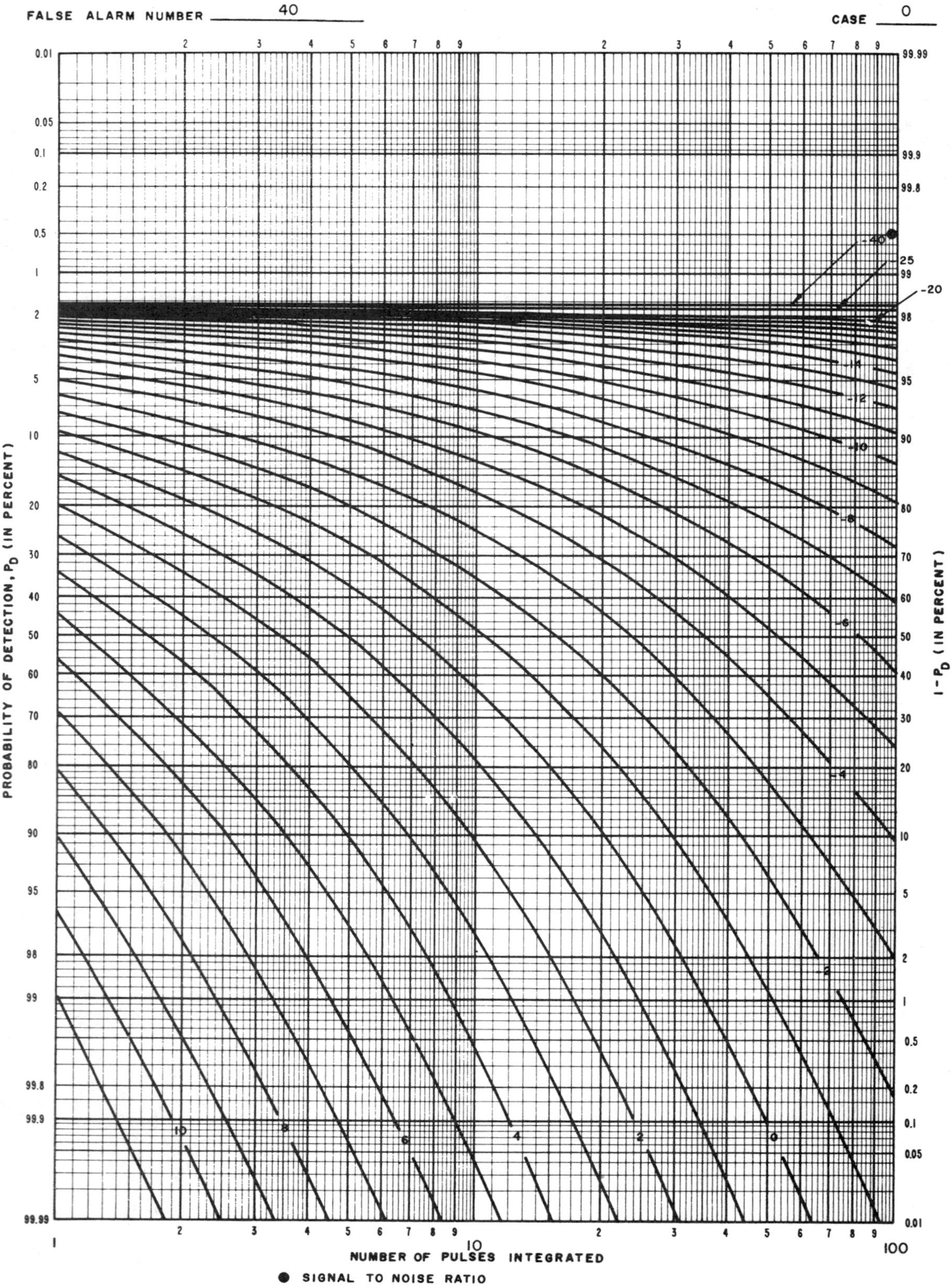
FALSE ALARM NUMBER 40
CASE 0
PROBABILITY OF DETECTION, P_D (IN PERCENT)
$1-P_D$ (IN PERCENT)
NUMBER OF PULSES INTEGRATED
● SIGNAL TO NOISE RATIO

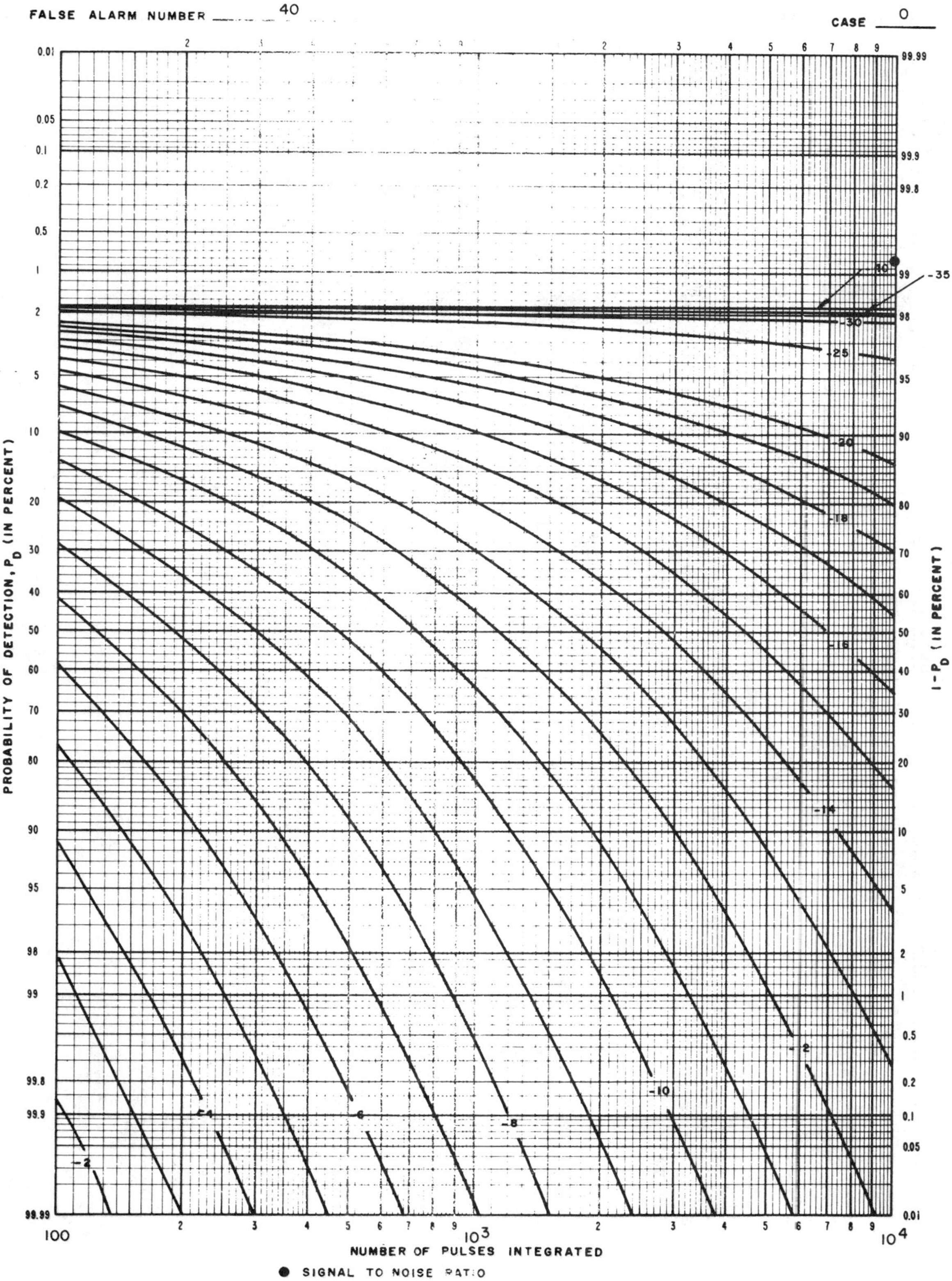
FALSE ALARM NUMBER 40
CASE 0
PROBABILITY OF DETECTION, P_D (IN PERCENT)
$1-P_D$ (IN PERCENT)
NUMBER OF PULSES INTEGRATED
● SIGNAL TO NOISE RATIO
100
10^3
10^4
-35
-30
-25
-20
-18
-16
-14
-12
-10
-8
-6
-4
-2

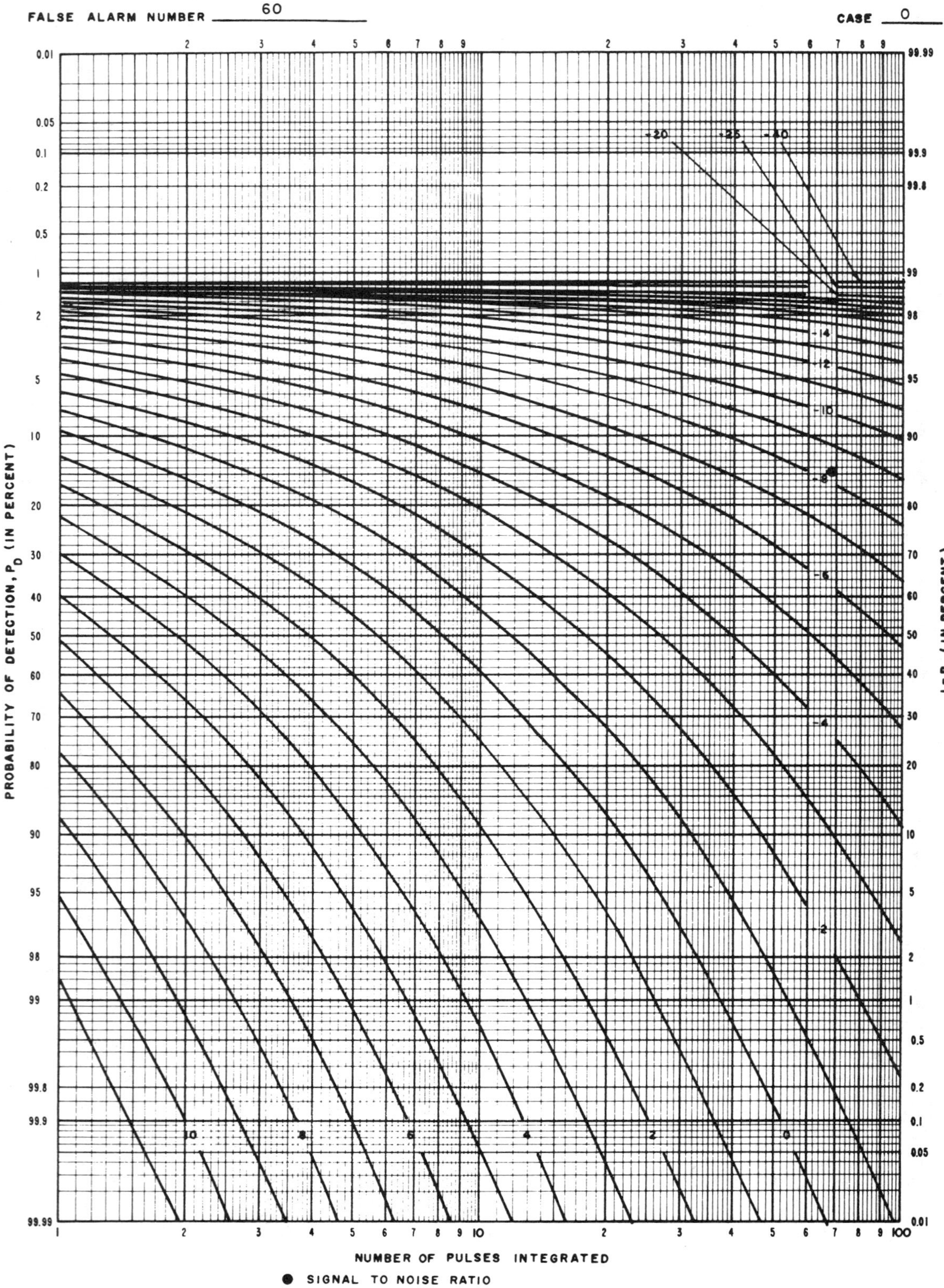
FALSE ALARM NUMBER 60
CASE 0
PROBABILITY OF DETECTION, P_D (IN PERCENT)
$1-P_D$ (IN PERCENT)
NUMBER OF PULSES INTEGRATED
● SIGNAL TO NOISE RATIO

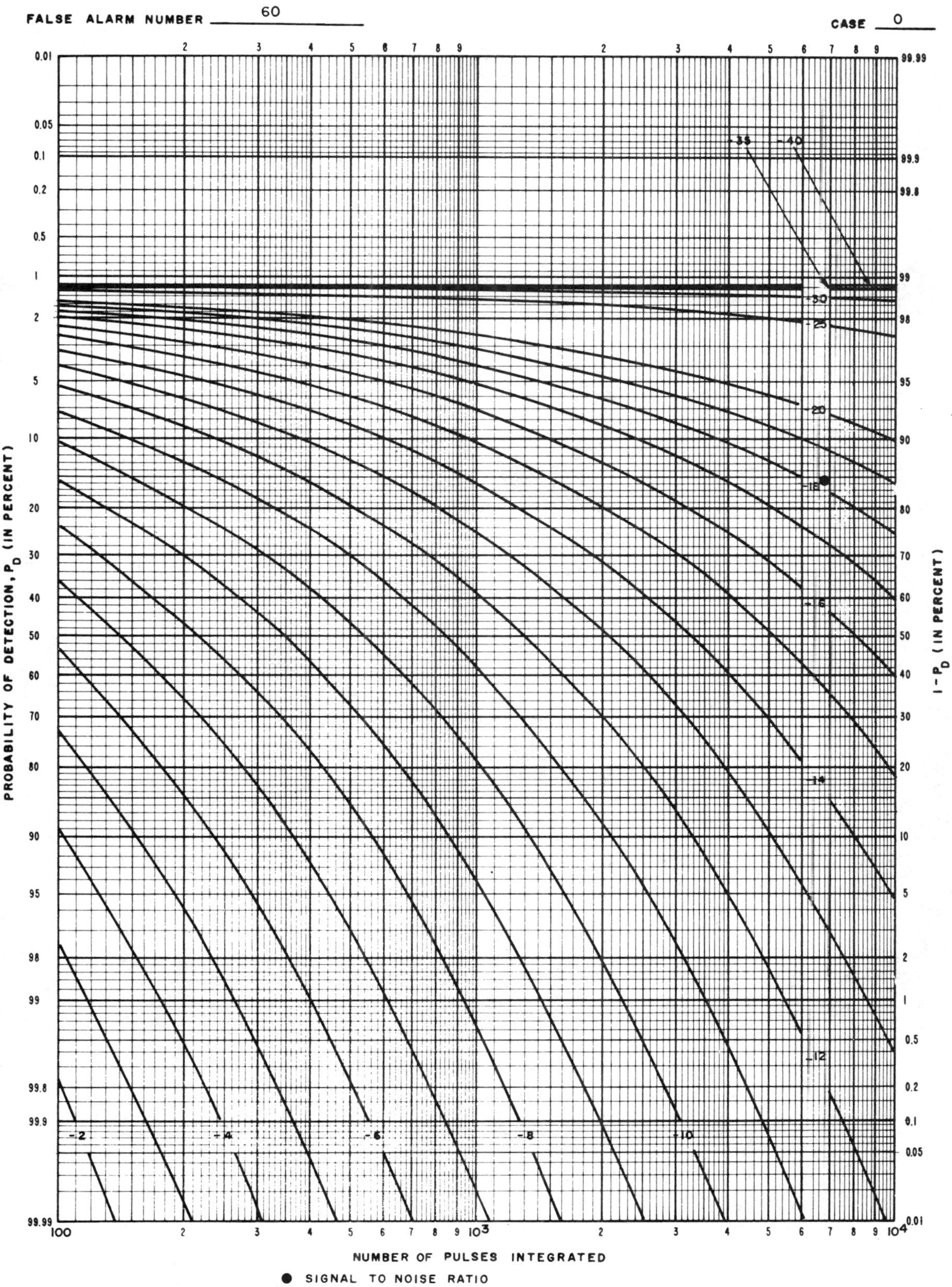
FALSE ALARM NUMBER 60
CASE 0
PROBABILITY OF DETECTION, P_D (IN PERCENT)
$1-P_D$ (IN PERCENT)
-35
-40
-30
-25
-20
-18
-16
-14
-12
-2
-4
-6
-8
-10
NUMBER OF PULSES INTEGRATED
● SIGNAL TO NOISE RATIO

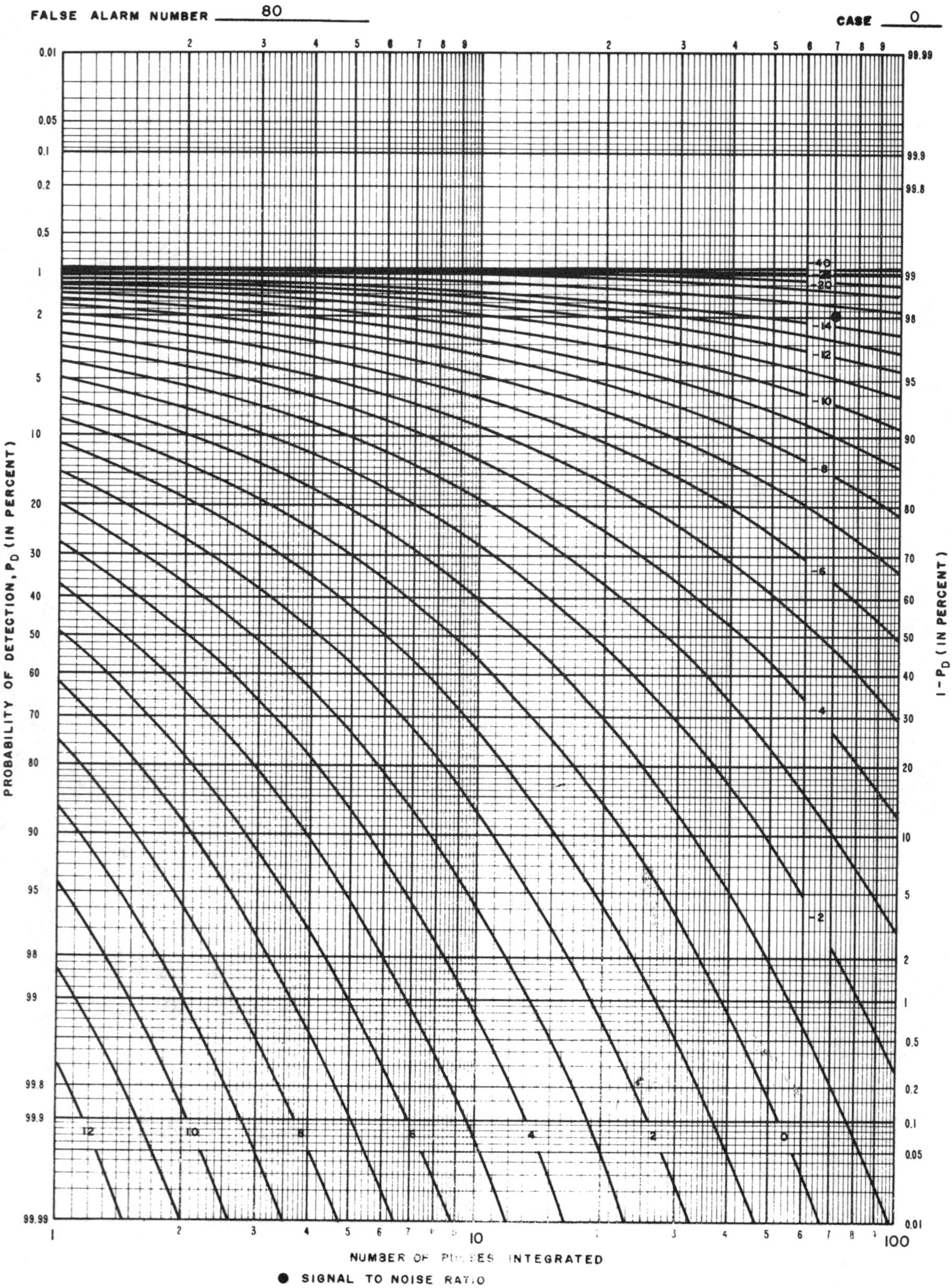
FALSE ALARM NUMBER 80
CASE 0
PROBABILITY OF DETECTION, P_D (IN PERCENT)
$1 - P_D$ (IN PERCENT)
NUMBER OF PULSES INTEGRATED
● SIGNAL TO NOISE RATIO

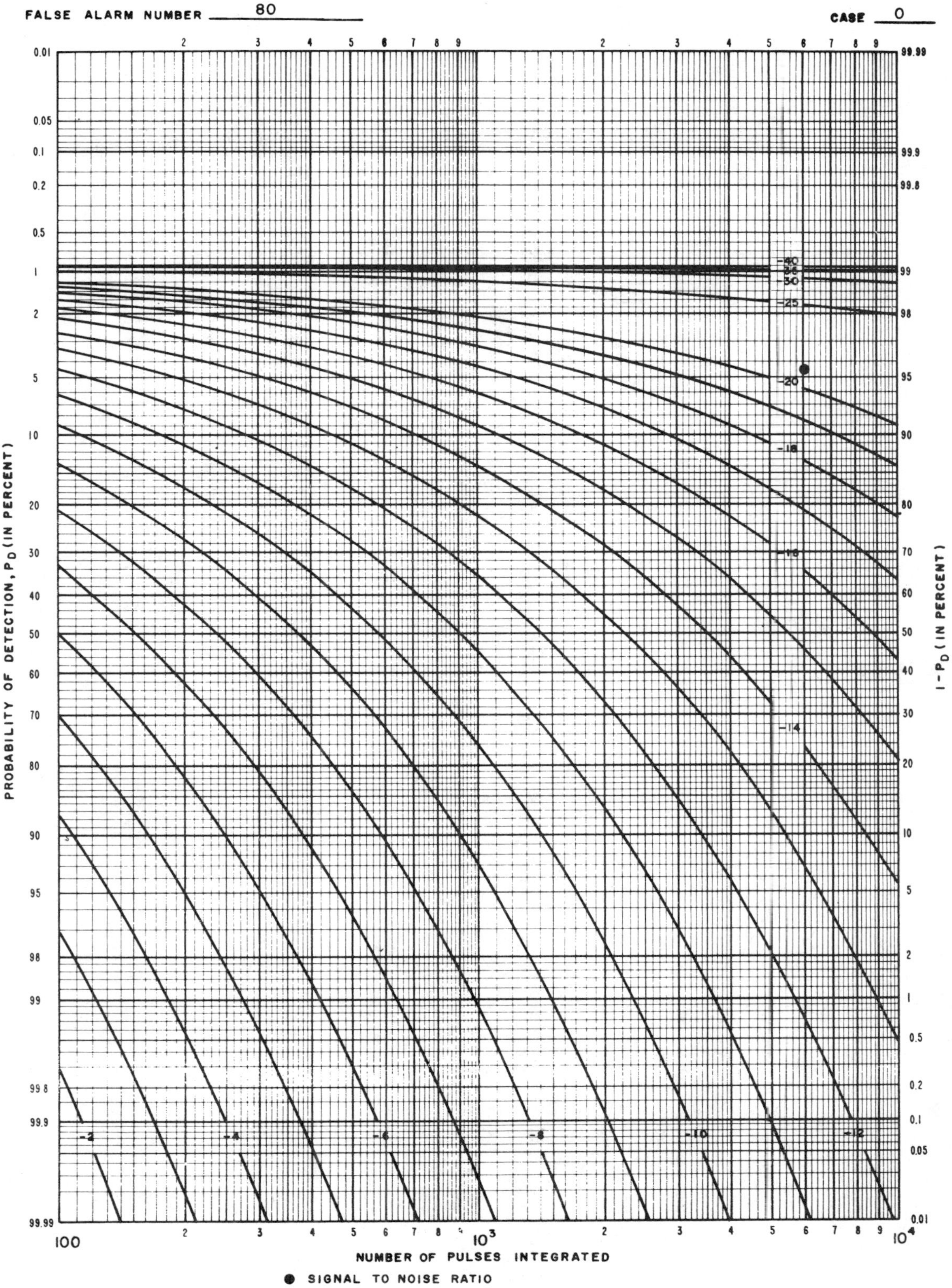
FALSE ALARM NUMBER 80
CASE 0
PROBABILITY OF DETECTION, P_D (IN PERCENT)
1 - P_D (IN PERCENT)
NUMBER OF PULSES INTEGRATED
● SIGNAL TO NOISE RATIO
-40
-35
-30
-25
-20
-18
-16
-14
-12
-10
-8
-6
-4
-2

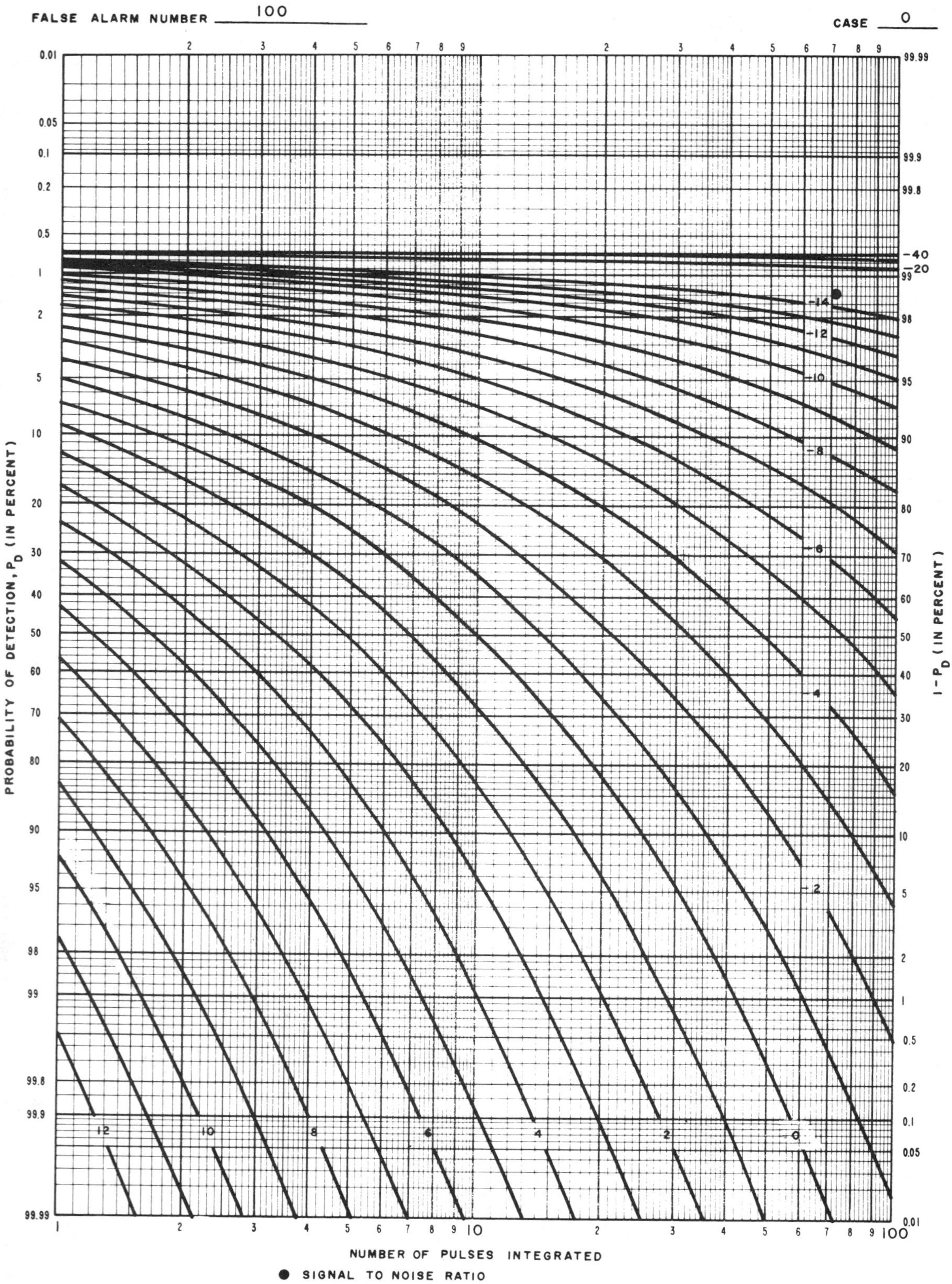
FALSE ALARM NUMBER 100
CASE 0
PROBABILITY OF DETECTION, P_D (IN PERCENT)
$1-P_D$ (IN PERCENT)
NUMBER OF PULSES INTEGRATED
● SIGNAL TO NOISE RATIO
-40
-20
-14
-12
-10
-8
-6
-4
-2
0
2
4
6
8
10
12

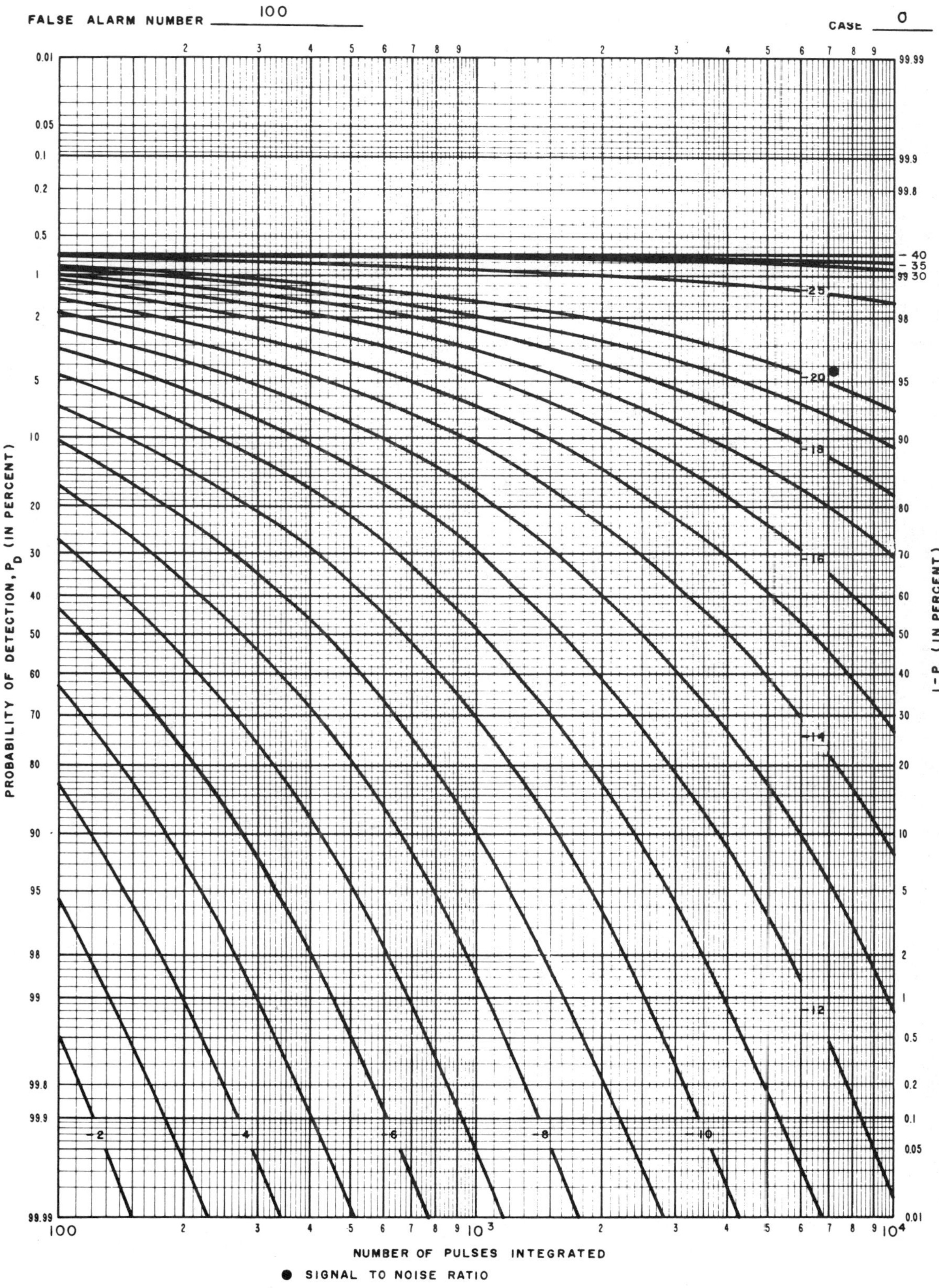
FALSE ALARM NUMBER 100
CASE 0
PROBABILITY OF DETECTION, P_D (IN PERCENT)
$1-P_D$ (IN PERCENT)
NUMBER OF PULSES INTEGRATED
● SIGNAL TO NOISE RATIO

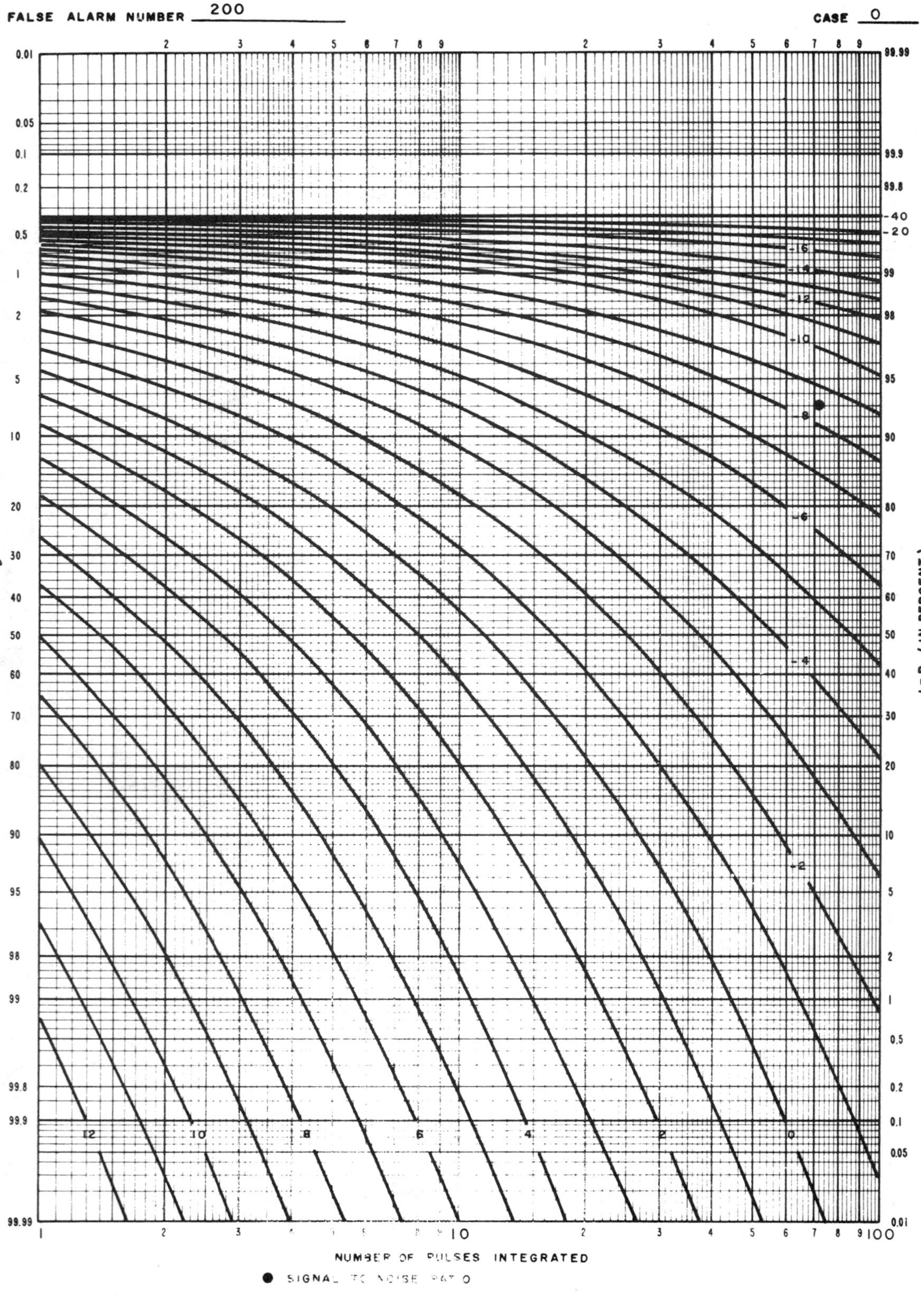
FALSE ALARM NUMBER 200
CASE 0
PROBABILITY OF DETECTION, P_D (IN PERCENT)
$1-P_D$ (IN PERCENT)
NUMBER OF PULSES INTEGRATED
● SIGNAL TO NOISE RATIO
-40
-20
-16
-14
-12
-10
-8
-6
-4
-2
0
2
4
6
8
10
12

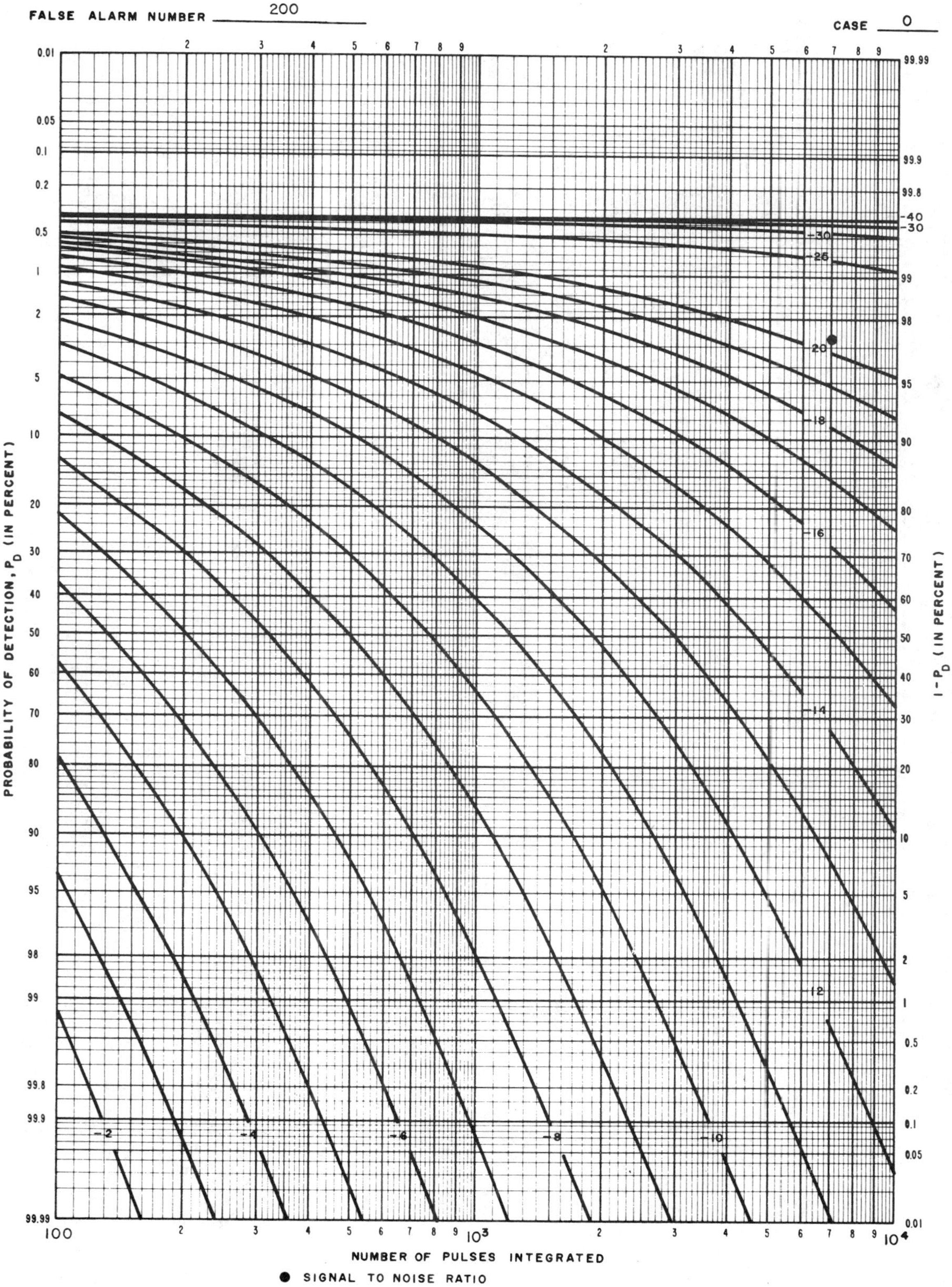
FALSE ALARM NUMBER 200
CASE 0
PROBABILITY OF DETECTION, P_D (IN PERCENT)
$1-P_D$ (IN PERCENT)
-40
-30
-25
-20
-18
-16
-14
-12
-10
-8
-6
-4
-2
100
10^3
10^4
NUMBER OF PULSES INTEGRATED
● SIGNAL TO NOISE RATIO

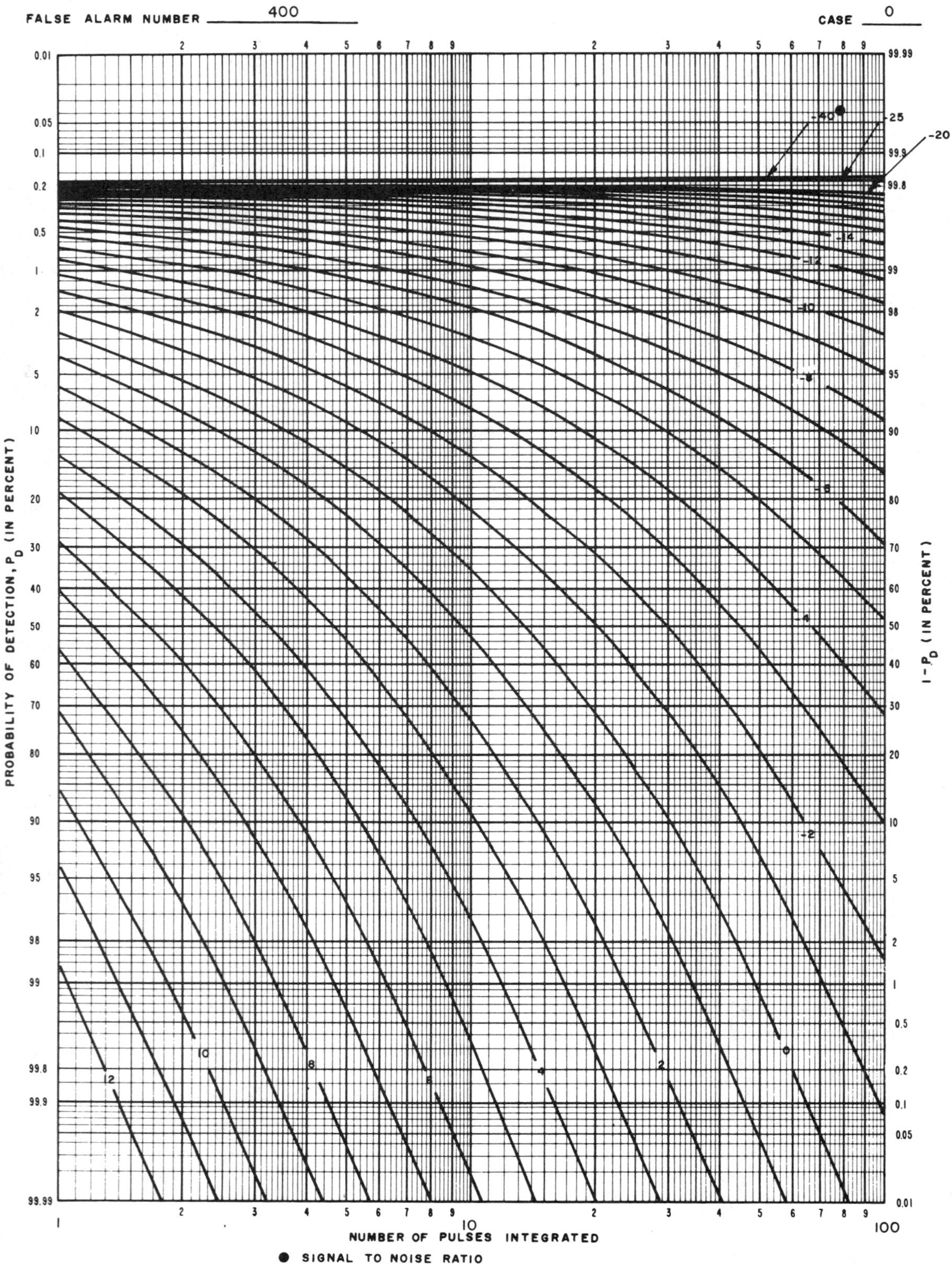
FALSE ALARM NUMBER 400
CASE 0
PROBABILITY OF DETECTION, P_D (IN PERCENT)
$1-P_D$ (IN PERCENT)
NUMBER OF PULSES INTEGRATED
● SIGNAL TO NOISE RATIO
-40
-25
-20
-14
-12
-10
-8
-6
-4
-2
0
2
4
6
8
10
12

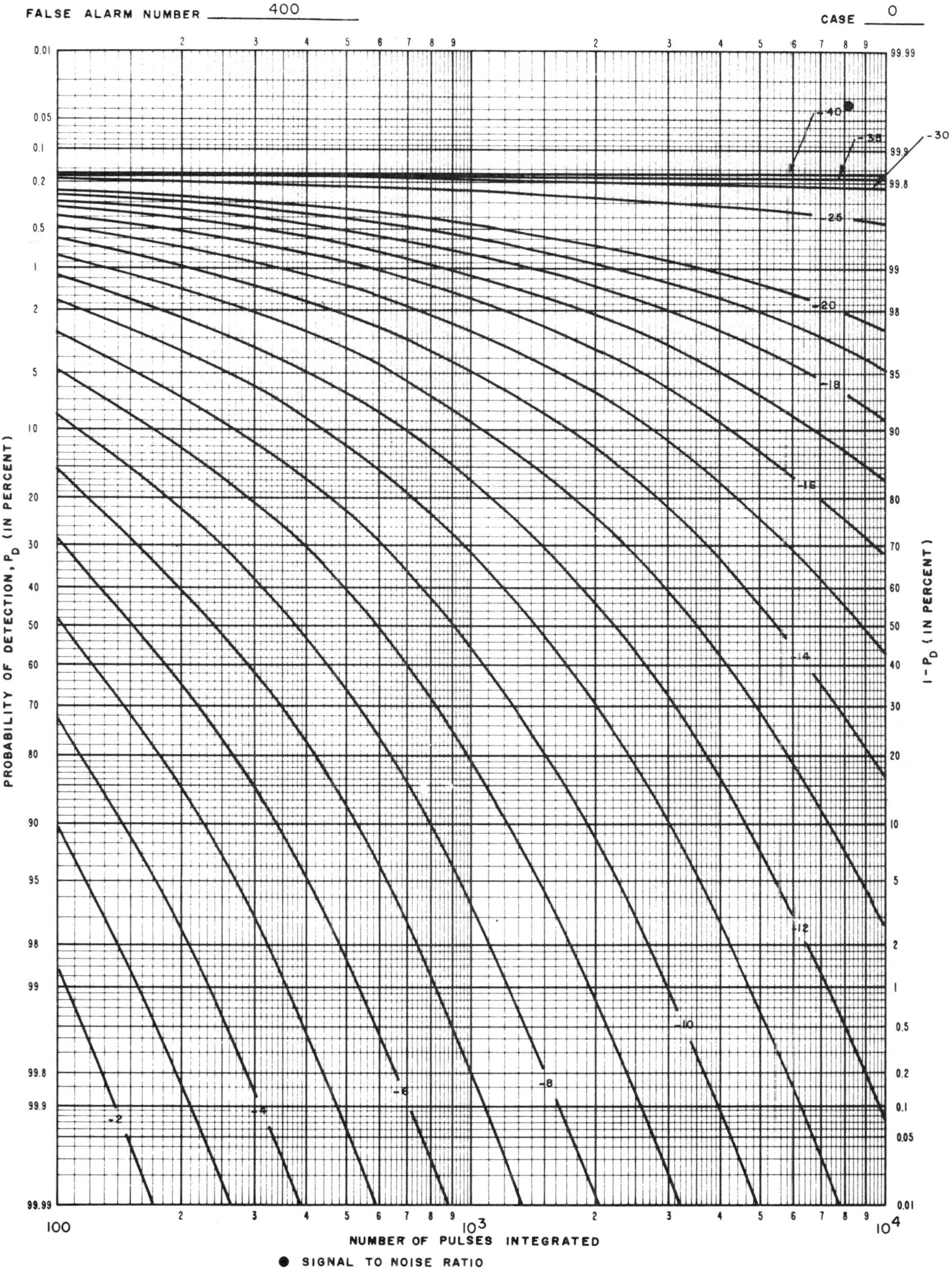
FALSE ALARM NUMBER 400
CASE 0
PROBABILITY OF DETECTION, P_D (IN PERCENT)
$1-P_D$ (IN PERCENT)
NUMBER OF PULSES INTEGRATED
100
10^3
10^4
-40
-35
-30
-25
-20
-18
-16
-14
-12
-10
-8
-6
-4
-2
● SIGNAL TO NOISE RATIO

FALSE ALARM NUMBER 600 CASE 0

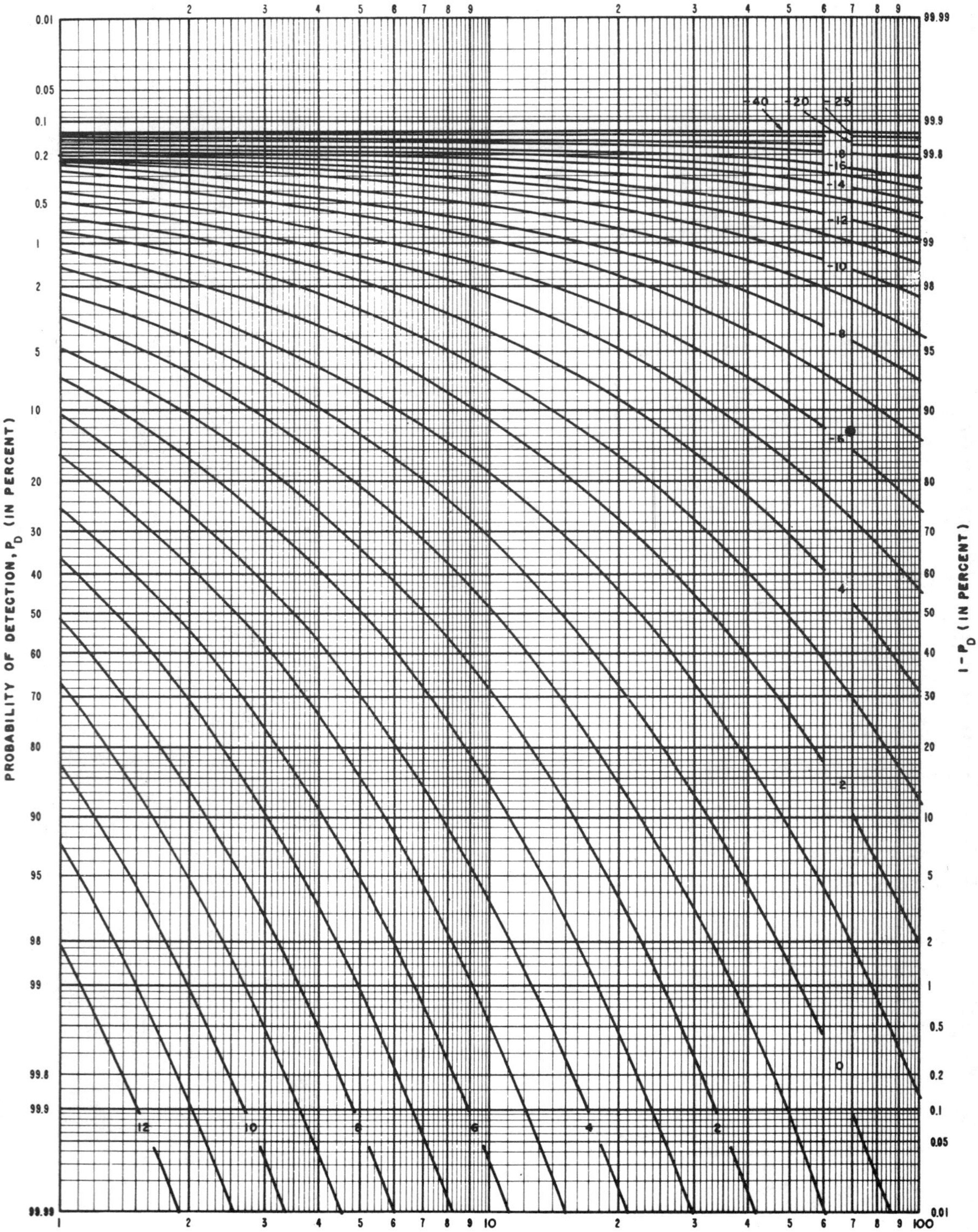

● SIGNAL TO NOISE RATIO

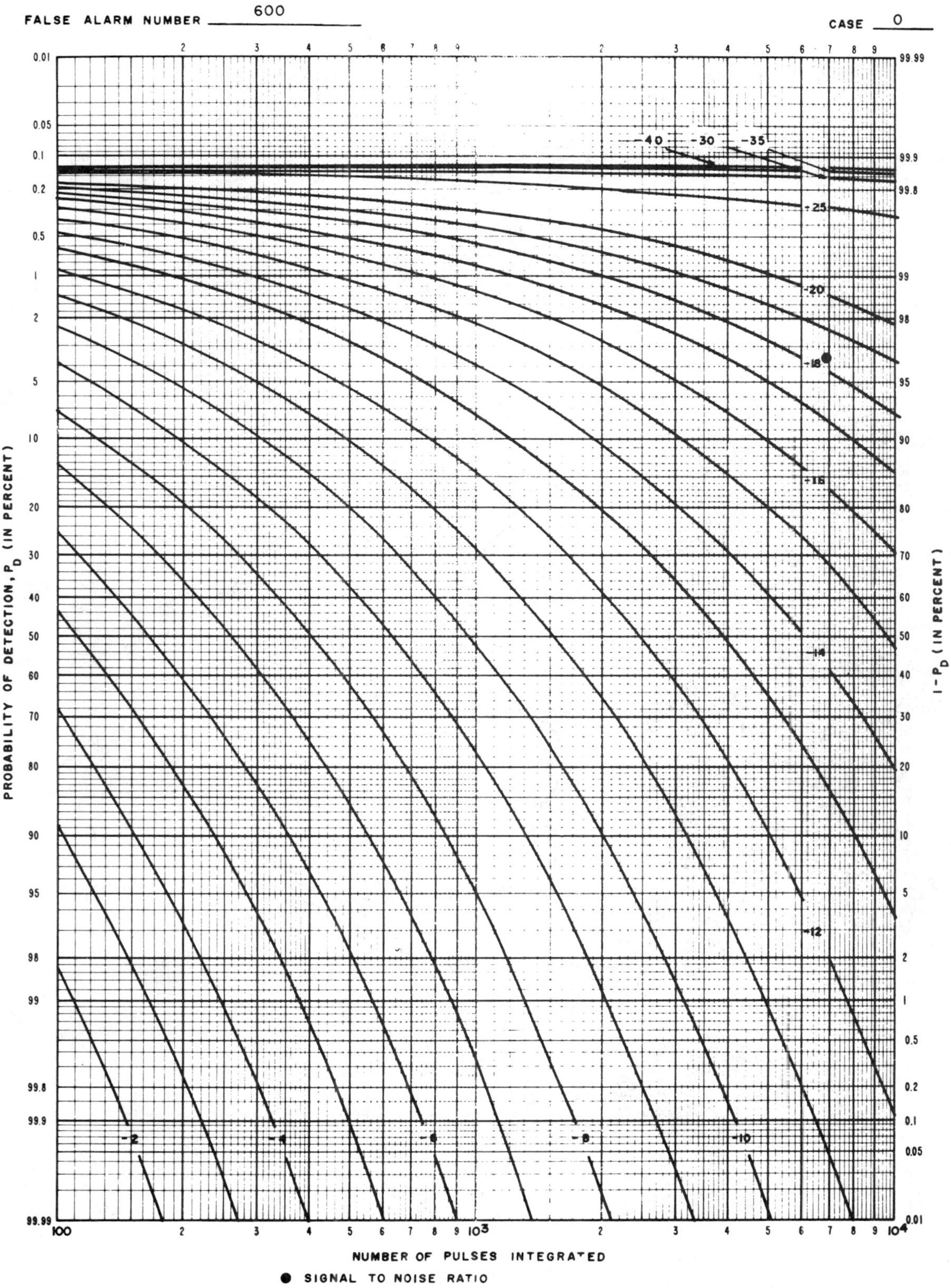
FALSE ALARM NUMBER 600
CASE 0
PROBABILITY OF DETECTION, P_D (IN PERCENT)
$1-P_D$ (IN PERCENT)
NUMBER OF PULSES INTEGRATED
● SIGNAL TO NOISE RATIO
-40
-30
-35
-25
-20
-18
-16
-14
-12
-10
-8
-6
-4
-2

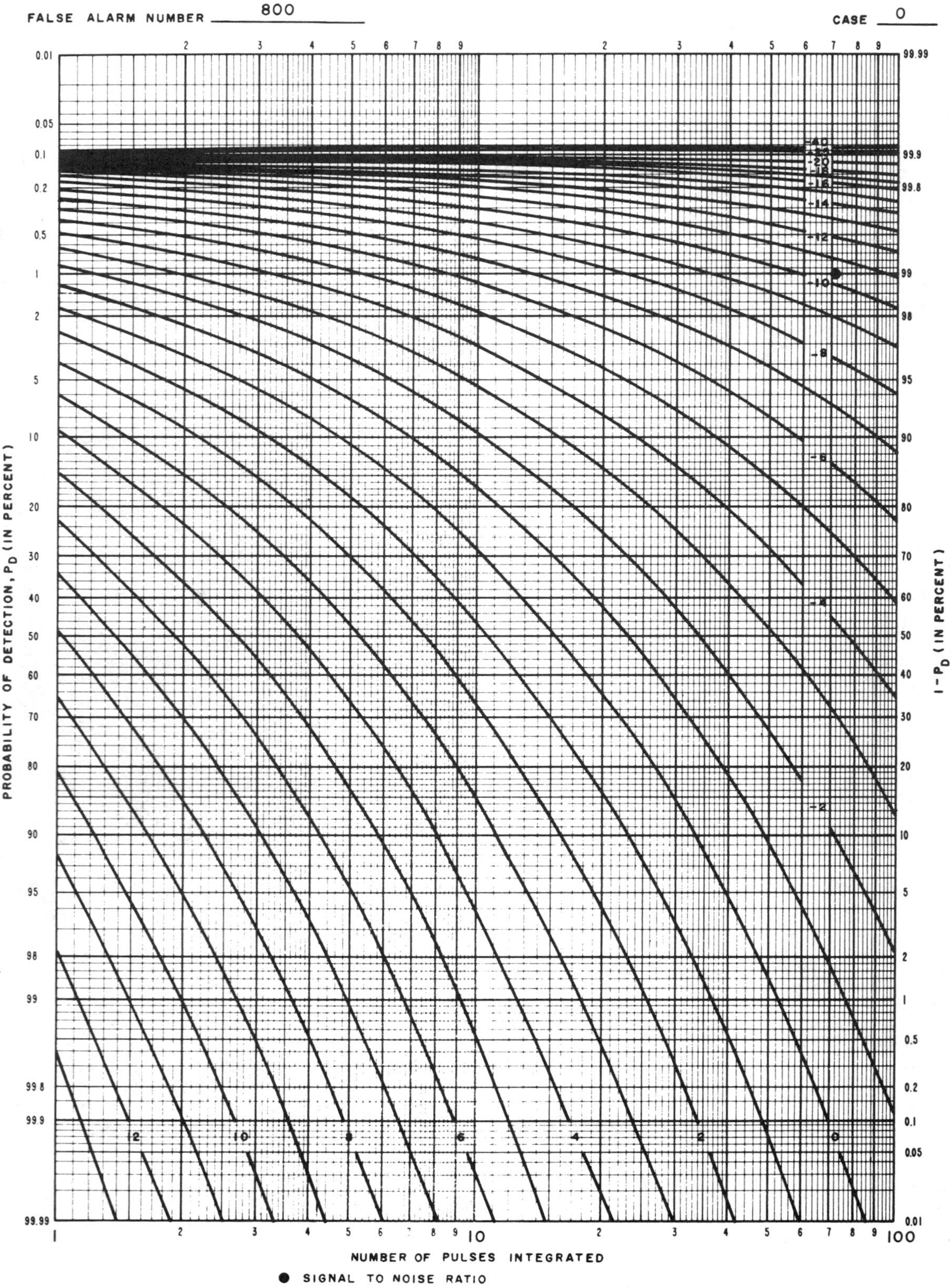
FALSE ALARM NUMBER 800
CASE 0
PROBABILITY OF DETECTION, P_D (IN PERCENT)
$1-P_D$ (IN PERCENT)
NUMBER OF PULSES INTEGRATED
● SIGNAL TO NOISE RATIO
-40
-20
-18
-16
-14
-12
-10
-8
-6
-4
-2
12
10
8
6
4
2
0

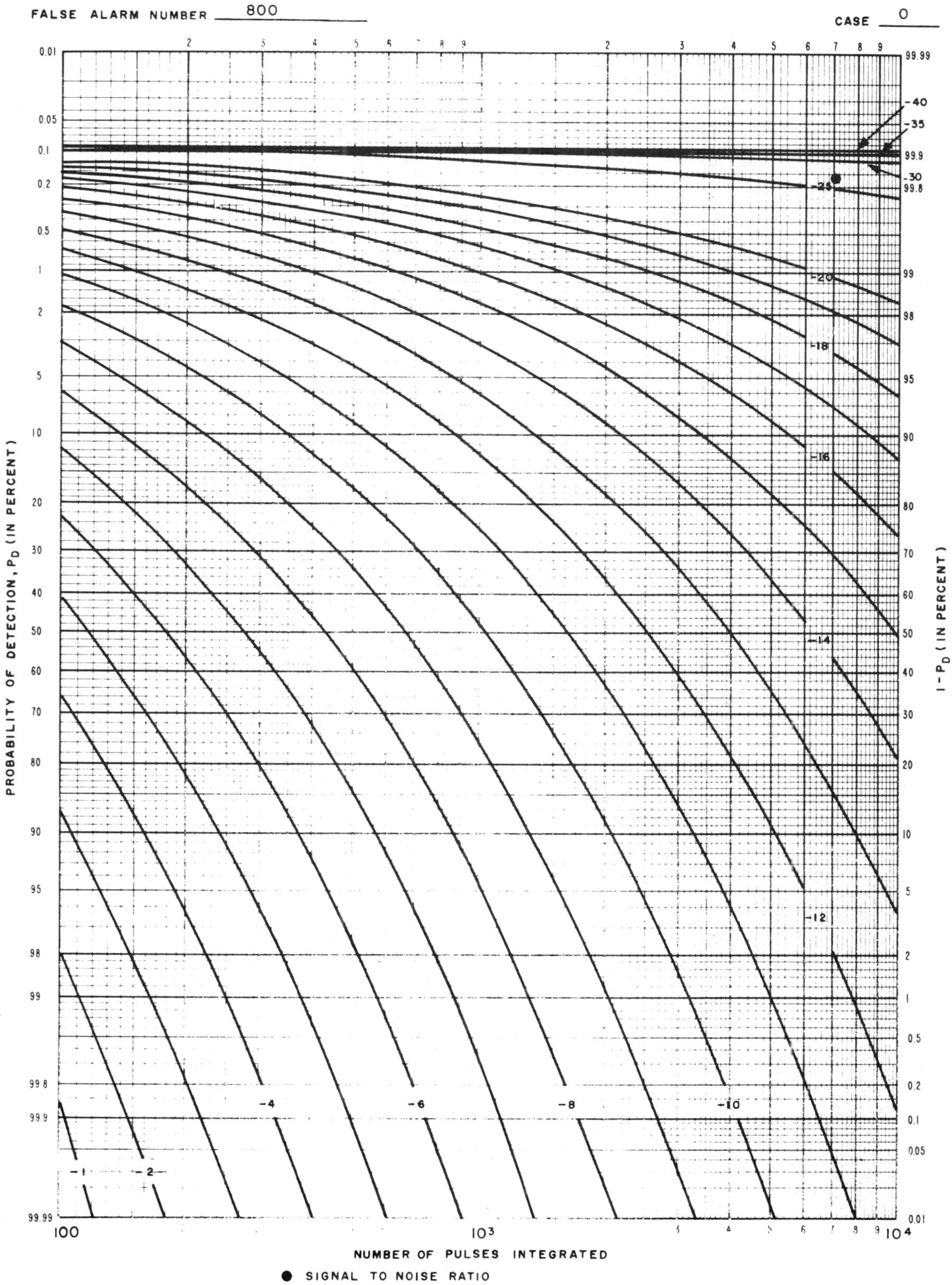
FALSE ALARM NUMBER 800
CASE 0
PROBABILITY OF DETECTION, P_D (IN PERCENT)
1 - P_D (IN PERCENT)
NUMBER OF PULSES INTEGRATED
-40
-35
-30
-25
-20
-18
-16
-14
-12
-10
-8
-6
-4
-2
-1
100
10^3
10^4
● SIGNAL TO NOISE RATIO

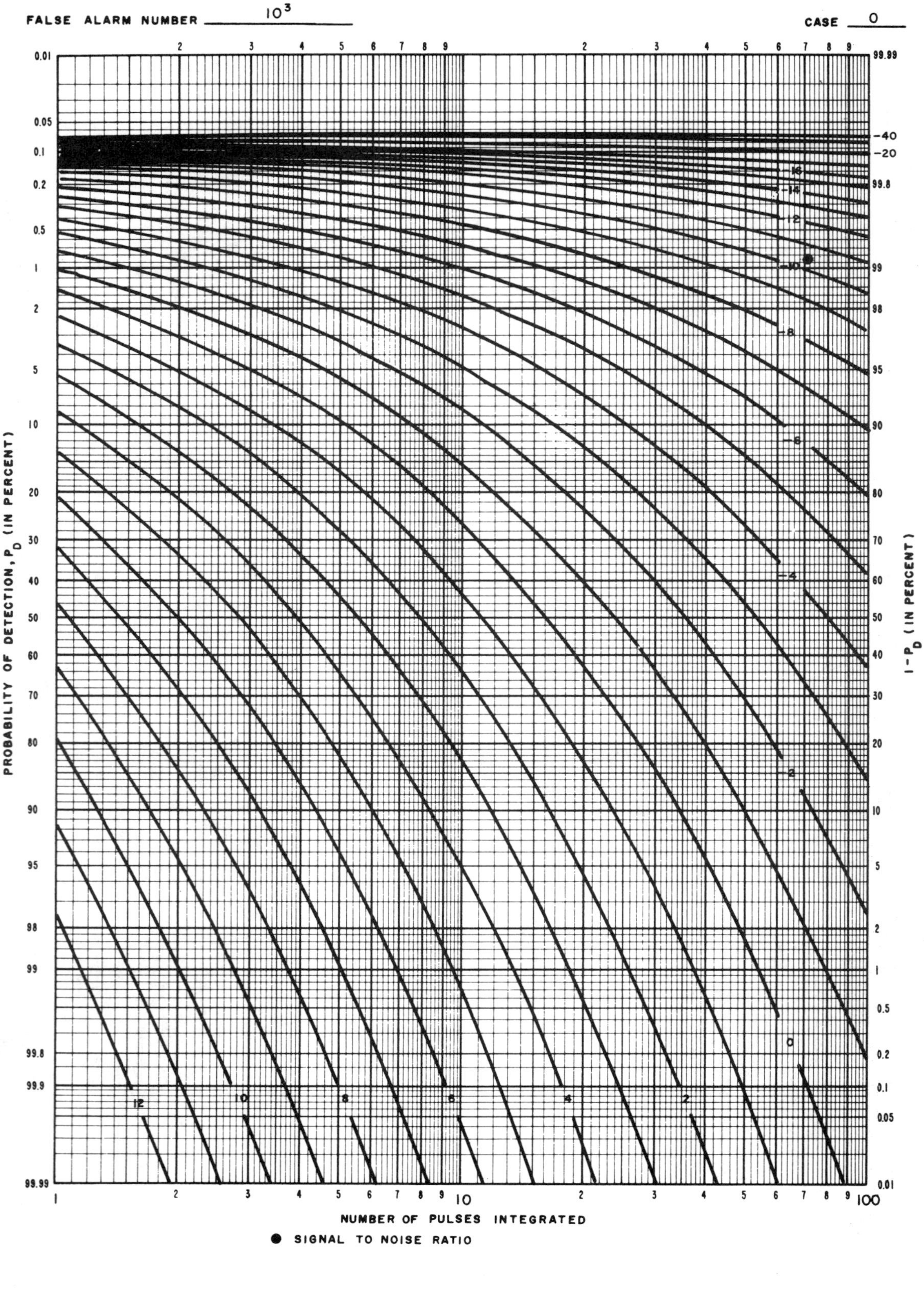
FALSE ALARM NUMBER 10^3
CASE 0
PROBABILITY OF DETECTION, P_D (IN PERCENT)
$1-P_D$ (IN PERCENT)
NUMBER OF PULSES INTEGRATED
● SIGNAL TO NOISE RATIO
-40
-20
-16
-14
-12
-10
-8
-6
-4
-2
0
2
4
6
8
10
12

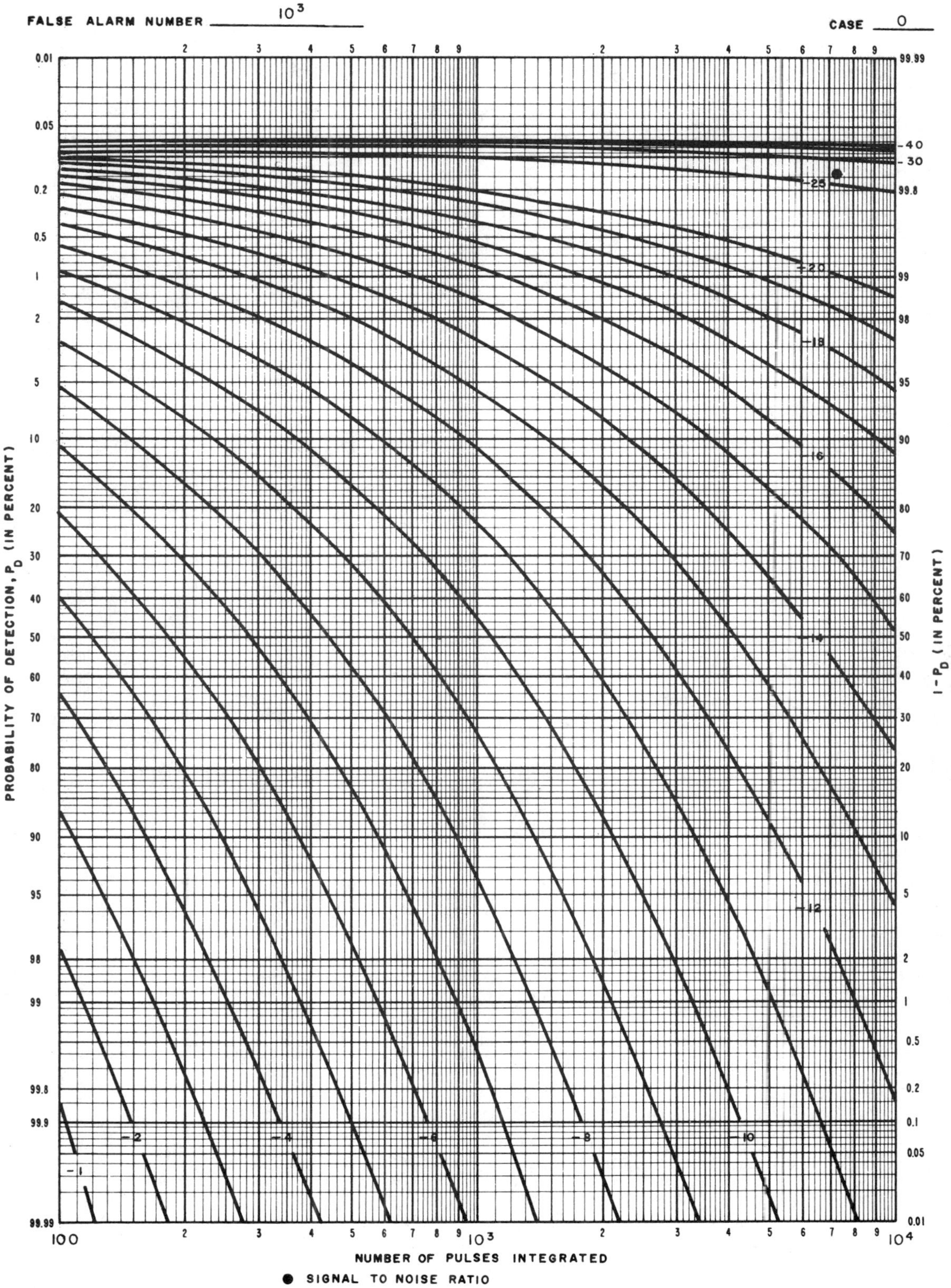
FALSE ALARM NUMBER 10^3
CASE 0
PROBABILITY OF DETECTION, P_D (IN PERCENT)
1-P_D (IN PERCENT)
NUMBER OF PULSES INTEGRATED
100
10^3
10^4
-40
-30
-25
-20
-18
-16
-14
-12
-10
-8
-6
-4
-2
-1
● SIGNAL TO NOISE RATIO

FALSE ALARM NUMBER 2×10^3 CASE 0

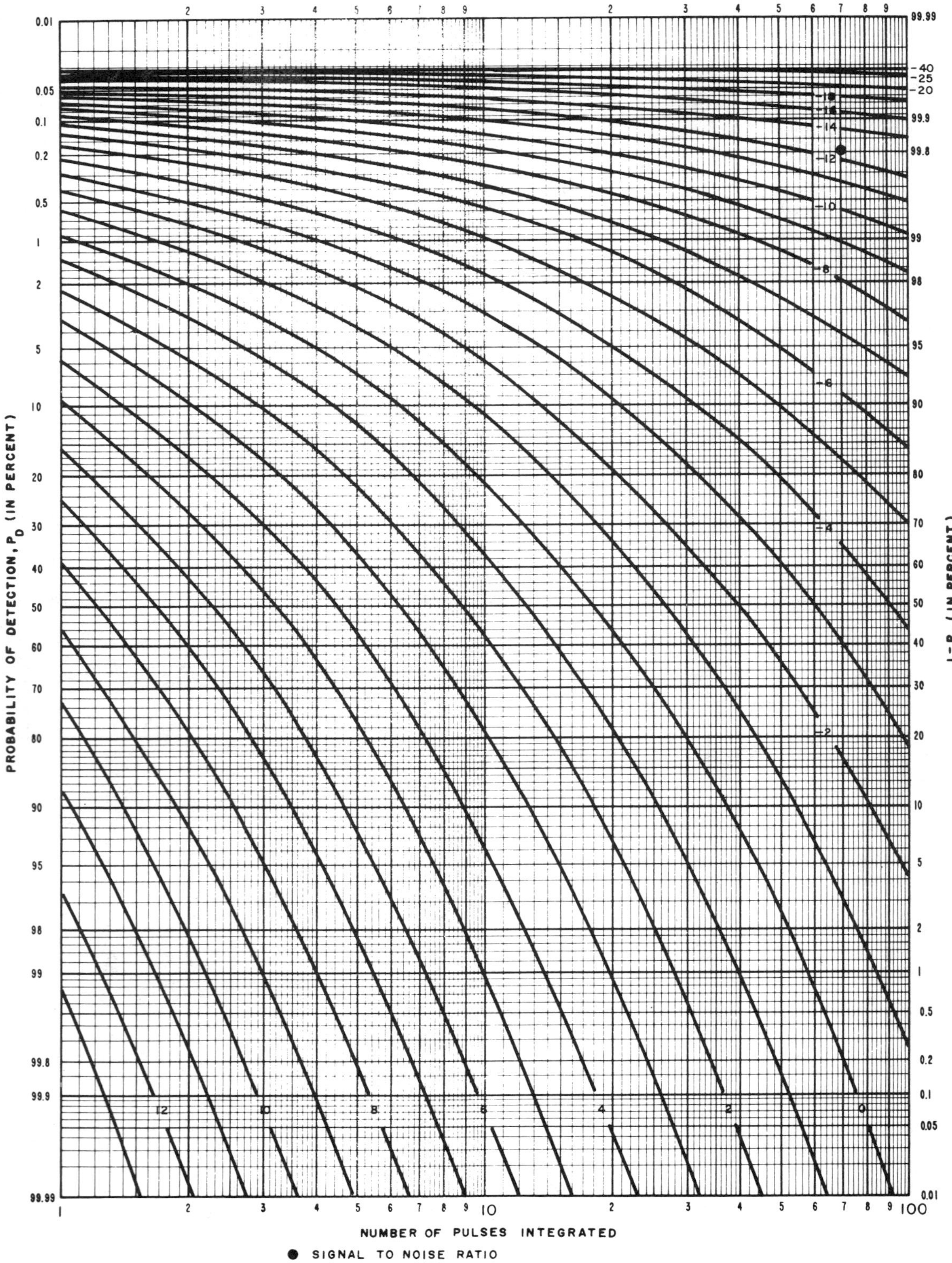

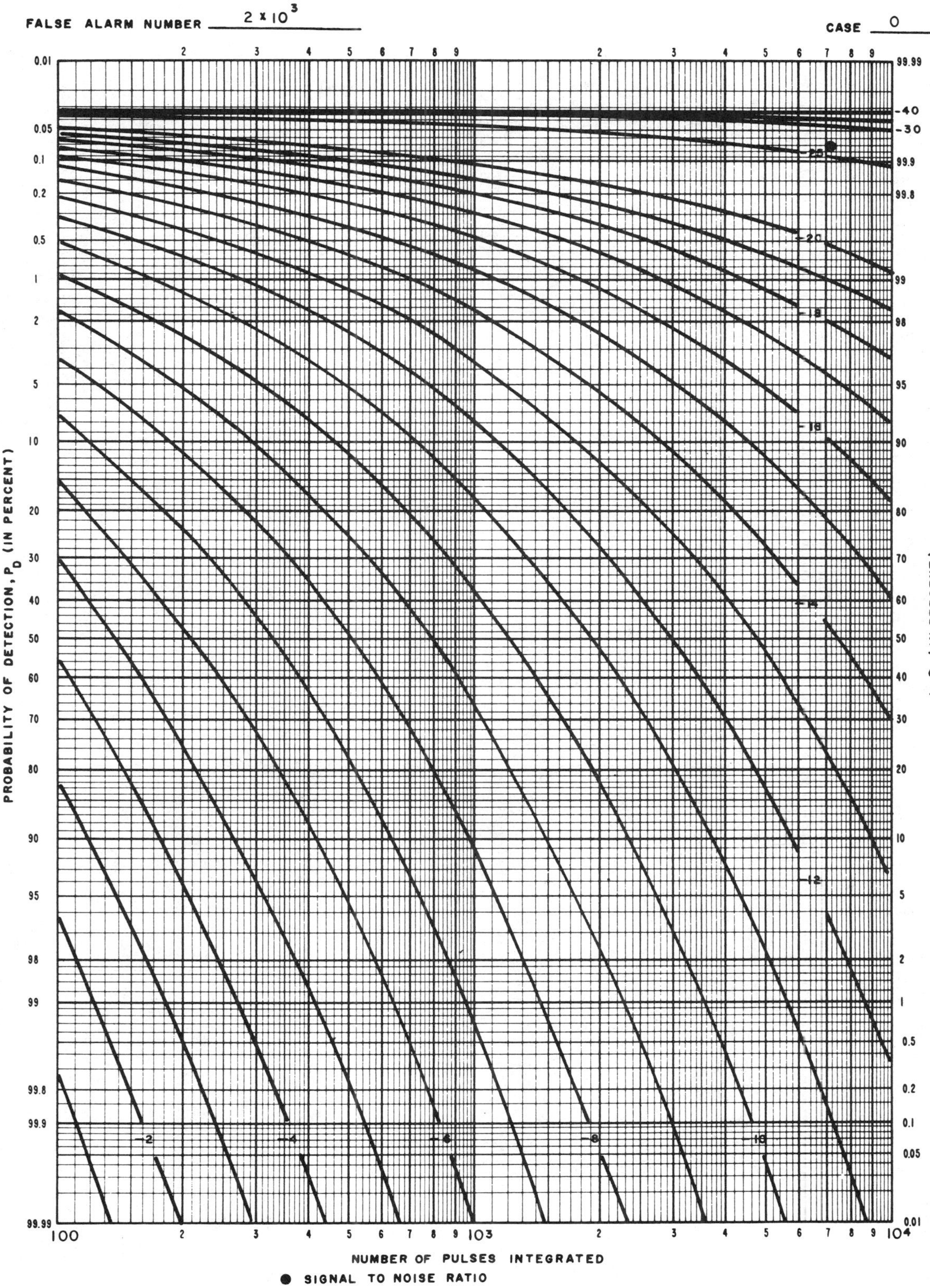
FALSE ALARM NUMBER 2 × 10³
CASE 0
PROBABILITY OF DETECTION, P_D (IN PERCENT)
1 - P_D (IN PERCENT)
0.01 0.05 0.1 0.2 0.5 1 2 5 10 20 30 40 50 60 70 80 90 95 98 99 99.8 99.9 99.99
99.99 99.9 99.8 99 98 95 90 80 70 60 50 40 30 20 10 5 2 1 0.5 0.2 0.1 0.05 0.01
100 2 3 4 5 6 7 8 9 10³ 2 3 4 5 6 7 8 9 10⁴
-2 -4 -6 -8 -10 -12 -14 -16 -18 -20 -25 -30 -40
NUMBER OF PULSES INTEGRATED
● SIGNAL TO NOISE RATIO

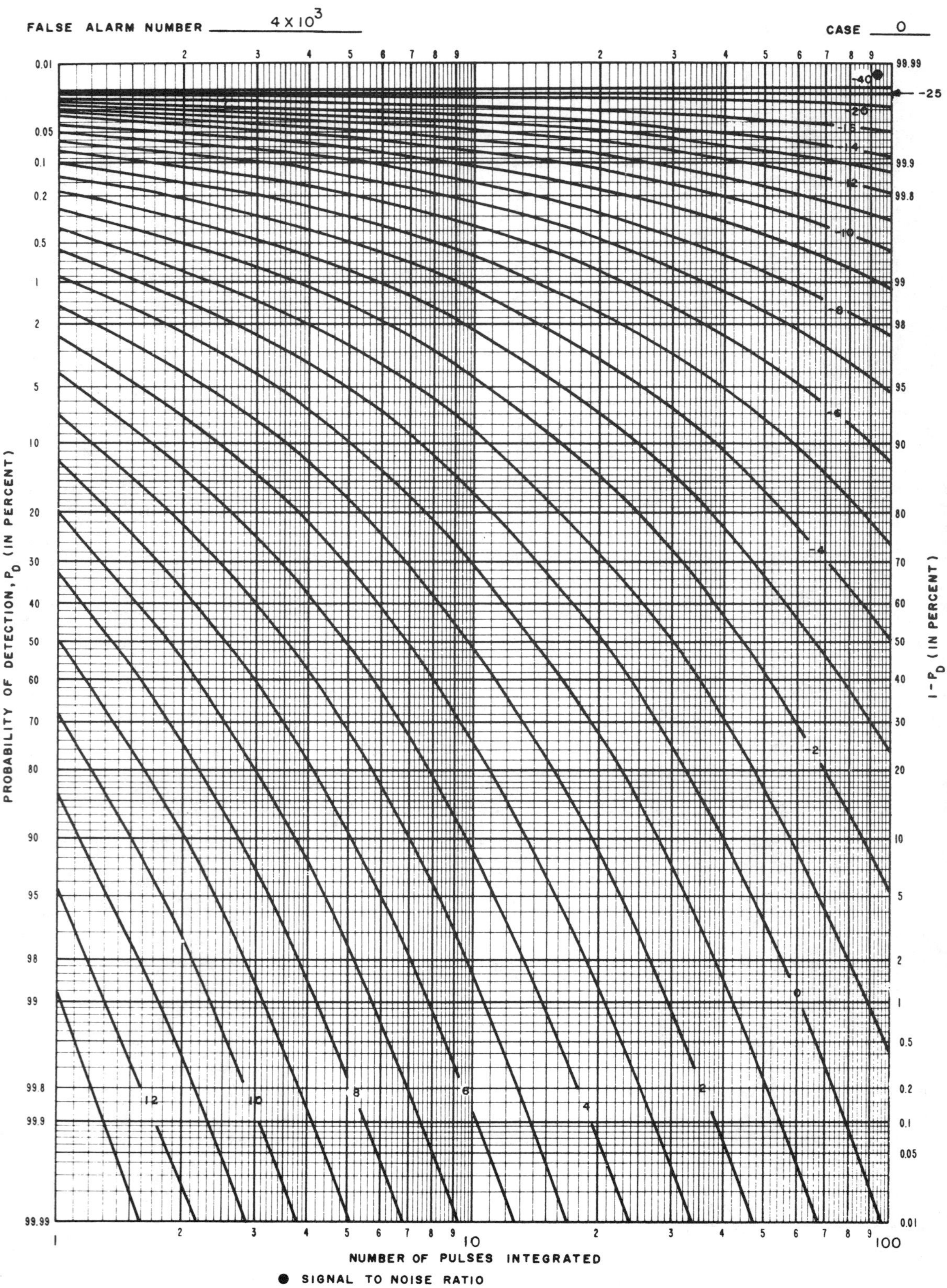
FALSE ALARM NUMBER 4×10^3
CASE 0
PROBABILITY OF DETECTION, P_D (IN PERCENT)
$1-P_D$ (IN PERCENT)
NUMBER OF PULSES INTEGRATED
● SIGNAL TO NOISE RATIO
-40
-25
-20
-15
-14
-12
-10
-8
-6
-4
-2
0
2
4
6
8
10
12

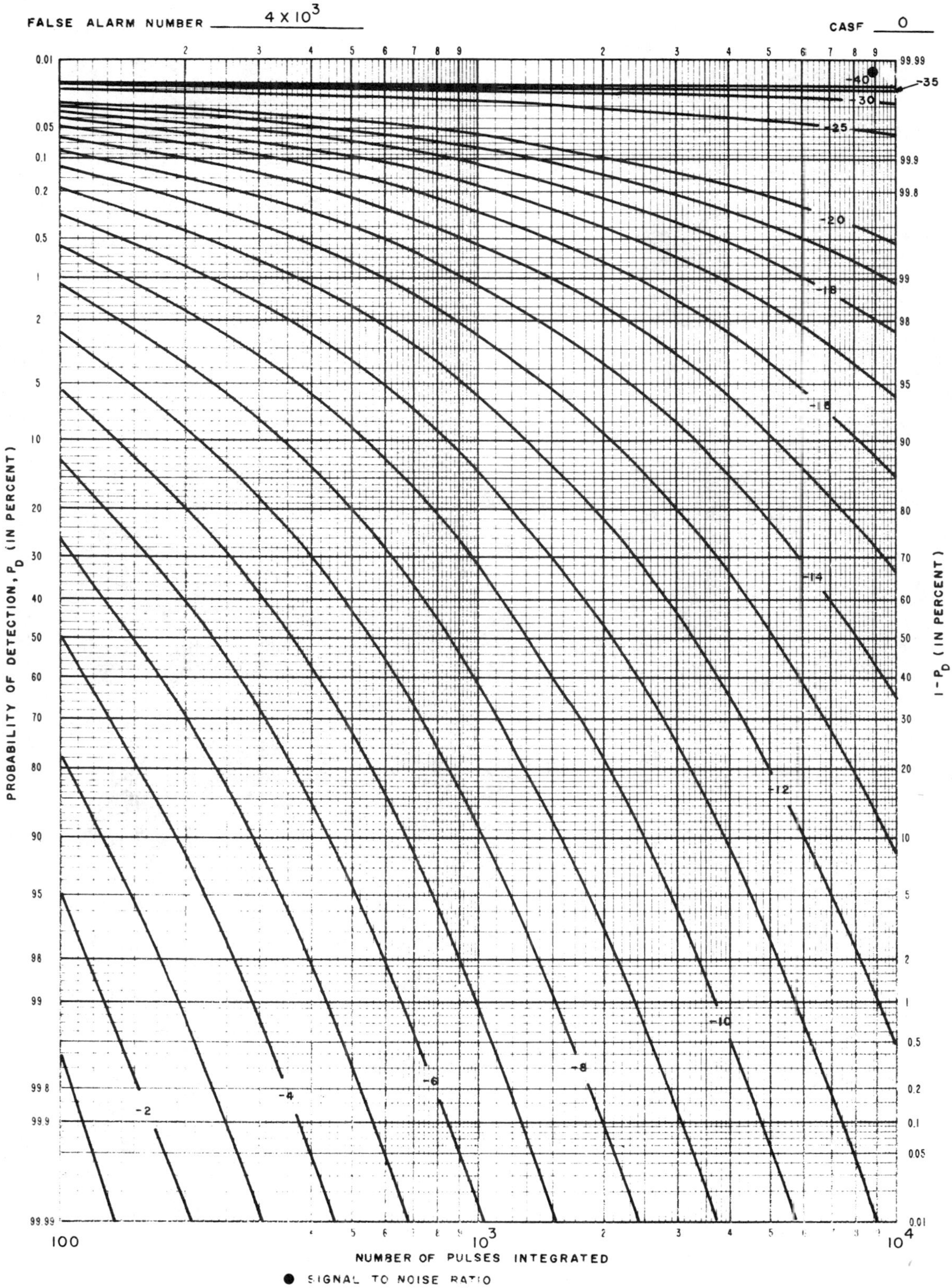
FALSE ALARM NUMBER 4 X 10^3
CASE 0
PROBABILITY OF DETECTION, P_D (IN PERCENT)
1 - P_D (IN PERCENT)
NUMBER OF PULSES INTEGRATED
100
10^3
10^4
-40
-35
-30
-25
-20
-18
-16
-14
-12
-10
-8
-6
-4
-2
● SIGNAL TO NOISE RATIO

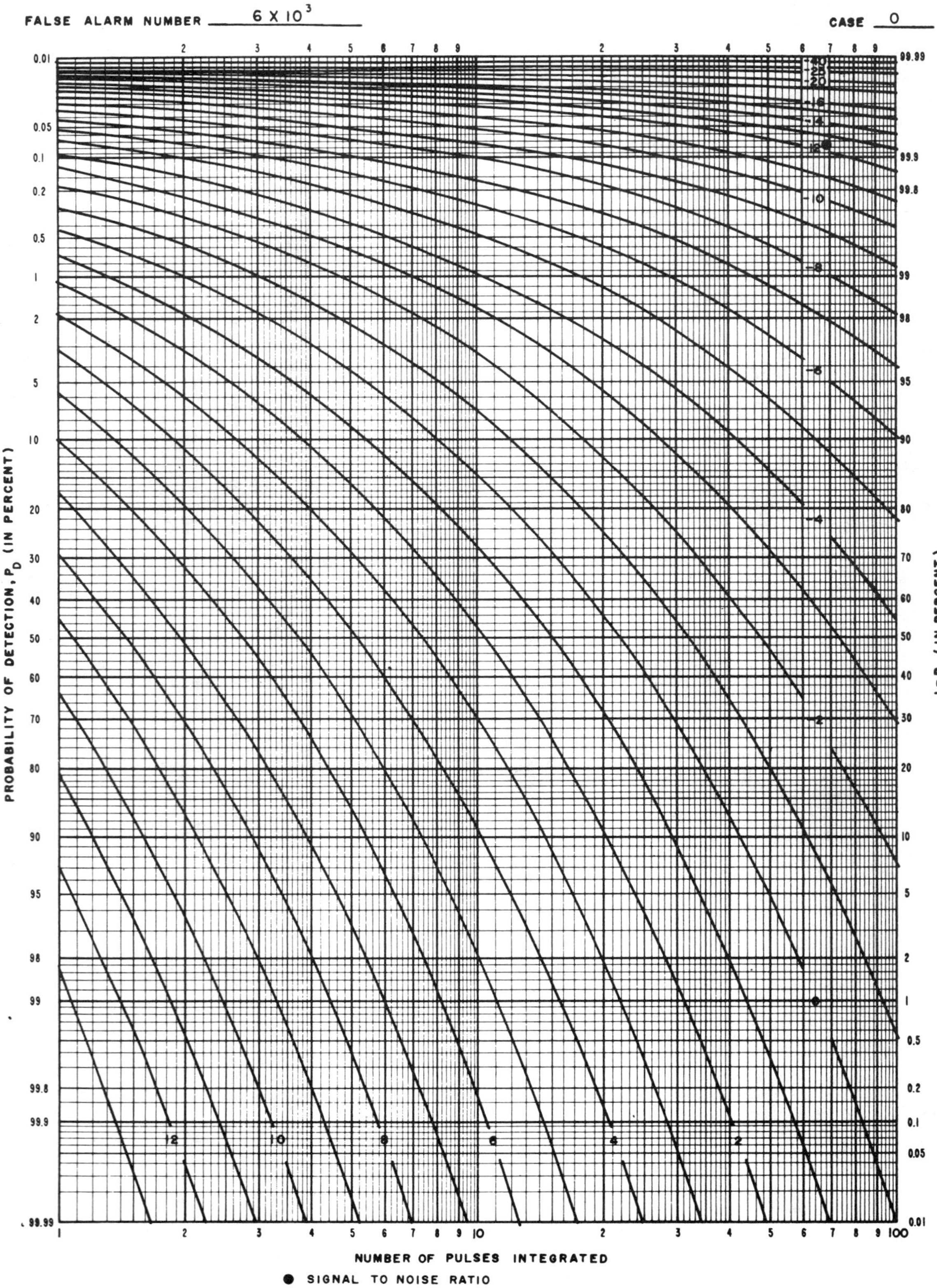
FALSE ALARM NUMBER 6 X 10^3
CASE 0
PROBABILITY OF DETECTION, P_D (IN PERCENT)
1 - P_D (IN PERCENT)
NUMBER OF PULSES INTEGRATED
● SIGNAL TO NOISE RATIO
-40
-25
-20
-16
-14
-12
-10
-8
-6
-4
-2
0
2
4
6
8
10
12

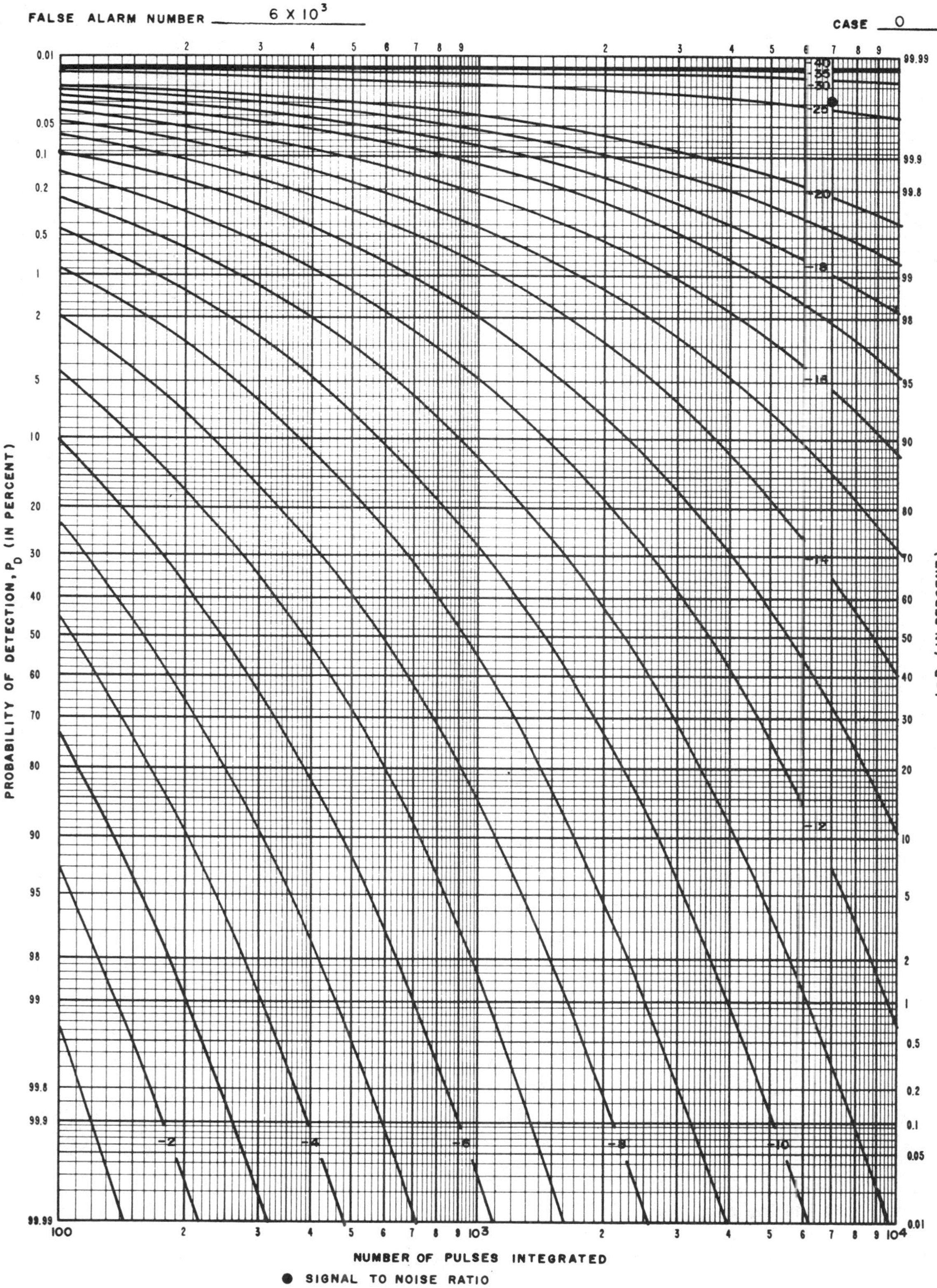
FALSE ALARM NUMBER 6 X 10^3
CASE 0
PROBABILITY OF DETECTION, P_D (IN PERCENT)
1 - P_D (IN PERCENT)
NUMBER OF PULSES INTEGRATED
● SIGNAL TO NOISE RATIO
40
35
30
25
20
18
16
14
12
-2
-4
-6
-8
-10

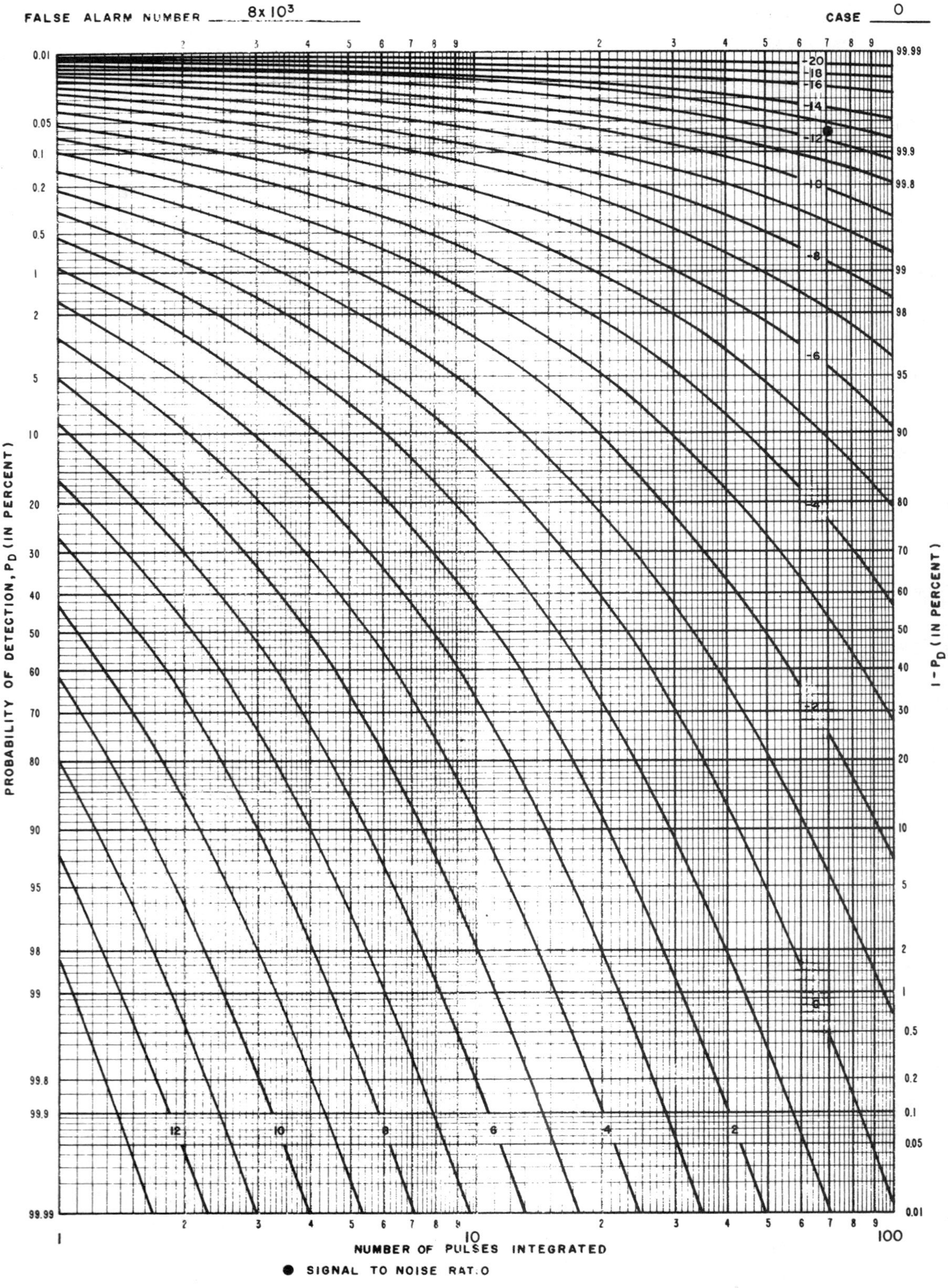
FALSE ALARM NUMBER 8x10³
CASE 0
PROBABILITY OF DETECTION, P_D (IN PERCENT)
1 - P_D (IN PERCENT)
NUMBER OF PULSES INTEGRATED
SIGNAL TO NOISE RATIO
-20
-18
-16
-14
-12
-10
-8
-6
-4
-2
0
2
4
6
8
10
12
1
10
100

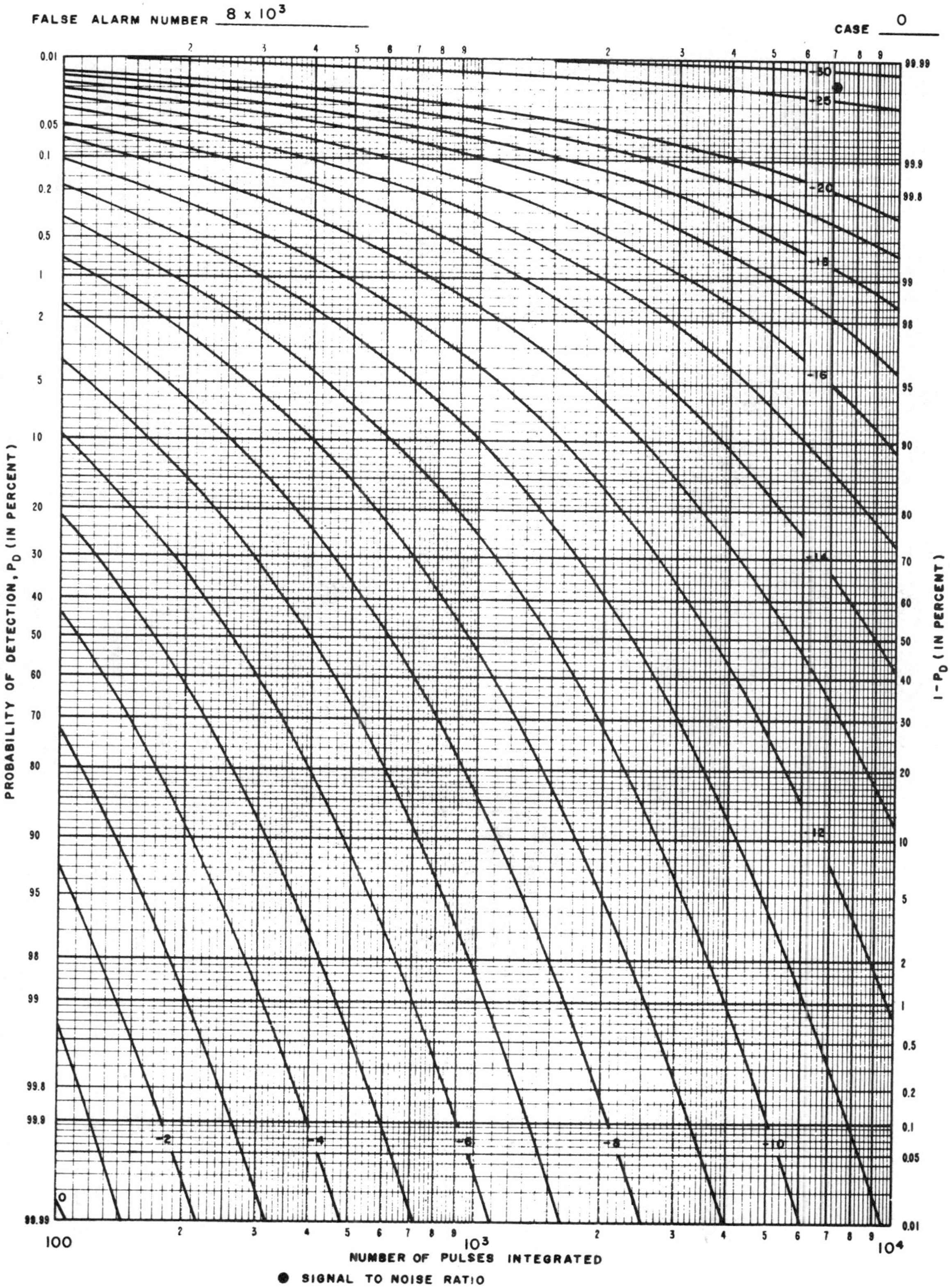
FALSE ALARM NUMBER 8 x 10³
CASE 0
PROBABILITY OF DETECTION, P_D (IN PERCENT)
1 - P_D (IN PERCENT)
NUMBER OF PULSES INTEGRATED
● SIGNAL TO NOISE RATIO
100
10³
10⁴
0
-2
-4
-6
-8
-10
-12
-14
-16
-18
-20
-25
-30

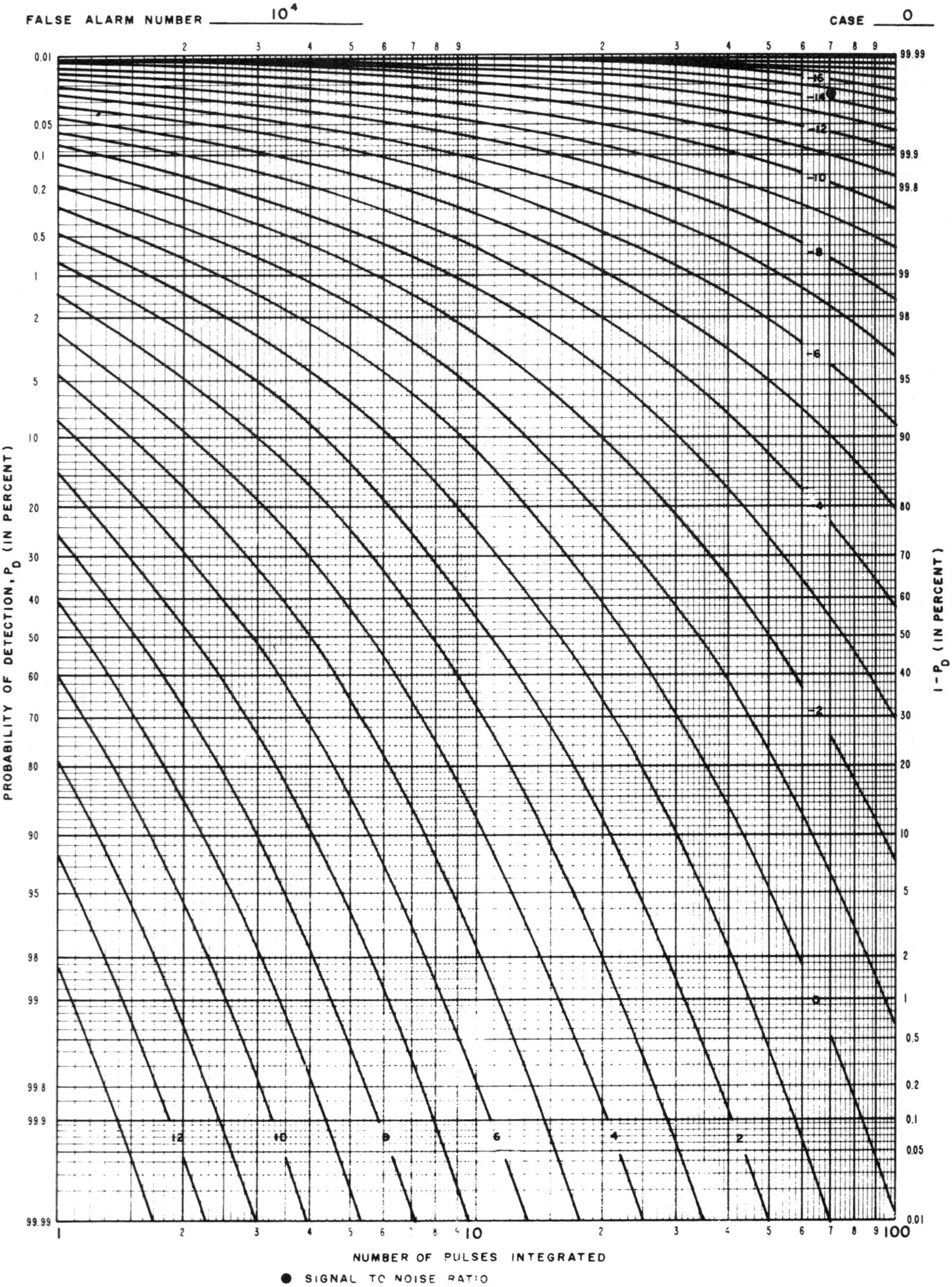

FALSE ALARM NUMBER 10^4
CASE 0
PROBABILITY OF DETECTION, P_D (IN PERCENT)
$1 - P_D$ (IN PERCENT)
NUMBER OF PULSES INTEGRATED
● SIGNAL TO NOISE RATIO

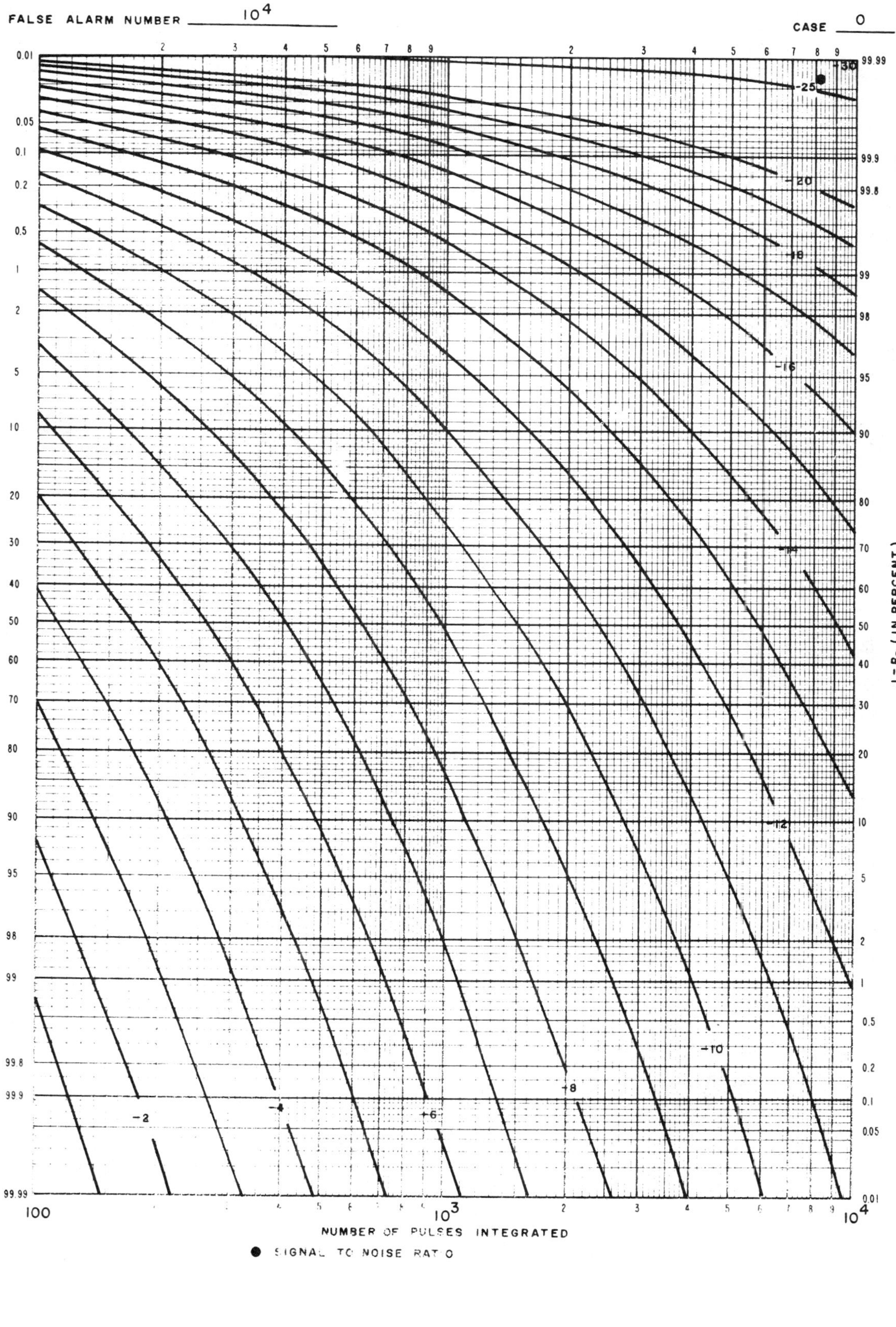
FALSE ALARM NUMBER 10^4
CASE 0
PROBABILITY OF DETECTION, P_D (IN PERCENT)
1 - P_D (IN PERCENT)
NUMBER OF PULSES INTEGRATED
100
10^3
10^4
-30
-25
-20
-18
-16
-14
-12
-10
+8
+6
-4
-2
● SIGNAL TO NOISE RATIO

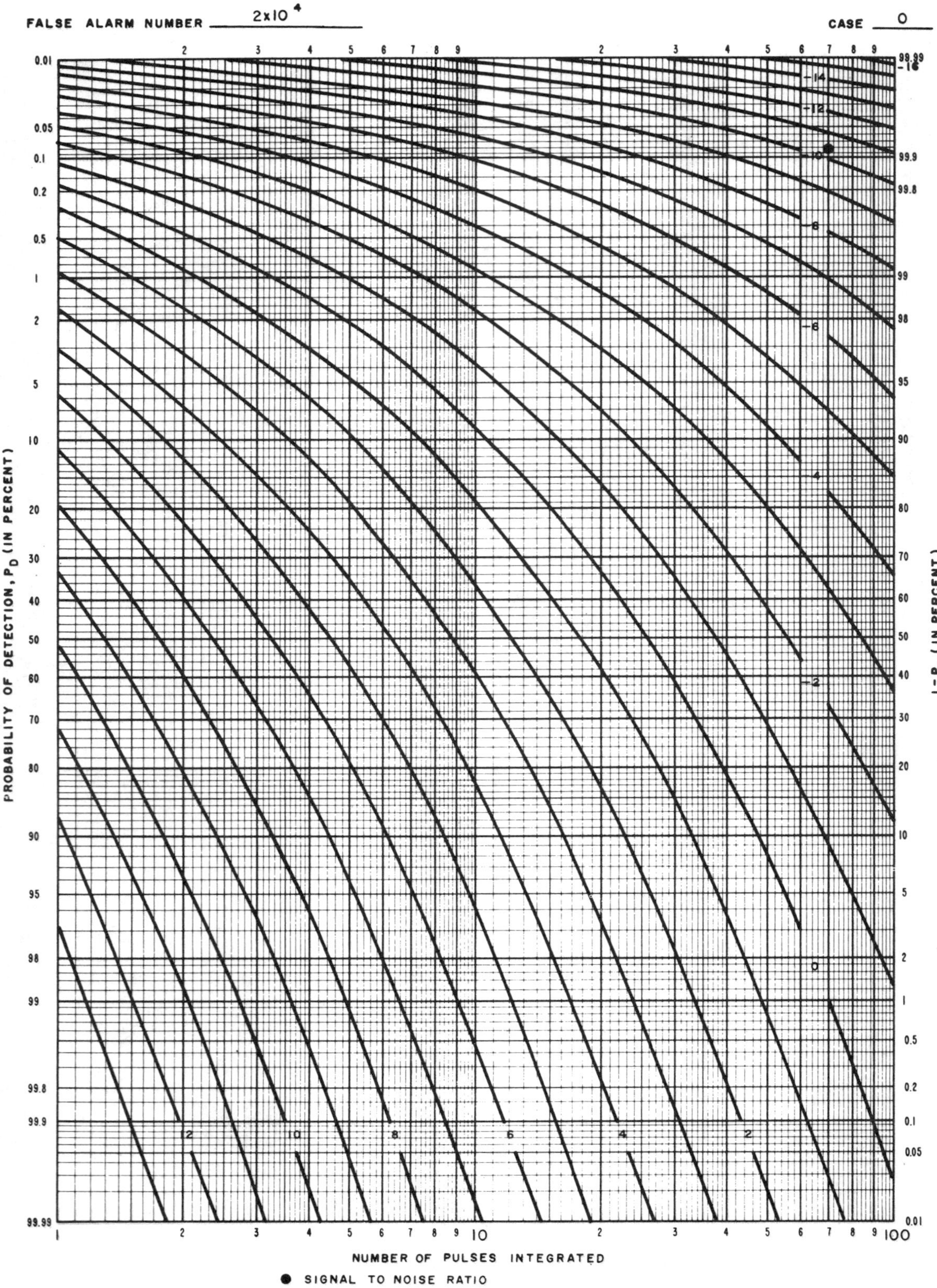
FALSE ALARM NUMBER 2x10^4
CASE 0
PROBABILITY OF DETECTION, P_D (IN PERCENT)
1-P_D (IN PERCENT)
NUMBER OF PULSES INTEGRATED
● SIGNAL TO NOISE RATIO

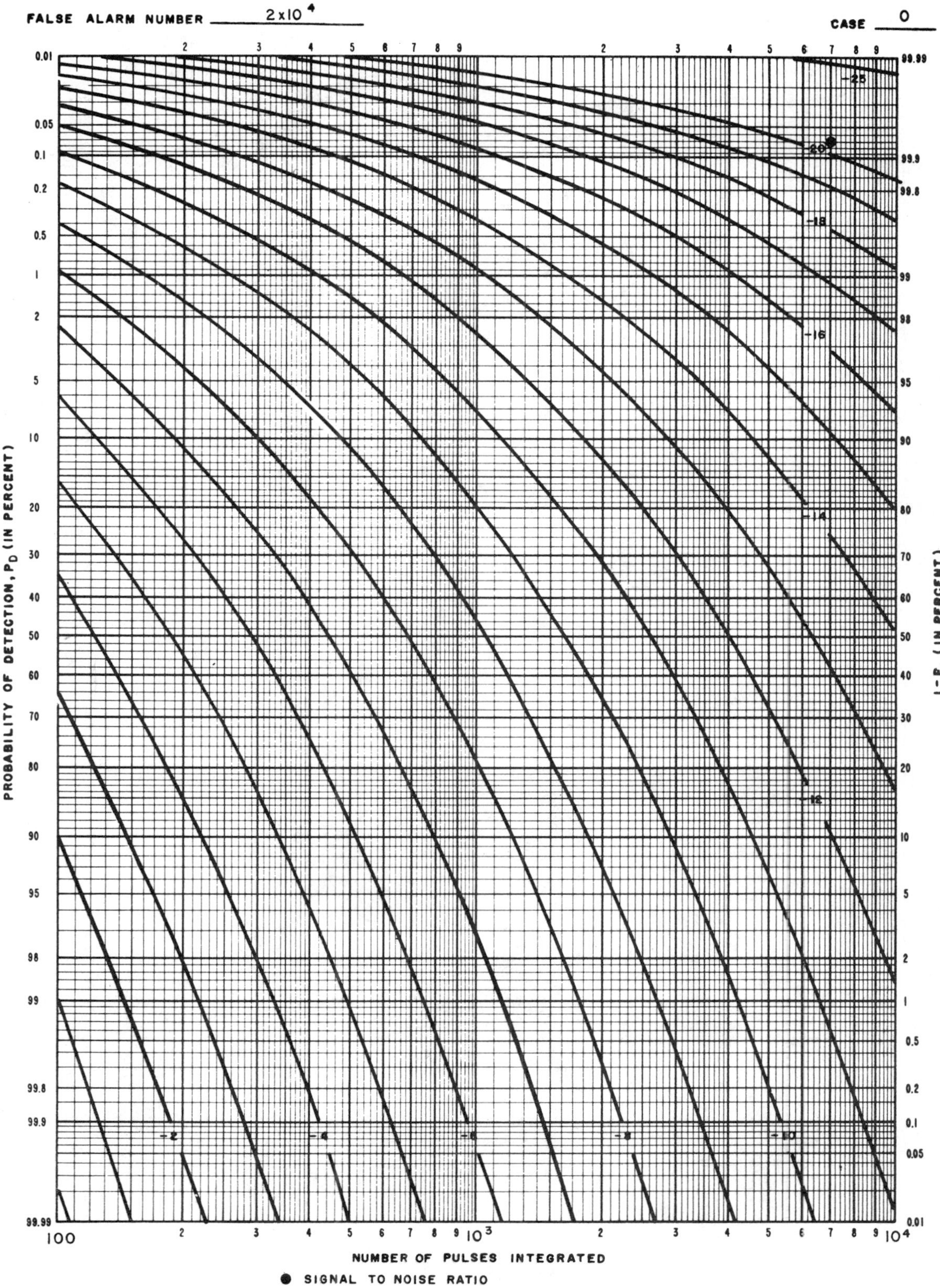
FALSE ALARM NUMBER 2x10^4
CASE 0
PROBABILITY OF DETECTION, P_D (IN PERCENT)
1 - P_D (IN PERCENT)
NUMBER OF PULSES INTEGRATED
● SIGNAL TO NOISE RATIO
100
10^3
10^4
-2
-4
-6
-8
-10
-12
-14
-16
-18
-20
-25

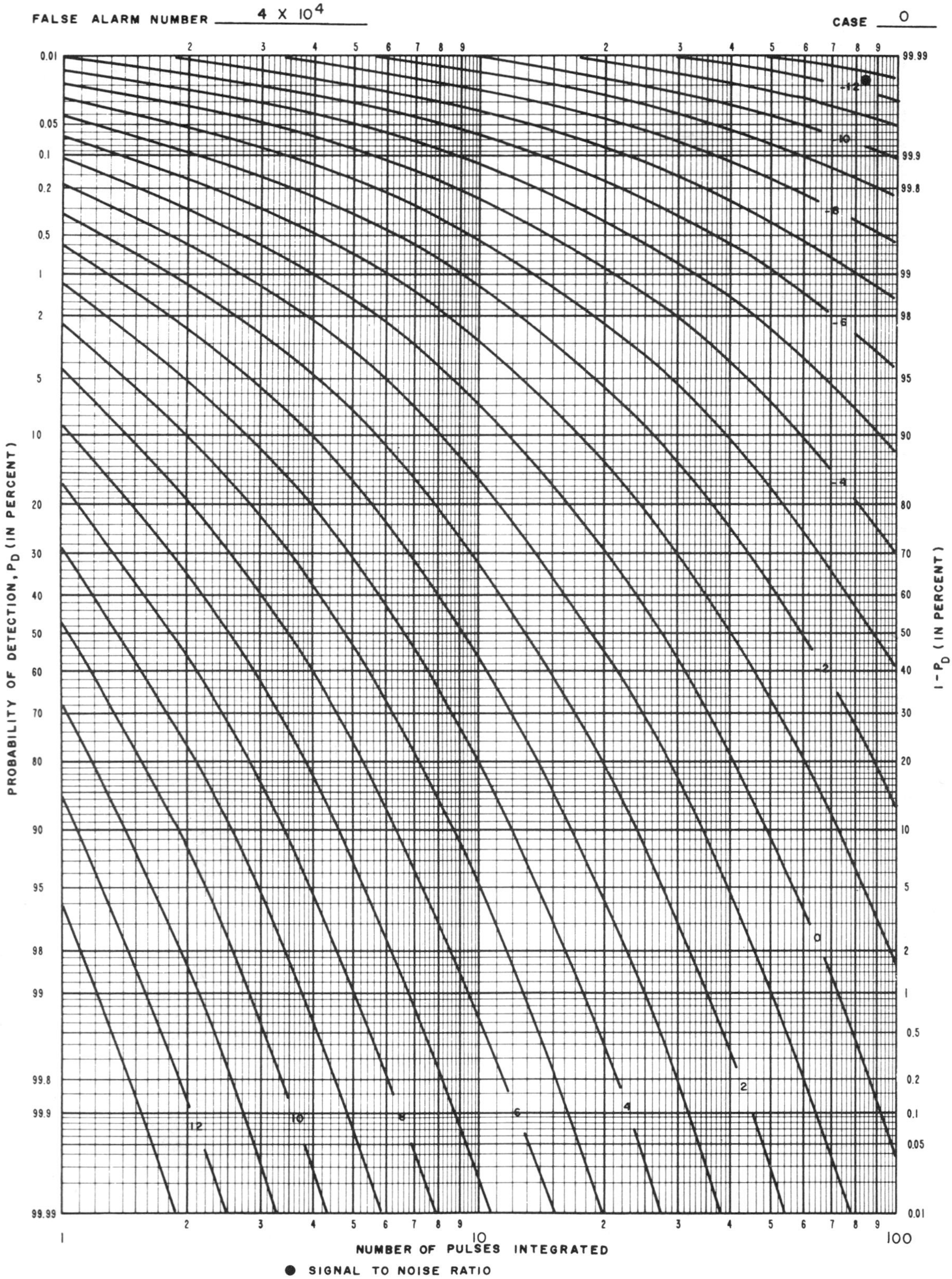
FALSE ALARM NUMBER 4 X 10^4
CASE 0
PROBABILITY OF DETECTION, P_D (IN PERCENT)
1 - P_D (IN PERCENT)
NUMBER OF PULSES INTEGRATED
● SIGNAL TO NOISE RATIO
-12
-10
-8
-6
-4
-2
0
2
4
6
8
10
12

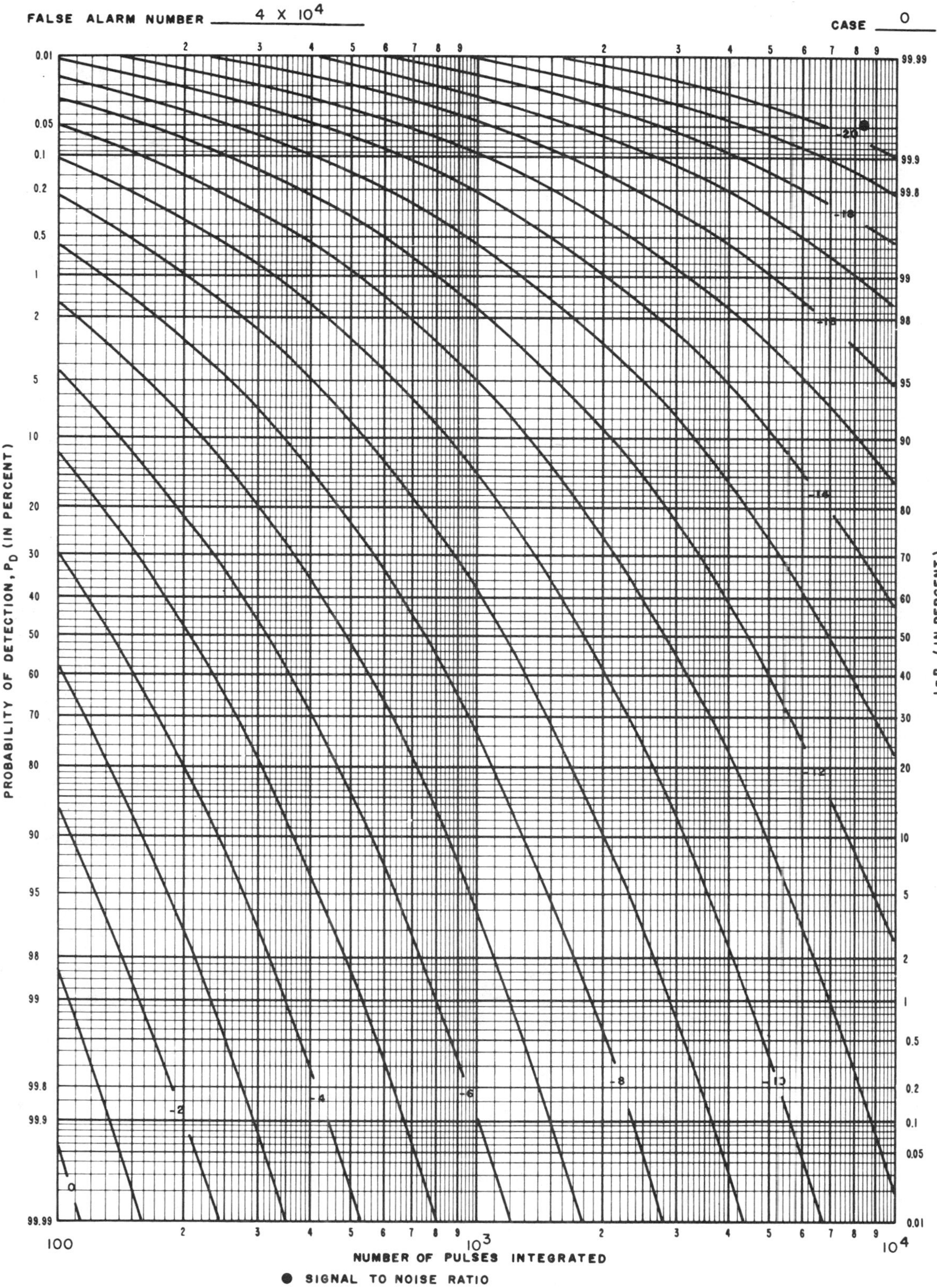
FALSE ALARM NUMBER 4 X 10^4
CASE 0
PROBABILITY OF DETECTION, P_D (IN PERCENT)
1 - P_D (IN PERCENT)
NUMBER OF PULSES INTEGRATED
100
10^3
10^4
0
-2
-4
-6
-8
-10
-12
-14
-16
-18
-20
● SIGNAL TO NOISE RATIO

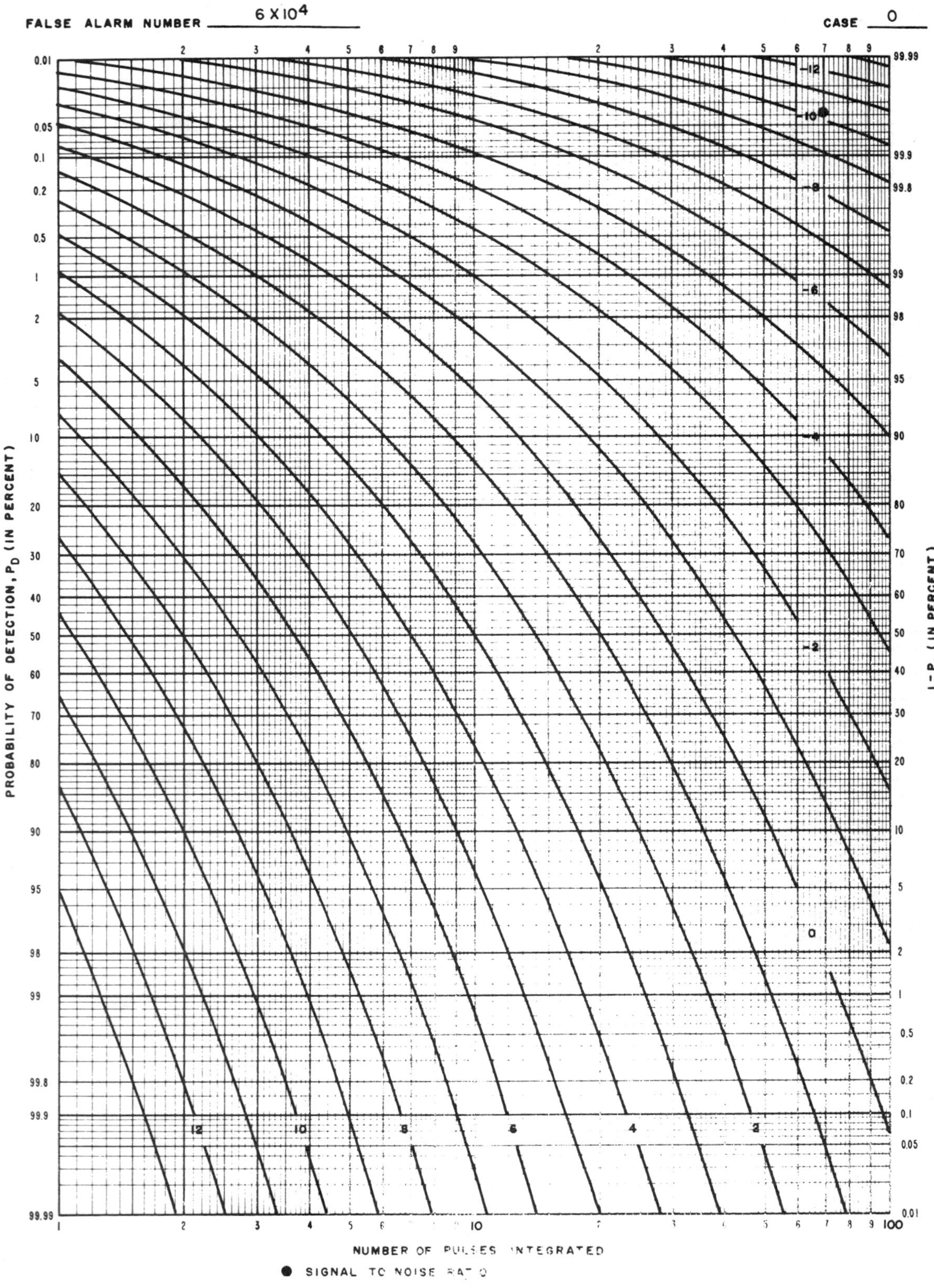
FALSE ALARM NUMBER 6 X 10^4
CASE 0
PROBABILITY OF DETECTION, P_D (IN PERCENT)
1 - P_D (IN PERCENT)
NUMBER OF PULSES INTEGRATED
SIGNAL TO NOISE RATIO
-12
-10
-8
-6
-4
-2
0
12
10
8
6
4
2

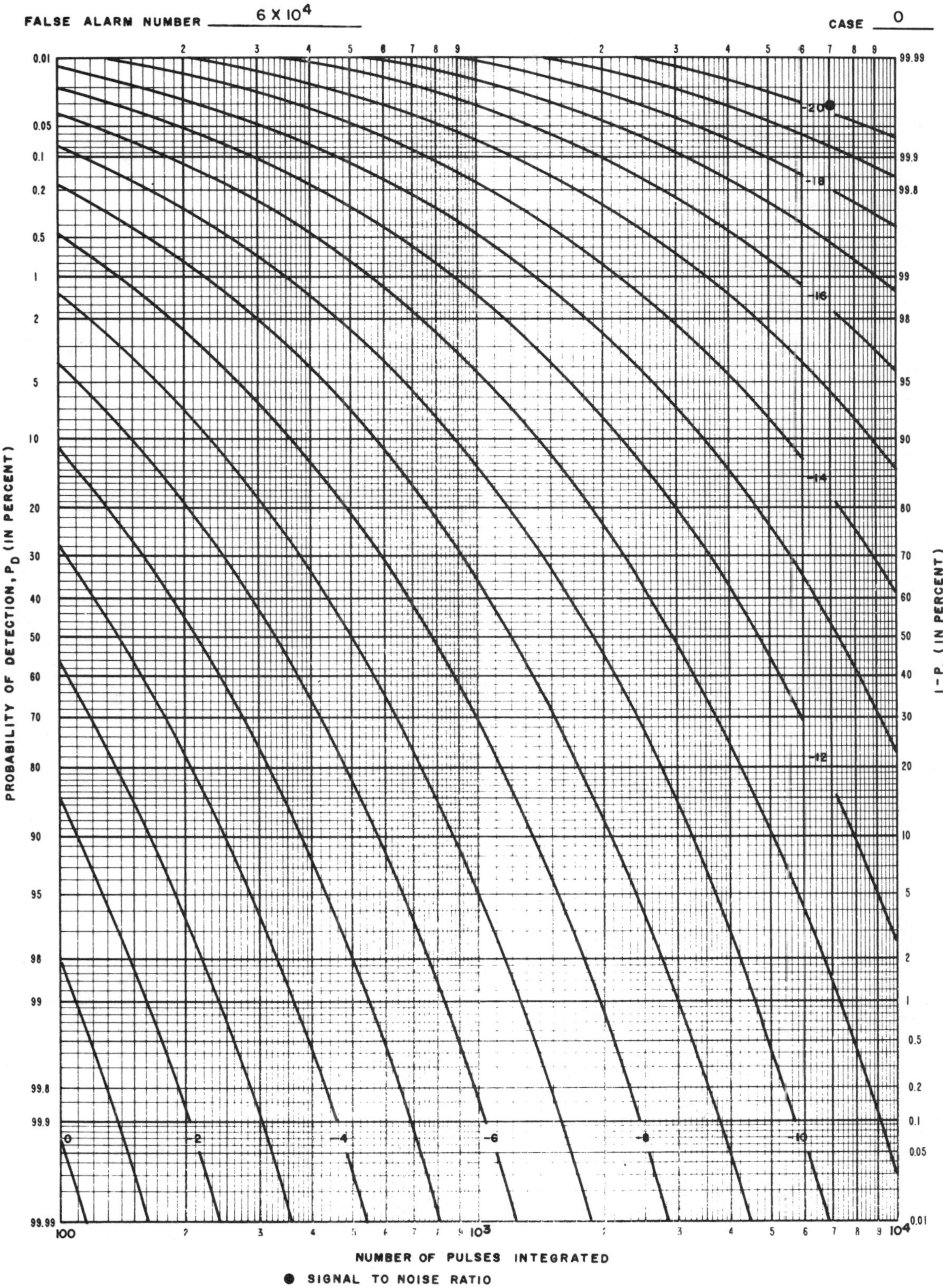
FALSE ALARM NUMBER 6 X 10^4
CASE 0
PROBABILITY OF DETECTION, P_D (IN PERCENT)
1 - P_D (IN PERCENT)
NUMBER OF PULSES INTEGRATED
● SIGNAL TO NOISE RATIO
100
10^3
10^4
0
-2
-4
-6
-8
-10
-12
-14
-16
-18
-20

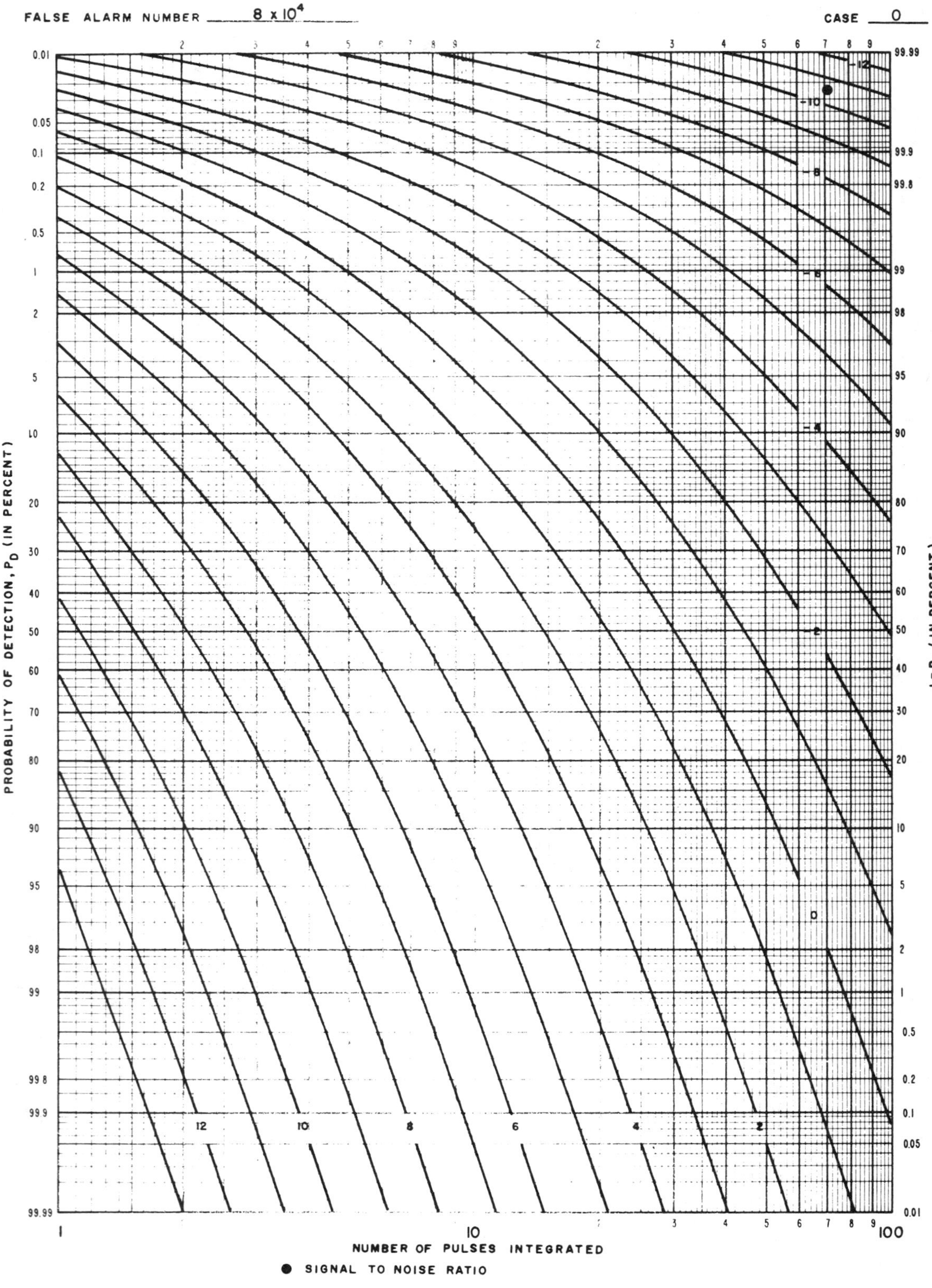
FALSE ALARM NUMBER 8 x 10^4
CASE 0
PROBABILITY OF DETECTION, P_D (IN PERCENT)
1 - P_D (IN PERCENT)
12
10
8
6
4
2
0
-2
-4
-6
-8
-10
-12
1
10
100
NUMBER OF PULSES INTEGRATED
● SIGNAL TO NOISE RATIO

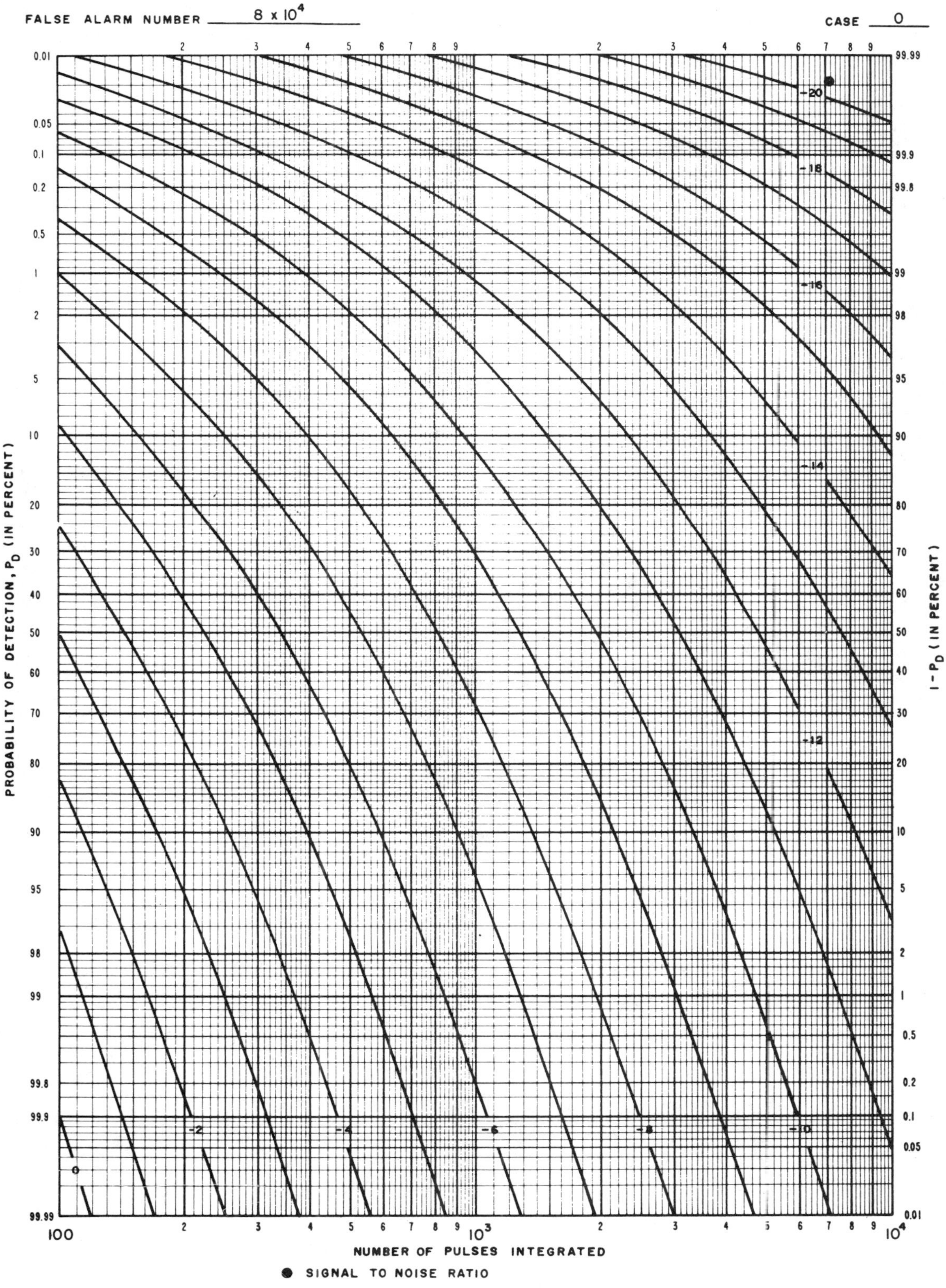
FALSE ALARM NUMBER 8 x 10^4
CASE 0
PROBABILITY OF DETECTION, P_D (IN PERCENT)
1 - P_D (IN PERCENT)
NUMBER OF PULSES INTEGRATED
100
10^3
10^4
0
-2
-4
-6
-8
-10
-12
-14
-16
-18
-20
● SIGNAL TO NOISE RATIO

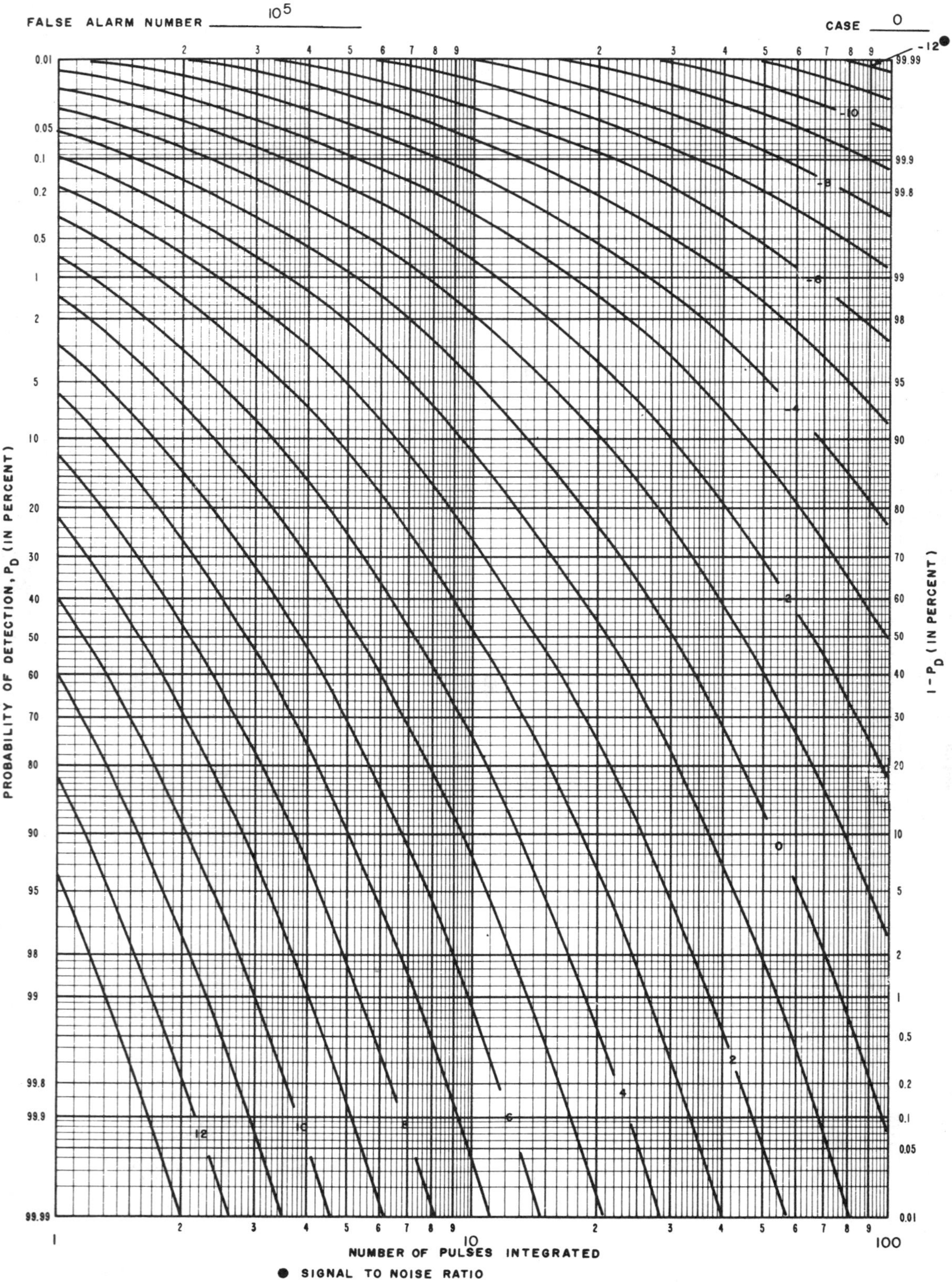
FALSE ALARM NUMBER 10^5
CASE 0
PROBABILITY OF DETECTION, P_D (IN PERCENT)
$1-P_D$ (IN PERCENT)
NUMBER OF PULSES INTEGRATED
● SIGNAL TO NOISE RATIO
-12
-10
-8
-6
-4
-2
0
2
4
6
8
10
12

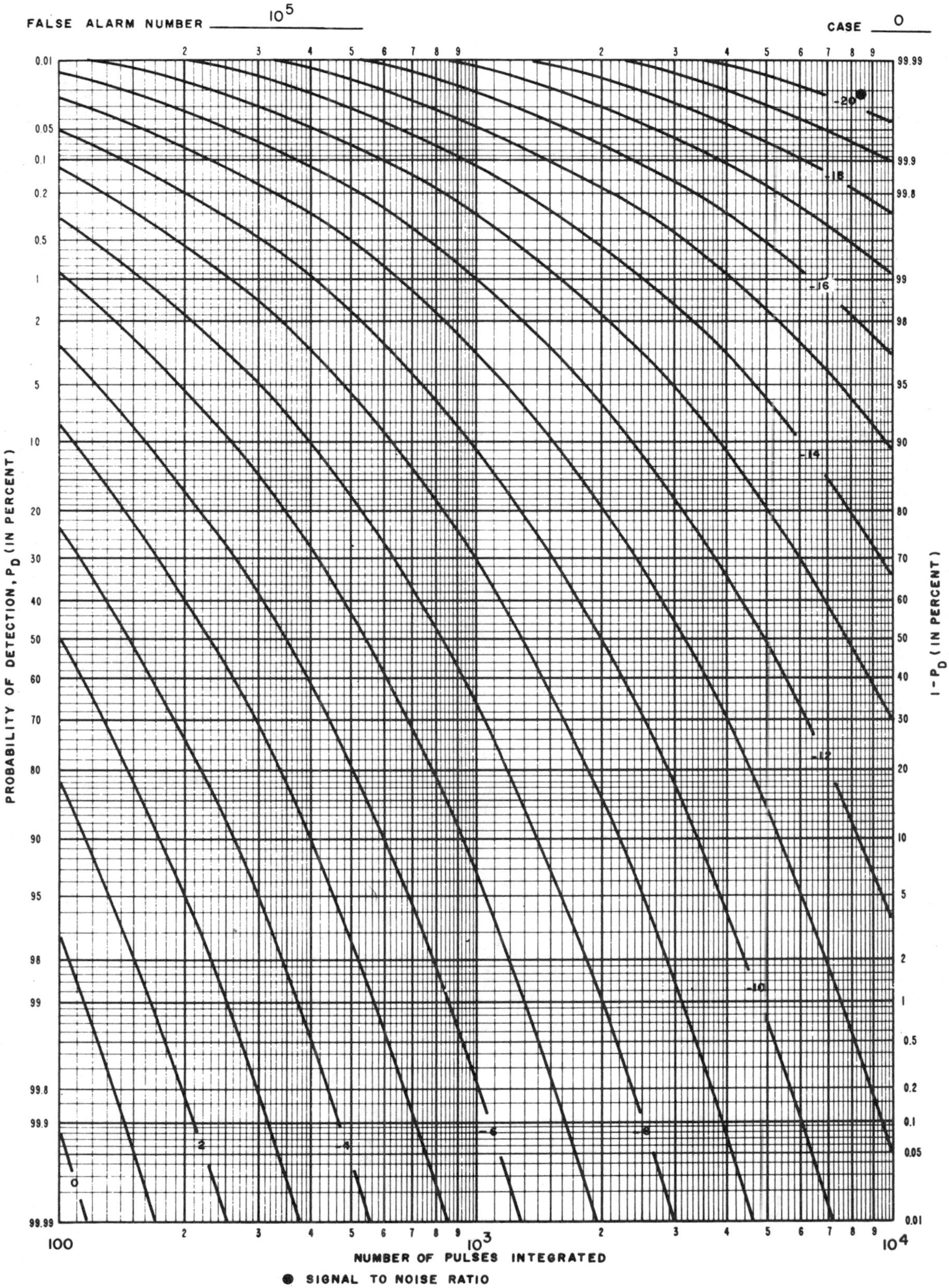
FALSE ALARM NUMBER 10^5
CASE 0
PROBABILITY OF DETECTION, P_D (IN PERCENT)
$1-P_D$ (IN PERCENT)
NUMBER OF PULSES INTEGRATED
100
10^3
10^4
-20
-18
-16
-14
-12
-10
-8
-6
-4
2
0
● SIGNAL TO NOISE RATIO

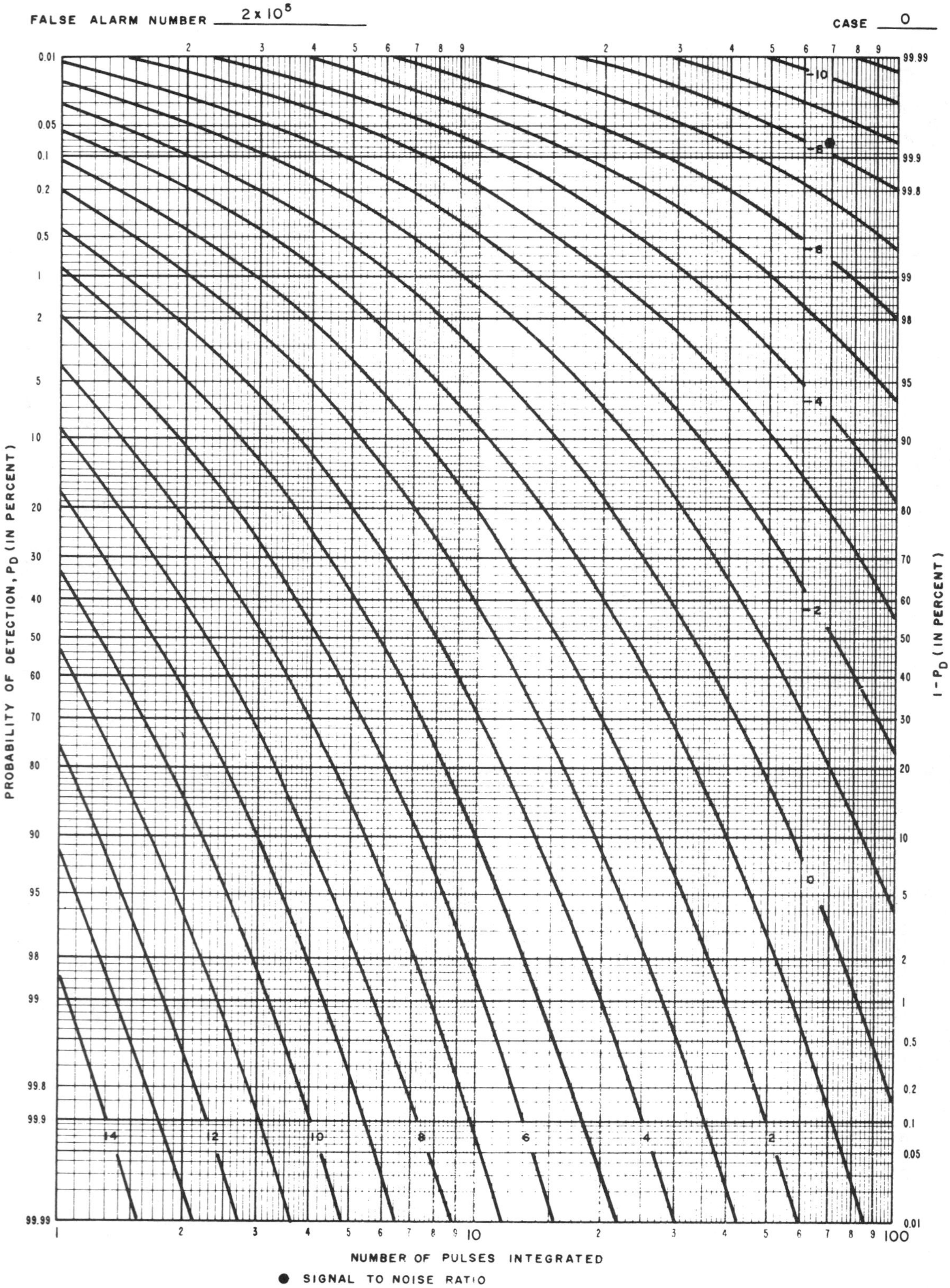
FALSE ALARM NUMBER 2 x 10^5
CASE 0
PROBABILITY OF DETECTION, P_D (IN PERCENT)
1 - P_D (IN PERCENT)
NUMBER OF PULSES INTEGRATED
● SIGNAL TO NOISE RATIO
-10
-8
-6
-4
-2
0
2
4
6
8
10
12
14

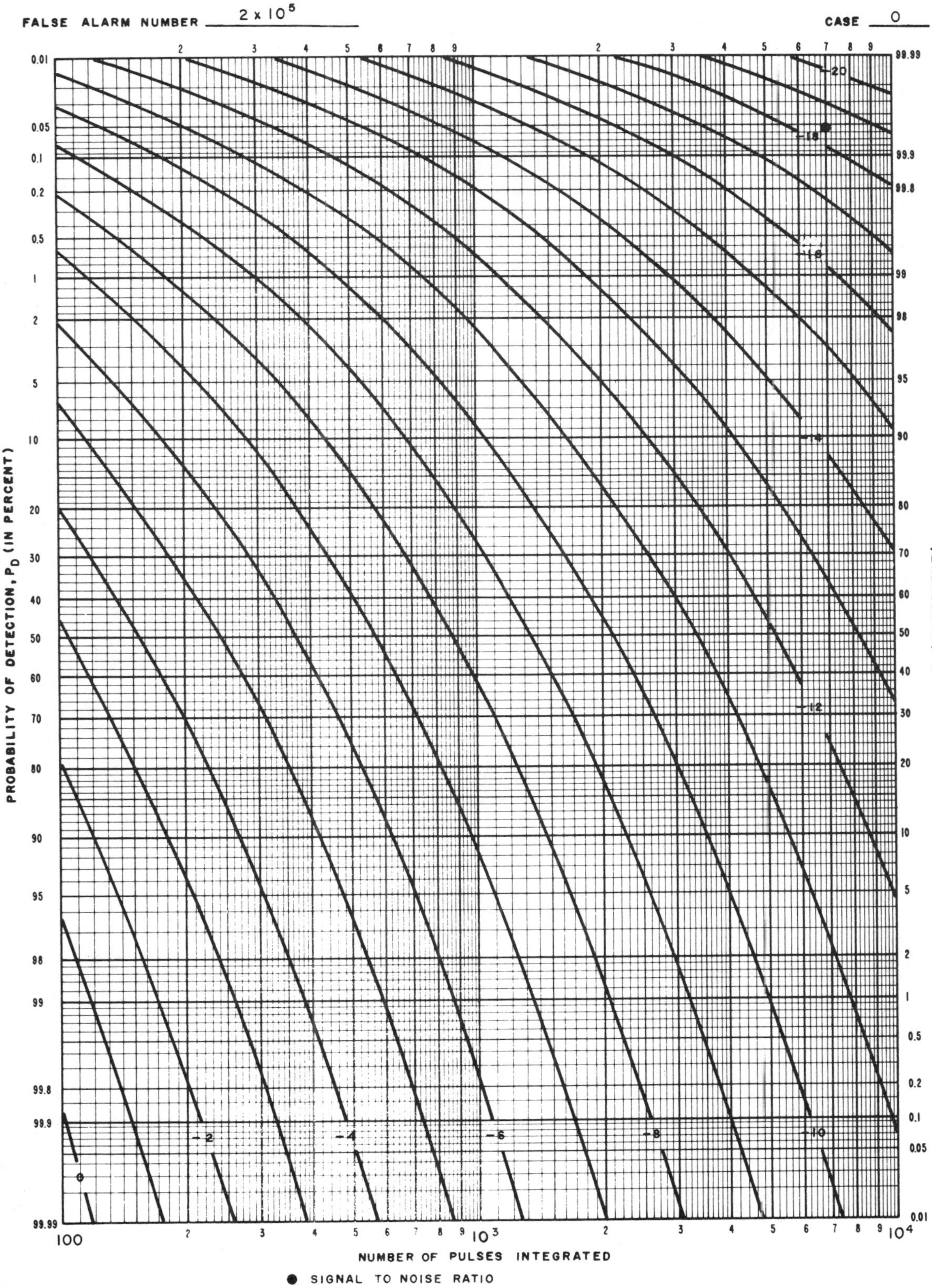
FALSE ALARM NUMBER 2×10^5
CASE 0
PROBABILITY OF DETECTION, P_D (IN PERCENT)
$1-P_D$ (IN PERCENT)
NUMBER OF PULSES INTEGRATED
● SIGNAL TO NOISE RATIO
0
-2
-4
-6
-8
-10
-12
-14
-16
-18
-20
100
10^3
10^4

FALSE ALARM NUMBER 4 X 10^5 CASE 0

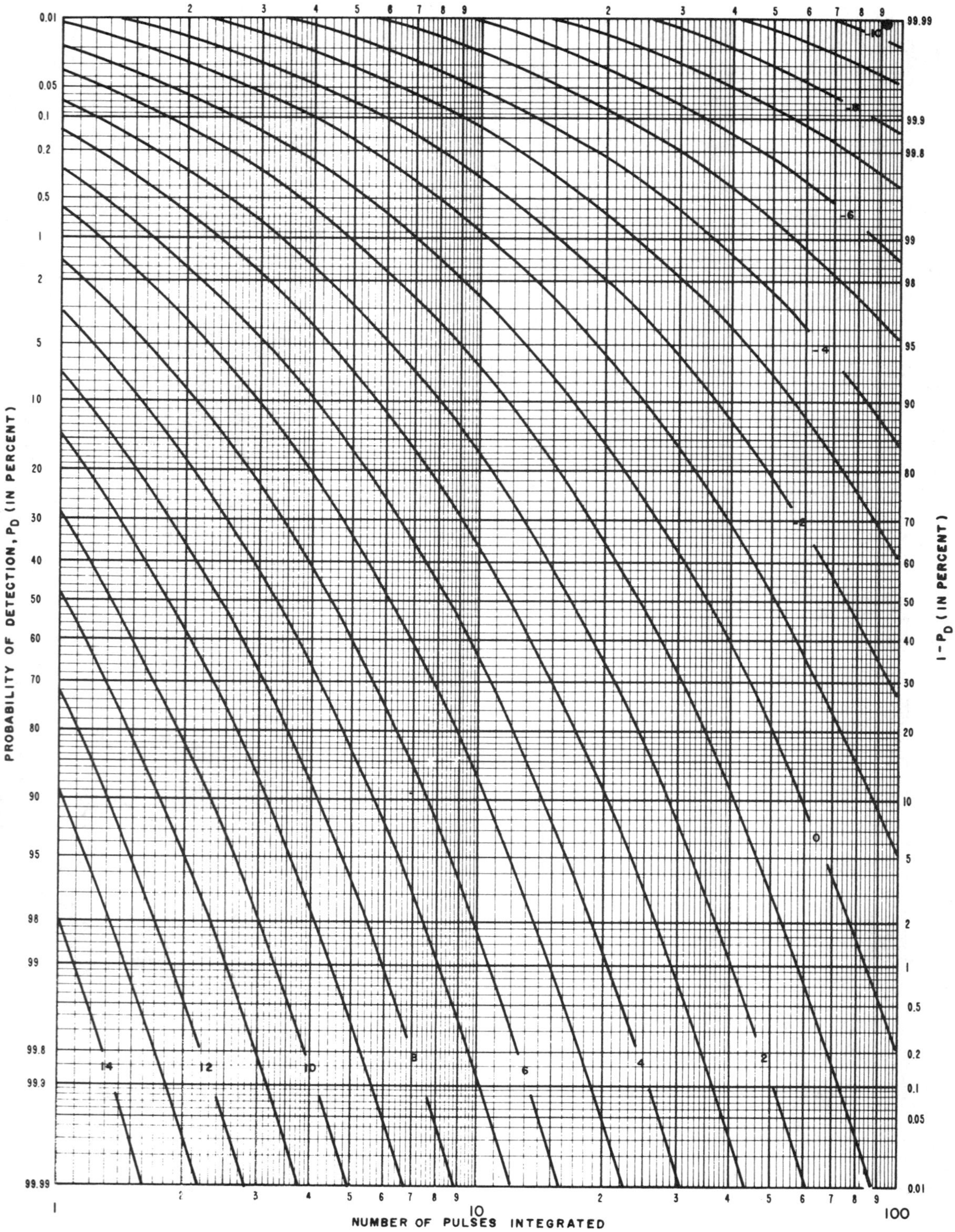

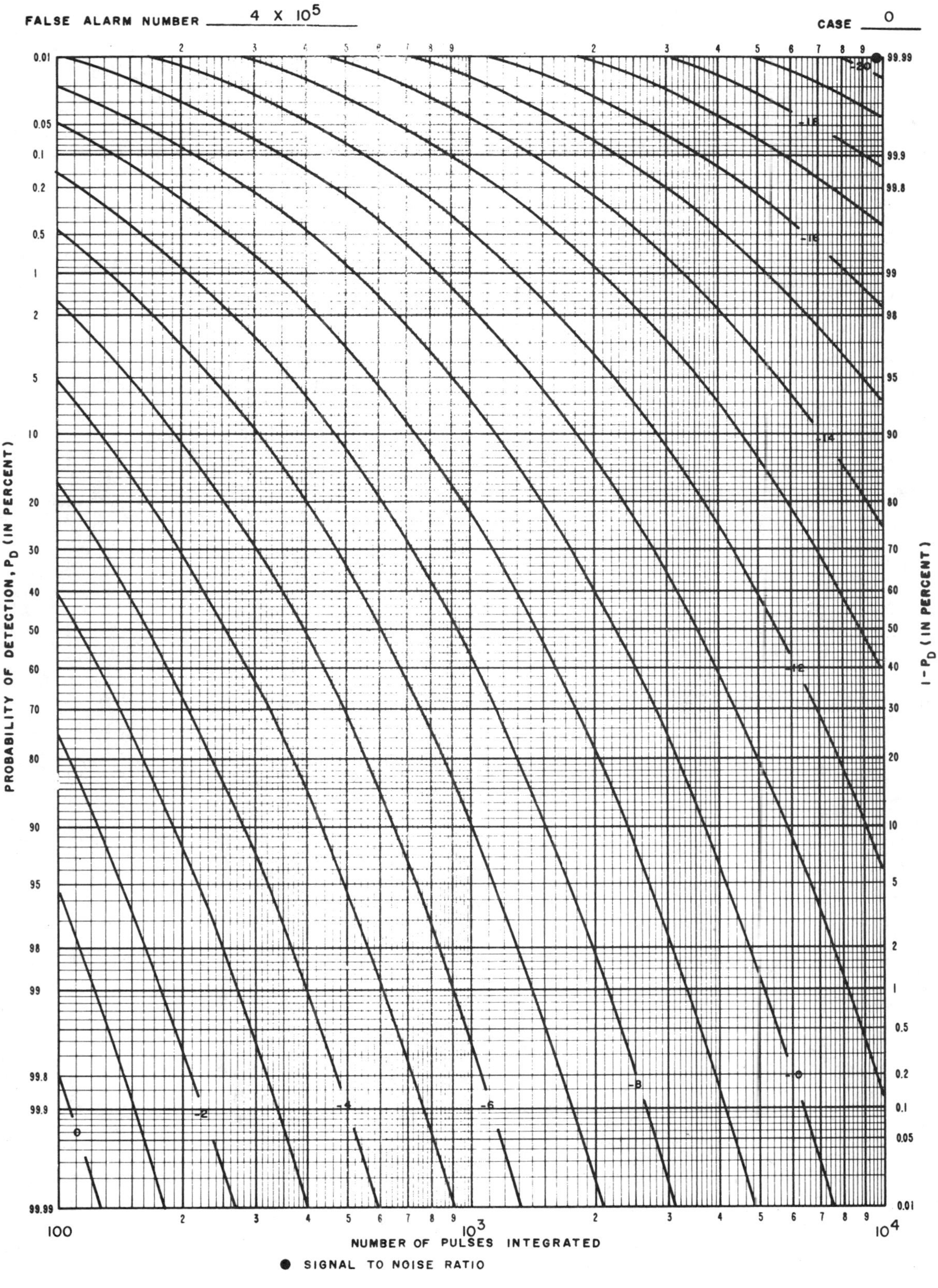
FALSE ALARM NUMBER 4 X 10^5
CASE 0
PROBABILITY OF DETECTION, P_D (IN PERCENT)
1 - P_D (IN PERCENT)
0 -2 -4 -6 -8 -10 -12 -14 -16 -18 -20
100
10^3
10^4
NUMBER OF PULSES INTEGRATED
● SIGNAL TO NOISE RATIO

FALSE ALARM NUMBER 6×10^5 CASE 0

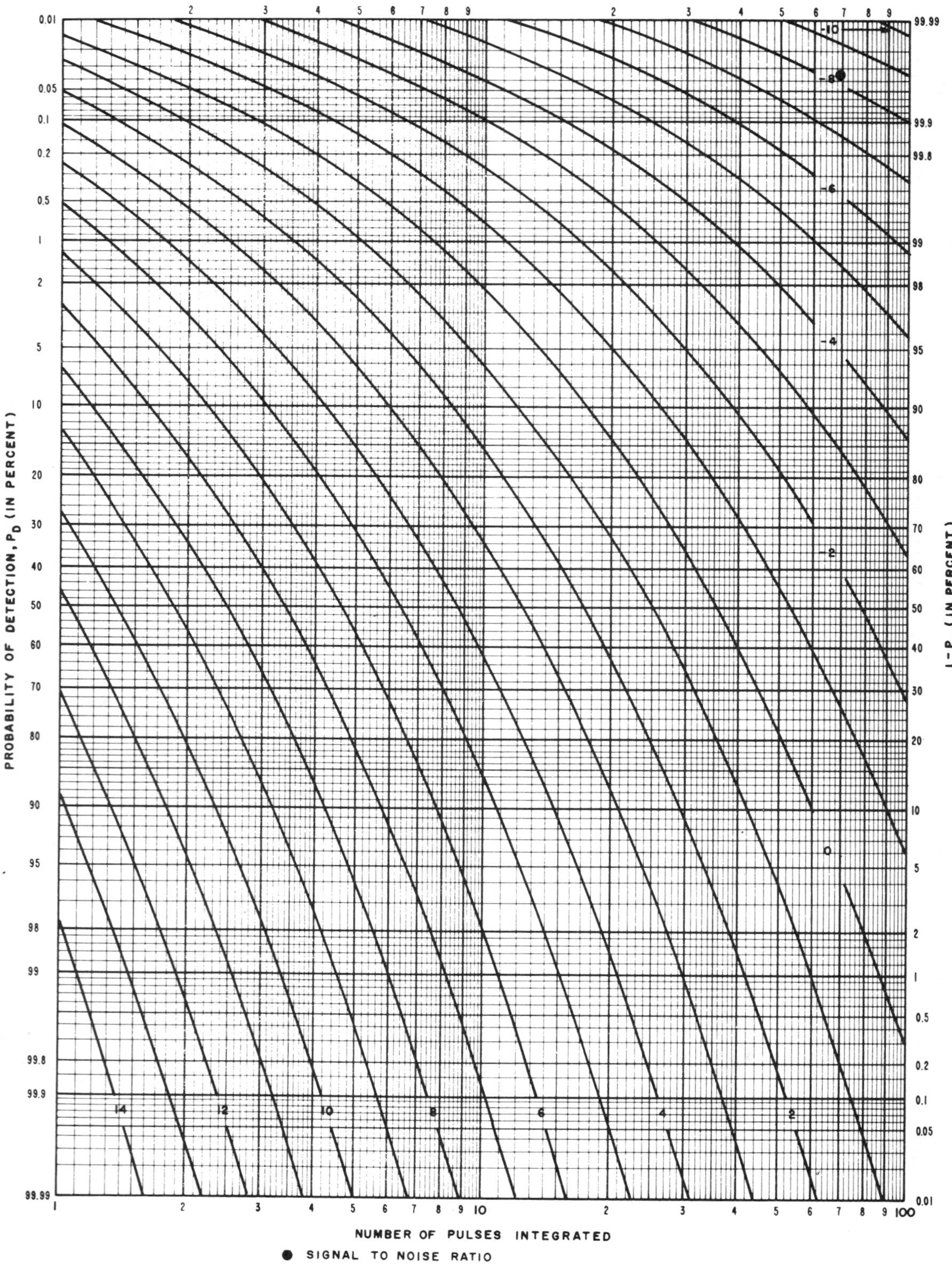

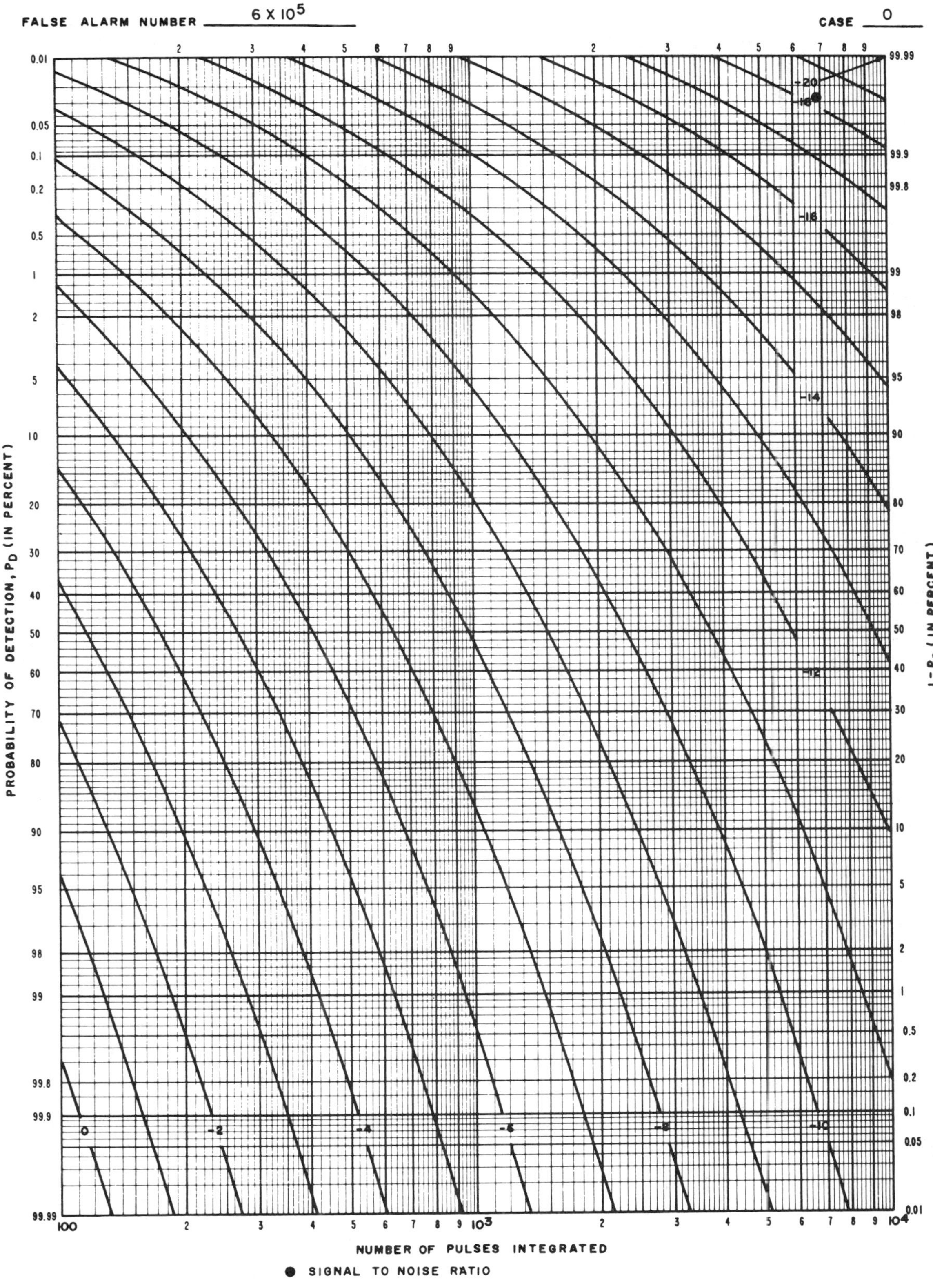
FALSE ALARM NUMBER 6 X 10^5
CASE 0
PROBABILITY OF DETECTION, P_D (IN PERCENT)
1 - P_D (IN PERCENT)
NUMBER OF PULSES INTEGRATED
● SIGNAL TO NOISE RATIO
0
-2
-4
-6
-8
-10
-12
-14
-16
-18
-20

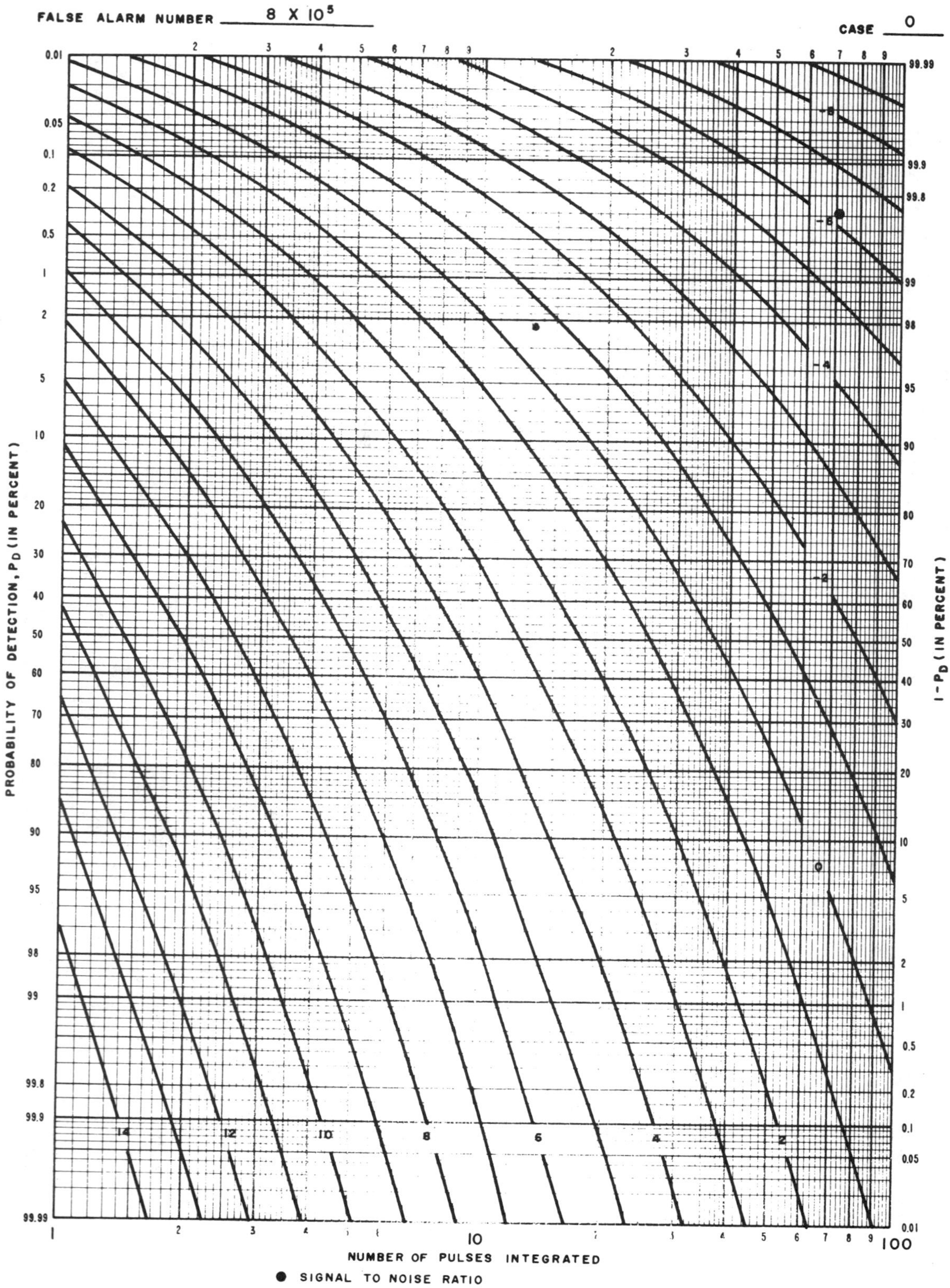
FALSE ALARM NUMBER 8 X 10^5
CASE 0
PROBABILITY OF DETECTION, P_D (IN PERCENT)
1-P_D (IN PERCENT)
NUMBER OF PULSES INTEGRATED
● SIGNAL TO NOISE RATIO
14
12
10
8
6
4
2
0
-2
-4
-6
-8
1
10
100

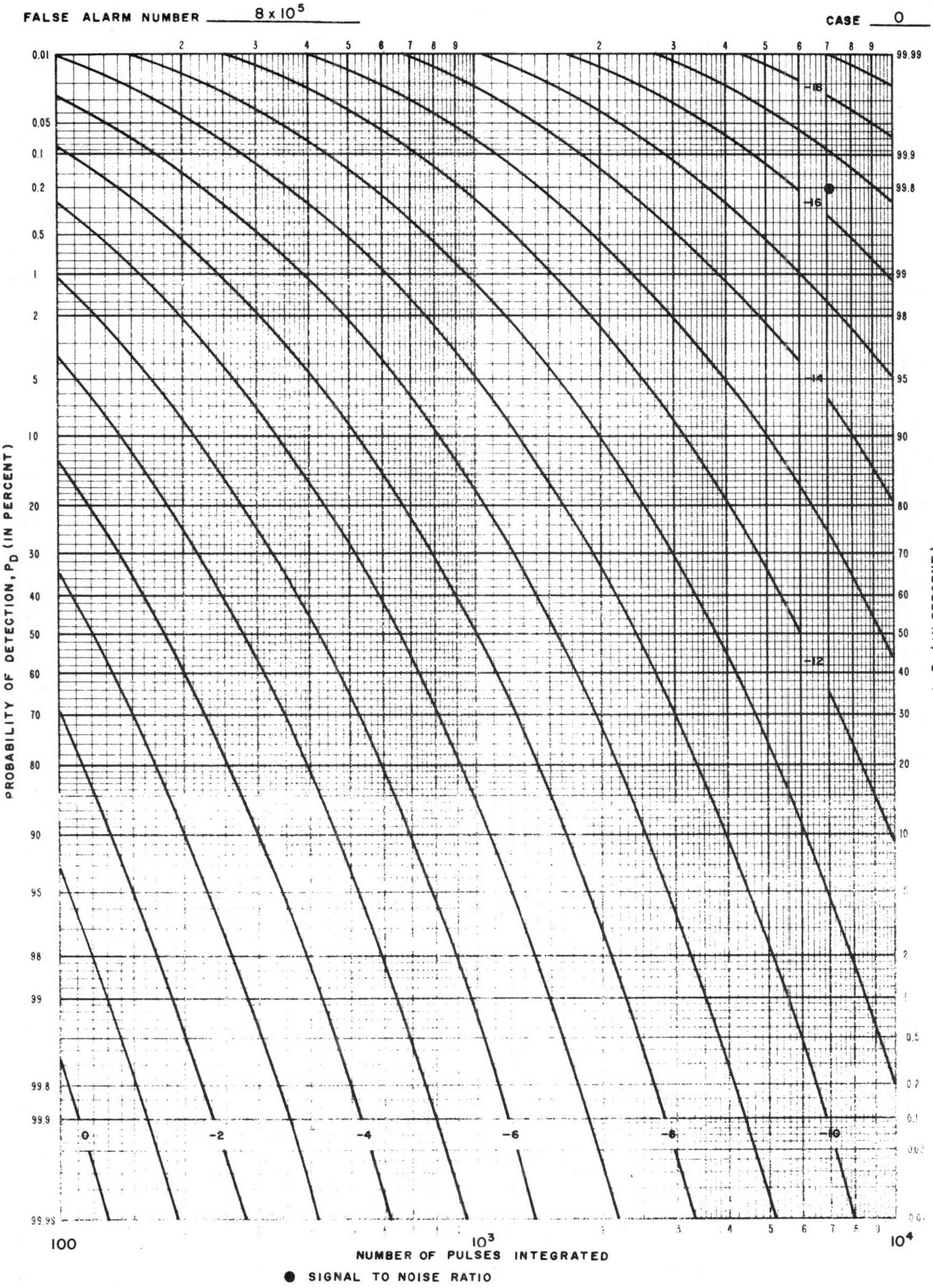
FALSE ALARM NUMBER 8 x 10^5
CASE 0
PROBABILITY OF DETECTION, P_D (IN PERCENT)
1 - P_D (IN PERCENT)
NUMBER OF PULSES INTEGRATED
● SIGNAL TO NOISE RATIO
0
-2
-4
-6
-8
-10
-12
-14
-16
-18
100
10^3
10^4

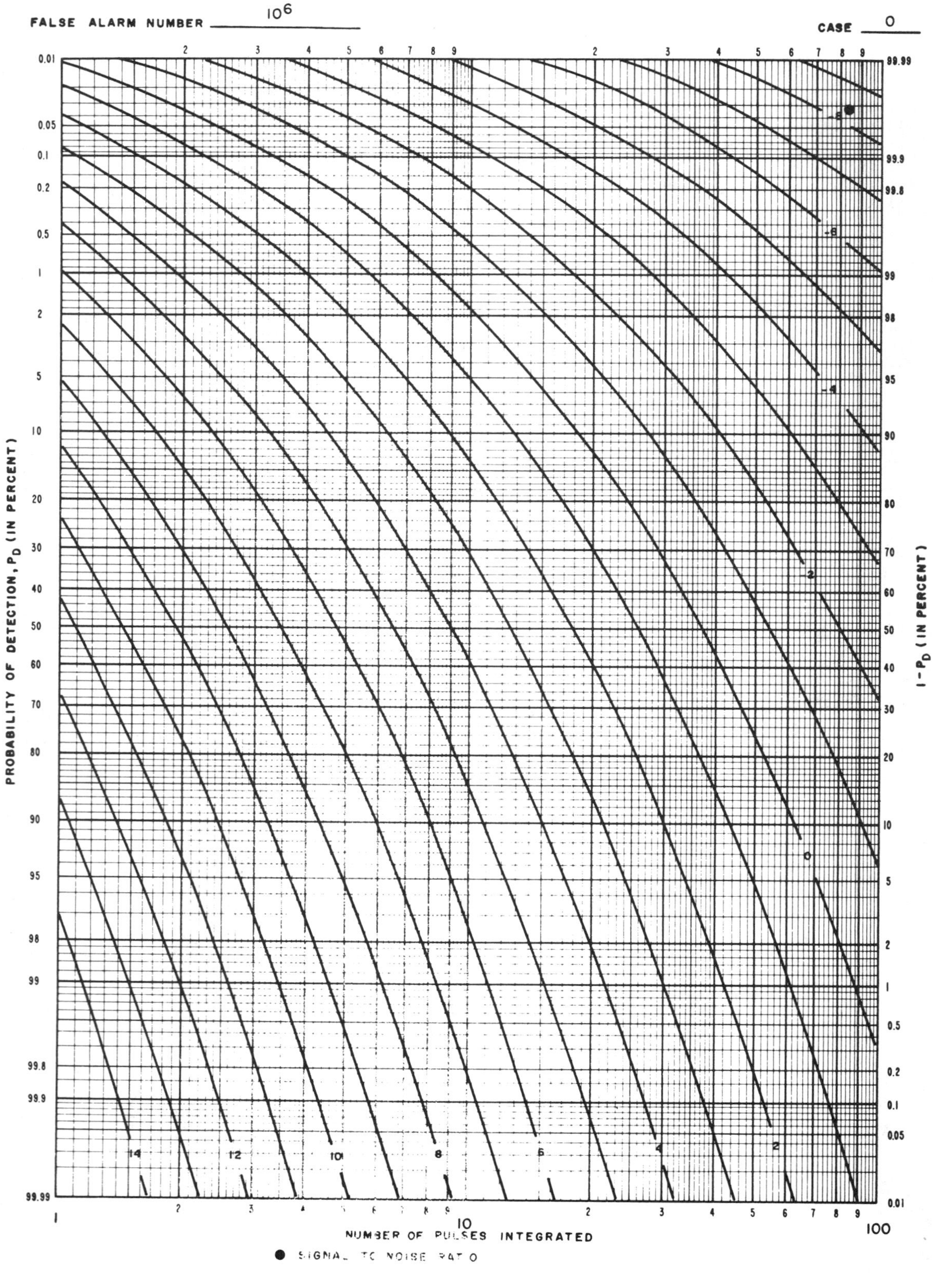

FALSE ALARM NUMBER 10^6
CASE 0
PROBABILITY OF DETECTION, P_D (IN PERCENT)
1 - P_D (IN PERCENT)
NUMBER OF PULSES INTEGRATED
SIGNAL TO NOISE RATIO
-8
-6
-4
-2
0
2
4
6
8
10
12
14

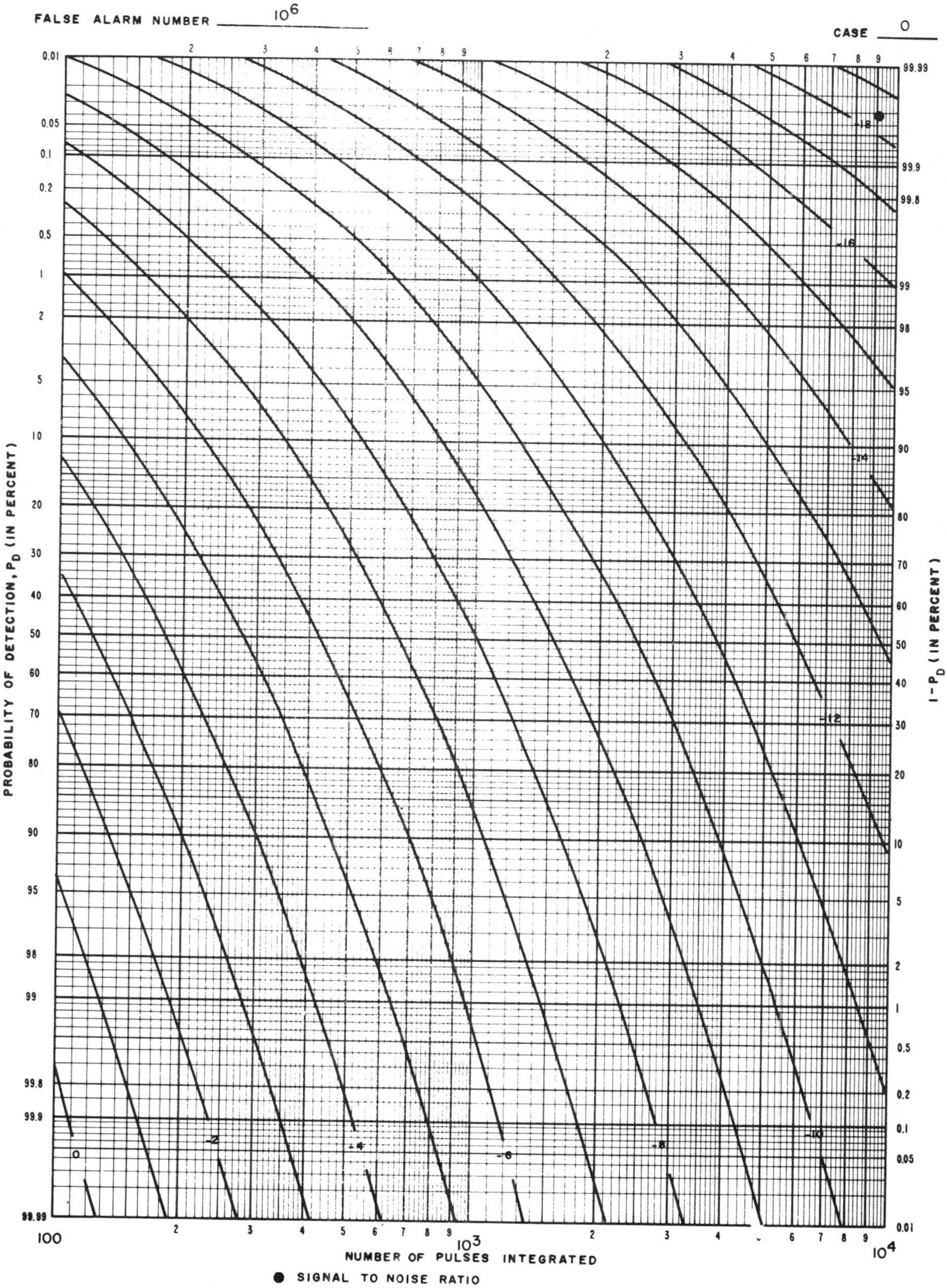
FALSE ALARM NUMBER 10^6
CASE 0
PROBABILITY OF DETECTION, P_D (IN PERCENT)
1-P_D (IN PERCENT)
NUMBER OF PULSES INTEGRATED
100
10^3
10^4
0
-2
-4
-6
-8
-10
-12
-14
-16
-18
● SIGNAL TO NOISE RATIO

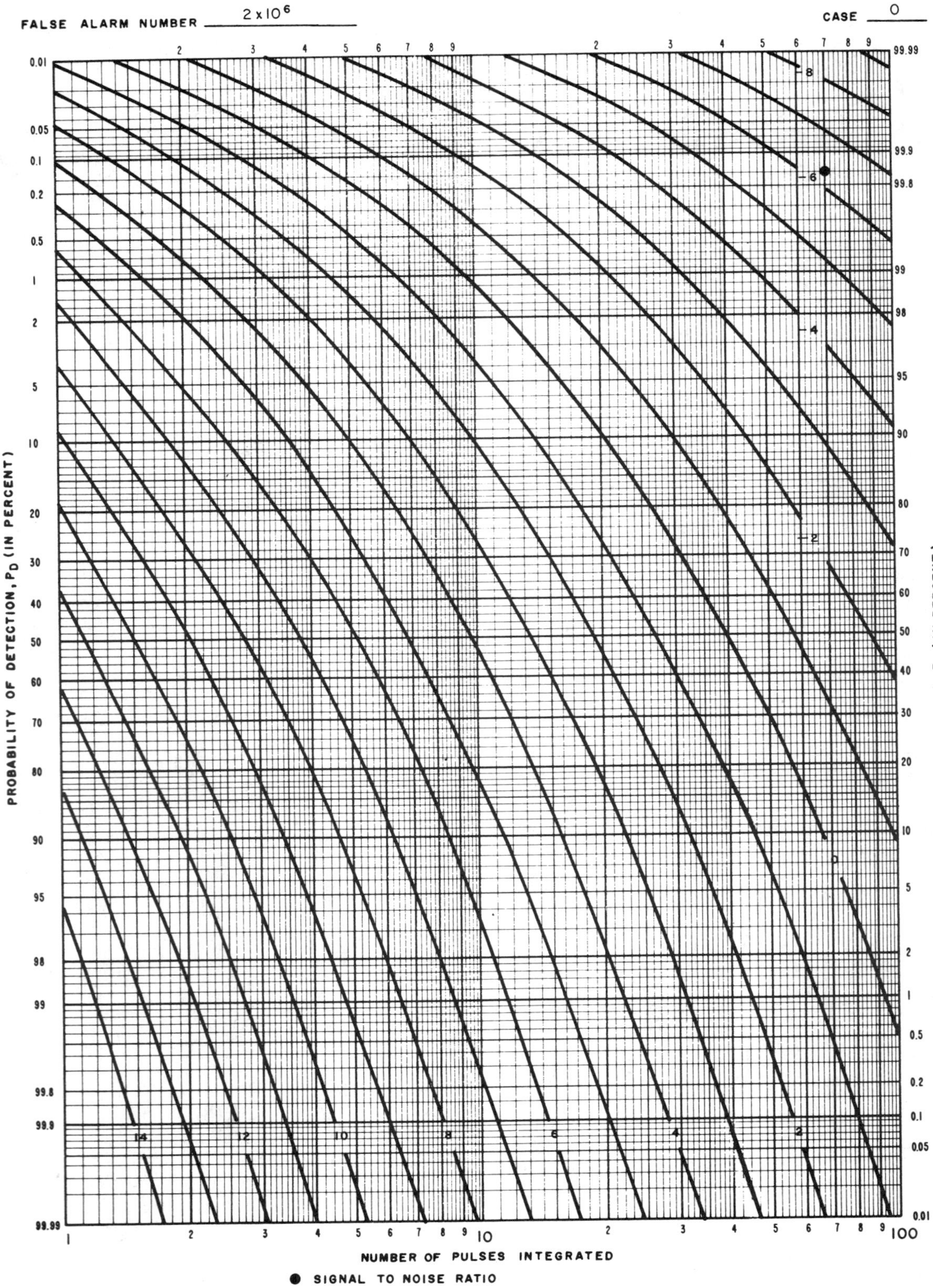
FALSE ALARM NUMBER 2 x 10^6
CASE 0
PROBABILITY OF DETECTION, P_D (IN PERCENT)
1 - P_D (IN PERCENT)
NUMBER OF PULSES INTEGRATED
● SIGNAL TO NOISE RATIO
14
12
10
8
6
4
2
0

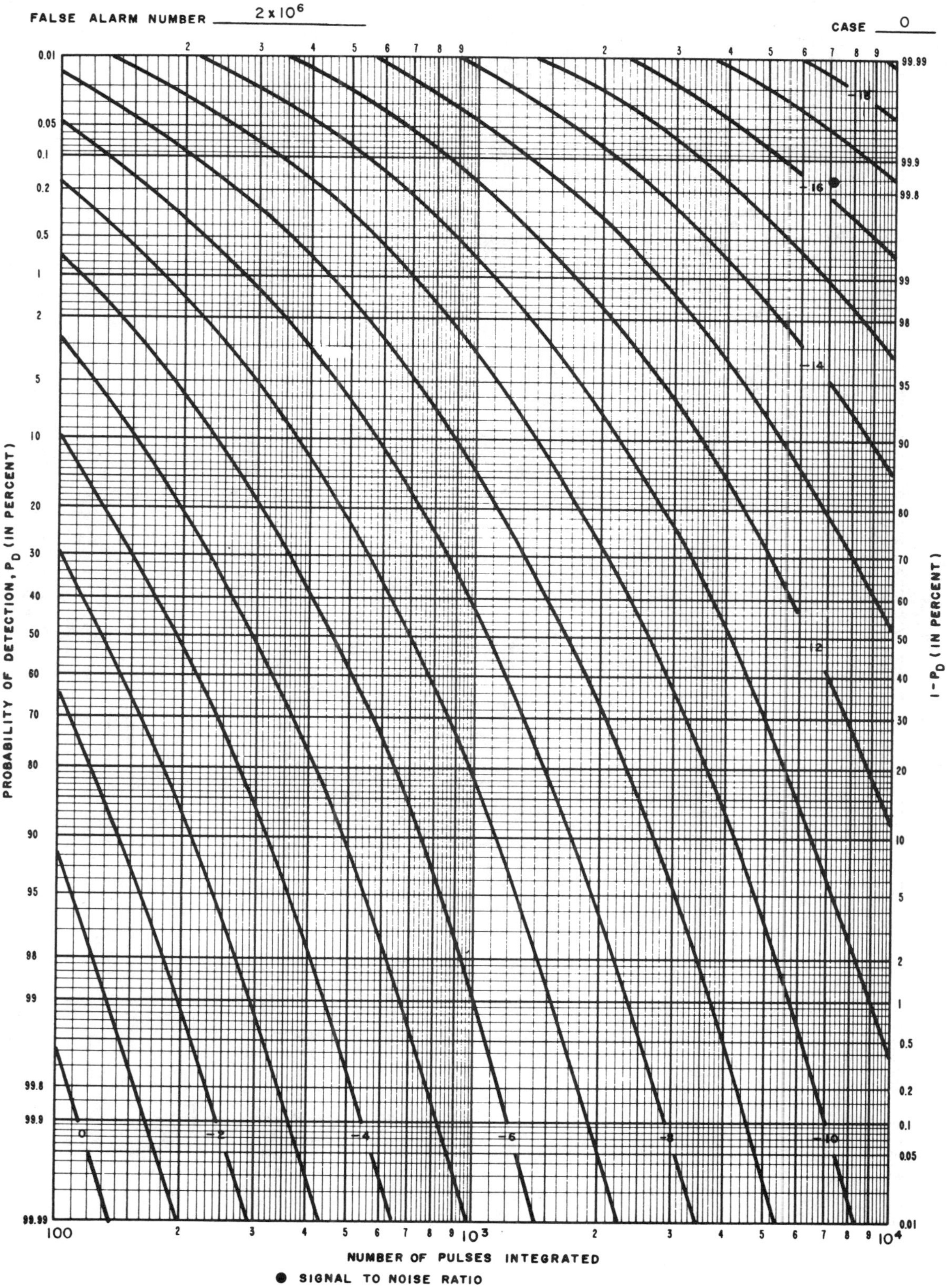
FALSE ALARM NUMBER 2 x 10^6
CASE 0
PROBABILITY OF DETECTION, P_D (IN PERCENT)
1 - P_D (IN PERCENT)
NUMBER OF PULSES INTEGRATED
100
10^3
10^4
0
-2
-4
-6
-8
-10
-12
-14
-16
-18
● SIGNAL TO NOISE RATIO

FALSE ALARM NUMBER 4 X 10^6 CASE 0

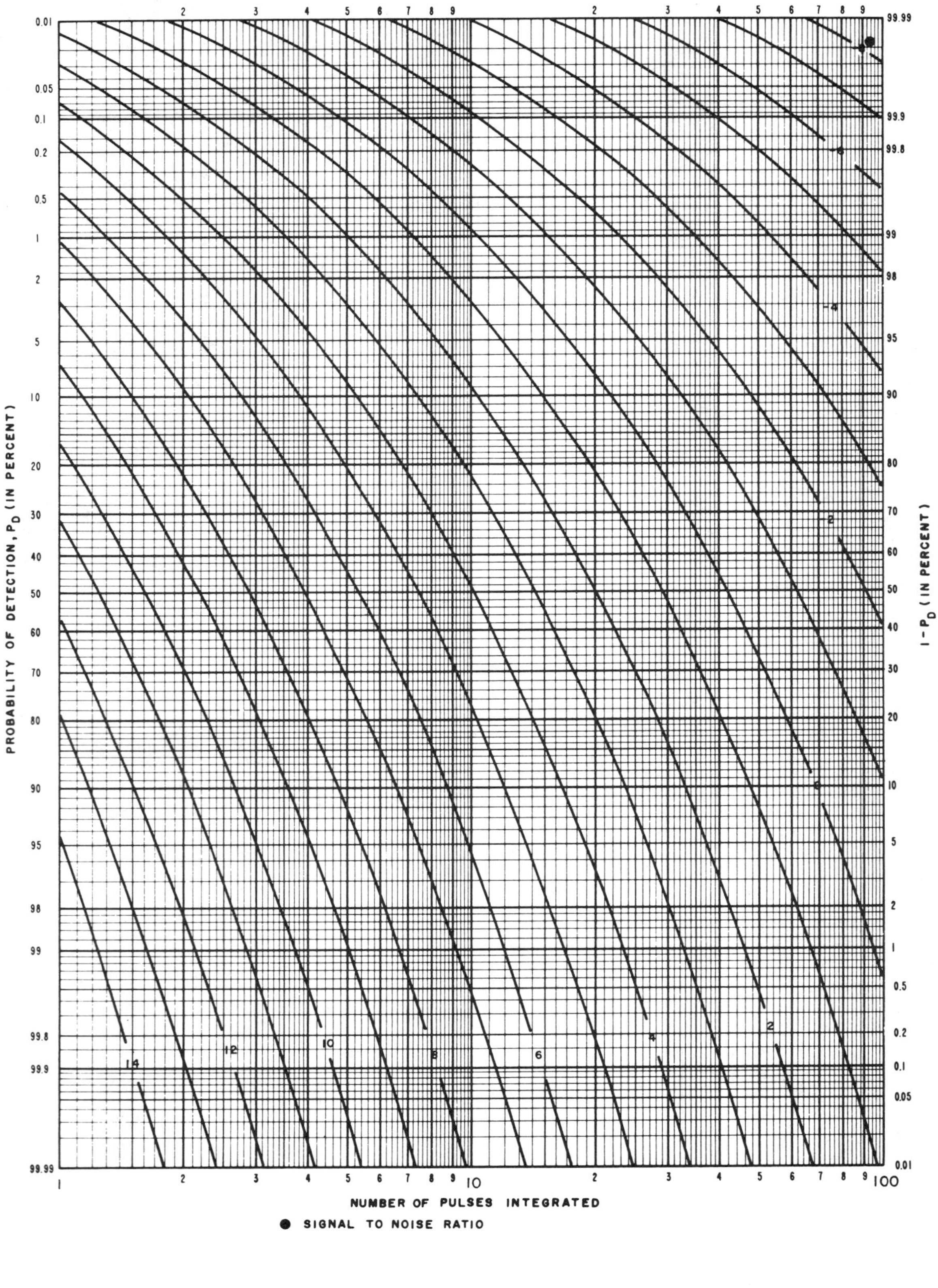

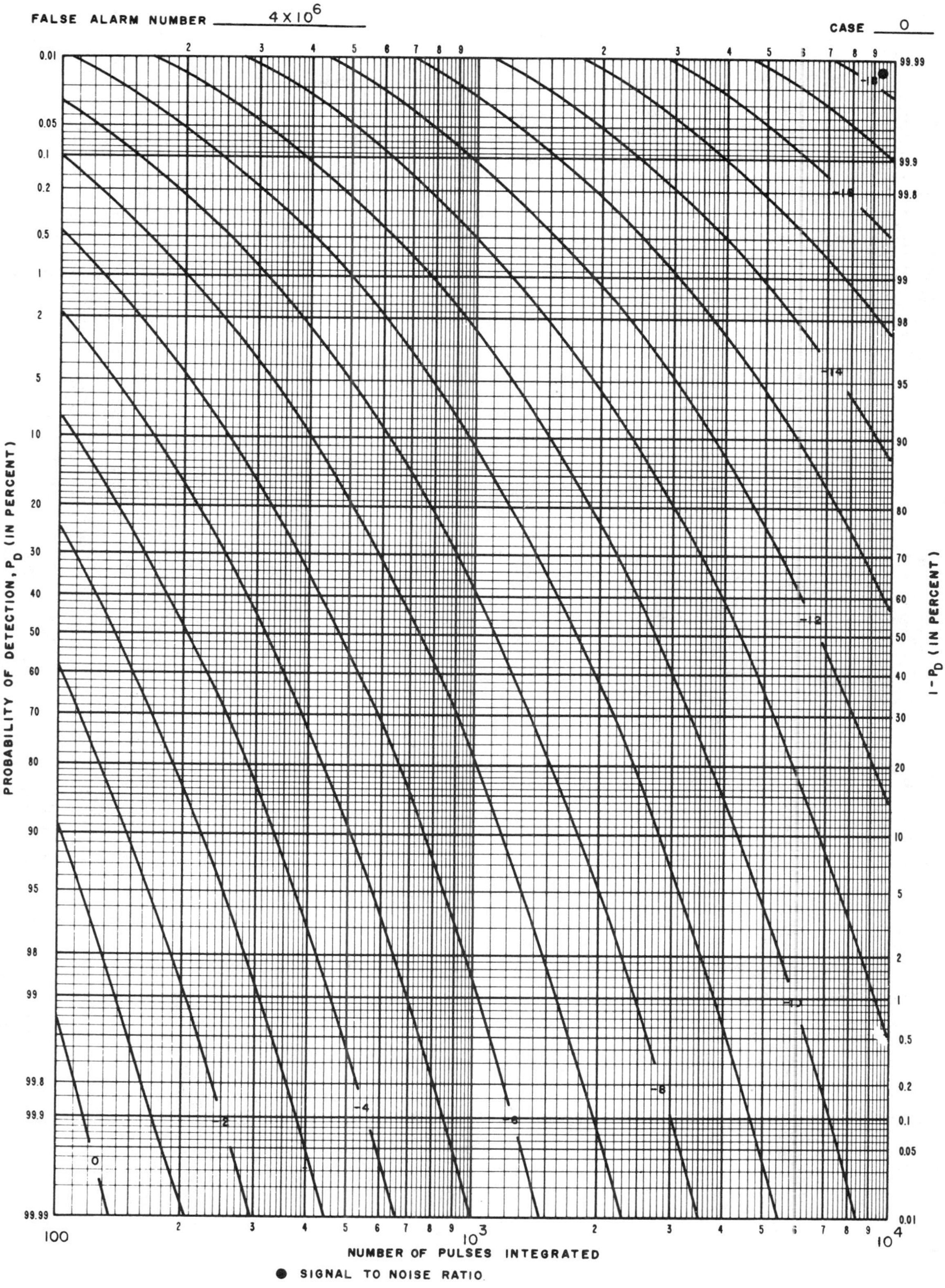
FALSE ALARM NUMBER 4 X 10^6
CASE 0
PROBABILITY OF DETECTION, P_D (IN PERCENT)
1 - P_D (IN PERCENT)
0
-2
-4
-6
-8
-12
-14
100
10^3
10^4
NUMBER OF PULSES INTEGRATED
● SIGNAL TO NOISE RATIO

FALSE ALARM NUMBER 6 X 10^6 CASE 0

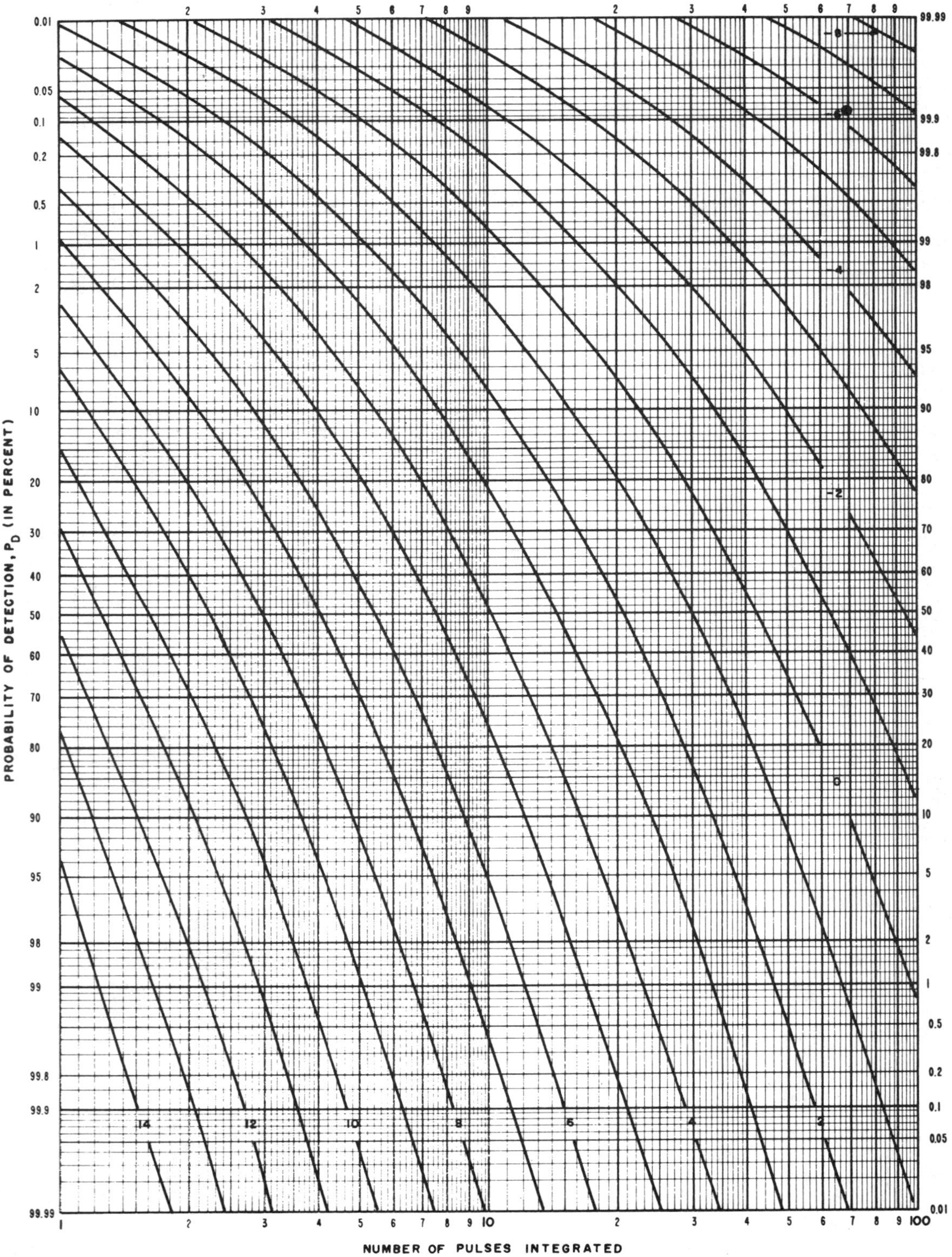

● SIGNAL TO NOISE RATIO

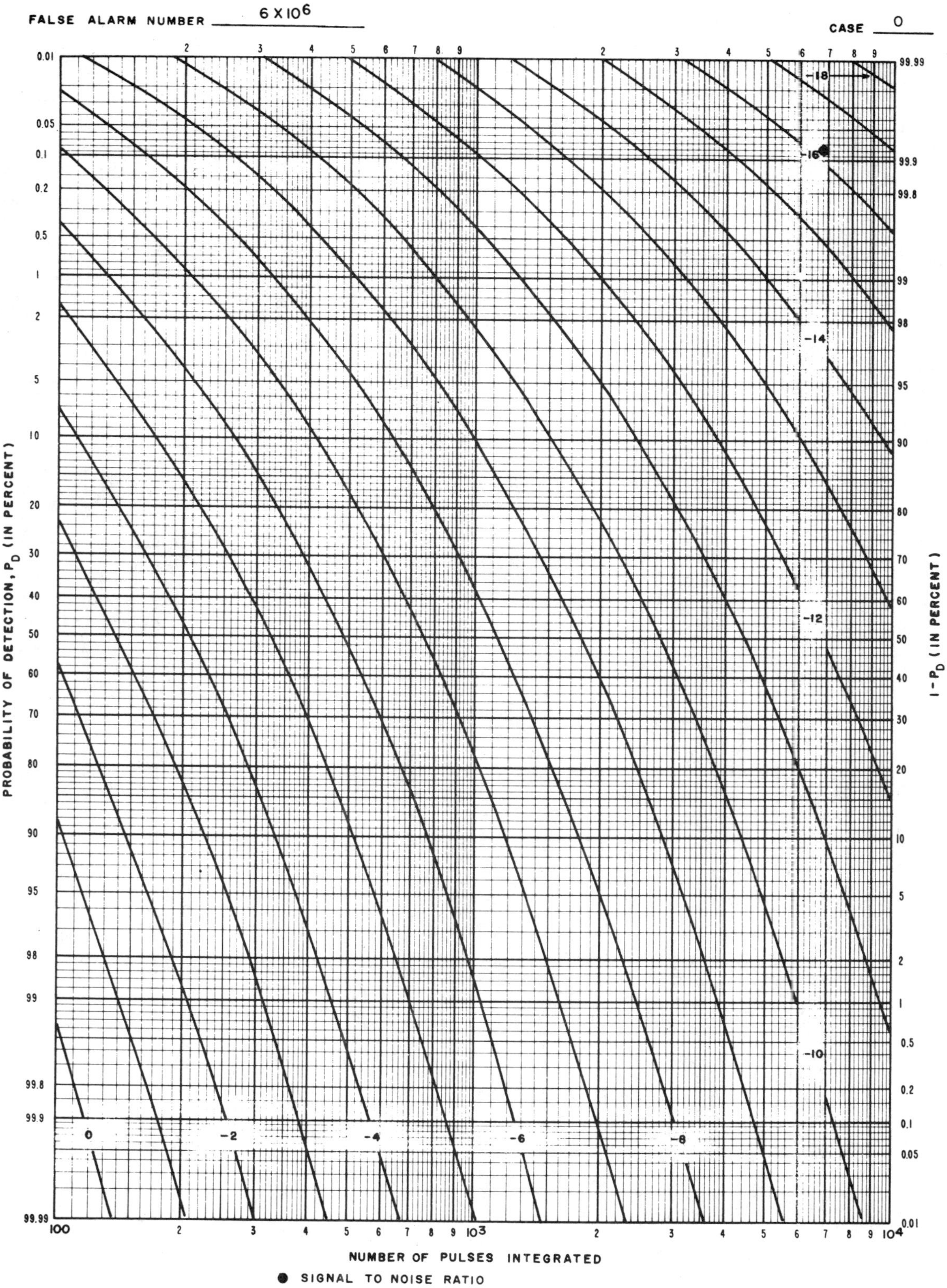
FALSE ALARM NUMBER 6 X 10^6
CASE 0
PROBABILITY OF DETECTION, P_D (IN PERCENT)
1 - P_D (IN PERCENT)
NUMBER OF PULSES INTEGRATED
● SIGNAL TO NOISE RATIO
0
-2
-4
-6
-8
-10
-12
-14
-16
-18
100
10^3
10^4

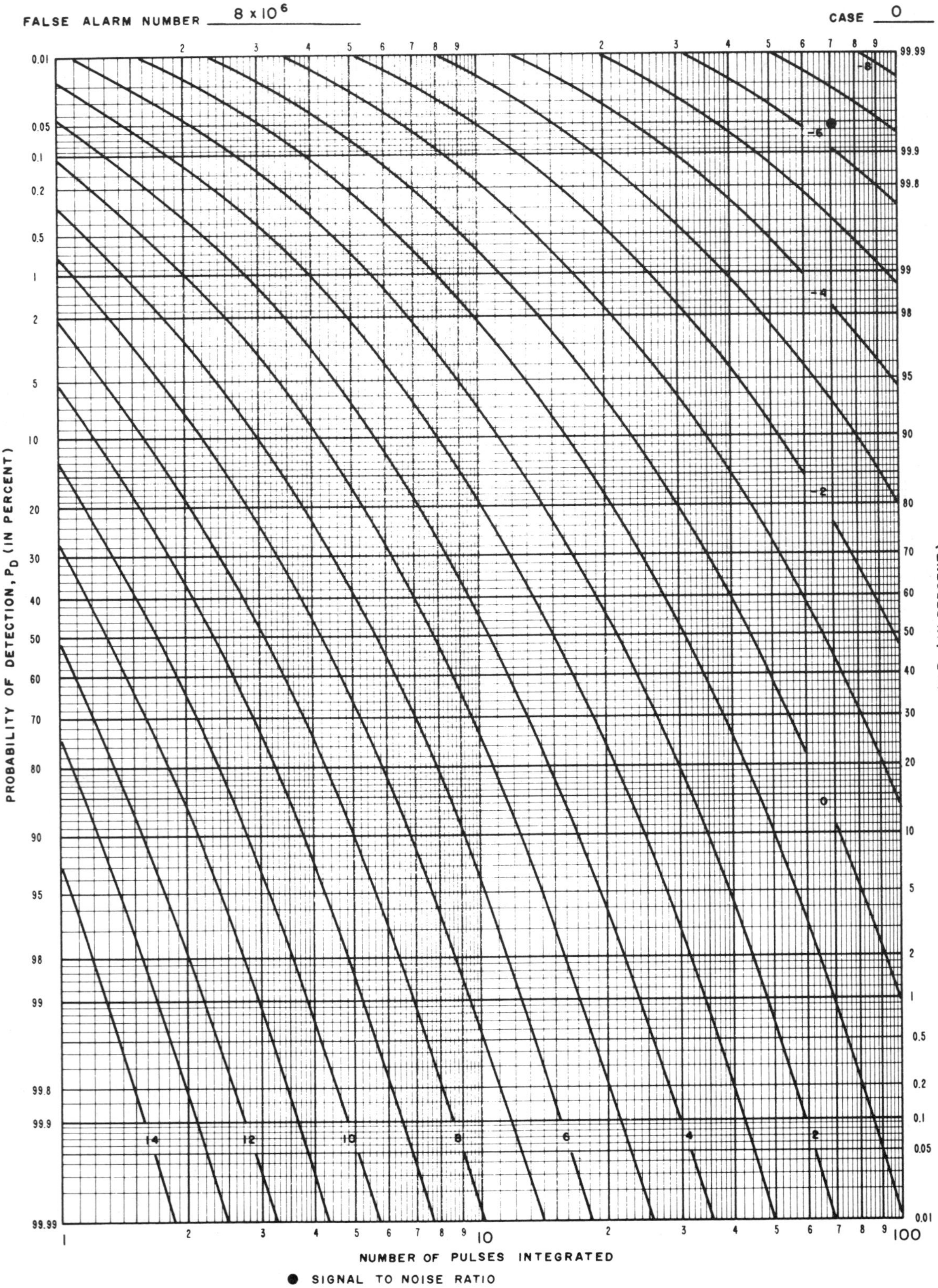
FALSE ALARM NUMBER 8 x 10^6
CASE 0
PROBABILITY OF DETECTION, P_D (IN PERCENT)
1 - P_D (IN PERCENT)
-8
-6
-4
-2
0
2
4
6
8
10
12
14
NUMBER OF PULSES INTEGRATED
● SIGNAL TO NOISE RATIO

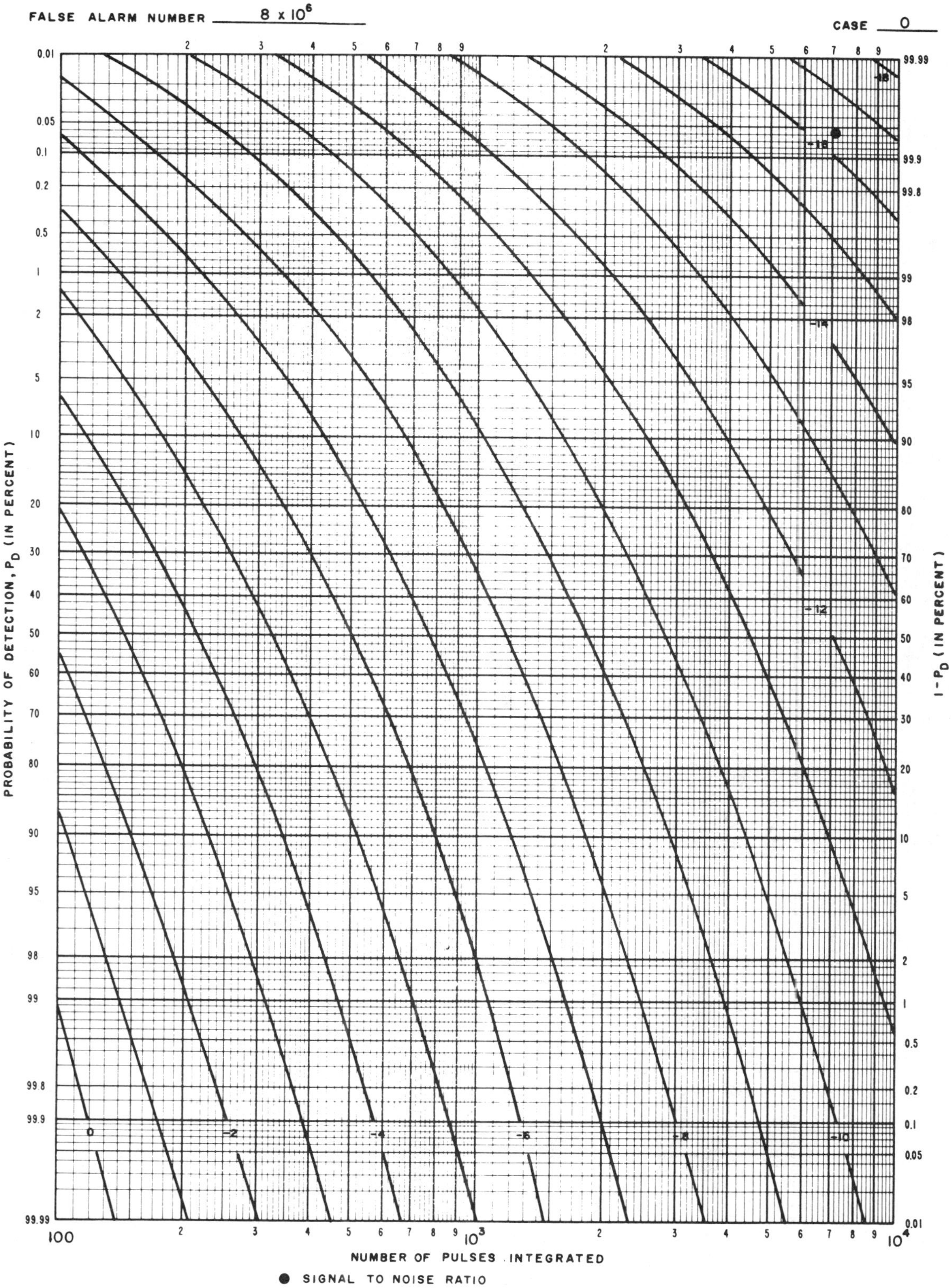
FALSE ALARM NUMBER 8 x 10^6
CASE 0
PROBABILITY OF DETECTION, P_D (IN PERCENT)
1-P_D (IN PERCENT)
NUMBER OF PULSES INTEGRATED
● SIGNAL TO NOISE RATIO
0
-2
-4
-6
-8
-10
-12
-14
-16
-18

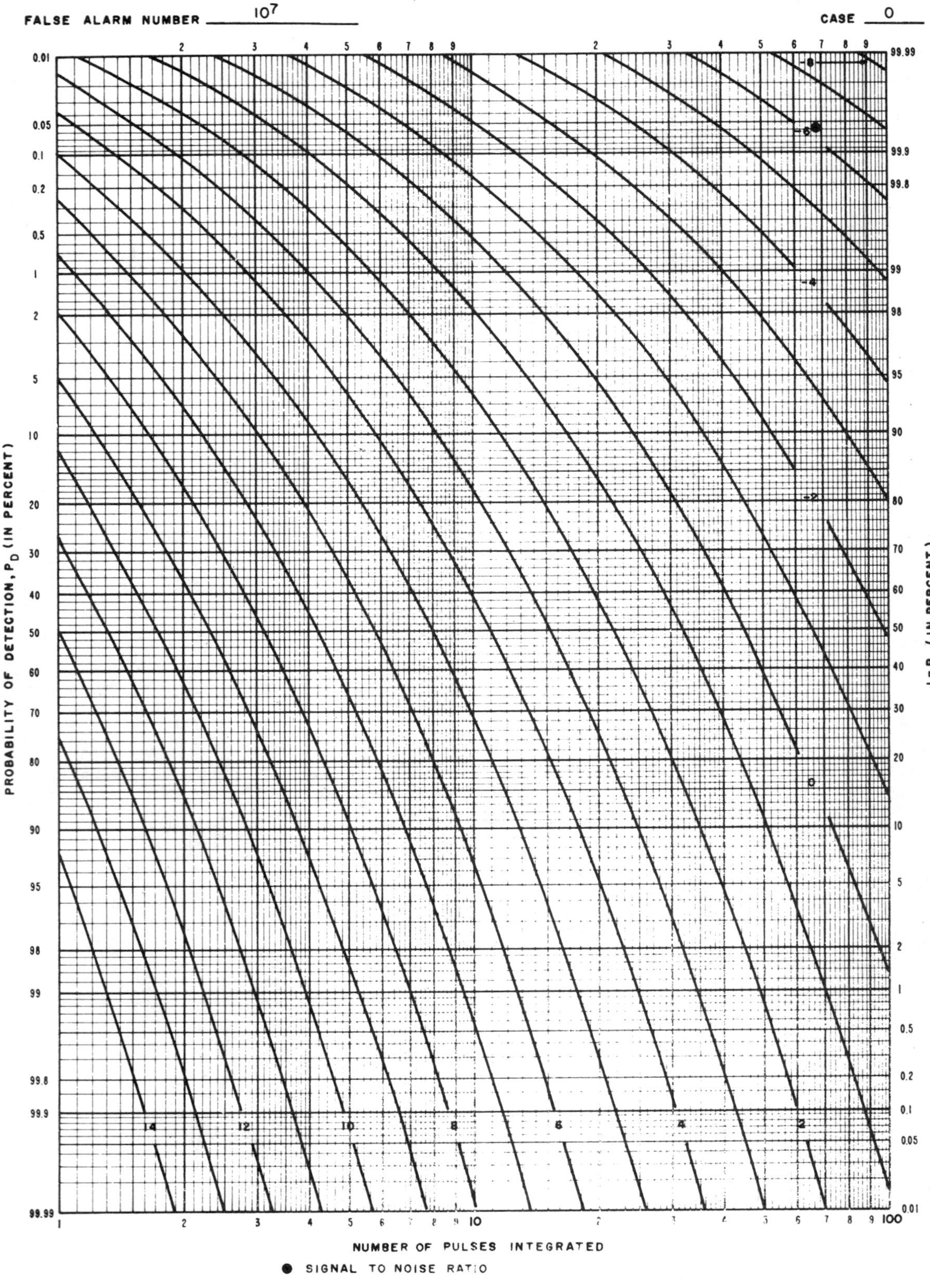
FALSE ALARM NUMBER 10^7
CASE 0
PROBABILITY OF DETECTION, P_D (IN PERCENT)
$1-P_D$ (IN PERCENT)
NUMBER OF PULSES INTEGRATED
● SIGNAL TO NOISE RATIO
-8
-6
-4
-2
0
2
4
6
8
10
12
14

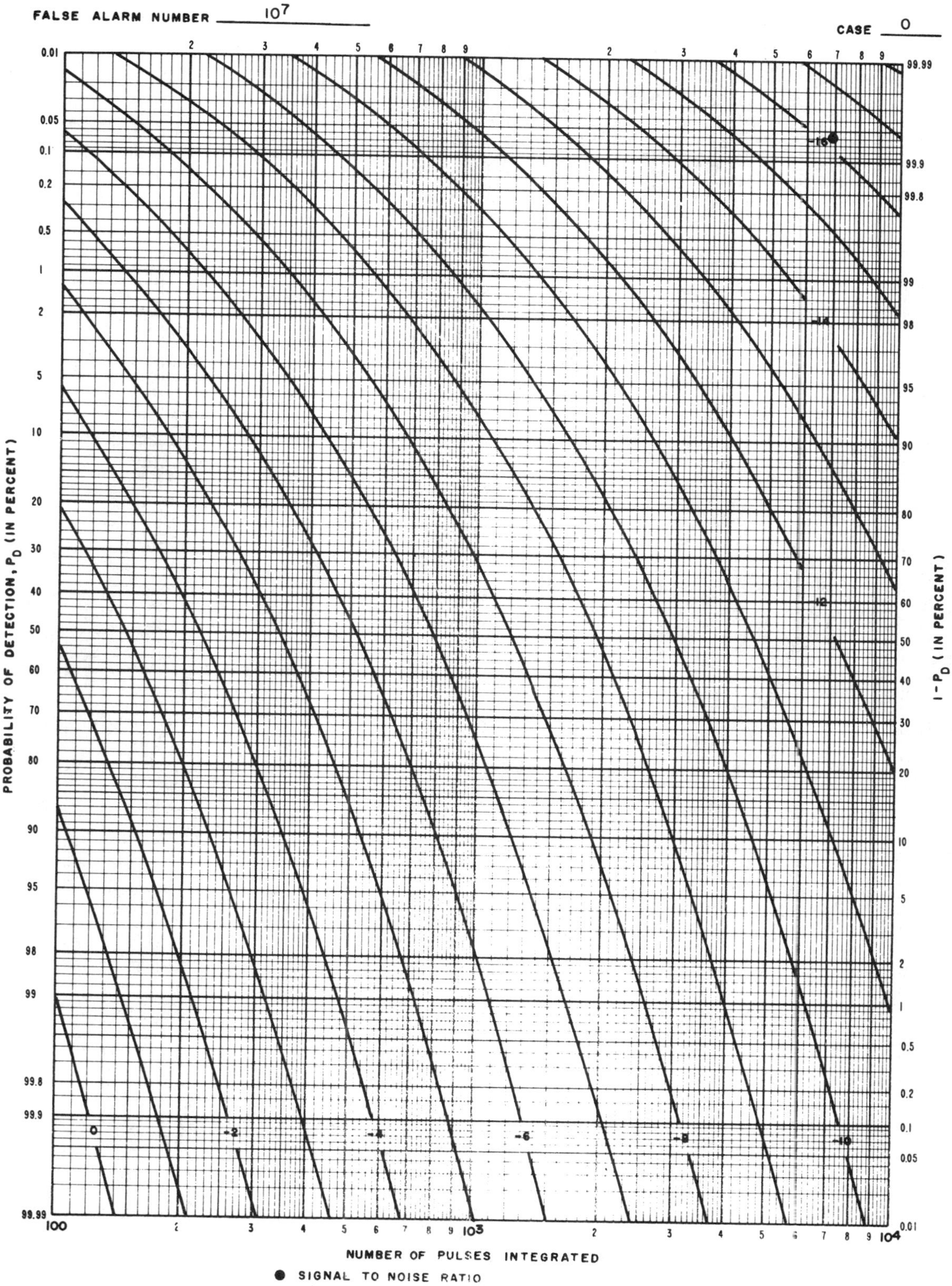
FALSE ALARM NUMBER 10^7
CASE 0
PROBABILITY OF DETECTION, P_D (IN PERCENT)
1 - P_D (IN PERCENT)
NUMBER OF PULSES INTEGRATED
● SIGNAL TO NOISE RATIO
0
-2
-4
-6
-8
-10
-12
-14
-16
100
10^3
10^4

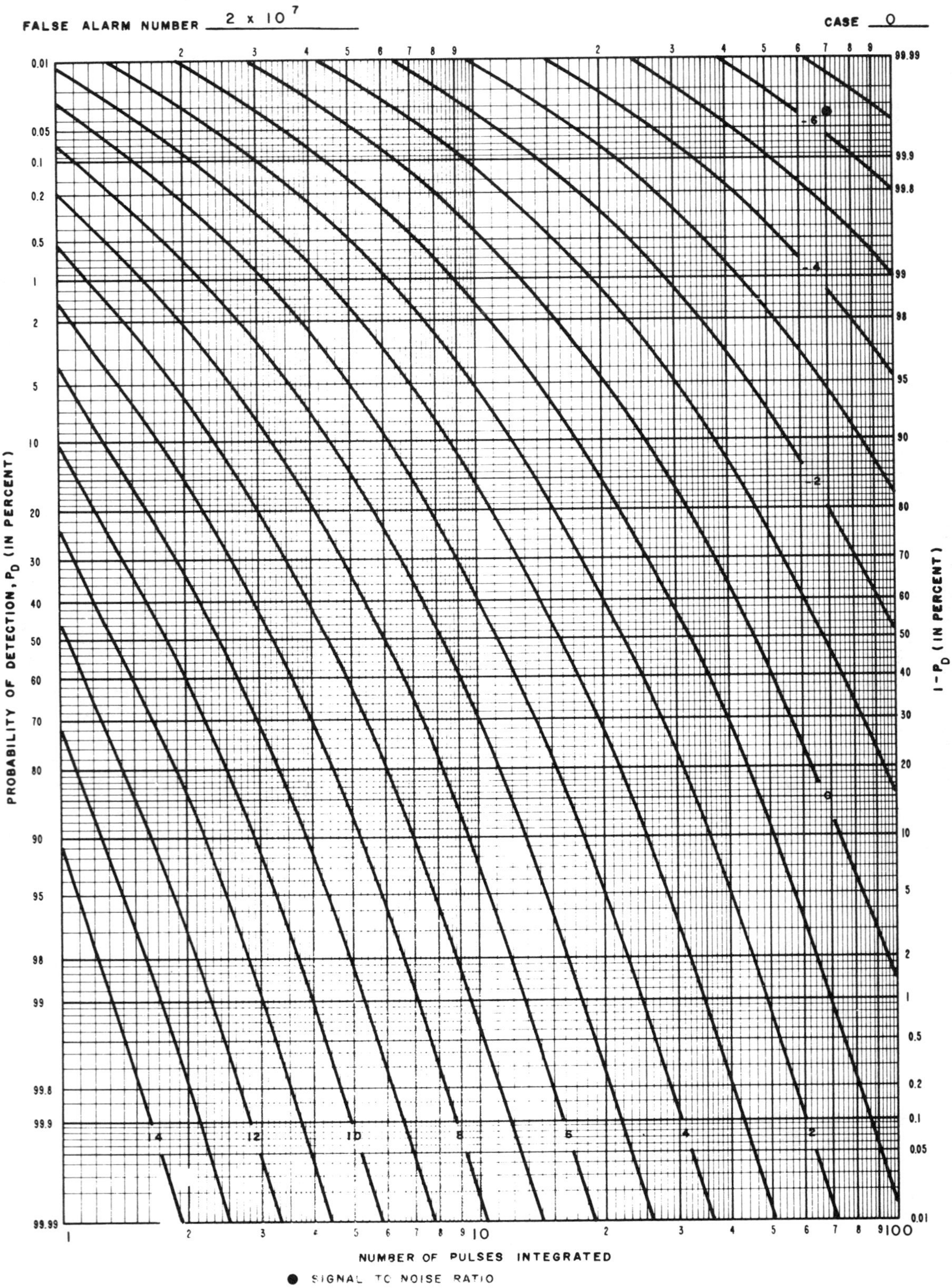
FALSE ALARM NUMBER 2 x 10^7
CASE 0
PROBABILITY OF DETECTION, P_D (IN PERCENT)
1-P_D (IN PERCENT)
NUMBER OF PULSES INTEGRATED
● SIGNAL TO NOISE RATIO
14
12
10
8
6
4
2
0
-2
-4
-6

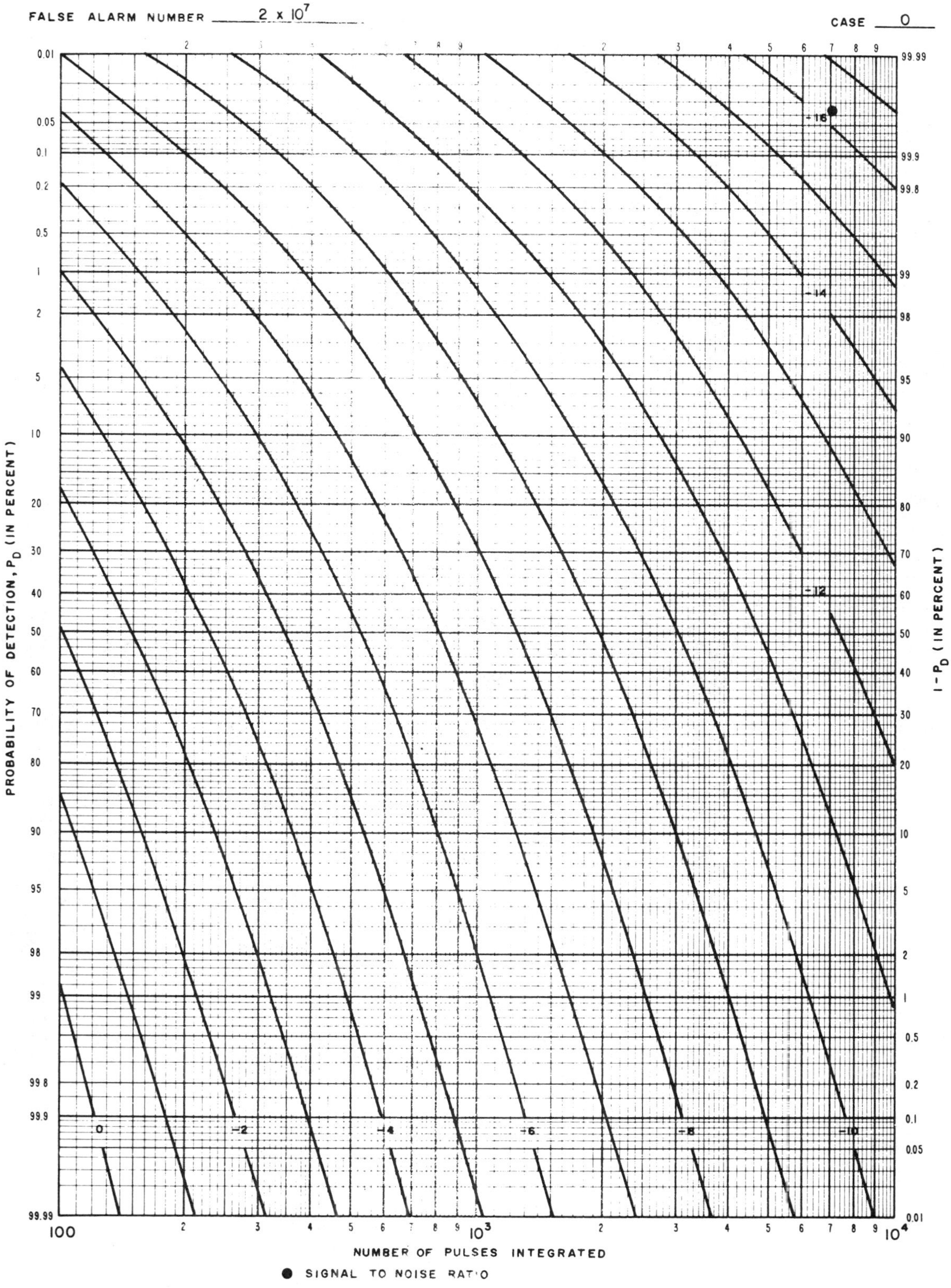
FALSE ALARM NUMBER 2 x 10^7
CASE 0
PROBABILITY OF DETECTION, P_D (IN PERCENT)
1 - P_D (IN PERCENT)
NUMBER OF PULSES INTEGRATED
SIGNAL TO NOISE RATIO
0
-2
-4
-6
-8
-10
-12
-14
-16
100
10^3
10^4

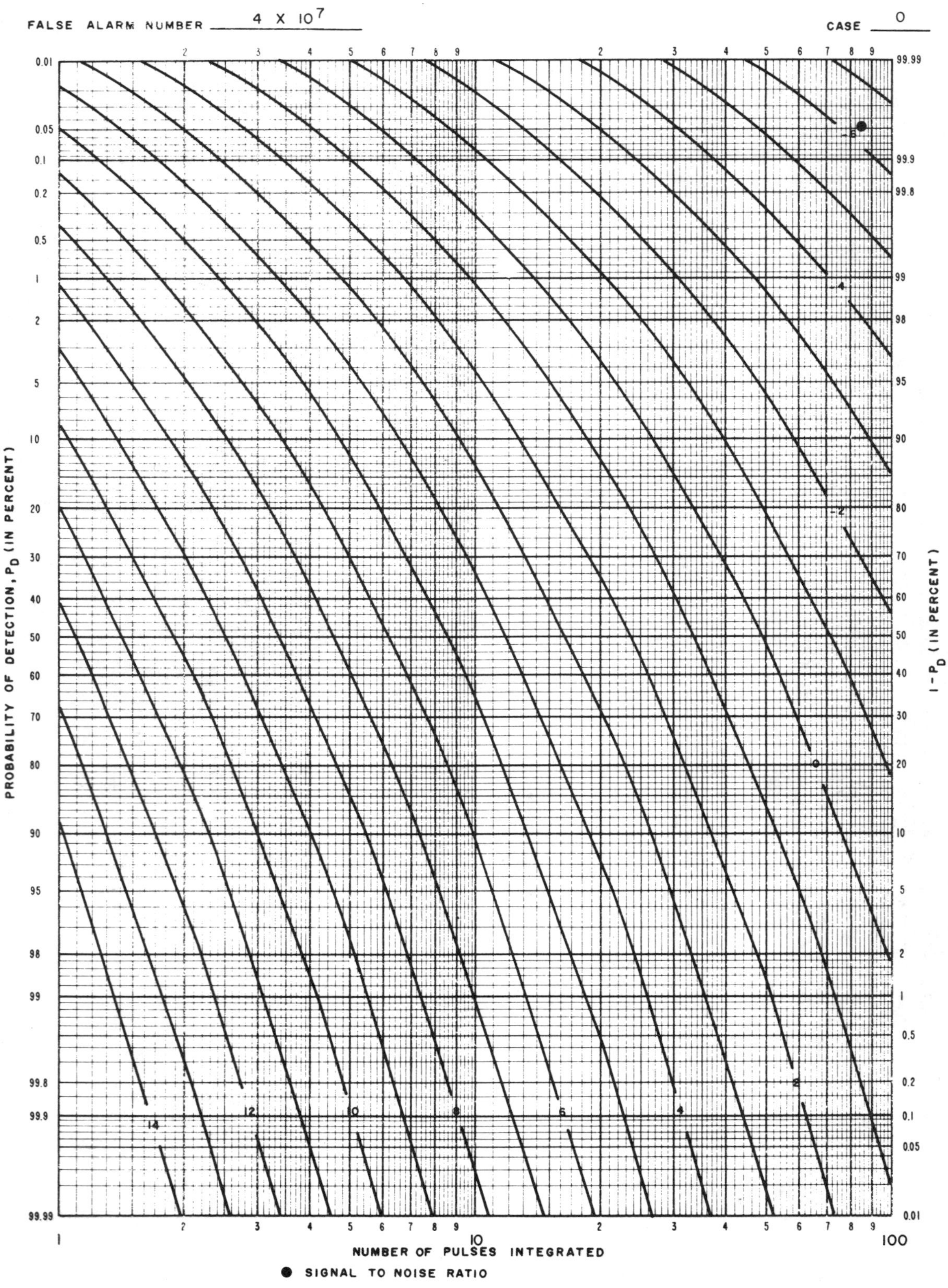
FALSE ALARM NUMBER 4 X 10^7
CASE 0
PROBABILITY OF DETECTION, P_D (IN PERCENT)
1 - P_D (IN PERCENT)
NUMBER OF PULSES INTEGRATED
● SIGNAL TO NOISE RATIO
-6
-4
-2
0
2
4
6
8
10
12
14
1
10
100

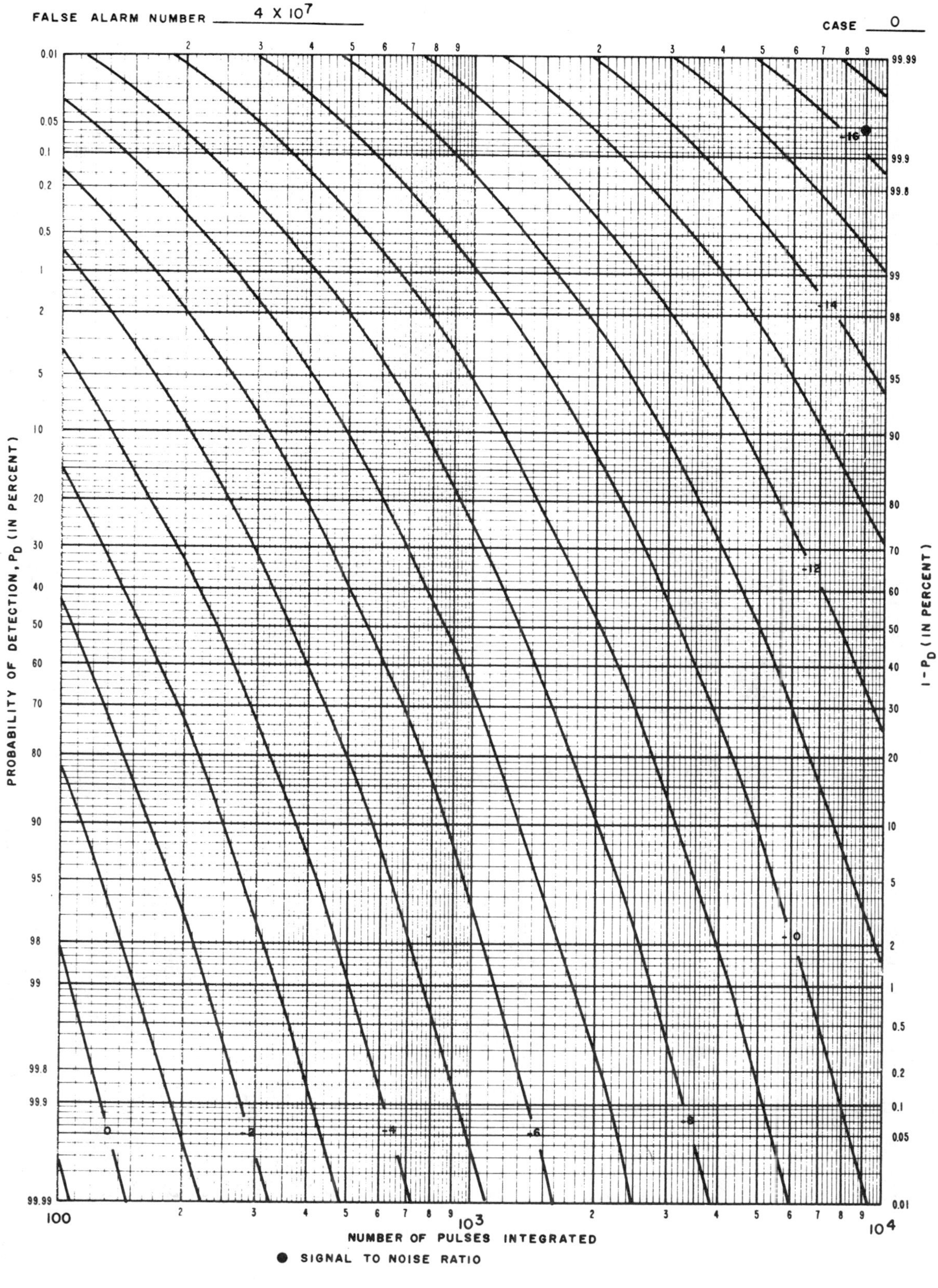
FALSE ALARM NUMBER 4 X 10^7
CASE 0
PROBABILITY OF DETECTION, P_D (IN PERCENT)
1 - P_D (IN PERCENT)
NUMBER OF PULSES INTEGRATED
100
10^3
10^4
0
+2
+4
+6
+8
+10
+12
+14
+16
● SIGNAL TO NOISE RATIO

FALSE ALARM NUMBER 6 X 10^7 CASE 0

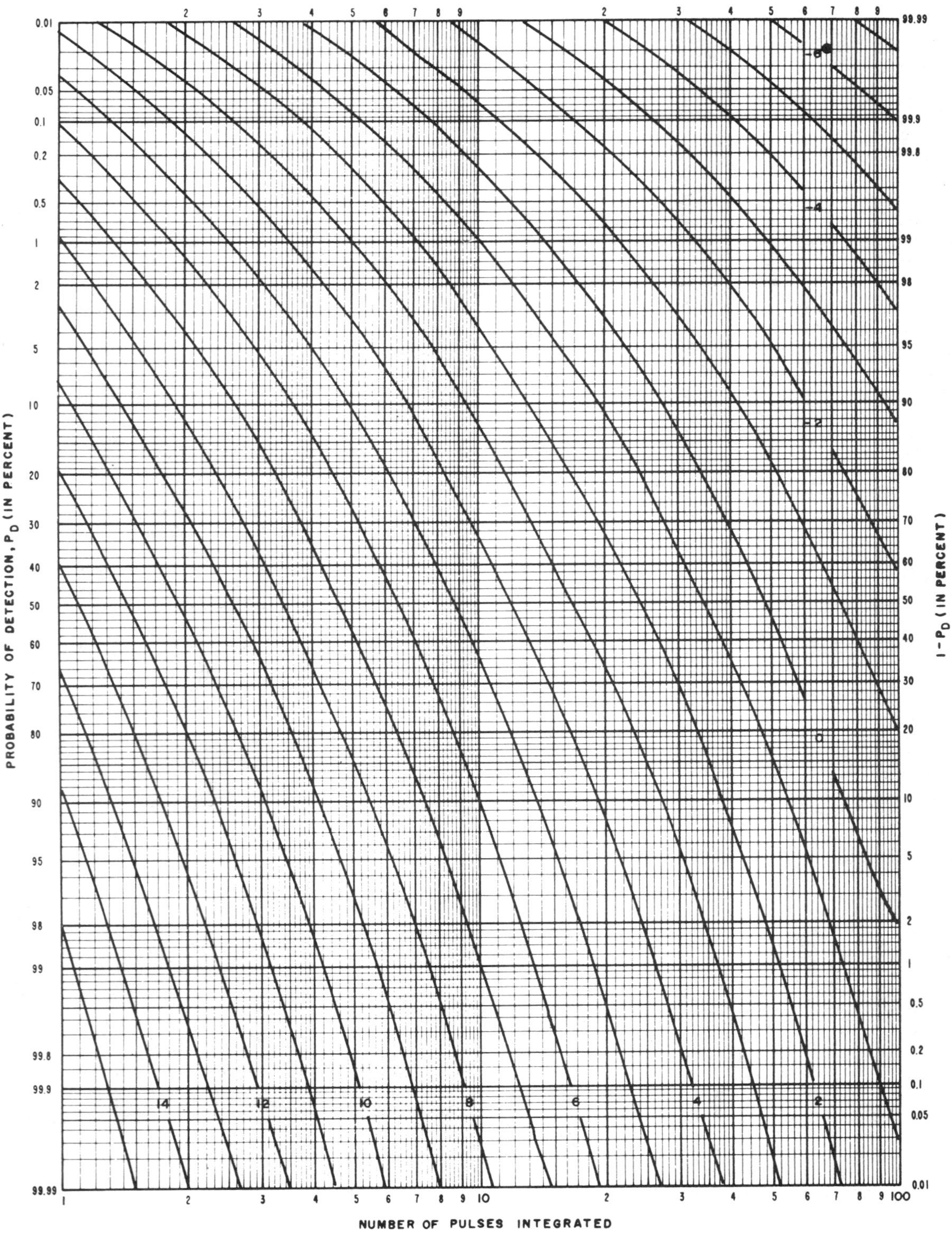

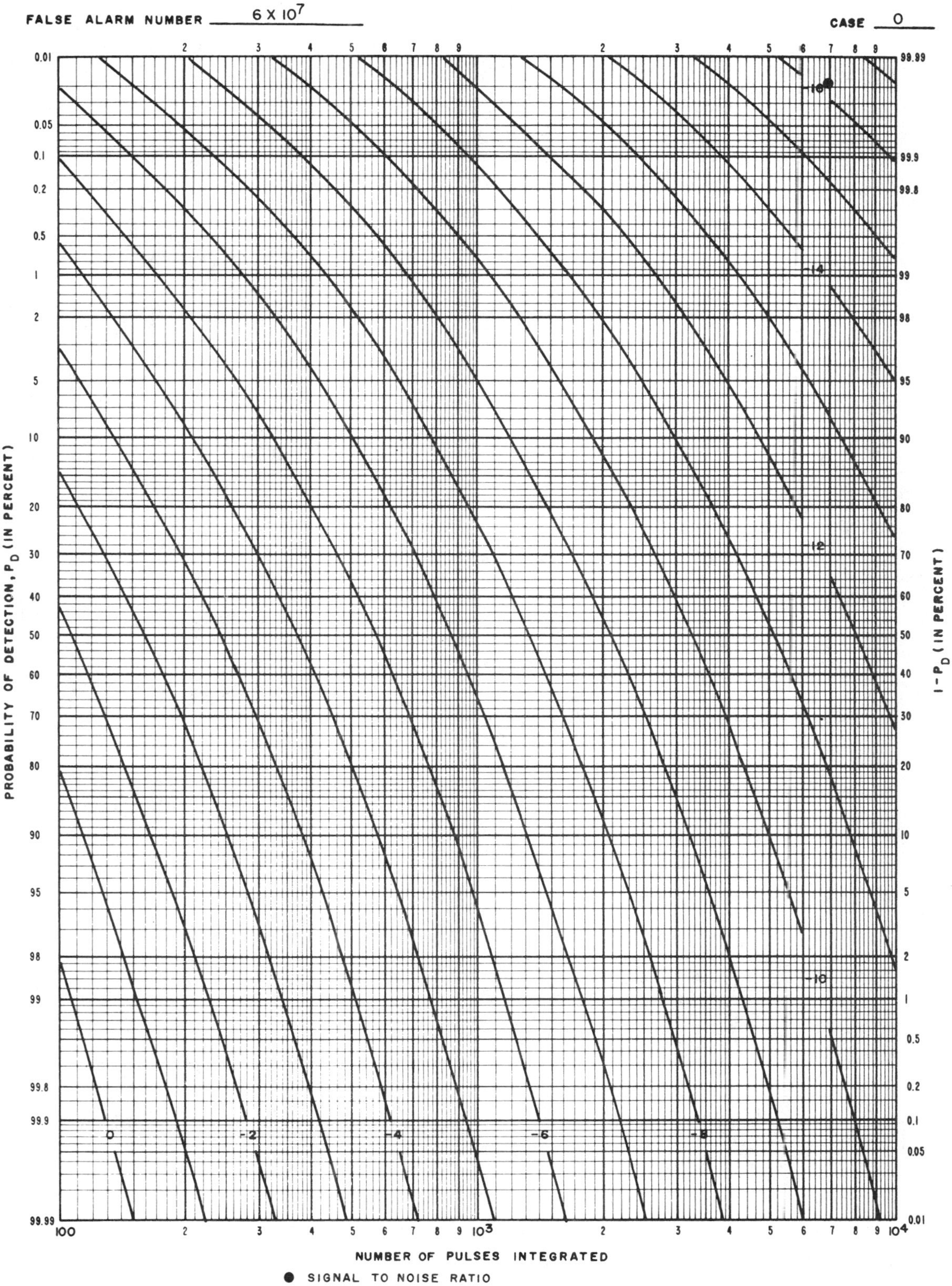
FALSE ALARM NUMBER 6 X 10^7
CASE 0
PROBABILITY OF DETECTION, P_D (IN PERCENT)
1 - P_D (IN PERCENT)
NUMBER OF PULSES INTEGRATED
● SIGNAL TO NOISE RATIO
0
-2
-4
-6
-8
-10
-12
-14
-16
100
10^3
10^4

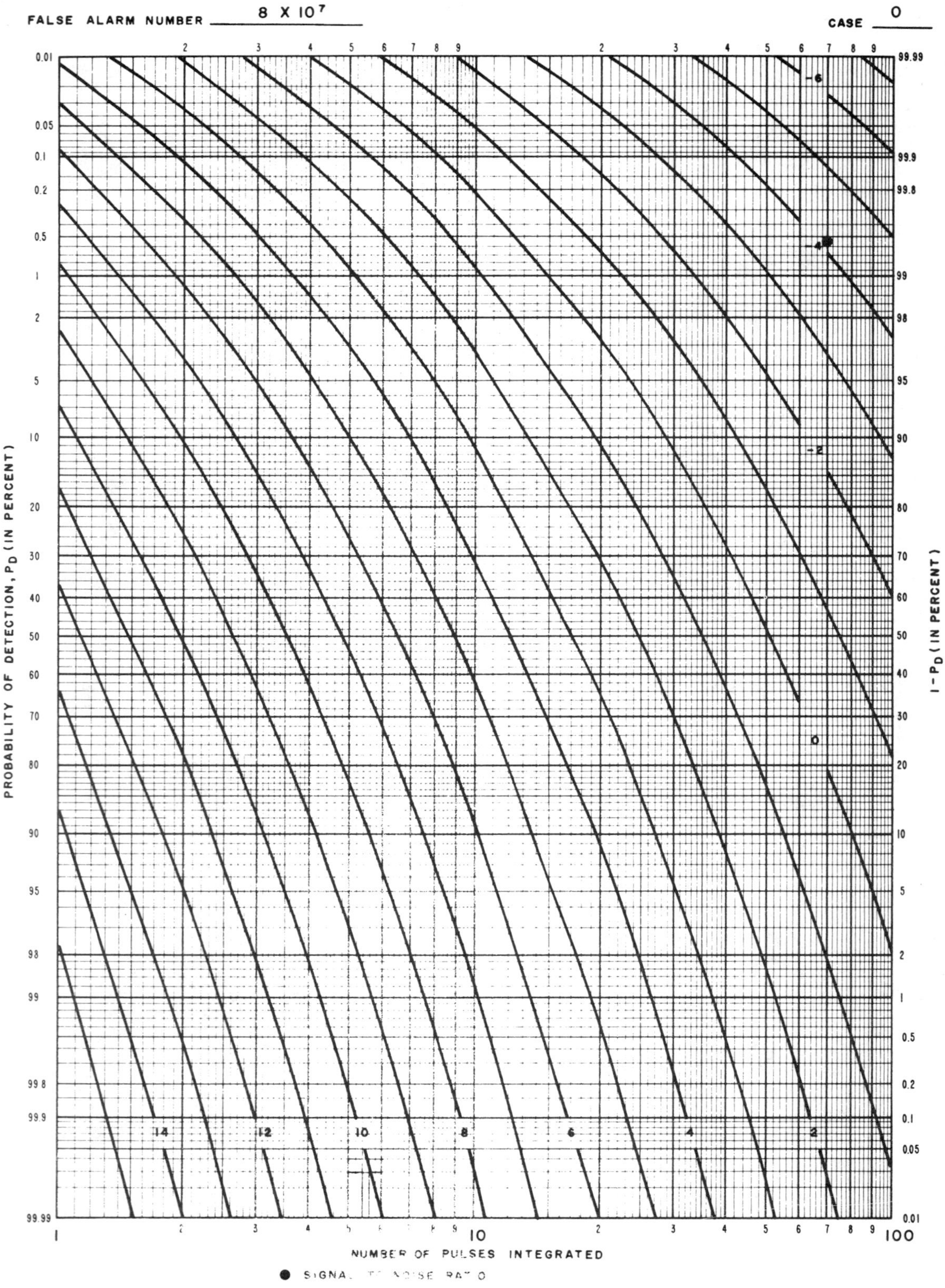

FALSE ALARM NUMBER 8 X 10^7
CASE 0
PROBABILITY OF DETECTION, P_D (IN PERCENT)
1 - P_D (IN PERCENT)
-6
-4
-2
0
2
4
6
8
10
12
14
NUMBER OF PULSES INTEGRATED
1
10
100

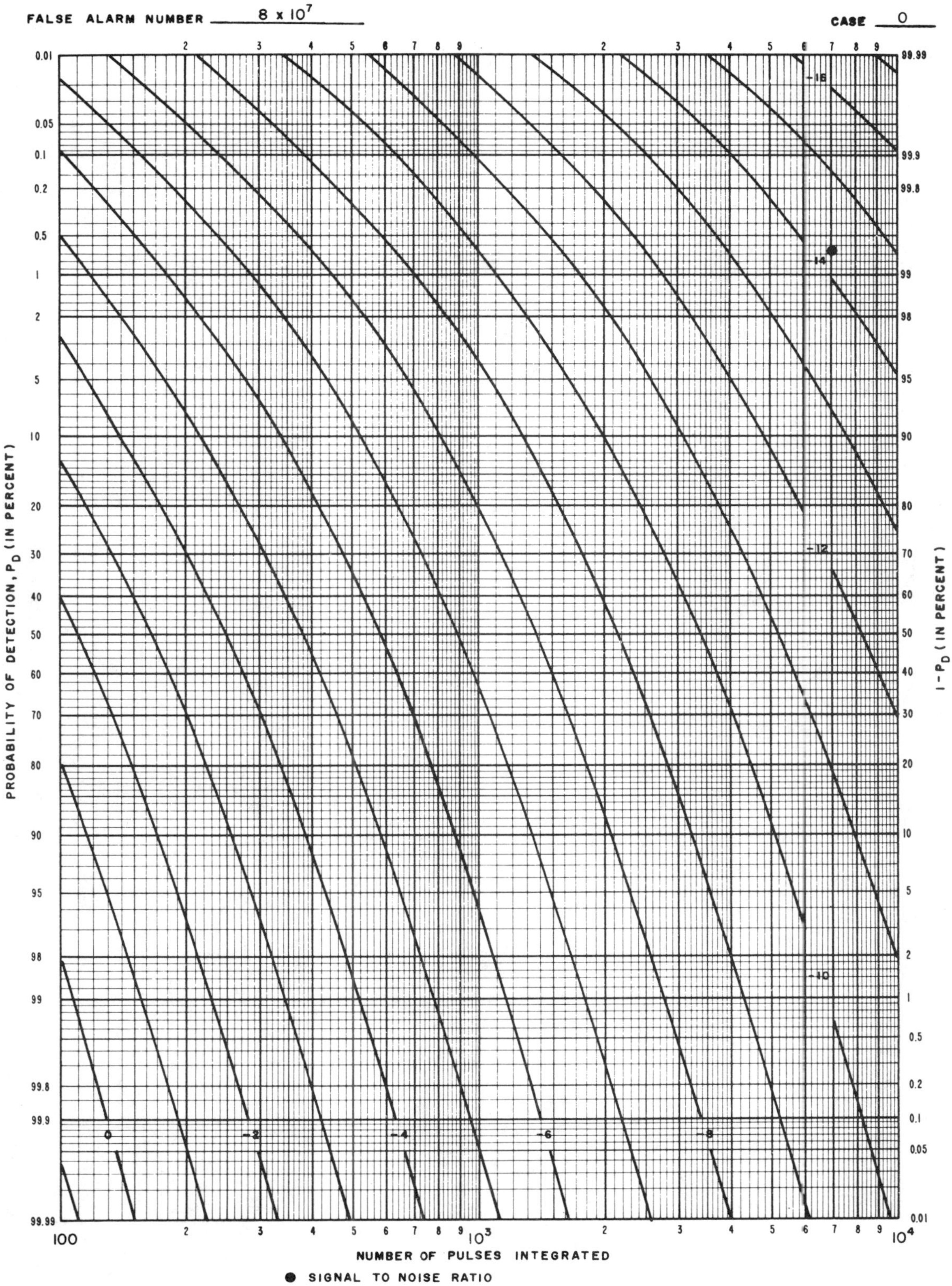
FALSE ALARM NUMBER 8 x 10^7
CASE 0
PROBABILITY OF DETECTION, P_D (IN PERCENT)
1 - P_D (IN PERCENT)
NUMBER OF PULSES INTEGRATED
● SIGNAL TO NOISE RATIO
100
10^3
10^4
0
-2
-4
-6
-8
-10
-12
-14
-16

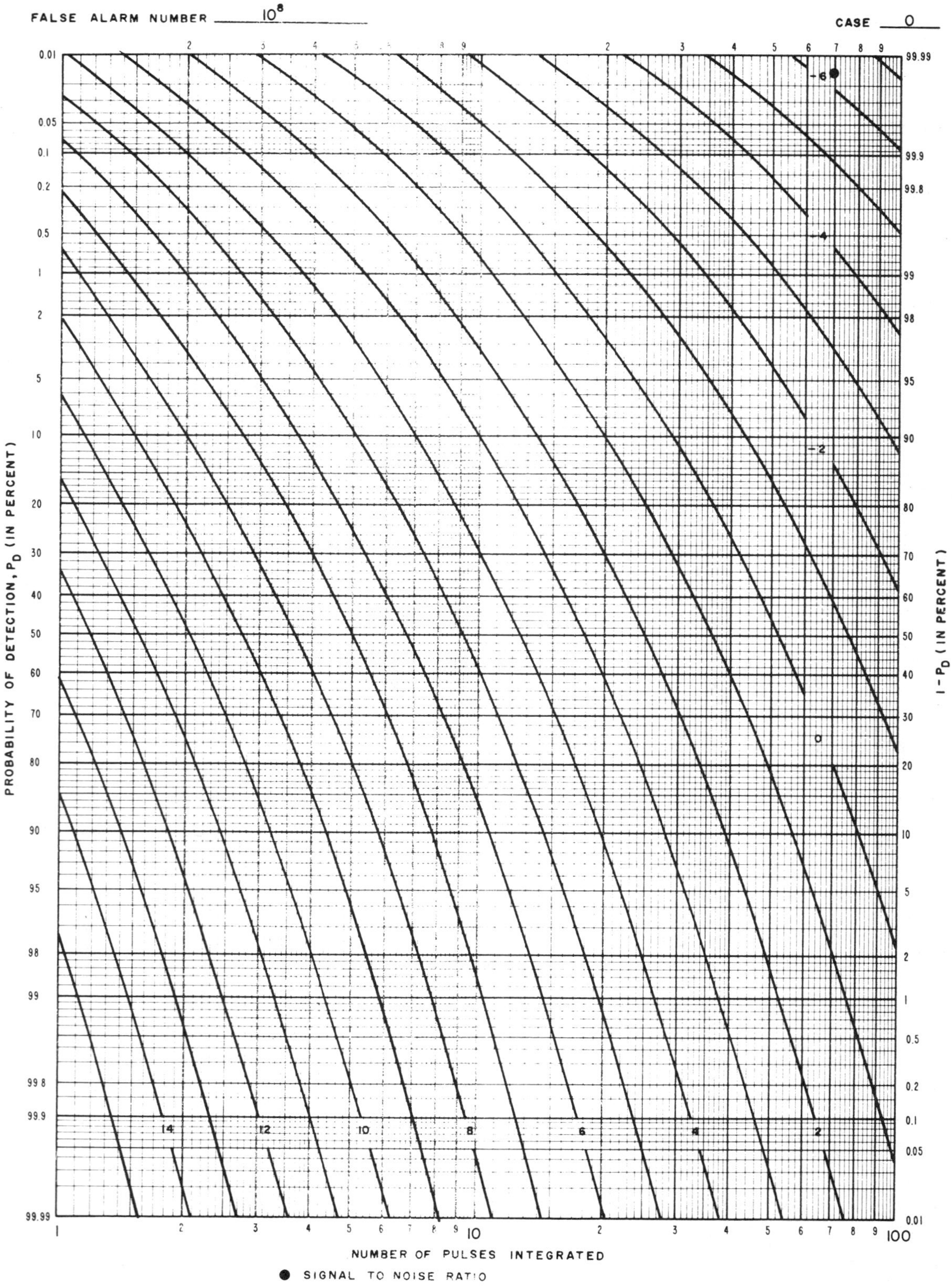

FALSE ALARM NUMBER 10^8
CASE 0
PROBABILITY OF DETECTION, P_D (IN PERCENT)
$1-P_D$ (IN PERCENT)
NUMBER OF PULSES INTEGRATED
-6
-4
-2
0
2
4
6
8
10
12
14
● SIGNAL TO NOISE RATIO

FALSE ALARM NUMBER 10^8 CASE 0

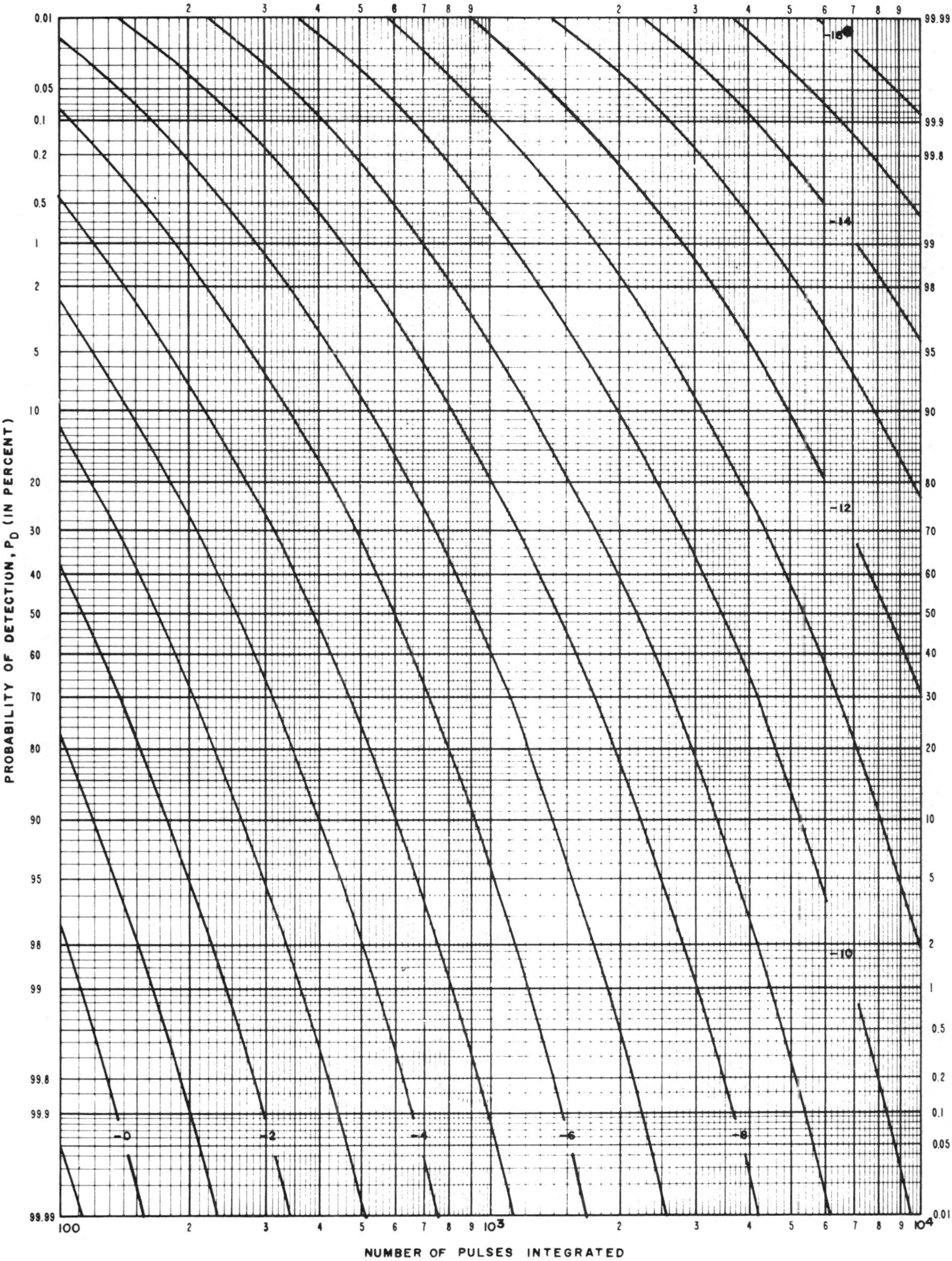

● SIGNAL TO NOISE RATIO

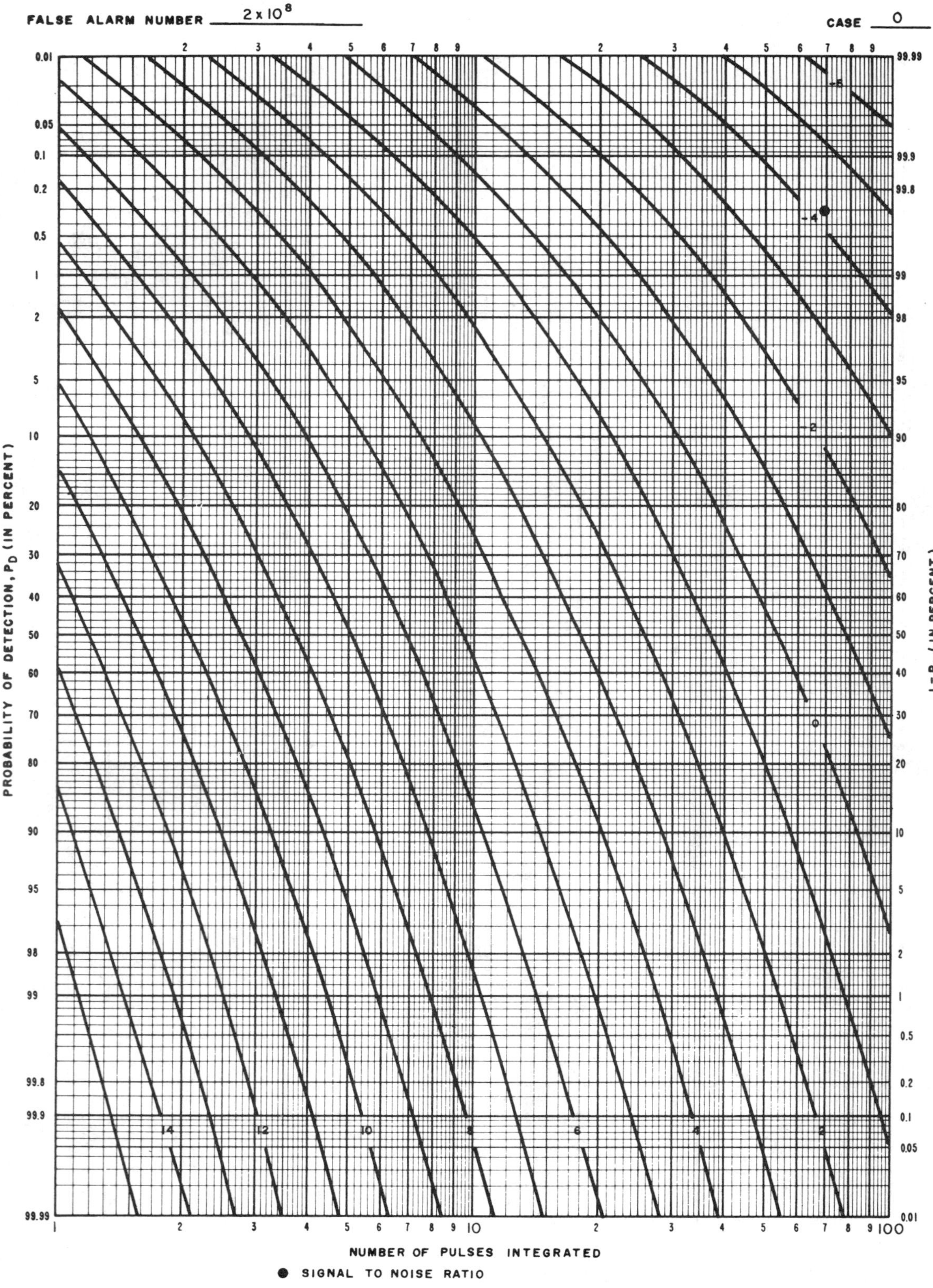
FALSE ALARM NUMBER 2 x 10^8
CASE 0
PROBABILITY OF DETECTION, P_D (IN PERCENT)
1 - P_D (IN PERCENT)
NUMBER OF PULSES INTEGRATED
● SIGNAL TO NOISE RATIO
14
12
10
8
6
4
2
0
-2
-4
-6

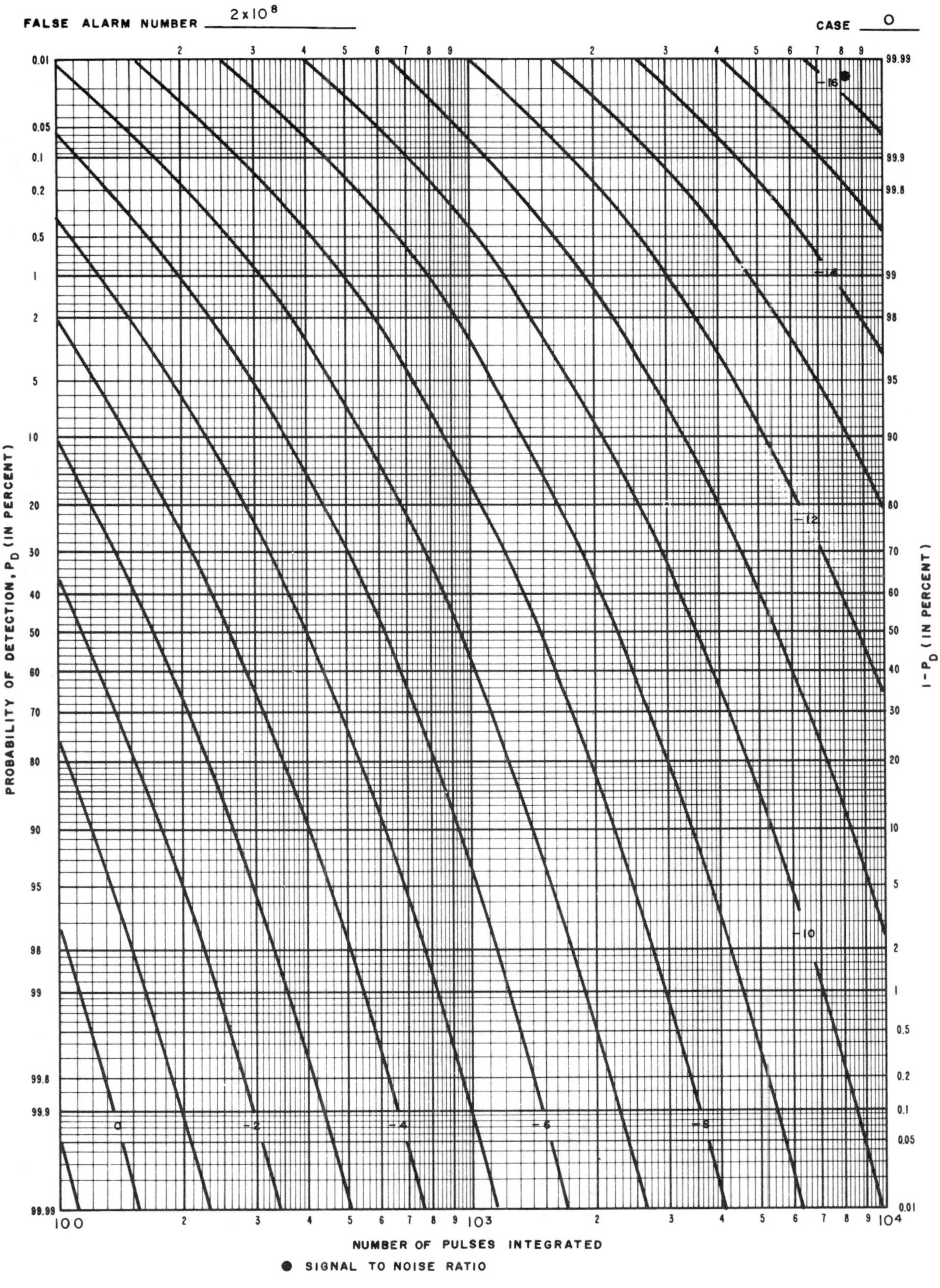
FALSE ALARM NUMBER 2x10^8
CASE 0
PROBABILITY OF DETECTION, P_D (IN PERCENT)
1-P_D (IN PERCENT)
NUMBER OF PULSES INTEGRATED
SIGNAL TO NOISE RATIO
0
-2
-4
-6
-8
-10
-12
-14
-16
100
10^3
10^4

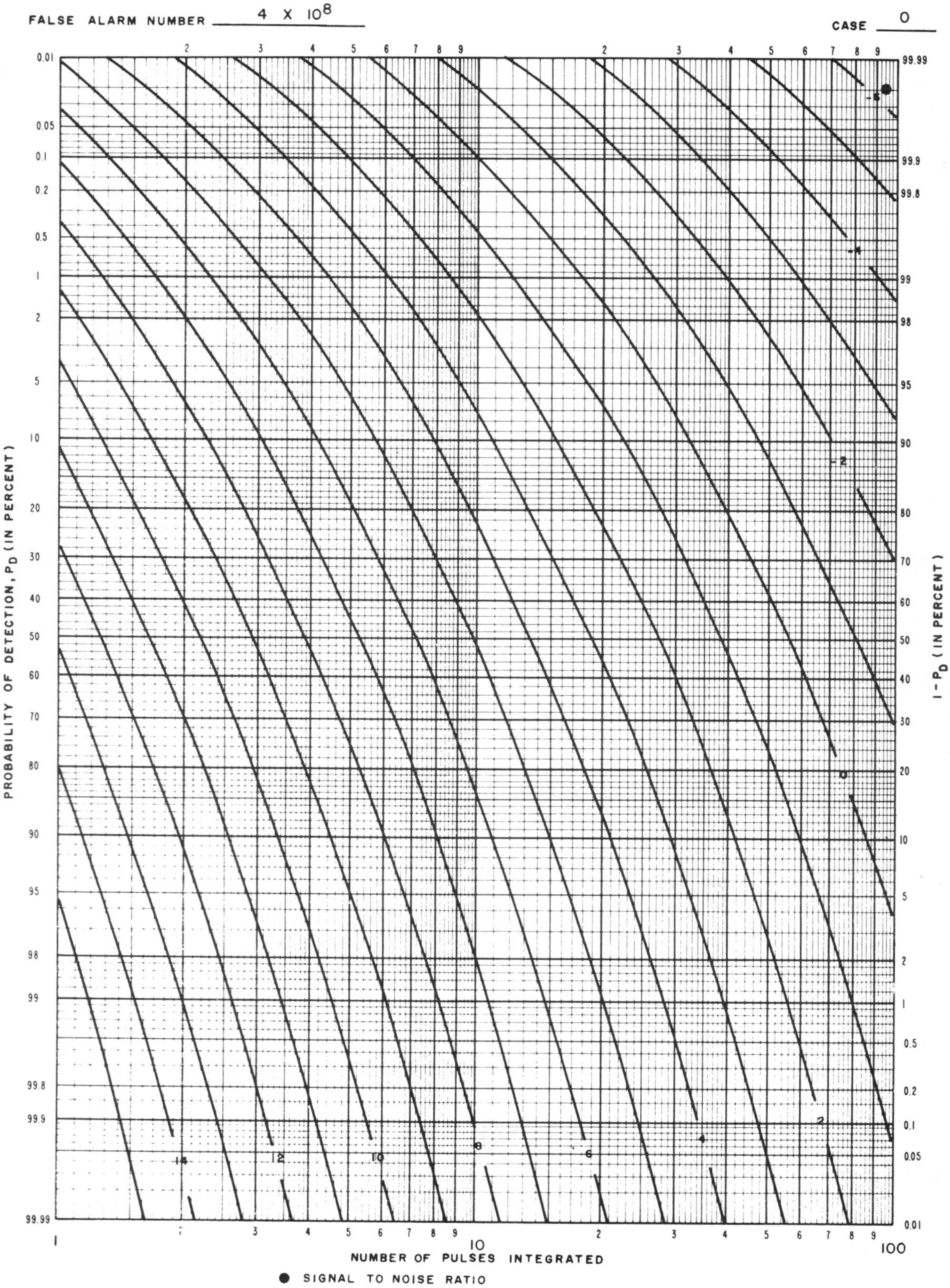
FALSE ALARM NUMBER 4 X 10^8
CASE 0
PROBABILITY OF DETECTION, P_D (IN PERCENT)
$1-P_D$ (IN PERCENT)
NUMBER OF PULSES INTEGRATED
● SIGNAL TO NOISE RATIO

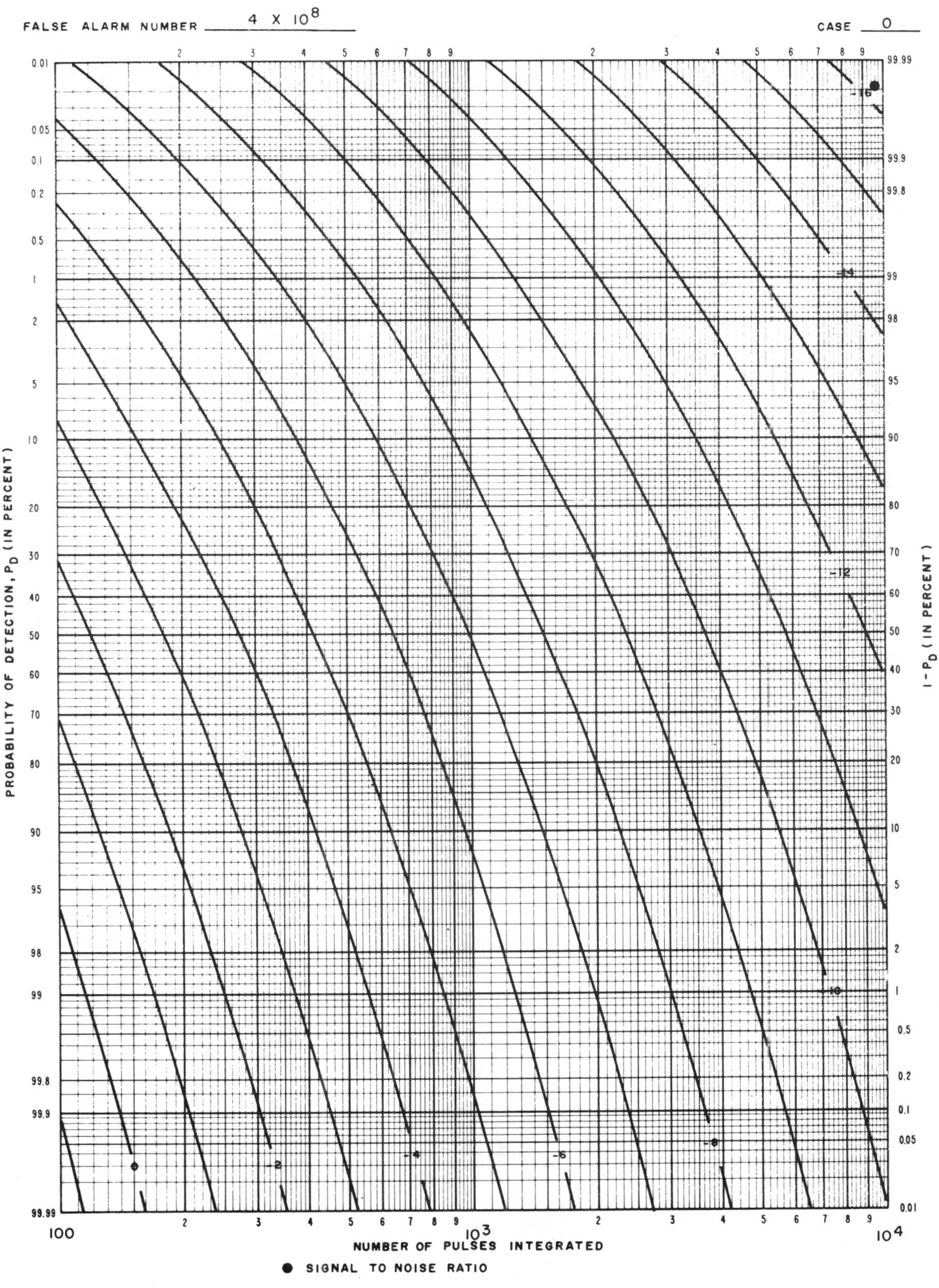
FALSE ALARM NUMBER 4 X 10^8
CASE 0
PROBABILITY OF DETECTION, P_D (IN PERCENT)
1 - P_D (IN PERCENT)
NUMBER OF PULSES INTEGRATED
100
10^3
10^4
● SIGNAL TO NOISE RATIO
-16
-14
-12
-10
-8
-6
-4
-2
0

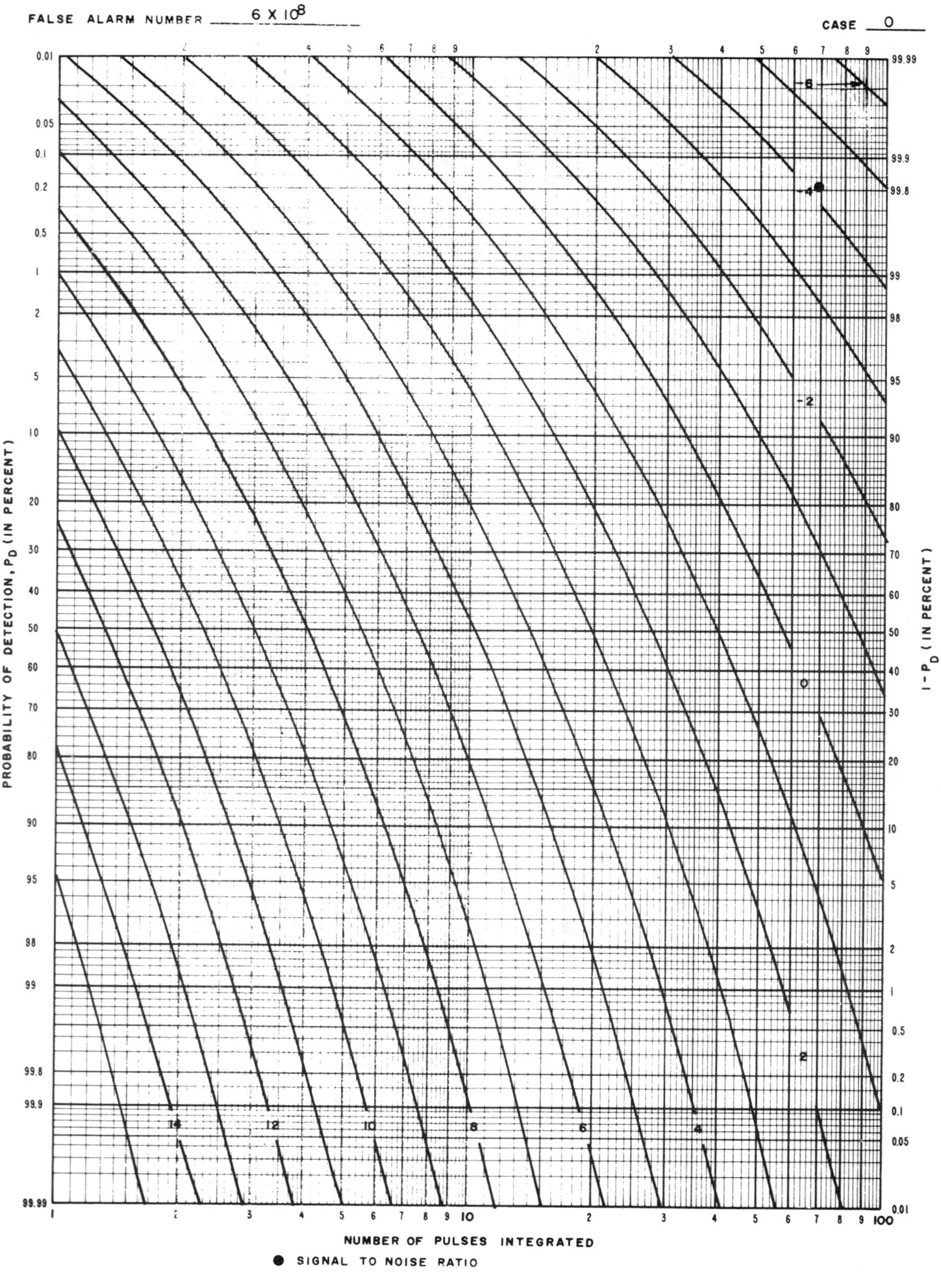
FALSE ALARM NUMBER 6 X 10^8
CASE 0
PROBABILITY OF DETECTION, P_D (IN PERCENT)
1 - P_D (IN PERCENT)
NUMBER OF PULSES INTEGRATED
● SIGNAL TO NOISE RATIO
-6
-4
-2
0
2
4
6
8
10
12
14

FALSE ALARM NUMBER 6 X 10^8 CASE 0

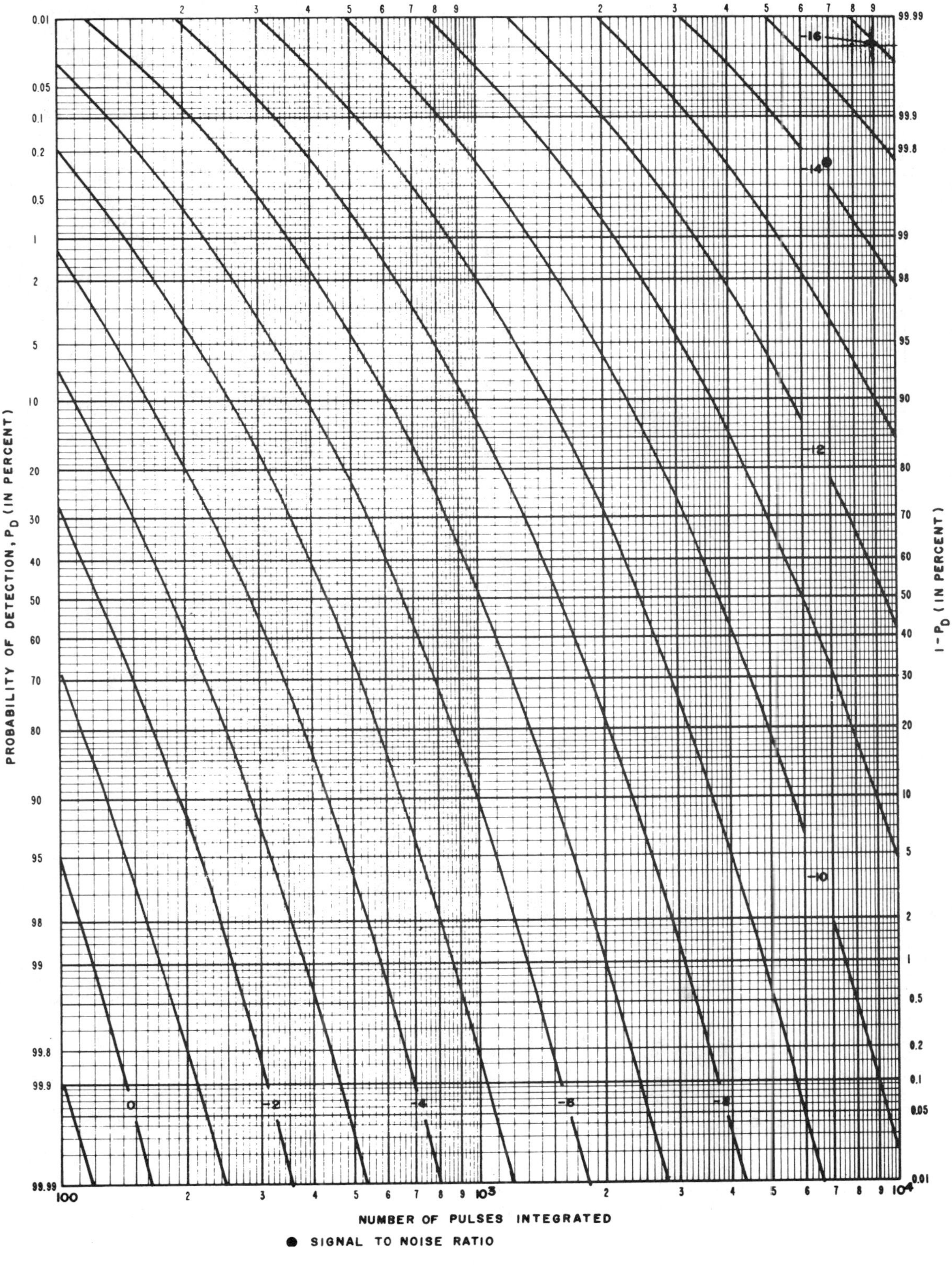

FALSE ALARM NUMBER 8 X 10^8

CASE 0

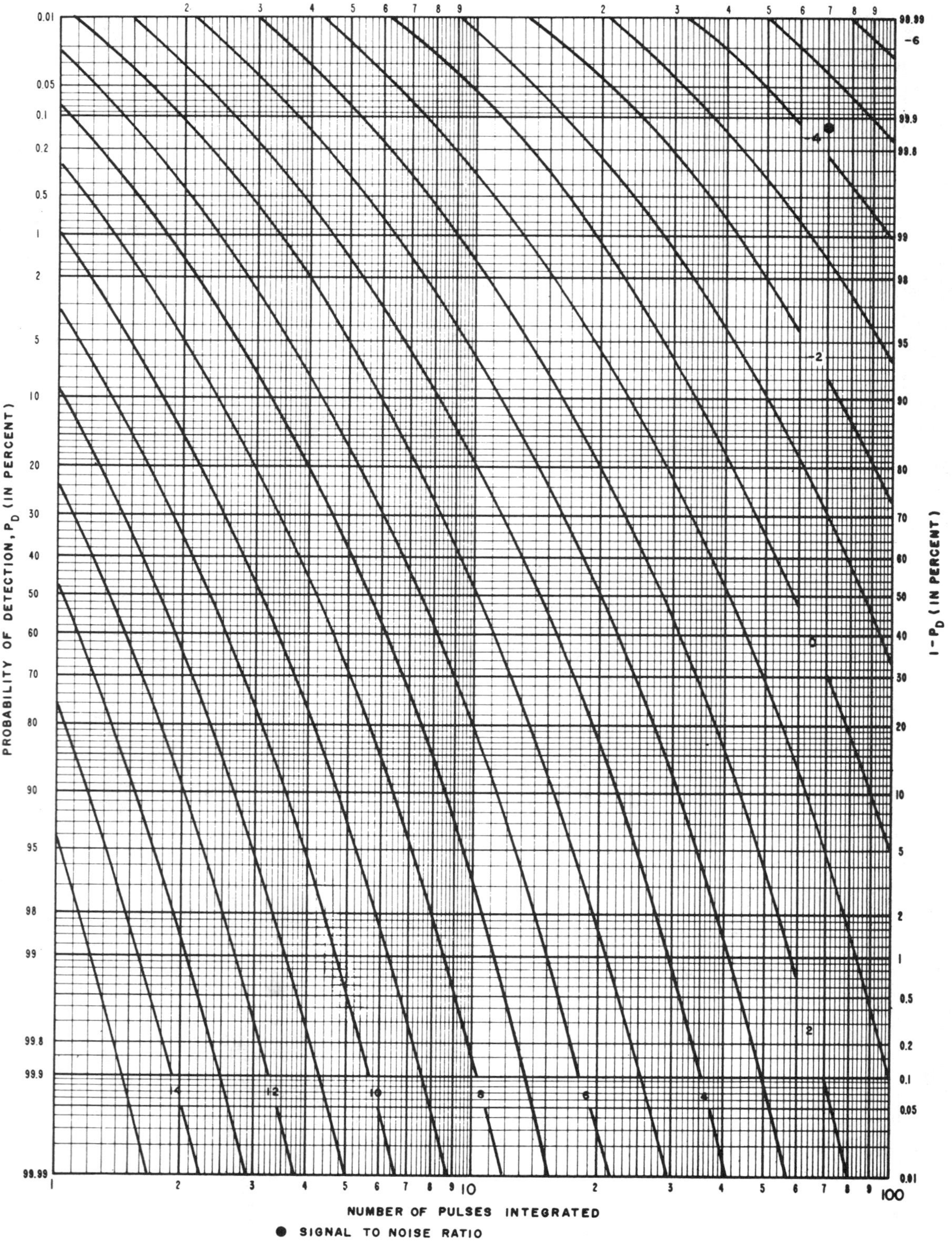

FALSE ALARM NUMBER 8 X 10⁸ CASE 0

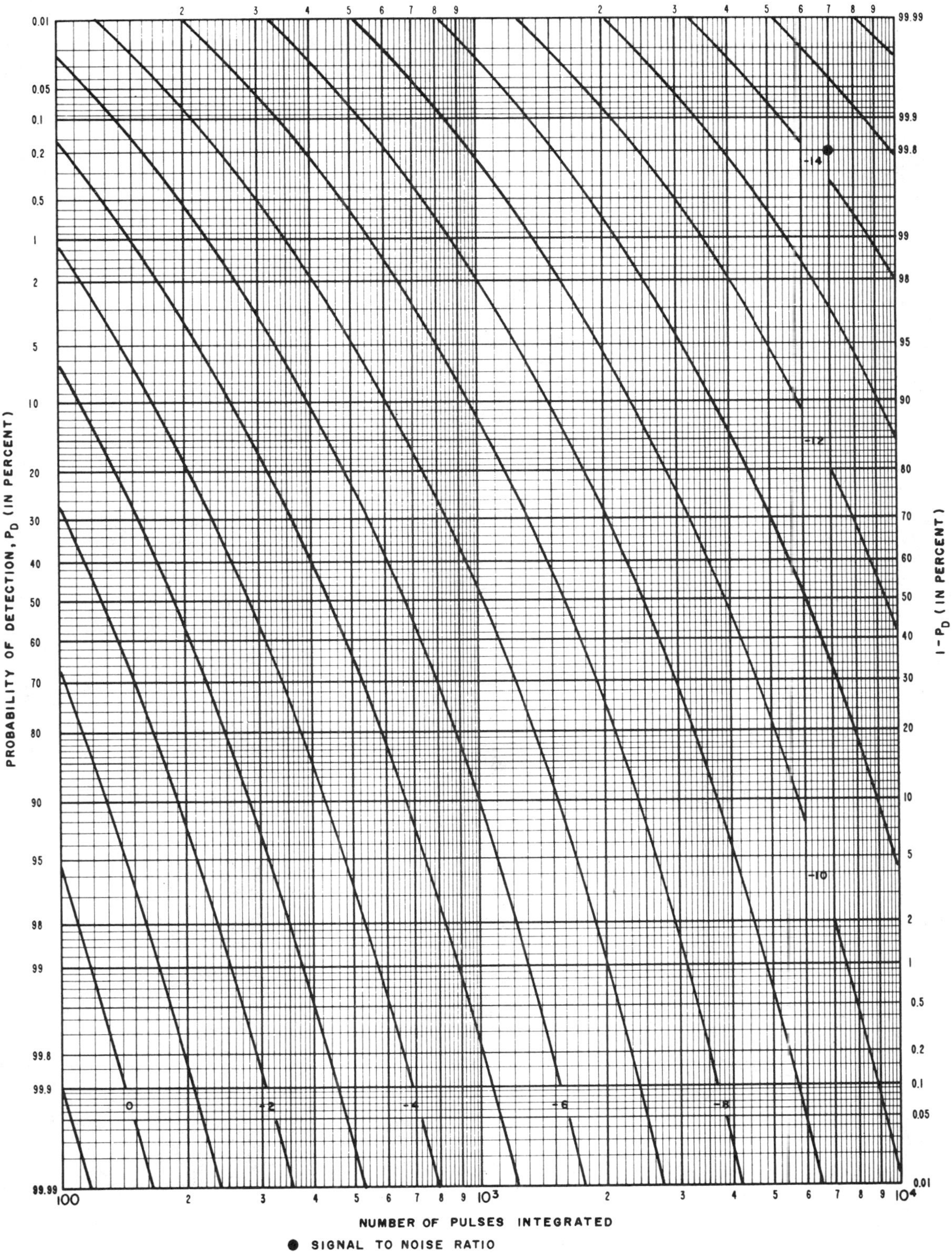

FALSE ALARM NUMBER 10^9 CASE 0

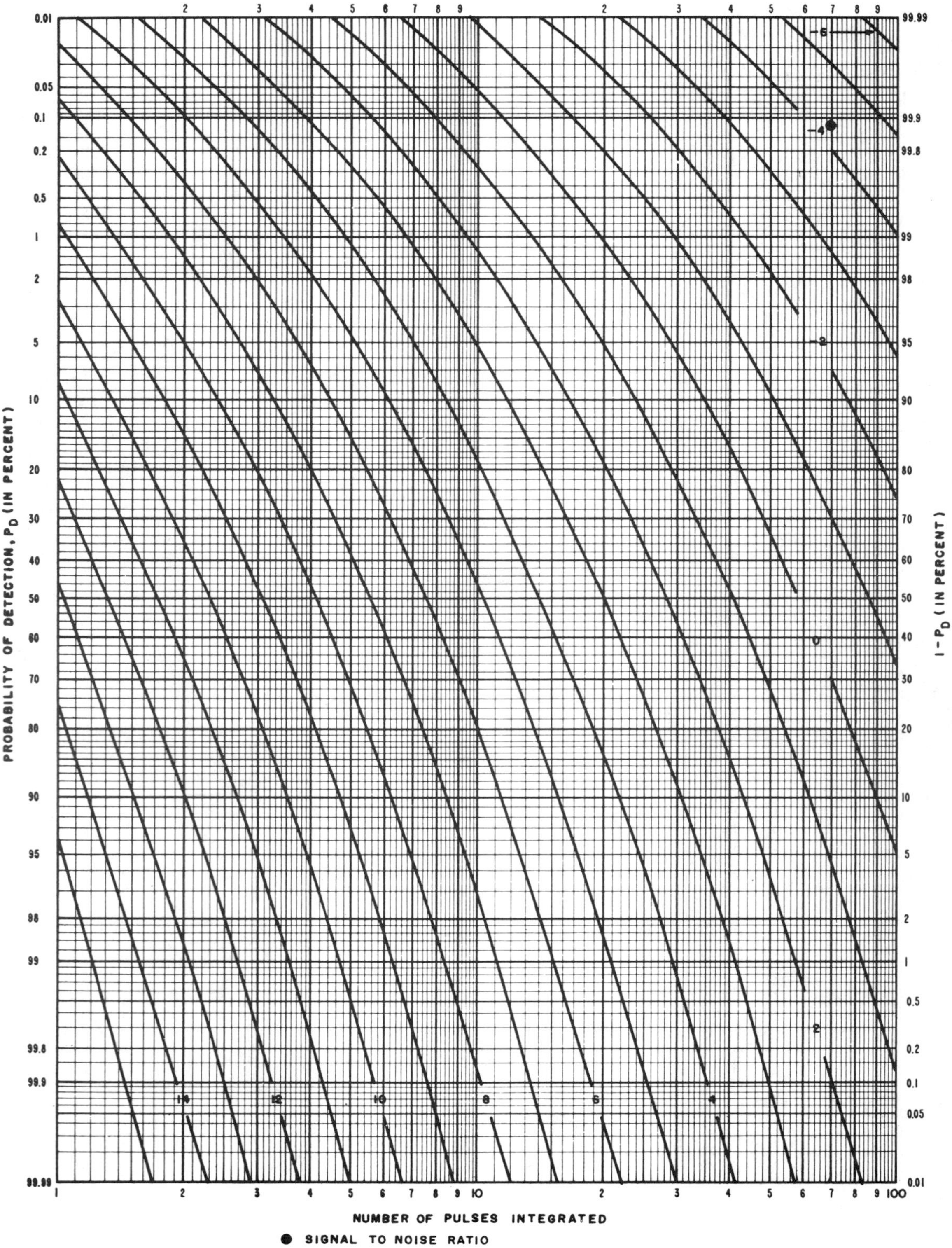

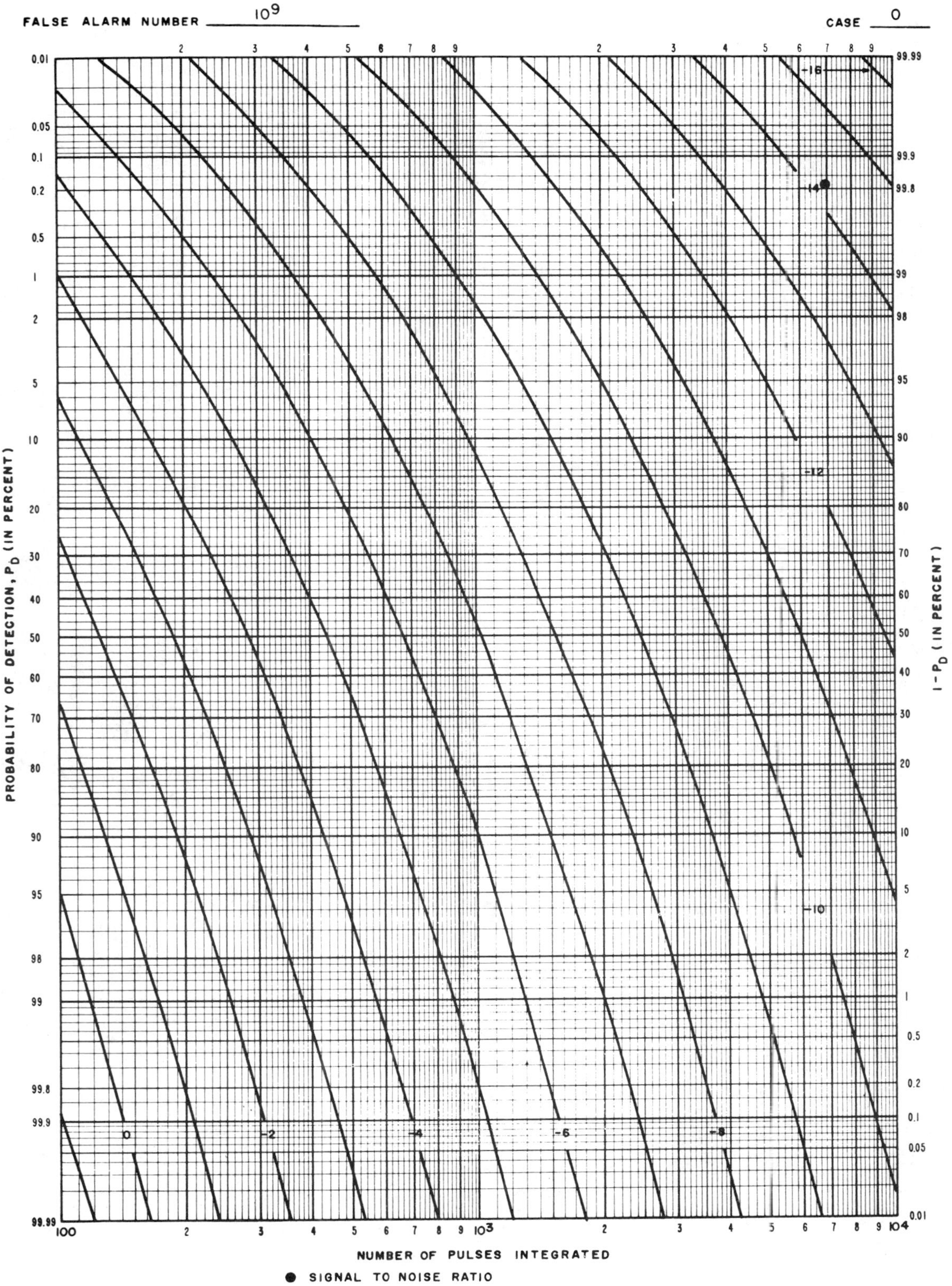
FALSE ALARM NUMBER 10^9
CASE 0
PROBABILITY OF DETECTION, P_D (IN PERCENT)
1 - P_D (IN PERCENT)
NUMBER OF PULSES INTEGRATED
● SIGNAL TO NOISE RATIO
0
-2
-4
-6
-8
-10
-12
-14
-16
100
10^3
10^4

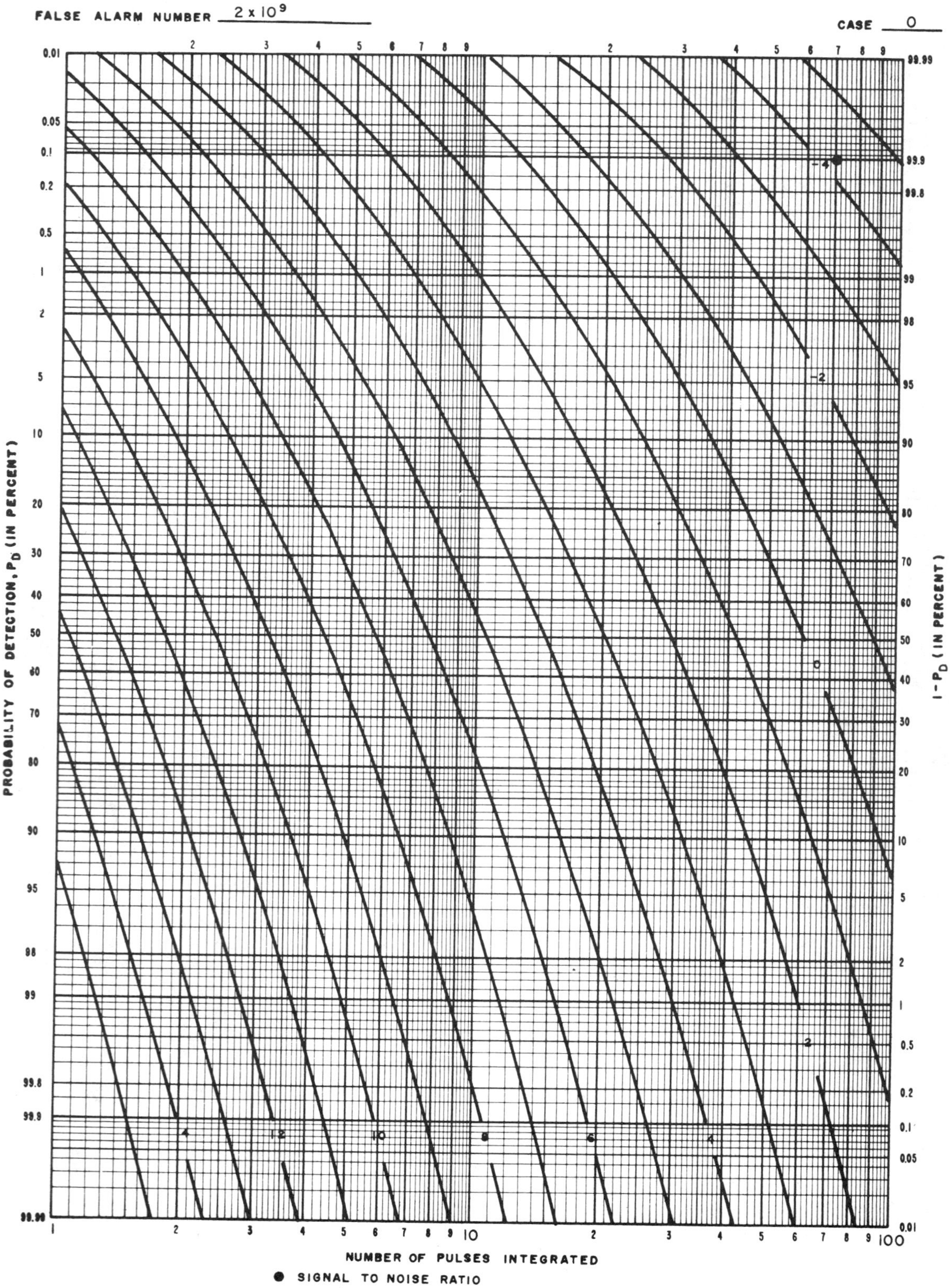
FALSE ALARM NUMBER 2 x 10^9
CASE 0
PROBABILITY OF DETECTION, P_D (IN PERCENT)
1 - P_D (IN PERCENT)
NUMBER OF PULSES INTEGRATED
● SIGNAL TO NOISE RATIO
14
12
10
8
6
4
2
0
-2
-4

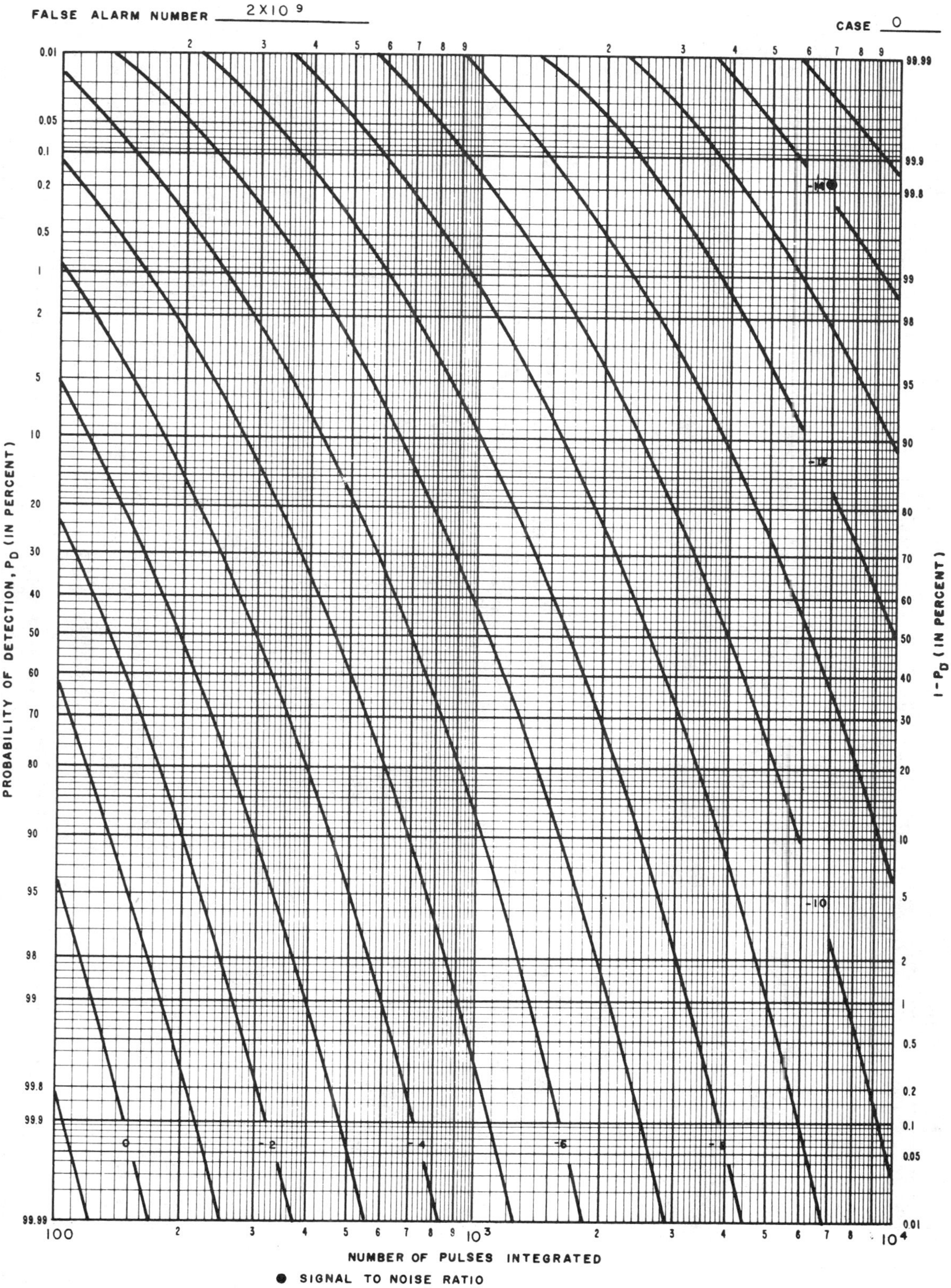
FALSE ALARM NUMBER 2X10 9
CASE 0
PROBABILITY OF DETECTION, P_D (IN PERCENT)
1-P_D (IN PERCENT)
0
-2
-4
-6
-8
-10
-12
-14
100
10 3
10 4
NUMBER OF PULSES INTEGRATED
● SIGNAL TO NOISE RATIO

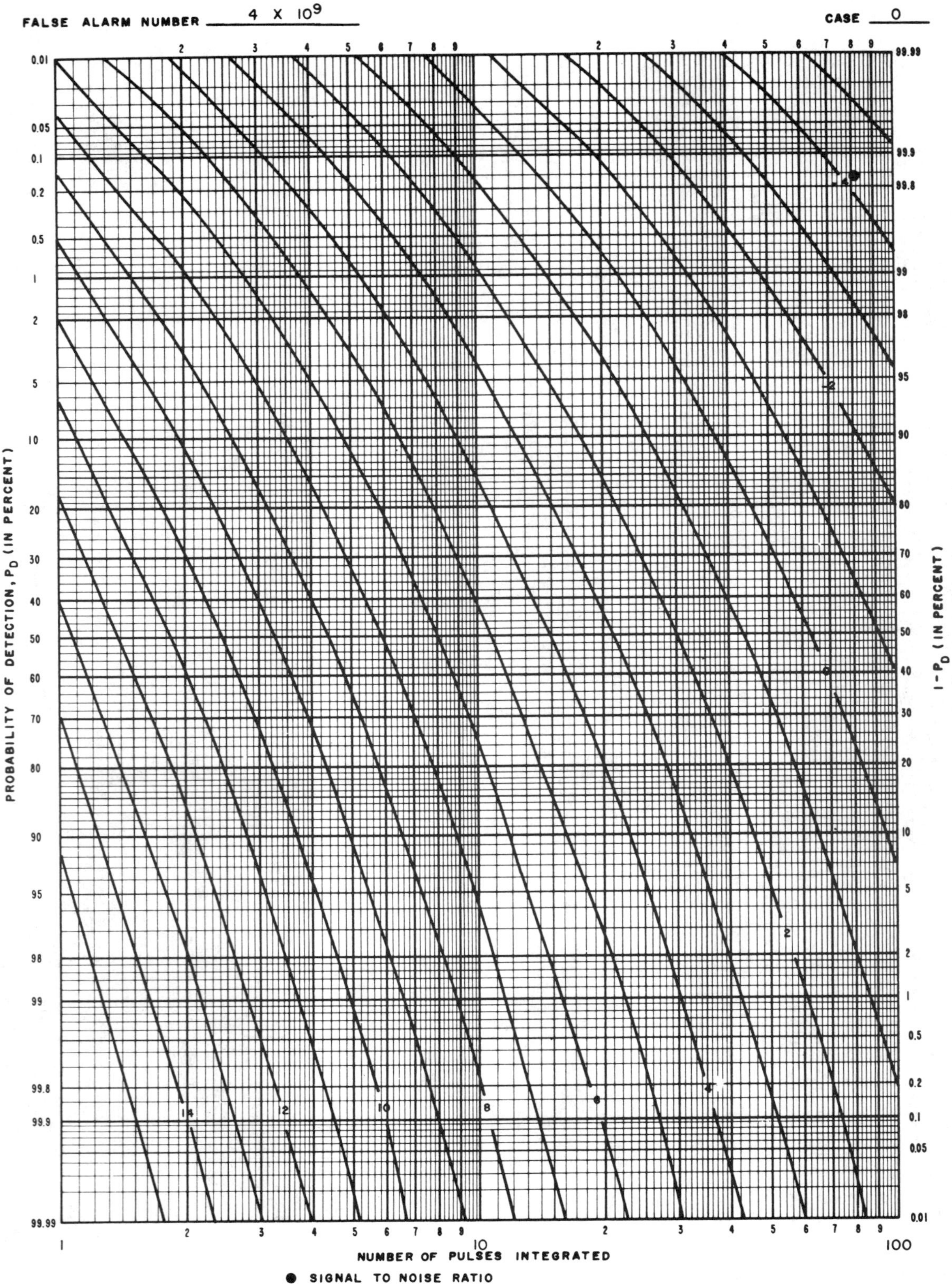
FALSE ALARM NUMBER 4 X 10^9
CASE 0
PROBABILITY OF DETECTION, P_D (IN PERCENT)
$1-P_D$ (IN PERCENT)
NUMBER OF PULSES INTEGRATED
● SIGNAL TO NOISE RATIO
14
12
10
8
6
4
2
0
-2
-4

FALSE ALARM NUMBER 4×10^9

CASE 0

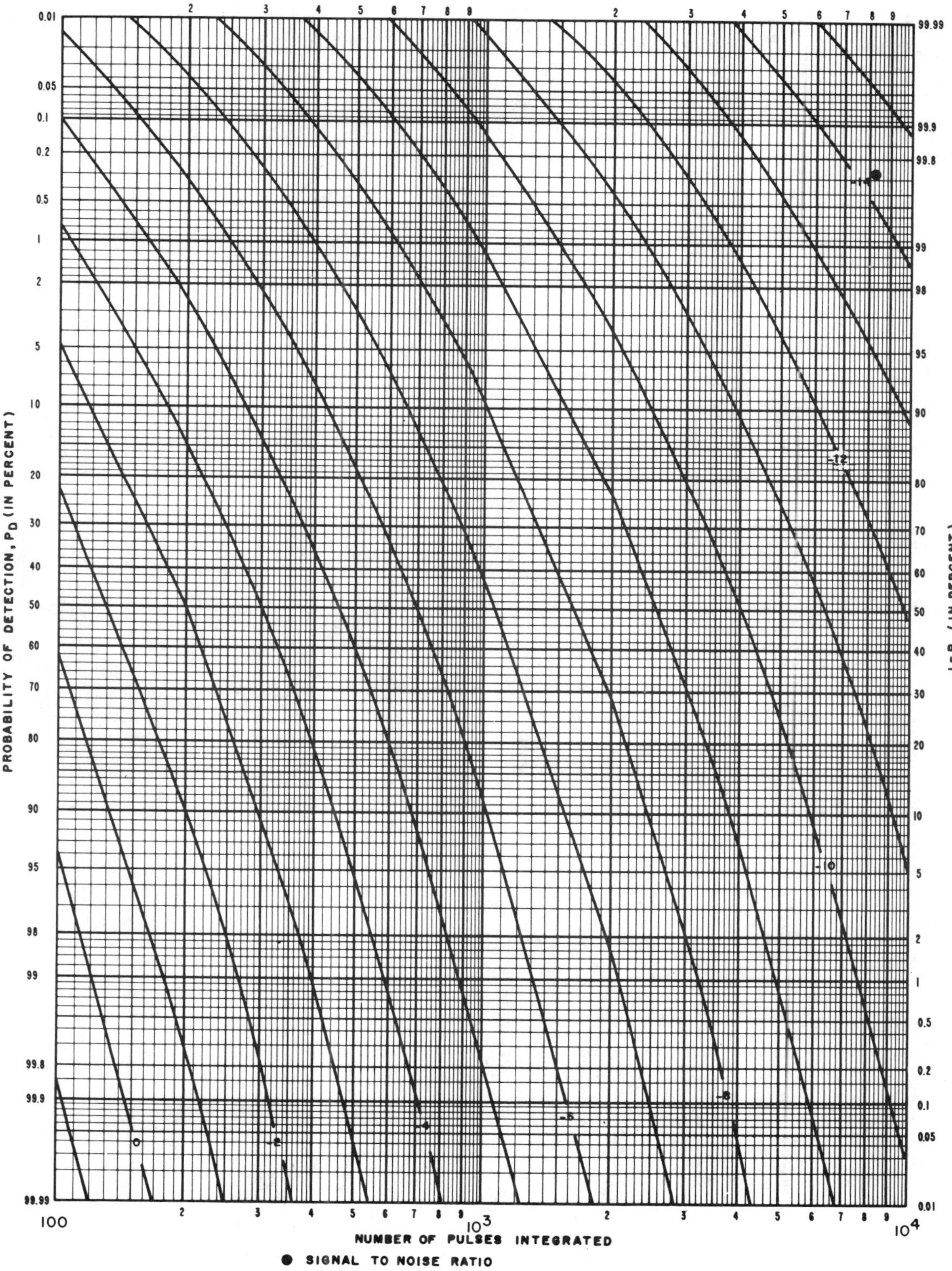

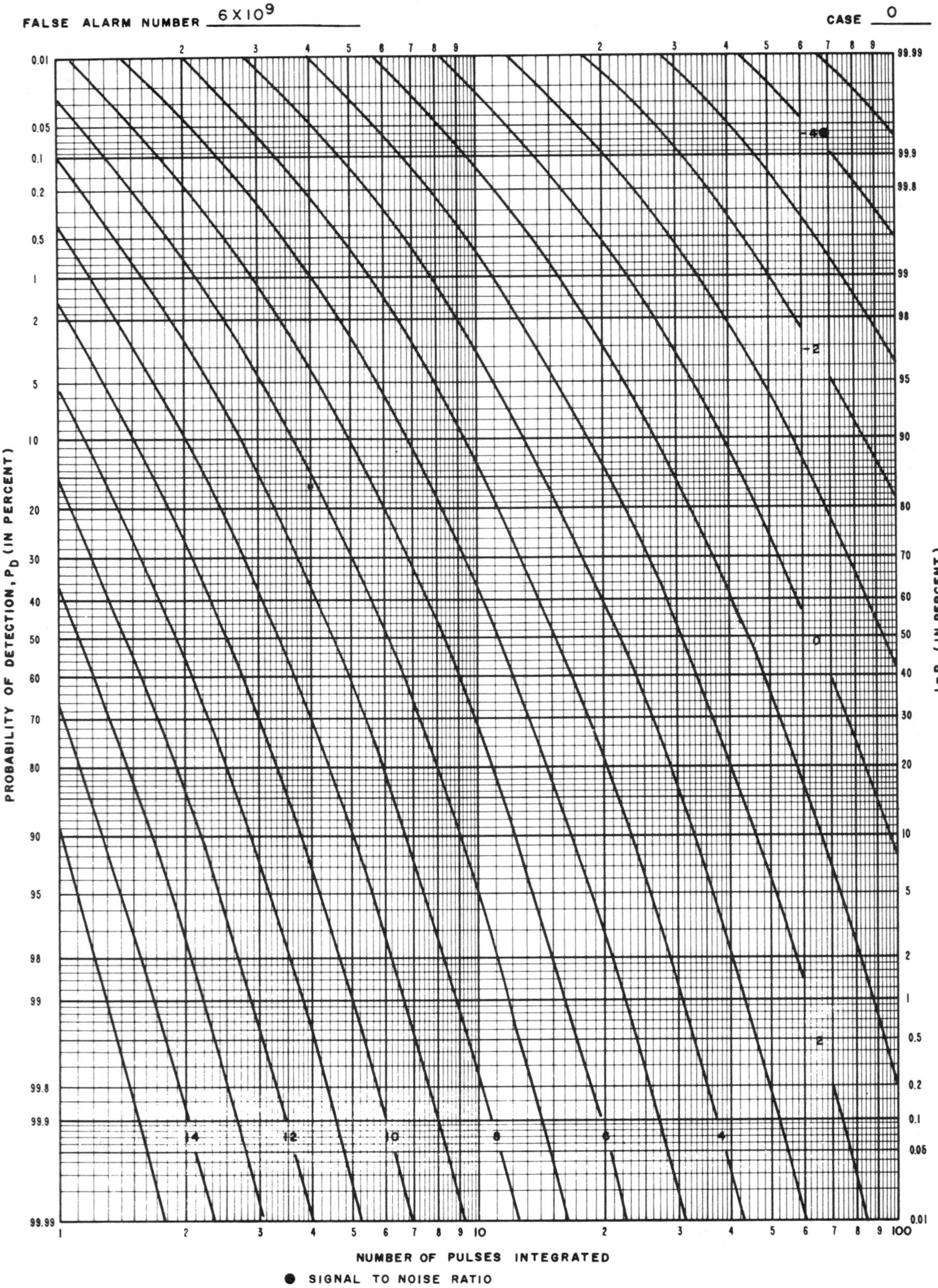
FALSE ALARM NUMBER 6X10^9
CASE 0
PROBABILITY OF DETECTION, P_D (IN PERCENT)
1 - P_D (IN PERCENT)
NUMBER OF PULSES INTEGRATED
● SIGNAL TO NOISE RATIO
14
12
10
8
6
4
2
0
-2
-4

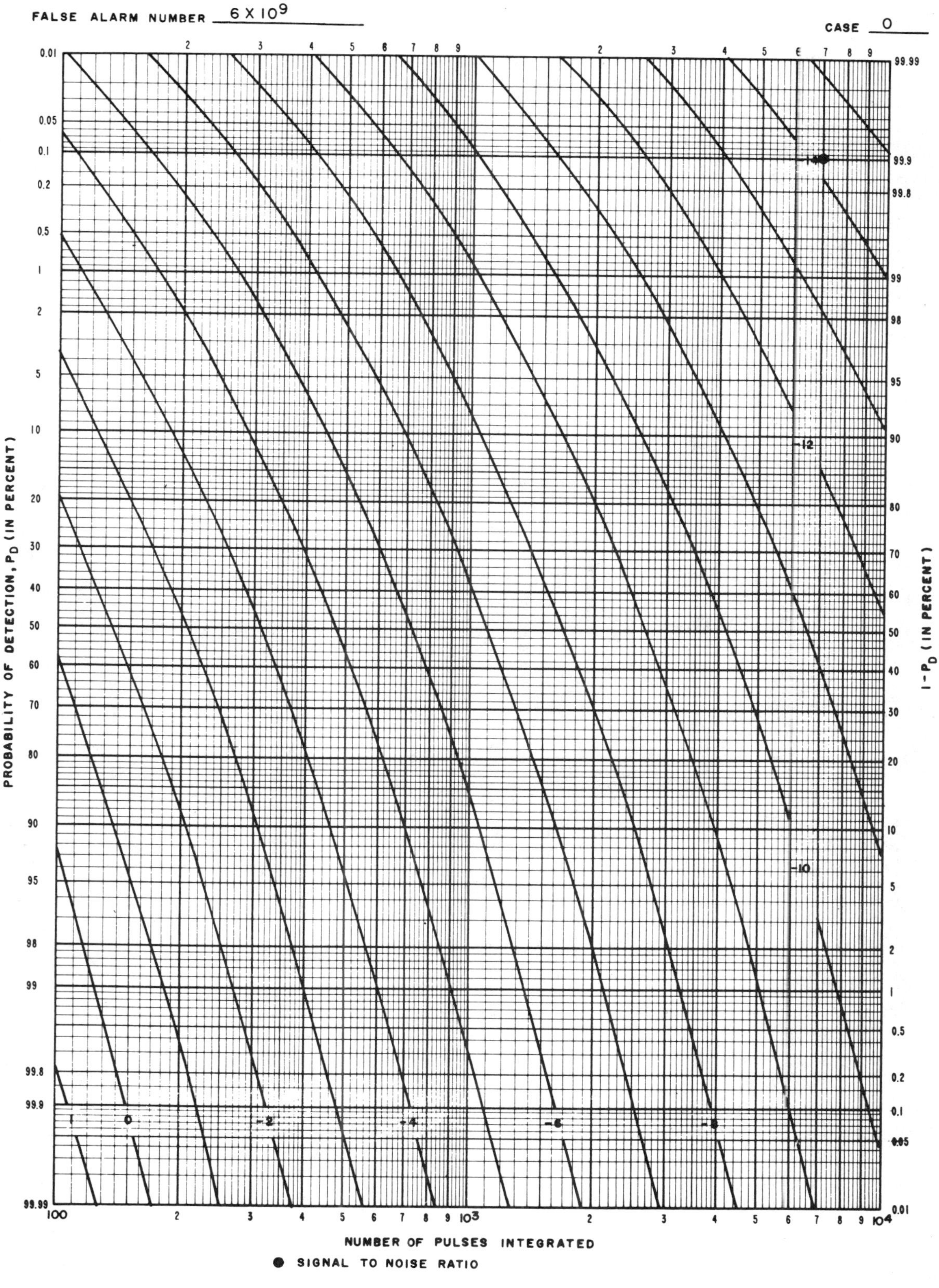
FALSE ALARM NUMBER 6 X 10^9
CASE 0
PROBABILITY OF DETECTION, P_D (IN PERCENT)
$1-P_D$ (IN PERCENT)
0
-2
-4
-6
-8
-10
-12
-14
NUMBER OF PULSES INTEGRATED
● SIGNAL TO NOISE RATIO

FALSE ALARM NUMBER 8 x 10^9 CASE 0

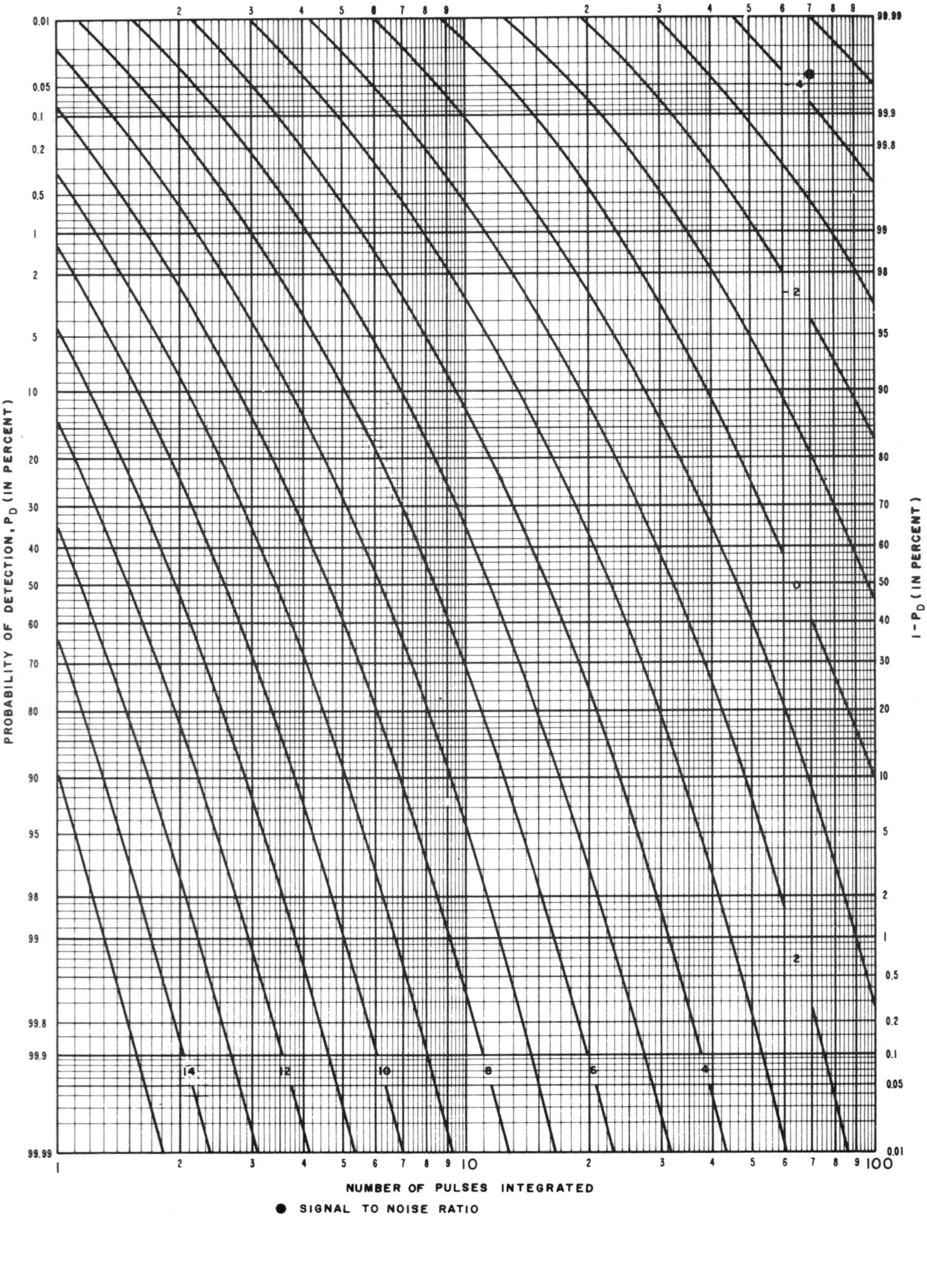

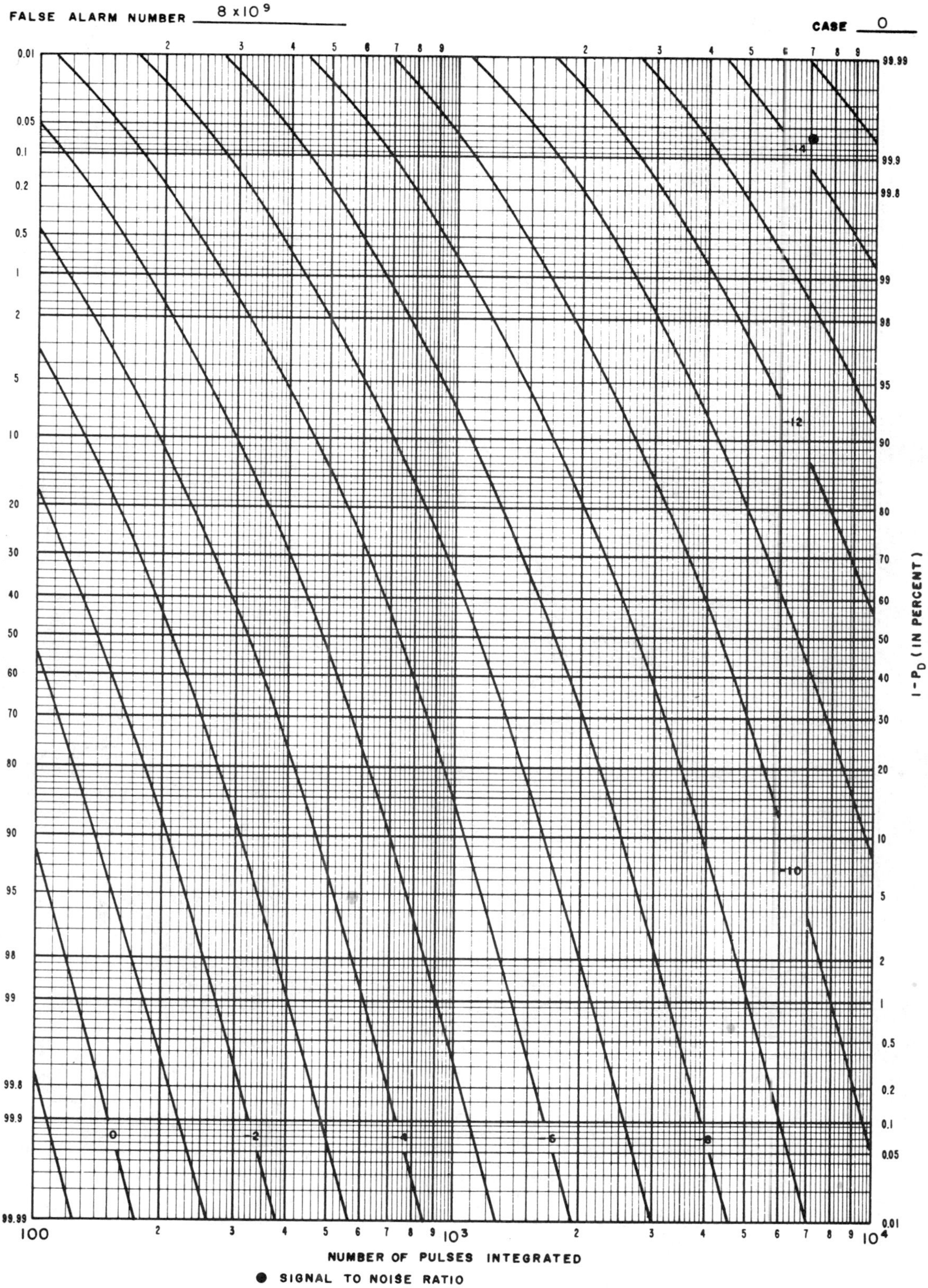
FALSE ALARM NUMBER 8 x 10⁹
CASE 0
PROBABILITY OF DETECTION, P_D (IN PERCENT)
1 - P_D (IN PERCENT)
0
-2
-4
-6
-8
-10
-12
-14
NUMBER OF PULSES INTEGRATED
● SIGNAL TO NOISE RATIO

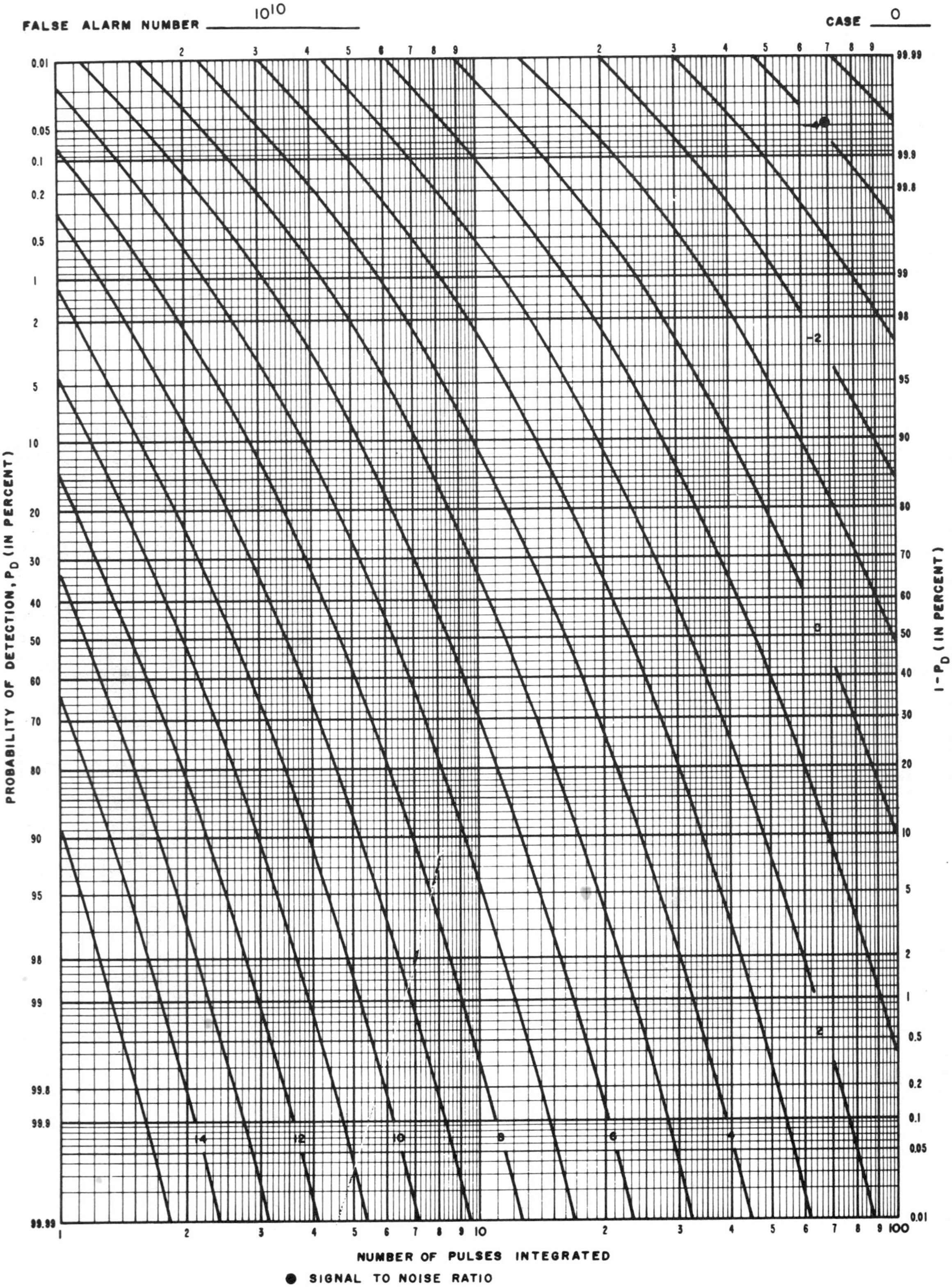
FALSE ALARM NUMBER 10^{10}
CASE 0
PROBABILITY OF DETECTION, P_D (IN PERCENT)
$1-P_D$ (IN PERCENT)
NUMBER OF PULSES INTEGRATED
● SIGNAL TO NOISE RATIO
14
12
10
8
6
4
2
0
-2
-4

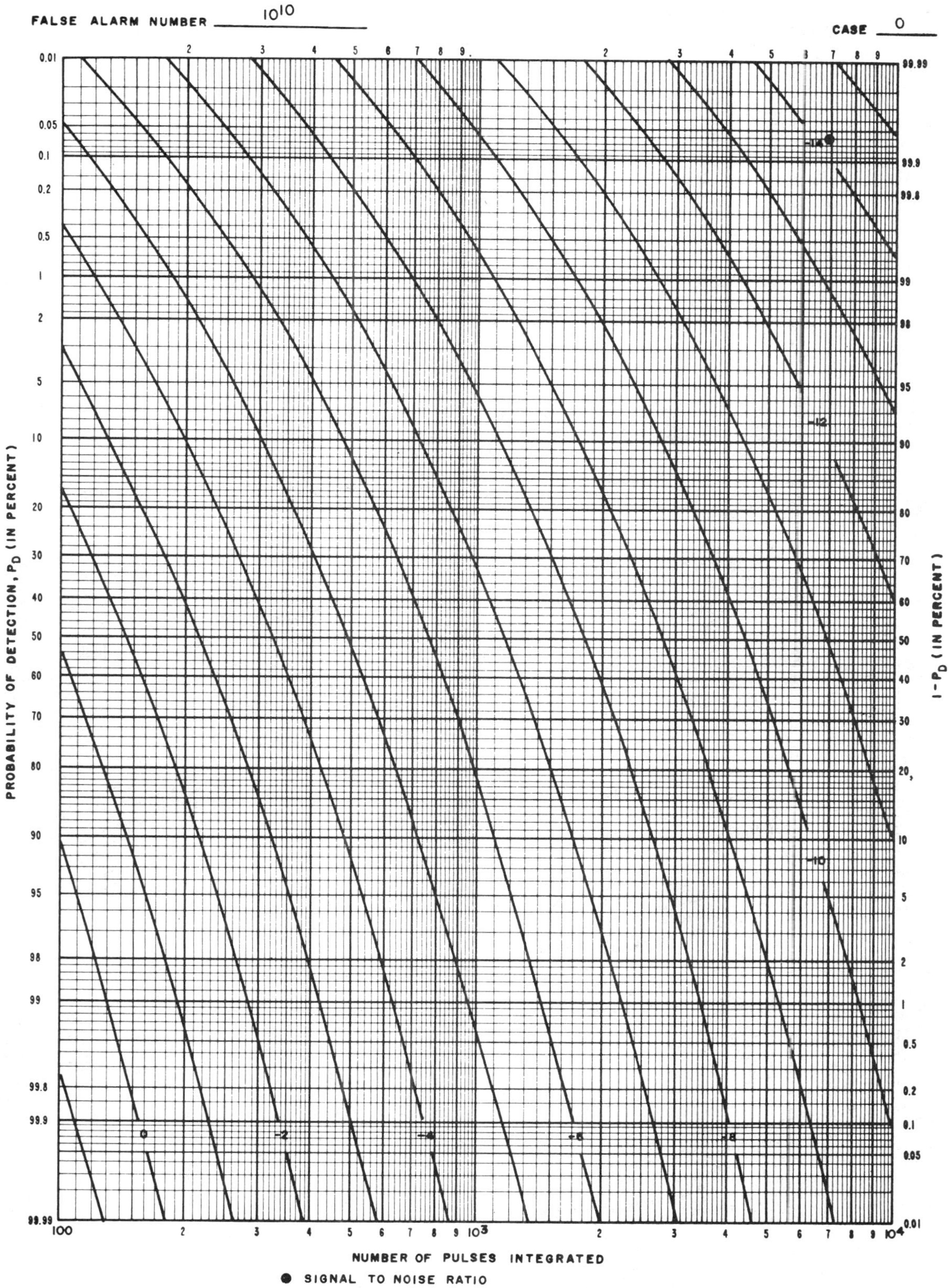
FALSE ALARM NUMBER 10^{10}
CASE 0
PROBABILITY OF DETECTION, P_D (IN PERCENT)
$1-P_D$ (IN PERCENT)
0
-2
-4
-6
-8
-10
-12
-14
NUMBER OF PULSES INTEGRATED
● SIGNAL TO NOISE RATIO

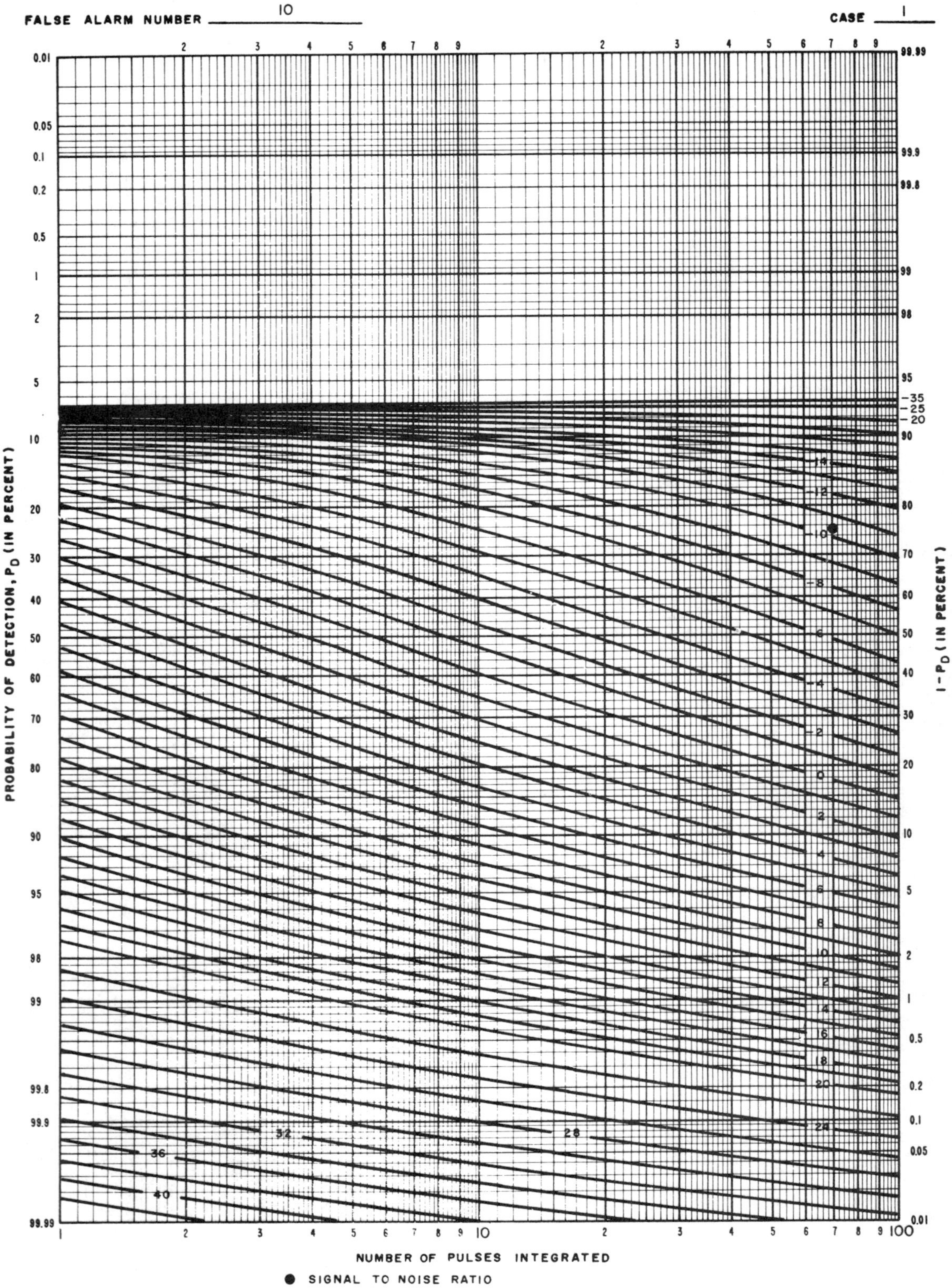
FALSE ALARM NUMBER 10
CASE 1
PROBABILITY OF DETECTION, P_D (IN PERCENT)
$1-P_D$ (IN PERCENT)
NUMBER OF PULSES INTEGRATED
● SIGNAL TO NOISE RATIO

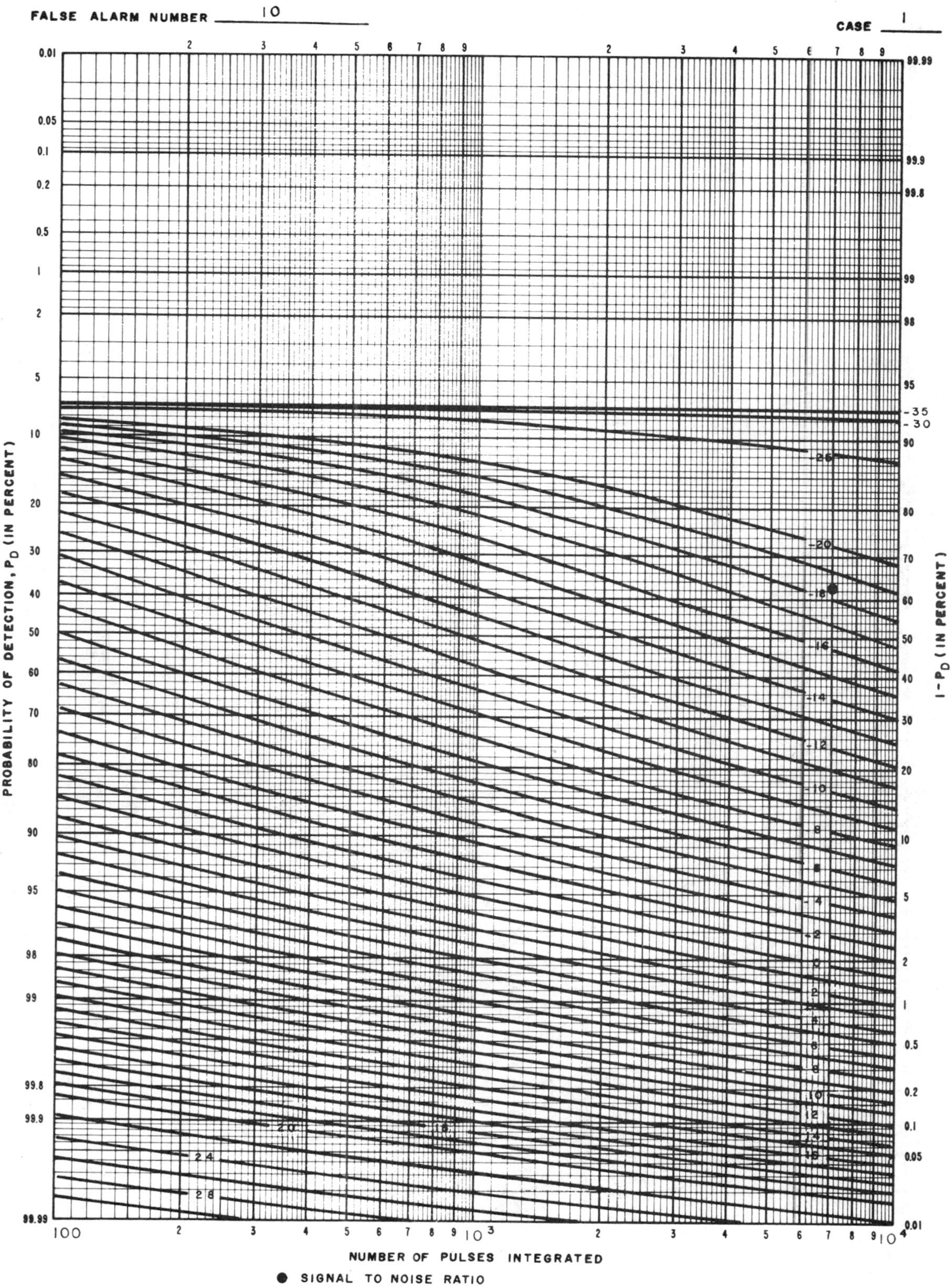
FALSE ALARM NUMBER 10
CASE 1
PROBABILITY OF DETECTION, P_D (IN PERCENT)
$1-P_D$ (IN PERCENT)
NUMBER OF PULSES INTEGRATED
● SIGNAL TO NOISE RATIO
0.01
0.05
0.1
0.2
0.5
1
2
5
10
20
30
40
50
60
70
80
90
95
98
99
99.8
99.9
99.99
100
10^3
10^4
-35
-30
-25
-20
-18
-16
-14
-12
-10
-8
-6
-4
-2
0
2
4
6
8
10
12
14
16
18
20
24
28

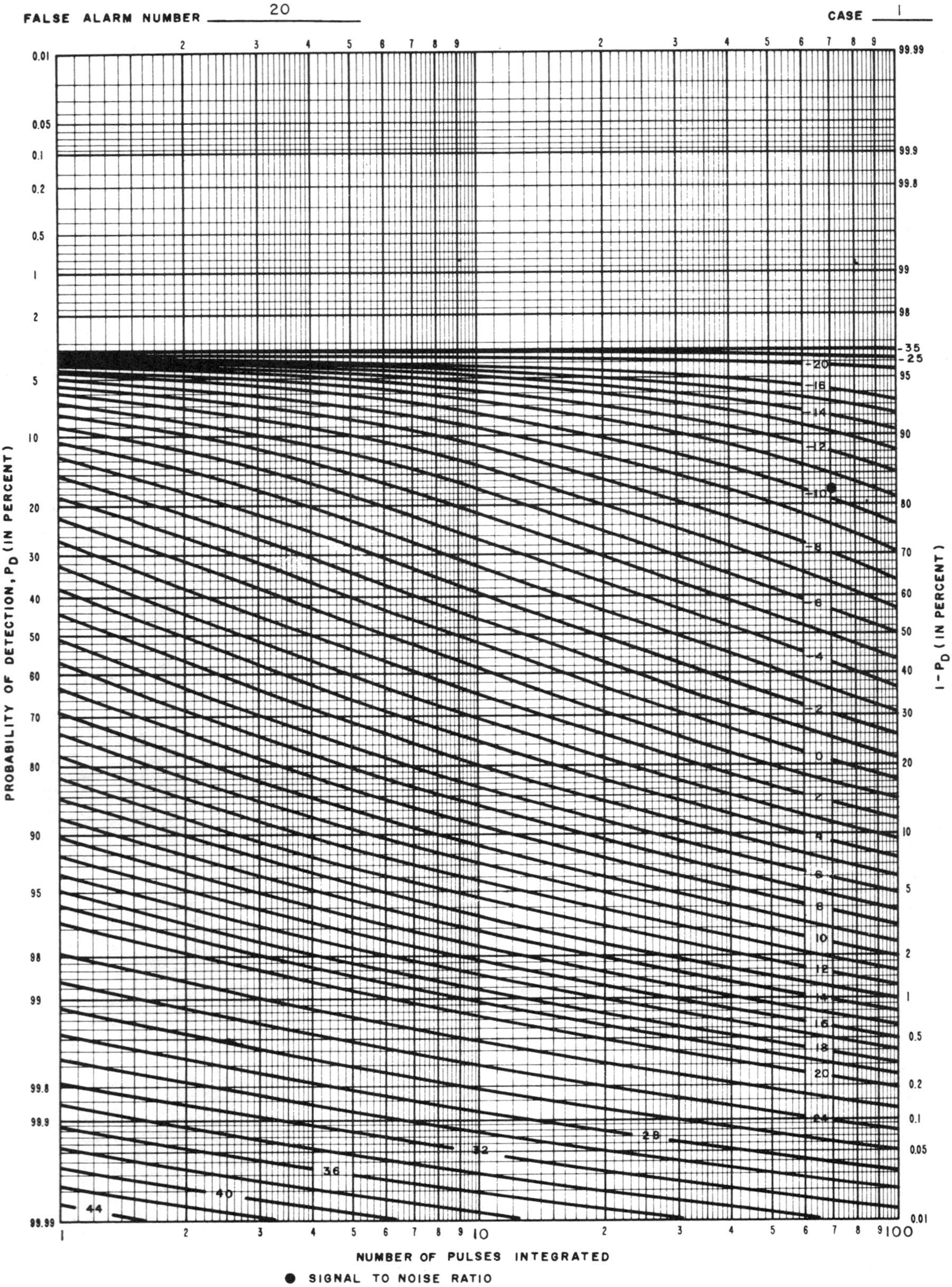
FALSE ALARM NUMBER 20
CASE I
PROBABILITY OF DETECTION, P_D (IN PERCENT)
$1 - P_D$ (IN PERCENT)
NUMBER OF PULSES INTEGRATED
● SIGNAL TO NOISE RATIO

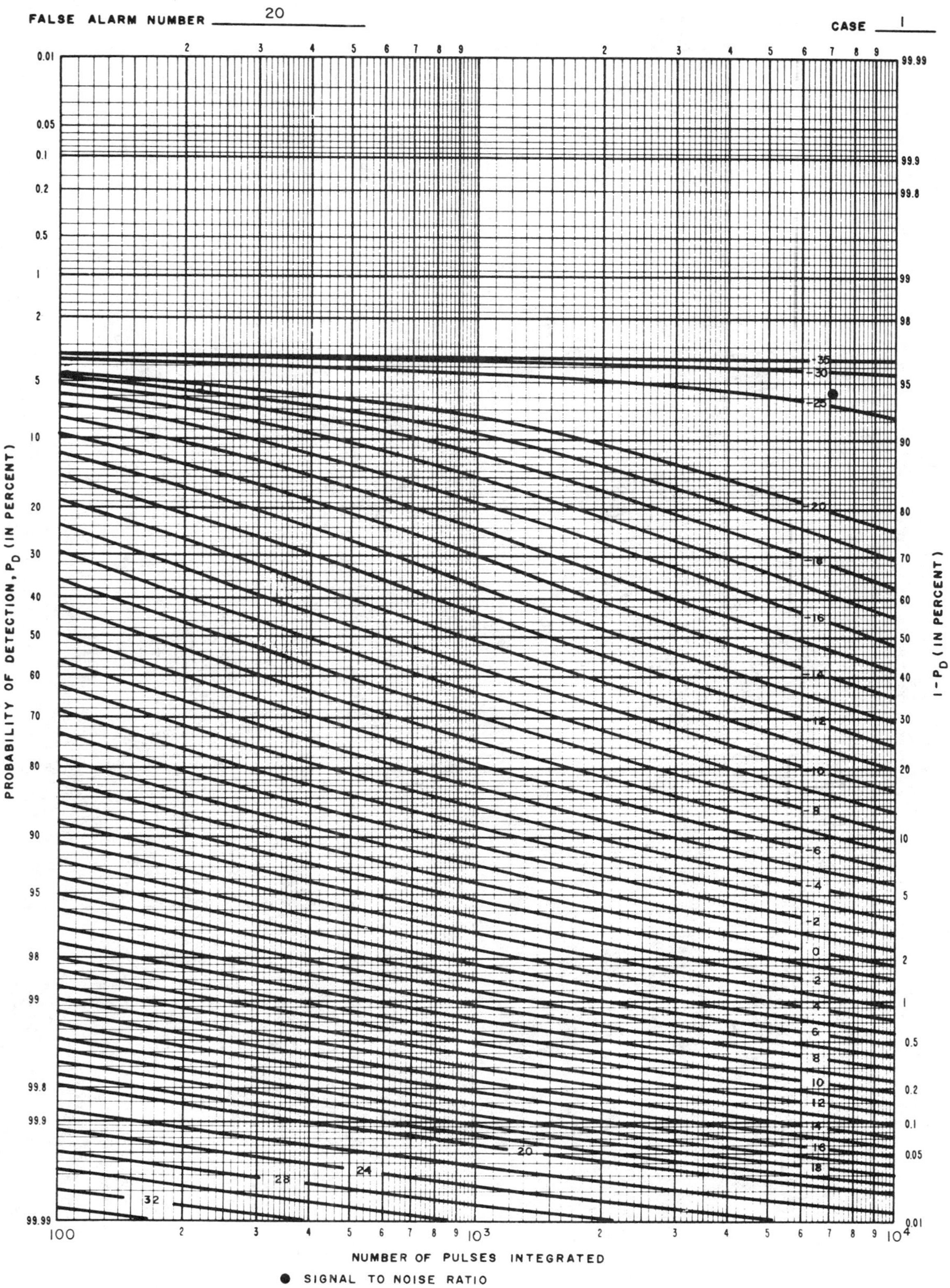
FALSE ALARM NUMBER 20
CASE I
PROBABILITY OF DETECTION, P_D (IN PERCENT)
$1-P_D$ (IN PERCENT)
NUMBER OF PULSES INTEGRATED
● SIGNAL TO NOISE RATIO

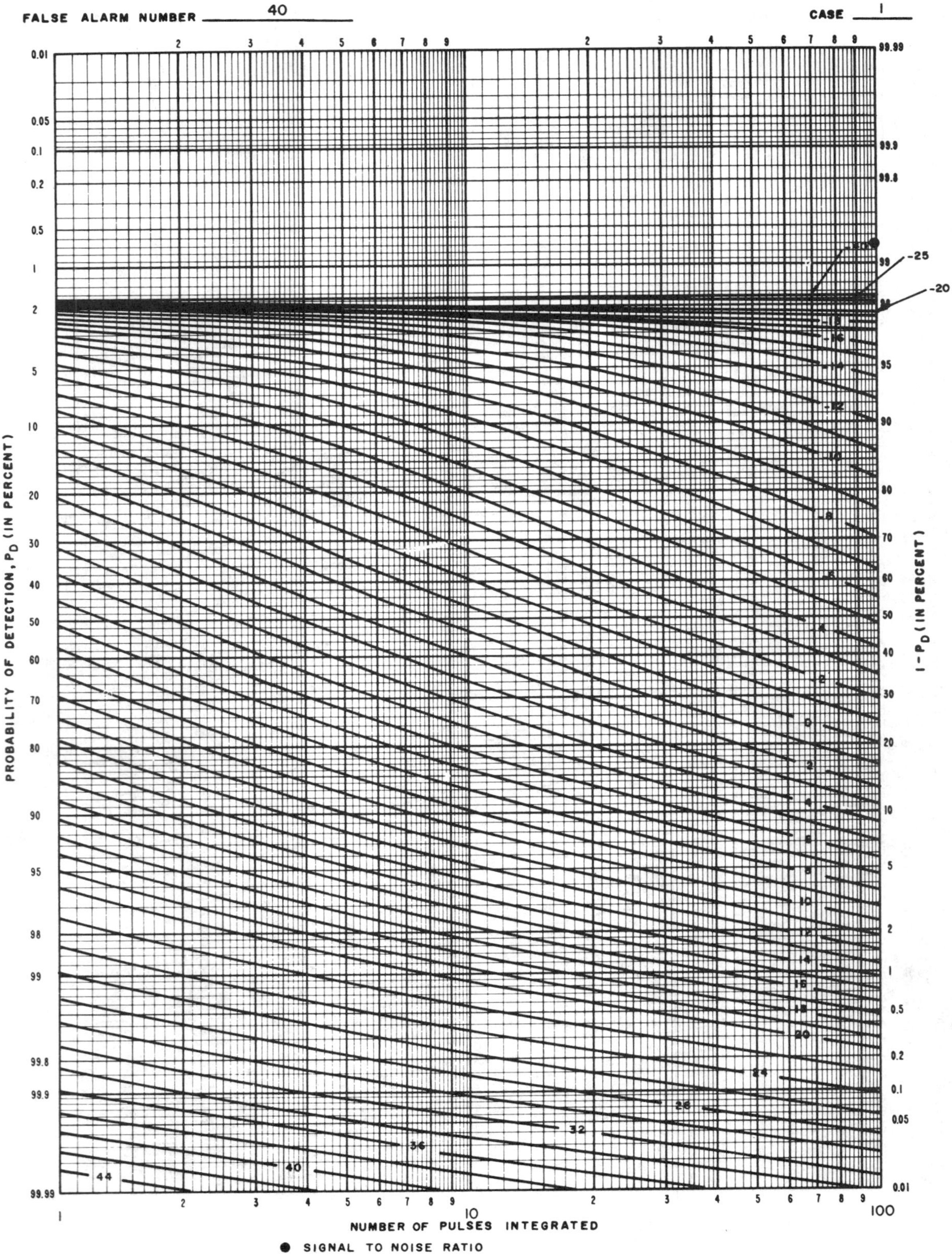
FALSE ALARM NUMBER 40
CASE 1
PROBABILITY OF DETECTION, P_D (IN PERCENT)
$1-P_D$ (IN PERCENT)
NUMBER OF PULSES INTEGRATED
● SIGNAL TO NOISE RATIO

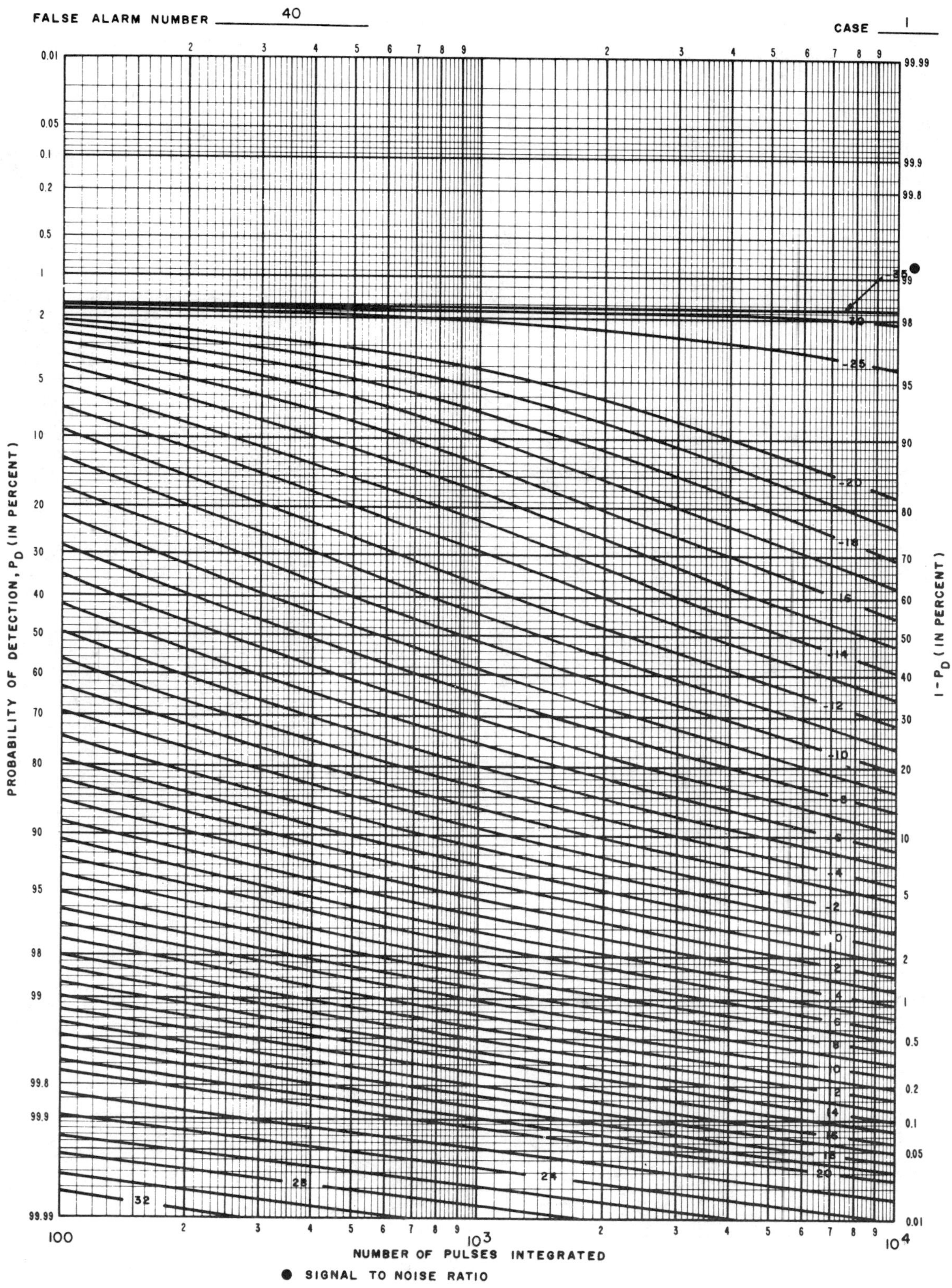
FALSE ALARM NUMBER 40
CASE 1
PROBABILITY OF DETECTION, P_D (IN PERCENT)
1 - P_D (IN PERCENT)
NUMBER OF PULSES INTEGRATED
● SIGNAL TO NOISE RATIO

FALSE ALARM NUMBER 60

CASE 1

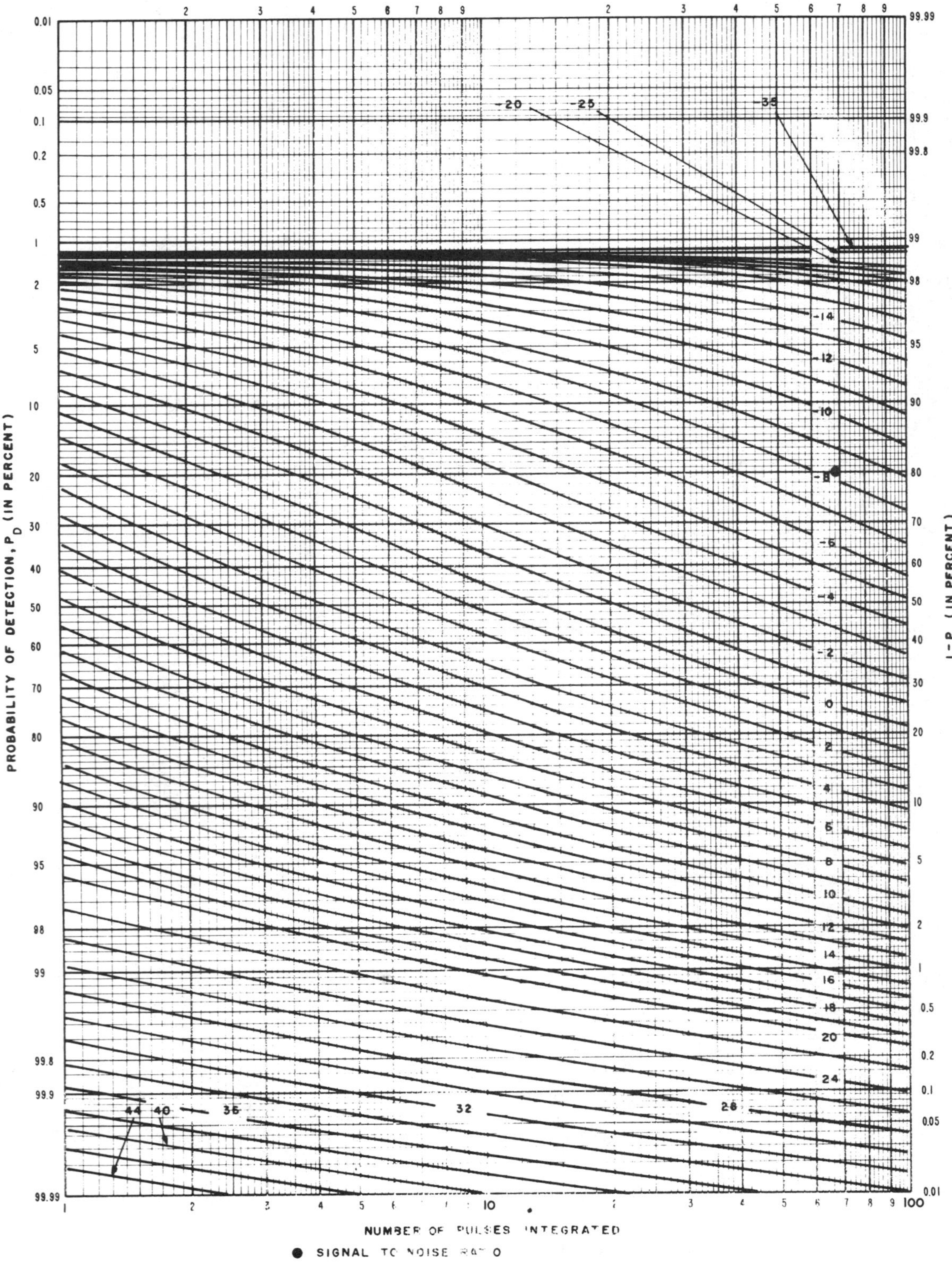

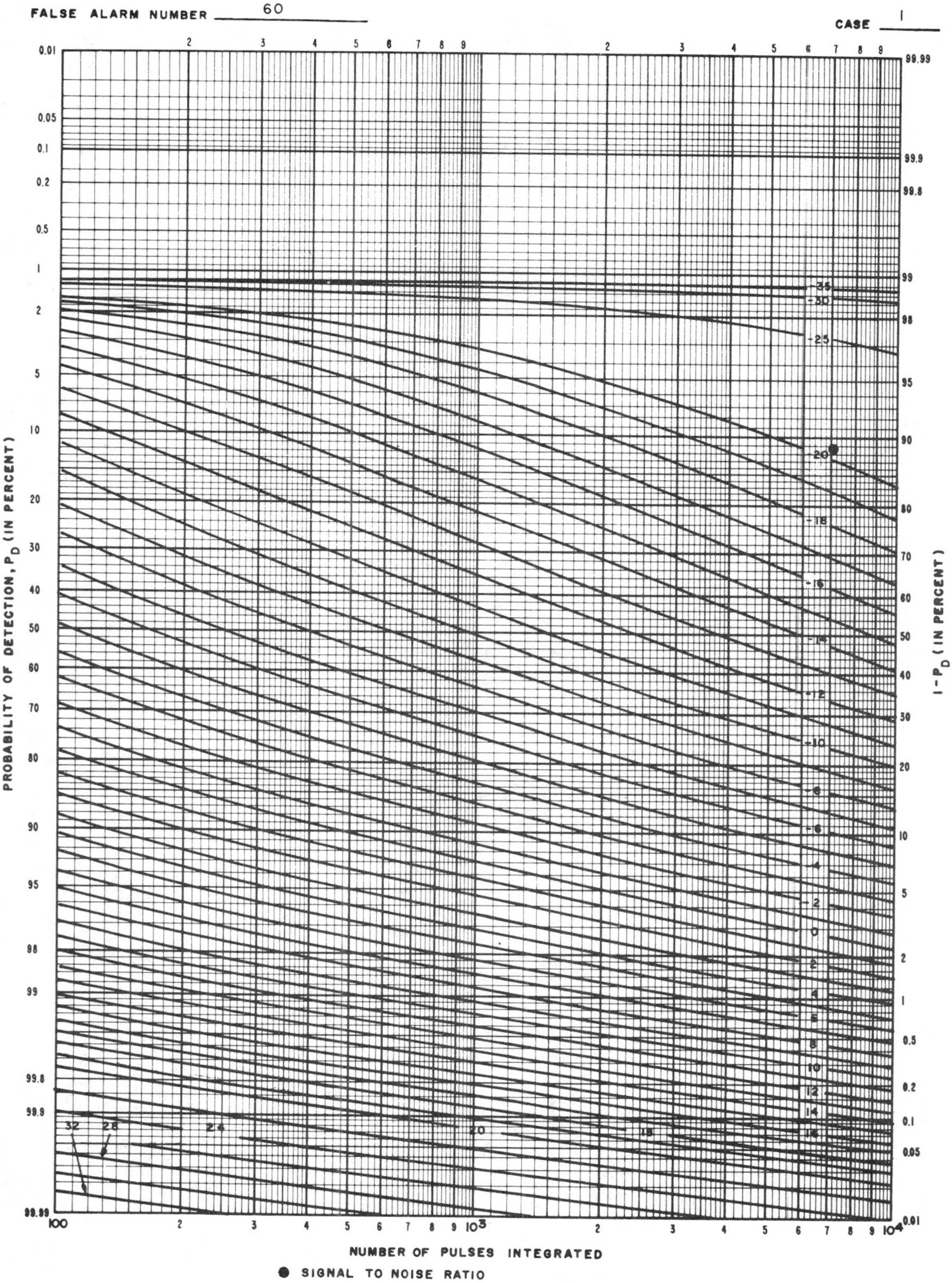
FALSE ALARM NUMBER 60
CASE 1
PROBABILITY OF DETECTION, P_D (IN PERCENT)
$1-P_D$ (IN PERCENT)
NUMBER OF PULSES INTEGRATED
● SIGNAL TO NOISE RATIO

FALSE ALARM NUMBER 80 CASE 1

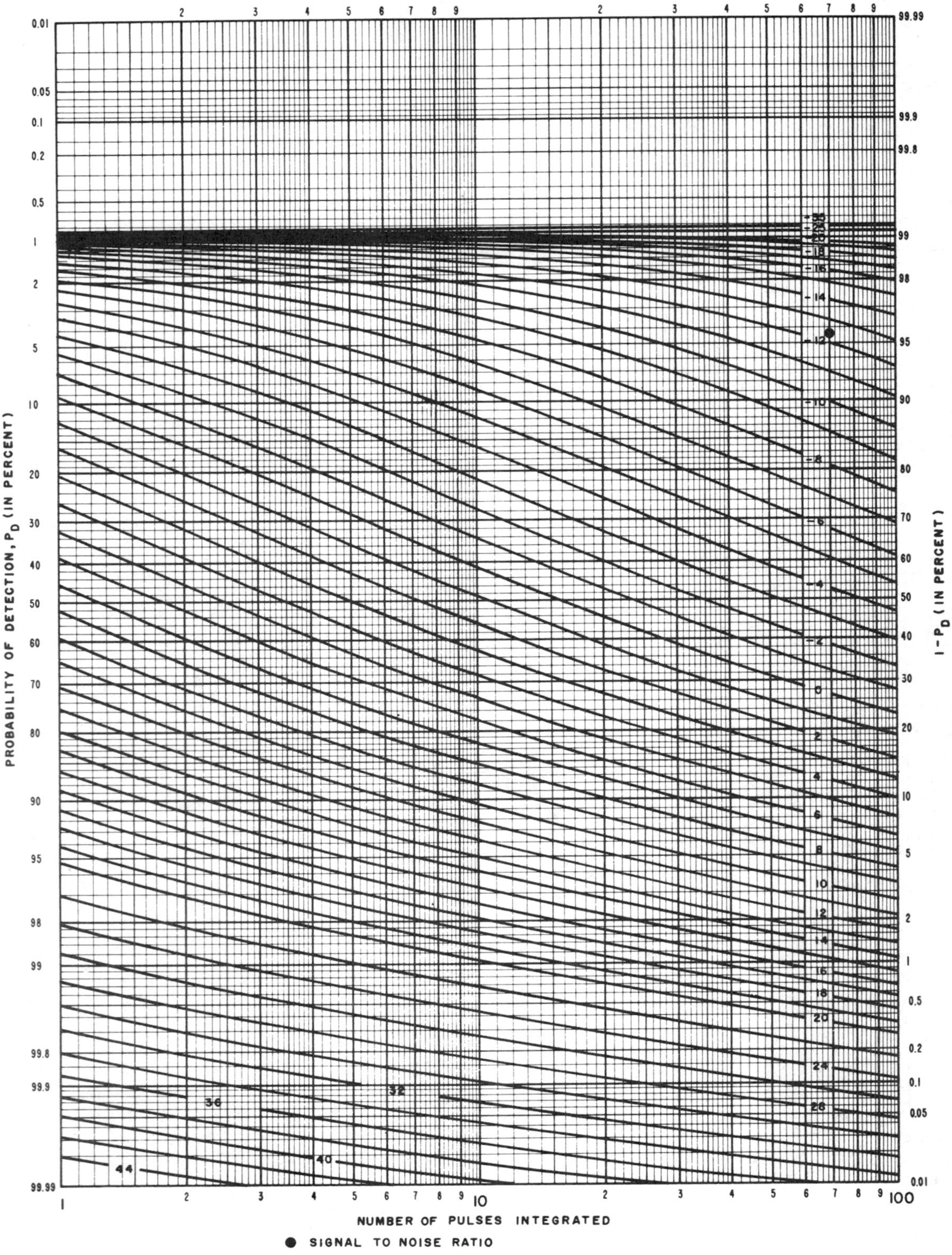

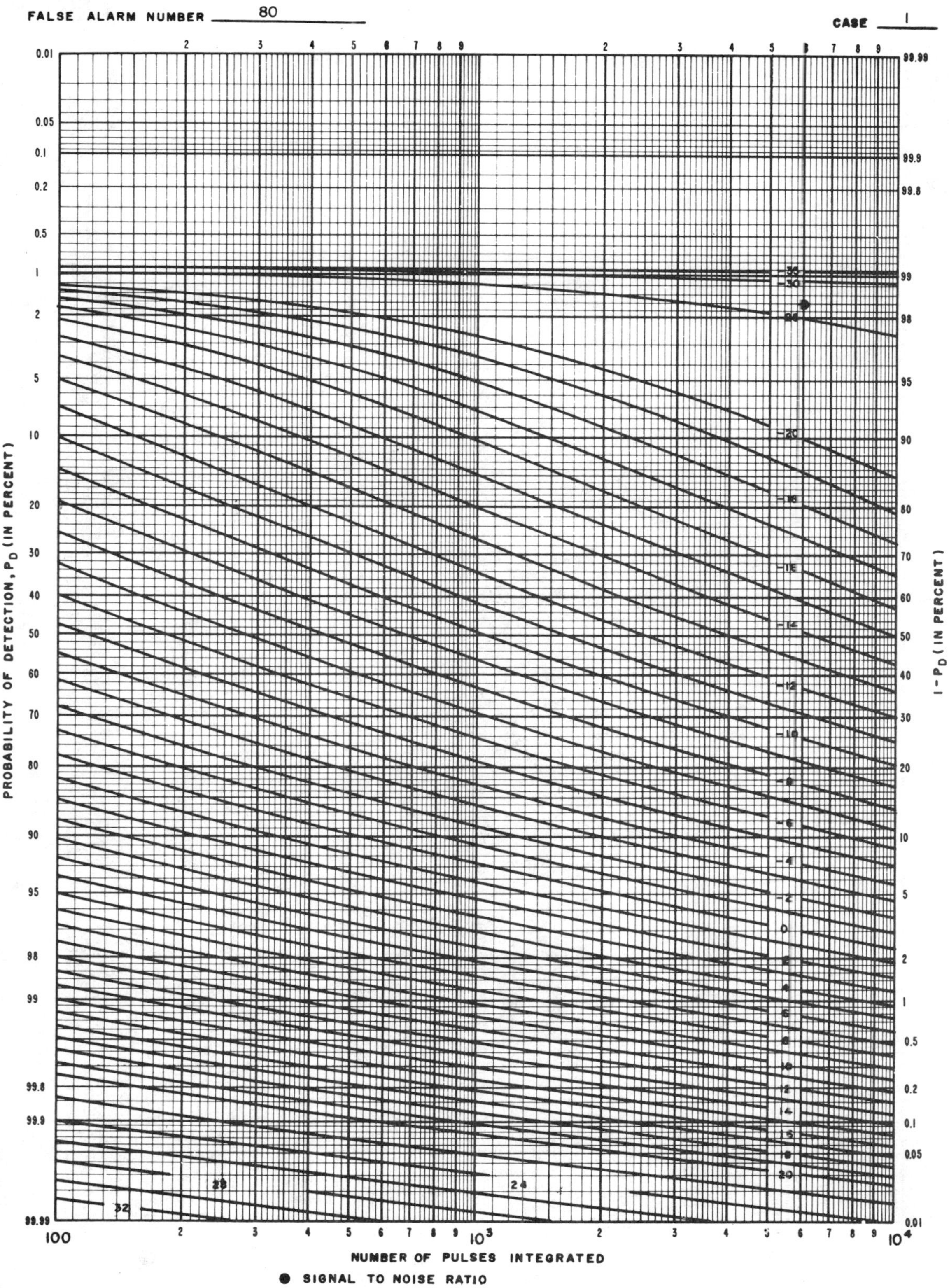
FALSE ALARM NUMBER 80
CASE 1
PROBABILITY OF DETECTION, P_D (IN PERCENT)
1 - P_D (IN PERCENT)
NUMBER OF PULSES INTEGRATED
● SIGNAL TO NOISE RATIO

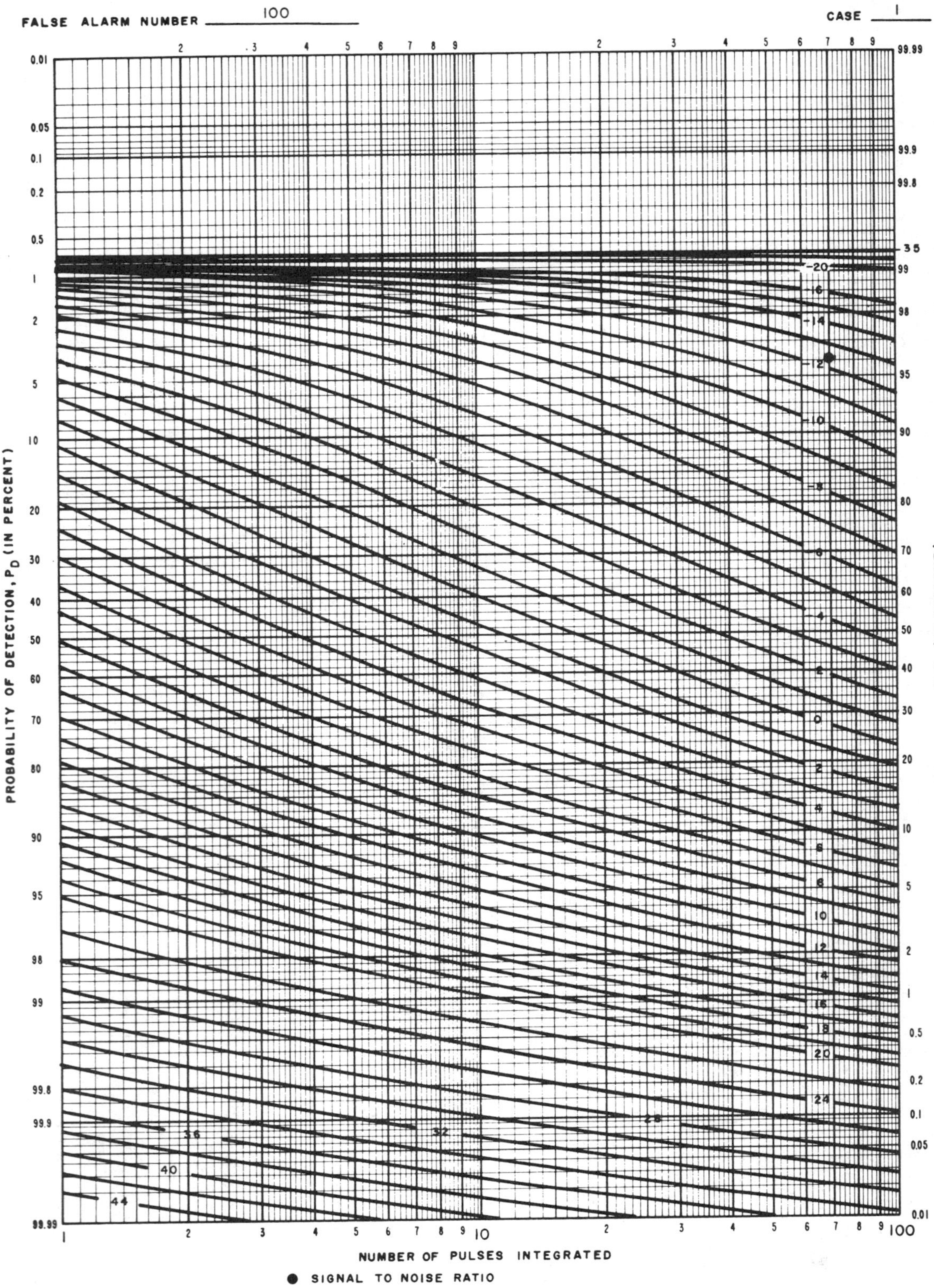
FALSE ALARM NUMBER 100
CASE 1
PROBABILITY OF DETECTION, P_D (IN PERCENT)
$1-P_D$ (IN PERCENT)
NUMBER OF PULSES INTEGRATED
● SIGNAL TO NOISE RATIO

FALSE ALARM NUMBER 100

CASE 1

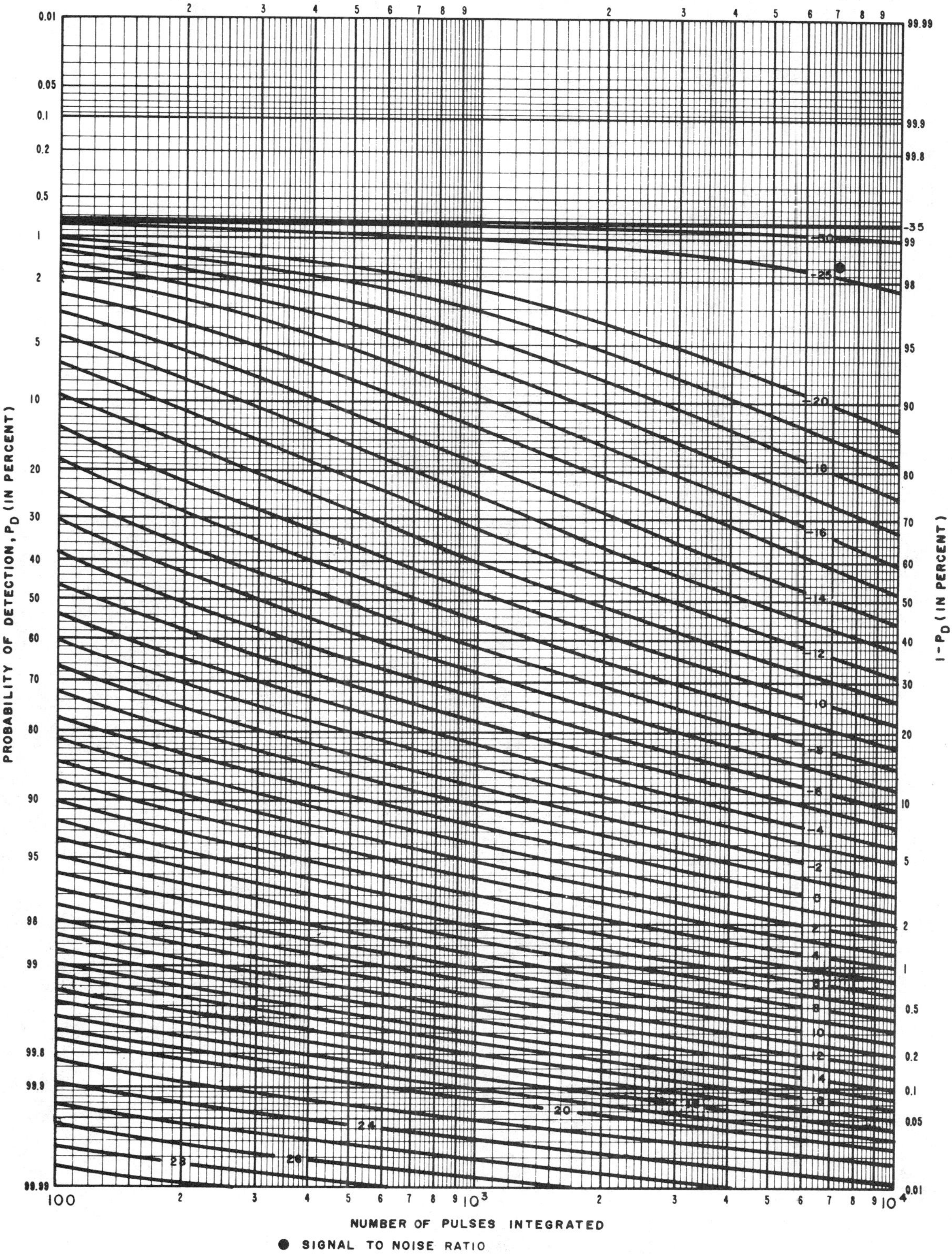

FALSE ALARM NUMBER 200

CASE 1

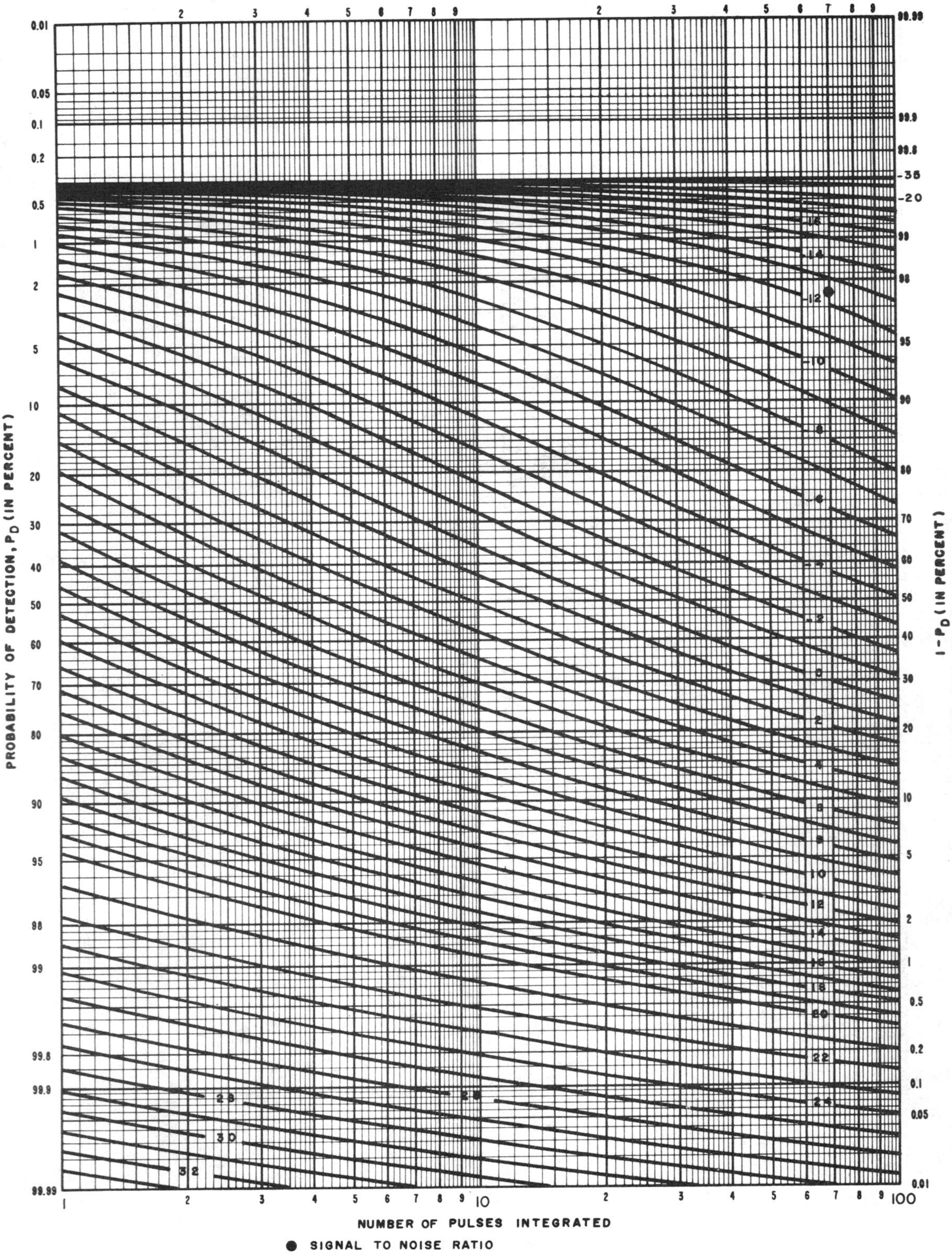

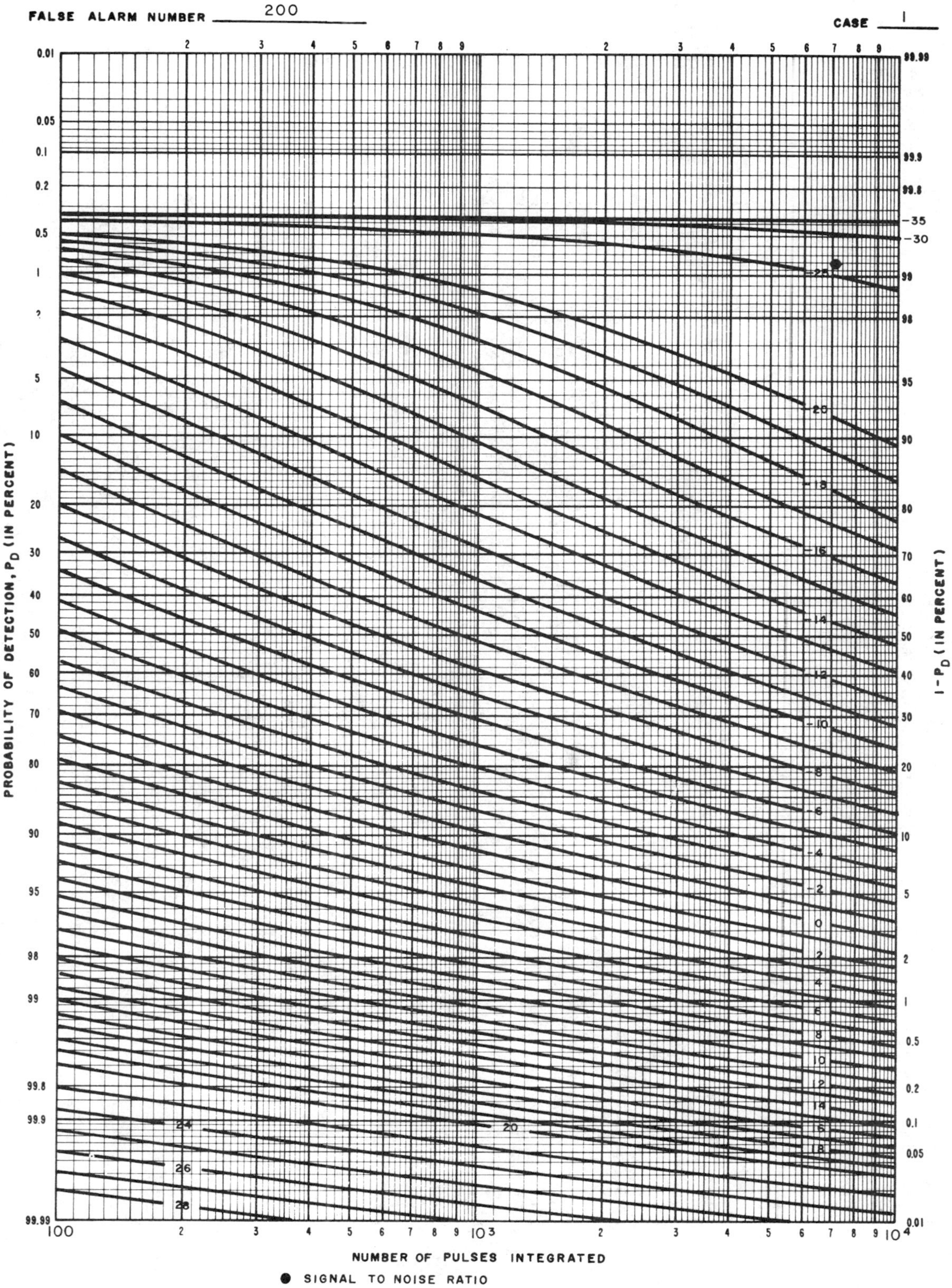
FALSE ALARM NUMBER 200
CASE 1
PROBABILITY OF DETECTION, P_D (IN PERCENT)
1 - P_D (IN PERCENT)
-35
-30
-25
-20
-18
-16
-14
-12
-10
-8
-6
-4
-2
0
2
4
6
8
10
12
14
16
18
20
24
26
28
100
10^3
10^4
NUMBER OF PULSES INTEGRATED
● SIGNAL TO NOISE RATIO

FALSE ALARM NUMBER 400 CASE 1

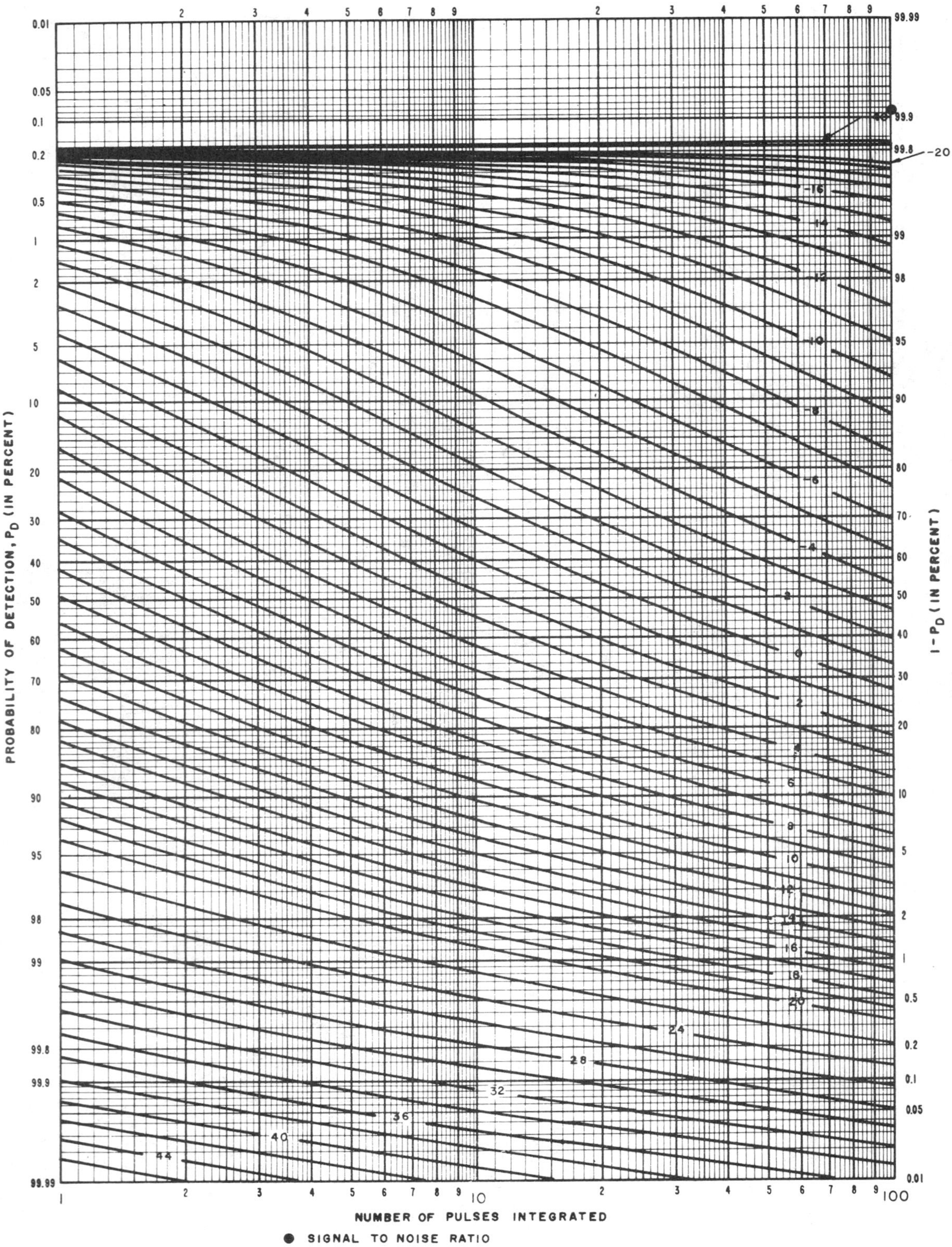

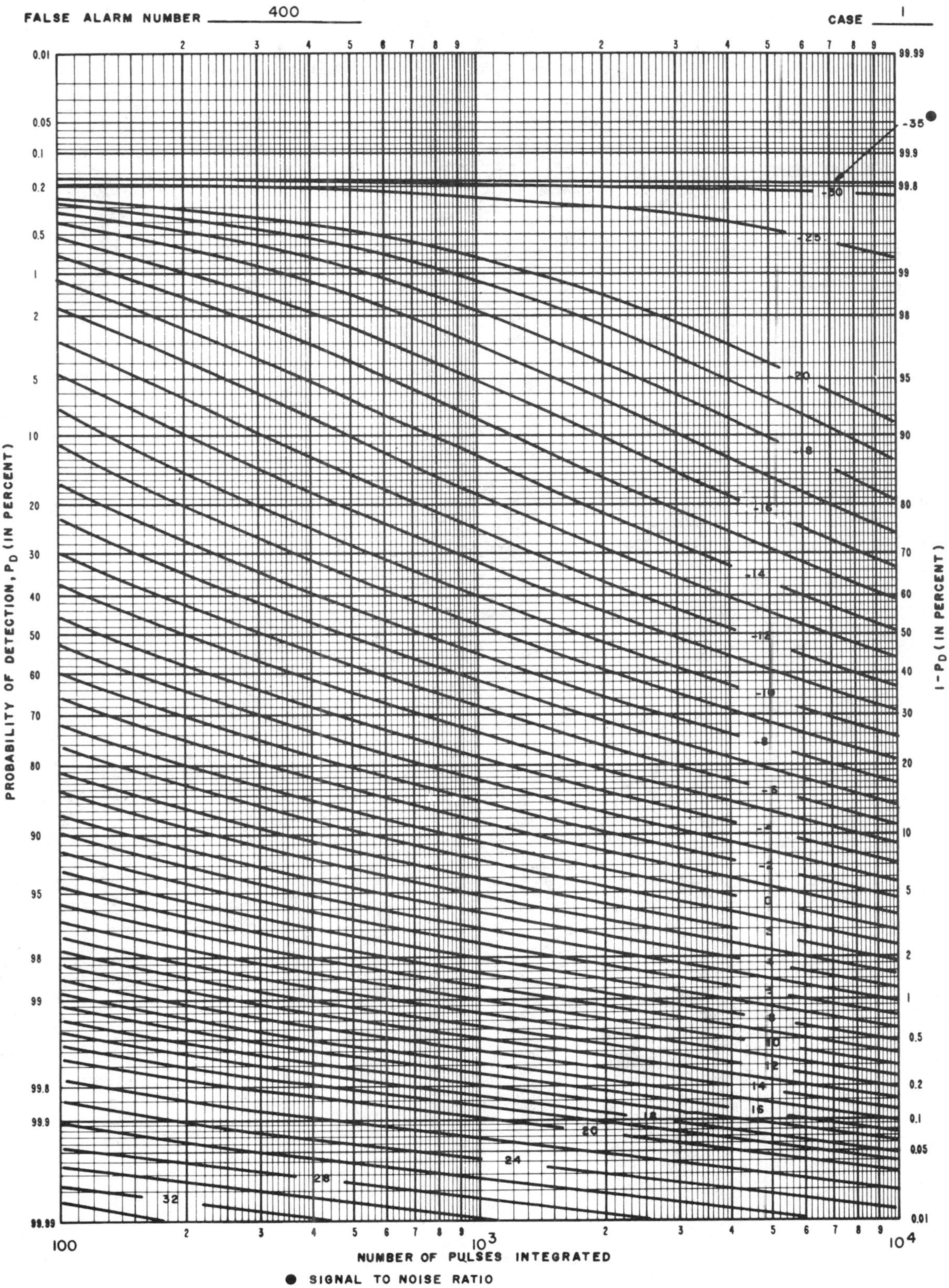
FALSE ALARM NUMBER 400
CASE 1
PROBABILITY OF DETECTION, P_D (IN PERCENT)
1 - P_D (IN PERCENT)
NUMBER OF PULSES INTEGRATED
● SIGNAL TO NOISE RATIO
100
10^3
10^4
-35
-30
-25
-20
-18
-16
-14
-12
-10
-8
-6
-4
-2
0
2
4
6
8
10
12
14
16
18
20
24
28
32

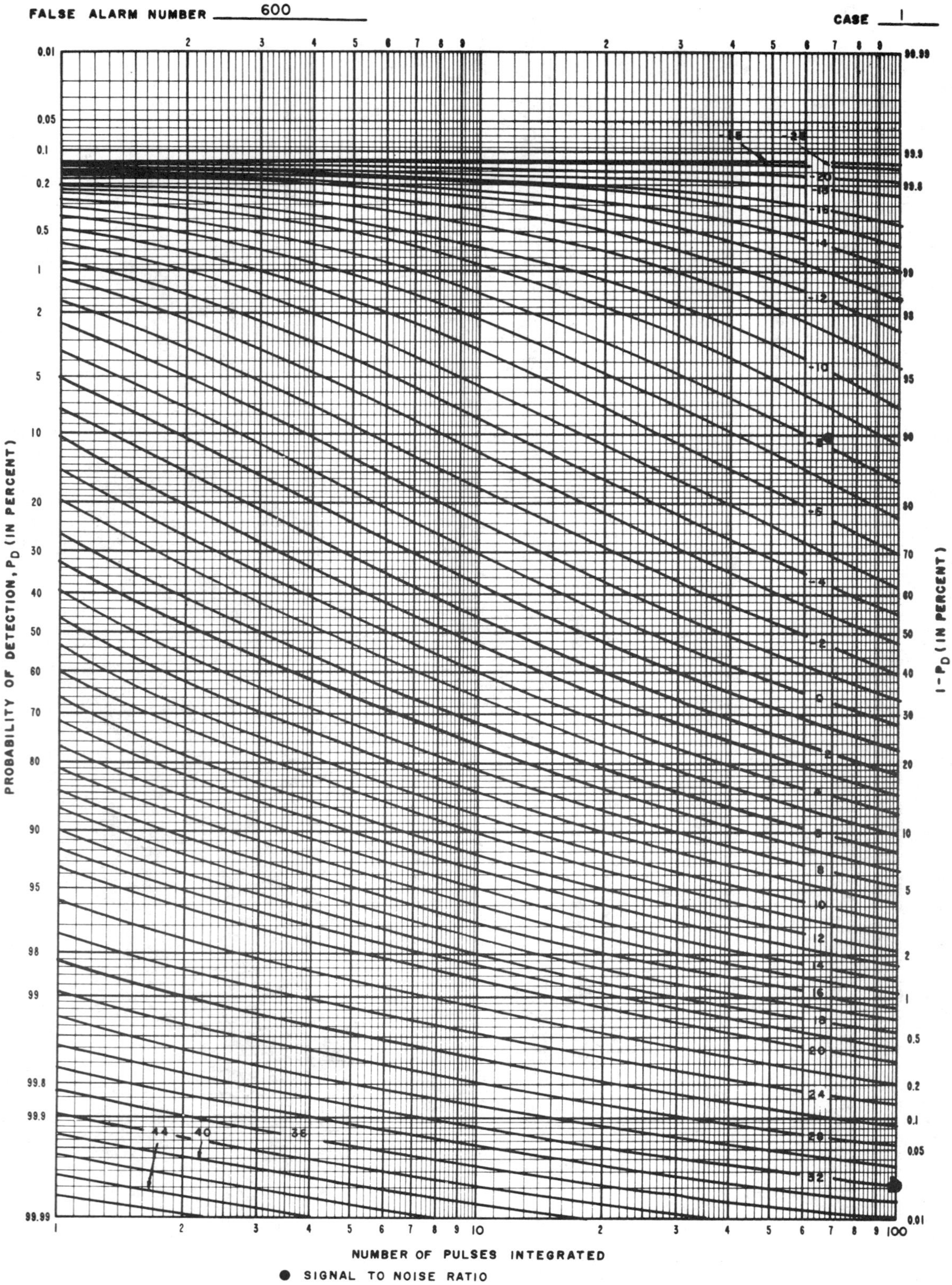
FALSE ALARM NUMBER 600
CASE I
PROBABILITY OF DETECTION, P_D (IN PERCENT)
$1-P_D$ (IN PERCENT)
NUMBER OF PULSES INTEGRATED
● SIGNAL TO NOISE RATIO

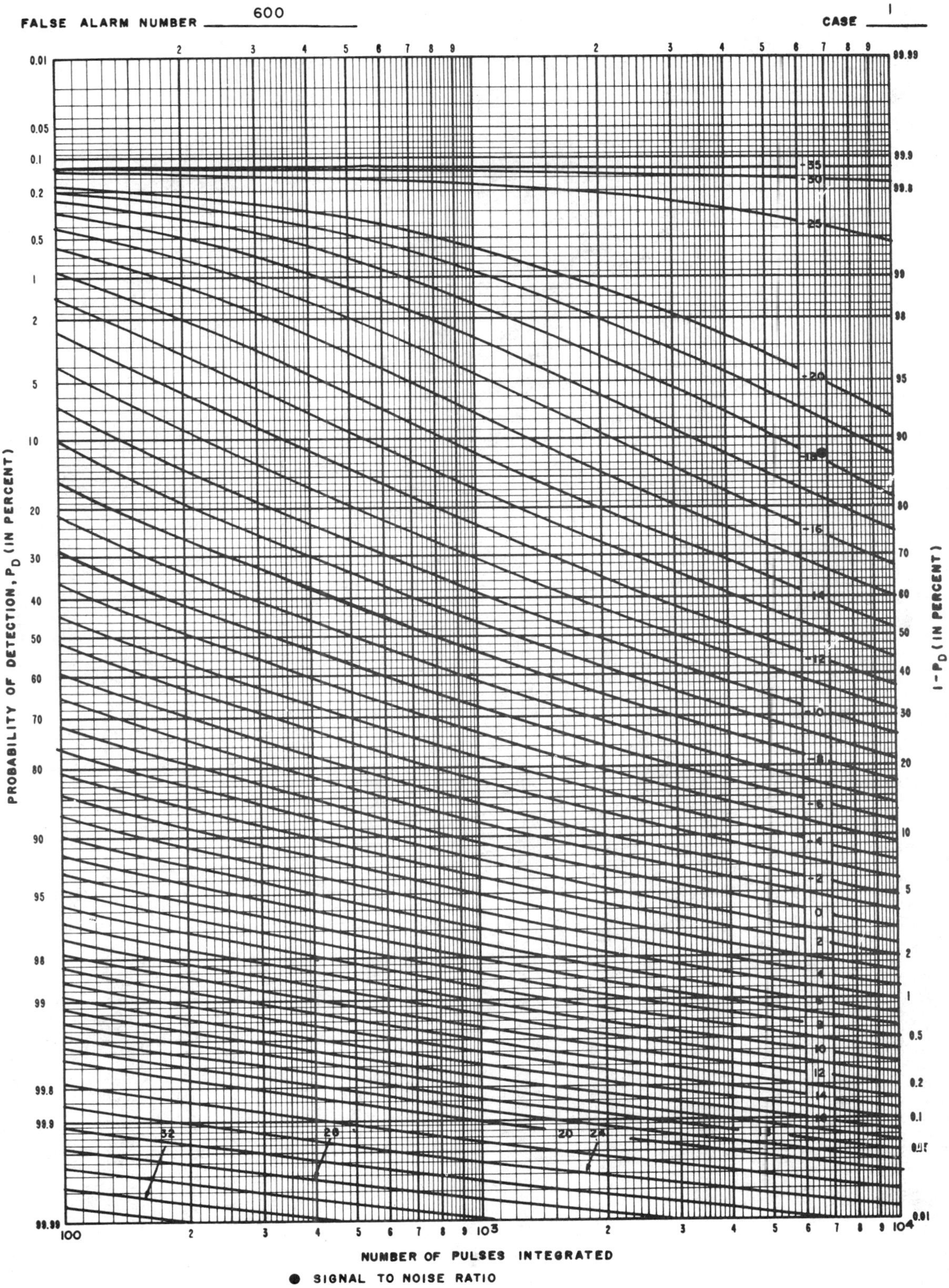
FALSE ALARM NUMBER 600
CASE 1
PROBABILITY OF DETECTION, P_D (IN PERCENT)
$1-P_D$ (IN PERCENT)
NUMBER OF PULSES INTEGRATED
● SIGNAL TO NOISE RATIO

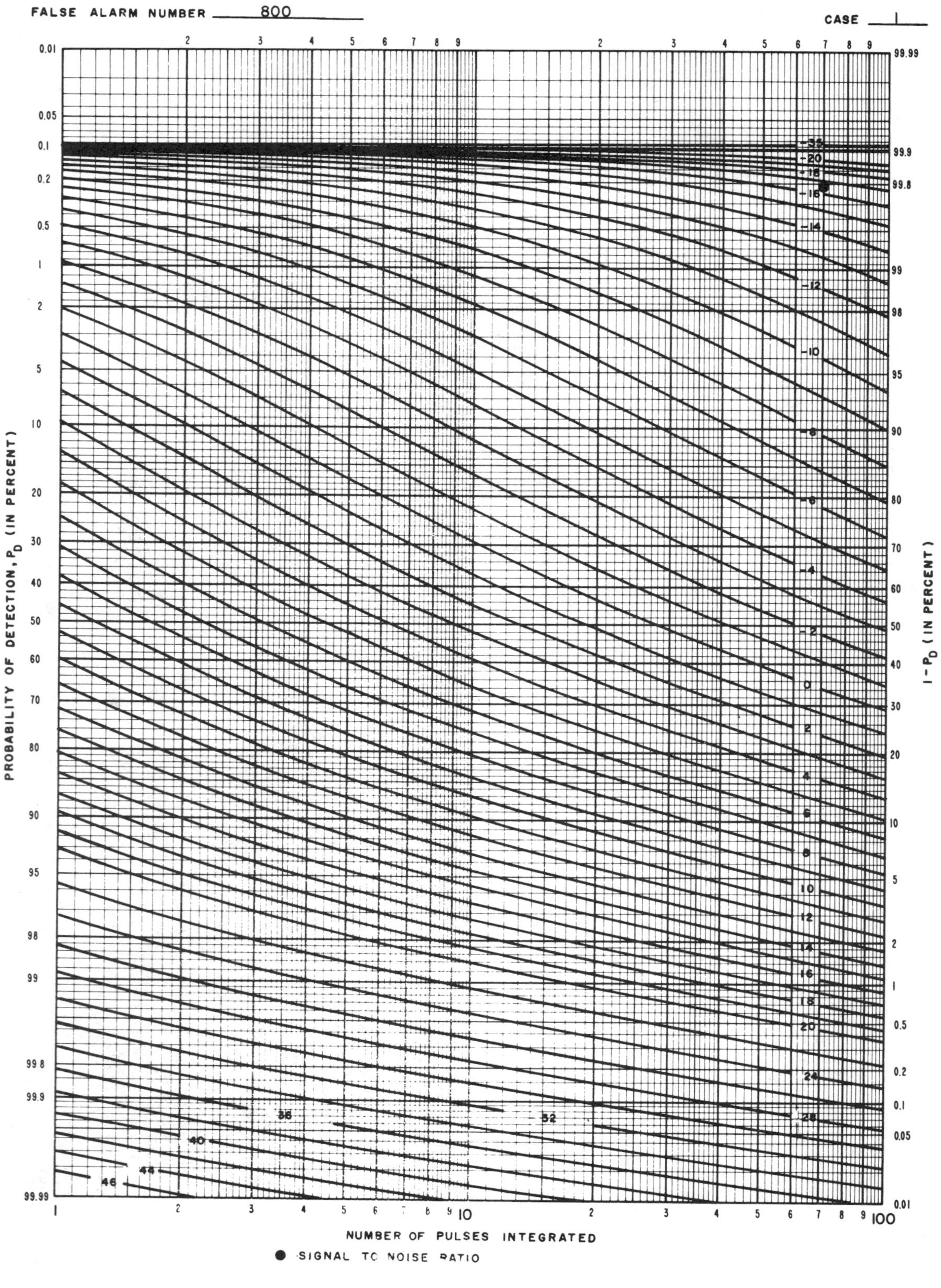
FALSE ALARM NUMBER 800
CASE I
PROBABILITY OF DETECTION, P_D (IN PERCENT)
$1-P_D$ (IN PERCENT)
NUMBER OF PULSES INTEGRATED
● SIGNAL TO NOISE RATIO

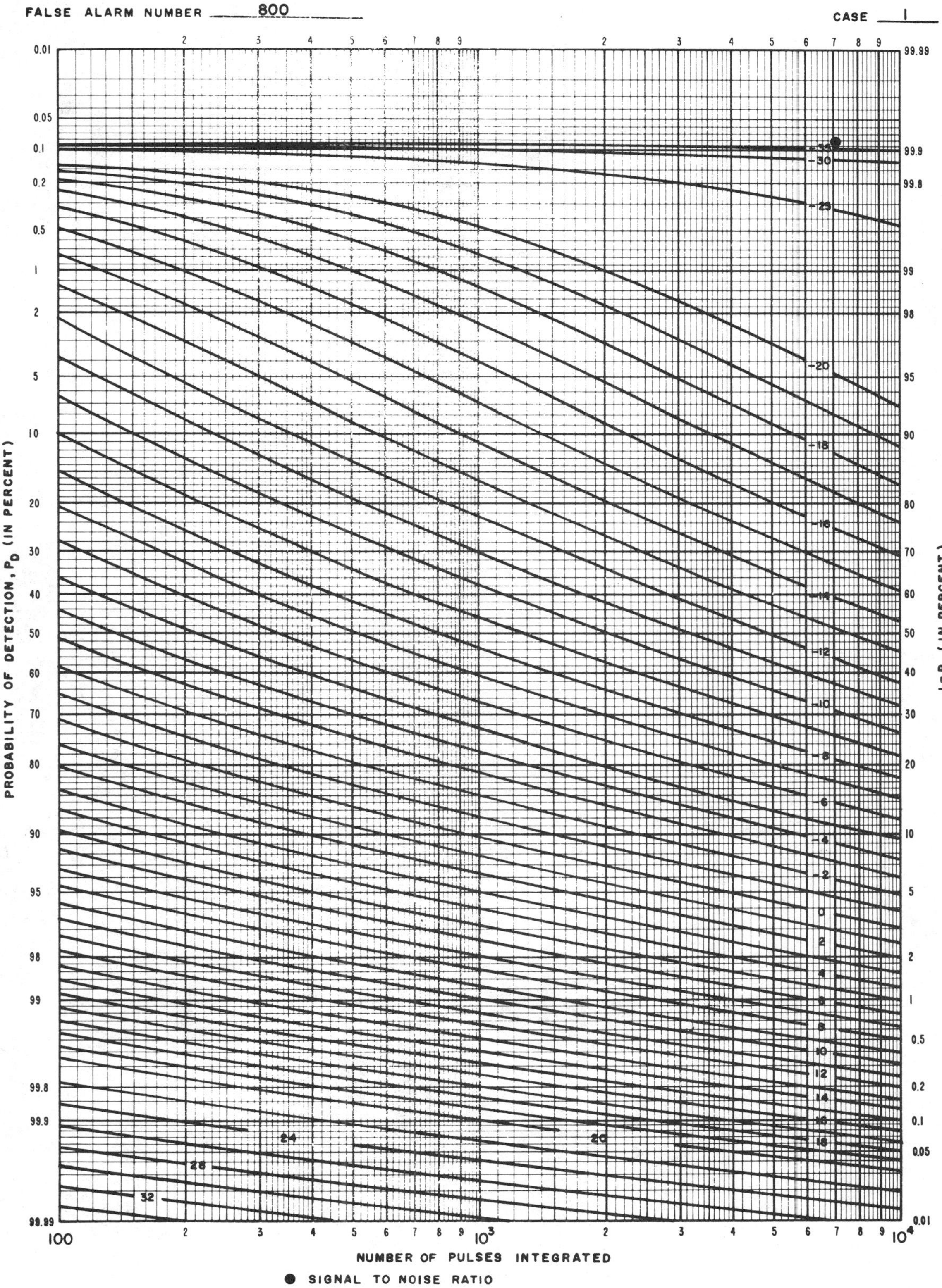
FALSE ALARM NUMBER 800
CASE 1
PROBABILITY OF DETECTION, P_D (IN PERCENT)
1 - P_D (IN PERCENT)
NUMBER OF PULSES INTEGRATED
● SIGNAL TO NOISE RATIO

FALSE ALARM NUMBER 10^3 CASE 1

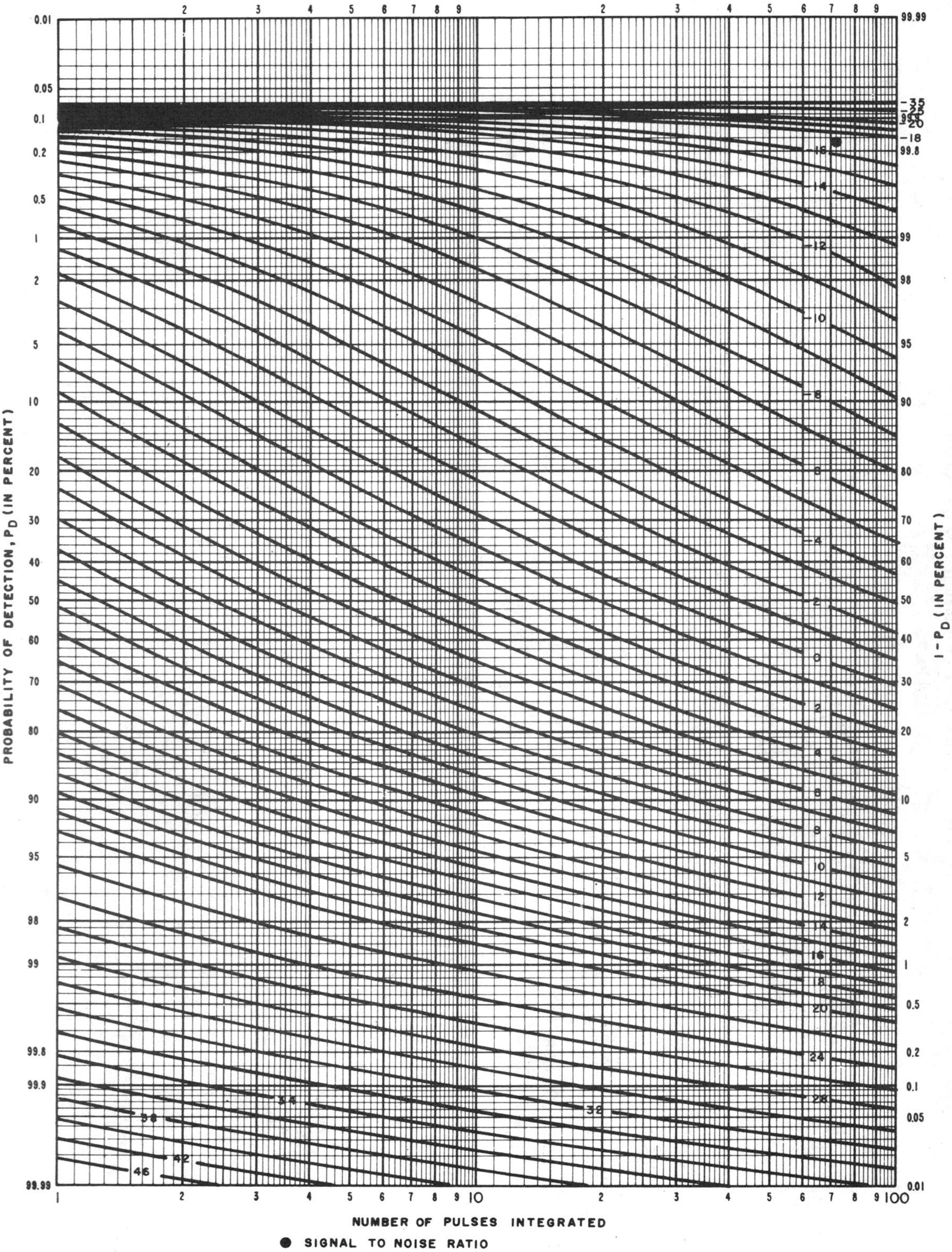

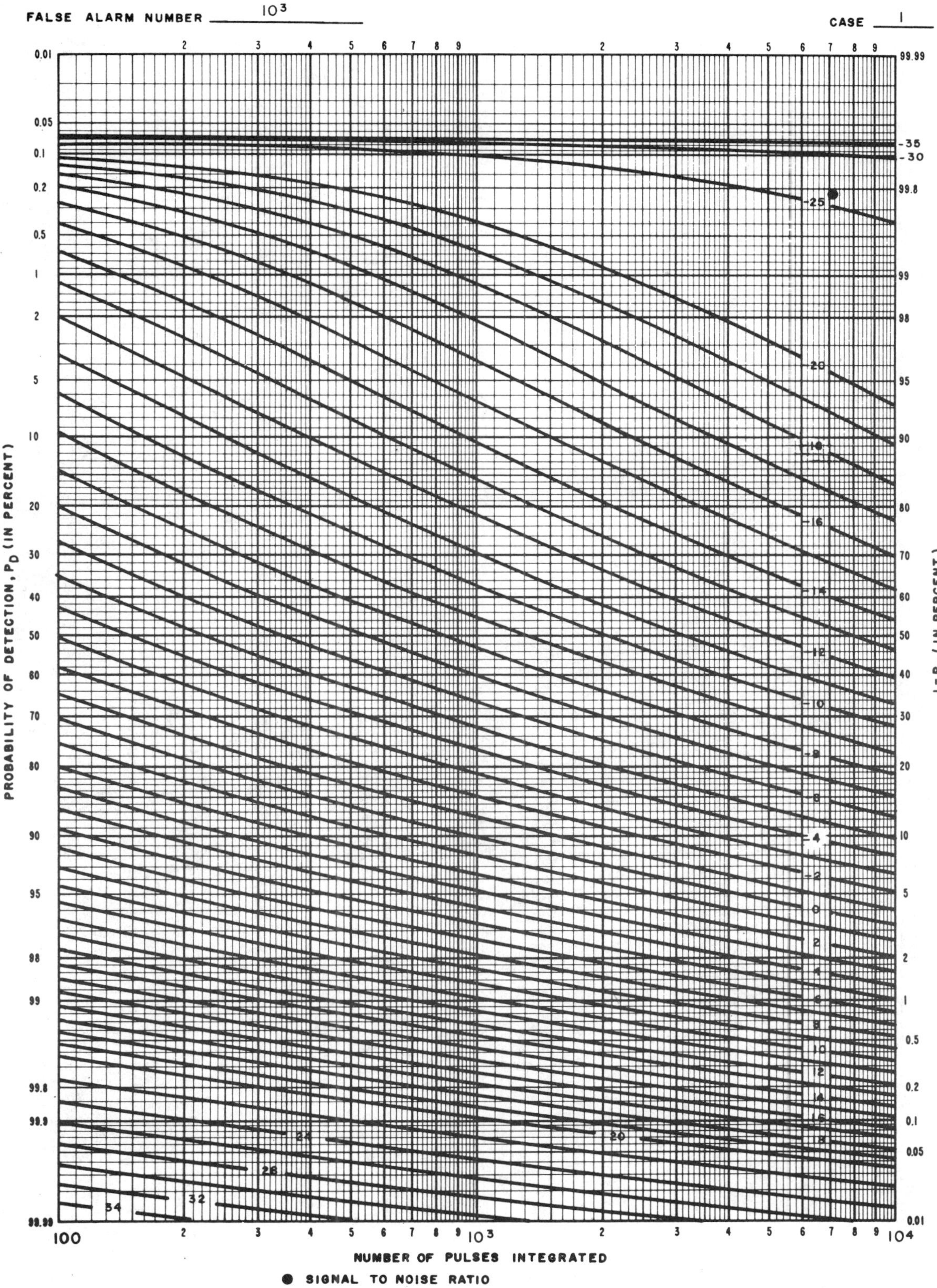
FALSE ALARM NUMBER 10^3
CASE 1
PROBABILITY OF DETECTION, P_D (IN PERCENT)
$1-P_D$ (IN PERCENT)
NUMBER OF PULSES INTEGRATED
● SIGNAL TO NOISE RATIO
100
10^3
10^4
-35
-30
-25
-20
-16
-14
-12
-10
-8
-4
-2
0
2
4
8
10
12
14
16
18
20
24
28
32

FALSE ALARM NUMBER 2×10^3 CASE 1

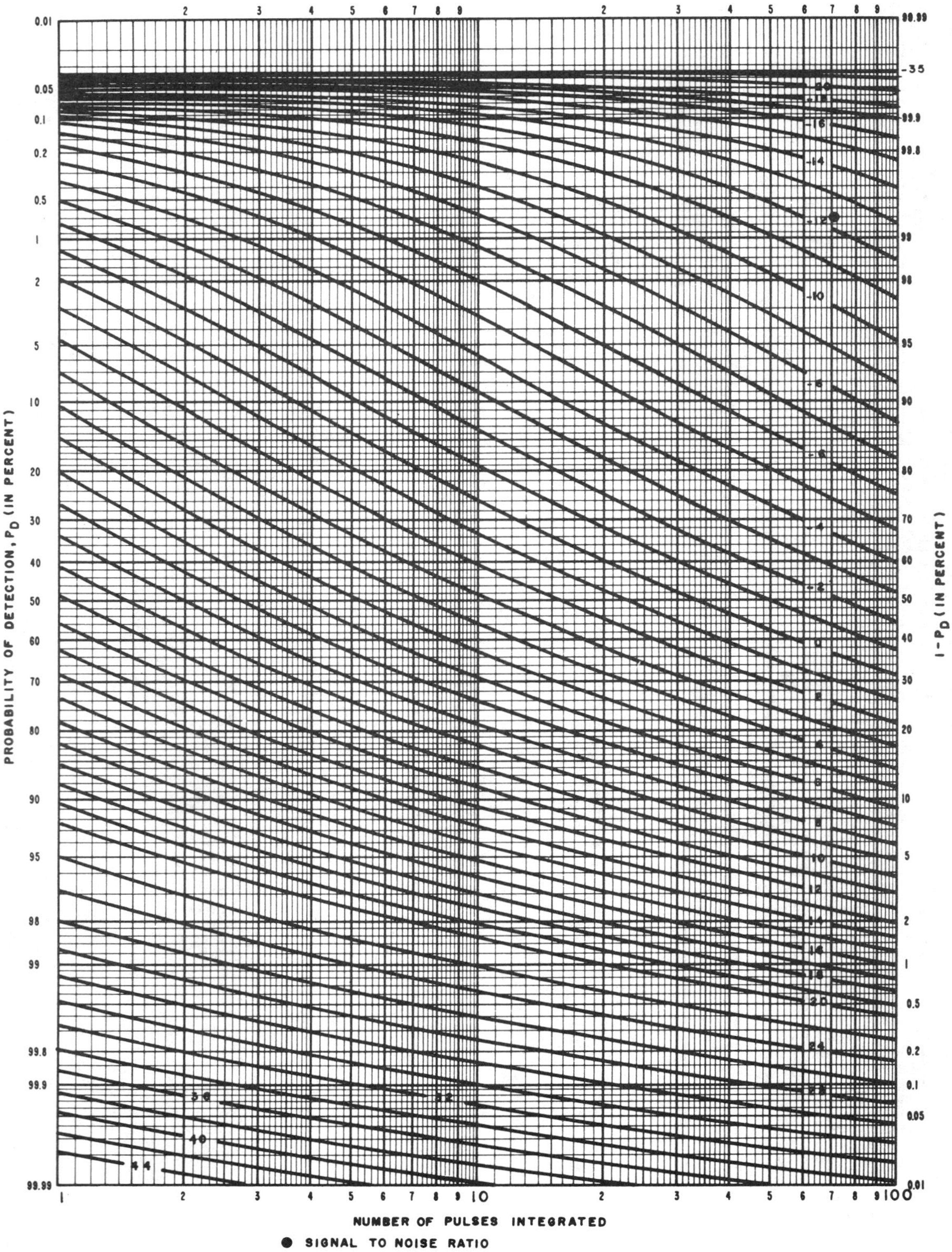

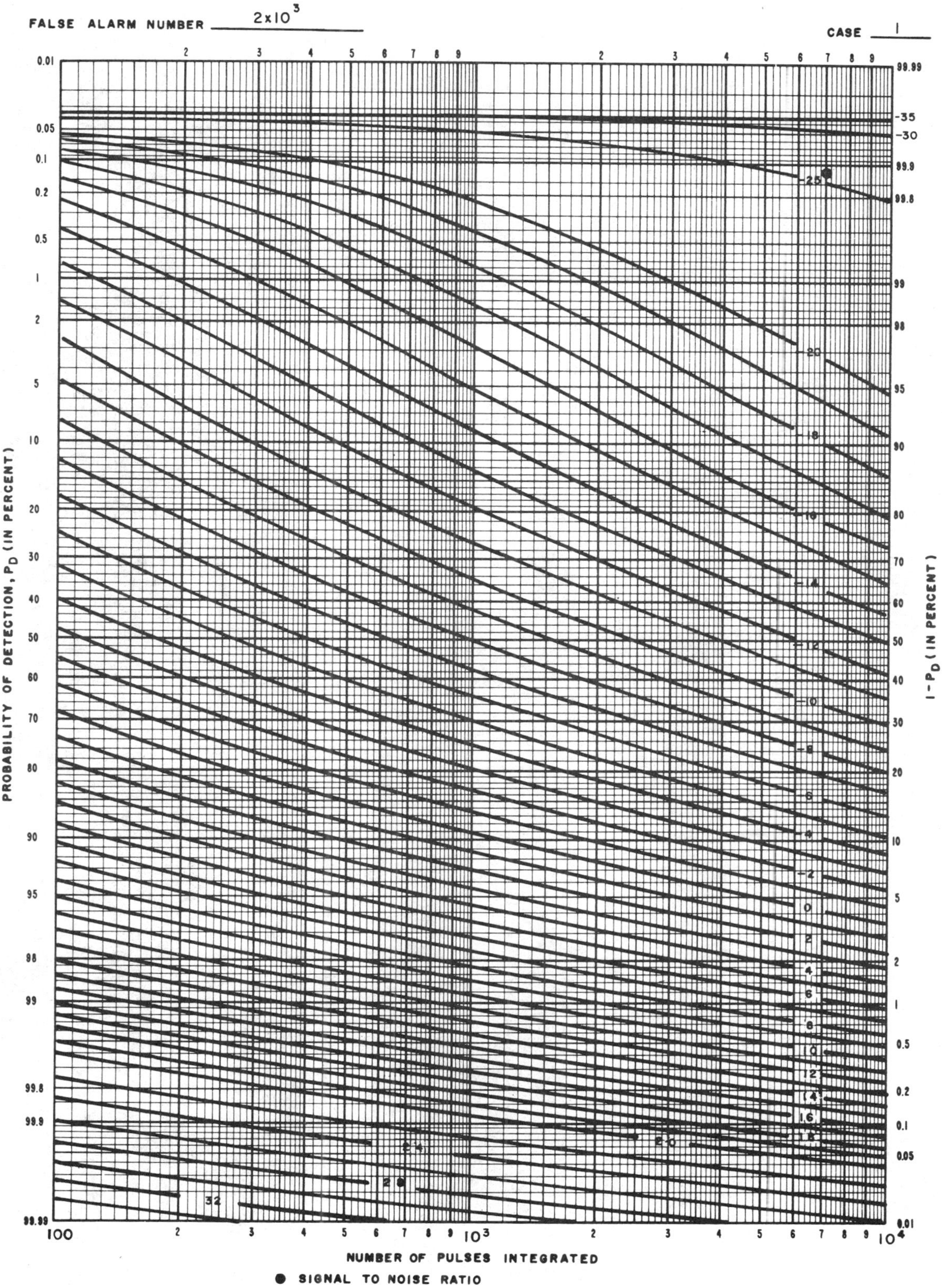
FALSE ALARM NUMBER 2x10^3
CASE 1
PROBABILITY OF DETECTION, P_D (IN PERCENT)
1-P_D (IN PERCENT)
-35
-30
-25
-20
-18
-16
-14
-12
-10
-8
-6
-4
-2
0
2
4
6
8
10
12
14
16
18
20
24
28
32
100
10^3
10^4
NUMBER OF PULSES INTEGRATED
● SIGNAL TO NOISE RATIO

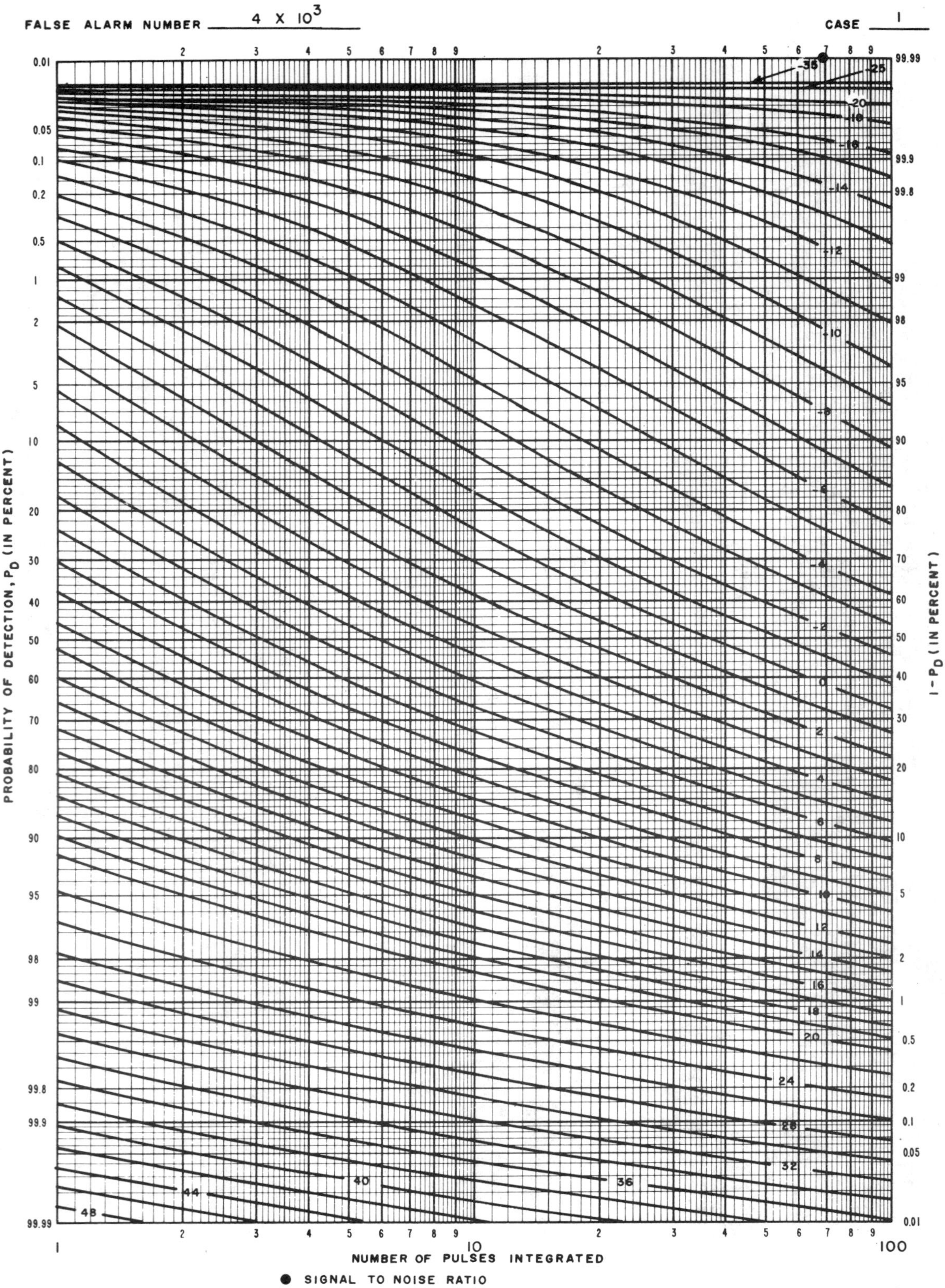
FALSE ALARM NUMBER 4 X 10^3
CASE I
PROBABILITY OF DETECTION, P_D (IN PERCENT)
1 - P_D (IN PERCENT)
NUMBER OF PULSES INTEGRATED
● SIGNAL TO NOISE RATIO
1
10
100

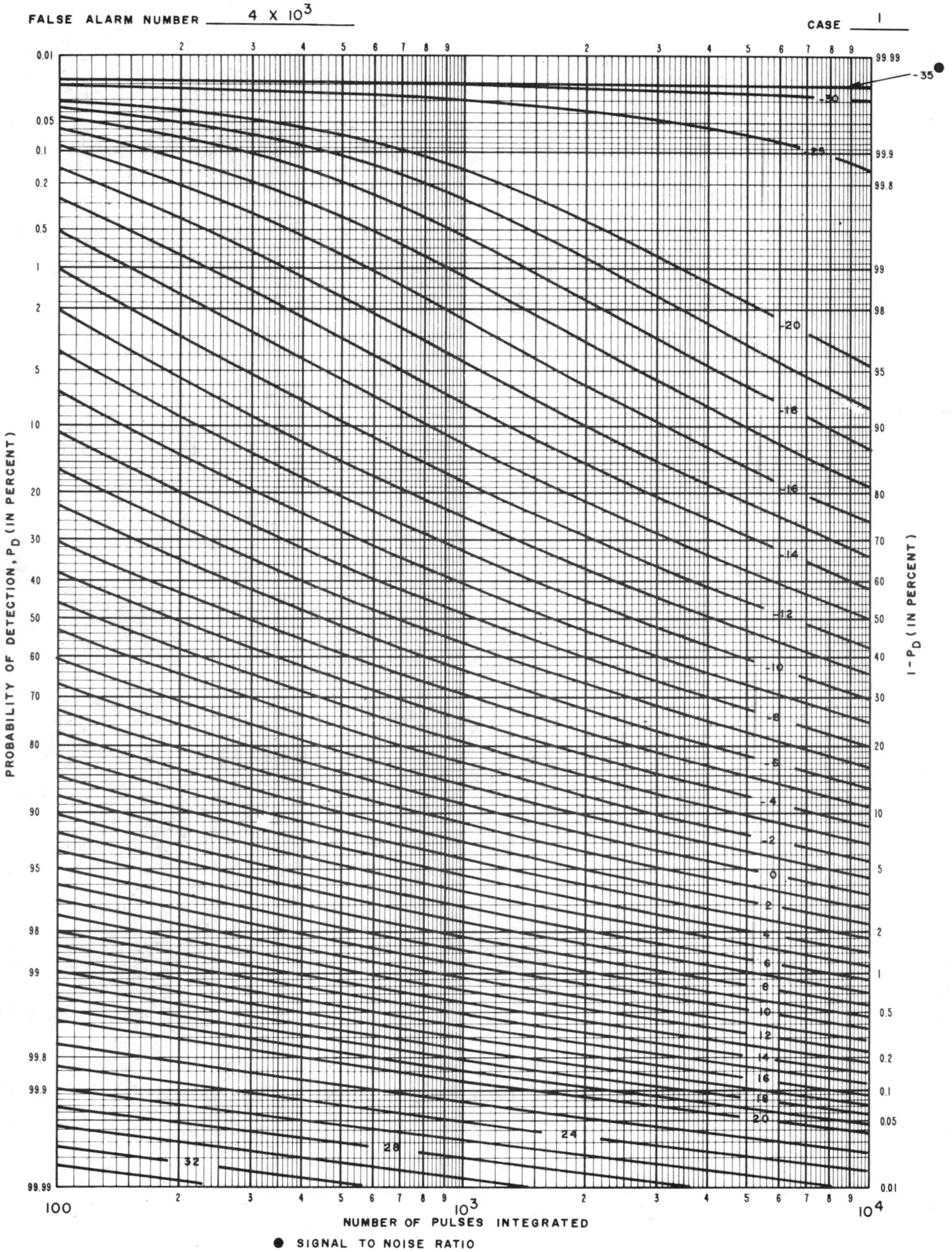
FALSE ALARM NUMBER 4 X 10^3
CASE I
PROBABILITY OF DETECTION, P_D (IN PERCENT)
1 - P_D (IN PERCENT)
NUMBER OF PULSES INTEGRATED
100
10^3
10^4
● SIGNAL TO NOISE RATIO
-35
-30
-25
-20
-18
-16
-14
-12
-10
-8
-6
-4
-2
0
2
4
6
8
10
12
14
16
18
20
24
28
32

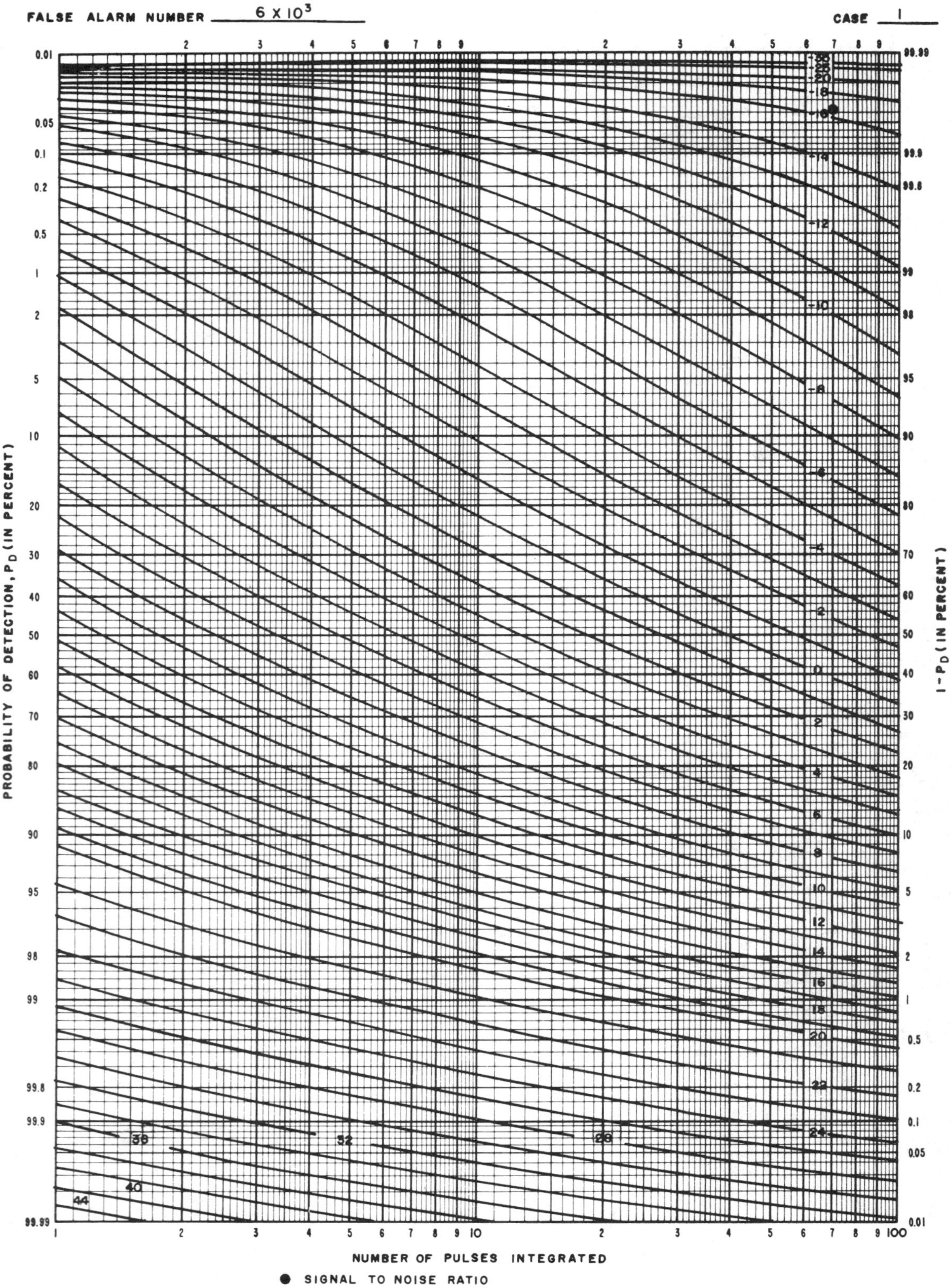
FALSE ALARM NUMBER 6 X 10^3
CASE 1
PROBABILITY OF DETECTION, P_D (IN PERCENT)
1 - P_D (IN PERCENT)
NUMBER OF PULSES INTEGRATED
● SIGNAL TO NOISE RATIO

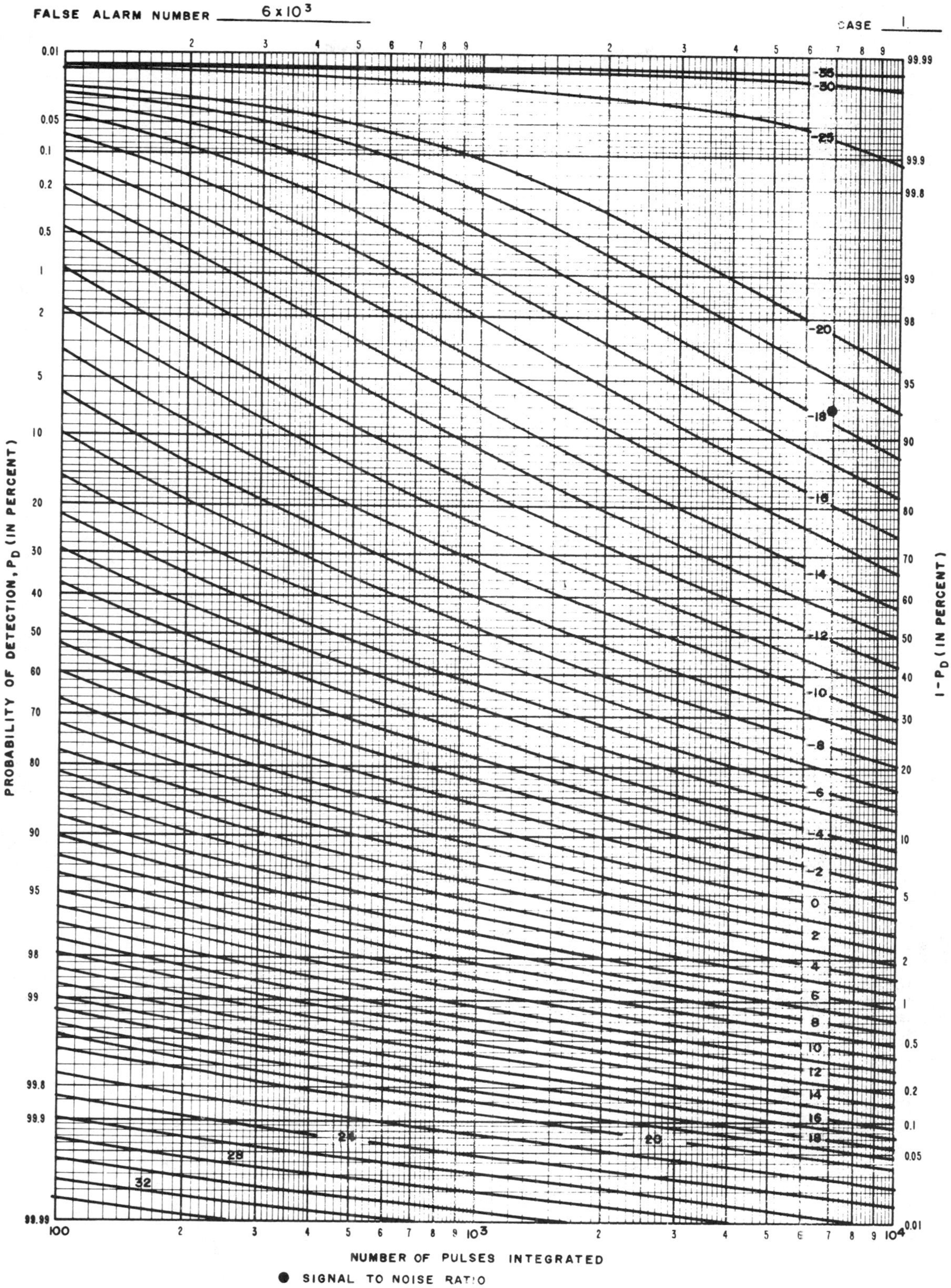
FALSE ALARM NUMBER 6 x 10^3
CASE 1
PROBABILITY OF DETECTION, P_D (IN PERCENT)
1 - P_D (IN PERCENT)
NUMBER OF PULSES INTEGRATED
● SIGNAL TO NOISE RATIO
-35
-30
-25
-20
-18
-16
-14
-12
-10
-8
-6
-4
-2
0
2
4
6
8
10
12
14
16
18
20
24
28
32

FALSE ALARM NUMBER 8×10^3 CASE 1

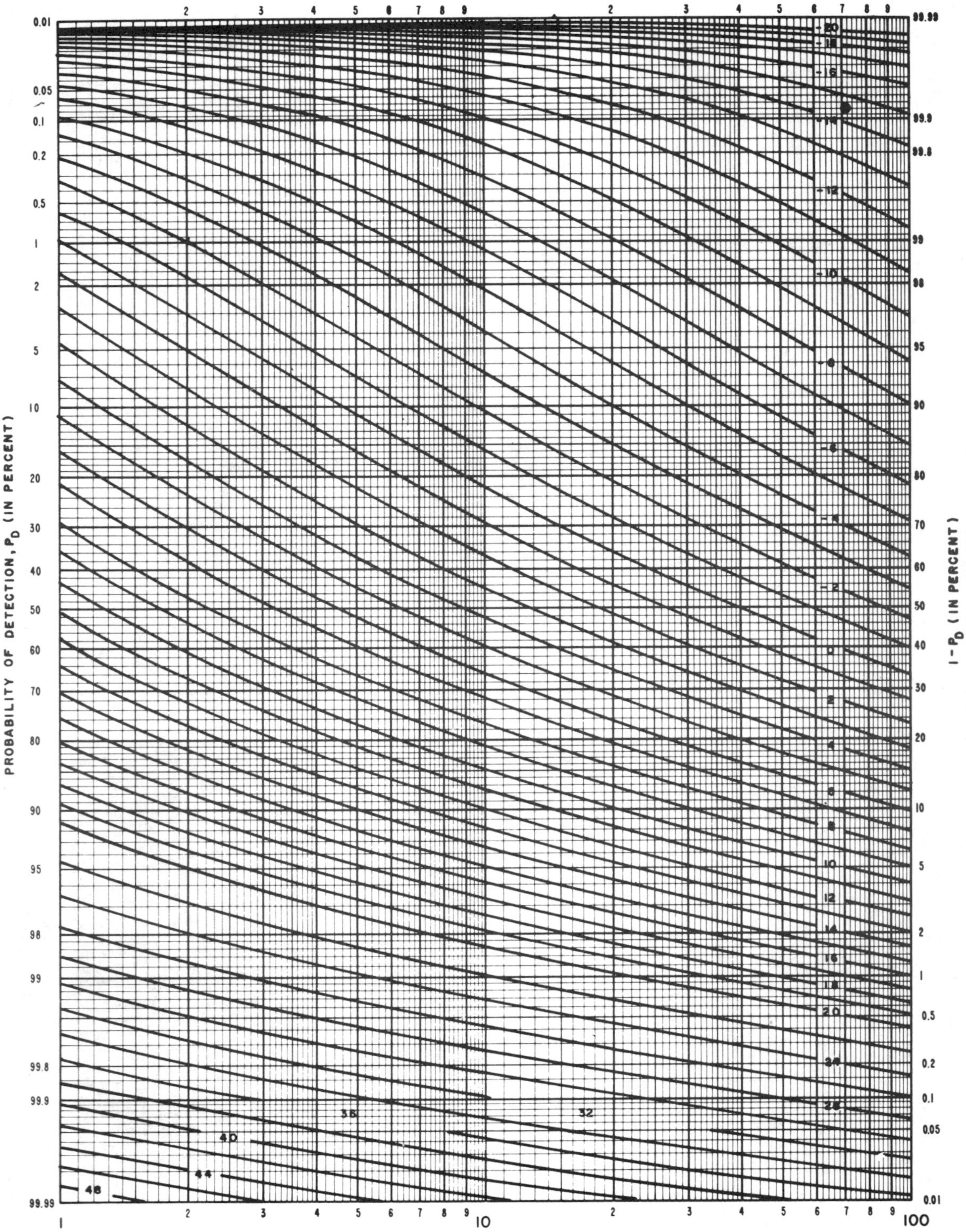

● SIGNAL TO NOISE RATIO

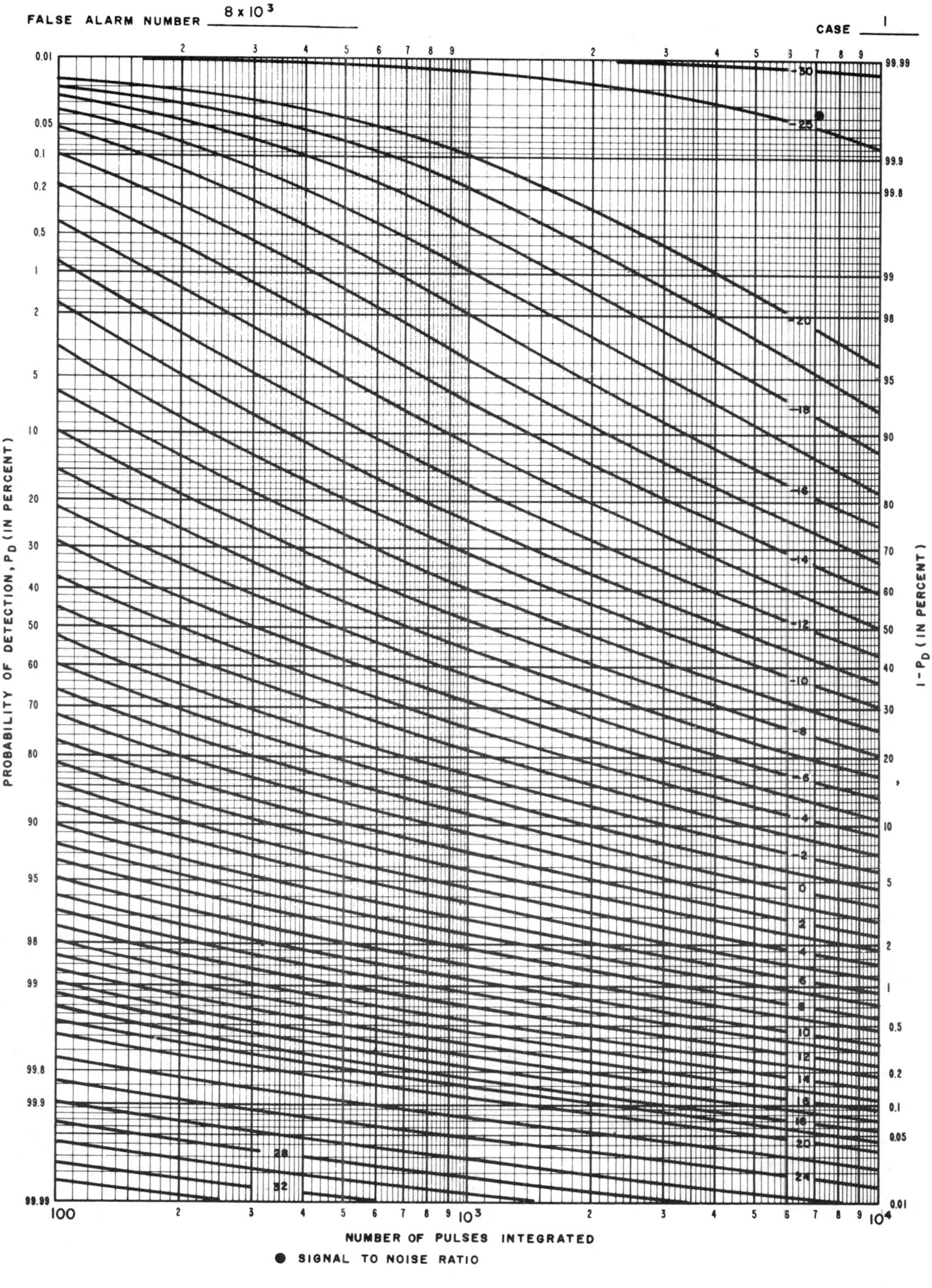
FALSE ALARM NUMBER 8 x 10^3
CASE 1
PROBABILITY OF DETECTION, P_D (IN PERCENT)
1 - P_D (IN PERCENT)
NUMBER OF PULSES INTEGRATED
● SIGNAL TO NOISE RATIO
100
10^3
10^4
-30
-25
-20
-18
-16
-14
-12
-10
-8
-6
-4
-2
0
2
4
6
8
10
12
14
16
18
20
24
28
32

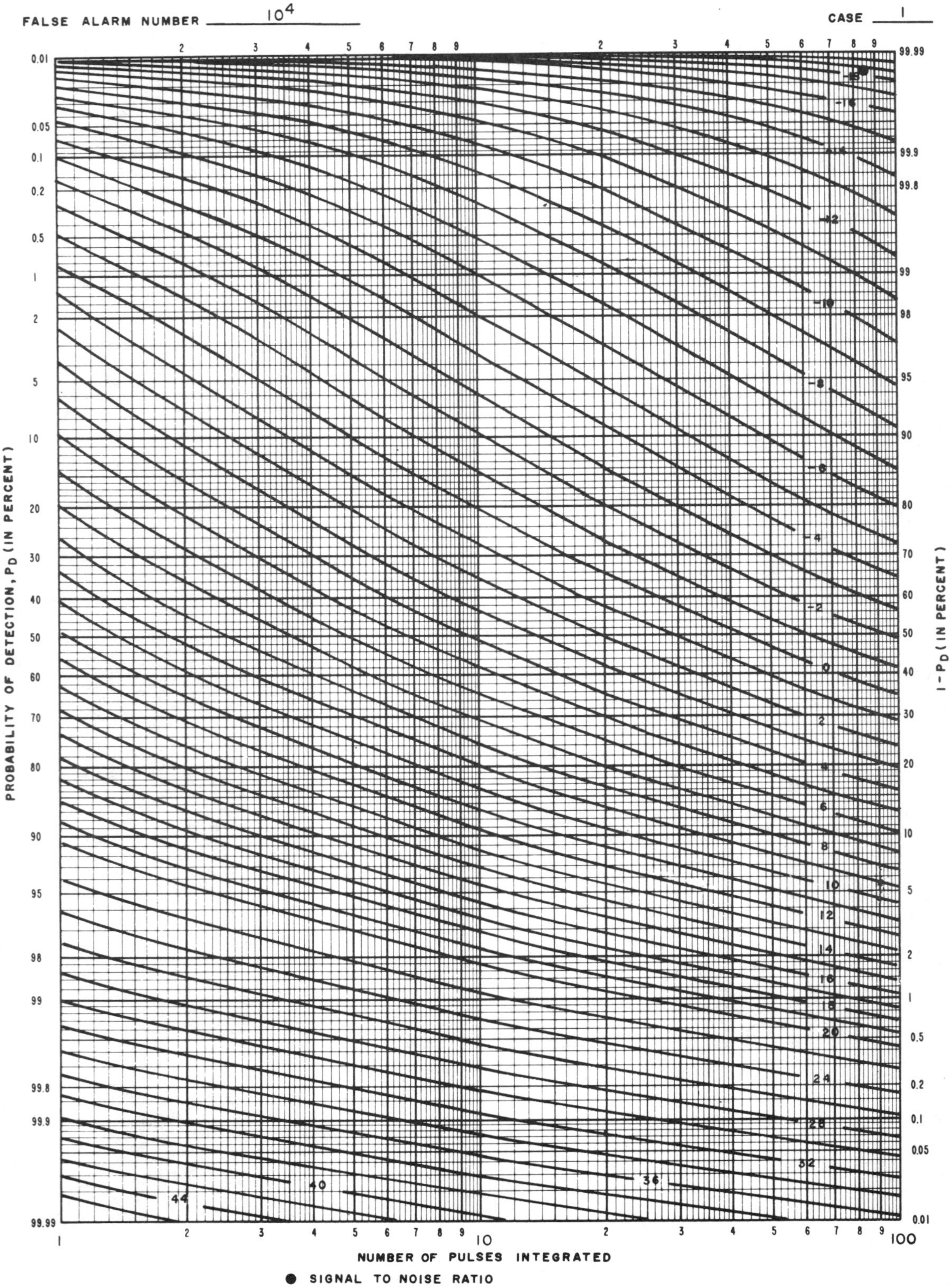
FALSE ALARM NUMBER 10^4
CASE 1
PROBABILITY OF DETECTION, P_D (IN PERCENT)
0.01 0.05 0.1 0.2 0.5 1 2 5 10 20 30 40 50 60 70 80 90 95 98 99 99.8 99.9 99.99
$1-P_D$ (IN PERCENT)
99.99 99.9 99.8 99 98 95 90 80 70 60 50 40 30 20 10 5 2 1 0.5 0.2 0.1 0.05 0.01
-18
-16
-14
-12
-10
-8
-6
-4
-2
0
2
4
6
8
10
12
14
16
18
20
24
28
32
36
40
44
1 2 3 4 5 6 7 8 9 10 2 3 4 5 6 7 8 9 100
NUMBER OF PULSES INTEGRATED
● SIGNAL TO NOISE RATIO

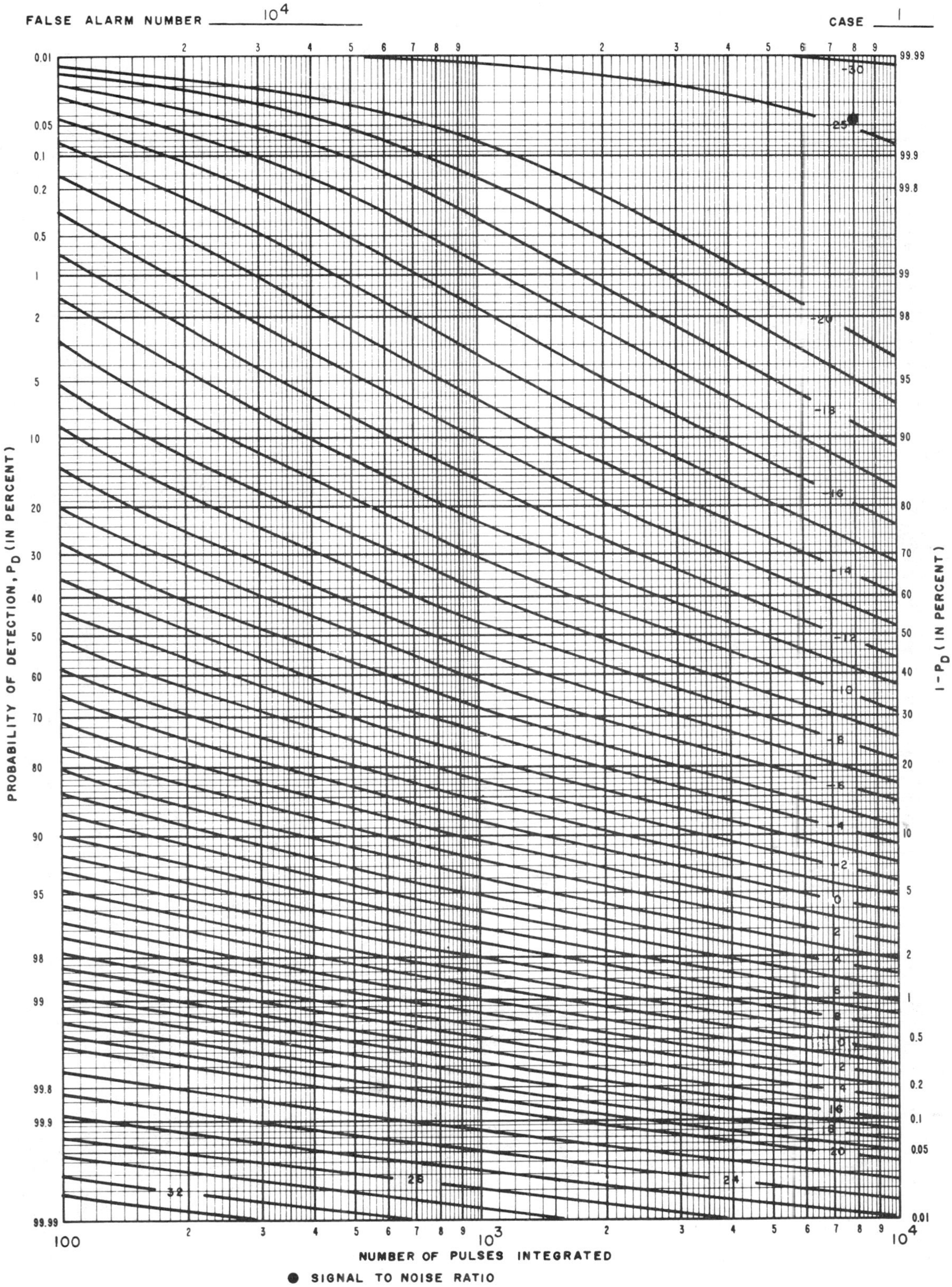
FALSE ALARM NUMBER 10^4
CASE 1
PROBABILITY OF DETECTION, P_D (IN PERCENT)
$1-P_D$ (IN PERCENT)
NUMBER OF PULSES INTEGRATED
● SIGNAL TO NOISE RATIO
-30
-25
-20
-18
-16
-14
-12
-10
-8
-6
-4
-2
0
2
4
6
8
10
12
14
16
18
20
24
28
32

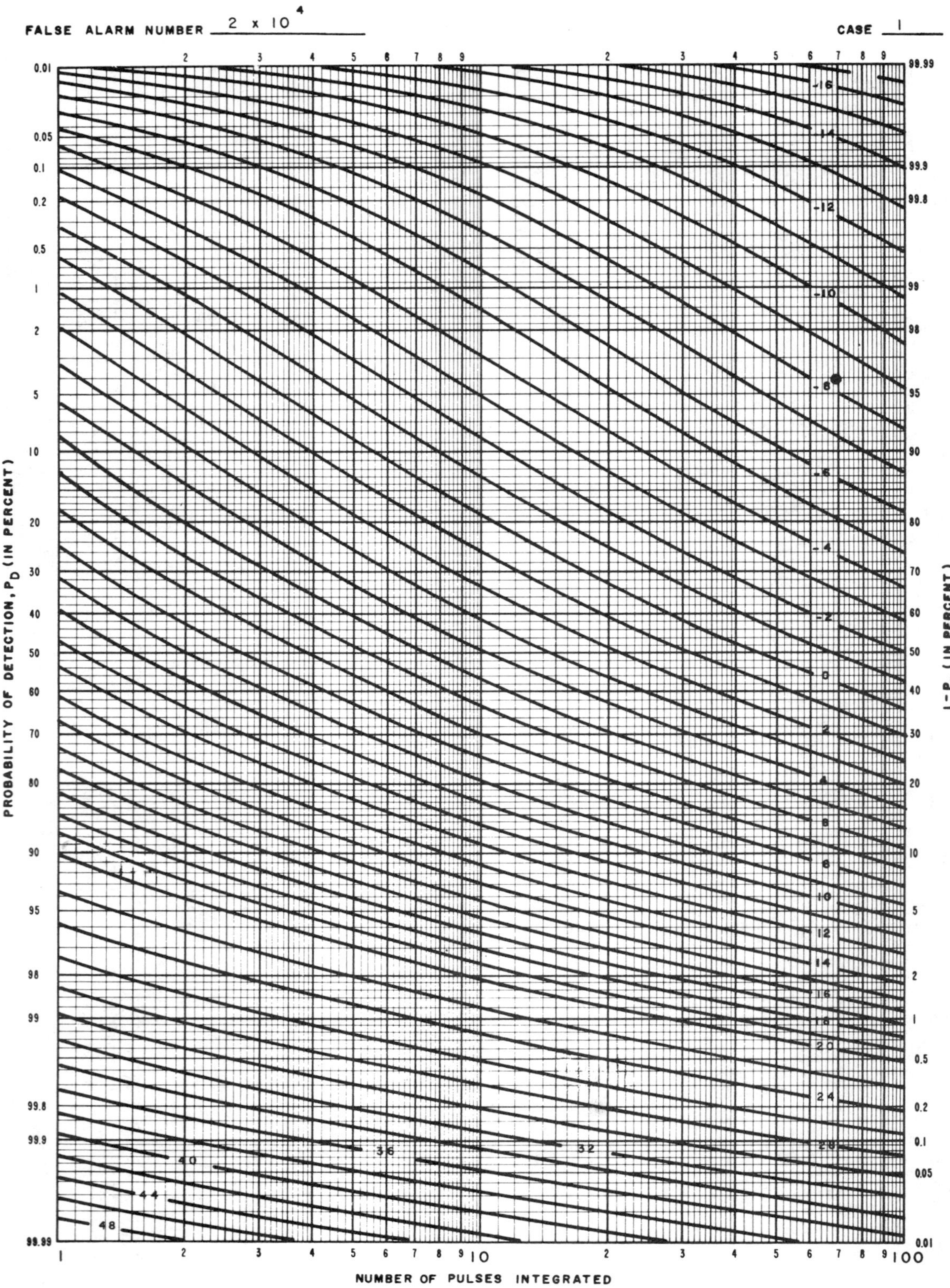

FALSE ALARM NUMBER 2 x 10^4
CASE I
PROBABILITY OF DETECTION, P_D (IN PERCENT)
1-P_D (IN PERCENT)
NUMBER OF PULSES INTEGRATED
● SIGNAL TO NOISE RATIO

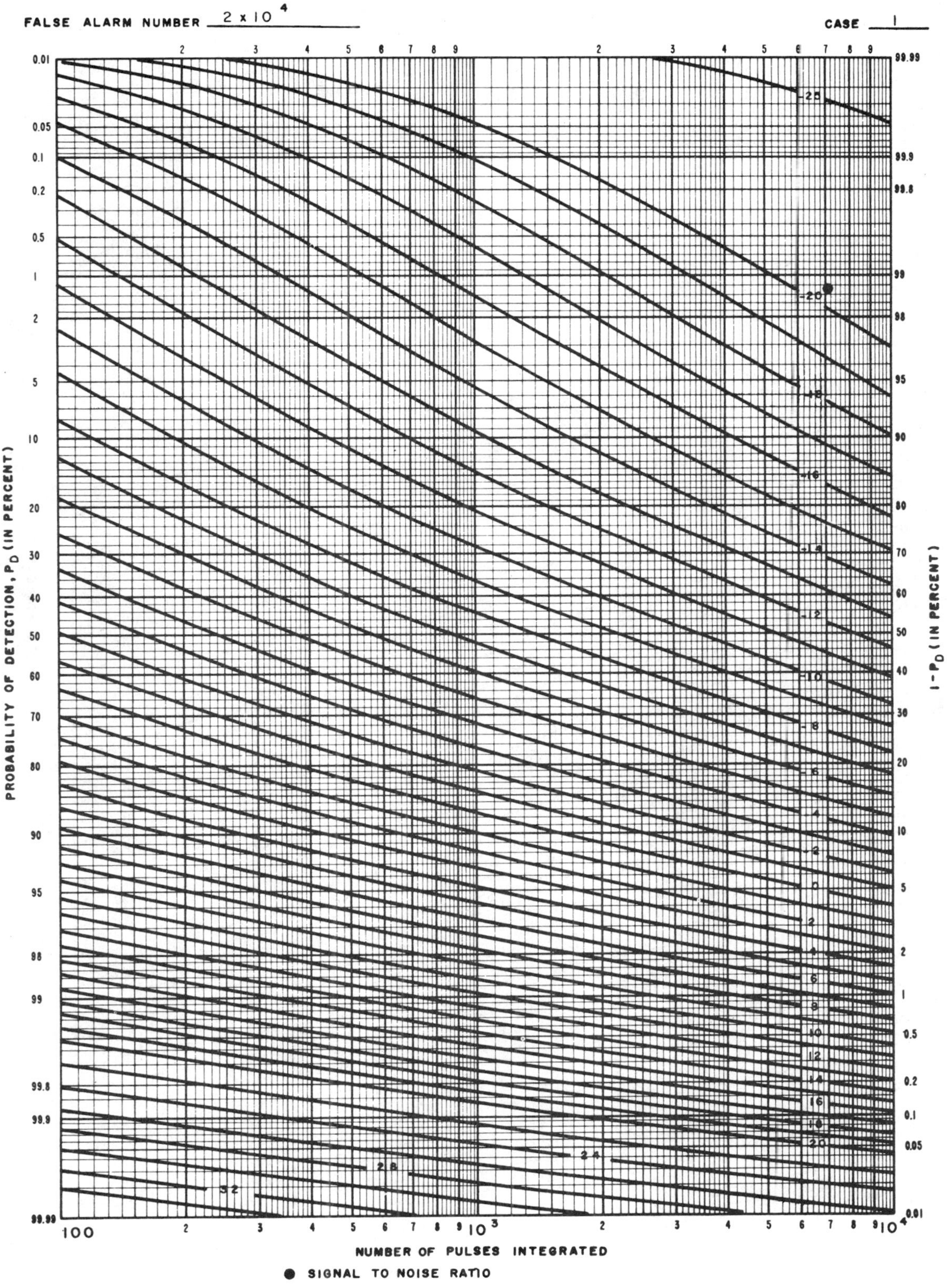
FALSE ALARM NUMBER 2 x 10^4
CASE 1
PROBABILITY OF DETECTION, P_D (IN PERCENT)
1 - P_D (IN PERCENT)
NUMBER OF PULSES INTEGRATED
● SIGNAL TO NOISE RATIO

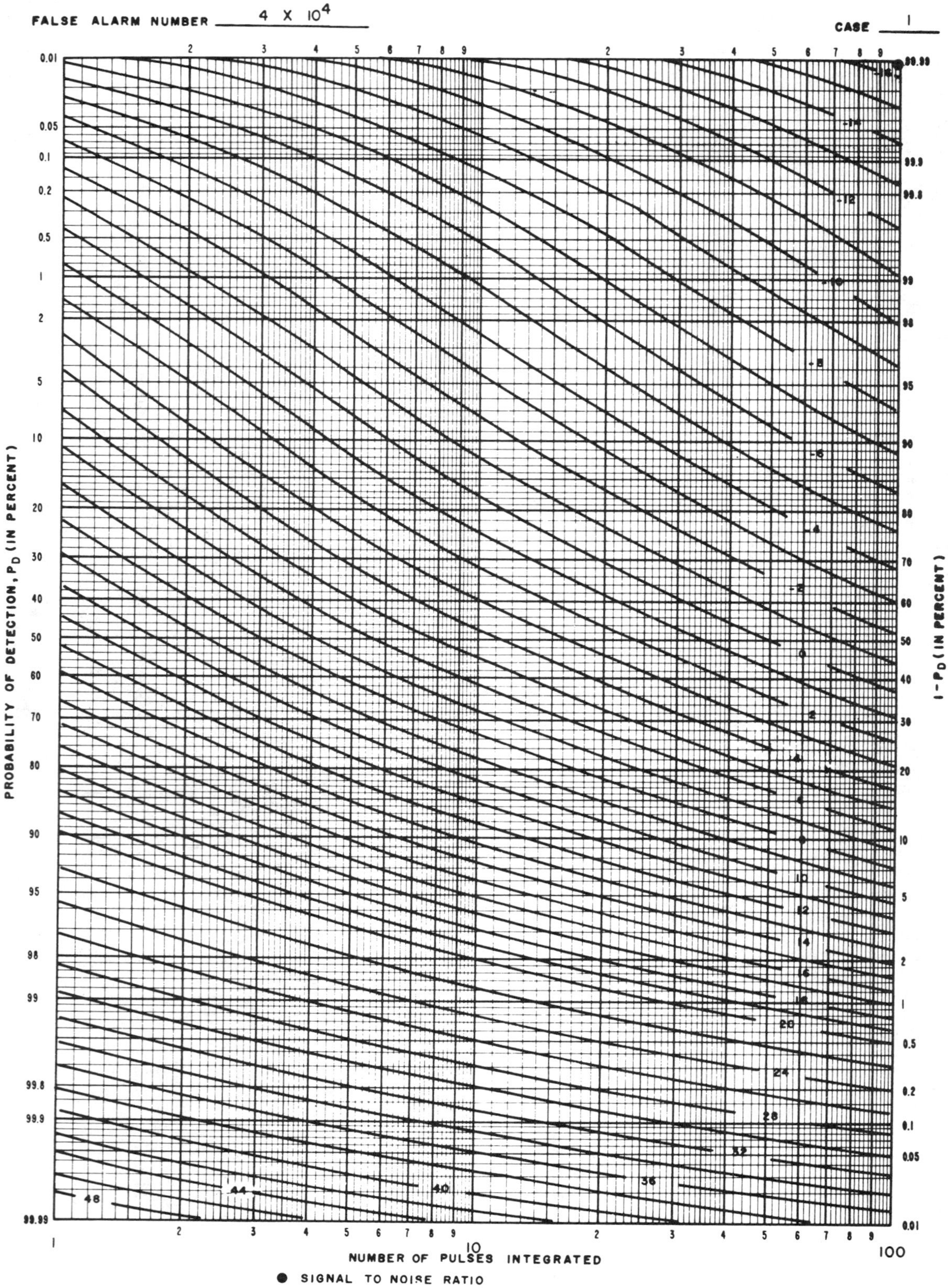

FALSE ALARM NUMBER 4 X 10^4
CASE I
PROBABILITY OF DETECTION, P_D (IN PERCENT)
1-P_D (IN PERCENT)
NUMBER OF PULSES INTEGRATED
● SIGNAL TO NOISE RATIO

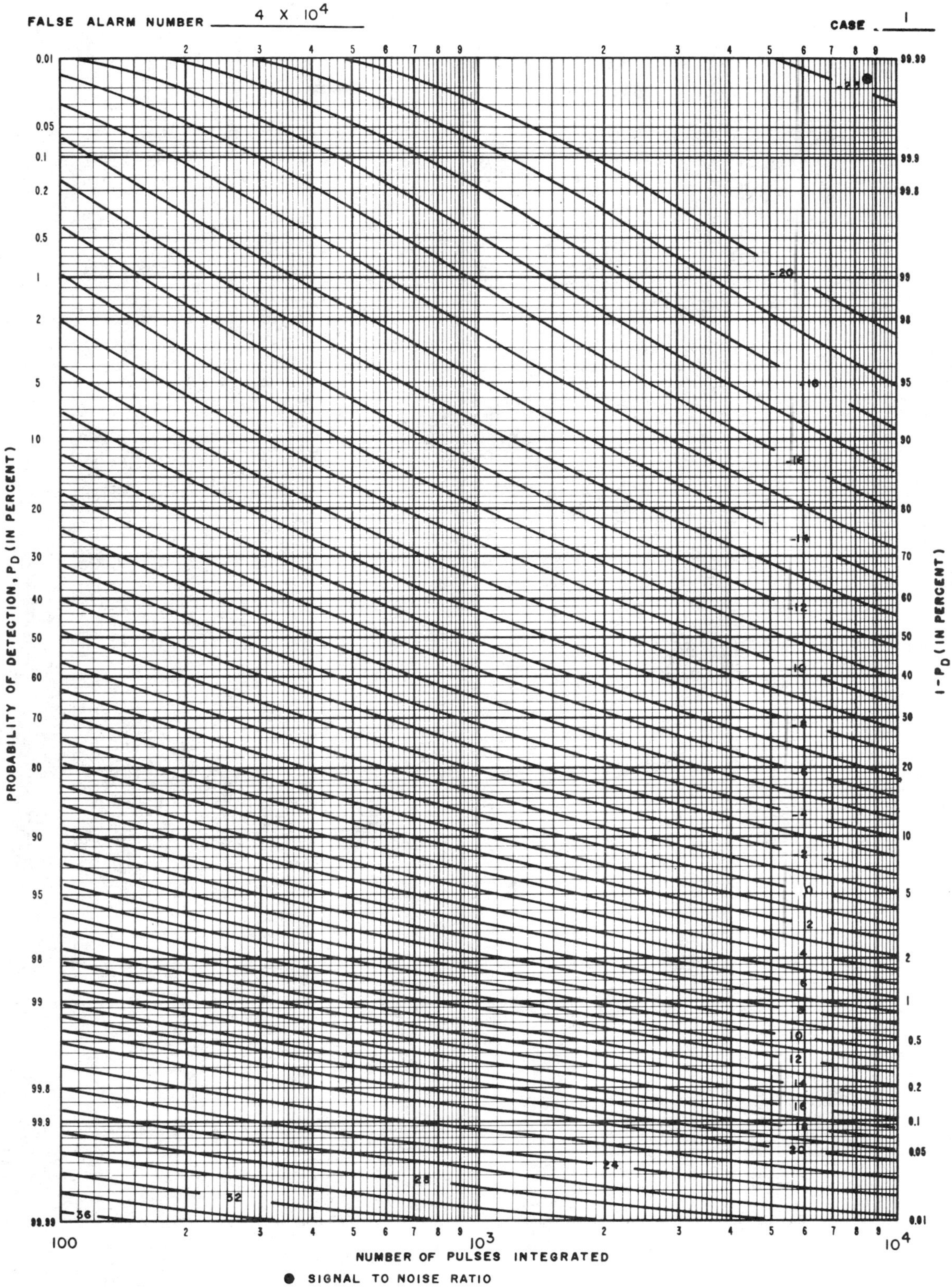
FALSE ALARM NUMBER 4 X 10^4
CASE 1
PROBABILITY OF DETECTION, P_D (IN PERCENT)
1 - P_D (IN PERCENT)
NUMBER OF PULSES INTEGRATED
100
10^3
10^4
● SIGNAL TO NOISE RATIO
-20
-18
-16
-14
-12
-10
-8
-6
-4
-2
0
2
4
6
8
10
12
14
16
18
20
24
28
32
36

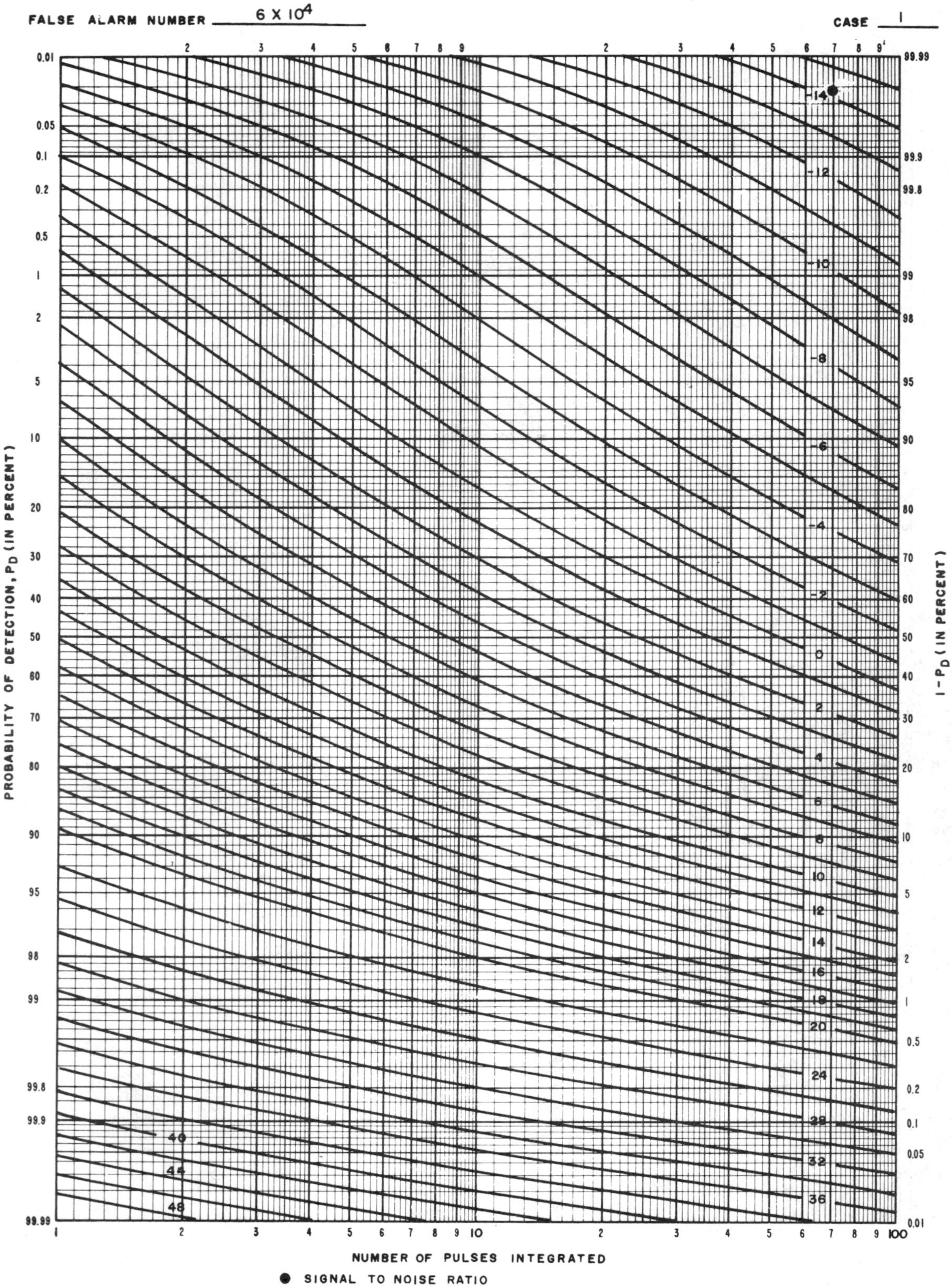
FALSE ALARM NUMBER 6 X 10^4
CASE 1
PROBABILITY OF DETECTION, P_D (IN PERCENT)
1 - P_D (IN PERCENT)
-14
-12
-10
-8
-6
-4
-2
0
2
4
6
8
10
12
14
16
18
20
24
28
32
36
40
44
48
NUMBER OF PULSES INTEGRATED
● SIGNAL TO NOISE RATIO

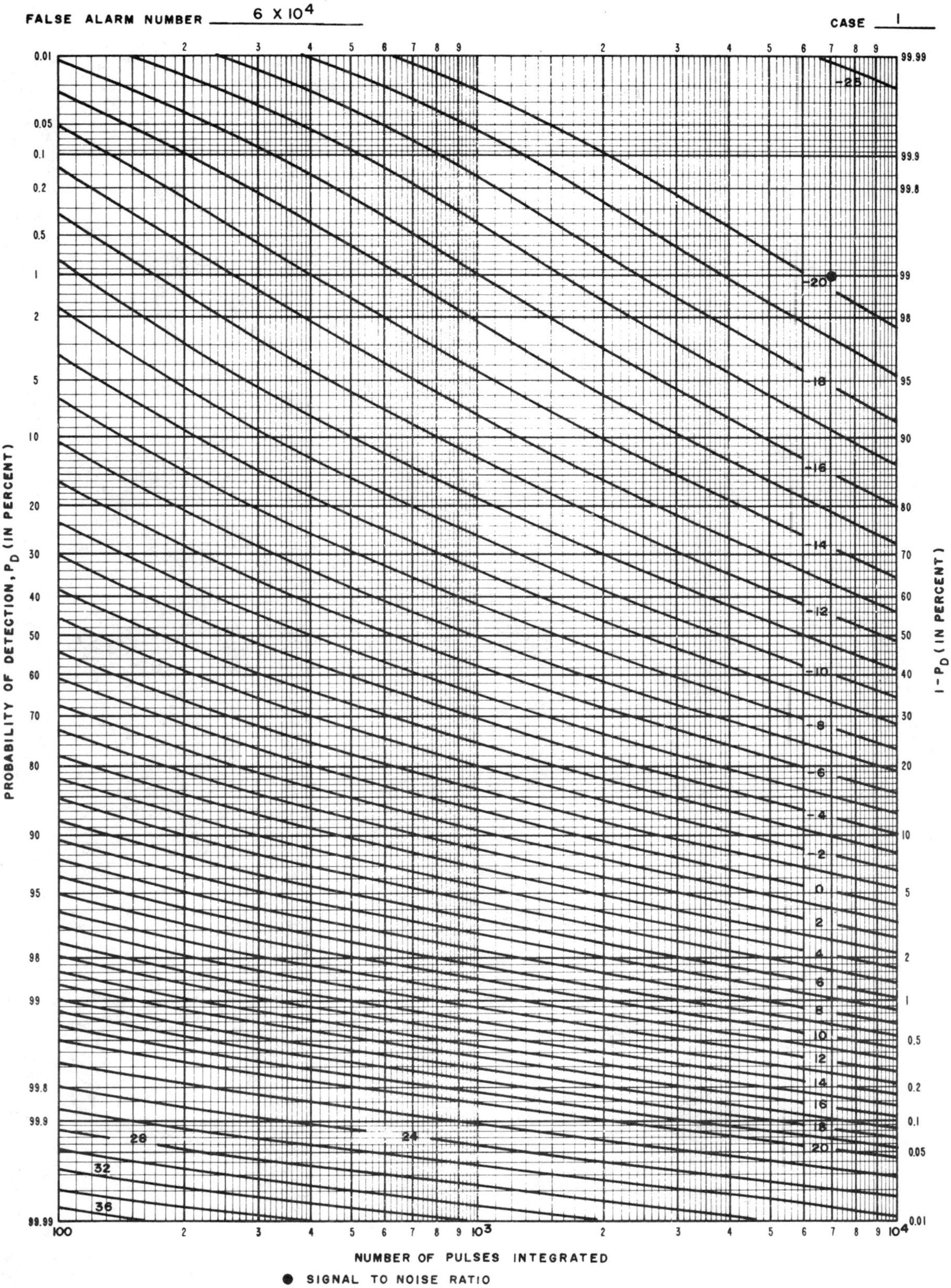
FALSE ALARM NUMBER 6 X 10^4
CASE I
PROBABILITY OF DETECTION, P_D (IN PERCENT)
1 - P_D (IN PERCENT)
NUMBER OF PULSES INTEGRATED
● SIGNAL TO NOISE RATIO

FALSE ALARM NUMBER 8×10^4

CASE I

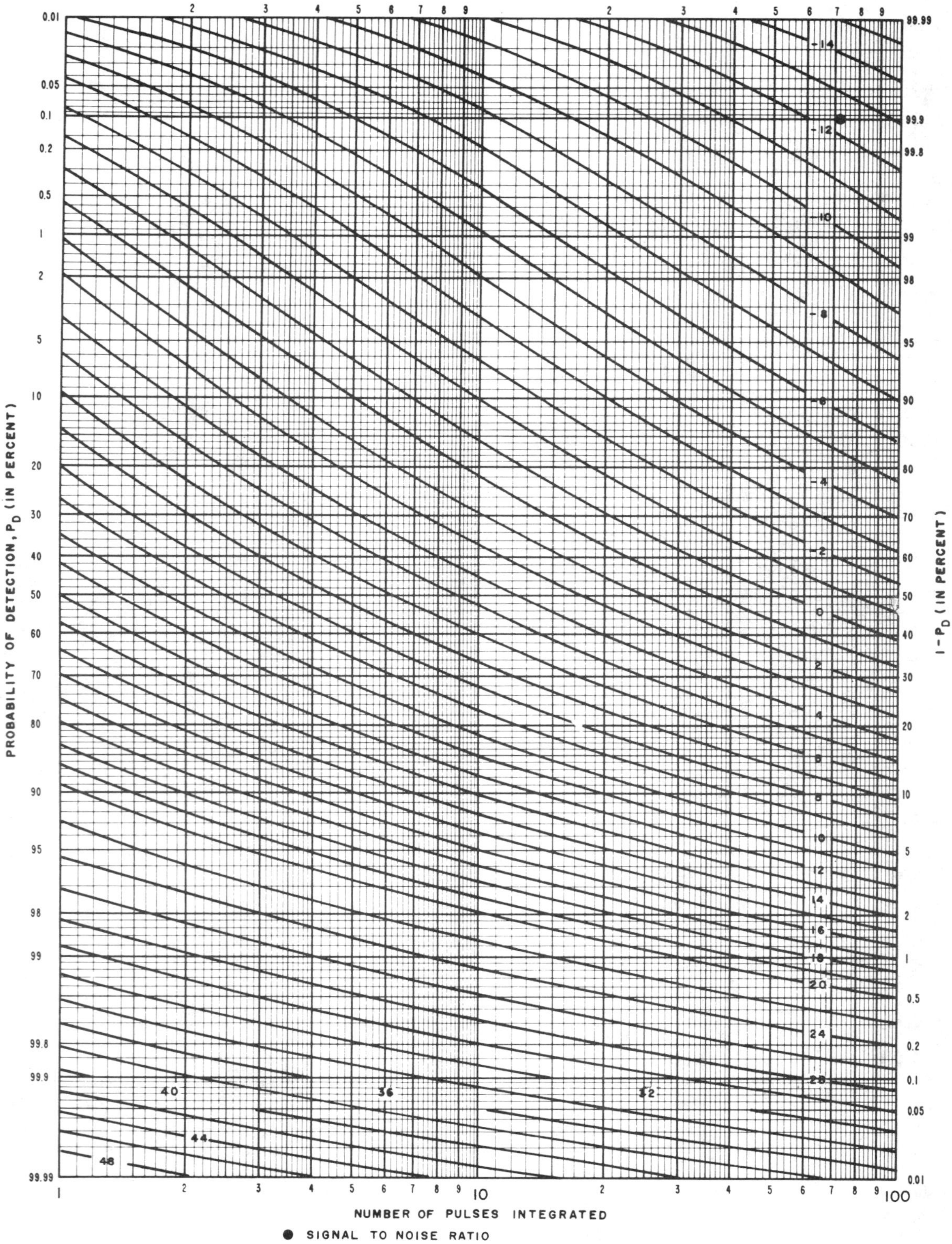

FALSE ALARM NUMBER 8×10^4 CASE 1

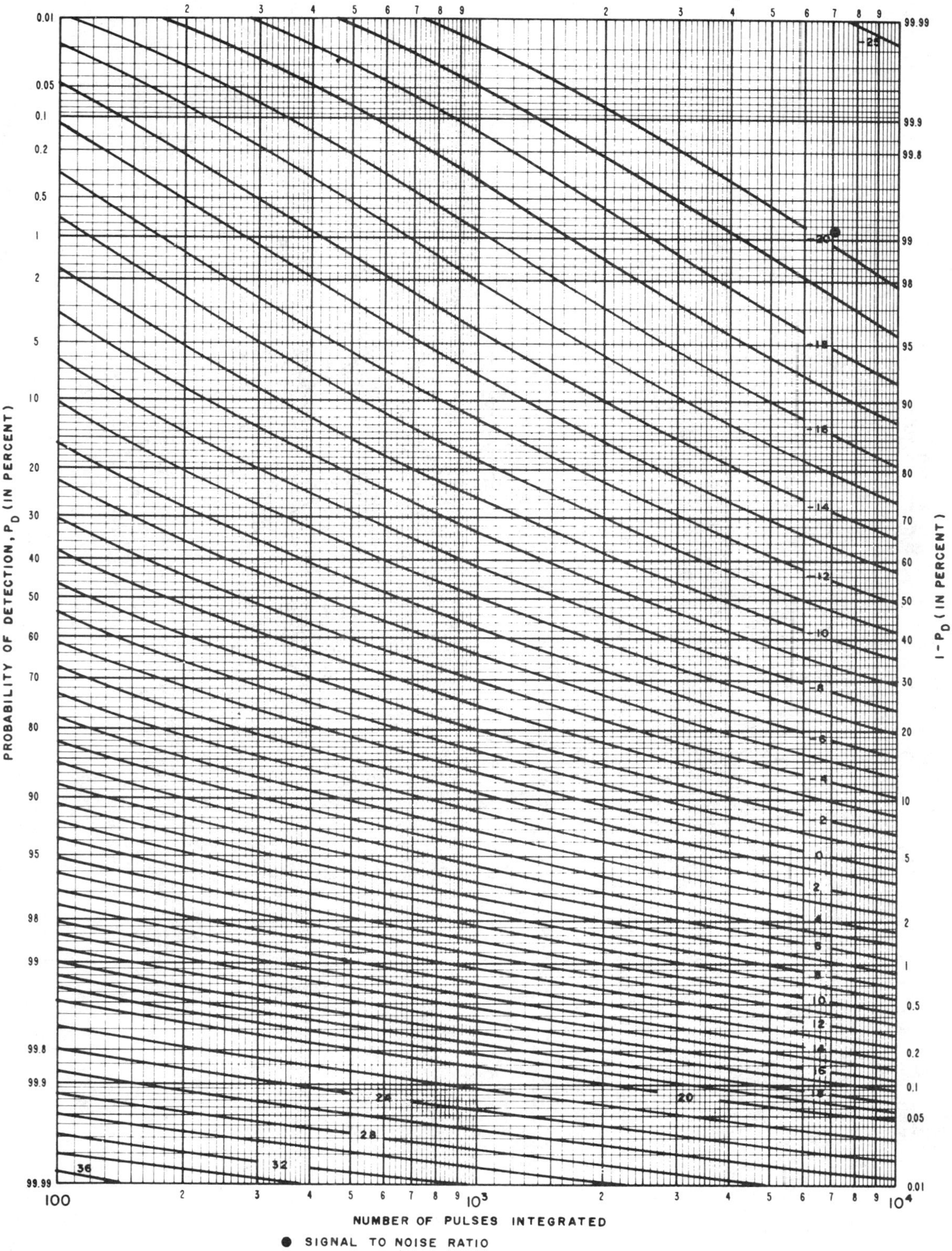

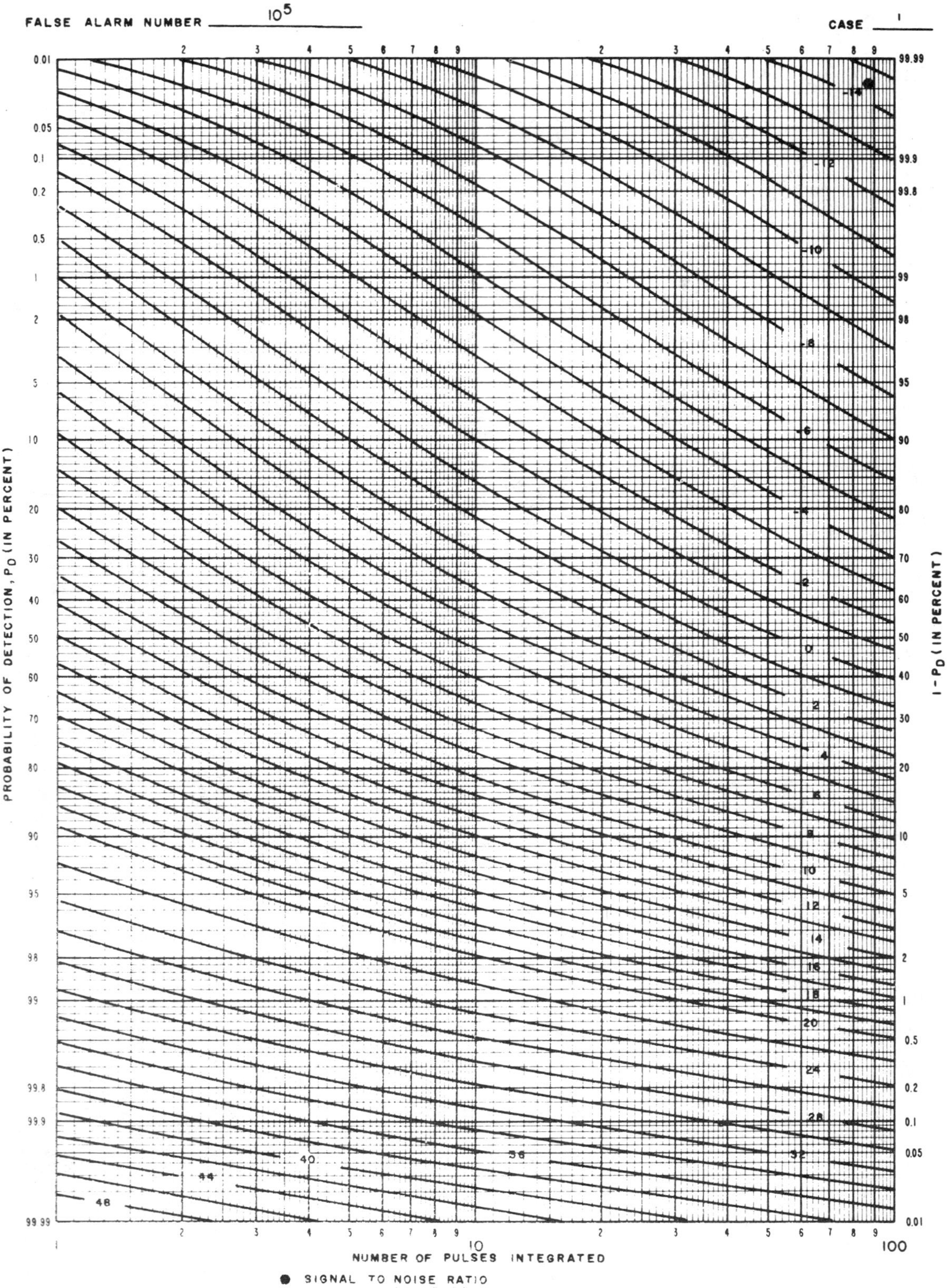
FALSE ALARM NUMBER 10^5
CASE 1
PROBABILITY OF DETECTION, P_D (IN PERCENT)
1 - P_D (IN PERCENT)
NUMBER OF PULSES INTEGRATED
● SIGNAL TO NOISE RATIO
-14
-12
-10
-8
-6
-4
-2
0
2
4
6
8
10
12
14
16
18
20
24
28
32
36
40
44
48

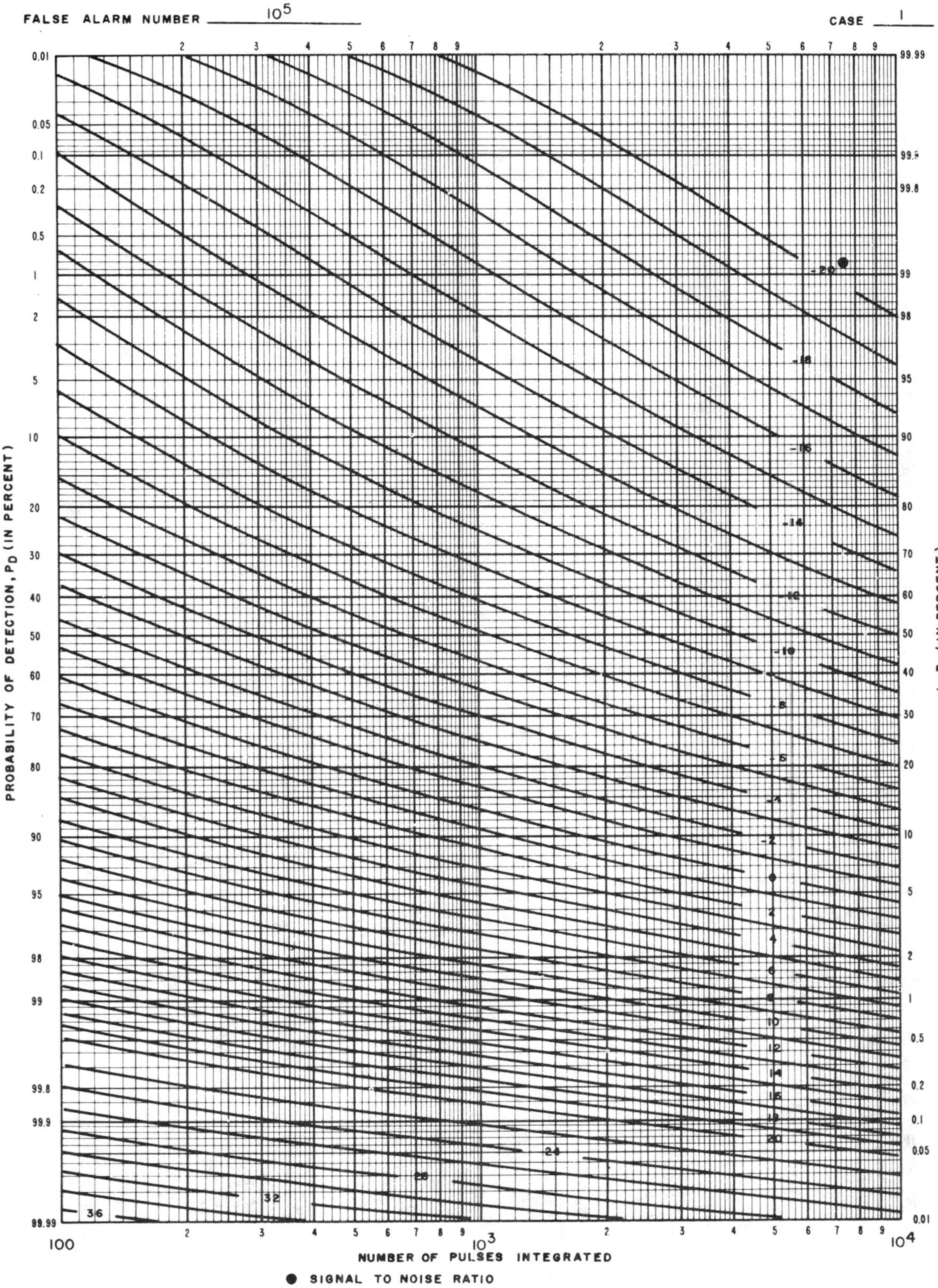
FALSE ALARM NUMBER 10^5
CASE 1
PROBABILITY OF DETECTION, P_D (IN PERCENT)
$1 - P_D$ (IN PERCENT)
NUMBER OF PULSES INTEGRATED
100
10^3
10^4
● SIGNAL TO NOISE RATIO

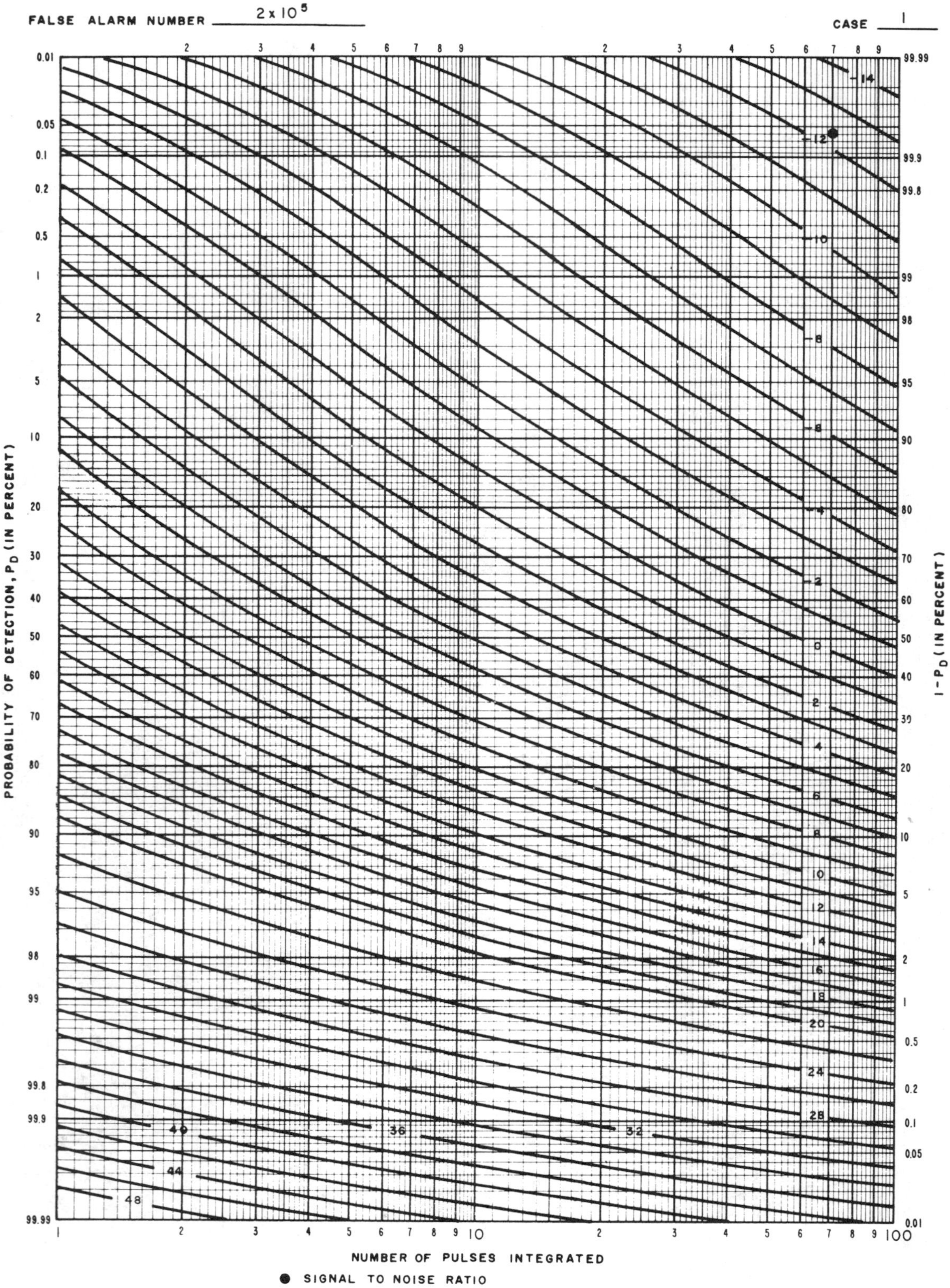
FALSE ALARM NUMBER 2 x 10^5
CASE 1
PROBABILITY OF DETECTION, P_D (IN PERCENT)
1 - P_D (IN PERCENT)
NUMBER OF PULSES INTEGRATED
● SIGNAL TO NOISE RATIO

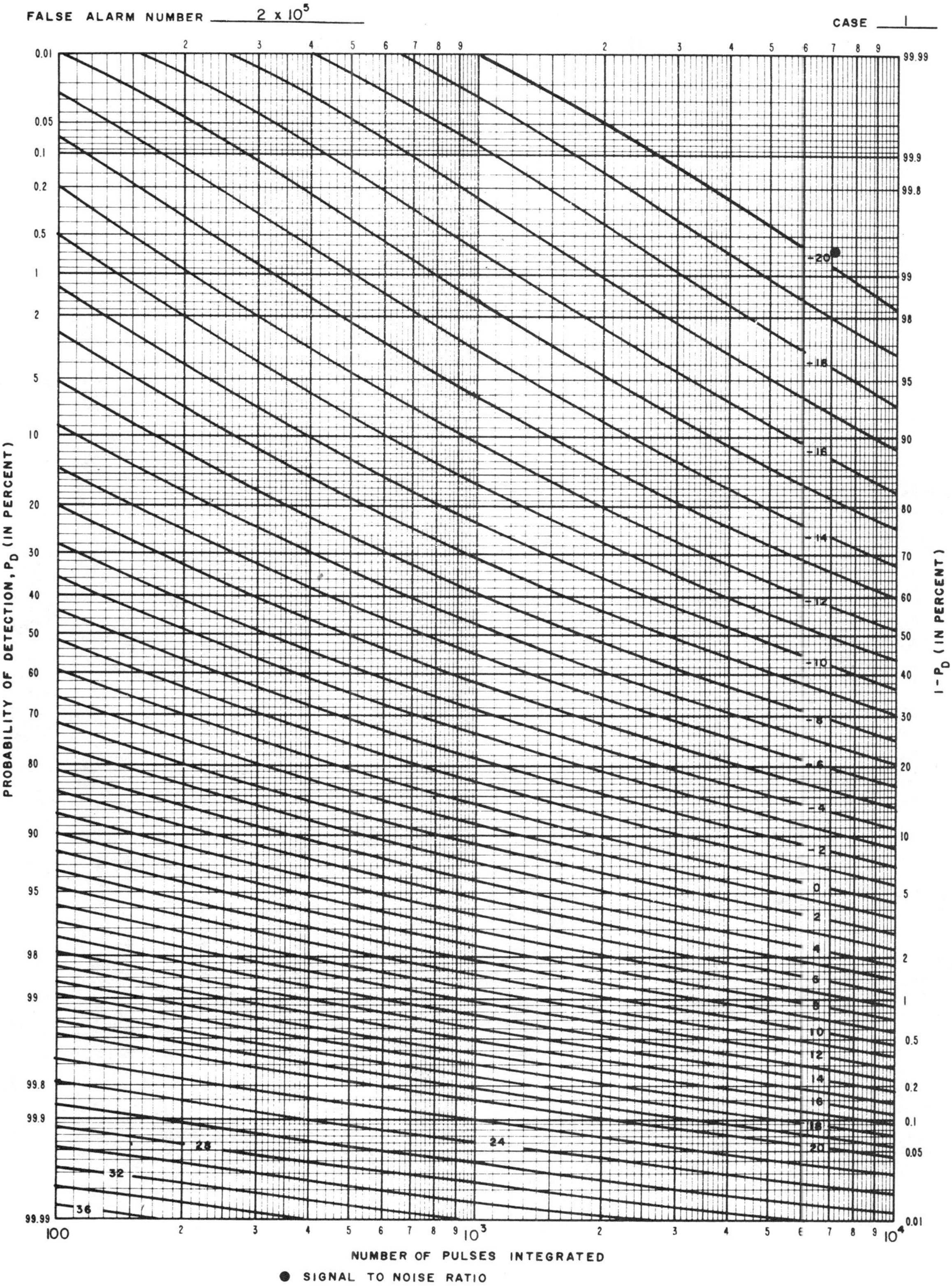
FALSE ALARM NUMBER 2 x 10^5
CASE 1
PROBABILITY OF DETECTION, P_D (IN PERCENT)
1-P_D (IN PERCENT)
-20
-18
-16
-14
-12
-10
-8
-6
-4
-2
0
2
4
6
8
10
12
14
16
18
20
24
28
32
36
100
10^3
10^4
NUMBER OF PULSES INTEGRATED
● SIGNAL TO NOISE RATIO

FALSE ALARM NUMBER 4×10^5 CASE 1

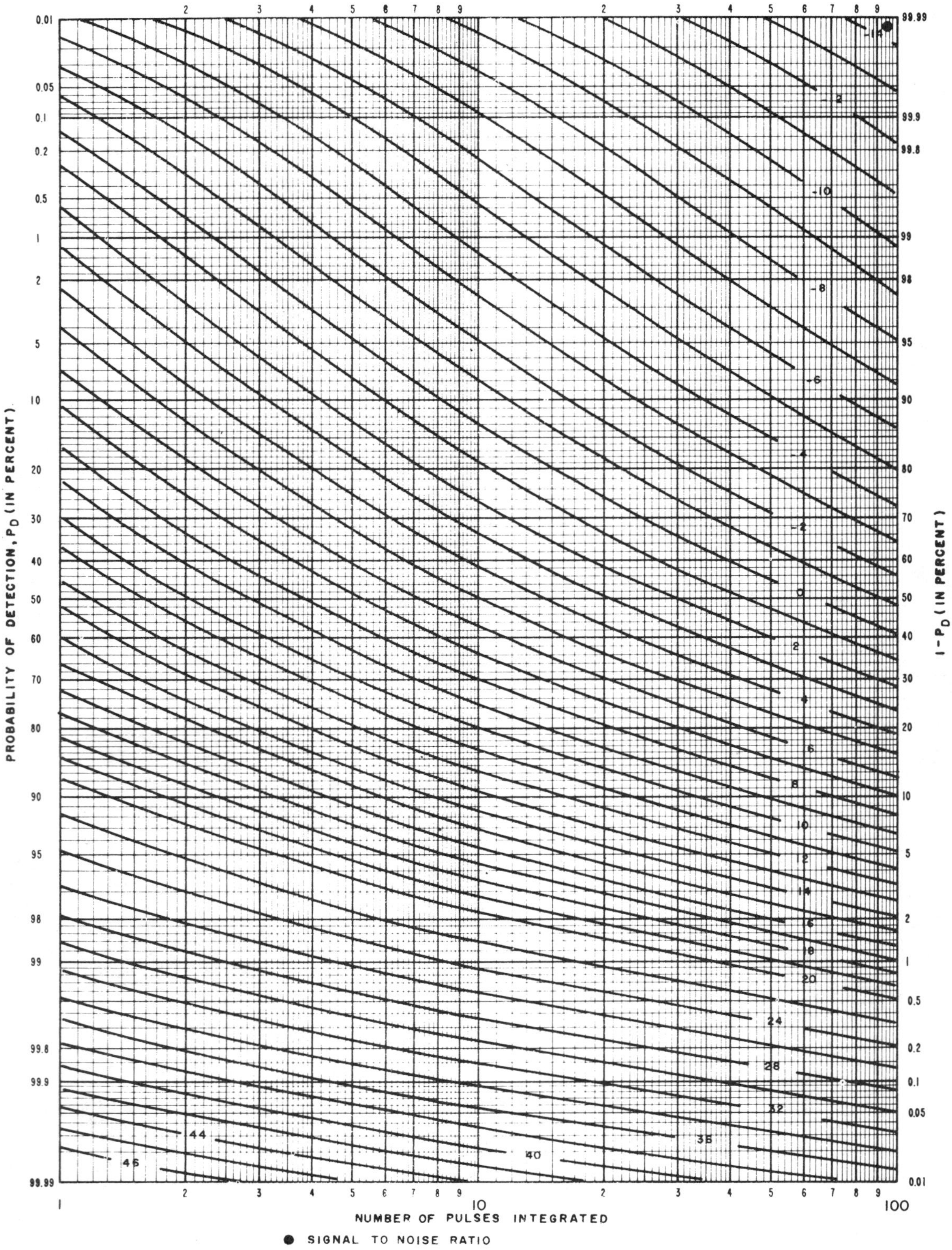

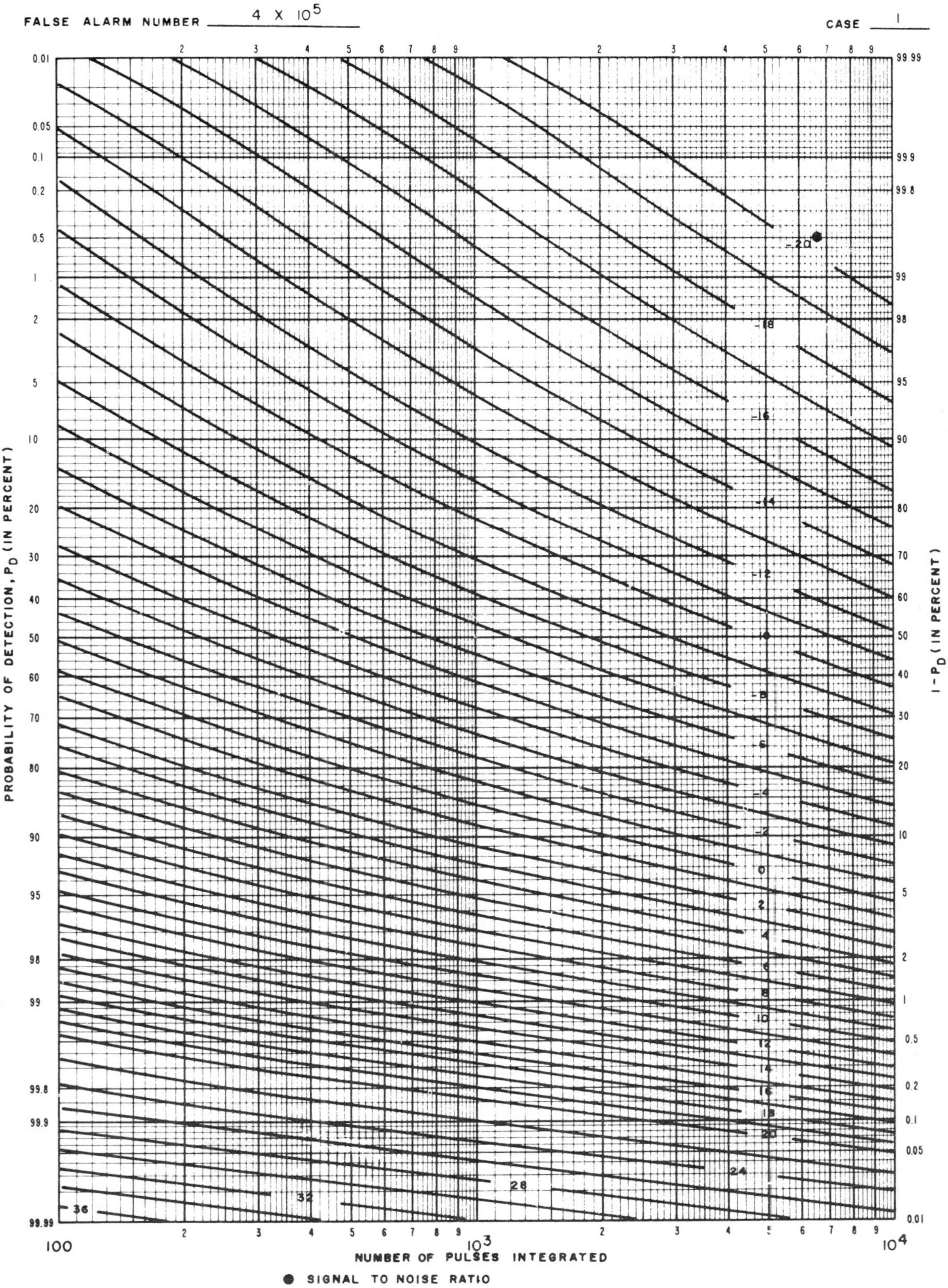
FALSE ALARM NUMBER 4 X 10^5
CASE I
PROBABILITY OF DETECTION, P_D (IN PERCENT)
1 - P_D (IN PERCENT)
NUMBER OF PULSES INTEGRATED
100
10^3
10^4
● SIGNAL TO NOISE RATIO
-20
-18
-16
-14
-12
-10
-8
-6
-4
-2
0
2
4
6
8
10
12
14
16
18
20
24
28
32
36

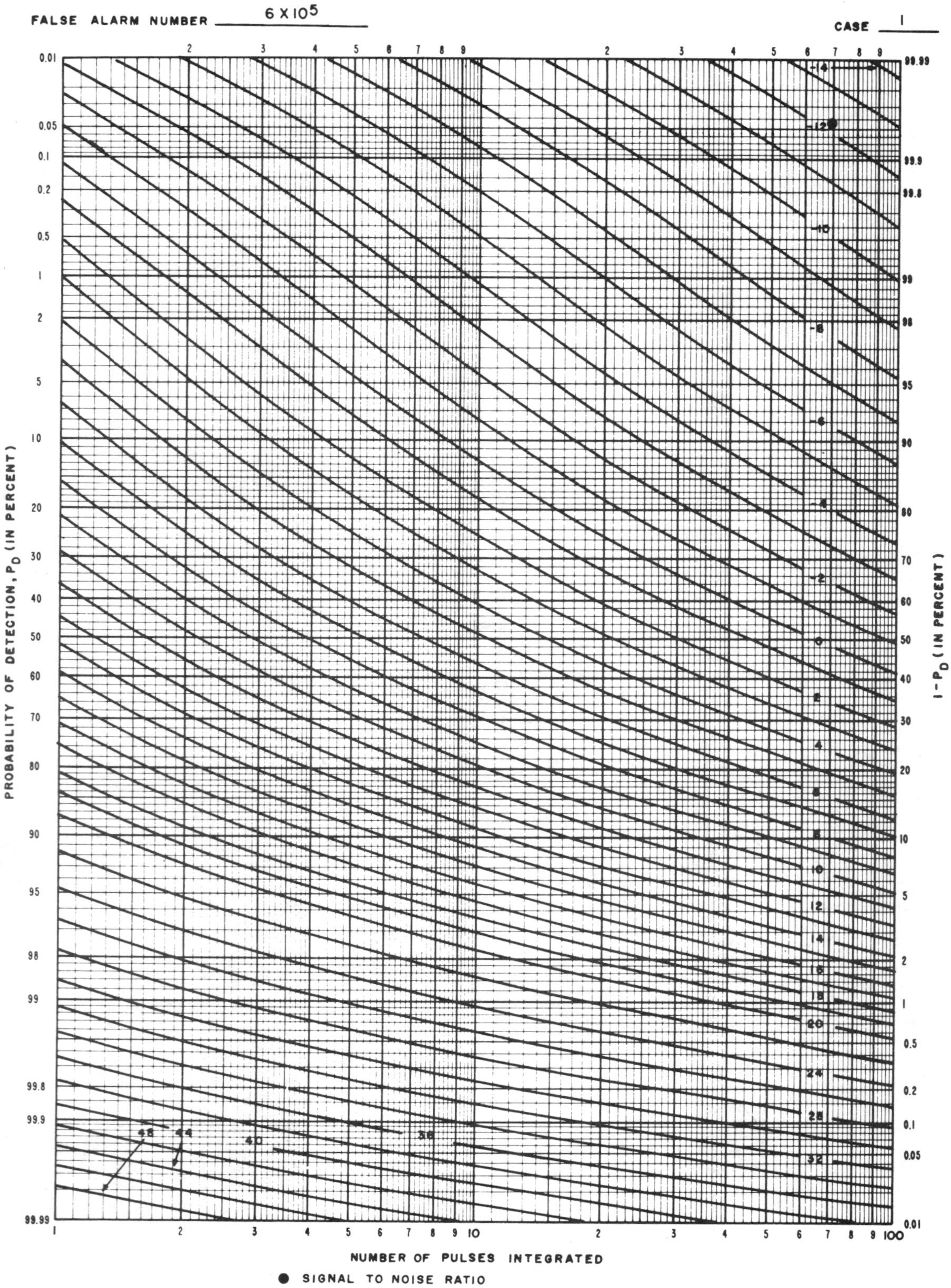
FALSE ALARM NUMBER 6 X 10^5
CASE 1
PROBABILITY OF DETECTION, P_D (IN PERCENT)
1 - P_D (IN PERCENT)
NUMBER OF PULSES INTEGRATED
● SIGNAL TO NOISE RATIO

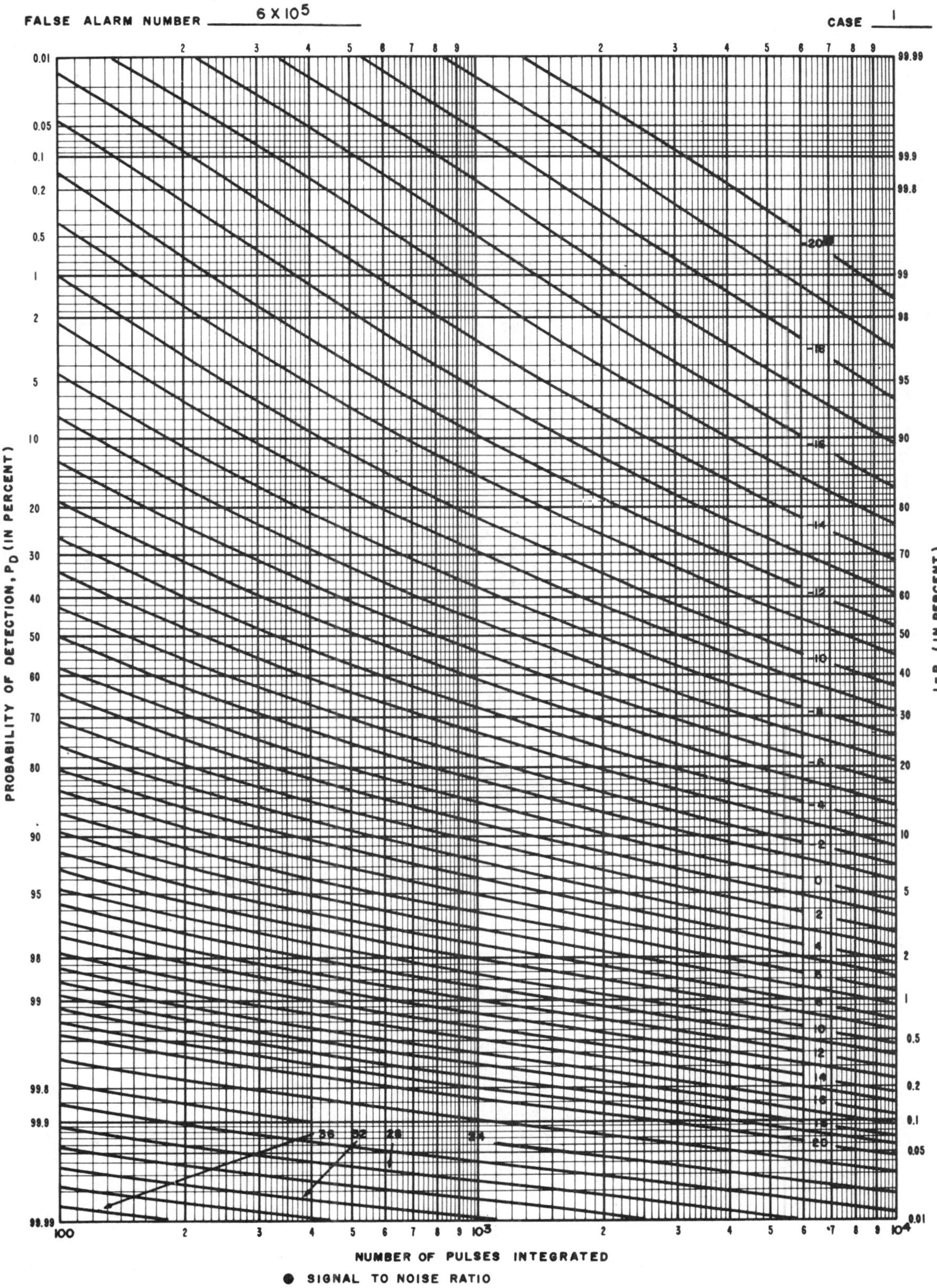
FALSE ALARM NUMBER 6 X 10^5
CASE 1
PROBABILITY OF DETECTION, P_D (IN PERCENT)
1-P_D (IN PERCENT)
NUMBER OF PULSES INTEGRATED
● SIGNAL TO NOISE RATIO

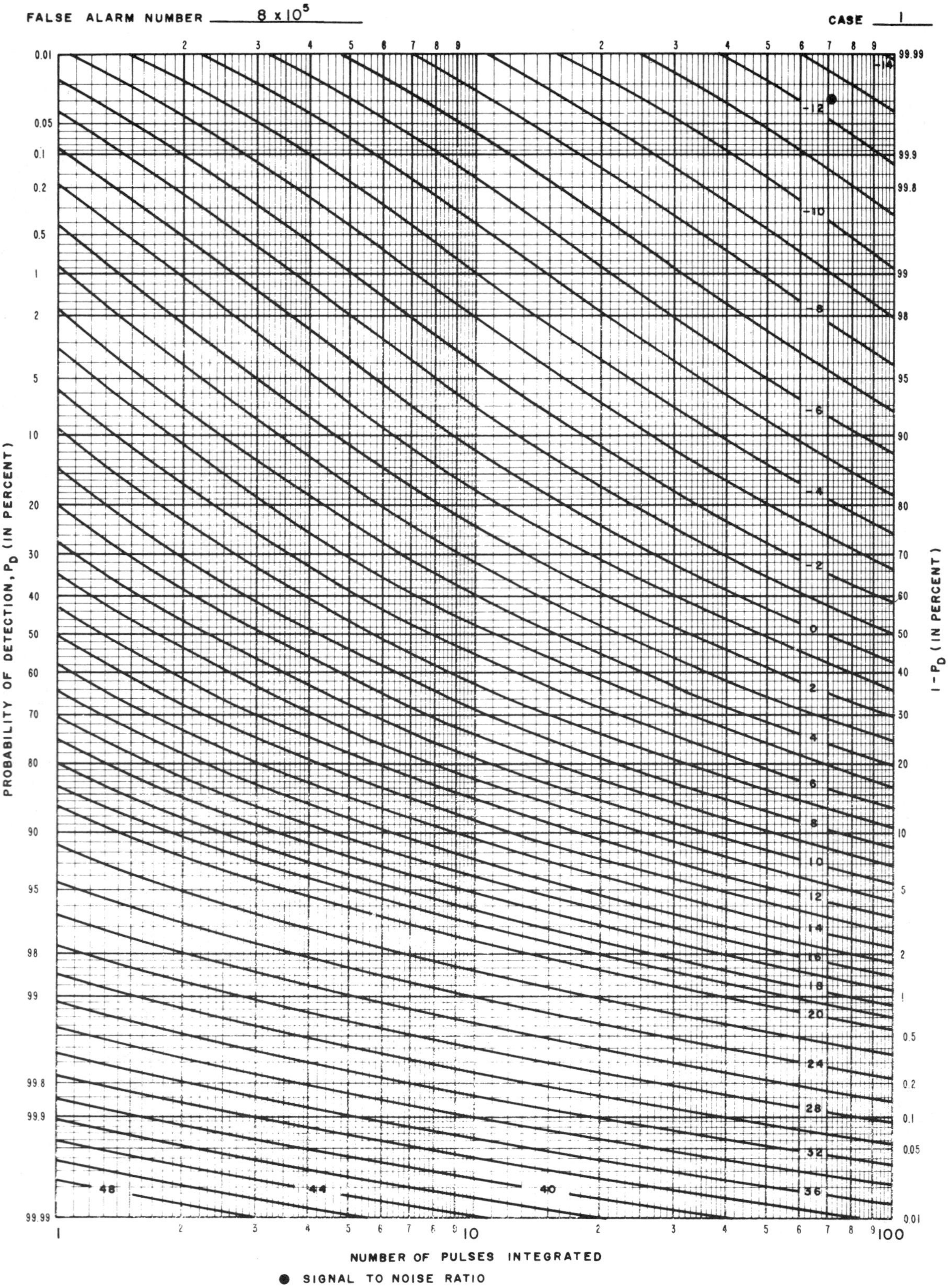
FALSE ALARM NUMBER 8 x 10^5
CASE 1
PROBABILITY OF DETECTION, P_D (IN PERCENT)
1 - P_D (IN PERCENT)
NUMBER OF PULSES INTEGRATED
● SIGNAL TO NOISE RATIO
-14
-12
-10
-8
-6
-4
-2
0
2
4
6
8
10
12
14
16
18
20
24
28
32
36
40
44
48

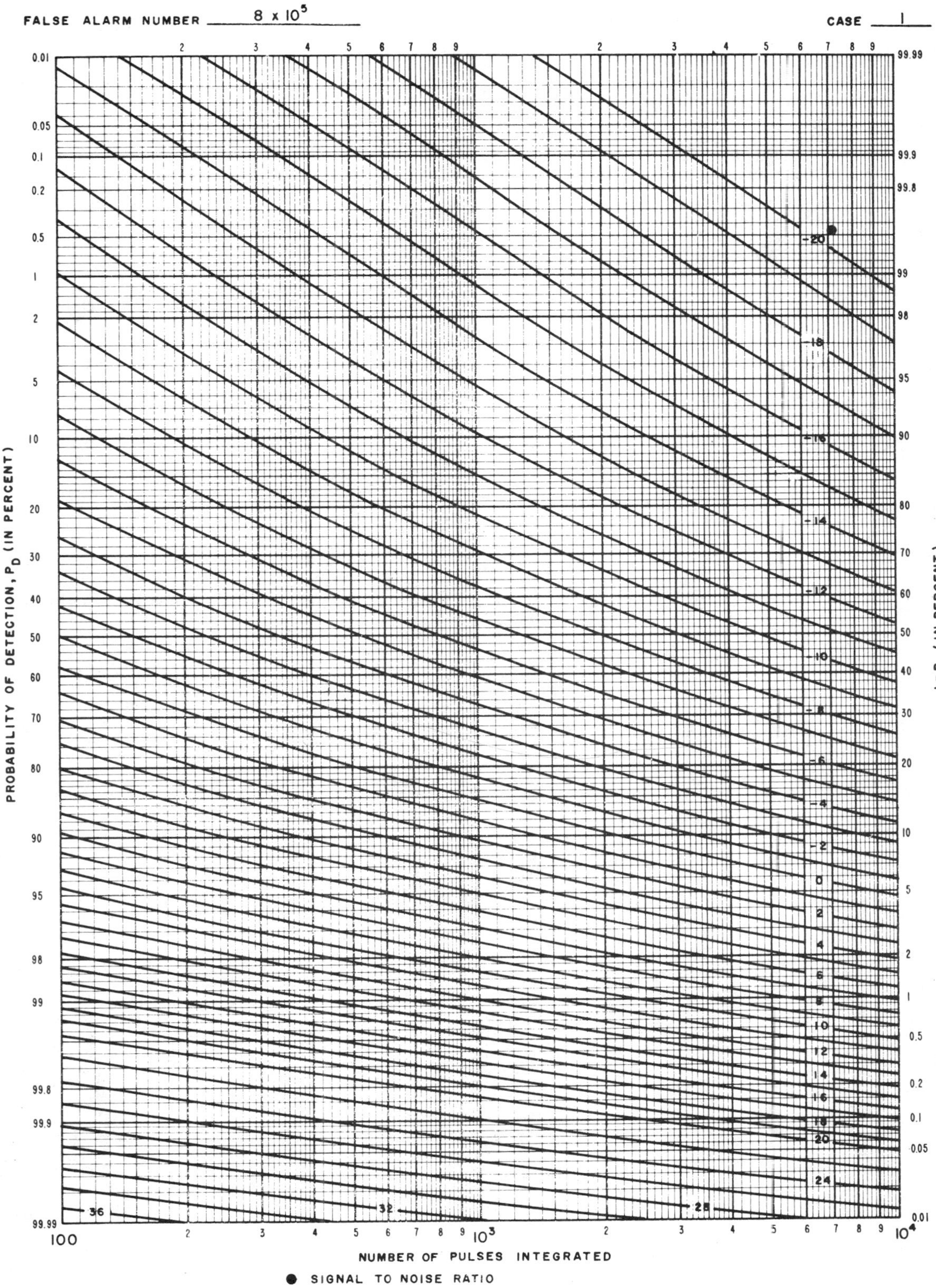
FALSE ALARM NUMBER 8 x 10^5
CASE 1
PROBABILITY OF DETECTION, P_D (IN PERCENT)
1 - P_D (IN PERCENT)
NUMBER OF PULSES INTEGRATED
100
10^3
10^4
● SIGNAL TO NOISE RATIO
-20
-18
-16
-14
-12
-10
-8
-6
-4
-2
0
2
4
6
8
10
12
14
16
18
20
24
28
32
36

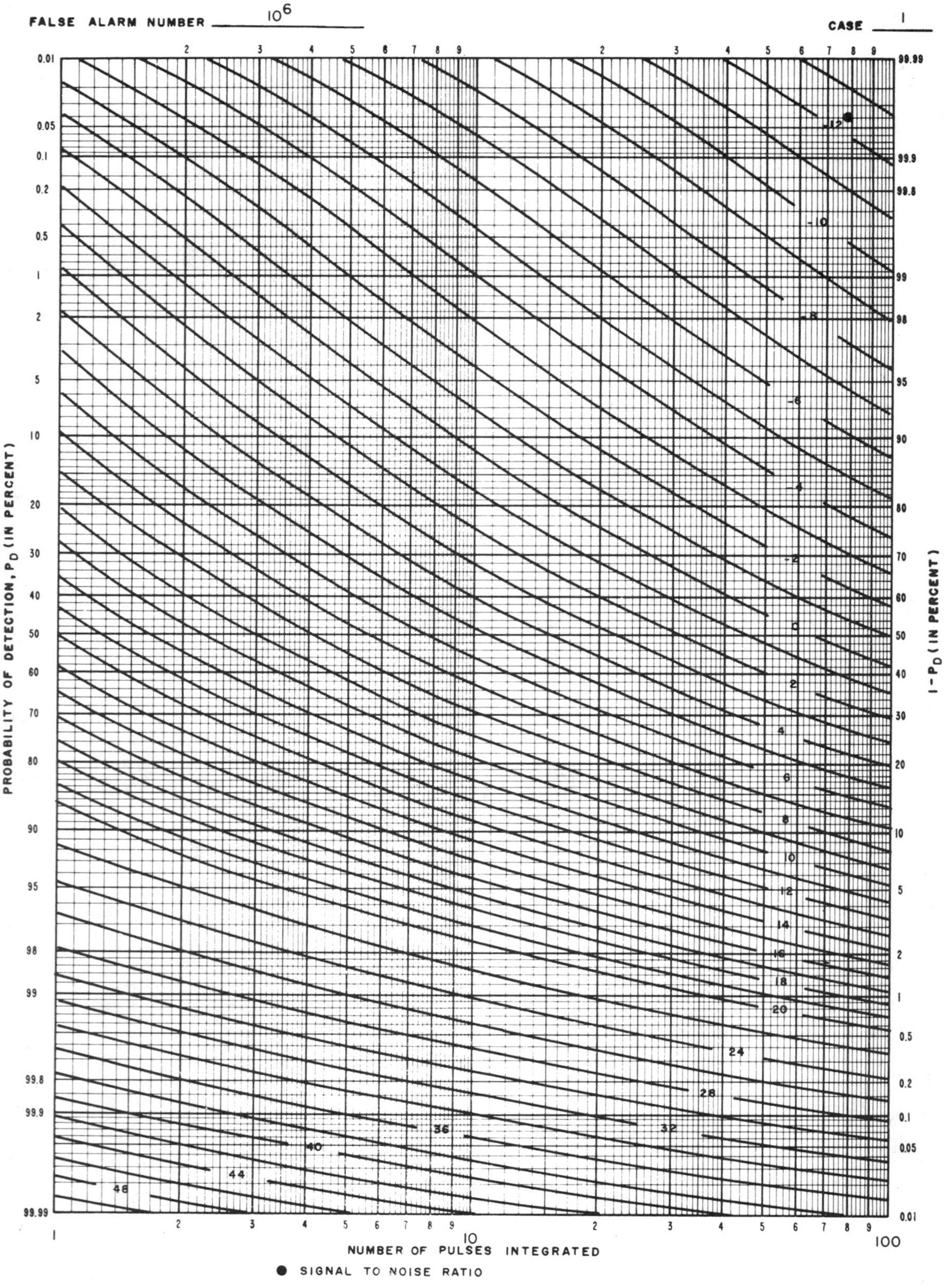
FALSE ALARM NUMBER 10^6
CASE 1
PROBABILITY OF DETECTION, P_D (IN PERCENT)
1 - P_D (IN PERCENT)
NUMBER OF PULSES INTEGRATED
● SIGNAL TO NOISE RATIO
-12
-10
-8
-6
-4
-2
0
2
4
6
8
10
12
14
16
18
20
24
28
32
36
40
44
48

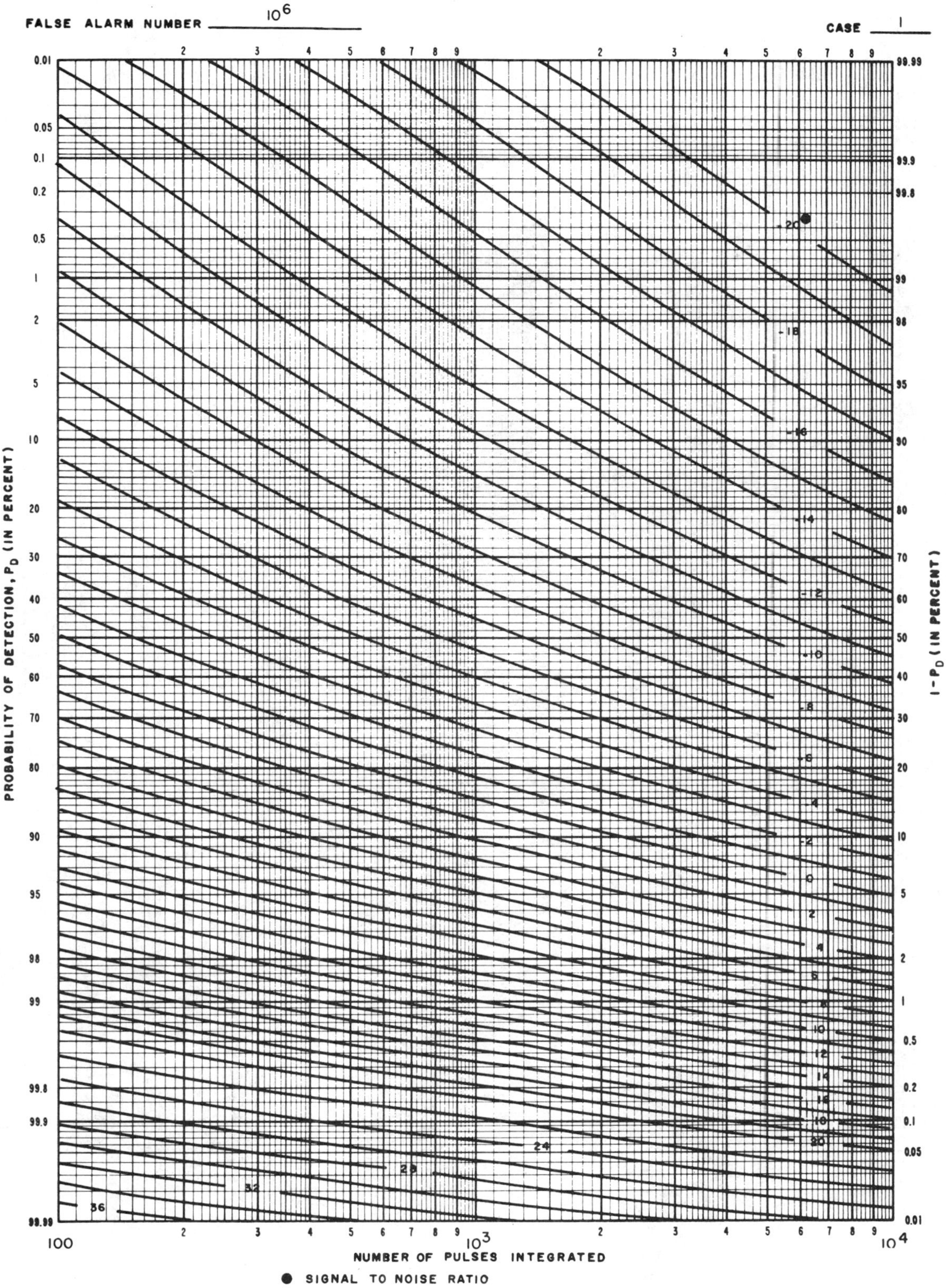
FALSE ALARM NUMBER 10^6
CASE 1
PROBABILITY OF DETECTION, P_D (IN PERCENT)
1 - P_D (IN PERCENT)
NUMBER OF PULSES INTEGRATED
● SIGNAL TO NOISE RATIO
100
10^3
10^4
-20
-18
-16
-14
-12
-10
-8
-6
-4
-2
0
2
4
6
8
10
12
14
16
18
20
24
28
32
36

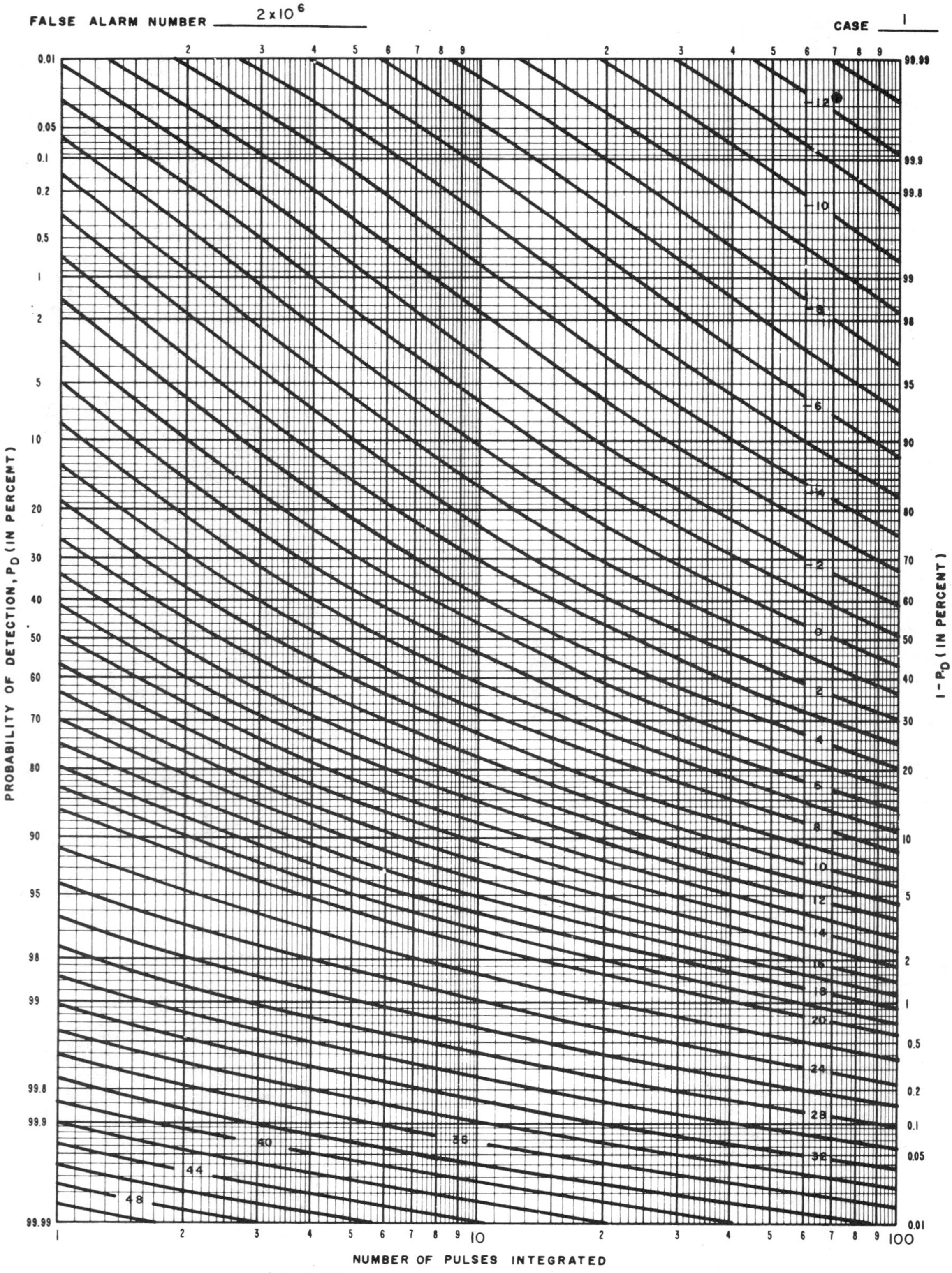
FALSE ALARM NUMBER 2x10^6
CASE I
PROBABILITY OF DETECTION, P_D (IN PERCENT)
1 - P_D (IN PERCENT)
NUMBER OF PULSES INTEGRATED
● SIGNAL TO NOISE RATIO

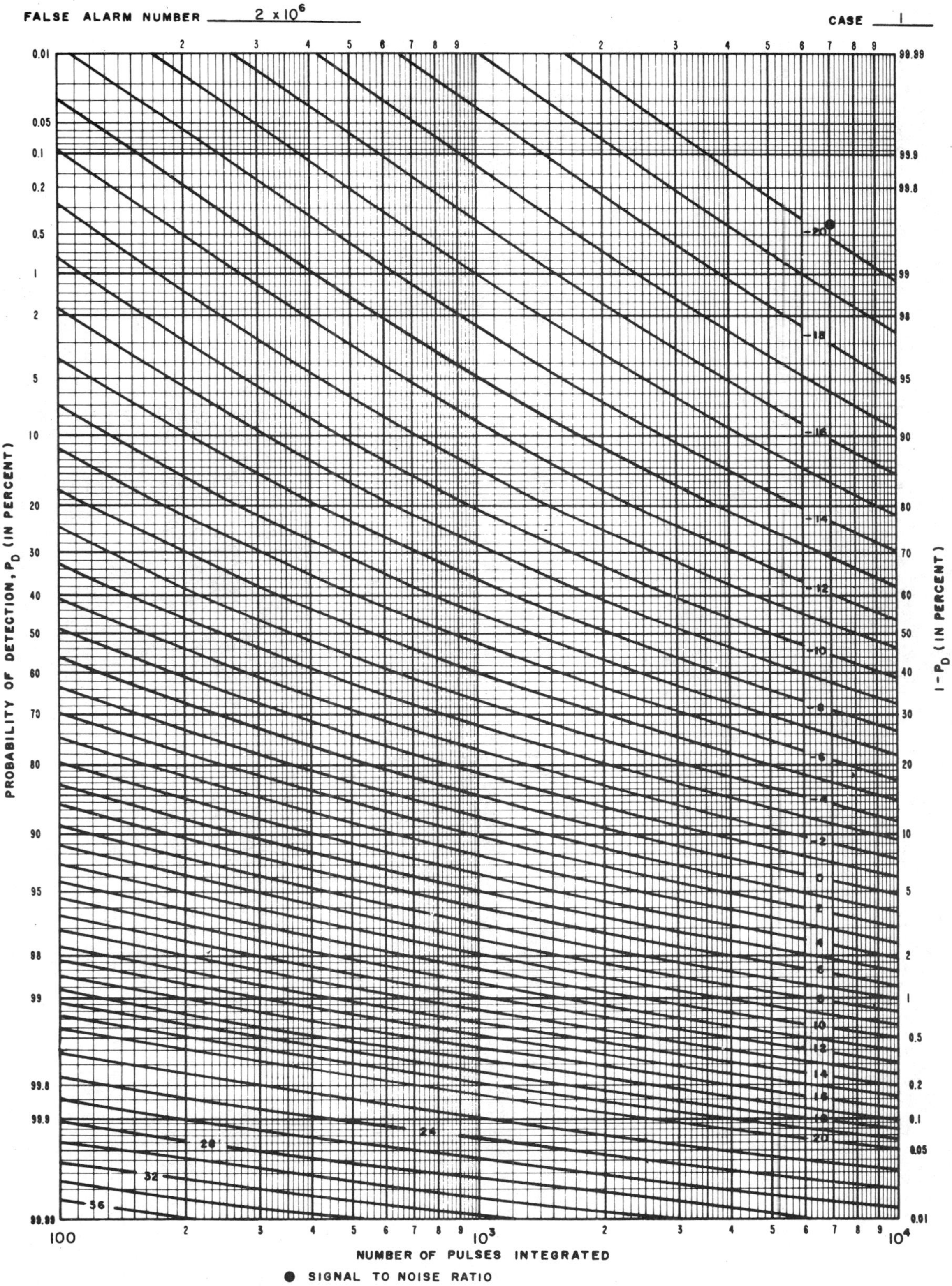
FALSE ALARM NUMBER 2 x 10^6
CASE 1
PROBABILITY OF DETECTION, P_D (IN PERCENT)
1 - P_D (IN PERCENT)
NUMBER OF PULSES INTEGRATED
● SIGNAL TO NOISE RATIO

FALSE ALARM NUMBER 4×10^6 CASE 1

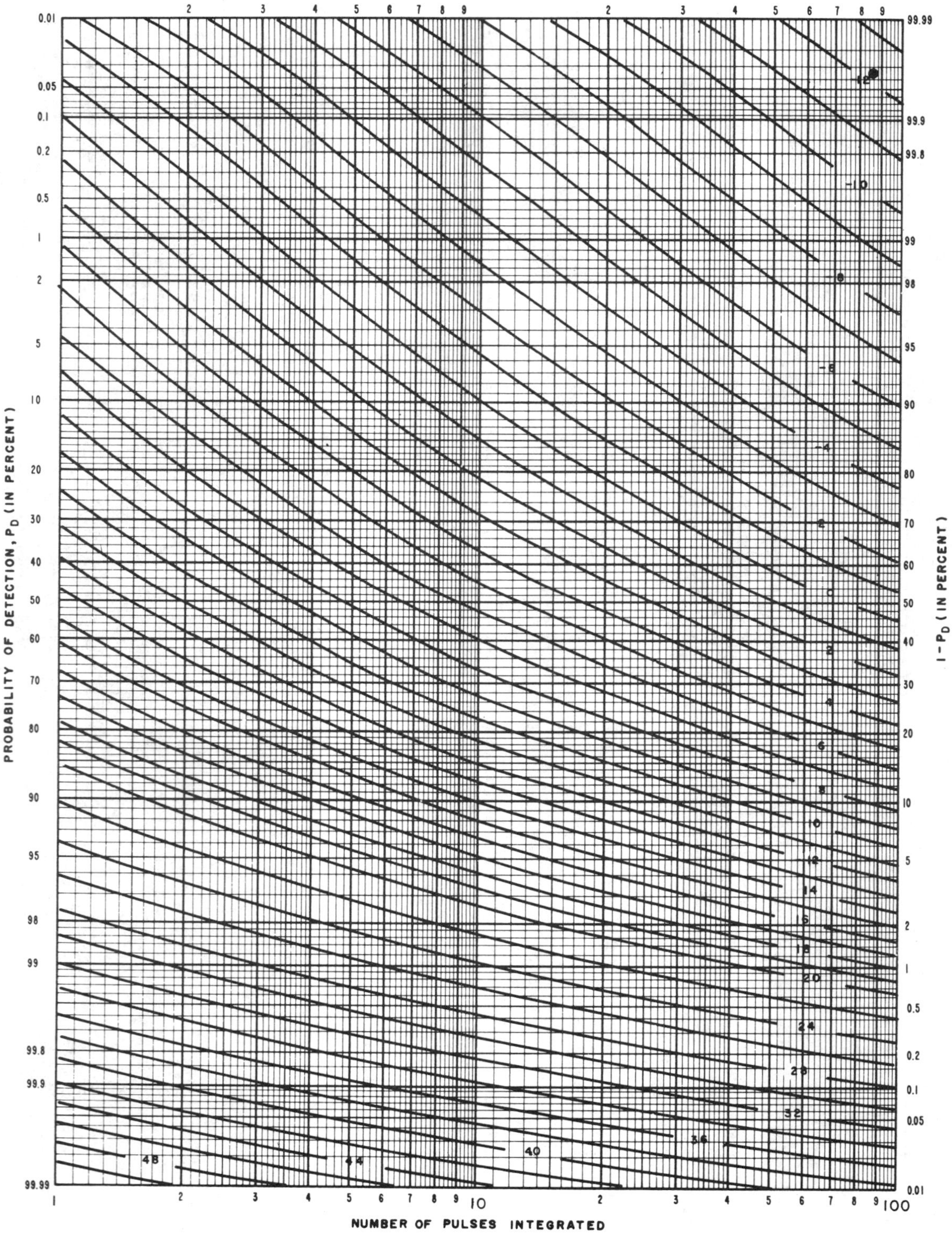

● SIGNAL TO NOISE RATIO

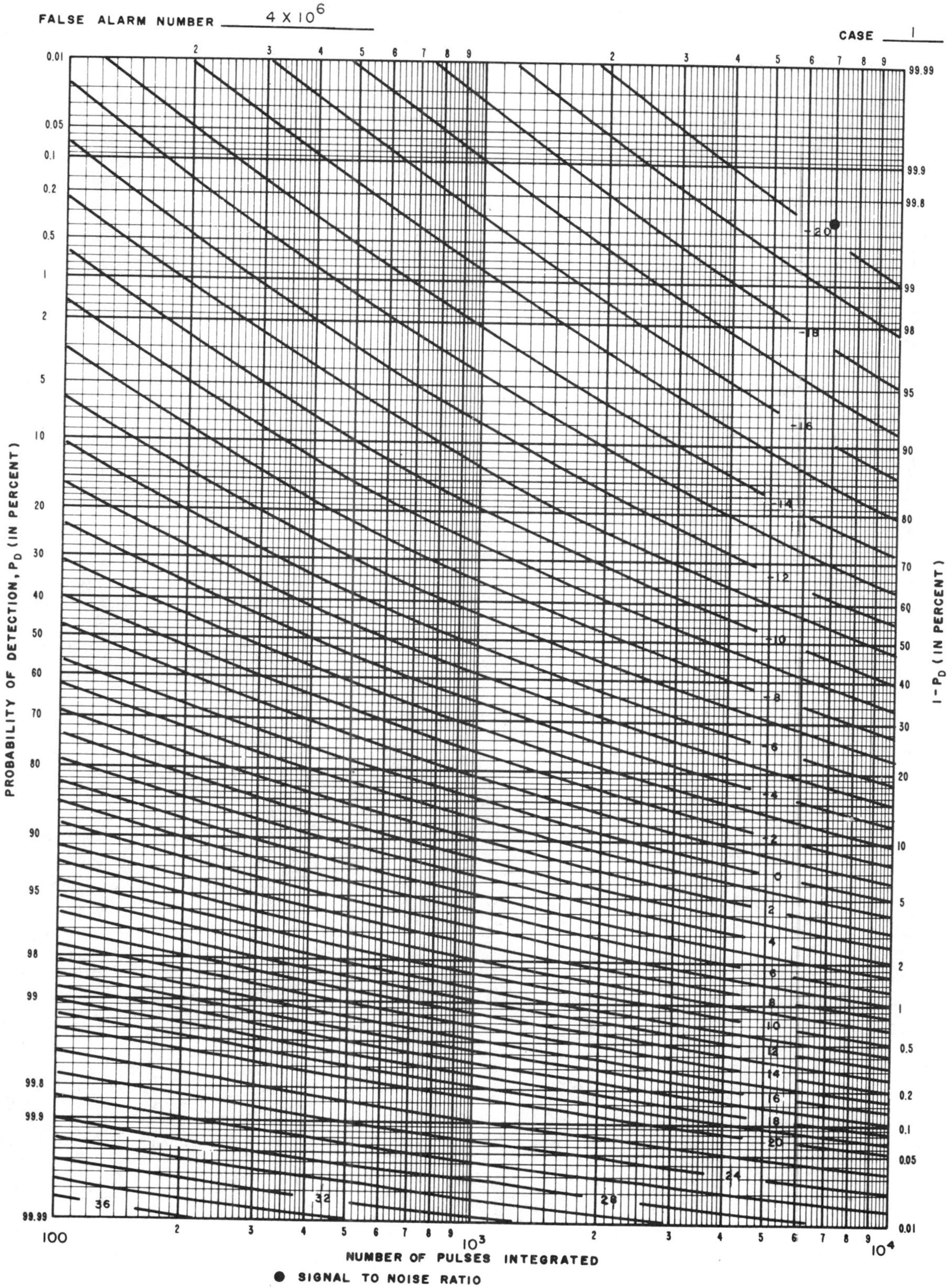
FALSE ALARM NUMBER 4 X 10^6
CASE 1
PROBABILITY OF DETECTION, P_D (IN PERCENT)
1 - P_D (IN PERCENT)
NUMBER OF PULSES INTEGRATED
● SIGNAL TO NOISE RATIO
-20
-18
-16
-14
-12
-10
-8
-6
-4
-2
0
2
4
6
8
10
12
14
16
18
20
24
28
32
36

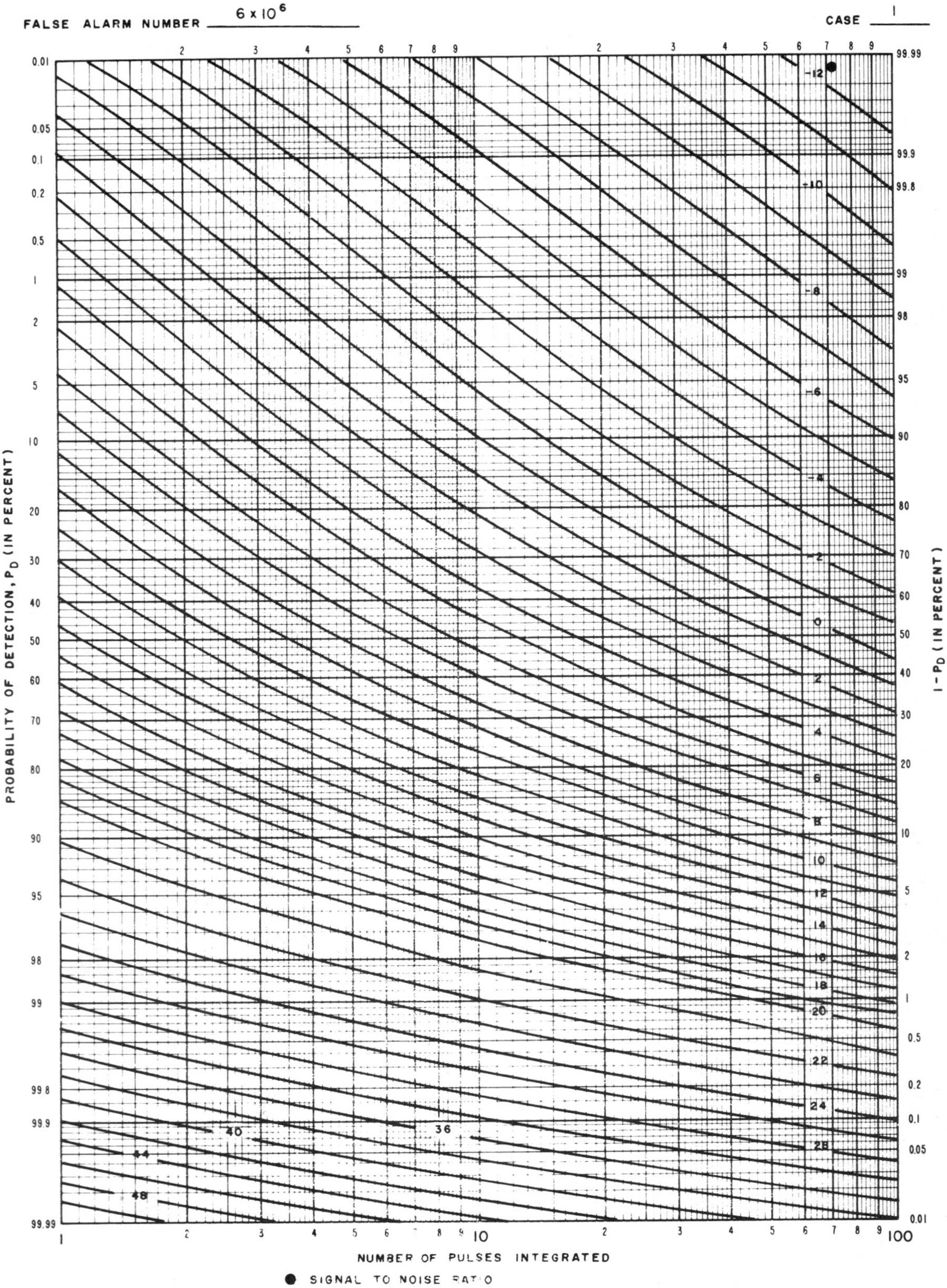
FALSE ALARM NUMBER 6 x 10^6
CASE 1
PROBABILITY OF DETECTION, P_D (IN PERCENT)
1 - P_D (IN PERCENT)
NUMBER OF PULSES INTEGRATED
● SIGNAL TO NOISE RATIO

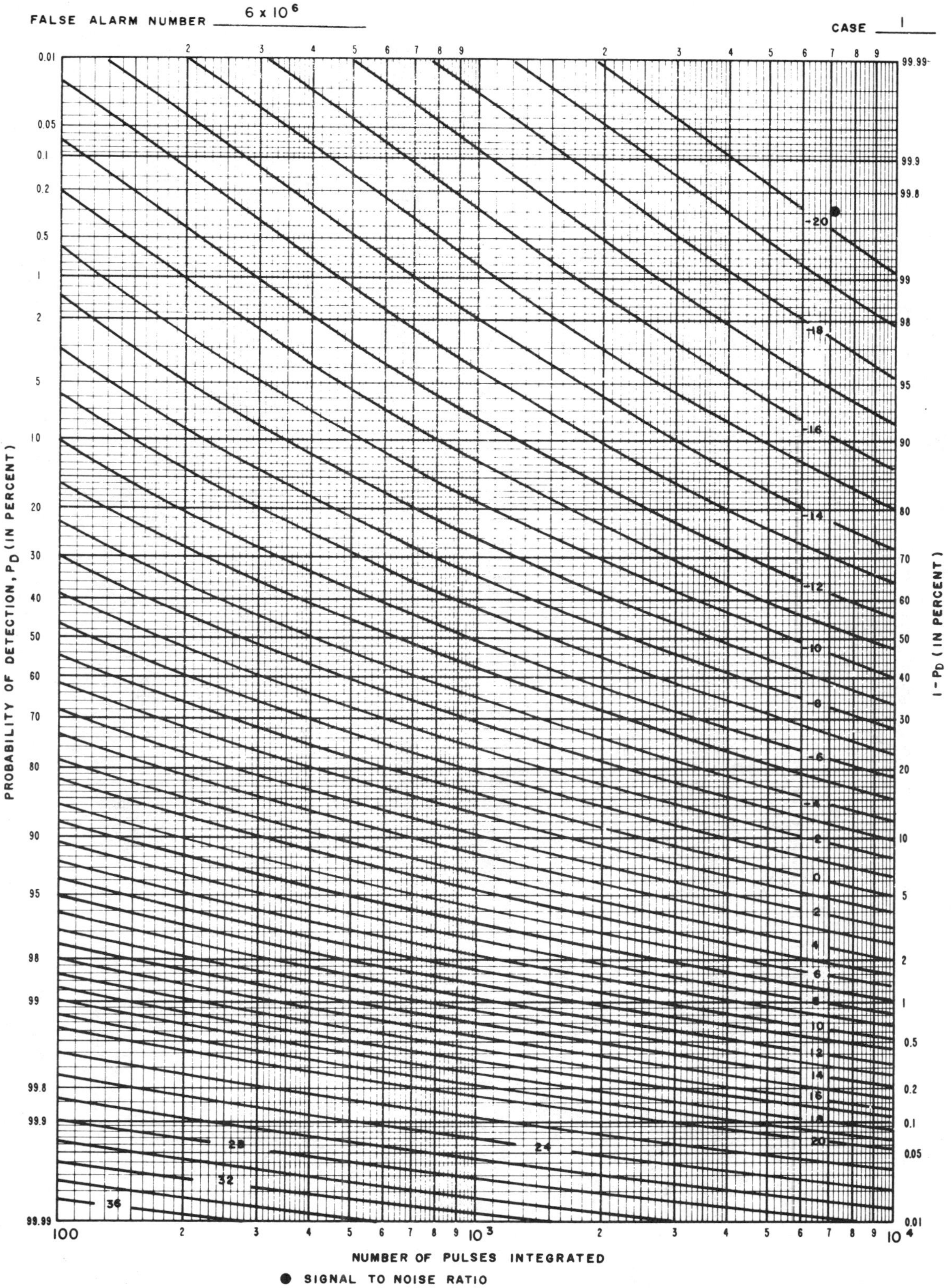
FALSE ALARM NUMBER 6 x 10^6
CASE 1
PROBABILITY OF DETECTION, P_D (IN PERCENT)
1 - P_D (IN PERCENT)
NUMBER OF PULSES INTEGRATED
100
10^3
10^4
-20
-18
-16
-14
-12
-10
-8
-6
-4
-2
0
2
4
6
8
10
12
14
16
18
20
24
28
32
36
SIGNAL TO NOISE RATIO

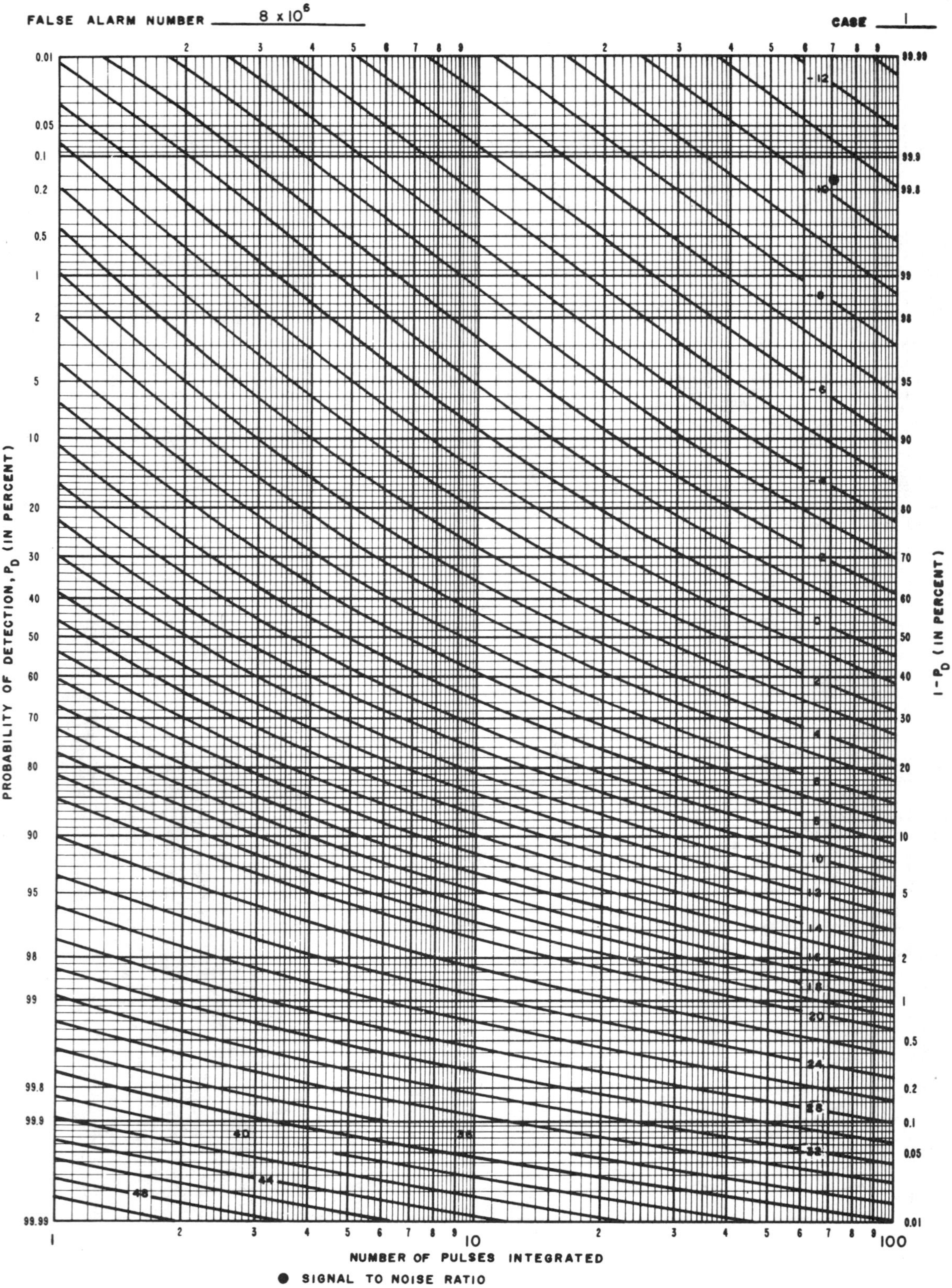
FALSE ALARM NUMBER 8 x 10^6
CASE 1
PROBABILITY OF DETECTION, P_D (IN PERCENT)
$1-P_D$ (IN PERCENT)
NUMBER OF PULSES INTEGRATED
● SIGNAL TO NOISE RATIO

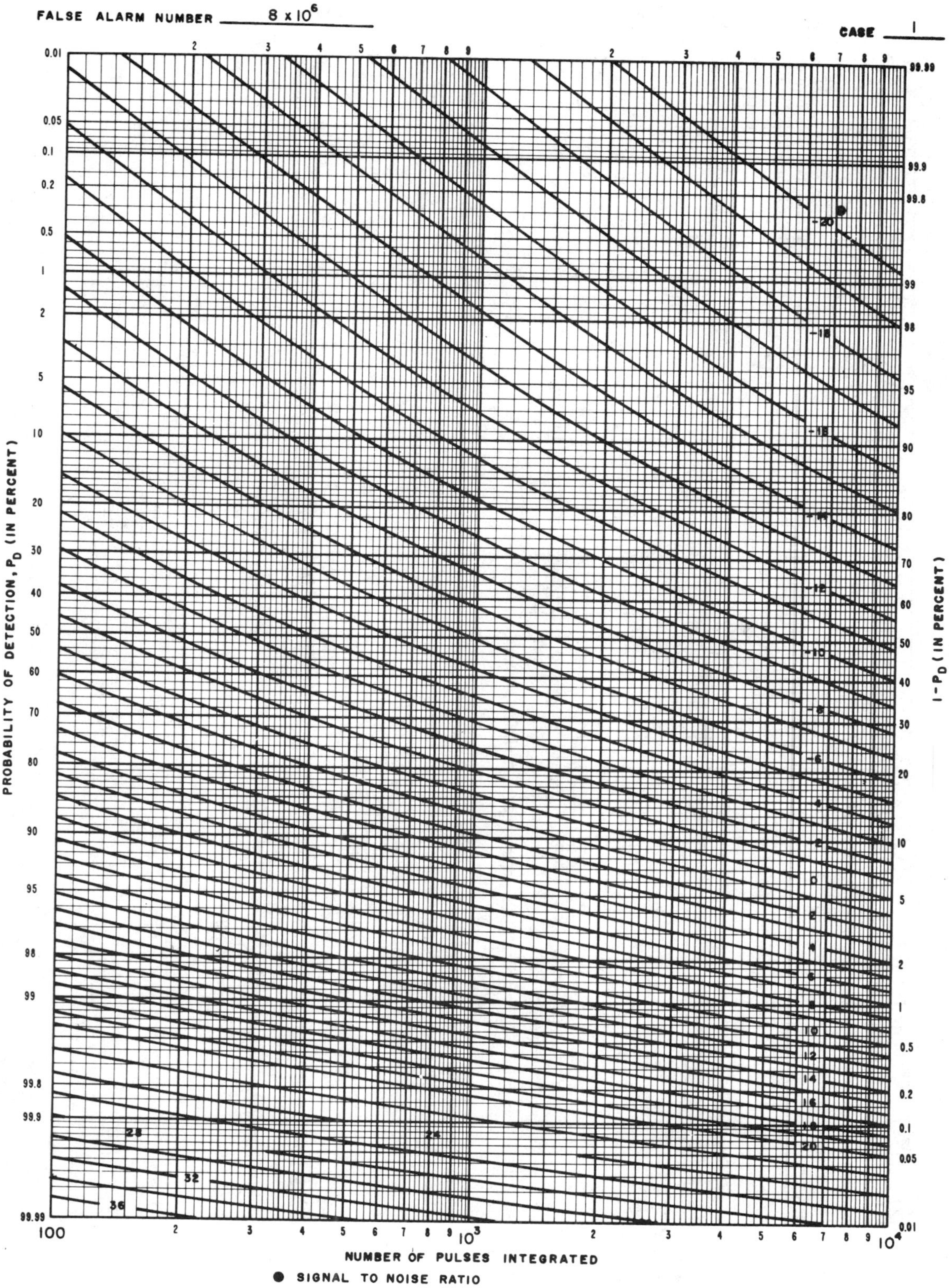
FALSE ALARM NUMBER 8 x 10^6
CASE 1
PROBABILITY OF DETECTION, P_D (IN PERCENT)
1 - P_D (IN PERCENT)
NUMBER OF PULSES INTEGRATED
● SIGNAL TO NOISE RATIO
100
10^3
10^4
-20
-18
-16
-12
-10
-8
-6
-4
-2
0
2
4
6
8
10
12
14
16
18
20
24
28
32
36

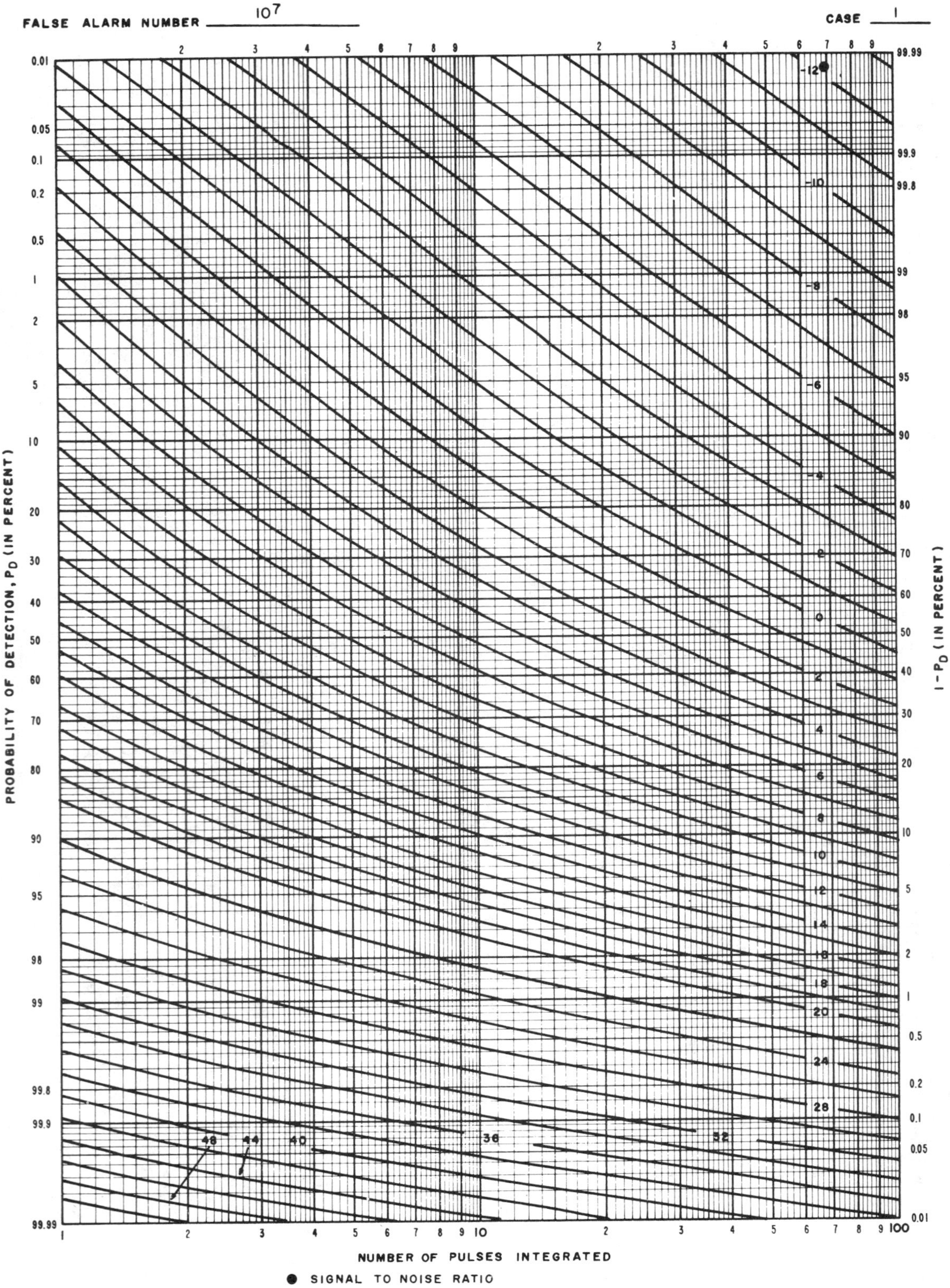
FALSE ALARM NUMBER 10^7
CASE 1
PROBABILITY OF DETECTION, P_D (IN PERCENT)
1 - P_D (IN PERCENT)
NUMBER OF PULSES INTEGRATED
● SIGNAL TO NOISE RATIO

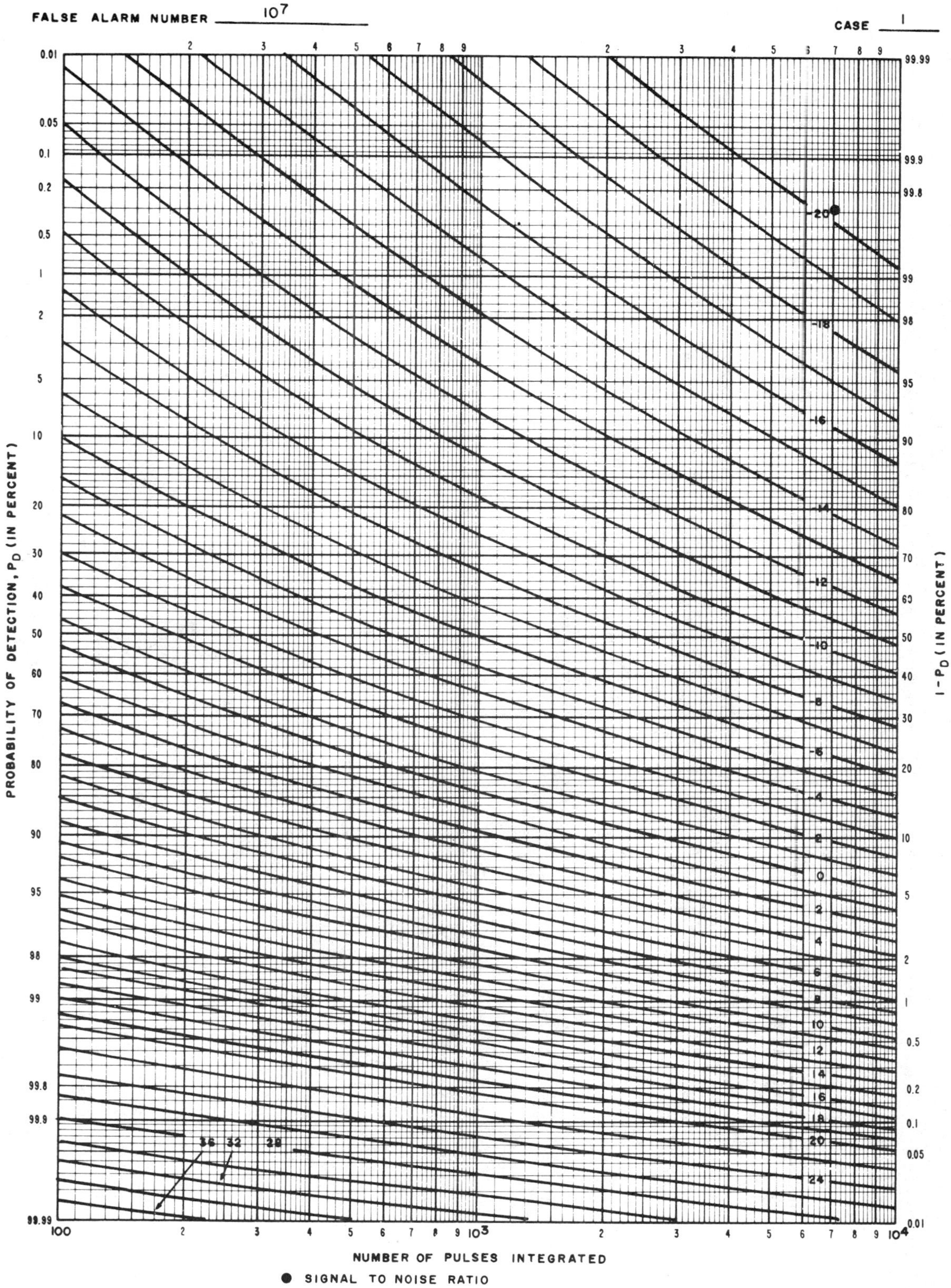
FALSE ALARM NUMBER 10^7
CASE 1
PROBABILITY OF DETECTION, P_D (IN PERCENT)
1 - P_D (IN PERCENT)
-20
-18
-16
-14
-12
-10
-8
-6
-4
-2
0
2
4
6
8
10
12
14
16
18
20
24
36
32
28
NUMBER OF PULSES INTEGRATED
● SIGNAL TO NOISE RATIO

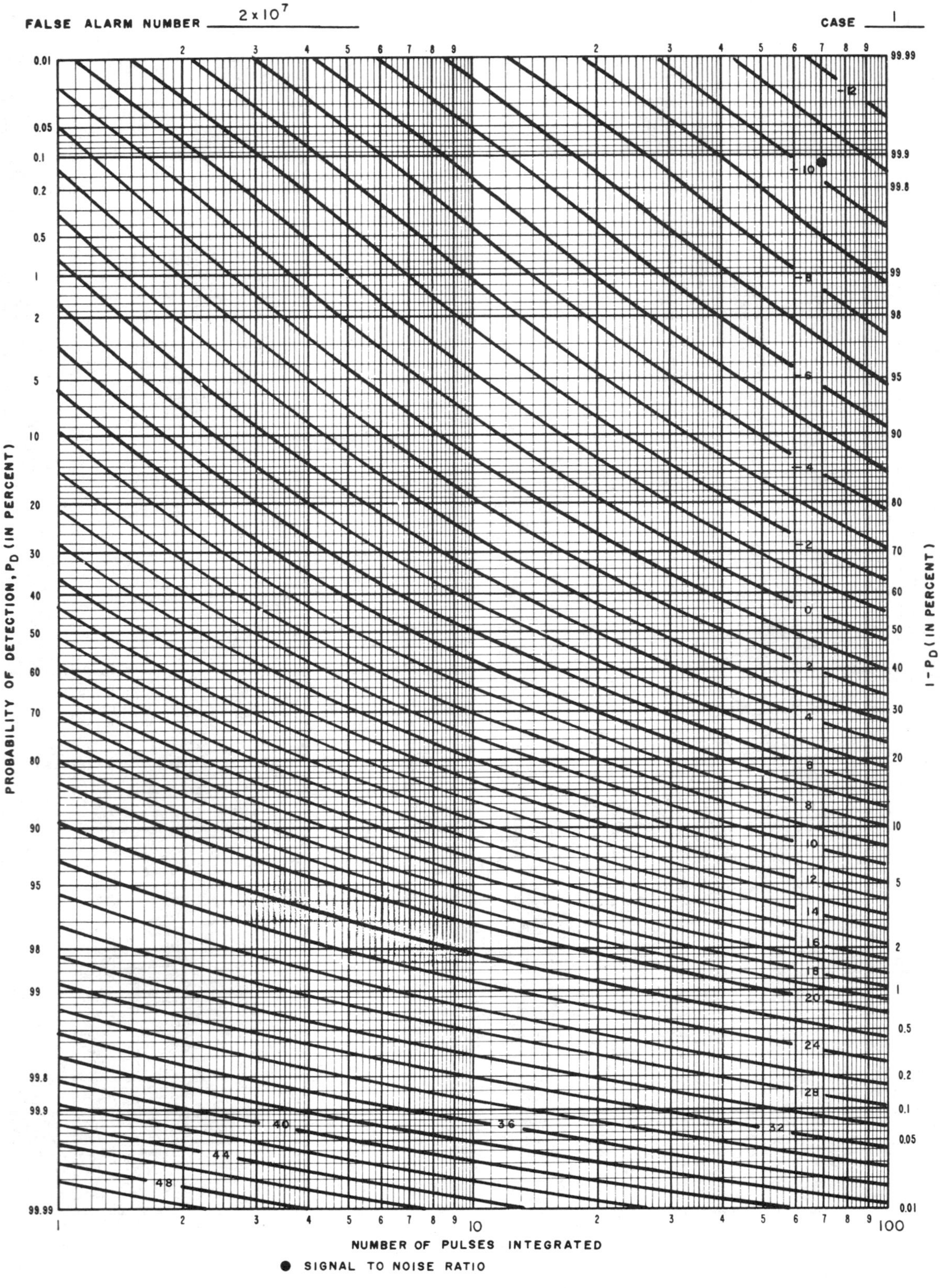
FALSE ALARM NUMBER 2 x 10^7
CASE 1
PROBABILITY OF DETECTION, P_D (IN PERCENT)
1 - P_D (IN PERCENT)
NUMBER OF PULSES INTEGRATED
● SIGNAL TO NOISE RATIO

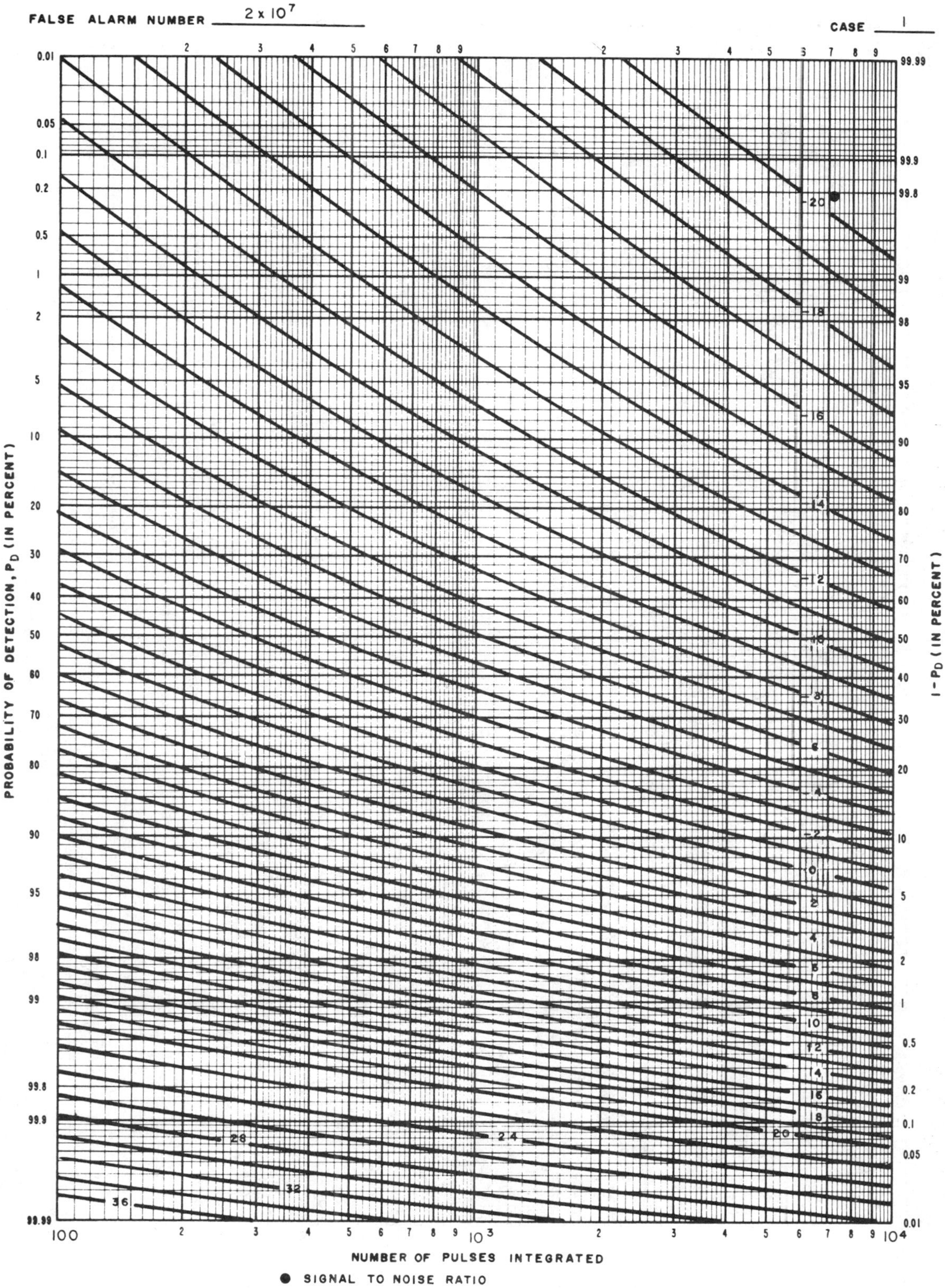
FALSE ALARM NUMBER 2 x 10^7
CASE 1
PROBABILITY OF DETECTION, P_D (IN PERCENT)
1 - P_D (IN PERCENT)
NUMBER OF PULSES INTEGRATED
● SIGNAL TO NOISE RATIO

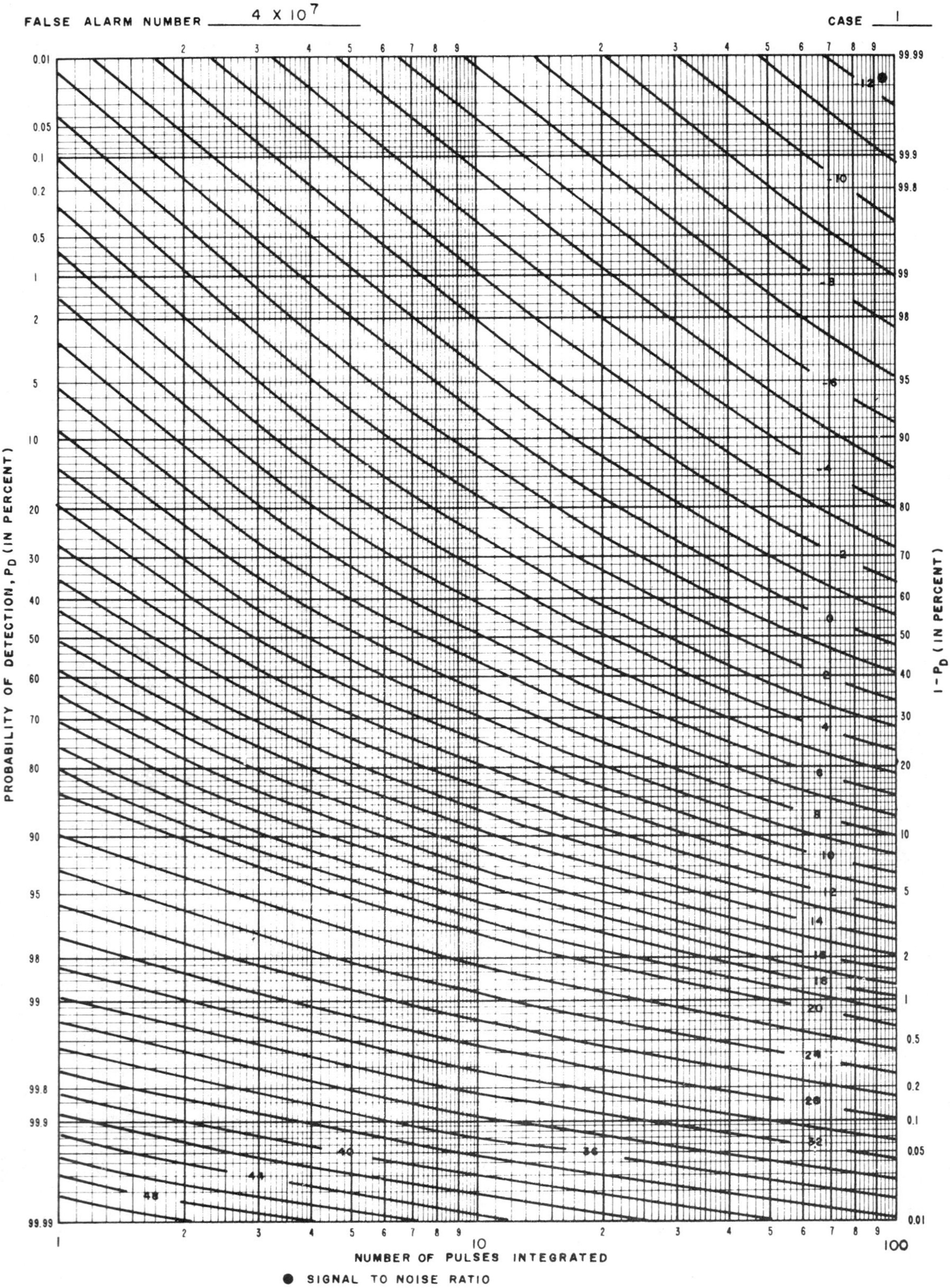
FALSE ALARM NUMBER 4 X 10^7
CASE 1
PROBABILITY OF DETECTION, P_D (IN PERCENT)
$1 - P_D$ (IN PERCENT)
NUMBER OF PULSES INTEGRATED
● SIGNAL TO NOISE RATIO

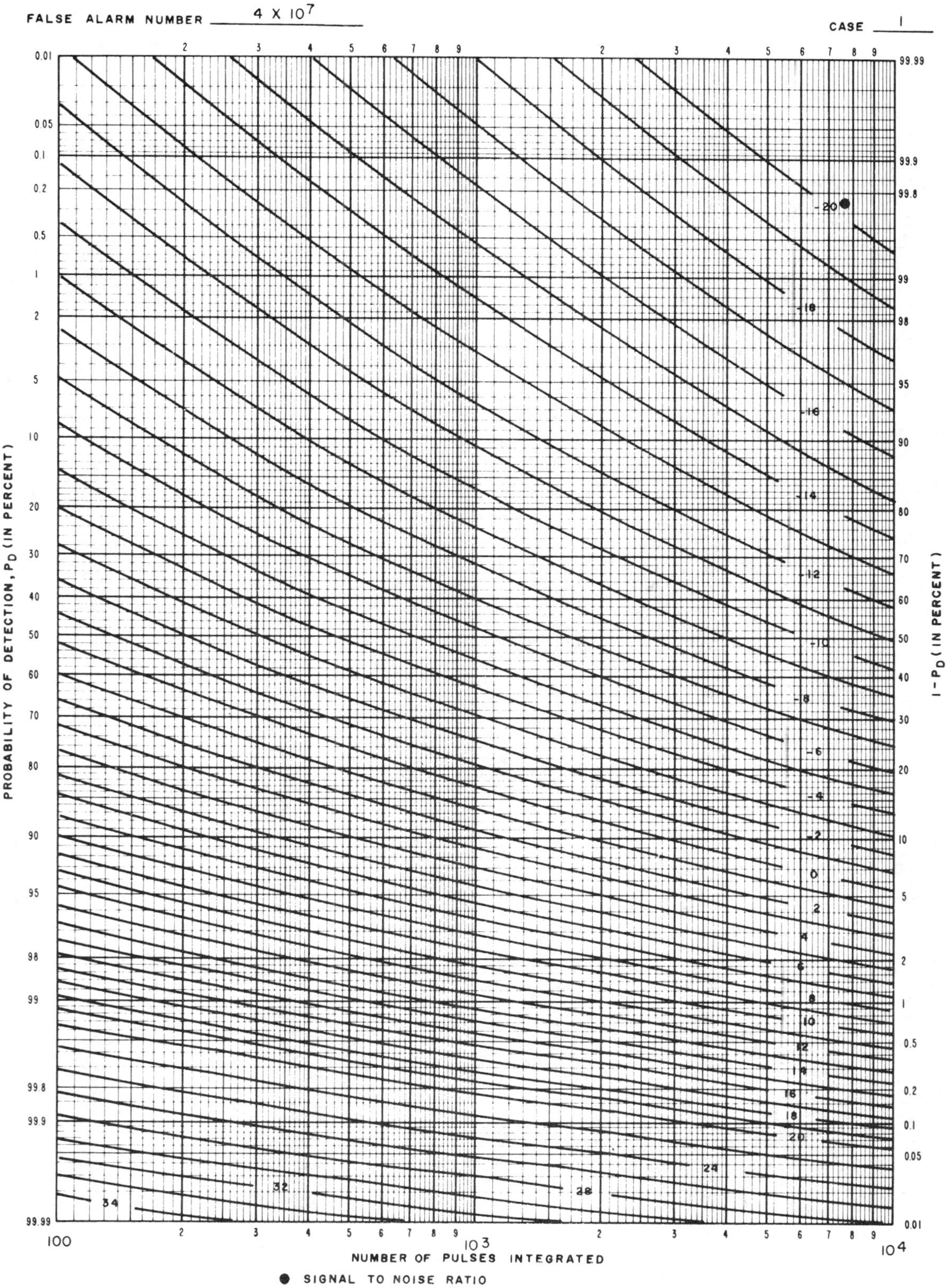
FALSE ALARM NUMBER 4 X 10^7
CASE 1
PROBABILITY OF DETECTION, P_D (IN PERCENT)
1 - P_D (IN PERCENT)
NUMBER OF PULSES INTEGRATED
● SIGNAL TO NOISE RATIO
100
10^3
10^4
-20
-18
-16
-14
-12
-10
-8
-6
-4
-2
0
2
4
6
8
10
12
14
16
18
20
24
28
32
34

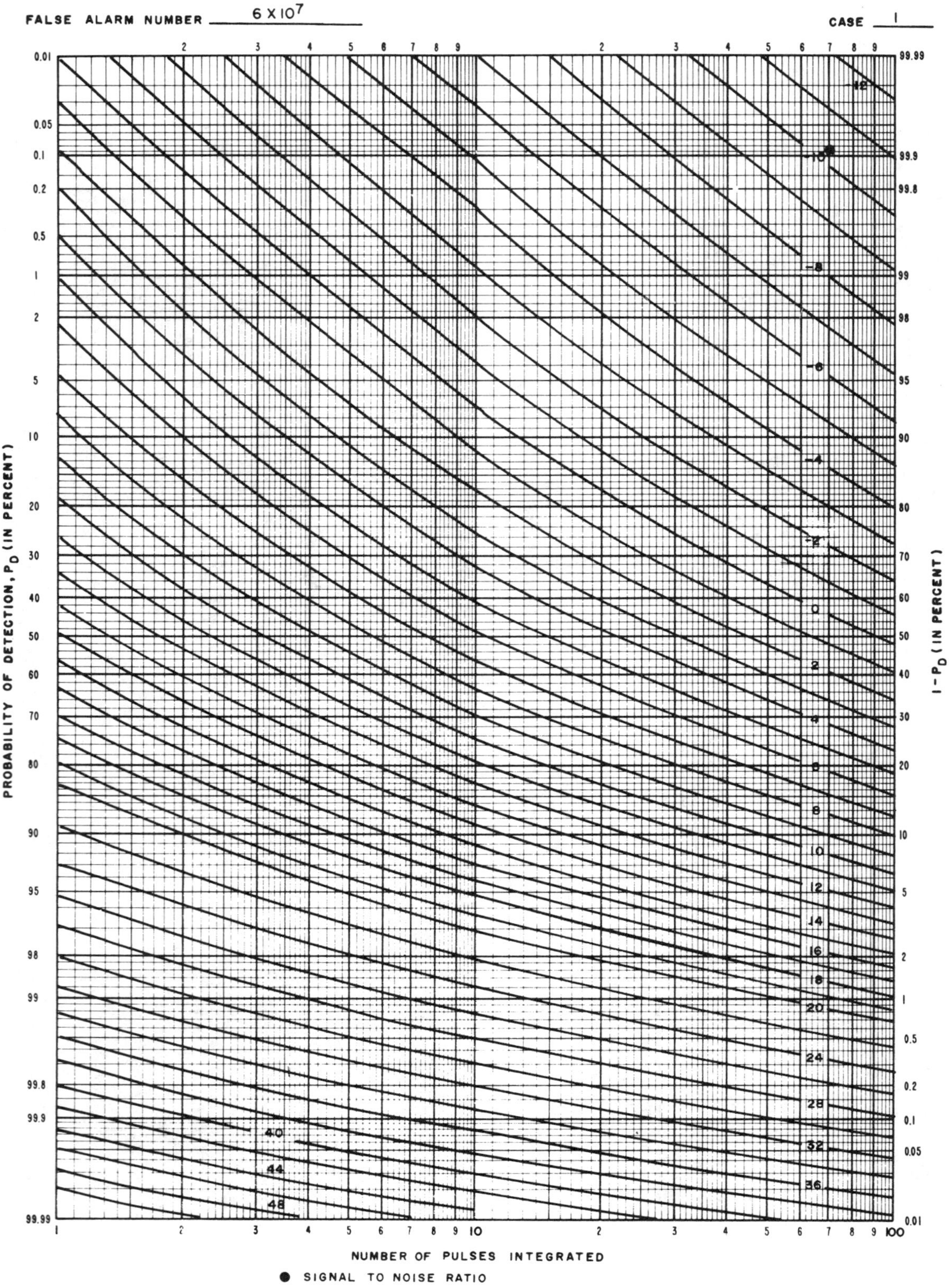
FALSE ALARM NUMBER 6 X 10^7
CASE 1
PROBABILITY OF DETECTION, P_D (IN PERCENT)
1 - P_D (IN PERCENT)
NUMBER OF PULSES INTEGRATED
● SIGNAL TO NOISE RATIO
-12
-10
-8
-6
-4
-2
0
2
4
6
8
10
12
14
16
18
20
24
28
32
36
40
44
48

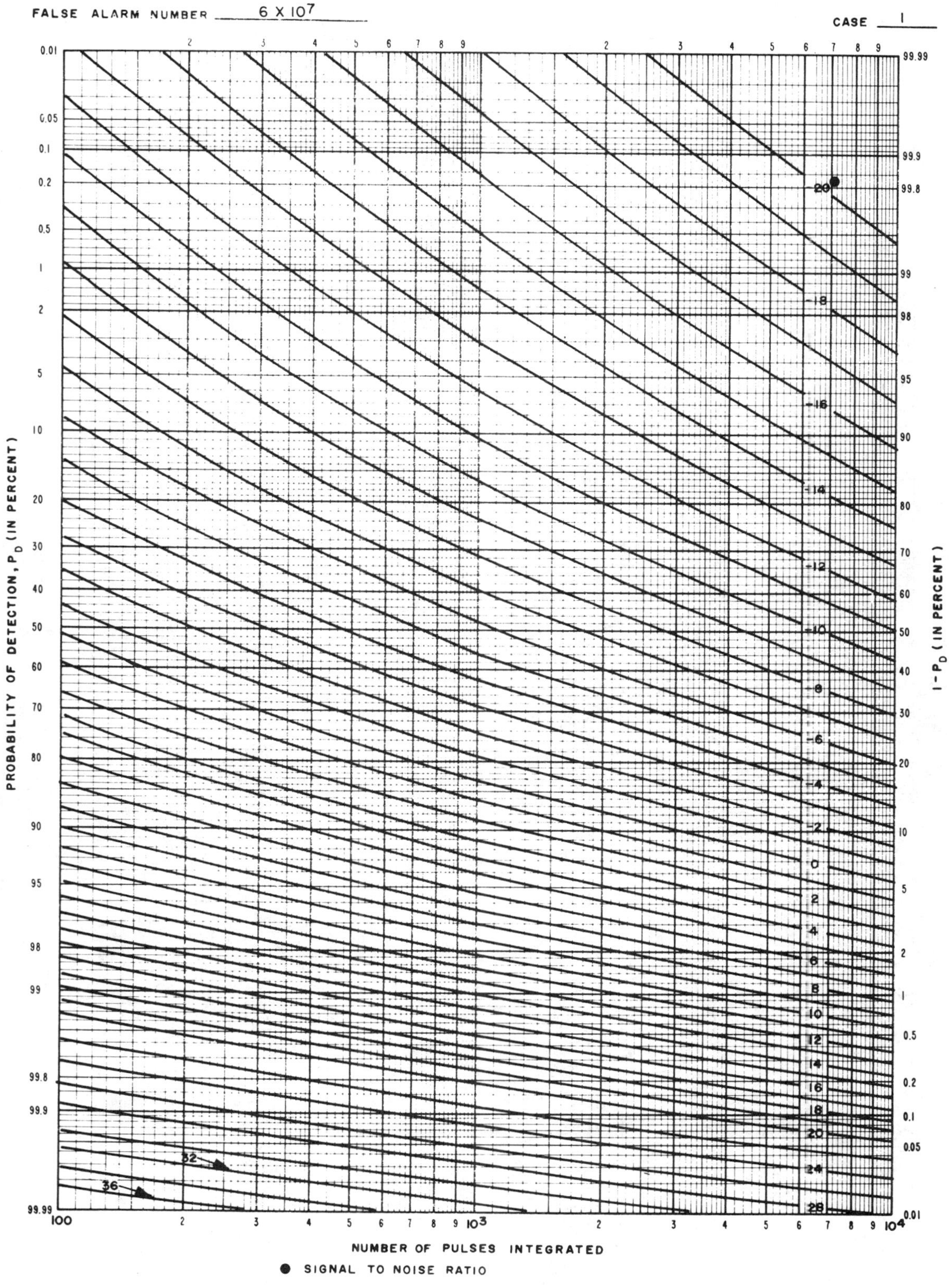
FALSE ALARM NUMBER 6 X 10^7
CASE 1
PROBABILITY OF DETECTION, P_D (IN PERCENT)
1 - P_D (IN PERCENT)
NUMBER OF PULSES INTEGRATED
● SIGNAL TO NOISE RATIO
20
18
16
14
12
10
8
6
4
2
0
2
4
6
8
10
12
14
16
18
20
24
28
32
36

FALSE ALARM NUMBER 8 X 10^7 CASE I

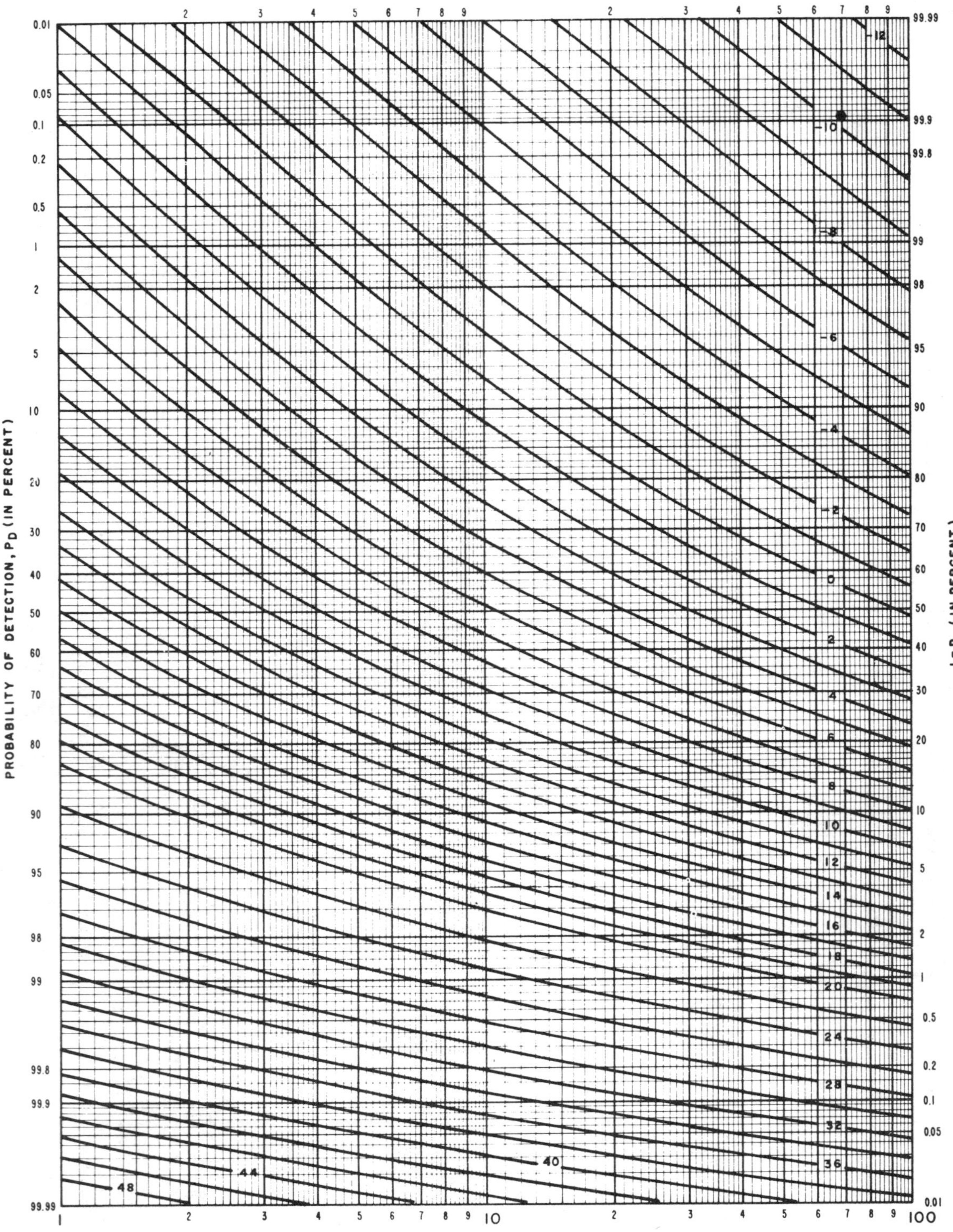

● SIGNAL TO NOISE RATIO

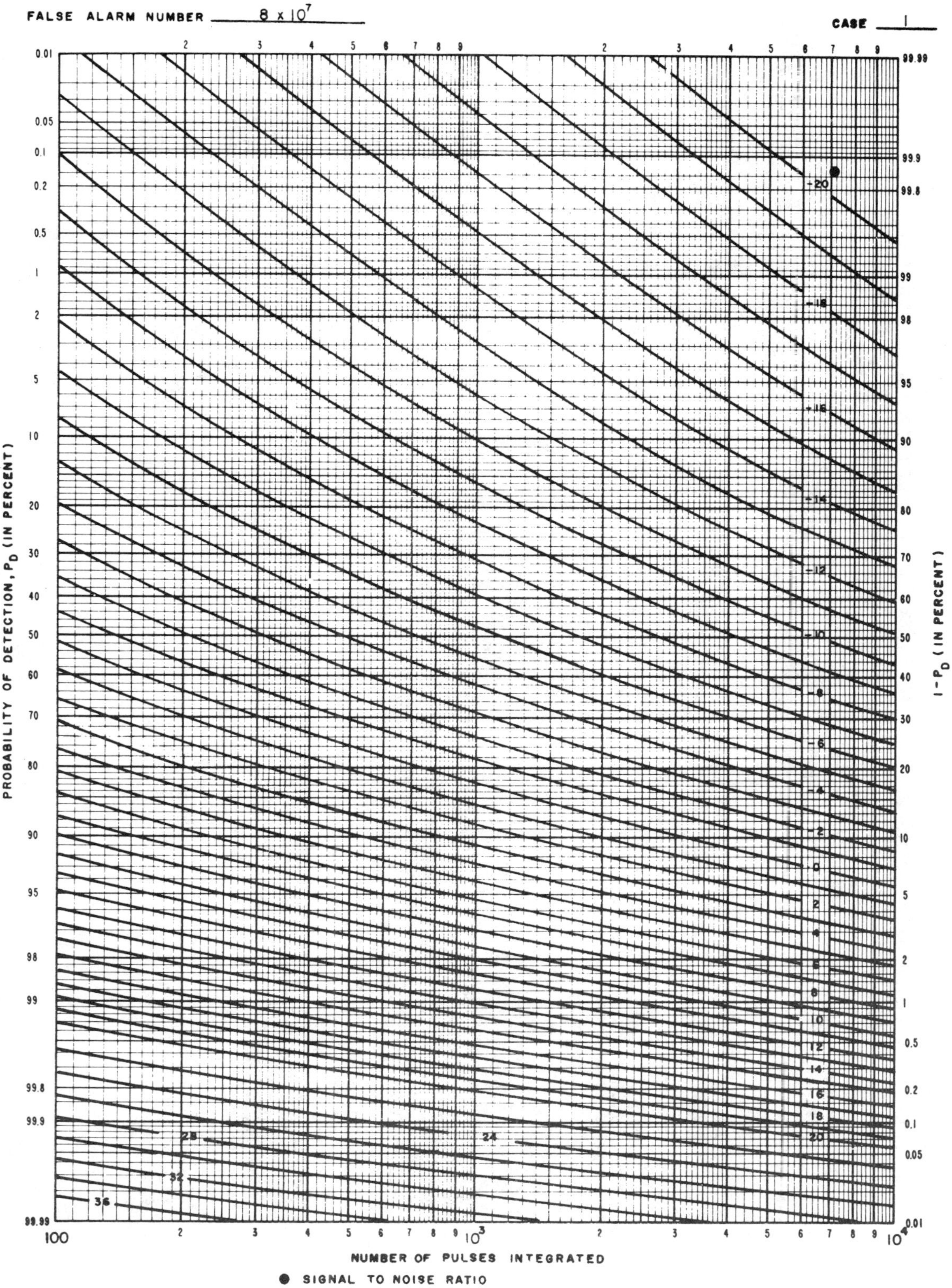
FALSE ALARM NUMBER 8 x 10^7
CASE 1
PROBABILITY OF DETECTION, P_D (IN PERCENT)
1 - P_D (IN PERCENT)
NUMBER OF PULSES INTEGRATED
● SIGNAL TO NOISE RATIO
-20
-18
-16
-14
-12
-10
-8
-6
-4
-2
0
2
4
6
8
10
12
14
16
18
20
24
28
32
36

FALSE ALARM NUMBER 10^8 CASE 1

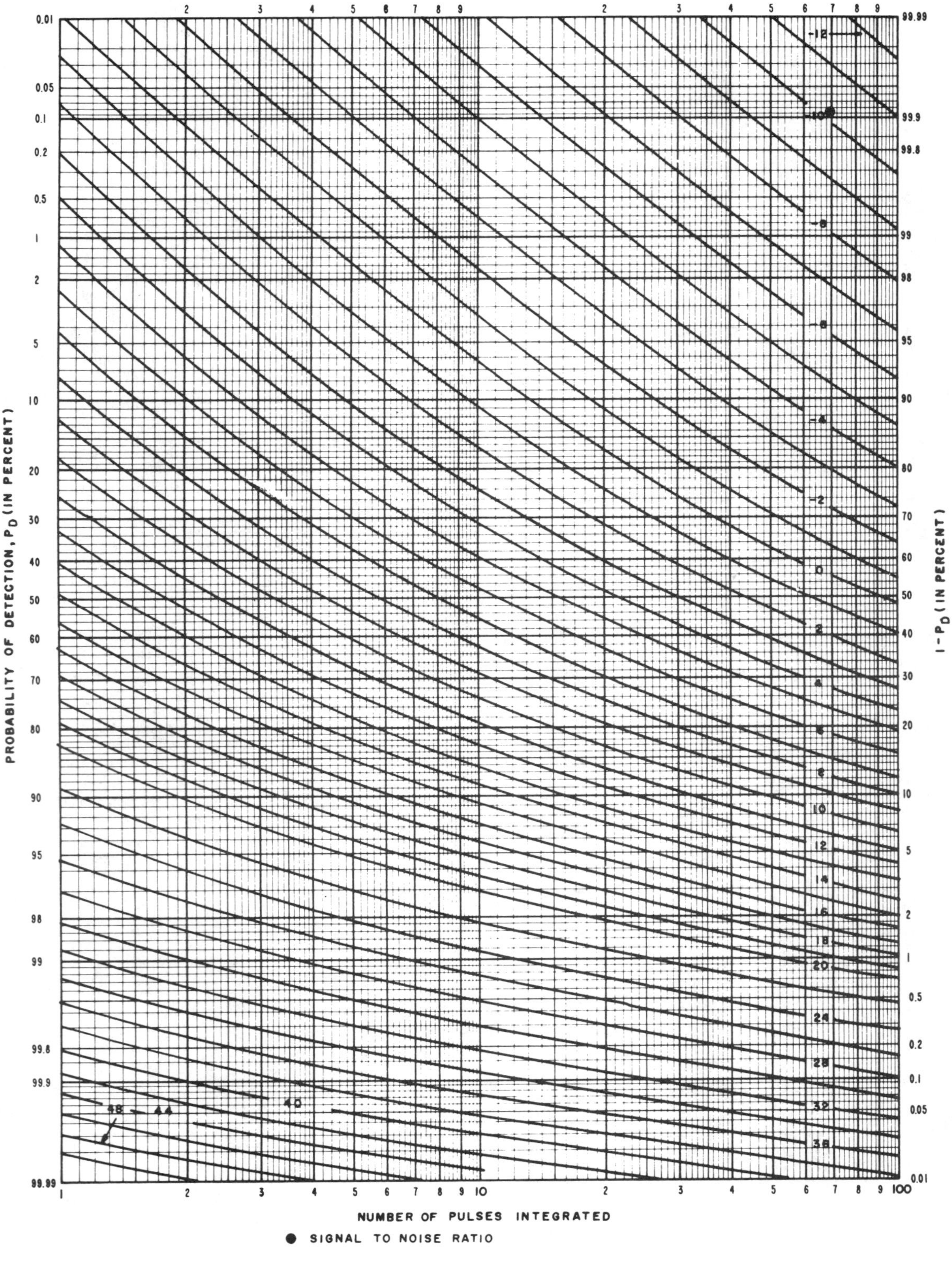

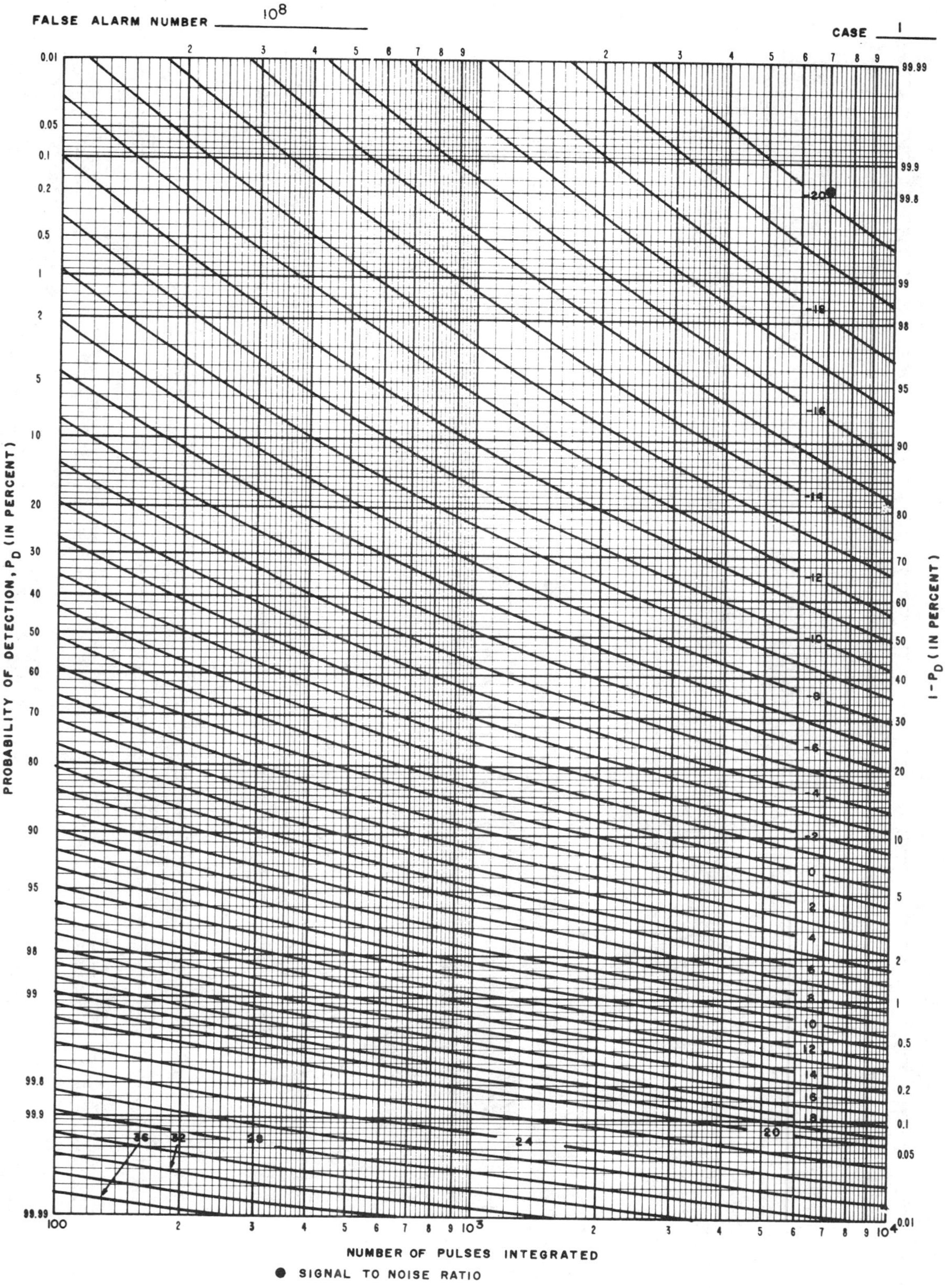
FALSE ALARM NUMBER 10^8
CASE 1
PROBABILITY OF DETECTION, P_D (IN PERCENT)
1 - P_D (IN PERCENT)
NUMBER OF PULSES INTEGRATED
● SIGNAL TO NOISE RATIO

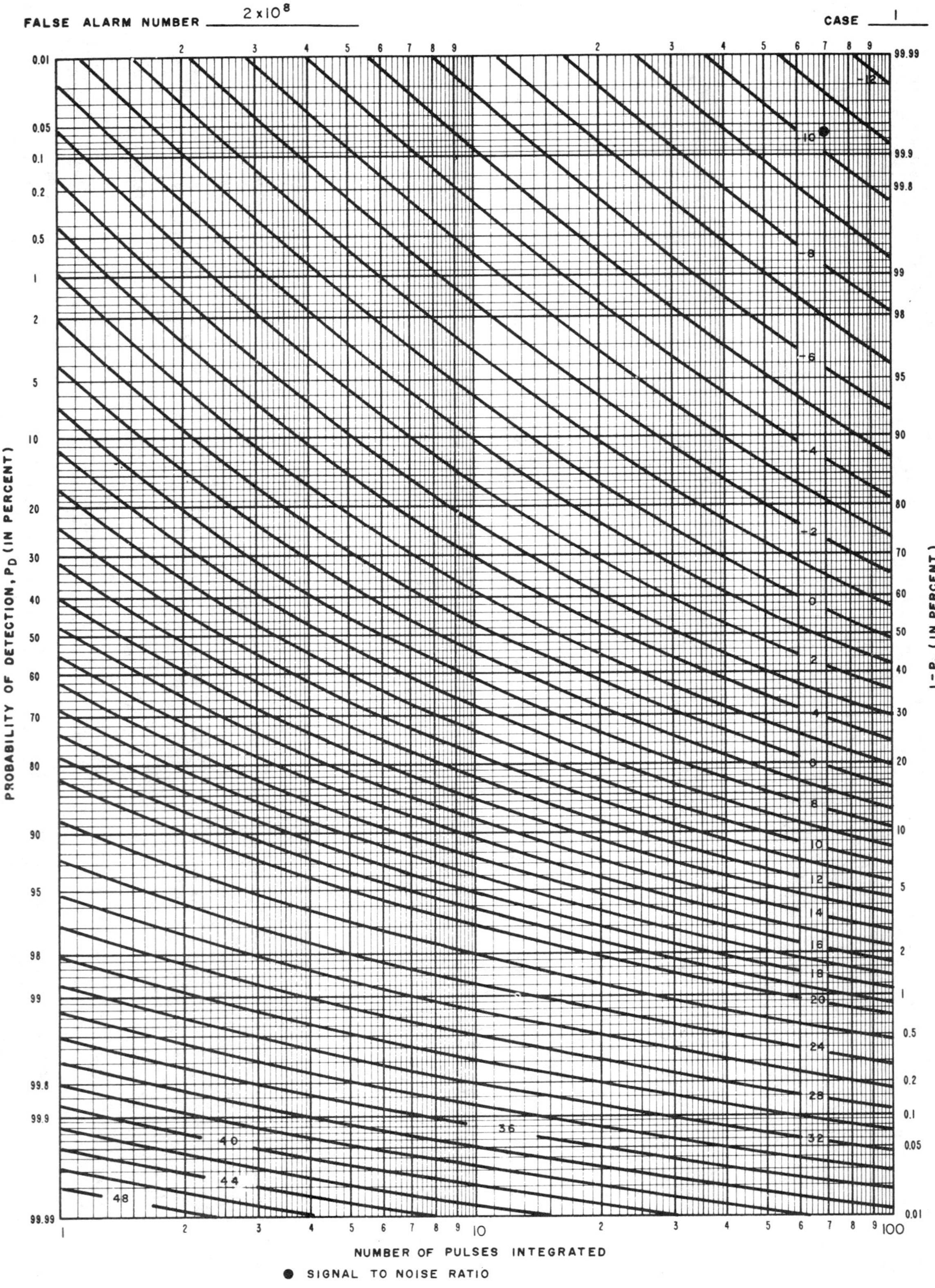
FALSE ALARM NUMBER 2 x 10⁸
CASE 1
PROBABILITY OF DETECTION, P_D (IN PERCENT)
1 - P_D (IN PERCENT)
NUMBER OF PULSES INTEGRATED
● SIGNAL TO NOISE RATIO
-12
-10
-8
-6
-4
-2
0
2
4
6
8
10
12
14
16
18
20
24
28
32
36
40
44
48

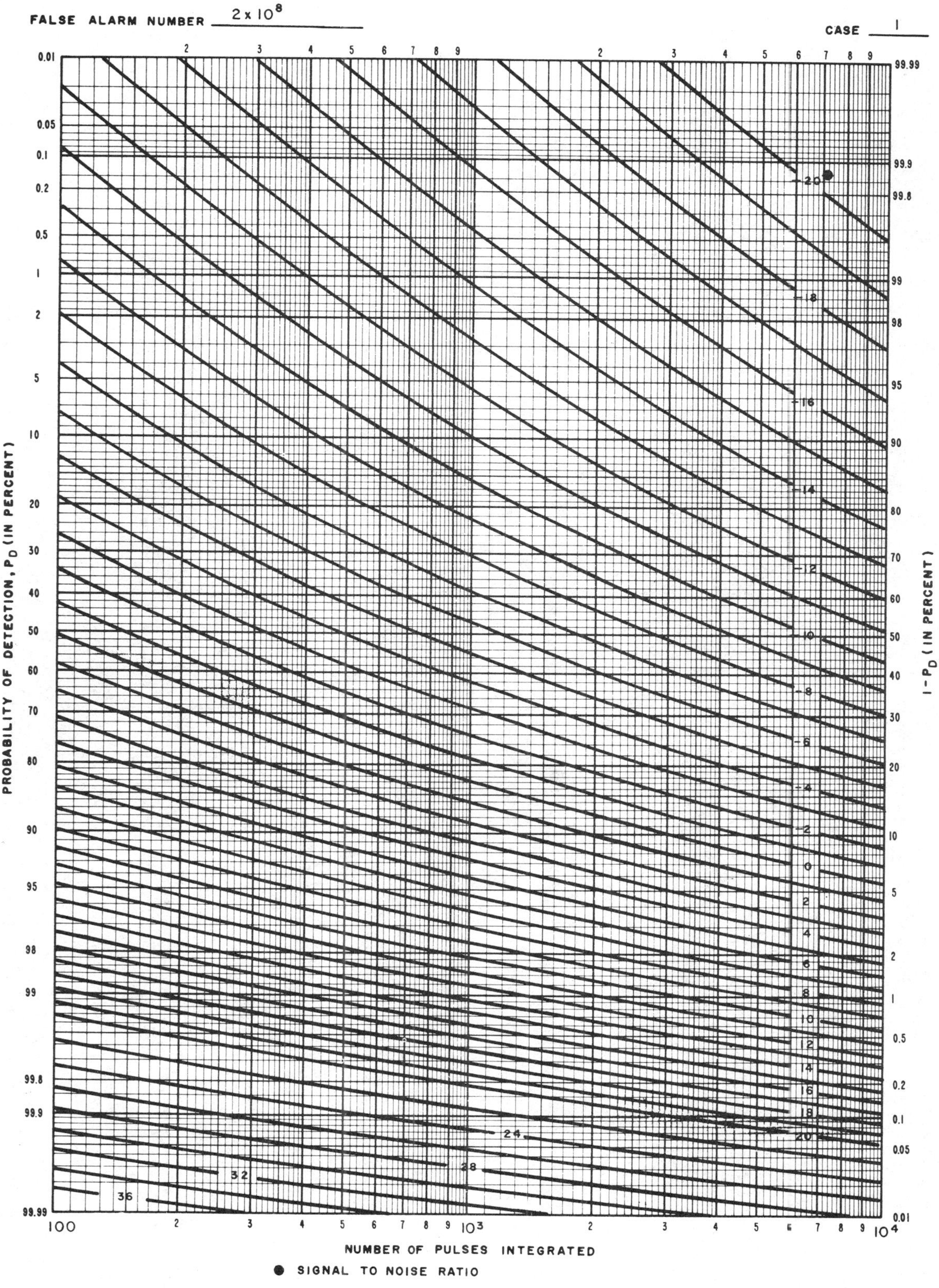
FALSE ALARM NUMBER 2 x 10^8
CASE 1
PROBABILITY OF DETECTION, P_D (IN PERCENT)
1 - P_D (IN PERCENT)
NUMBER OF PULSES INTEGRATED
● SIGNAL TO NOISE RATIO
100
10^3
10^4
0.01
0.05
0.1
0.2
0.5
1
2
5
10
20
30
40
50
60
70
80
90
95
98
99
99.8
99.9
99.99
-20
-18
-16
-14
-12
-10
-8
-6
-4
-2
0
2
4
6
8
10
12
14
16
18
20
24
28
32
36

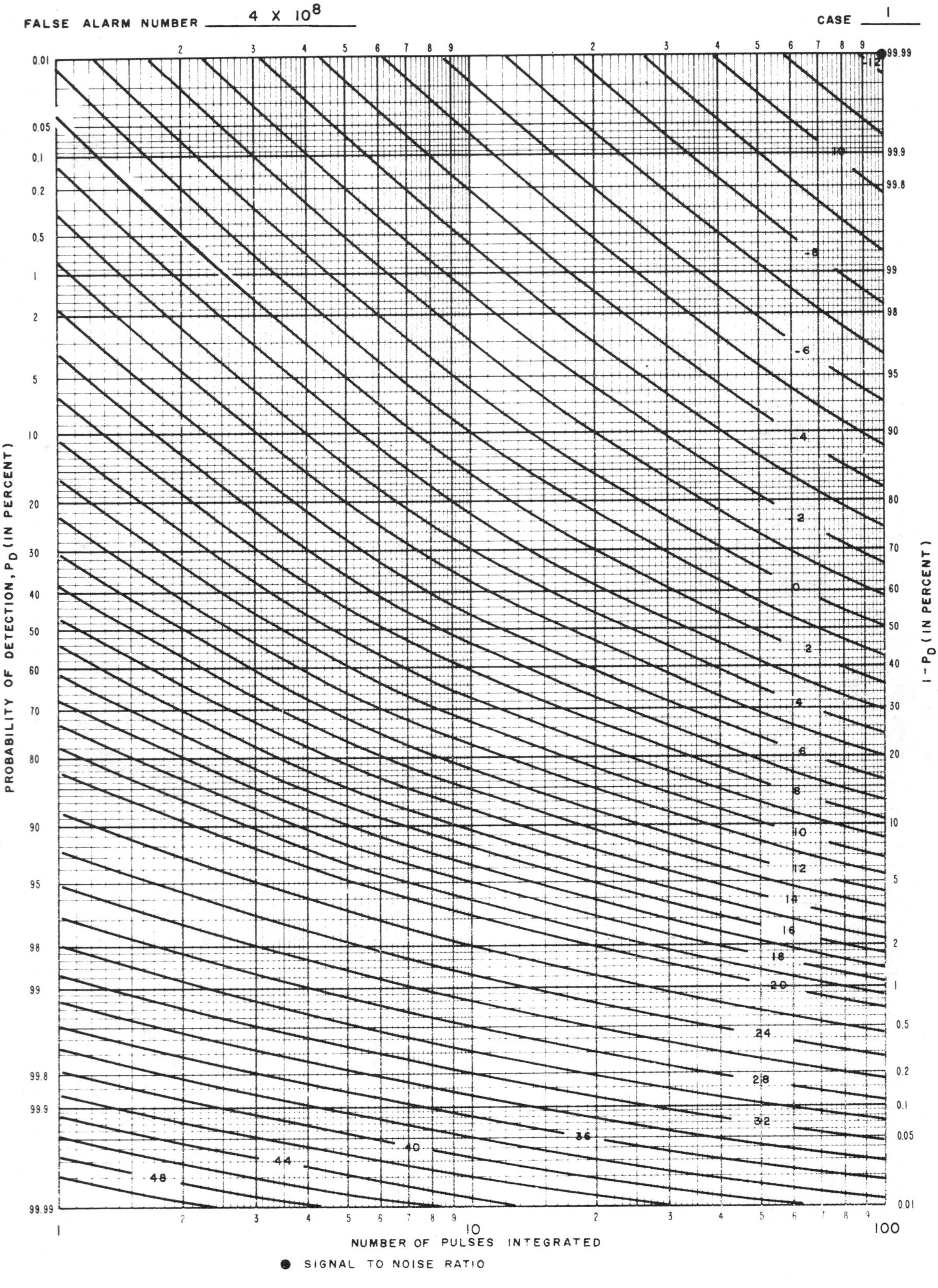
FALSE ALARM NUMBER 4 X 10^8
CASE 1
PROBABILITY OF DETECTION, P_D (IN PERCENT)
1 - P_D (IN PERCENT)
-12
-10
-8
-6
-4
-2
0
2
4
6
8
10
12
14
16
18
20
24
28
32
36
40
44
48
NUMBER OF PULSES INTEGRATED
● SIGNAL TO NOISE RATIO

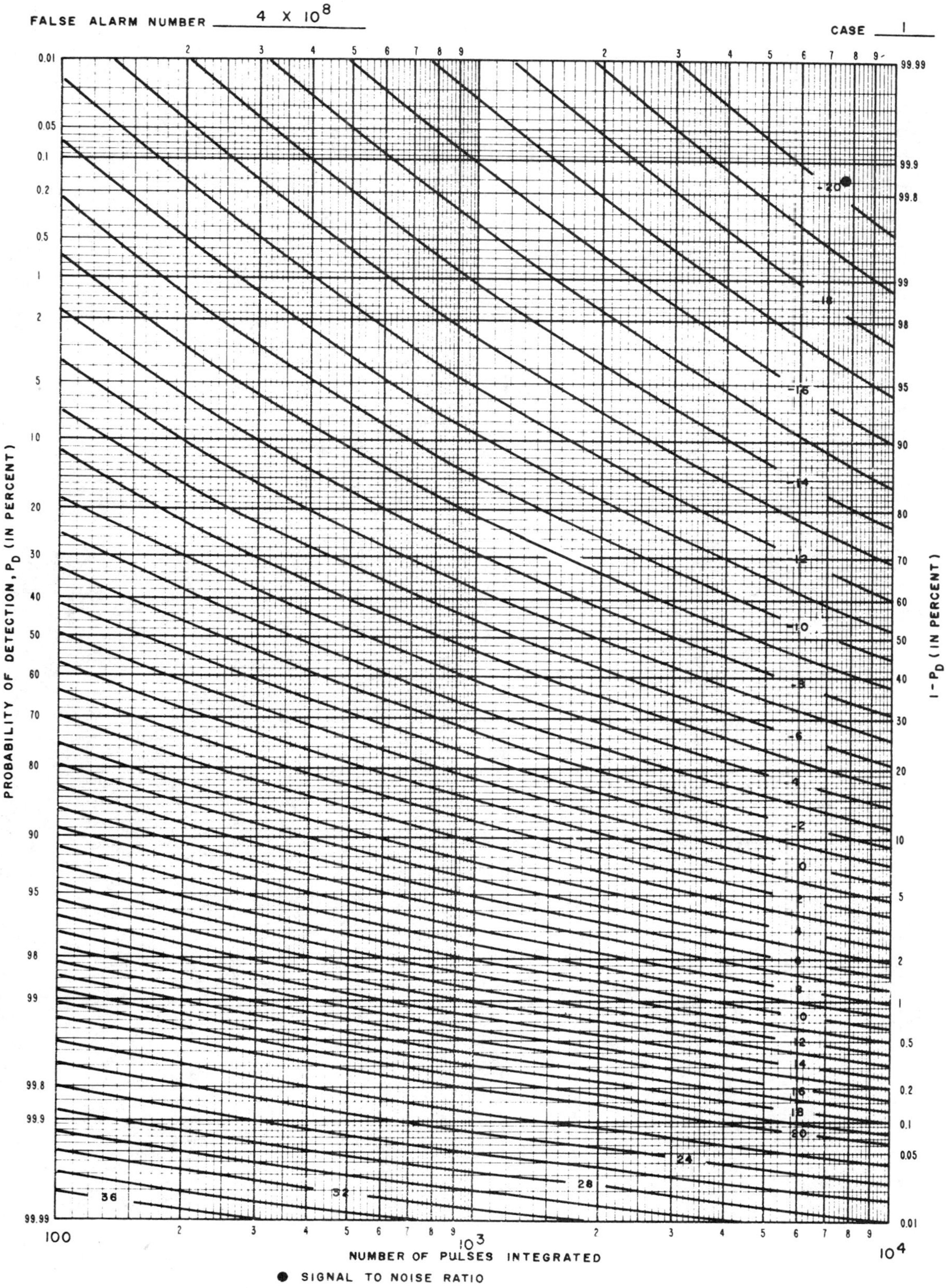
FALSE ALARM NUMBER 4 X 10^8
CASE 1
PROBABILITY OF DETECTION, P_D (IN PERCENT)
1 - P_D (IN PERCENT)
NUMBER OF PULSES INTEGRATED
100
10^3
10^4
● SIGNAL TO NOISE RATIO

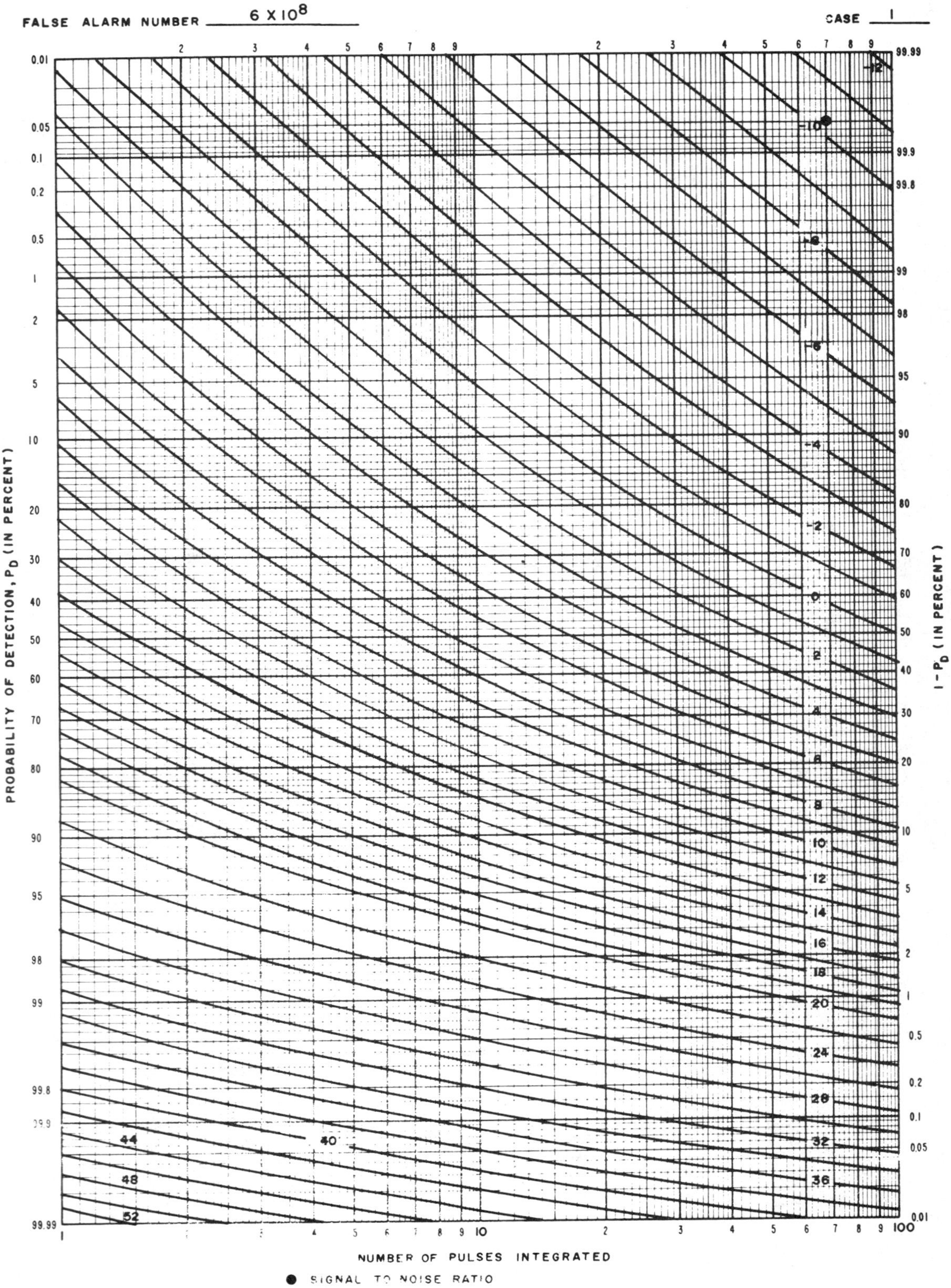
FALSE ALARM NUMBER 6 X 10^8
CASE 1
PROBABILITY OF DETECTION, P_D (IN PERCENT)
1 - P_D (IN PERCENT)
NUMBER OF PULSES INTEGRATED
● SIGNAL TO NOISE RATIO

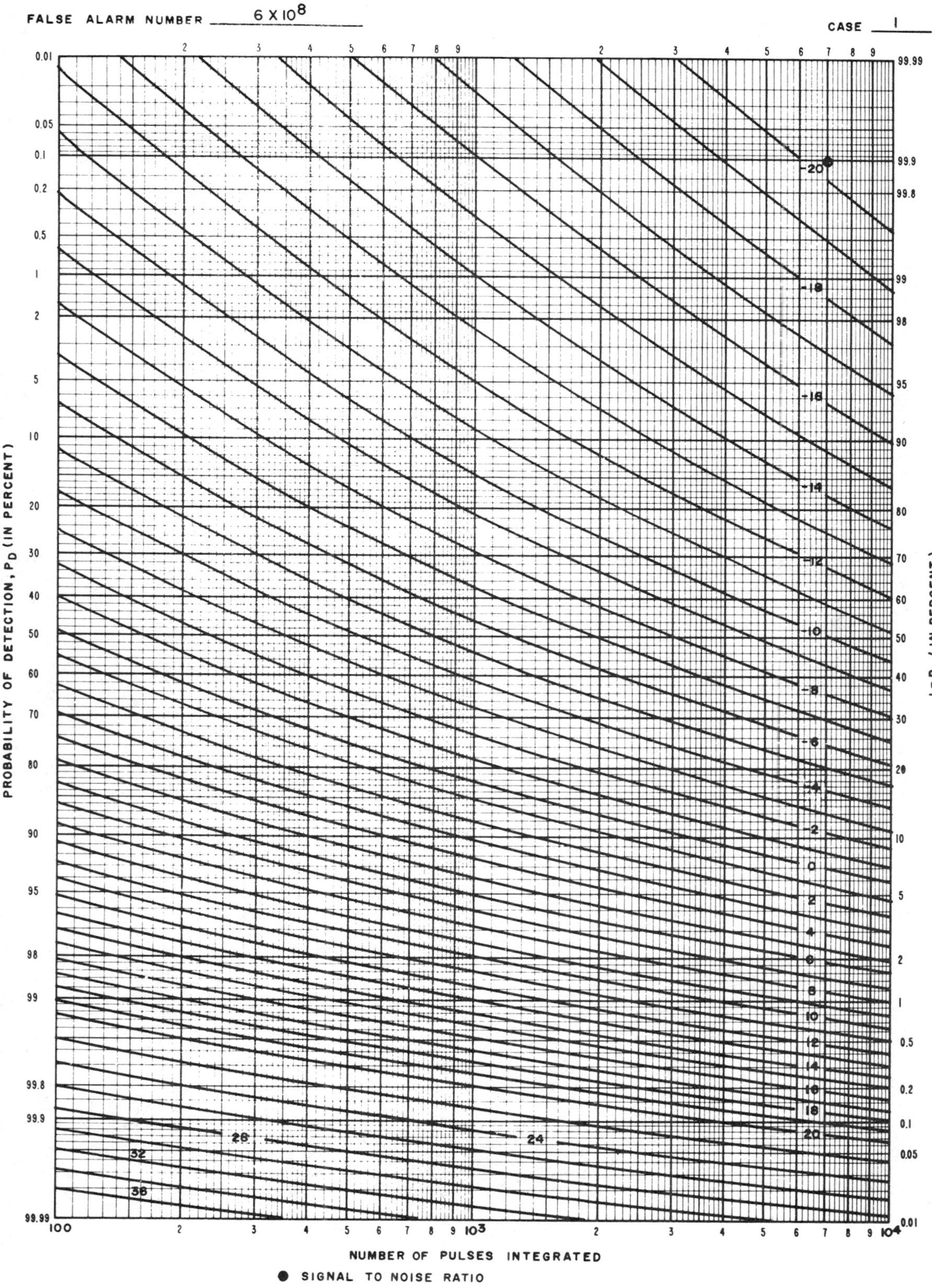
FALSE ALARM NUMBER 6 X 10^8
CASE I
PROBABILITY OF DETECTION, P_D (IN PERCENT)
1 - P_D (IN PERCENT)
NUMBER OF PULSES INTEGRATED
● SIGNAL TO NOISE RATIO

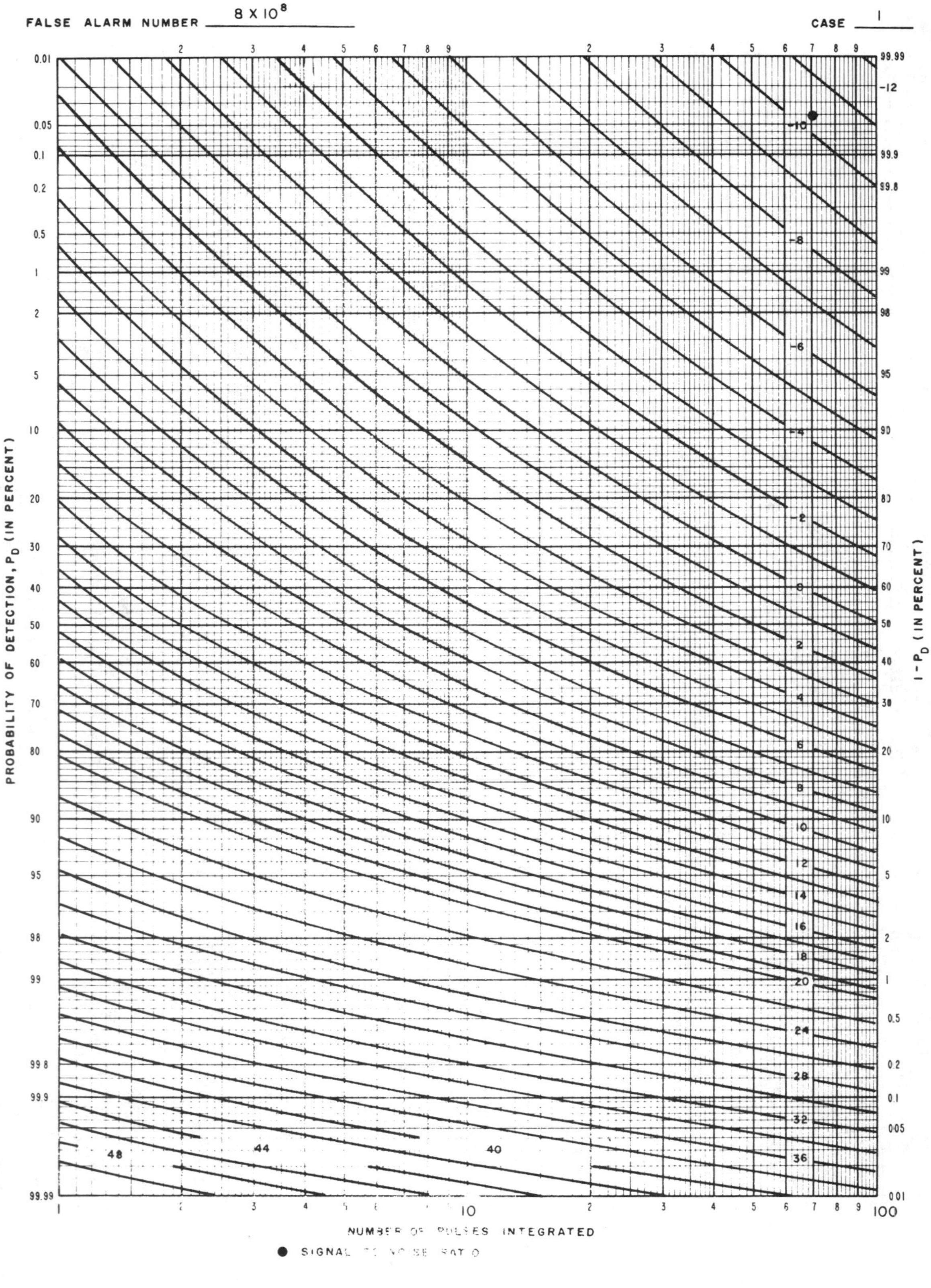
FALSE ALARM NUMBER 8 X 10^8
CASE 1
PROBABILITY OF DETECTION, P_D (IN PERCENT)
1 - P_D (IN PERCENT)
NUMBER OF PULSES INTEGRATED
SIGNAL TO NOISE RATIO
-12
-10
-8
-6
-4
-2
0
2
4
6
8
10
12
14
16
18
20
24
28
32
36
40
44
48

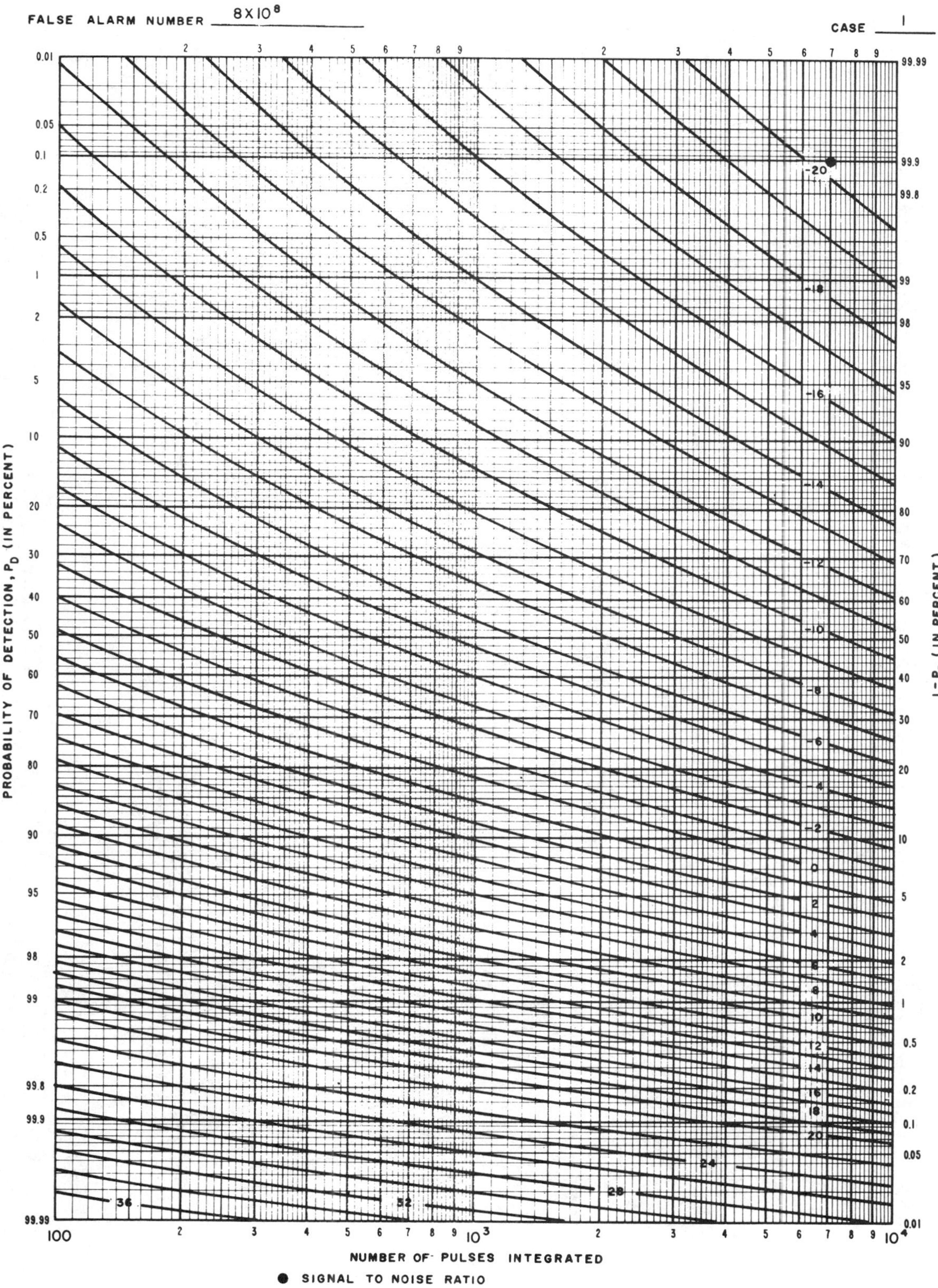
FALSE ALARM NUMBER 8X10^8
CASE 1
PROBABILITY OF DETECTION, P_D (IN PERCENT)
1-P_D (IN PERCENT)
NUMBER OF PULSES INTEGRATED
● SIGNAL TO NOISE RATIO
-20
-18
-16
-14
-12
-10
-8
-6
-4
-2
0
2
4
6
8
10
12
14
16
18
20
24
28
32
36

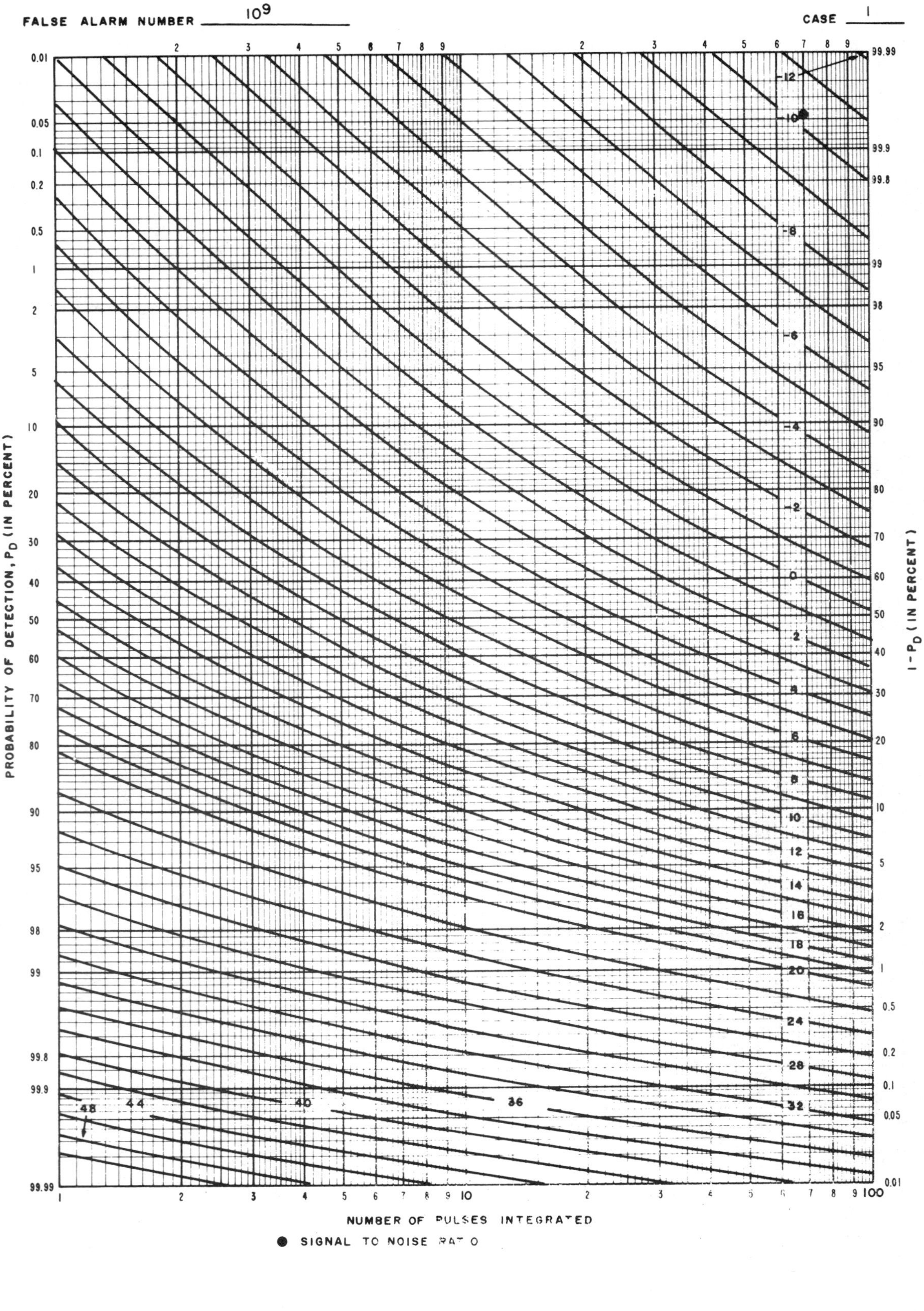
FALSE ALARM NUMBER 10^9
CASE 1
PROBABILITY OF DETECTION, P_D (IN PERCENT)
$1-P_D$ (IN PERCENT)
NUMBER OF PULSES INTEGRATED
● SIGNAL TO NOISE RATIO
-12
-10
-8
-6
-4
-2
0
2
4
6
8
10
12
14
16
18
20
24
28
32
36
40
44
48

FALSE ALARM NUMBER 10^9 CASE 1

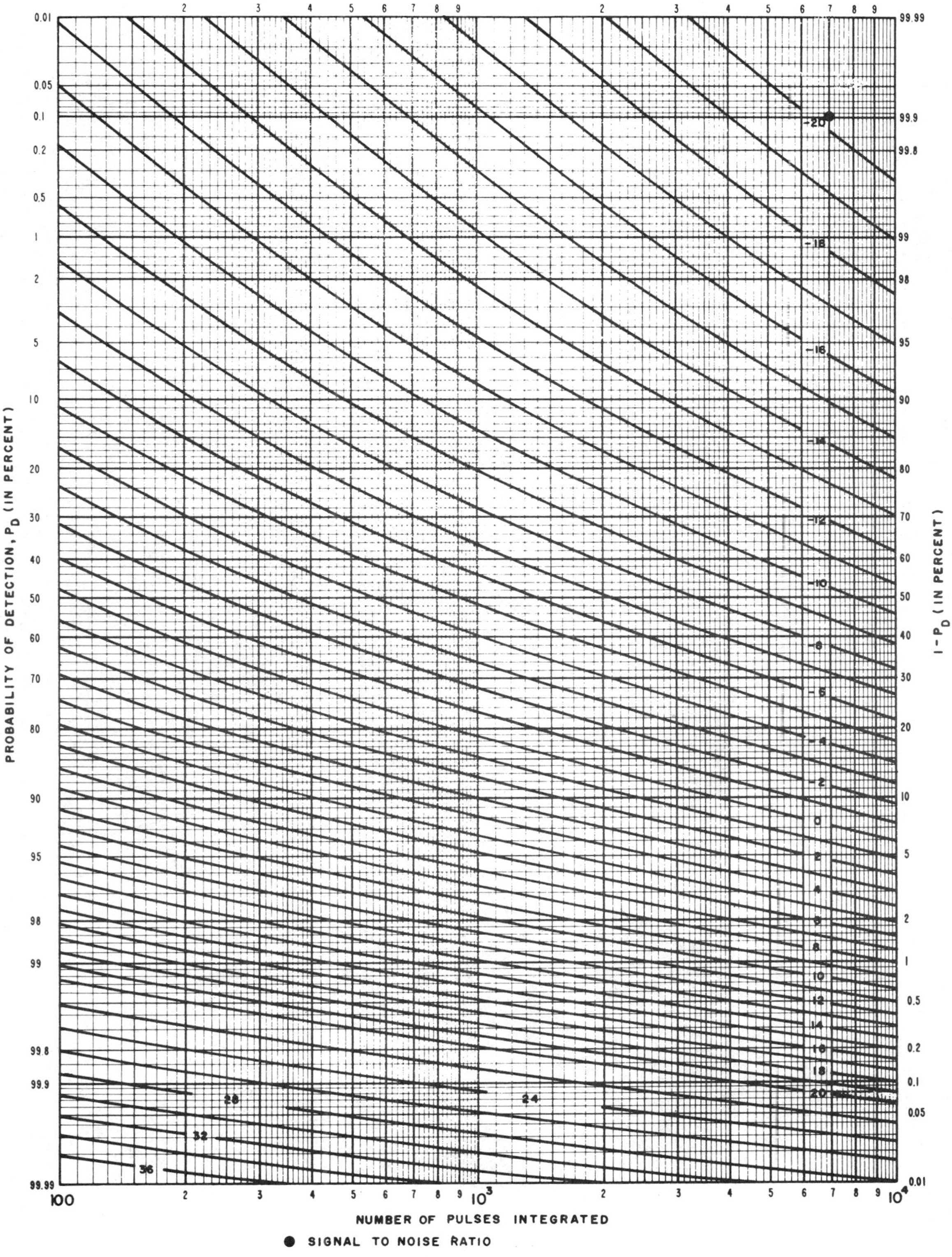

FALSE ALARM NUMBER 2 x 10⁹ CASE 1

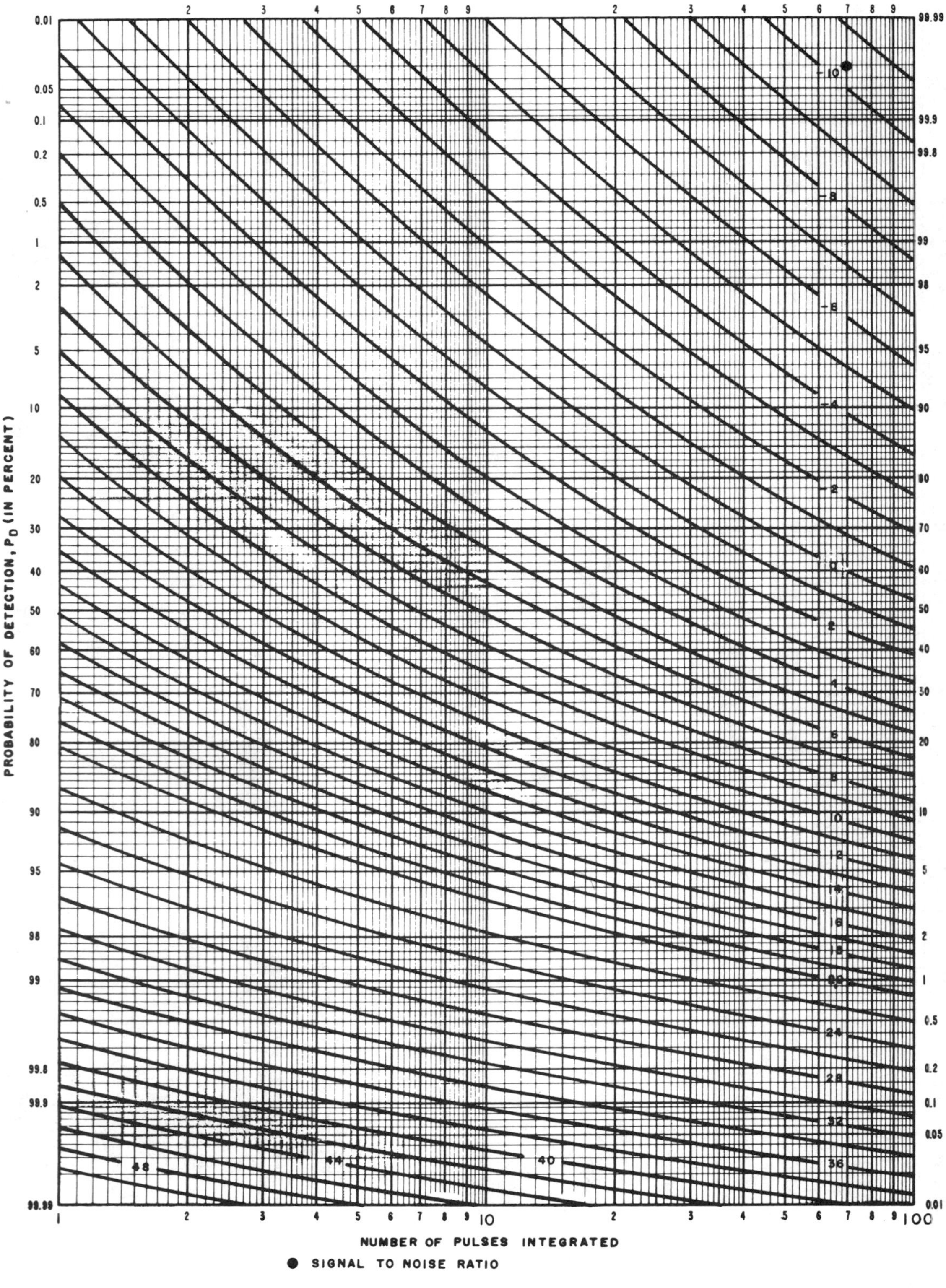

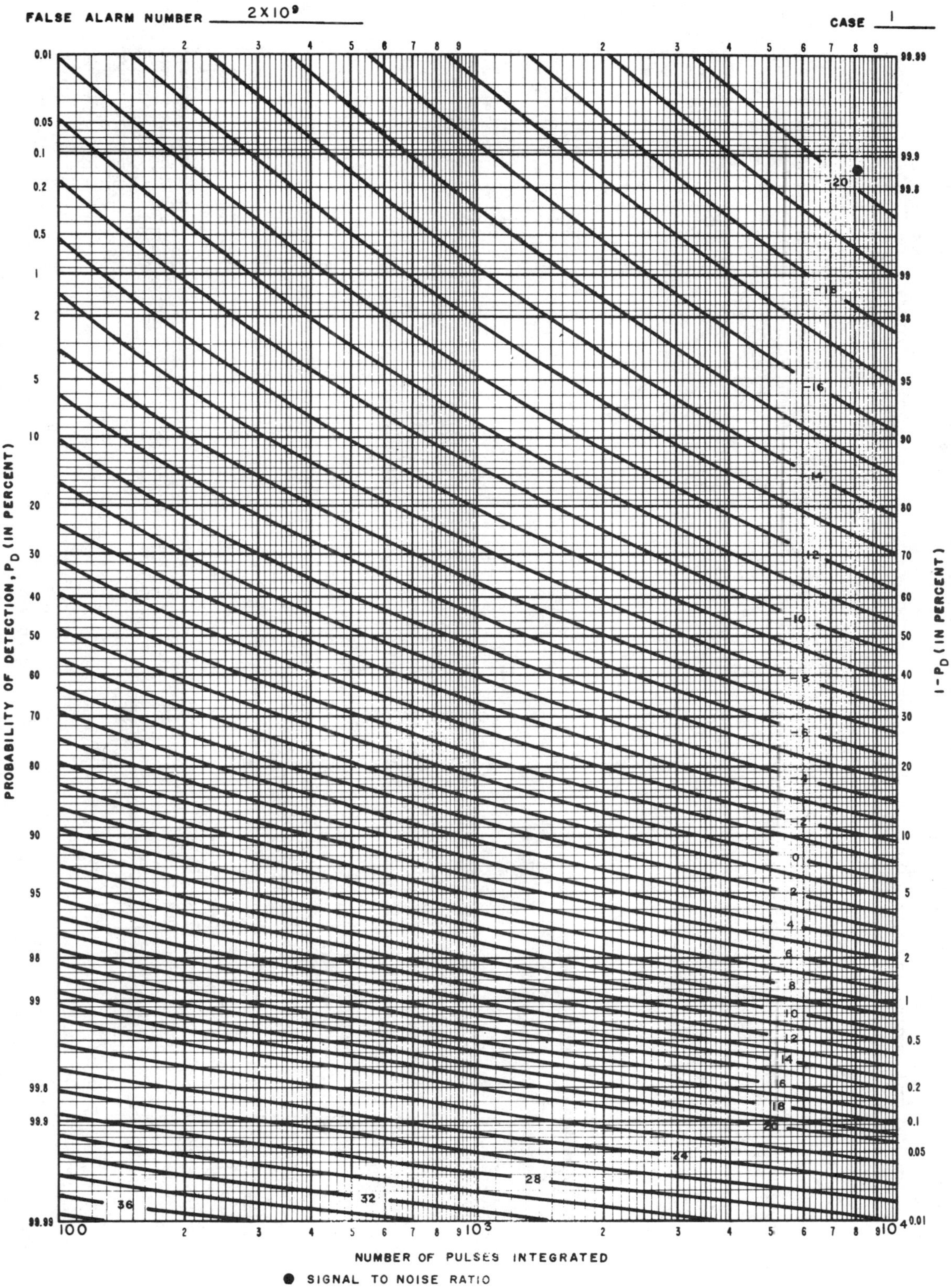
FALSE ALARM NUMBER 2X10⁹
CASE 1
PROBABILITY OF DETECTION, P_D (IN PERCENT)
$1-P_D$ (IN PERCENT)
NUMBER OF PULSES INTEGRATED
● SIGNAL TO NOISE RATIO
-20
-18
-16
-14
-12
-10
-8
-6
-4
-2
0
2
4
6
8
10
12
14
16
18
20
24
28
32
36

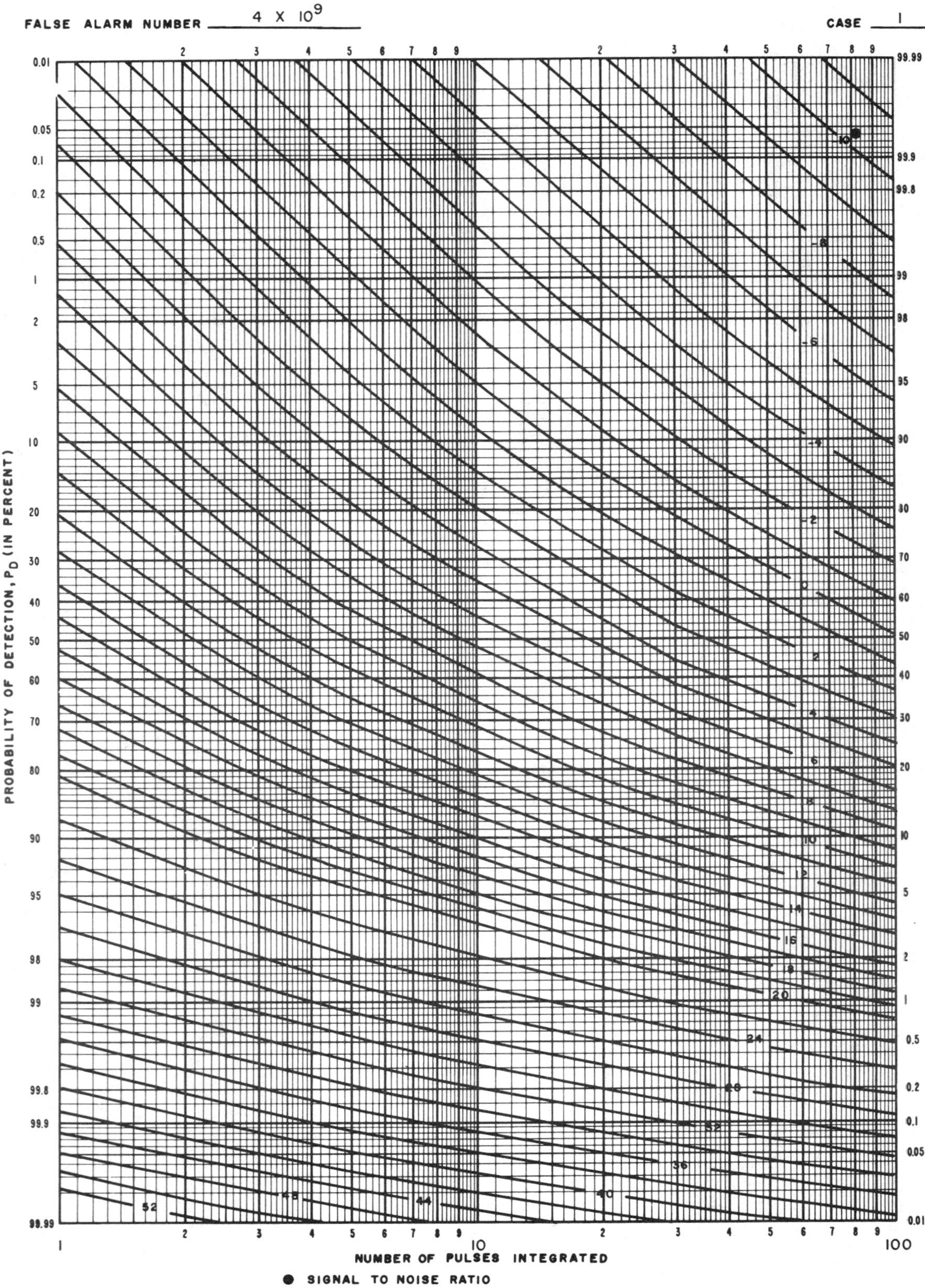
FALSE ALARM NUMBER 4 X 10⁹
CASE 1
PROBABILITY OF DETECTION, P_D (IN PERCENT)
1 - P_D (IN PERCENT)
NUMBER OF PULSES INTEGRATED
● SIGNAL TO NOISE RATIO

FALSE ALARM NUMBER 4 X 10^9 CASE 1

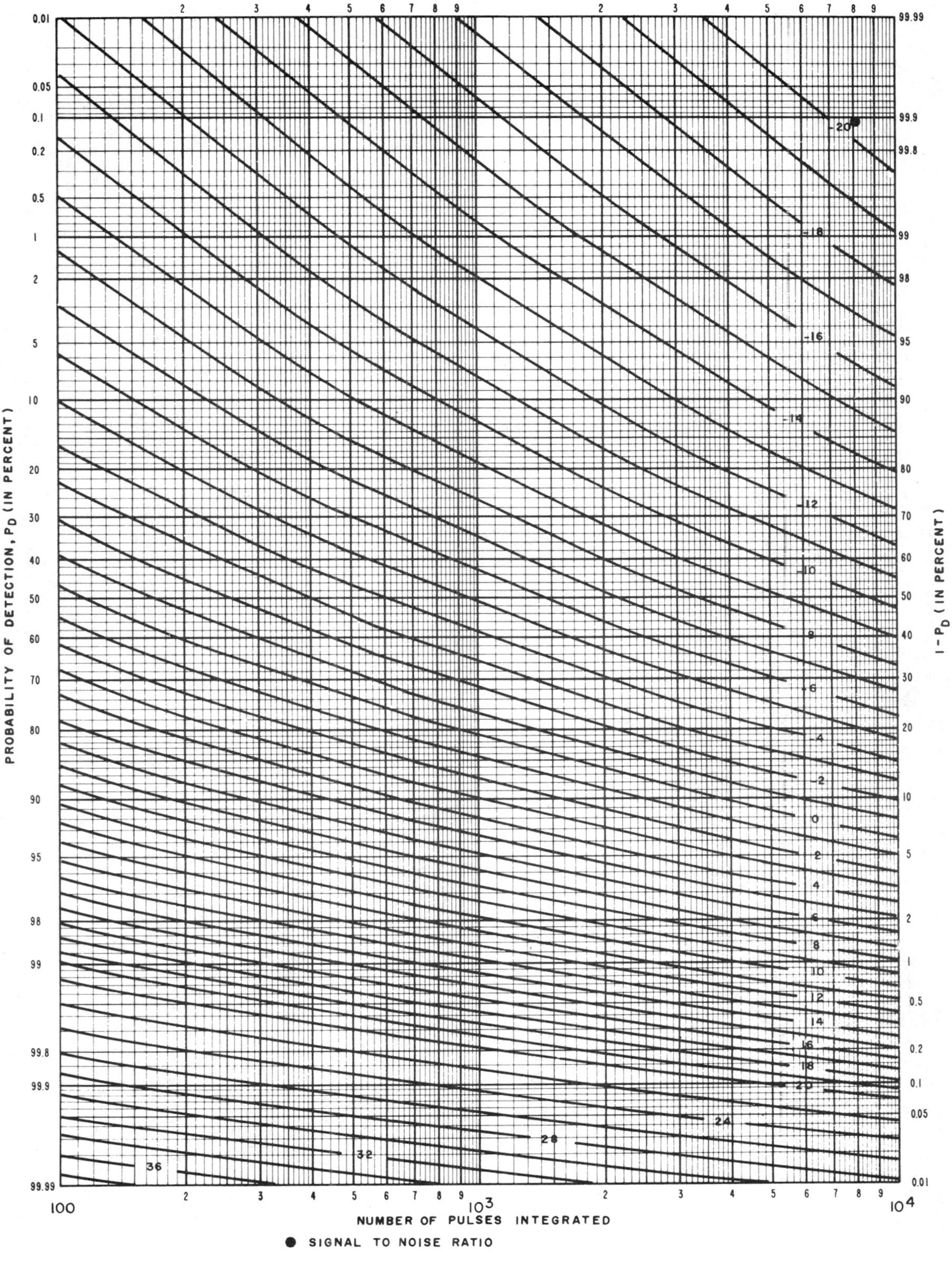

FALSE ALARM NUMBER 6×10^9 CASE 1

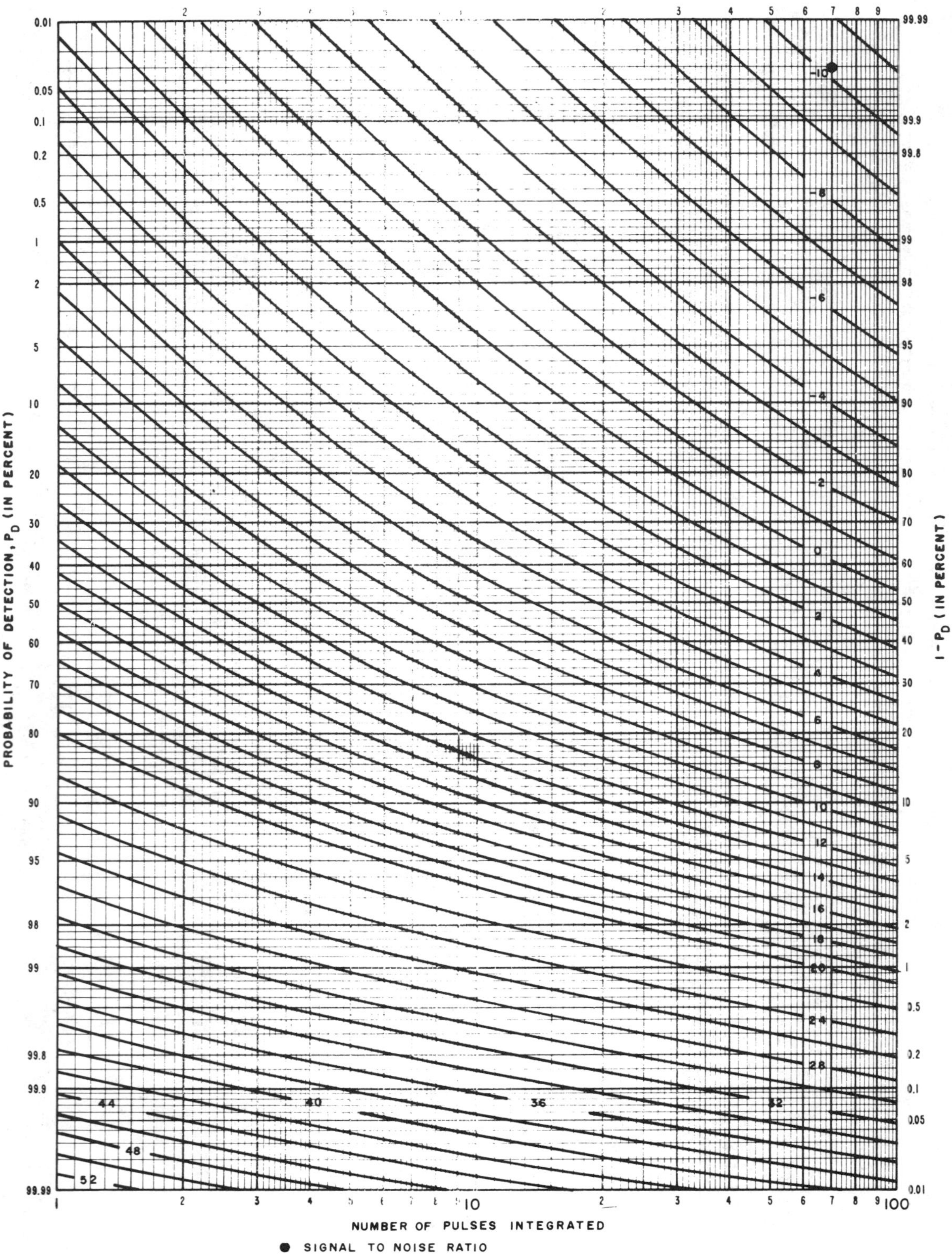

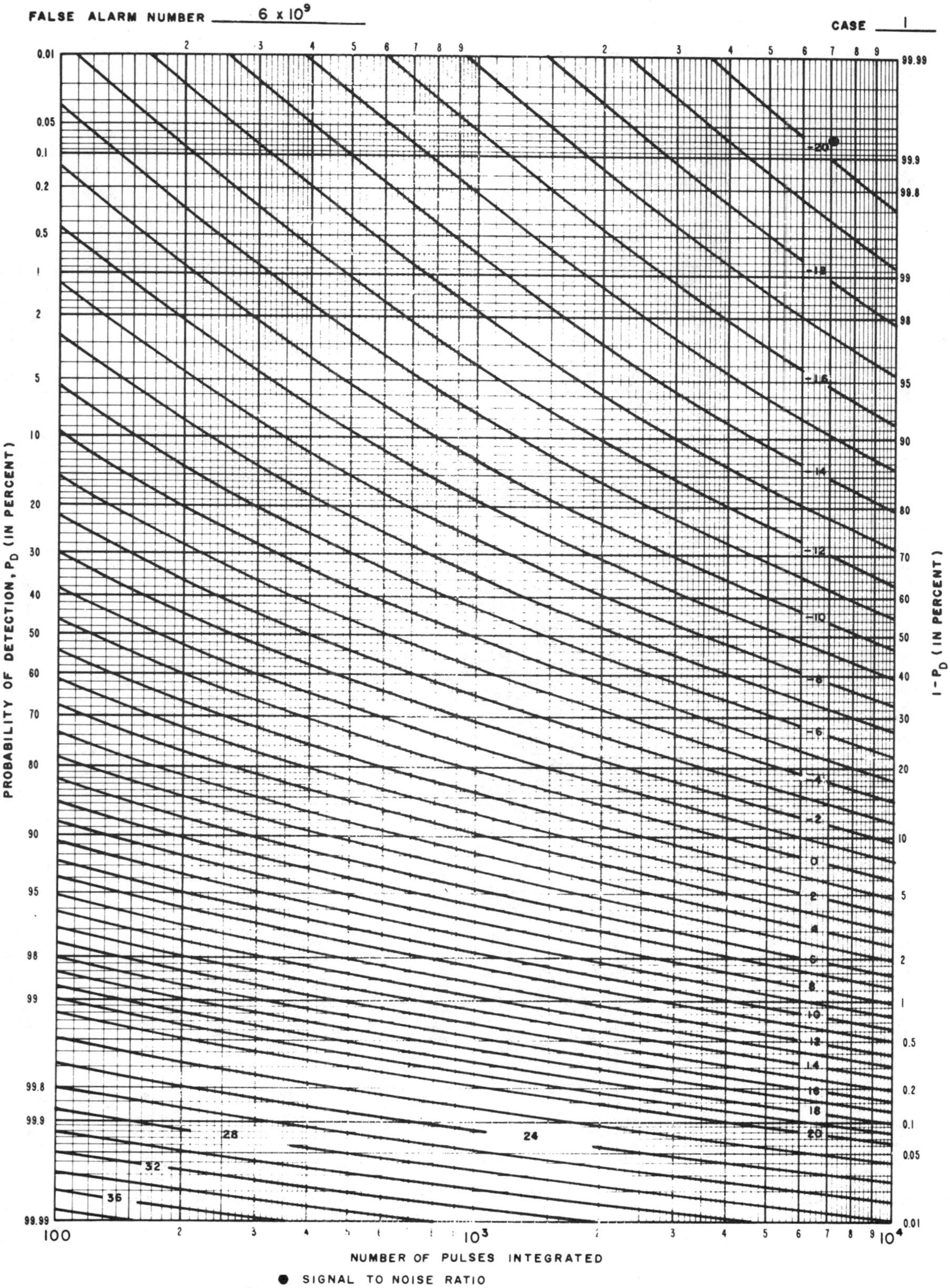
FALSE ALARM NUMBER 6 x 10^9
CASE 1
PROBABILITY OF DETECTION, P_D (IN PERCENT)
1 - P_D (IN PERCENT)
NUMBER OF PULSES INTEGRATED
● SIGNAL TO NOISE RATIO
-20
-18
-16
-14
-12
-10
-8
-6
-4
-2
0
2
4
6
8
10
12
14
16
18
20
24
28
32
36
100
10^3
10^4

FALSE ALARM NUMBER 8 x 10⁹ CASE 1

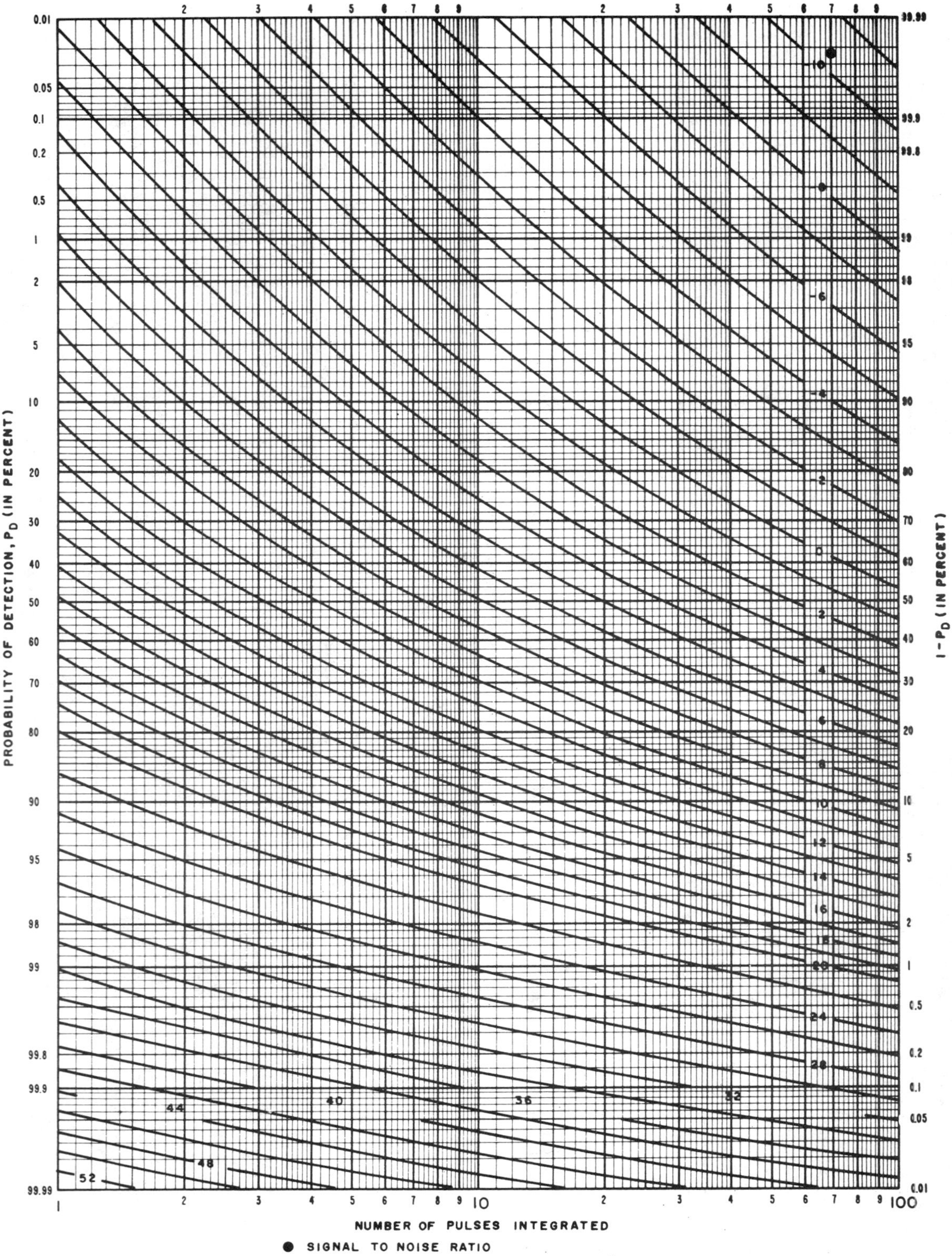

FALSE ALARM NUMBER 8 x 10⁹ CASE 1

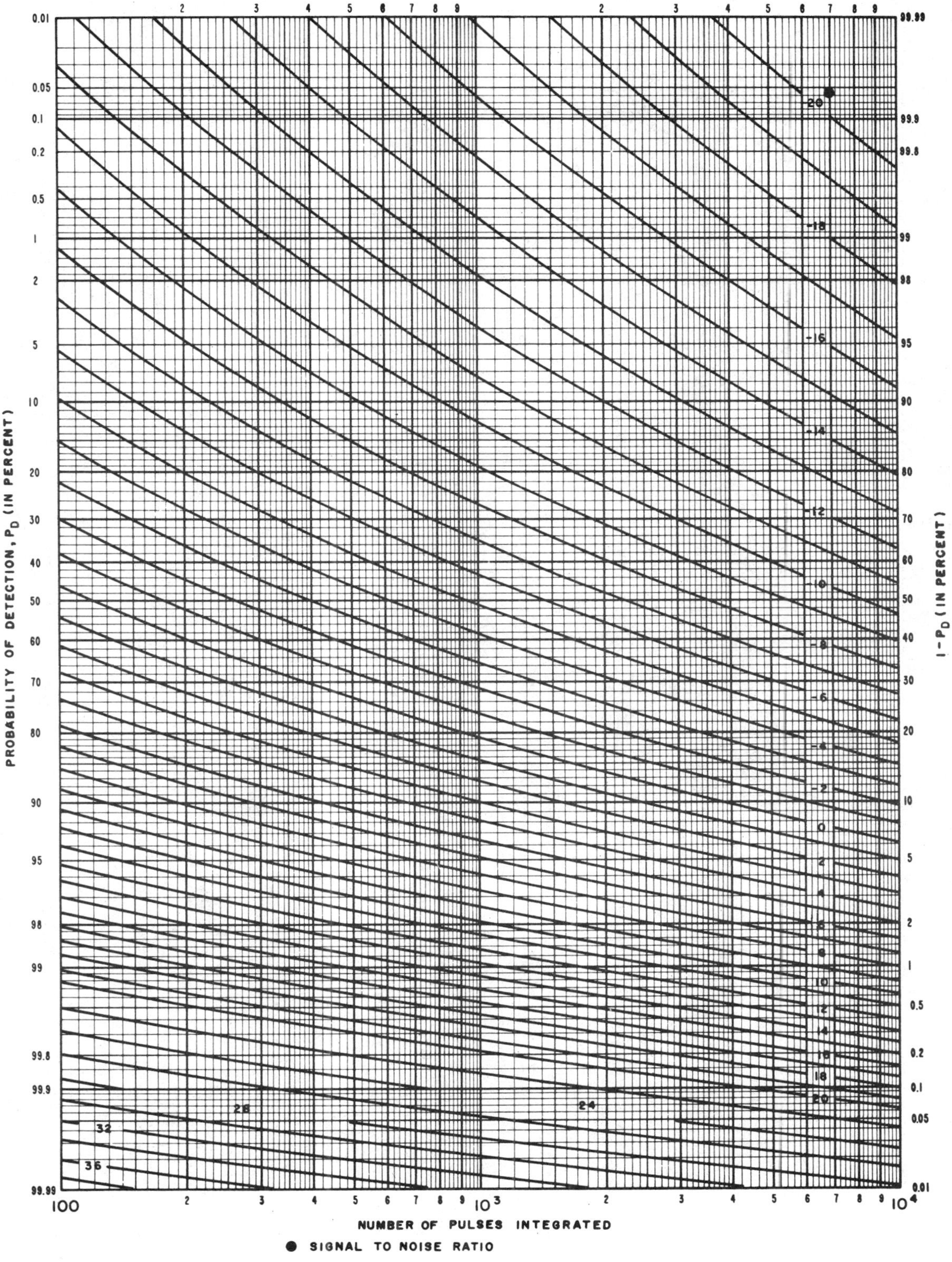

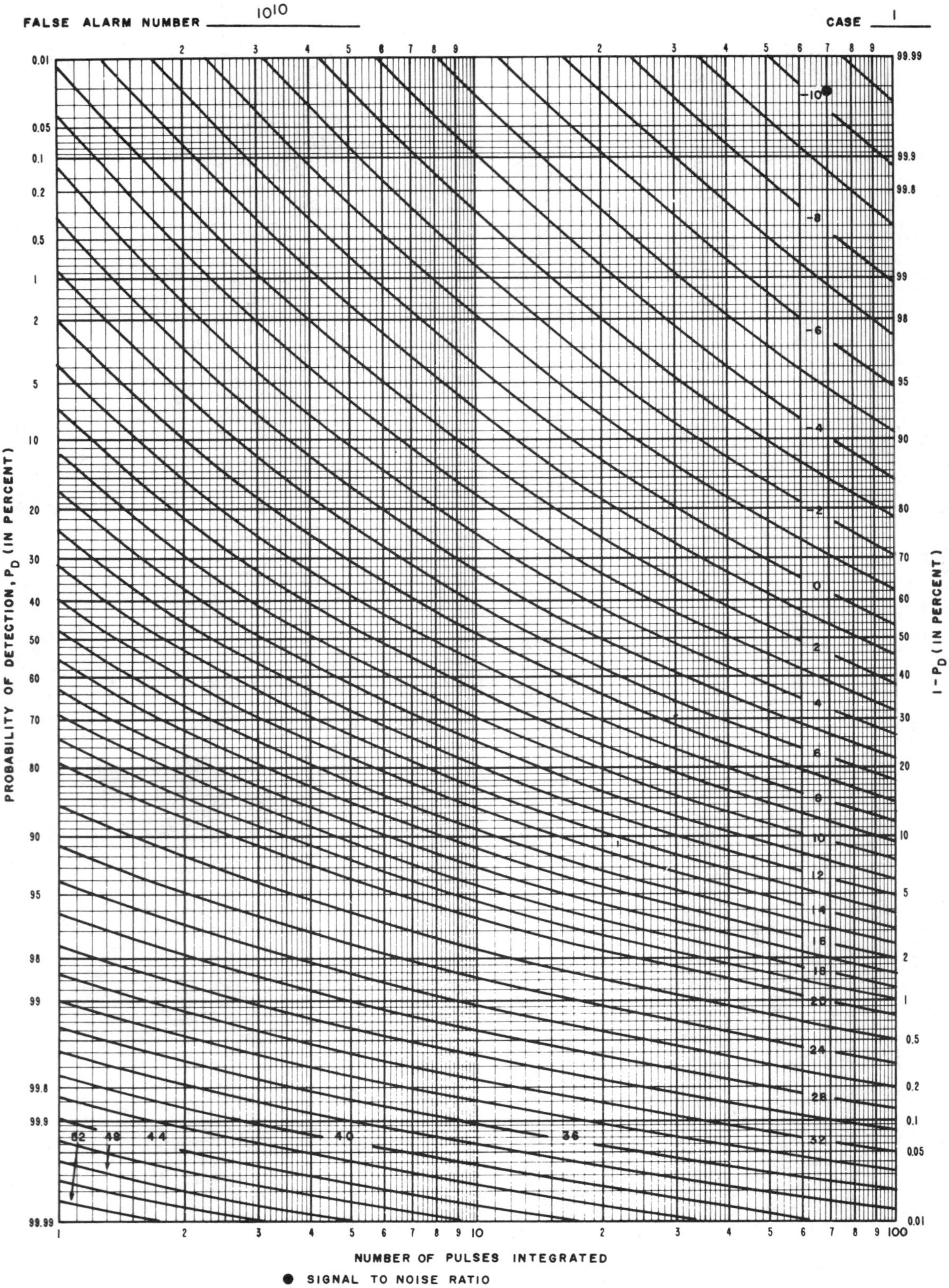
FALSE ALARM NUMBER 10^{10}
CASE 1
PROBABILITY OF DETECTION, P_D (IN PERCENT)
$1-P_D$ (IN PERCENT)
NUMBER OF PULSES INTEGRATED
● SIGNAL TO NOISE RATIO

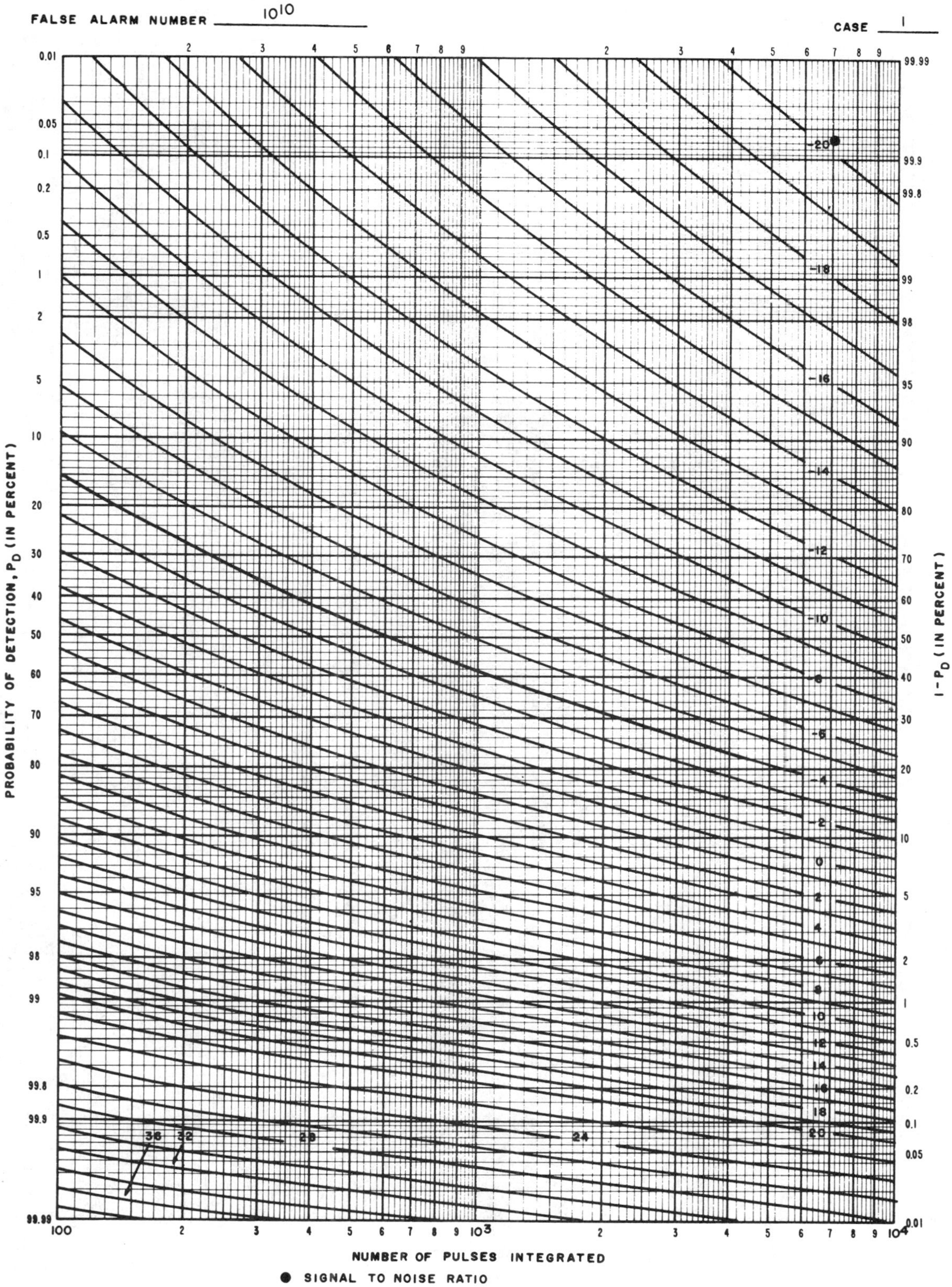
FALSE ALARM NUMBER 10^{10}
CASE 1
PROBABILITY OF DETECTION, P_D (IN PERCENT)
1 - P_D (IN PERCENT)
NUMBER OF PULSES INTEGRATED
● SIGNAL TO NOISE RATIO
-20
-18
-16
-14
-12
-10
-8
-6
-4
-2
0
2
4
6
8
10
12
14
16
18
20
24
28
32
36
100
10^3
10^4

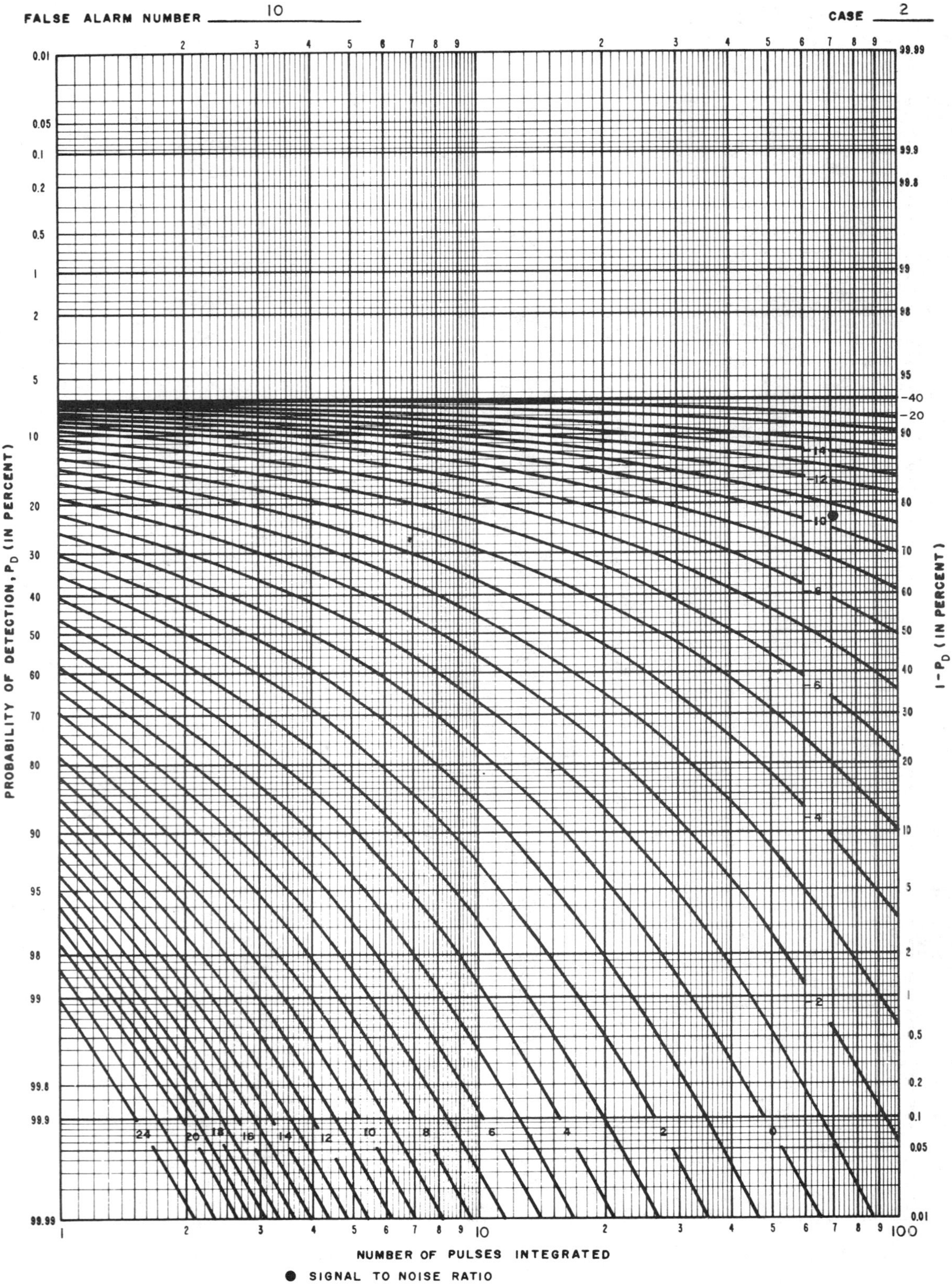
FALSE ALARM NUMBER 10
CASE 2
PROBABILITY OF DETECTION, P_D (IN PERCENT)
$1-P_D$ (IN PERCENT)
NUMBER OF PULSES INTEGRATED
● SIGNAL TO NOISE RATIO

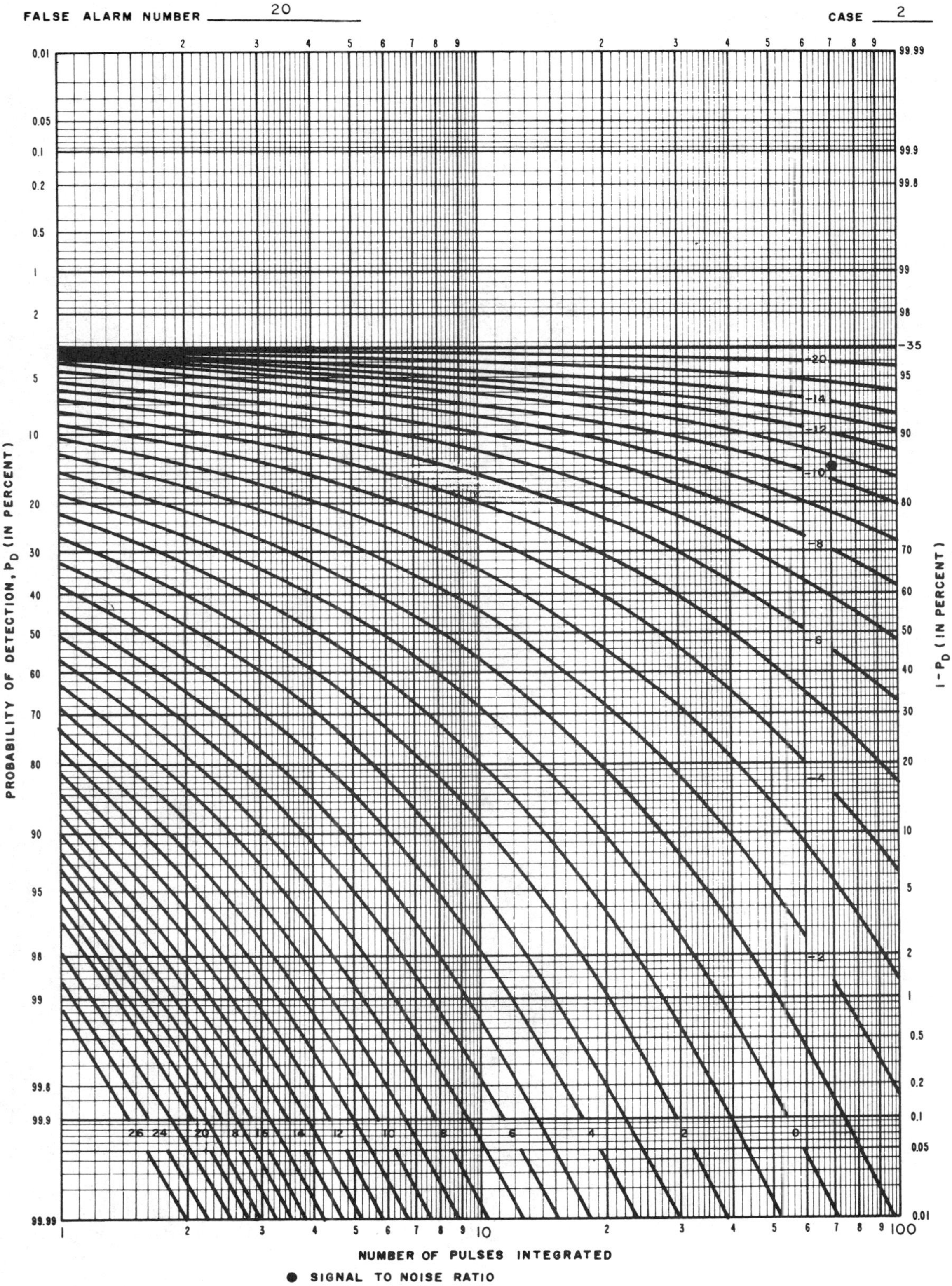
FALSE ALARM NUMBER 20
CASE 2
PROBABILITY OF DETECTION, P_D (IN PERCENT)
$1-P_D$ (IN PERCENT)
NUMBER OF PULSES INTEGRATED
● SIGNAL TO NOISE RATIO

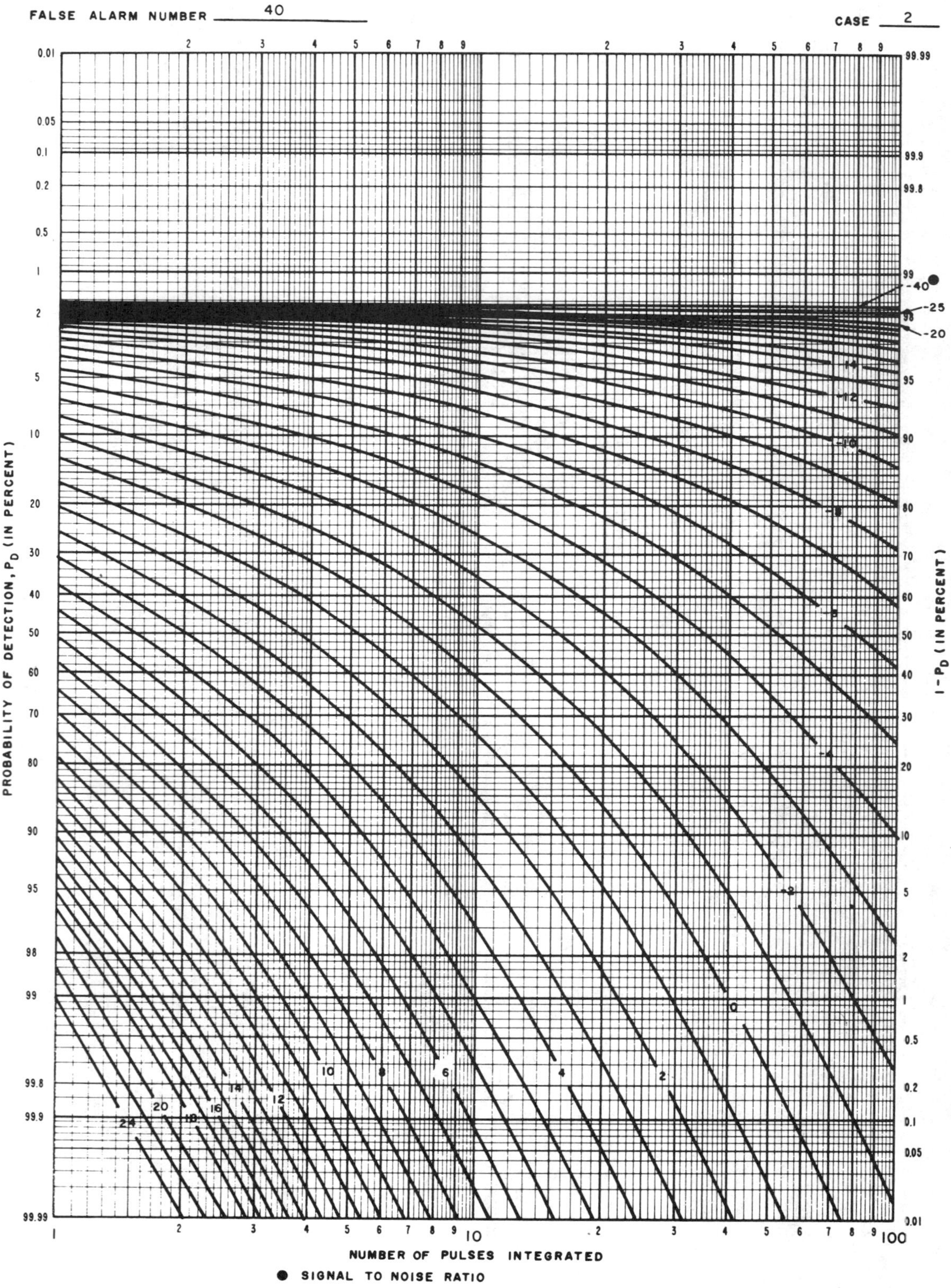
FALSE ALARM NUMBER 40
CASE 2
PROBABILITY OF DETECTION, P_D (IN PERCENT)
1 - P_D (IN PERCENT)
NUMBER OF PULSES INTEGRATED
● SIGNAL TO NOISE RATIO
-40
-25
-20
-14
-12
-10
-8
-6
-4
-2
0
2
4
6
8
10
12
14
16
18
20
24

FALSE ALARM NUMBER 60 CASE 2

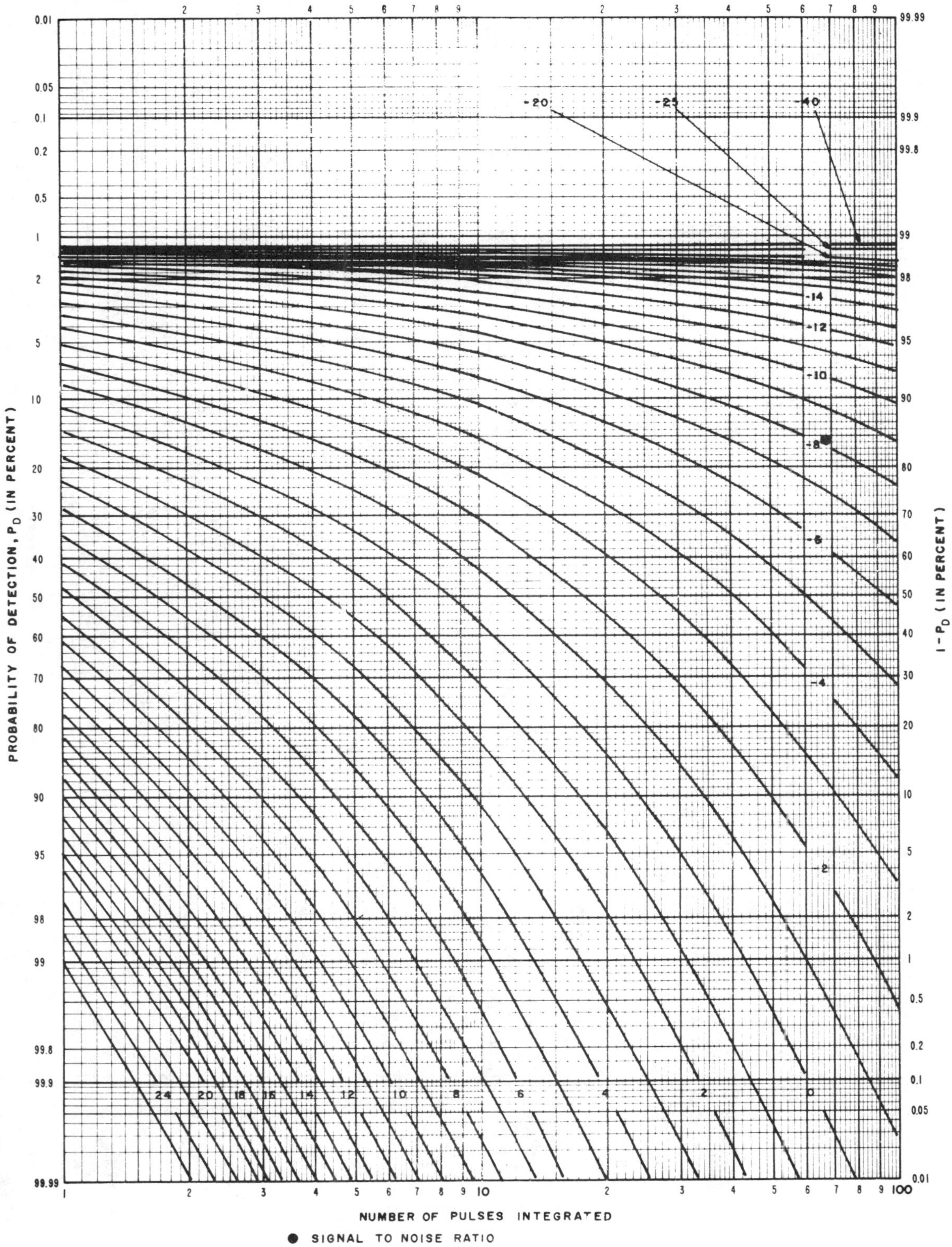

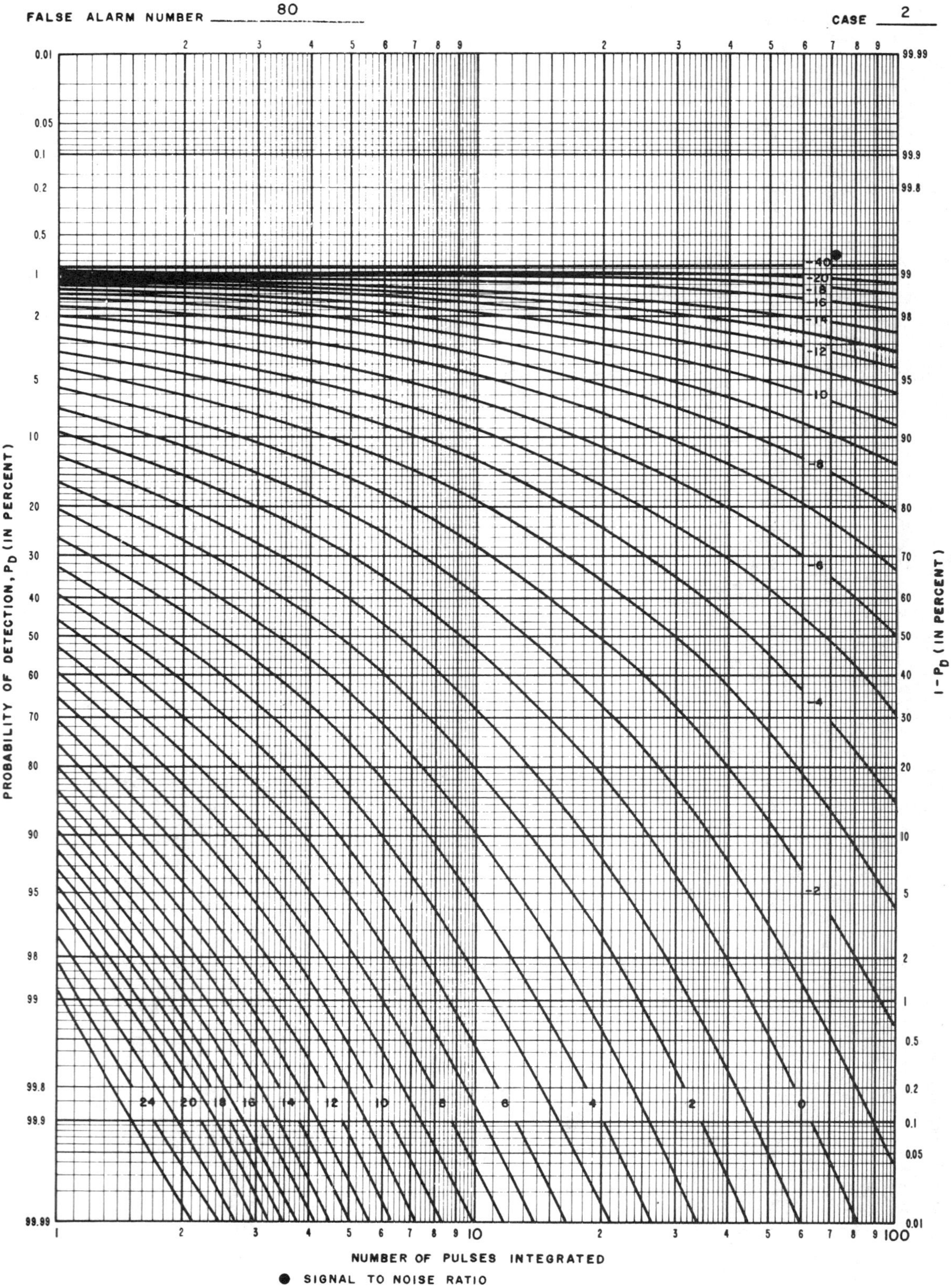
FALSE ALARM NUMBER 80
CASE 2
PROBABILITY OF DETECTION, P_D (IN PERCENT)
$1-P_D$ (IN PERCENT)
NUMBER OF PULSES INTEGRATED
● SIGNAL TO NOISE RATIO
-40
-20
-18
-16
-14
-12
-10
-8
-6
-4
-2
24
20
18
16
14
12
10
8
6
4
2
0

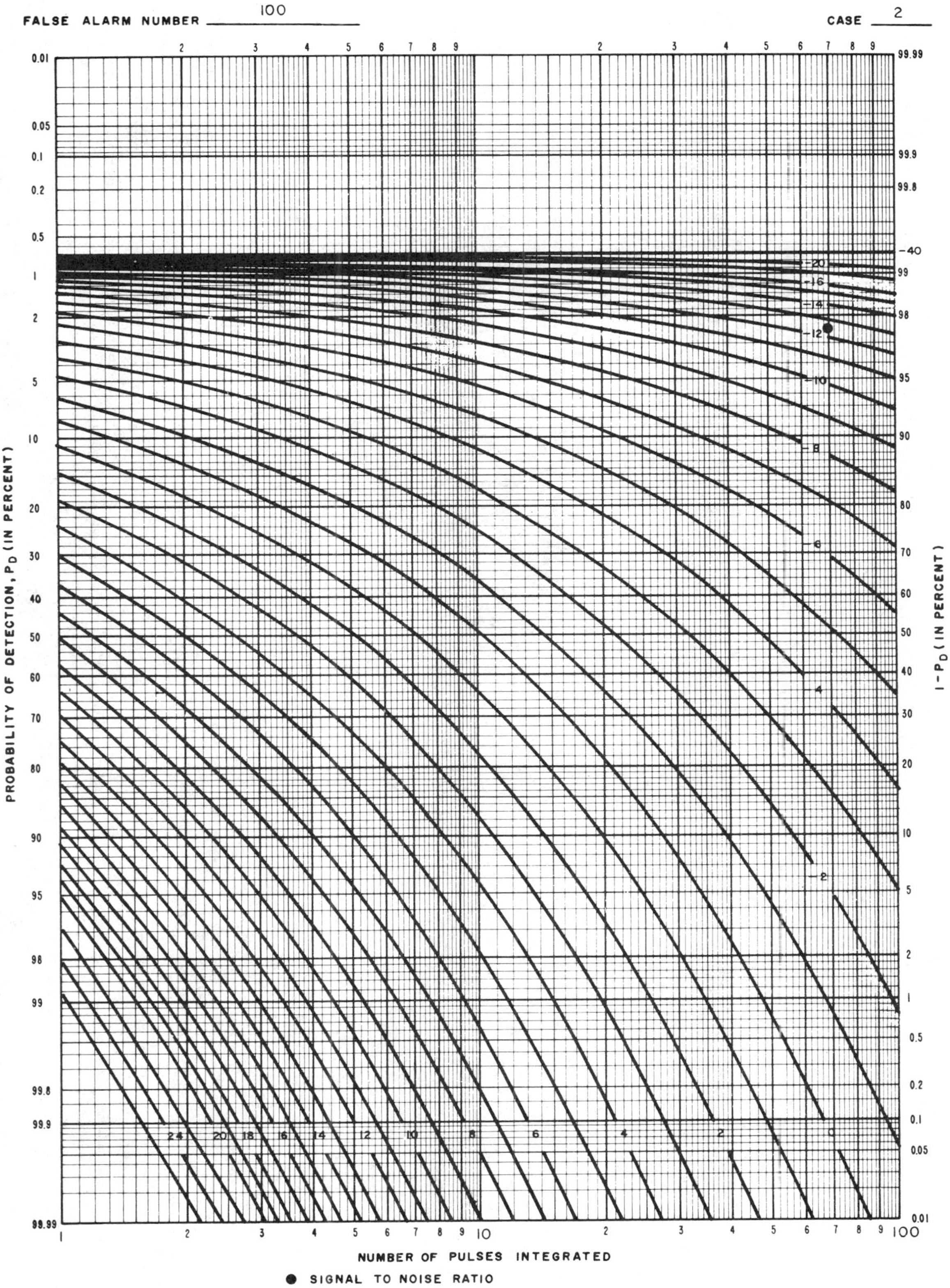
FALSE ALARM NUMBER 100
CASE 2
PROBABILITY OF DETECTION, P_D (IN PERCENT)
$1 - P_D$ (IN PERCENT)
NUMBER OF PULSES INTEGRATED
● SIGNAL TO NOISE RATIO
-40
-20
-16
-14
-12
-10
-8
-6
-4
-2
24 20 18 16 14 12 10 8 6 4 2 0

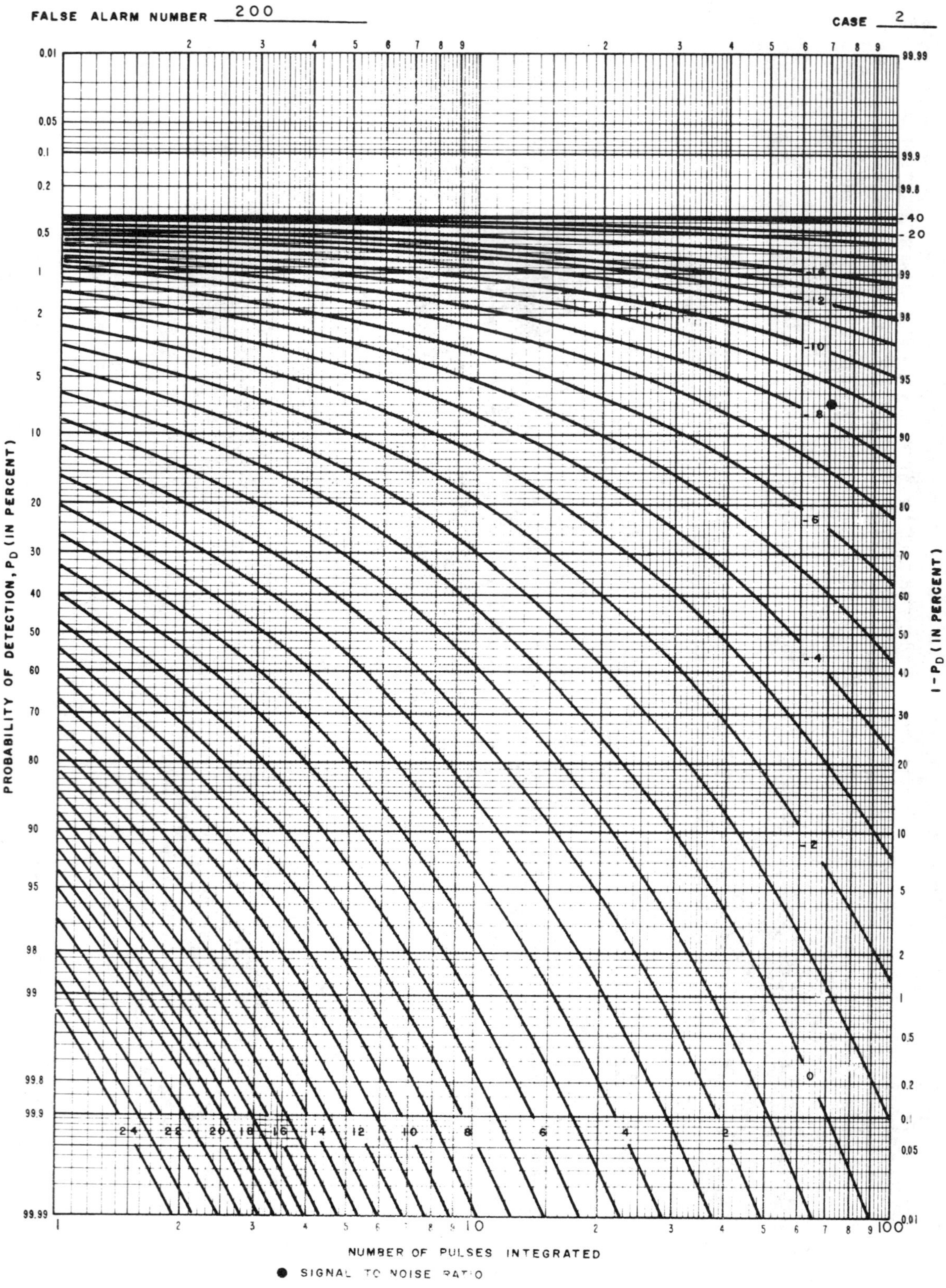
FALSE ALARM NUMBER 200
CASE 2
PROBABILITY OF DETECTION, P_D (IN PERCENT)
1 - P_D (IN PERCENT)
NUMBER OF PULSES INTEGRATED
● SIGNAL TO NOISE RATIO

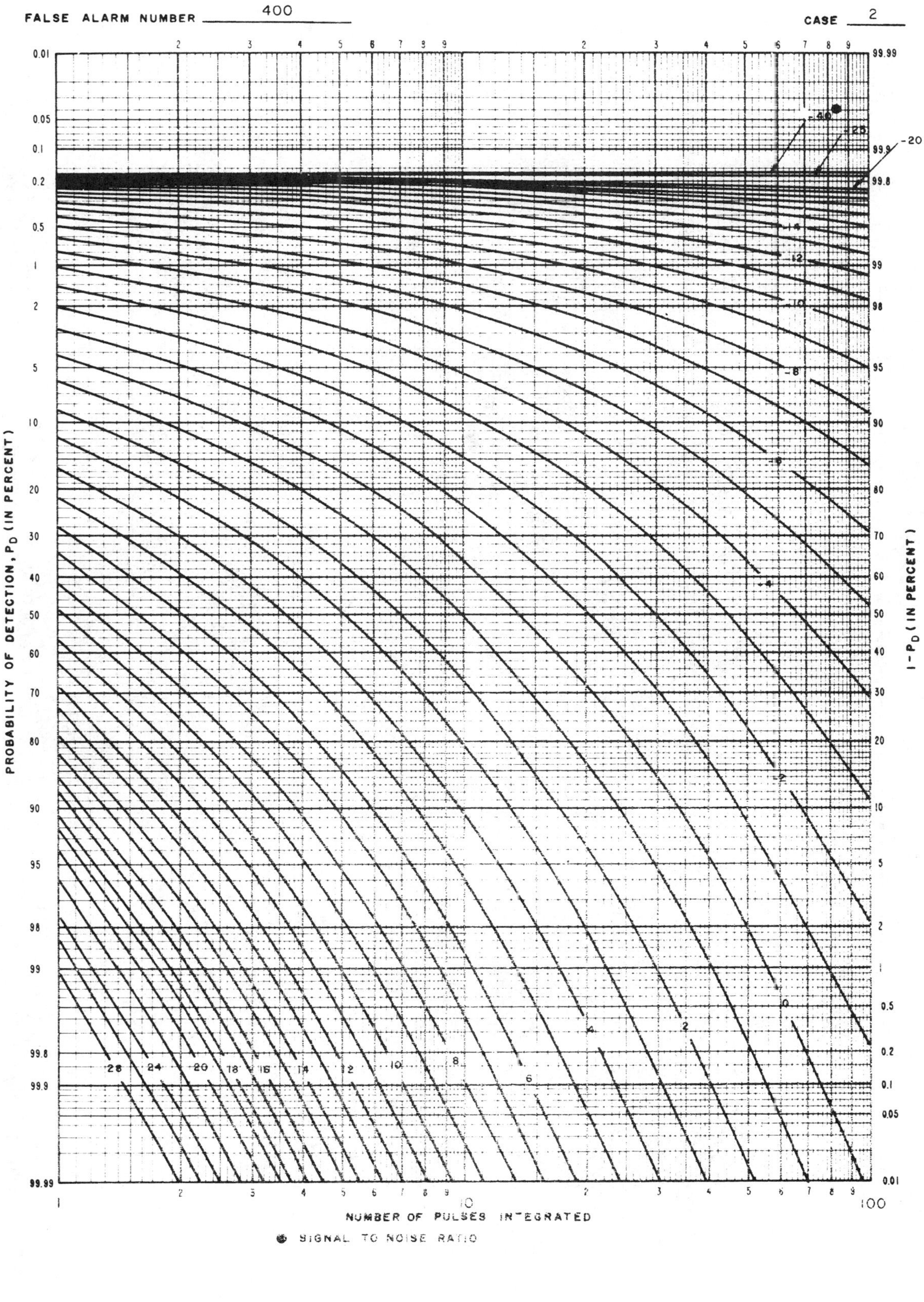
FALSE ALARM NUMBER 400
CASE 2
PROBABILITY OF DETECTION, P_D (IN PERCENT)
$1-P_D$ (IN PERCENT)
NUMBER OF PULSES INTEGRATED
● SIGNAL TO NOISE RATIO
-40
-25
-20
-14
-12
-10
-8
-6
-4
-2
0
2
4
6
8
10
12
14
16
18
20
24
26

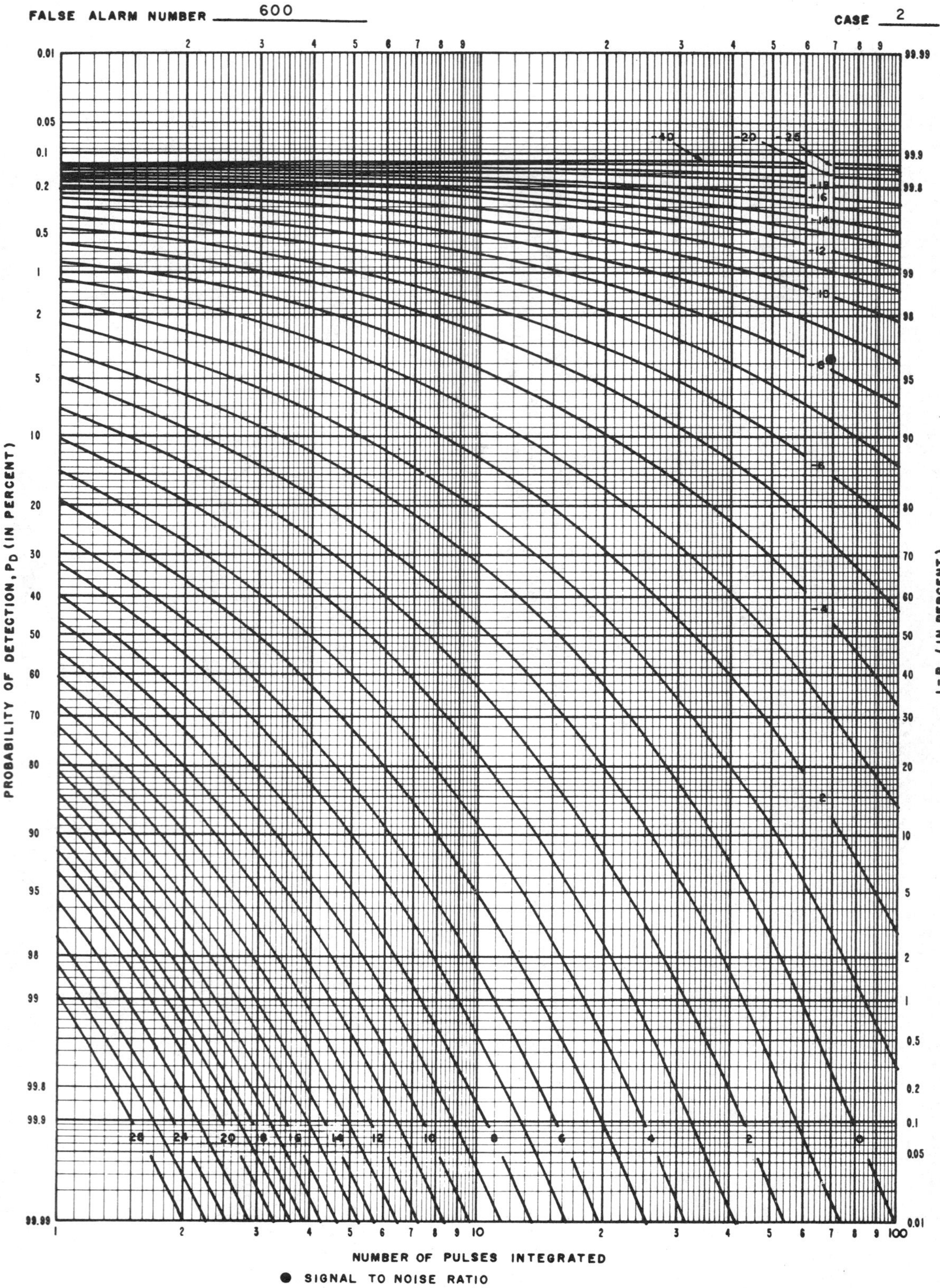

FALSE ALARM NUMBER 600
CASE 2
PROBABILITY OF DETECTION, P_D (IN PERCENT)
1 - P_D (IN PERCENT)
NUMBER OF PULSES INTEGRATED
● SIGNAL TO NOISE RATIO
-40
-20
-25
-18
-16
-14
-12
-10
-8
-6
-4
-2
26
24
20
18
16
14
12
10
8
6
4
2
0

FALSE ALARM NUMBER 800

CASE 2

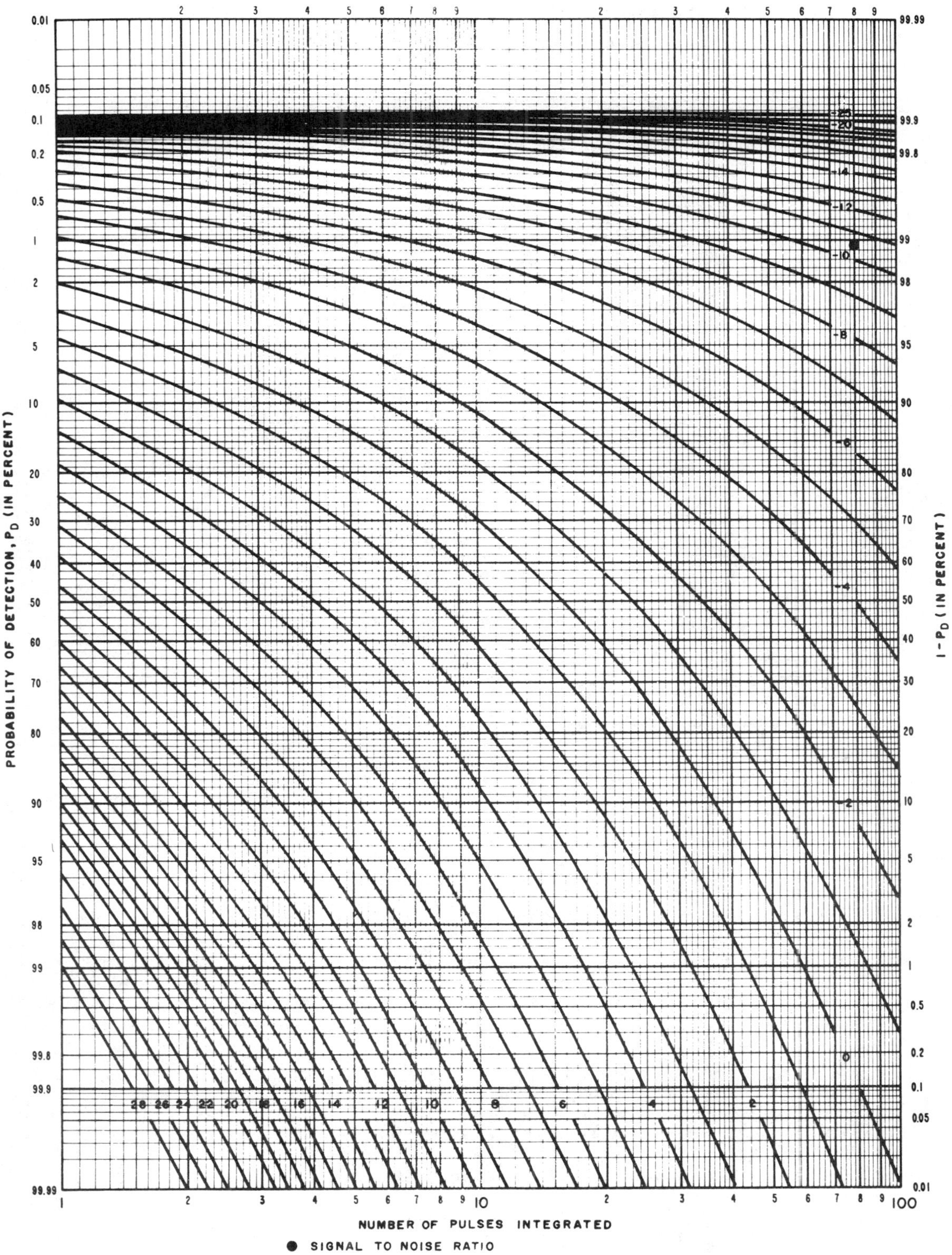

FALSE ALARM NUMBER 10^3 CASE 2

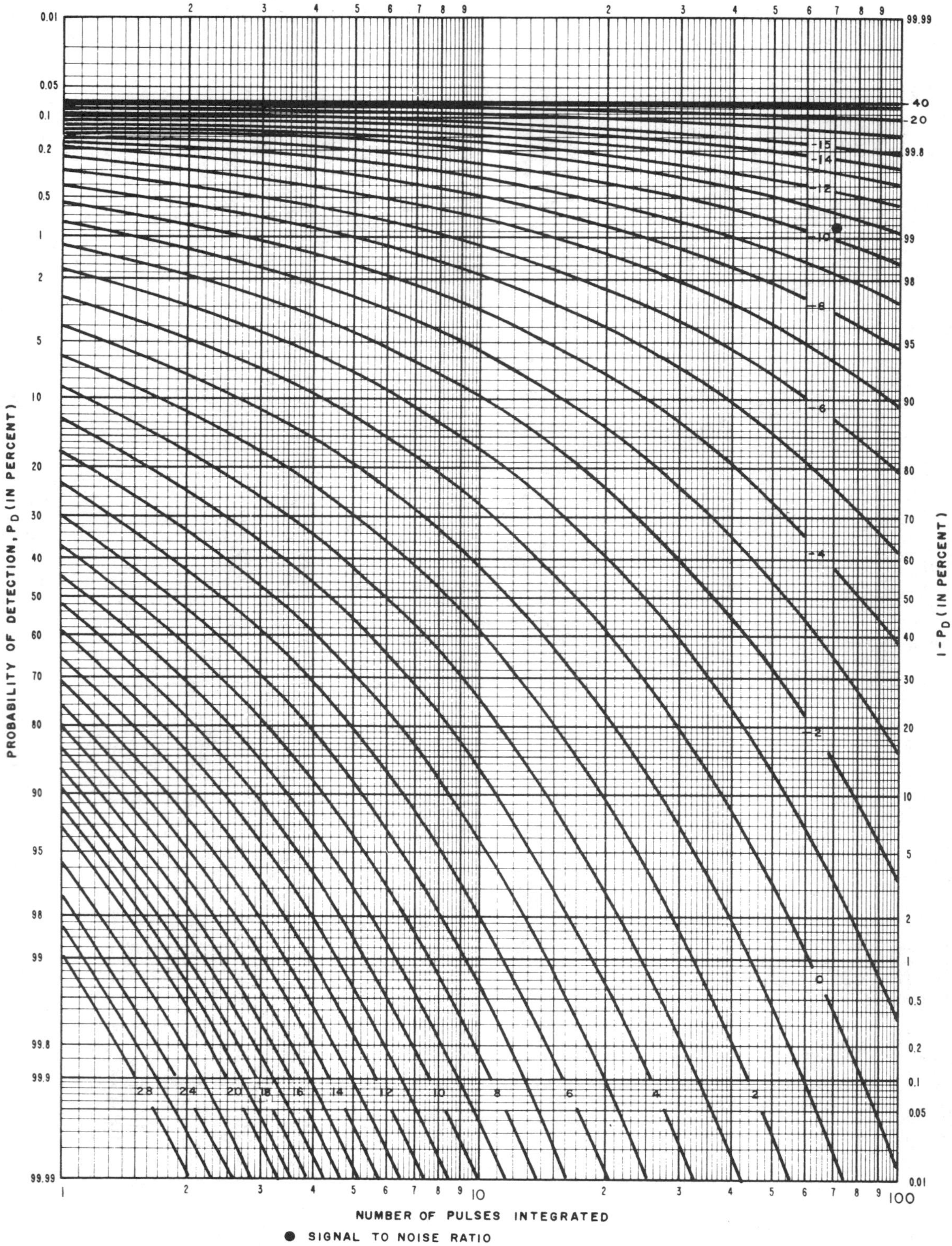

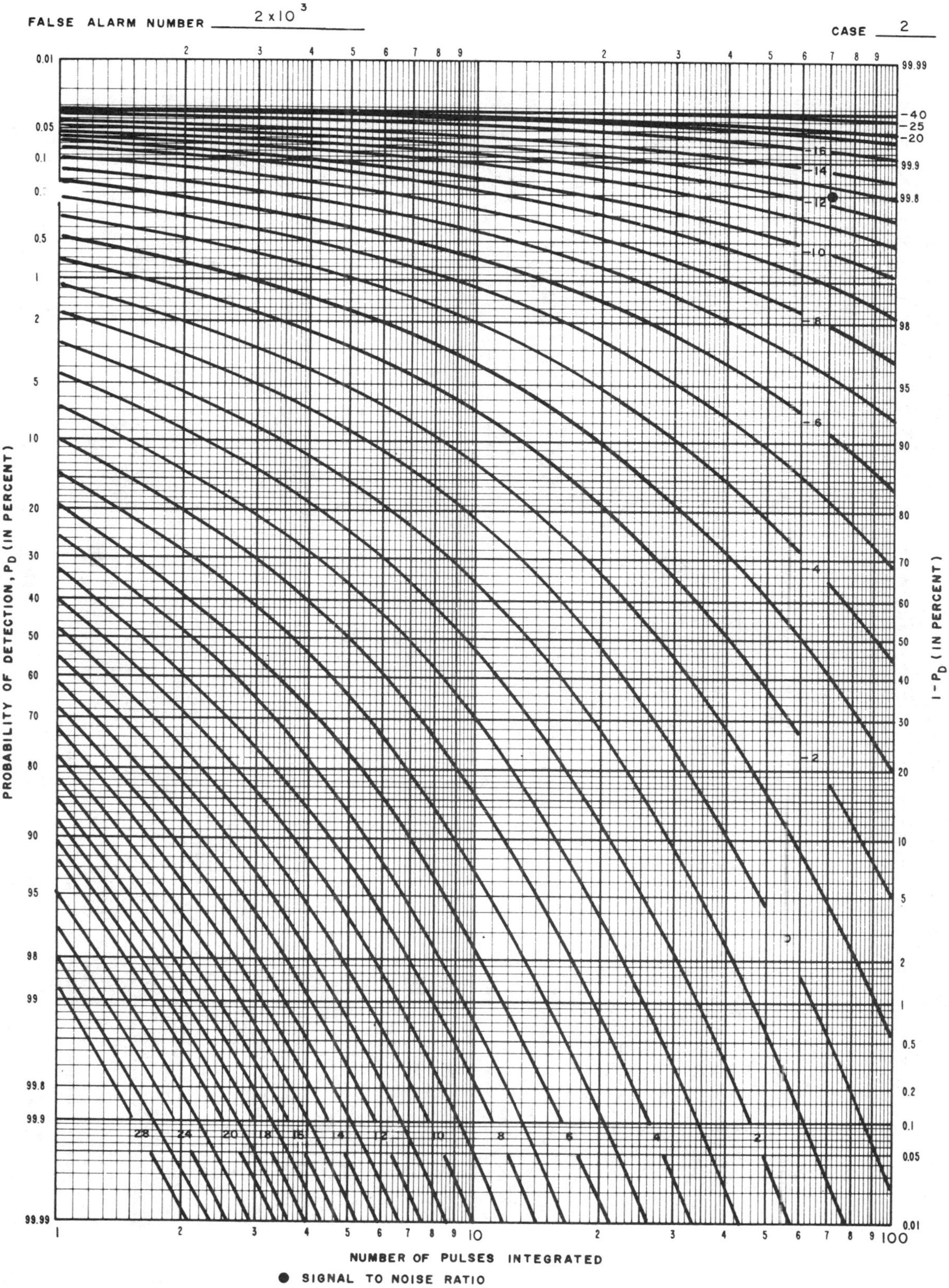
FALSE ALARM NUMBER 2 x 10^3
CASE 2
PROBABILITY OF DETECTION, P_D (IN PERCENT)
$1 - P_D$ (IN PERCENT)
NUMBER OF PULSES INTEGRATED
● SIGNAL TO NOISE RATIO
-40
-25
-20
16
14
12
10
8
6
4
2
28
24
20
18
16
14
12
10
8
6
4
2

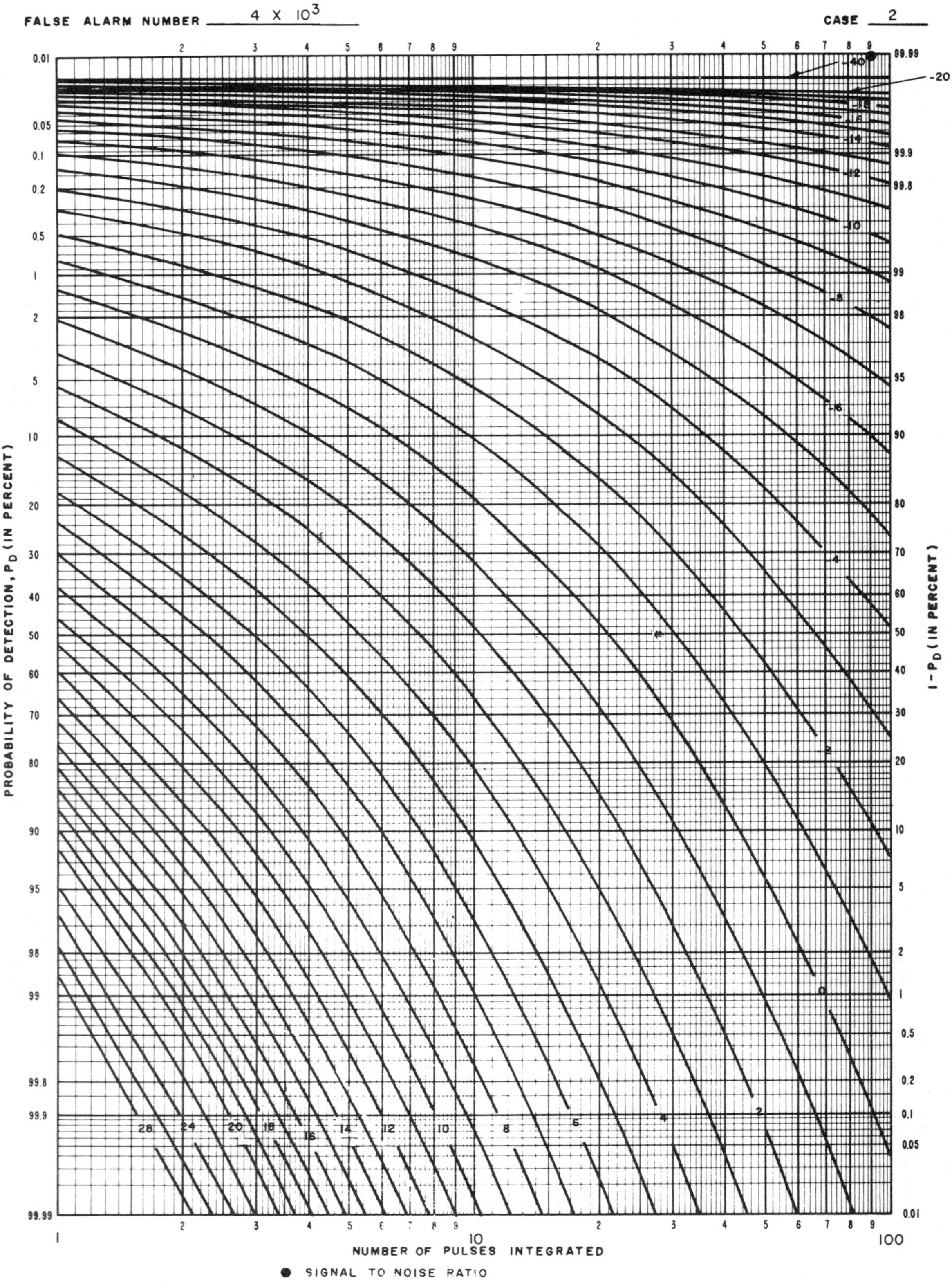
FALSE ALARM NUMBER 4 X 10³
CASE 2
PROBABILITY OF DETECTION, P_D (IN PERCENT)
1 - P_D (IN PERCENT)
NUMBER OF PULSES INTEGRATED
● SIGNAL TO NOISE RATIO
-40
-20
-18
-16
-14
-12
-10
-8
-6
-4
-2
0
2
4
6
8
10
12
14
16
18
20
24
28

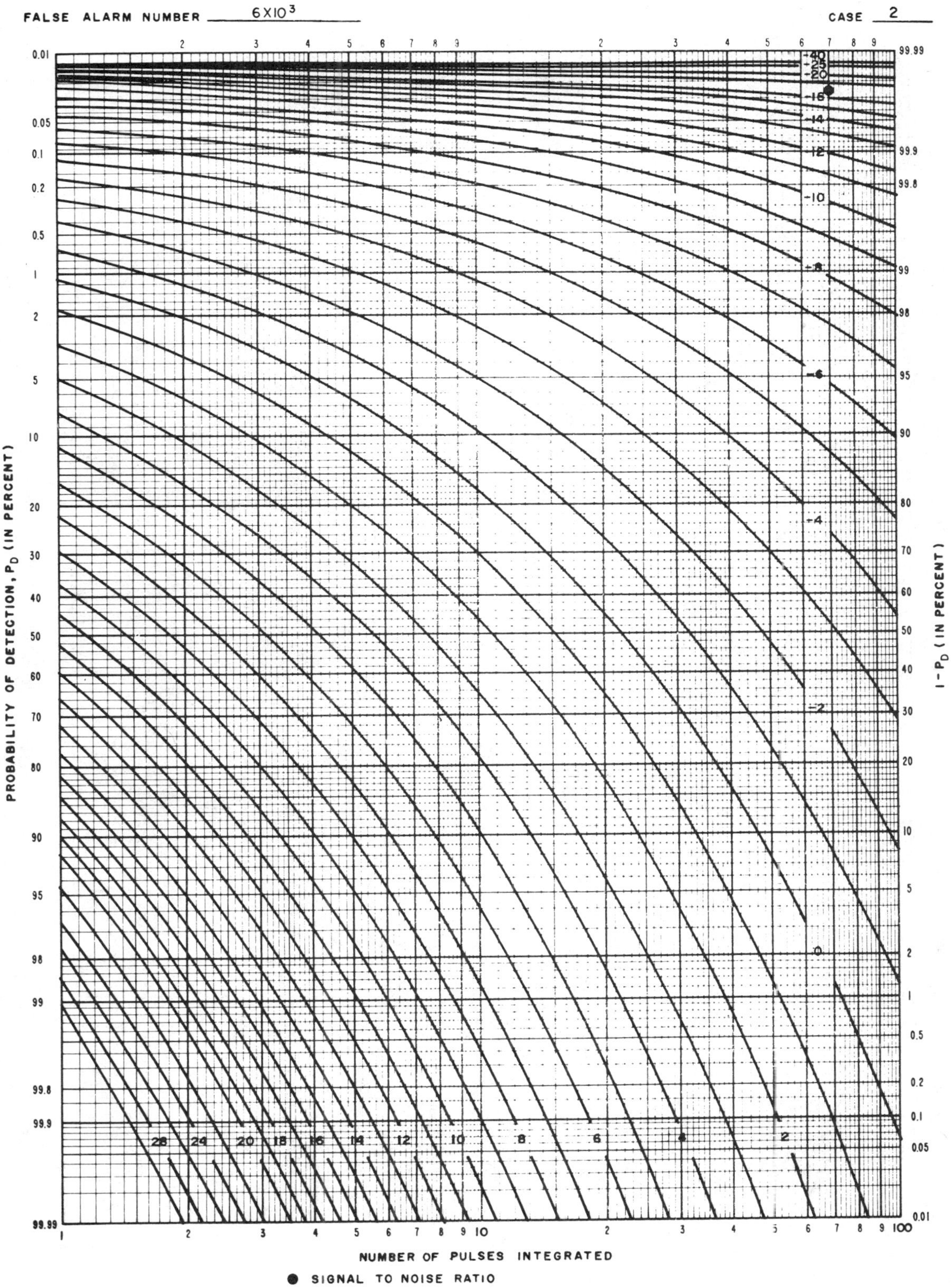
FALSE ALARM NUMBER 6X10³
CASE 2
PROBABILITY OF DETECTION, P_D (IN PERCENT)
1 - P_D (IN PERCENT)
NUMBER OF PULSES INTEGRATED
● SIGNAL TO NOISE RATIO
40
25
20
16
14
12
10
8
6
4
2
0
-2
-4
-6
-8
-10
-12
-14
-16
-20
-25
-40
28
24
20
18
16
14
12
10
8
6
4
2

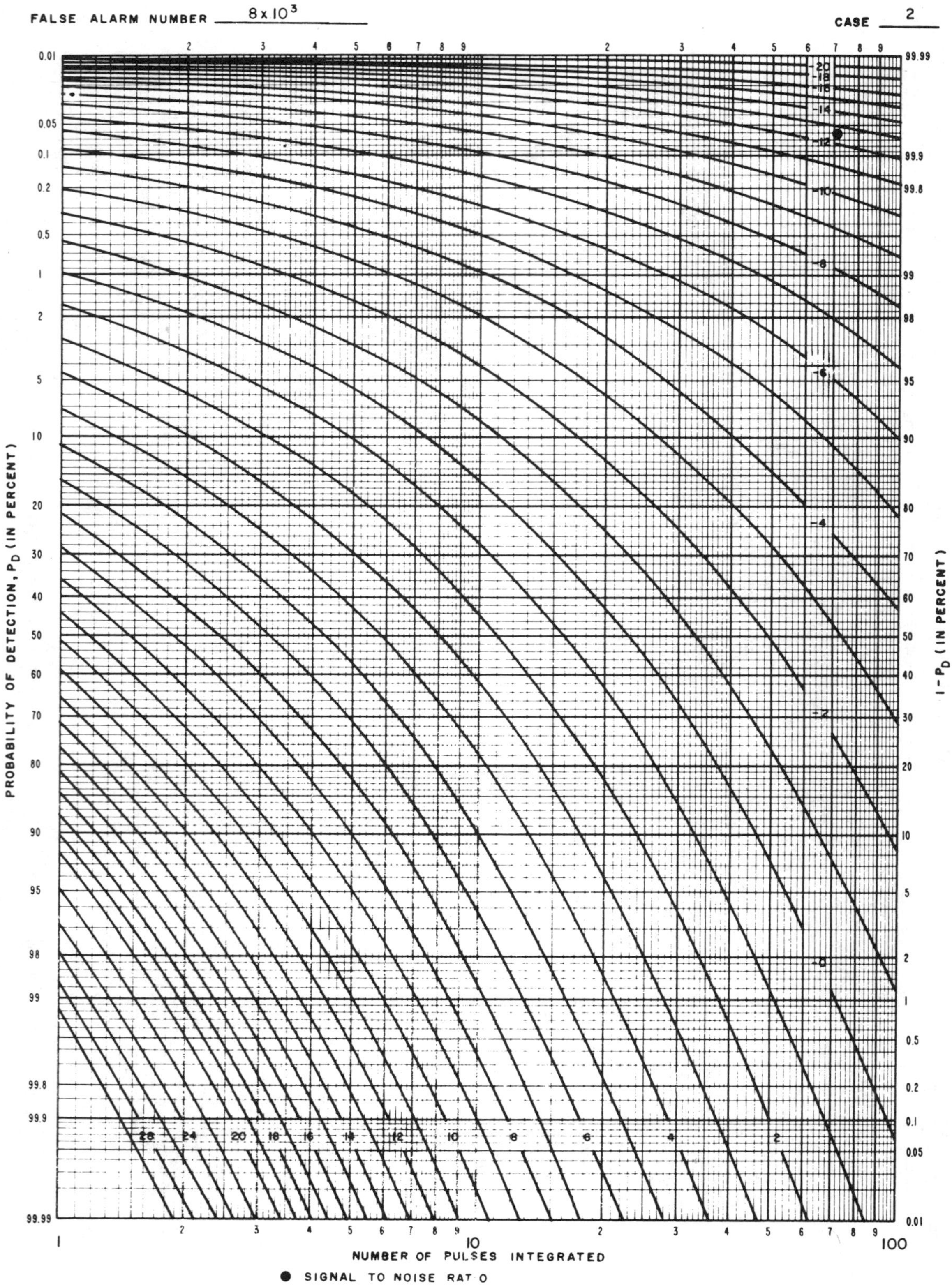

FALSE ALARM NUMBER 8 x 10³
CASE 2
PROBABILITY OF DETECTION, P_D (IN PERCENT)
1 - P_D (IN PERCENT)
NUMBER OF PULSES INTEGRATED
● SIGNAL TO NOISE RATIO

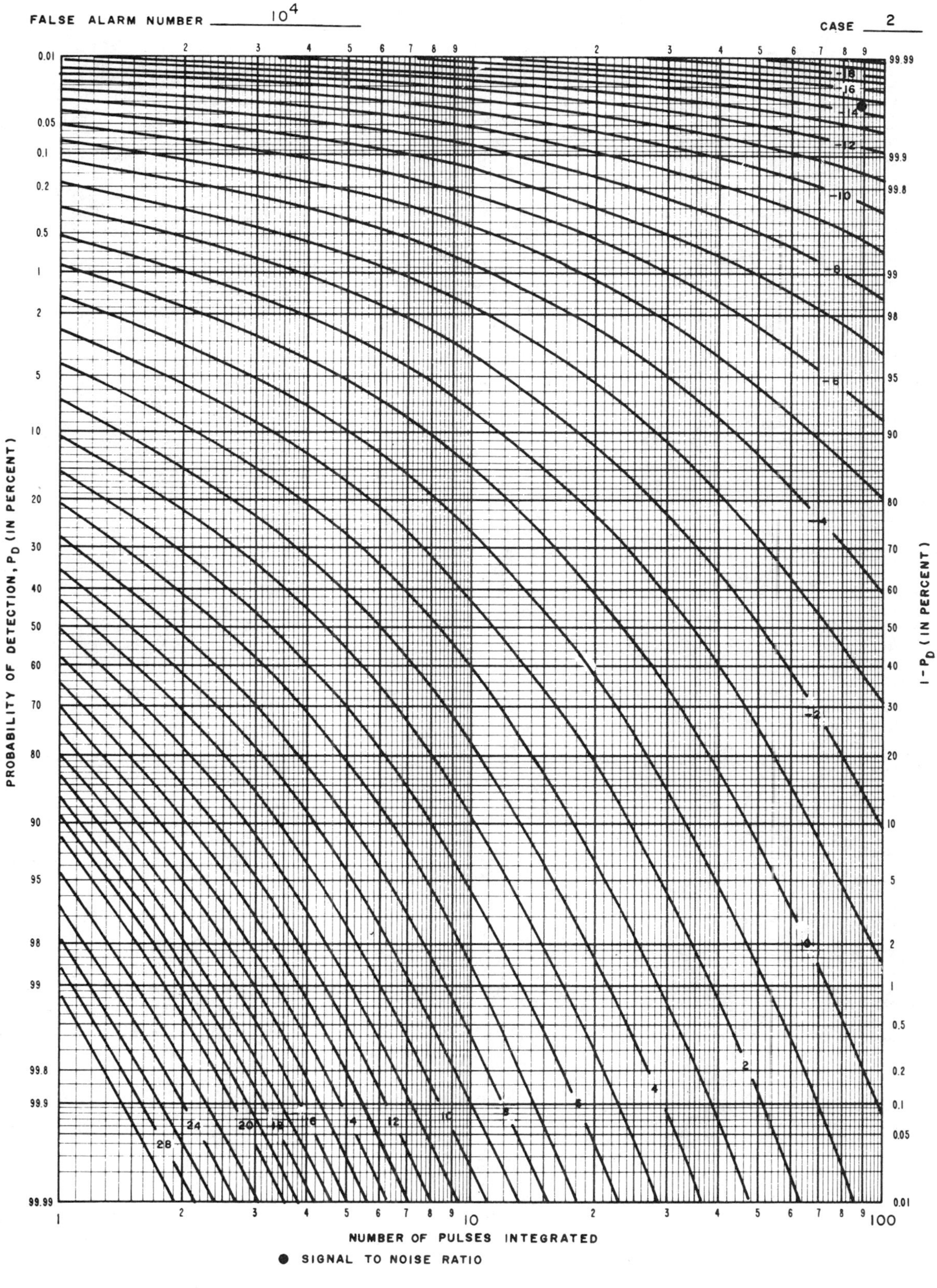
FALSE ALARM NUMBER 10^4
CASE 2
PROBABILITY OF DETECTION, P_D (IN PERCENT)
$1-P_D$ (IN PERCENT)
NUMBER OF PULSES INTEGRATED
● SIGNAL TO NOISE RATIO

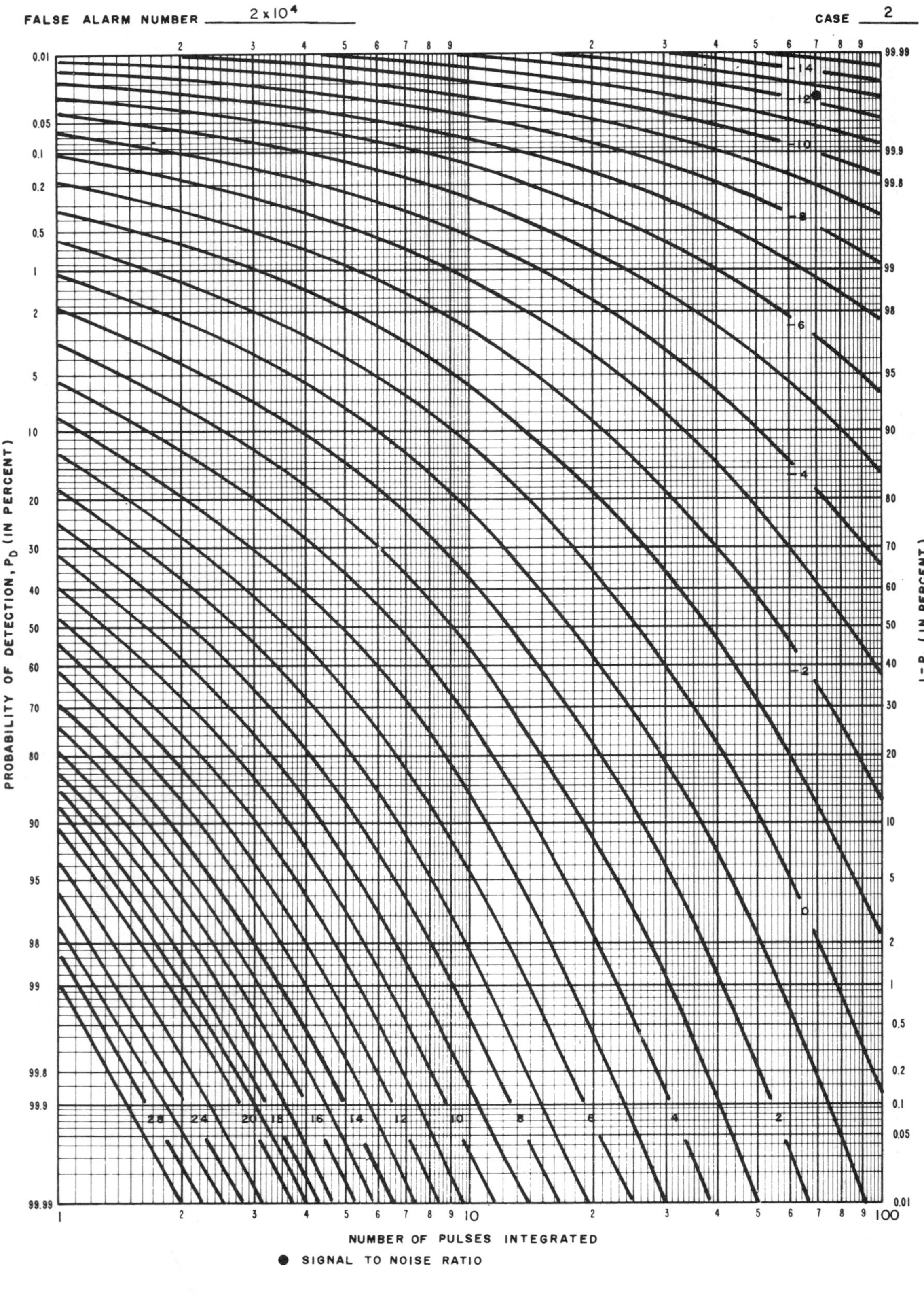
FALSE ALARM NUMBER 2 x 10⁴
CASE 2
PROBABILITY OF DETECTION, P_D (IN PERCENT)
1 - P_D (IN PERCENT)
NUMBER OF PULSES INTEGRATED
● SIGNAL TO NOISE RATIO
14
12
10
8
6
4
2
0
-2
28
24
20
18
16
14
12
10
8
6
4
2

FALSE ALARM NUMBER 4×10^4 CASE 2

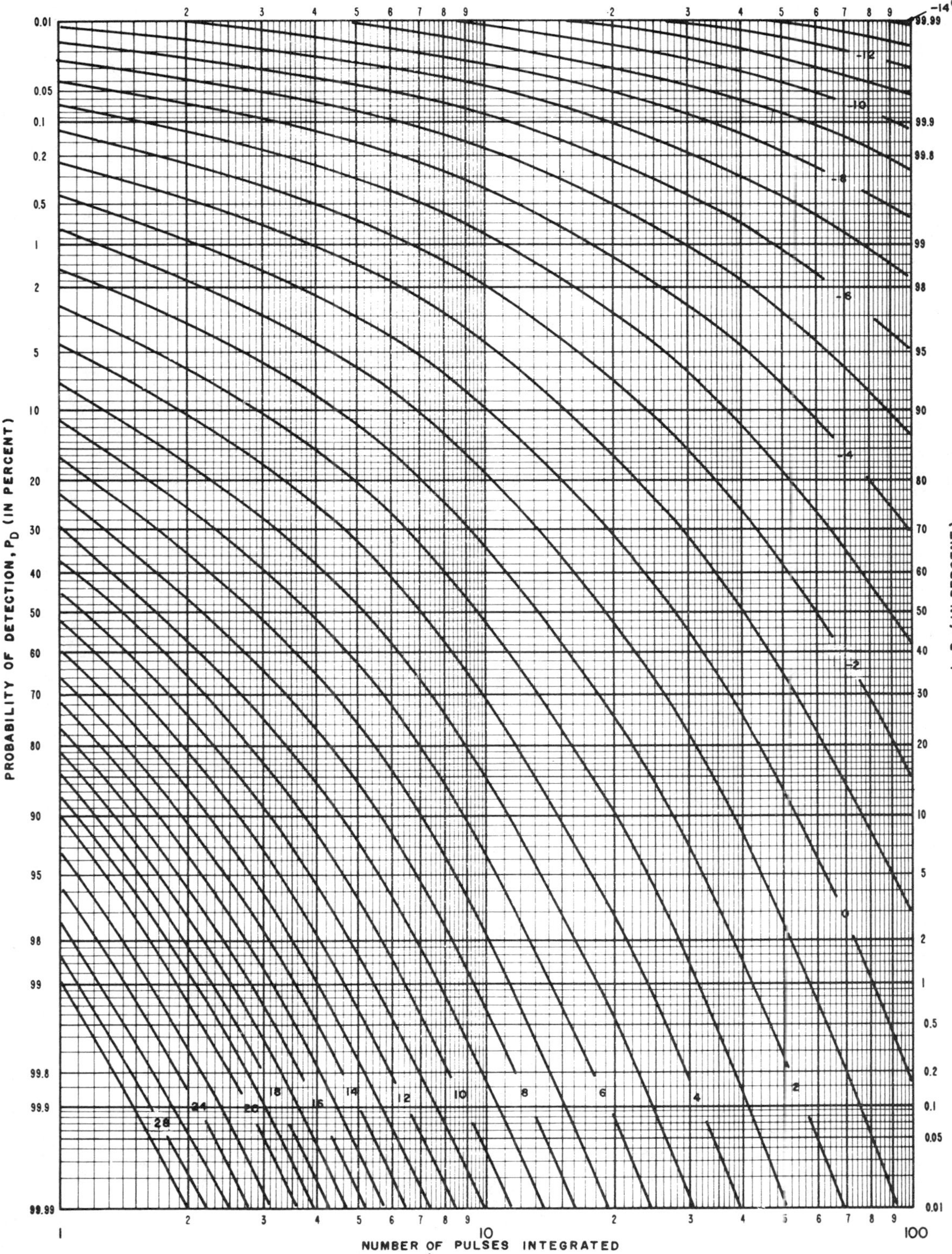

● SIGNAL TO NOISE RATIO

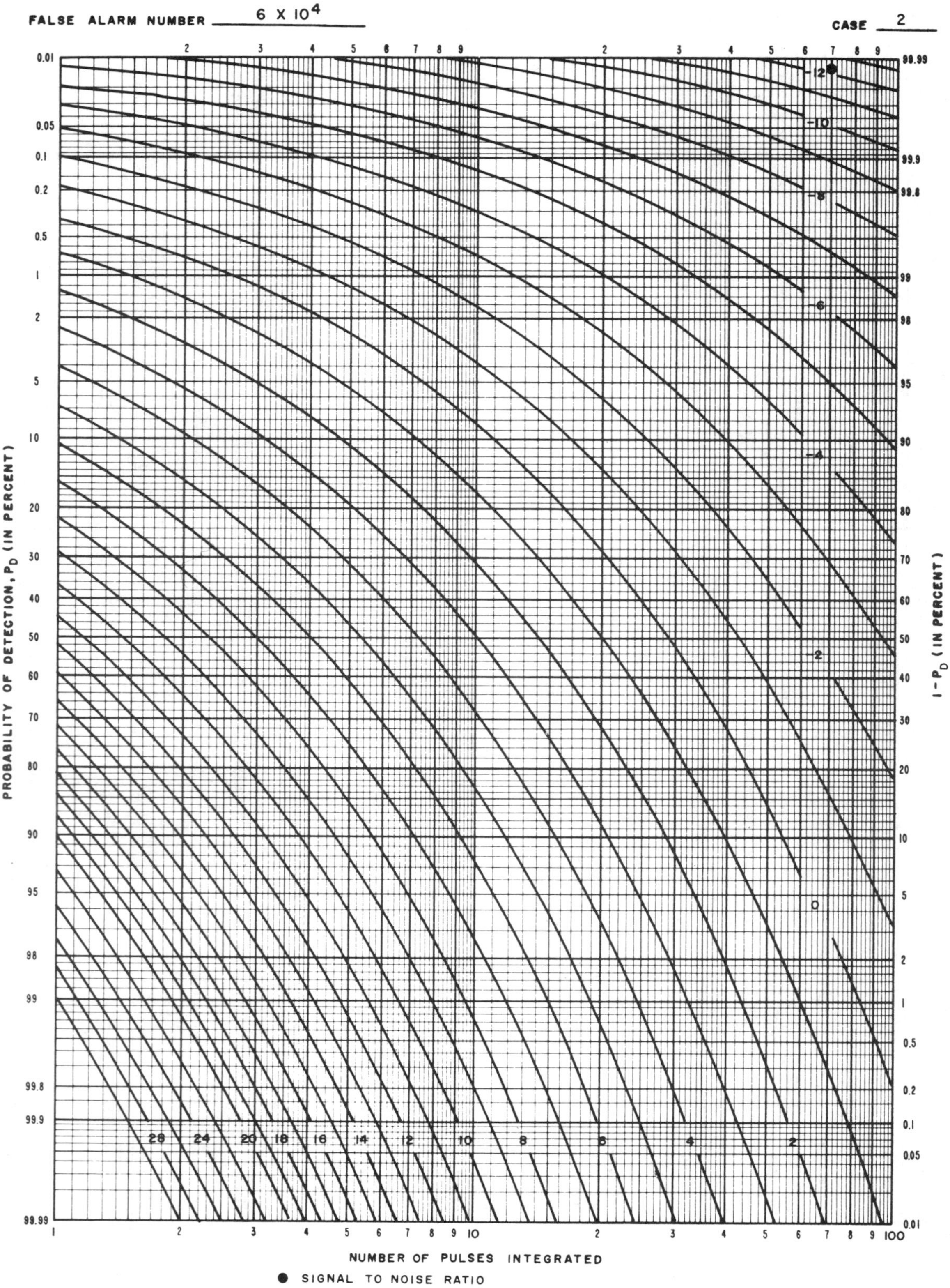
FALSE ALARM NUMBER 6 X 10^4
CASE 2
PROBABILITY OF DETECTION, P_D (IN PERCENT)
1 - P_D (IN PERCENT)
-12
-10
-8
-6
-4
-2
0
28
24
20
18
16
14
12
10
8
6
4
2
NUMBER OF PULSES INTEGRATED
● SIGNAL TO NOISE RATIO

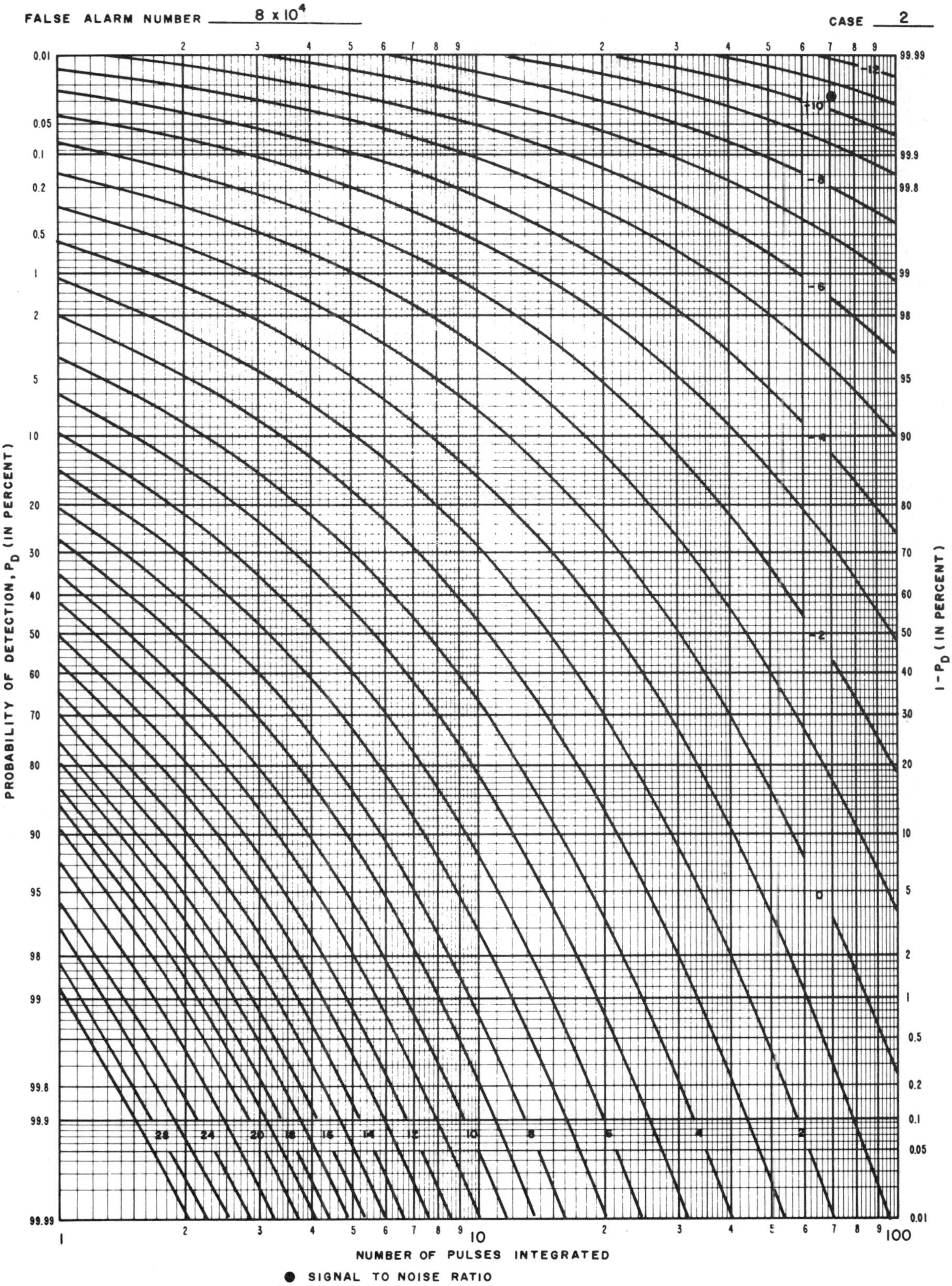
FALSE ALARM NUMBER 8 x 10^4
CASE 2
PROBABILITY OF DETECTION, P_D (IN PERCENT)
1 - P_D (IN PERCENT)
NUMBER OF PULSES INTEGRATED
● SIGNAL TO NOISE RATIO
28 24 20 18 16 14 12 10 8 6 4 2 0 -2 -4 -6 -8 -10 -12

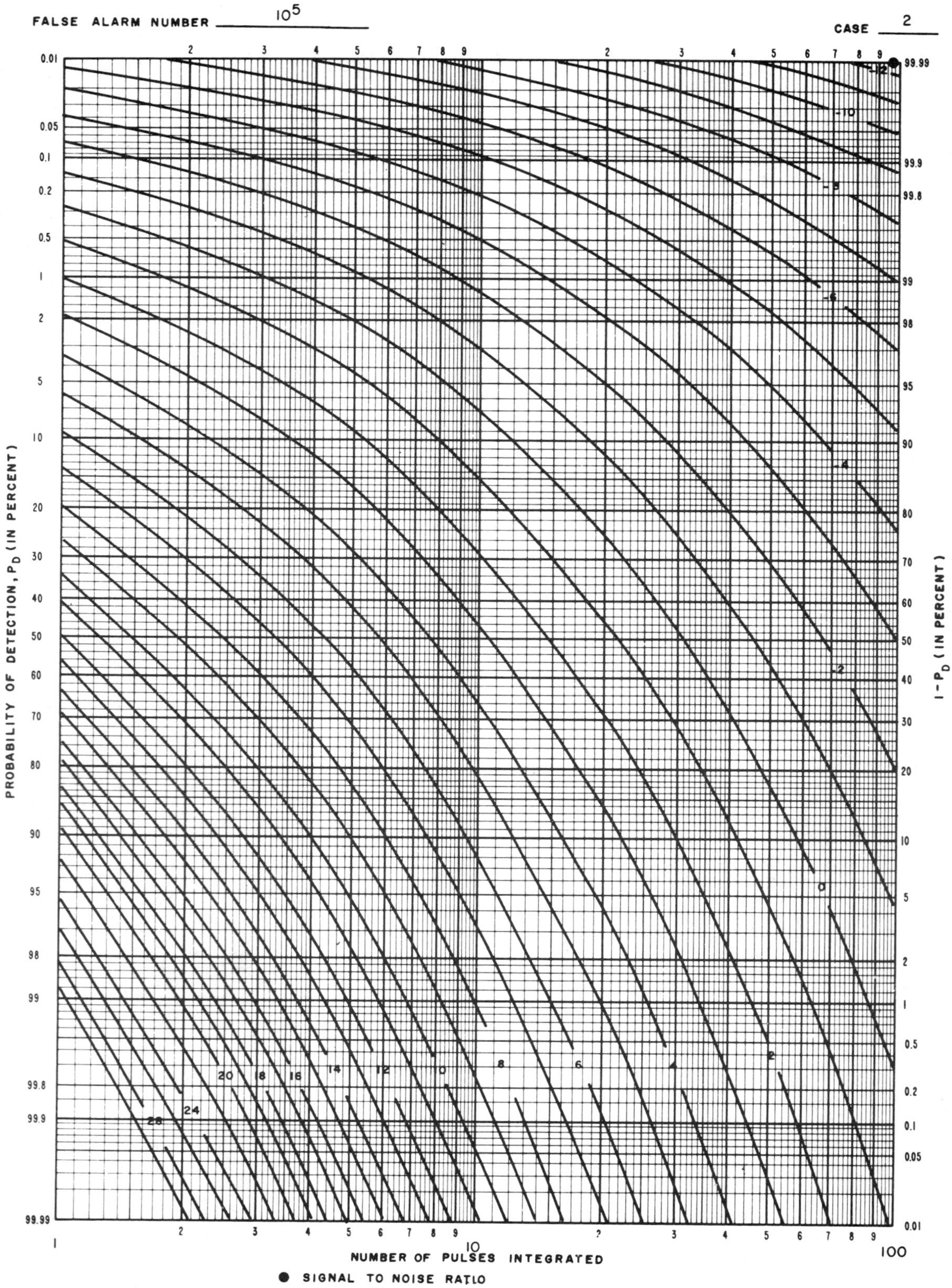
FALSE ALARM NUMBER 10^5
CASE 2
PROBABILITY OF DETECTION, P_D (IN PERCENT)
1 - P_D (IN PERCENT)
NUMBER OF PULSES INTEGRATED
● SIGNAL TO NOISE RATIO
-12
-10
-8
-6
-4
-2
0
2
4
6
8
10
12
14
16
18
20
24
28

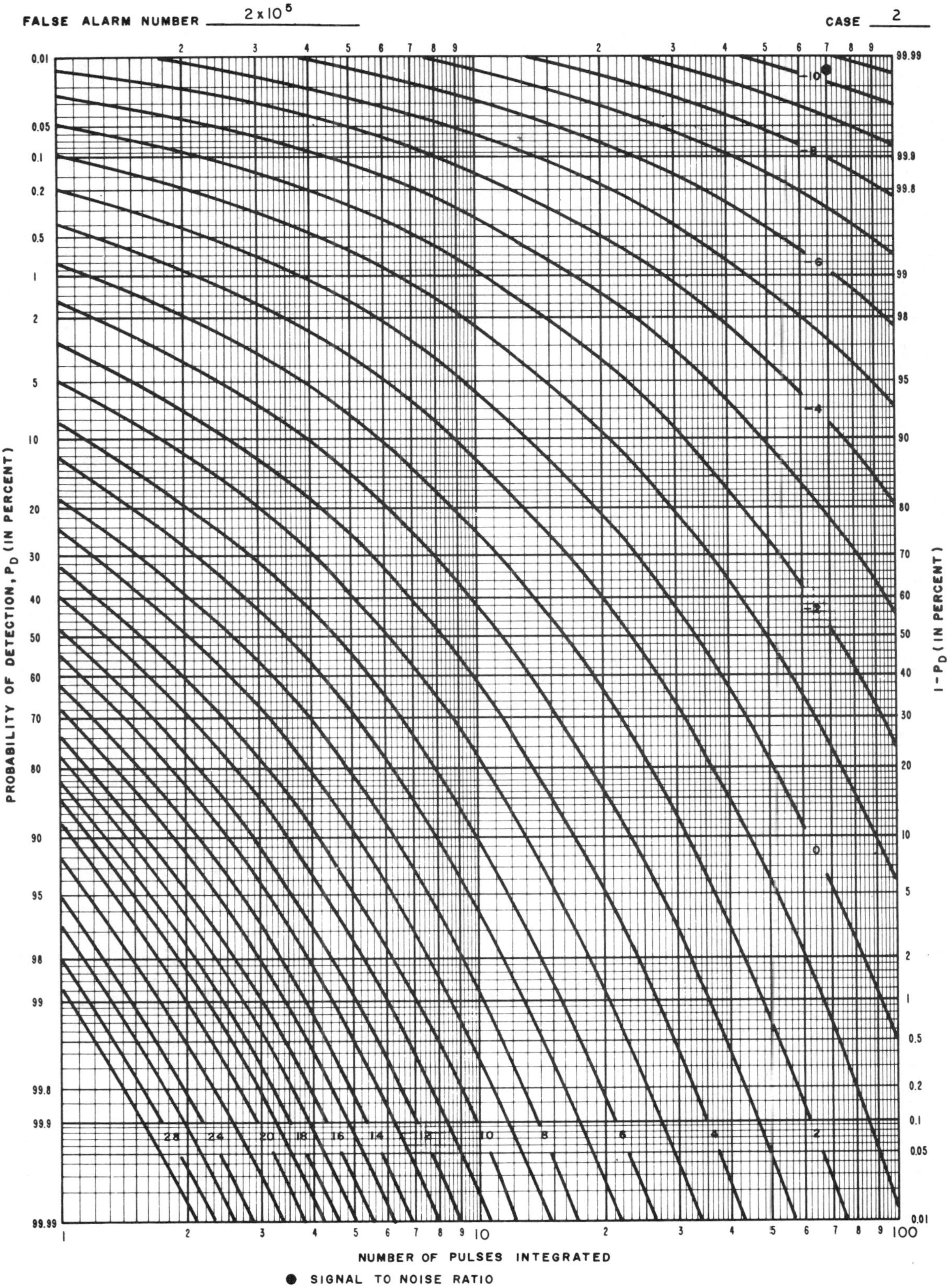
FALSE ALARM NUMBER 2 x 10^5
CASE 2
PROBABILITY OF DETECTION, P_D (IN PERCENT)
1 - P_D (IN PERCENT)
NUMBER OF PULSES INTEGRATED
SIGNAL TO NOISE RATIO

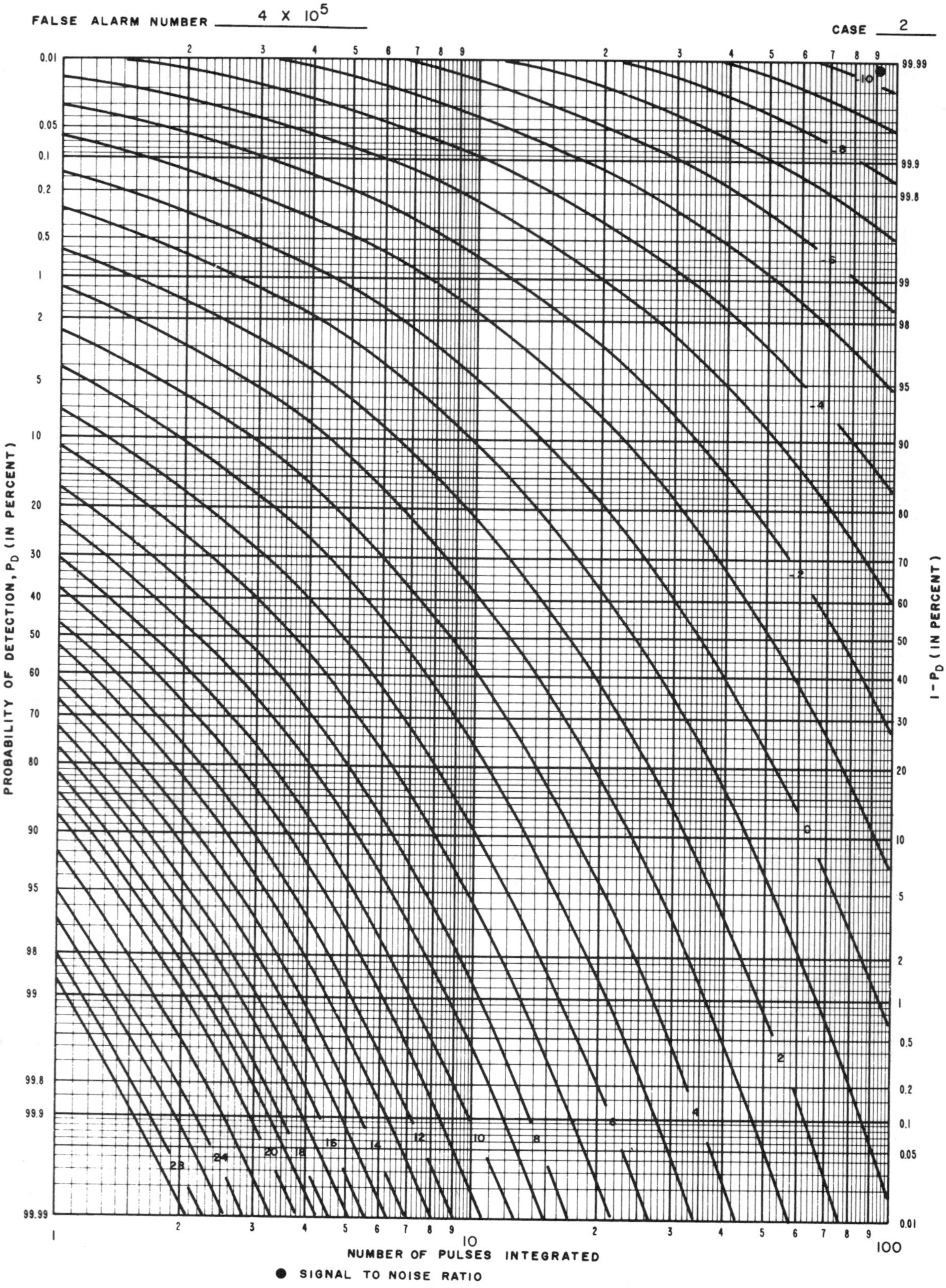
FALSE ALARM NUMBER 4 X 10^5
CASE 2
PROBABILITY OF DETECTION, P_D (IN PERCENT)
1 - P_D (IN PERCENT)
NUMBER OF PULSES INTEGRATED
● SIGNAL TO NOISE RATIO
0.01
0.05
0.1
0.2
0.5
1
2
5
10
20
30
40
50
60
70
80
90
95
98
99
99.8
99.9
99.99
1
10
100
28
24
20
18
16
14
12
10
8
6
4
2
0
-2
-4
-6
-8
-10

FALSE ALARM NUMBER 6 X 10^5 CASE 2

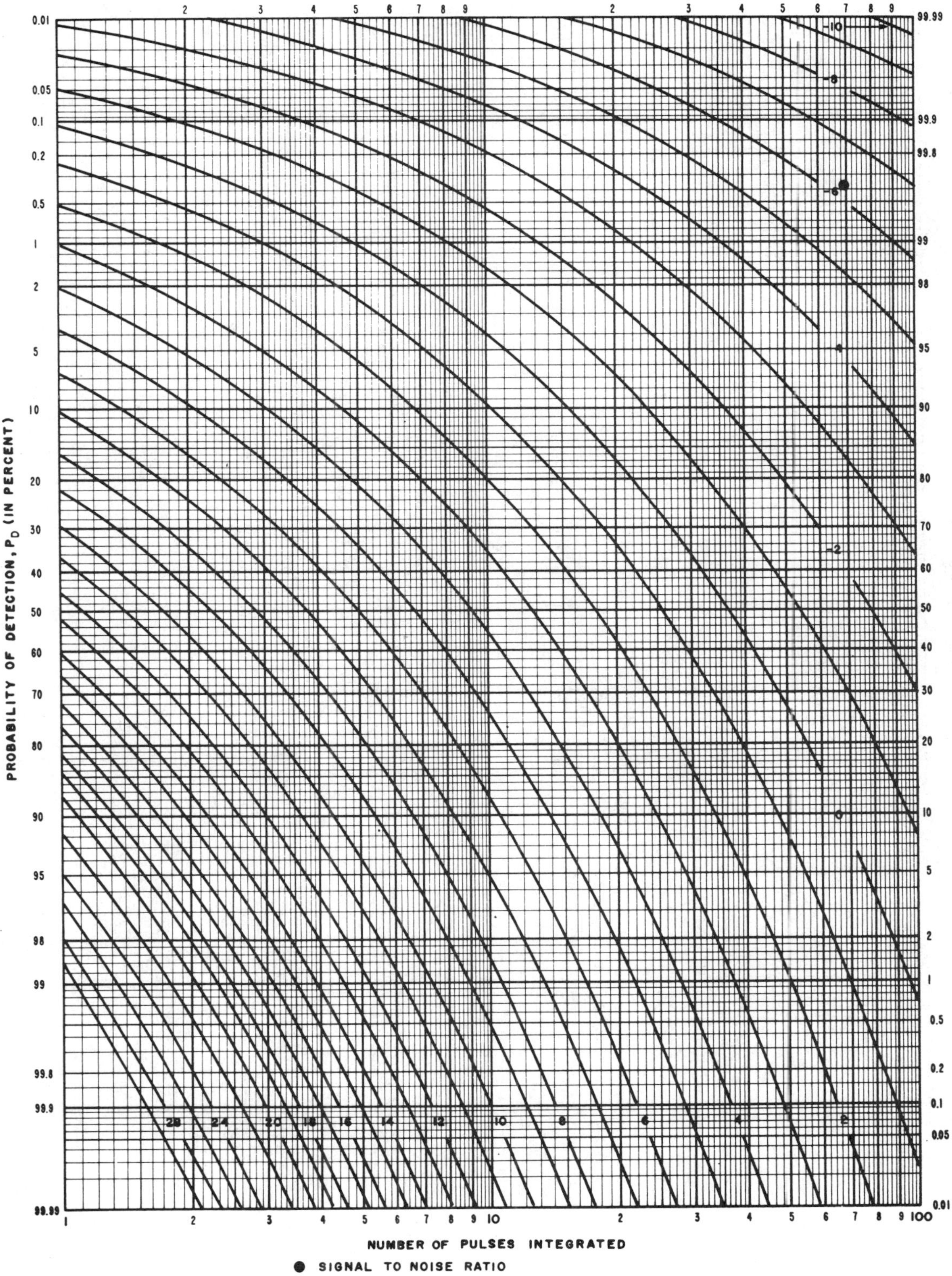

FALSE ALARM NUMBER 8×10^5

CASE 2

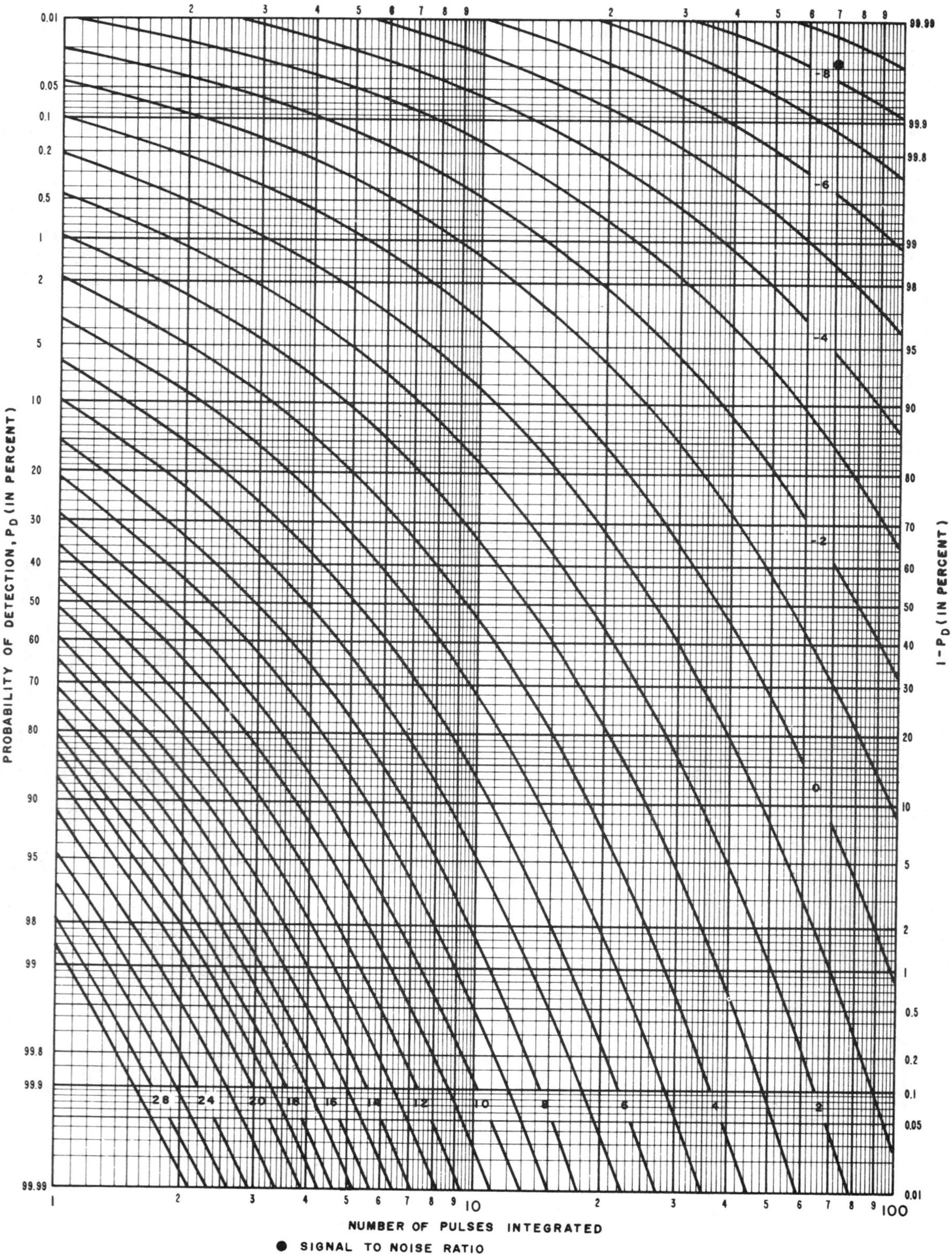

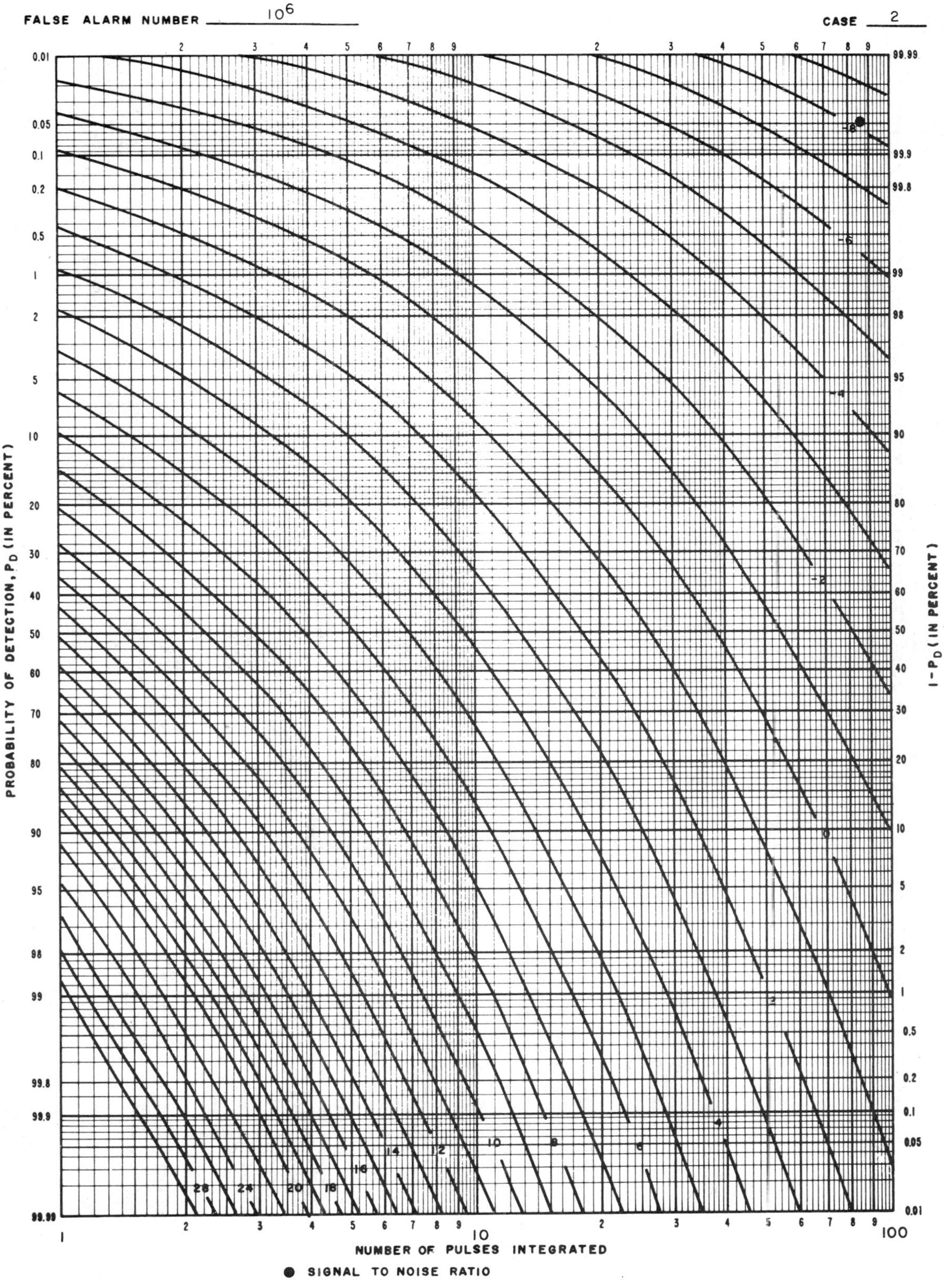
FALSE ALARM NUMBER 10^6
CASE 2
PROBABILITY OF DETECTION, P_D (IN PERCENT)
1-P_D (IN PERCENT)
NUMBER OF PULSES INTEGRATED
● SIGNAL TO NOISE RATIO
-8
-6
-4
-2
0
2
4
6
8
10
12
14
16
18
20
24
28

FALSE ALARM NUMBER 2×10^6 CASE 2

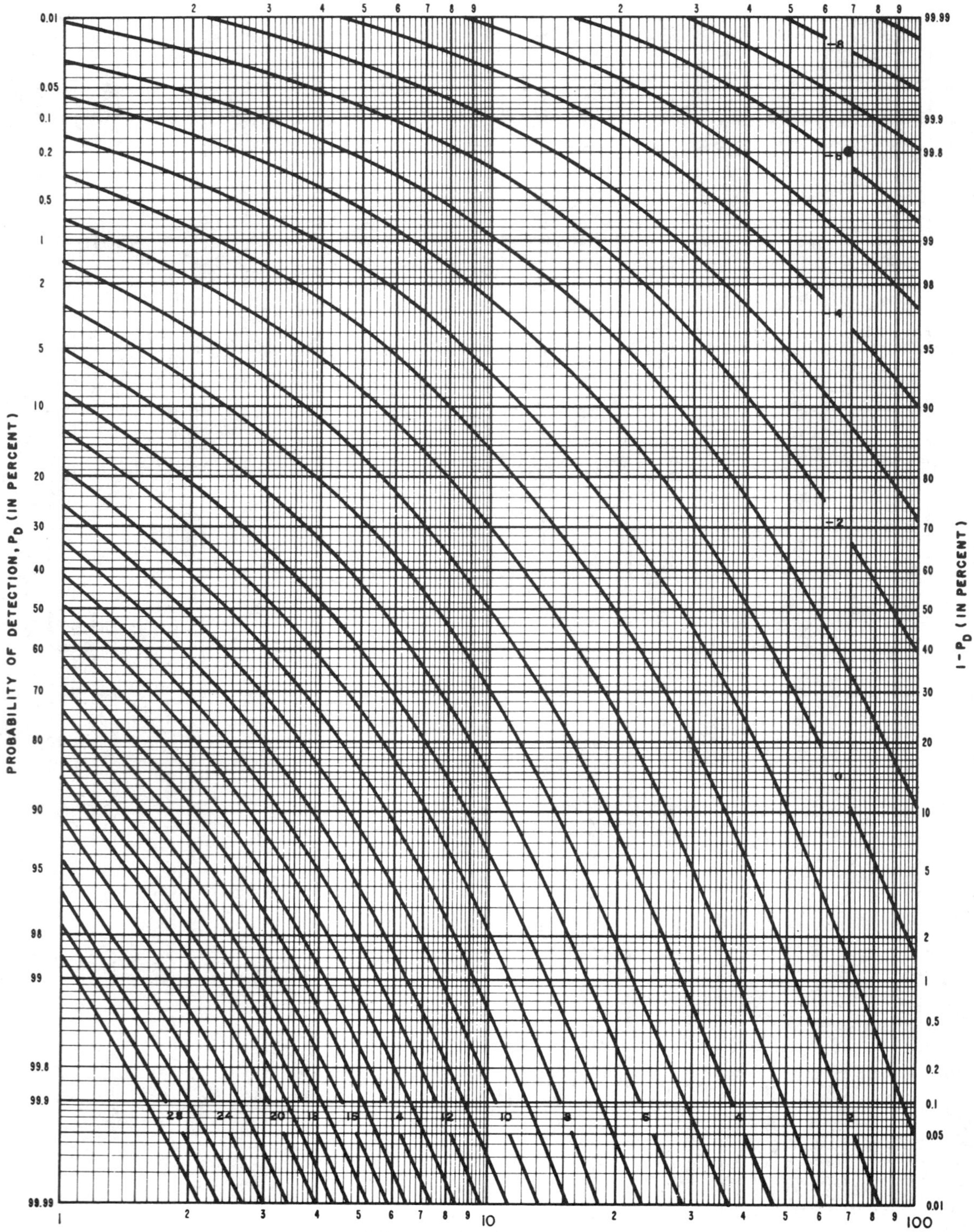

NUMBER OF PULSES INTEGRATED

● SIGNAL TO NOISE RATIO

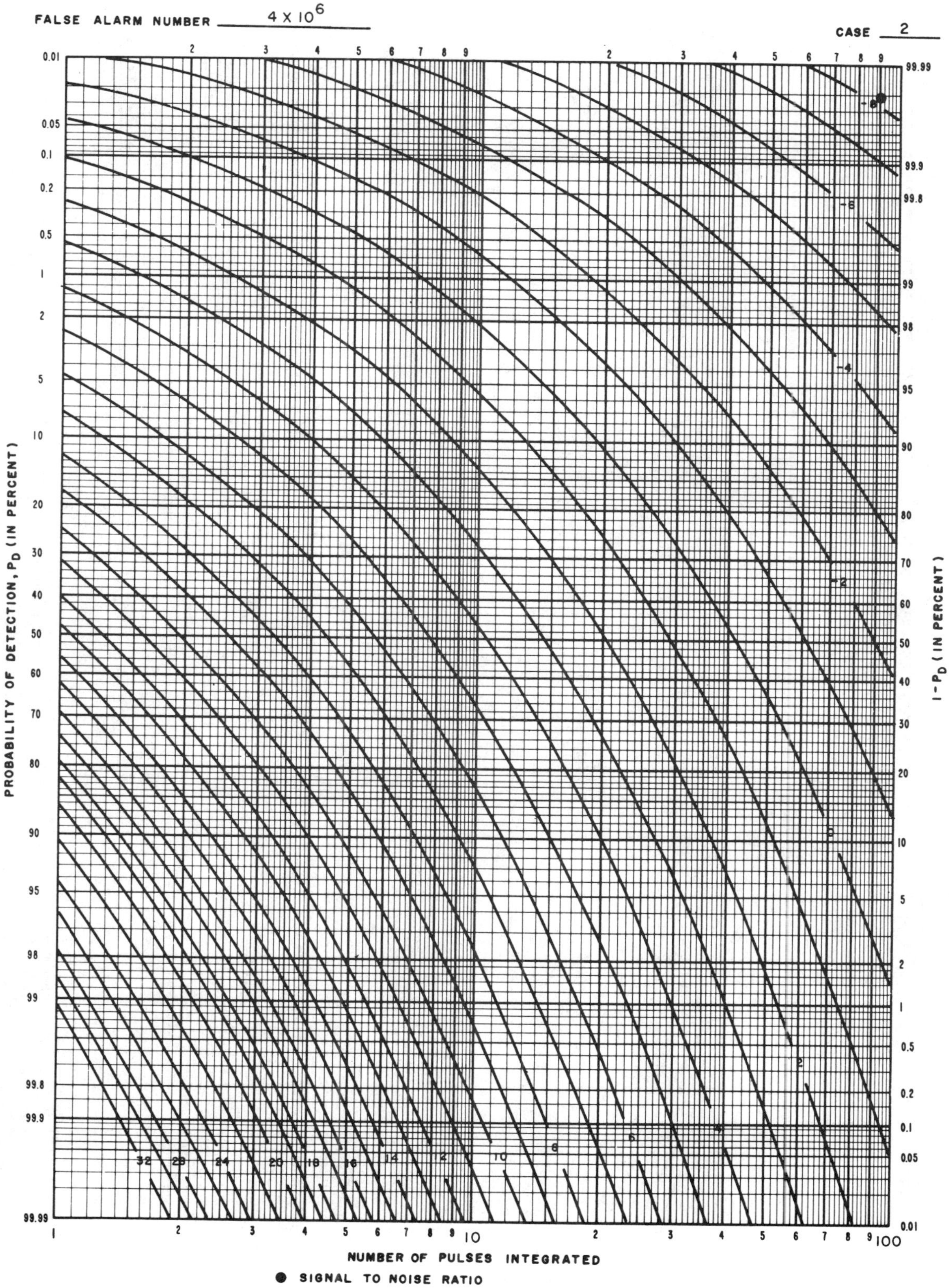
FALSE ALARM NUMBER 4 X 10^6
CASE 2
PROBABILITY OF DETECTION, P_D (IN PERCENT)
1 - P_D (IN PERCENT)
NUMBER OF PULSES INTEGRATED
● SIGNAL TO NOISE RATIO

FALSE ALARM NUMBER 6 X 10^6

CASE 2

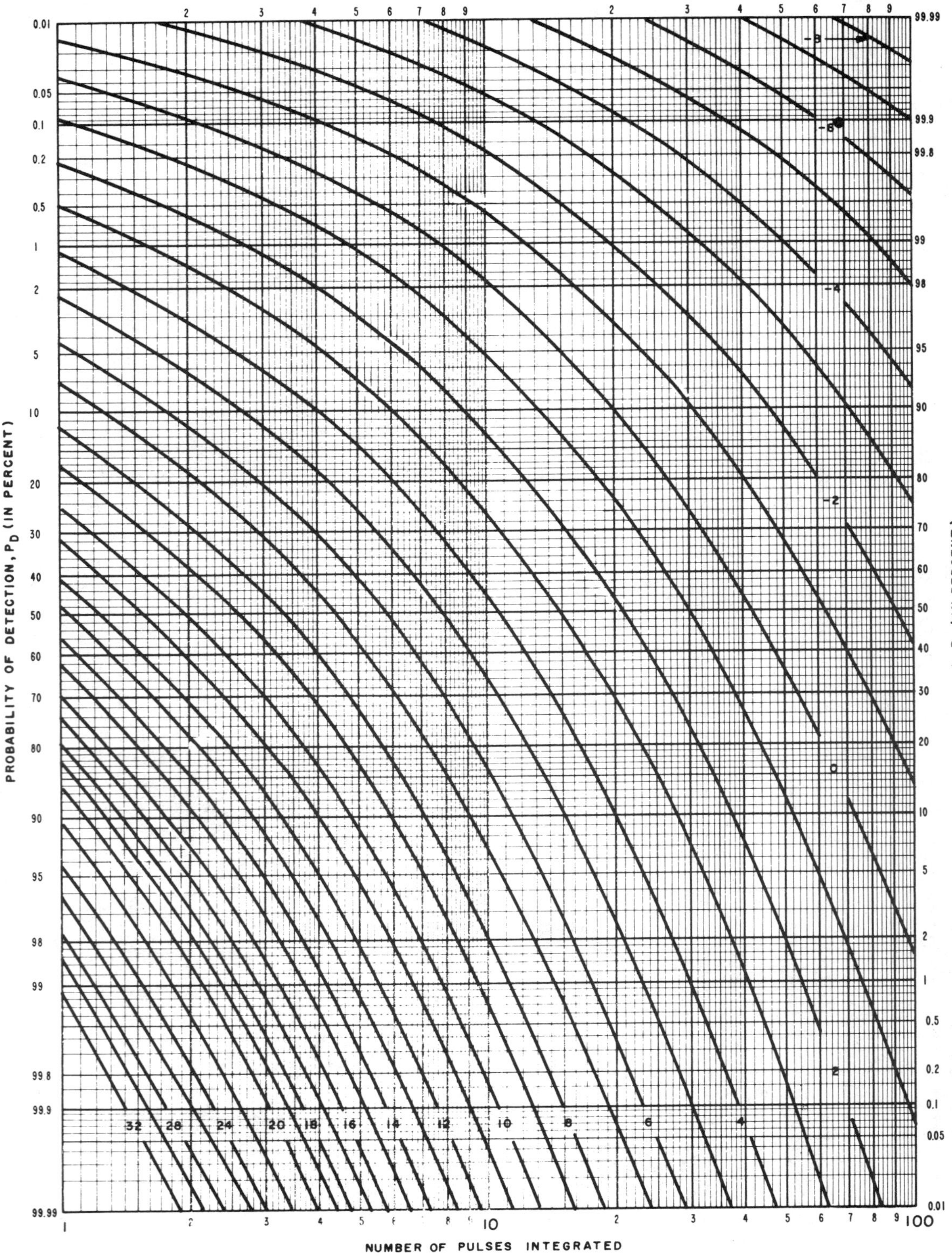

● SIGNAL TO NOISE RATIO

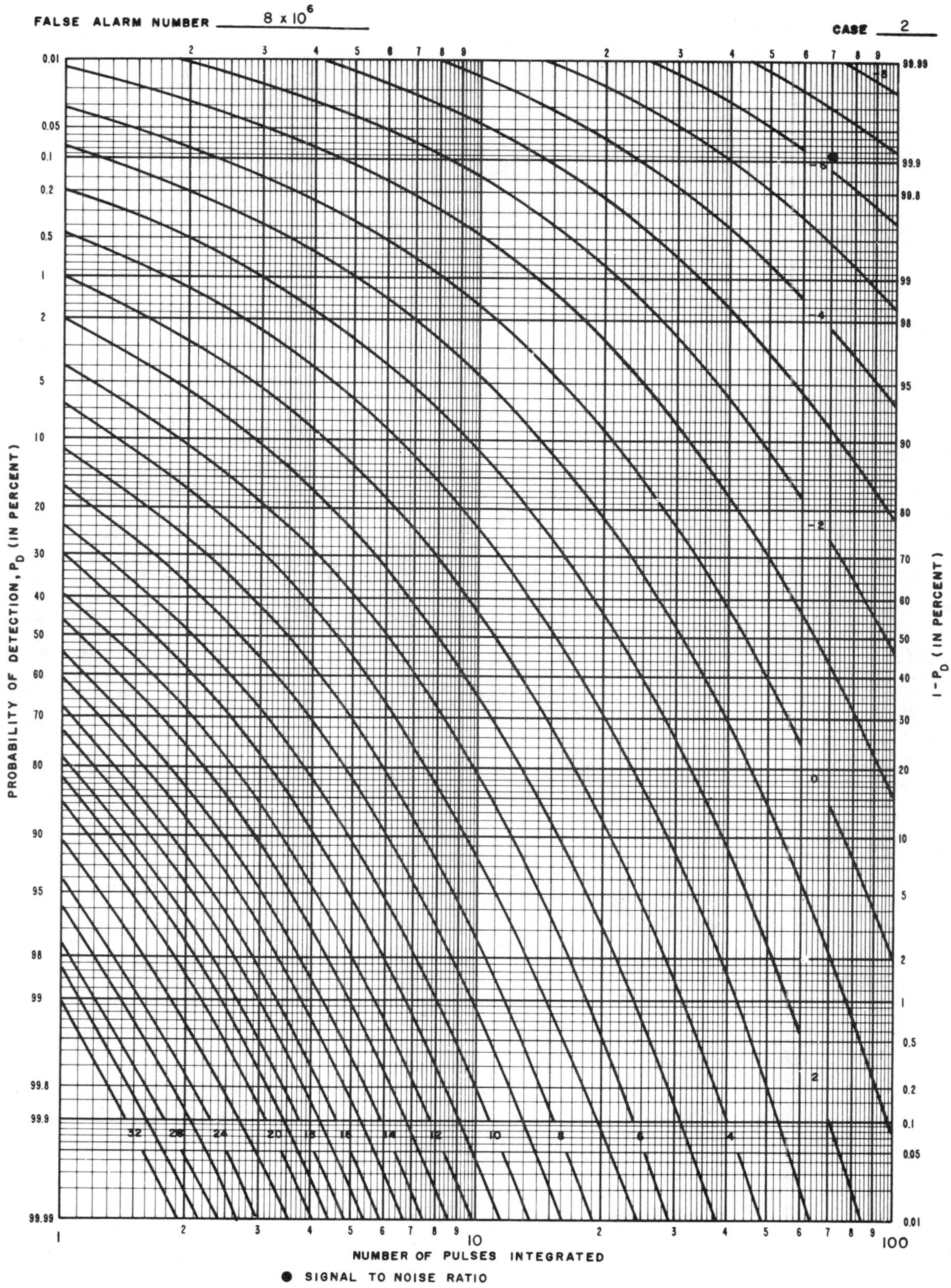
FALSE ALARM NUMBER 8 x 10^6
CASE 2
PROBABILITY OF DETECTION, P_D (IN PERCENT)
1 - P_D (IN PERCENT)
NUMBER OF PULSES INTEGRATED
● SIGNAL TO NOISE RATIO
32
28
24
20
18
16
14
12
10
8
6
4
2
0
-2
-4
-6
-8

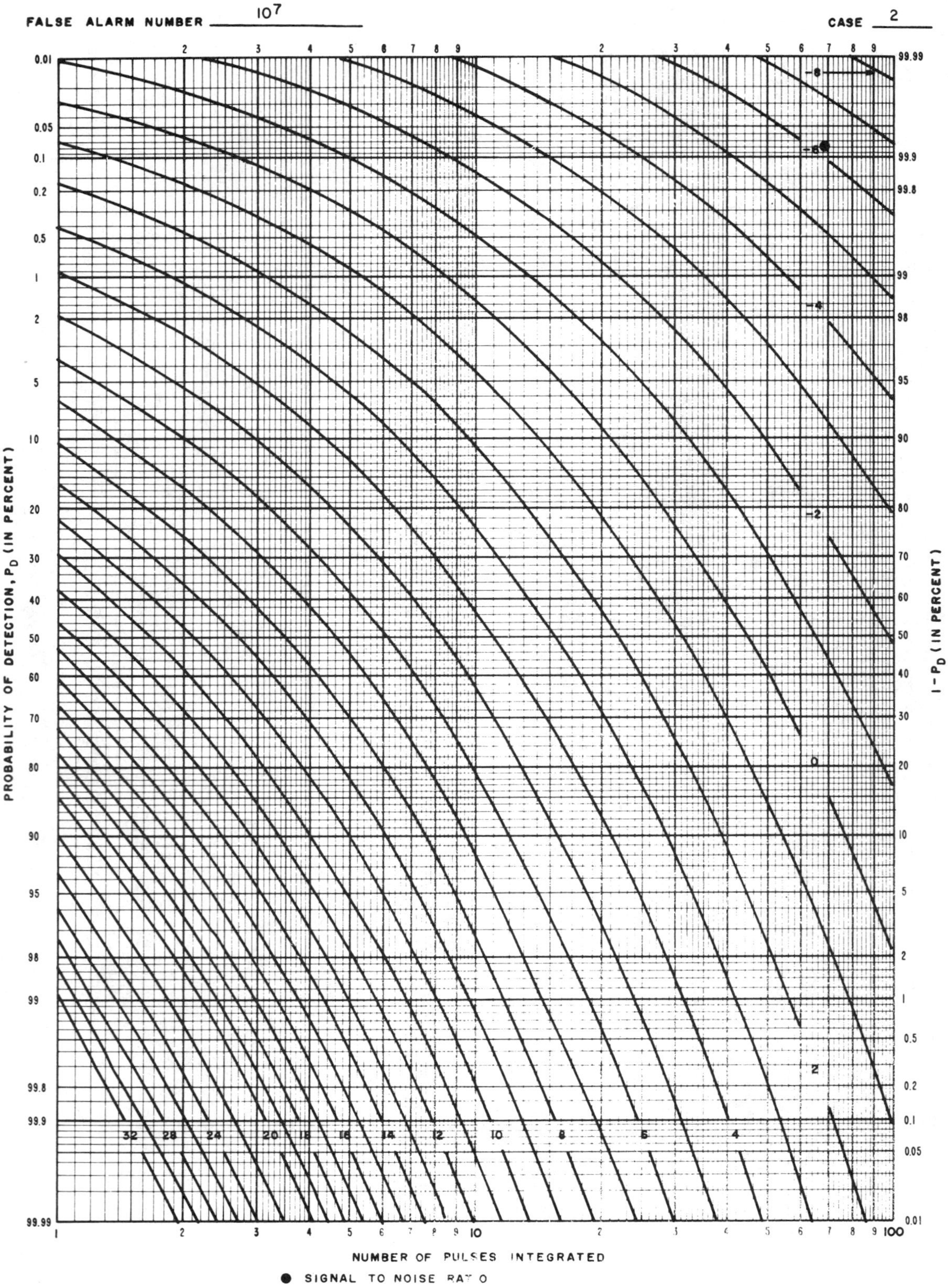
FALSE ALARM NUMBER 10^7
CASE 2
PROBABILITY OF DETECTION, P_D (IN PERCENT)
1 - P_D (IN PERCENT)
NUMBER OF PULSES INTEGRATED
● SIGNAL TO NOISE RATIO
-6
-4
-2
0
2
32
28
24
20
18
16
14
12
10
8
6
4

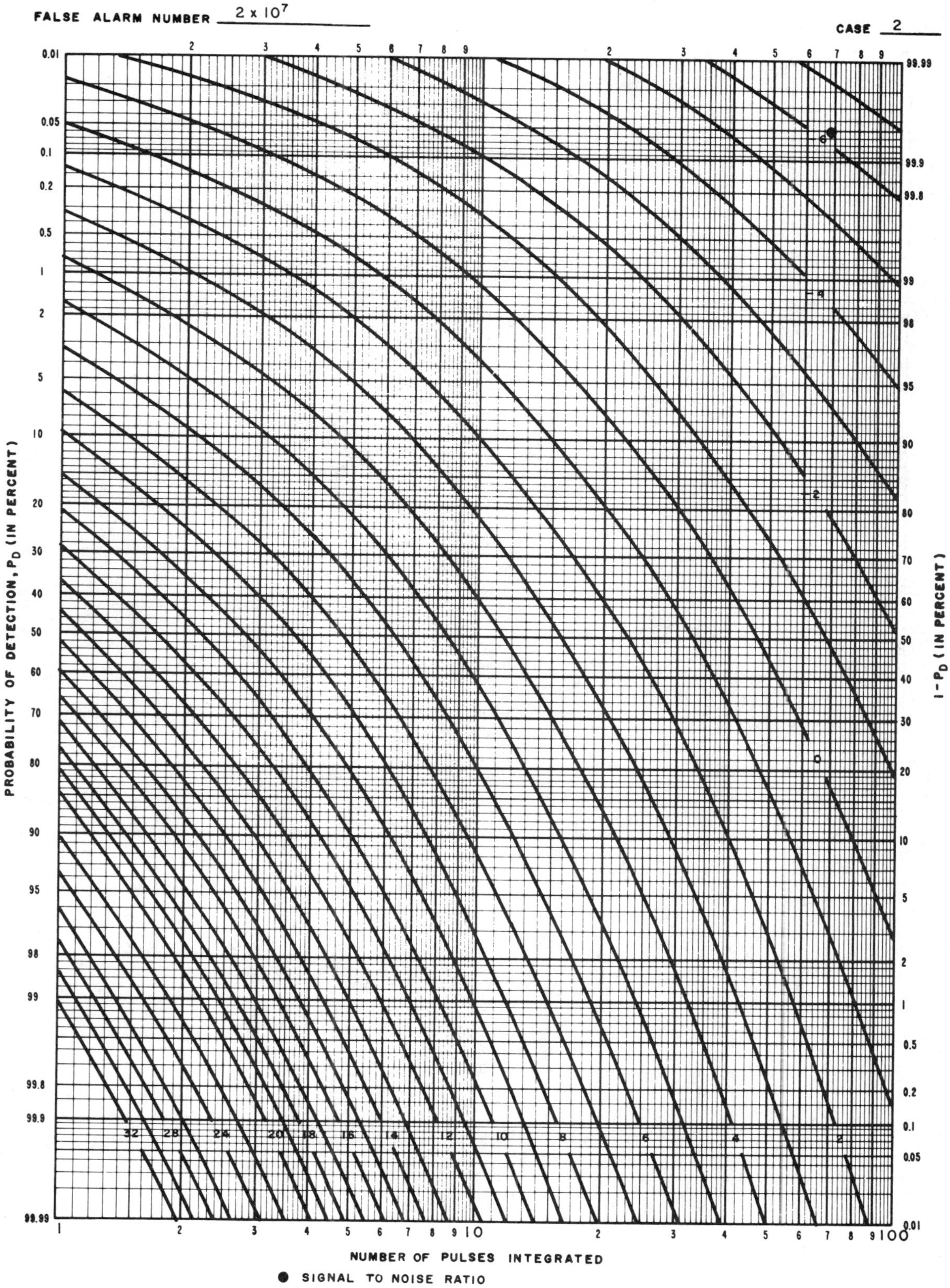
FALSE ALARM NUMBER 2 x 10^7
CASE 2
PROBABILITY OF DETECTION, P_D (IN PERCENT)
1-P_D (IN PERCENT)
NUMBER OF PULSES INTEGRATED
32
28
24
20
18
16
14
12
10
8
6
4
2
0
-2
-4
-6
● SIGNAL TO NOISE RATIO

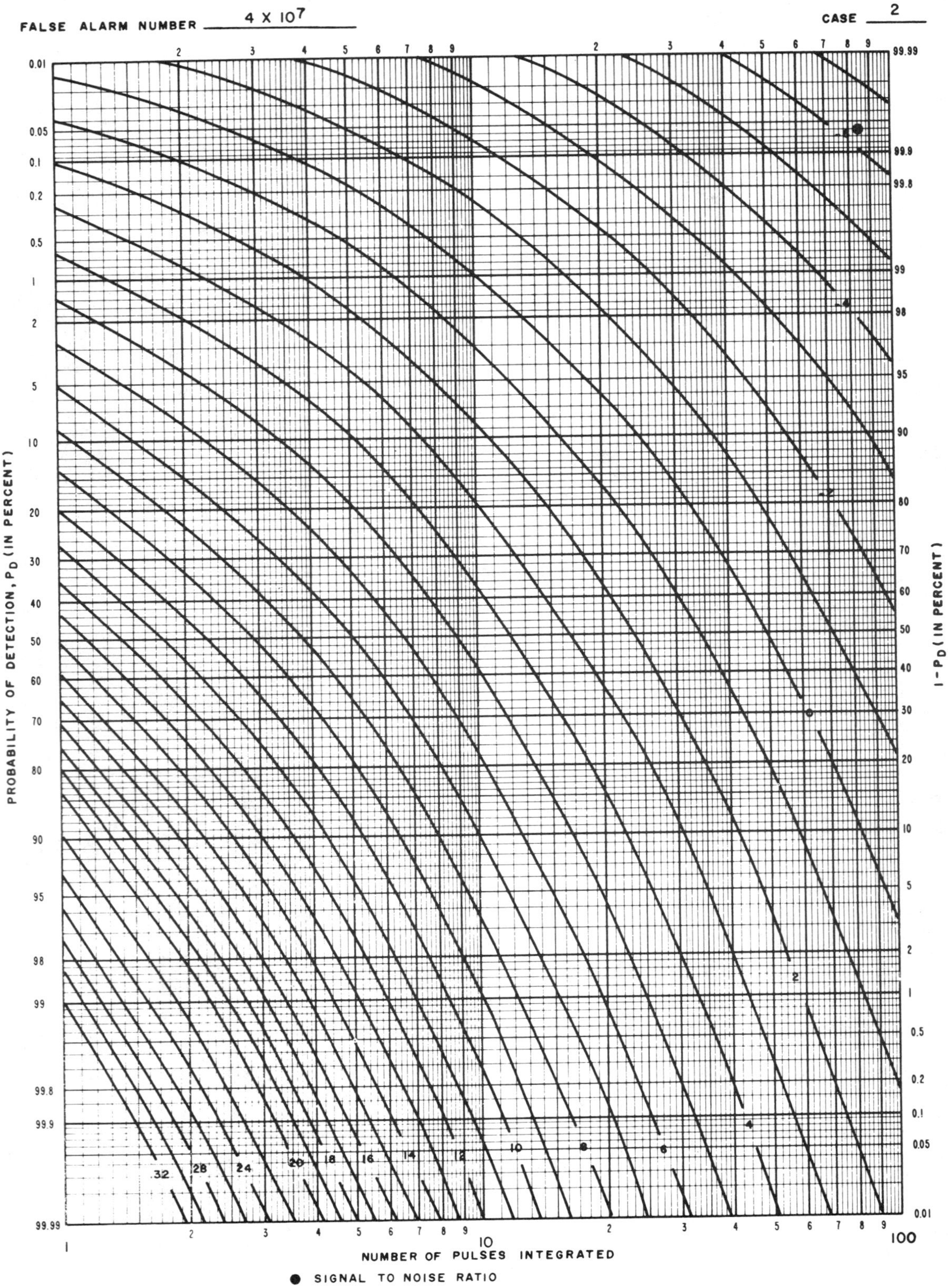

FALSE ALARM NUMBER 4 X 10^7
CASE 2
PROBABILITY OF DETECTION, P_D (IN PERCENT)
1 - P_D (IN PERCENT)
NUMBER OF PULSES INTEGRATED
● SIGNAL TO NOISE RATIO
32
28
24
20
18
16
14
12
10
8
6
4
2
0

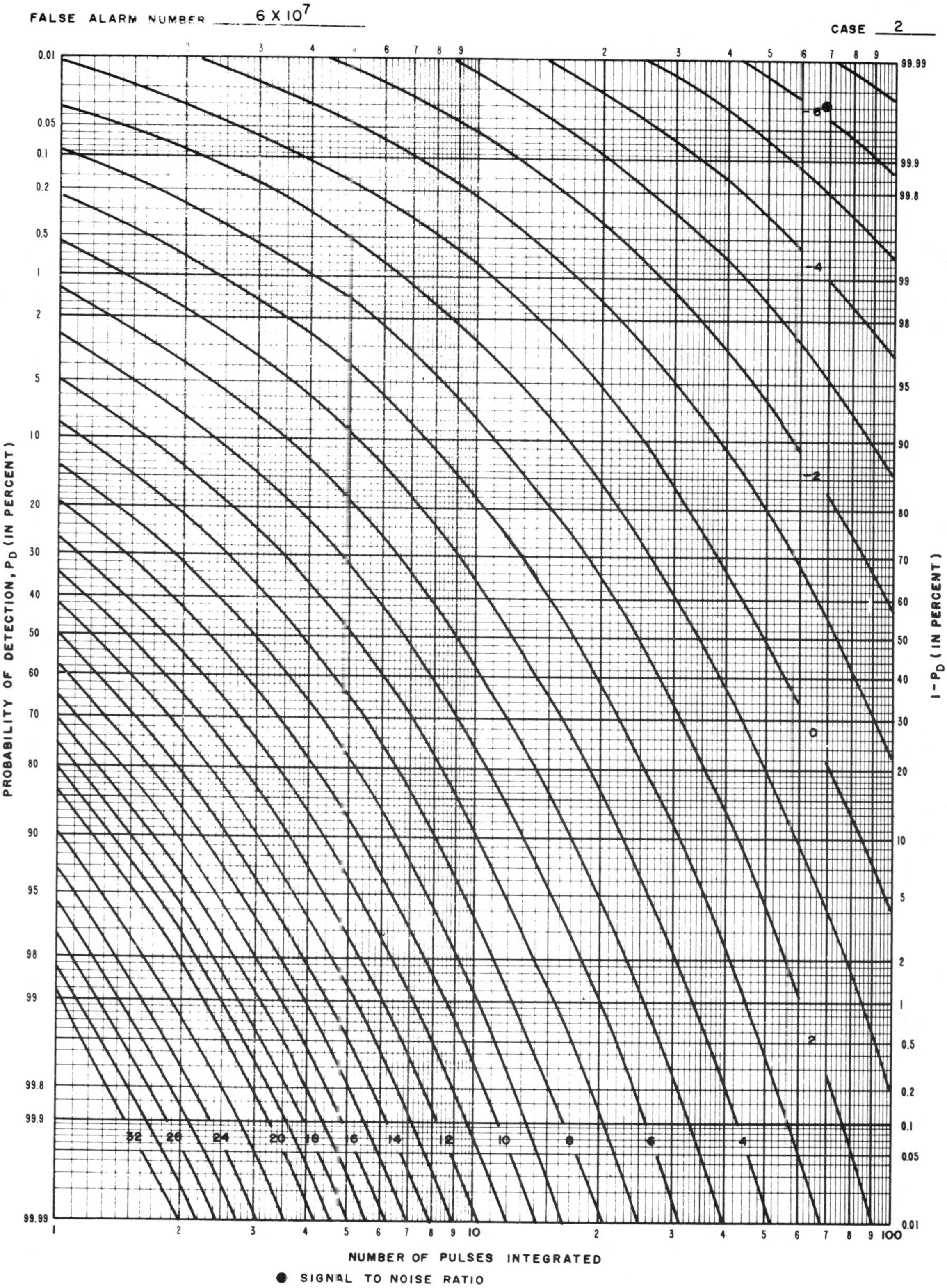
FALSE ALARM NUMBER 6 X 10^7
CASE 2
PROBABILITY OF DETECTION, P_D (IN PERCENT)
1 - P_D (IN PERCENT)
NUMBER OF PULSES INTEGRATED
● SIGNAL TO NOISE RATIO
32
28
24
20
18
16
14
12
10
8
6
4
2
0
-2
-4
-6

FALSE ALARM NUMBER 8×10^7 CASE 2

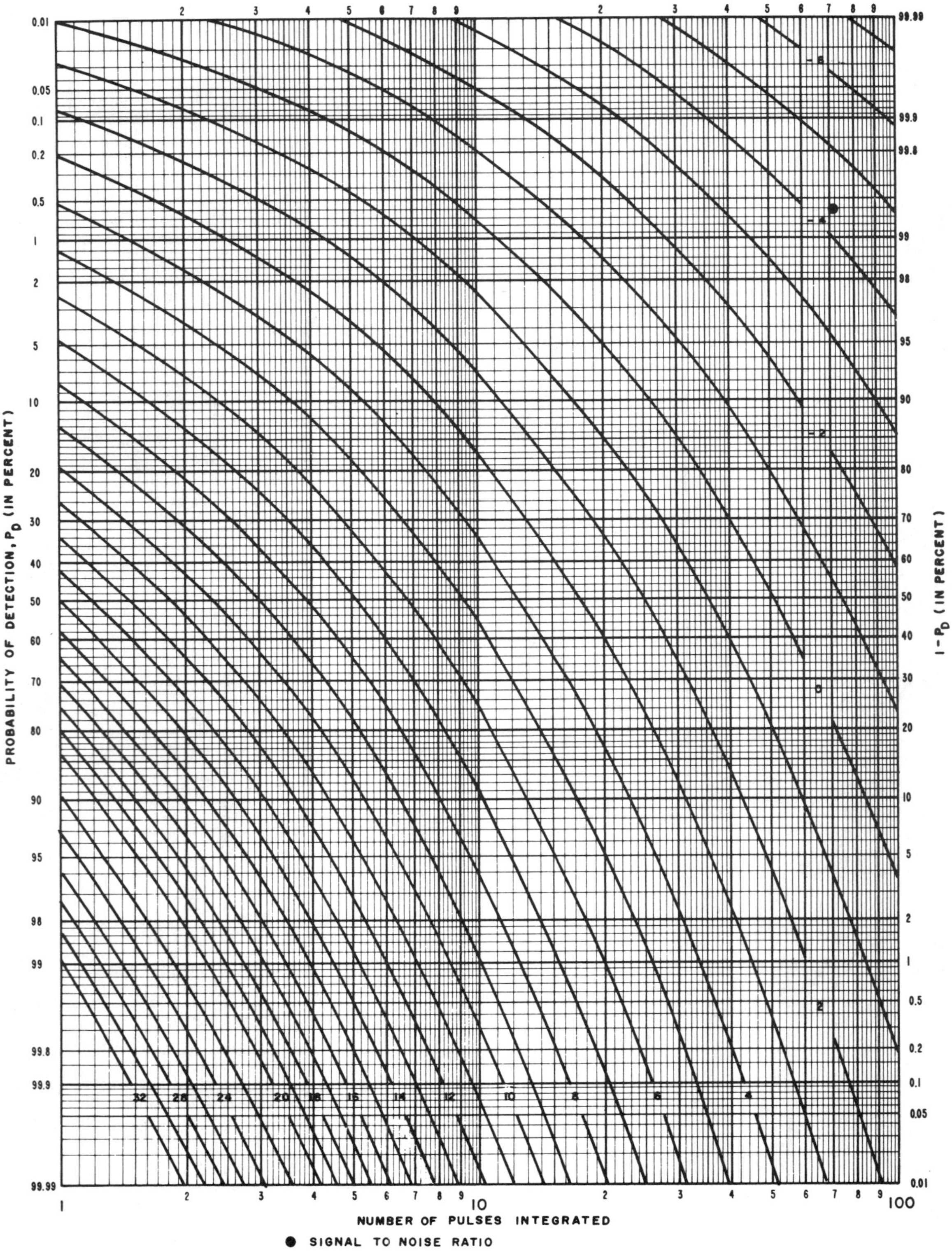

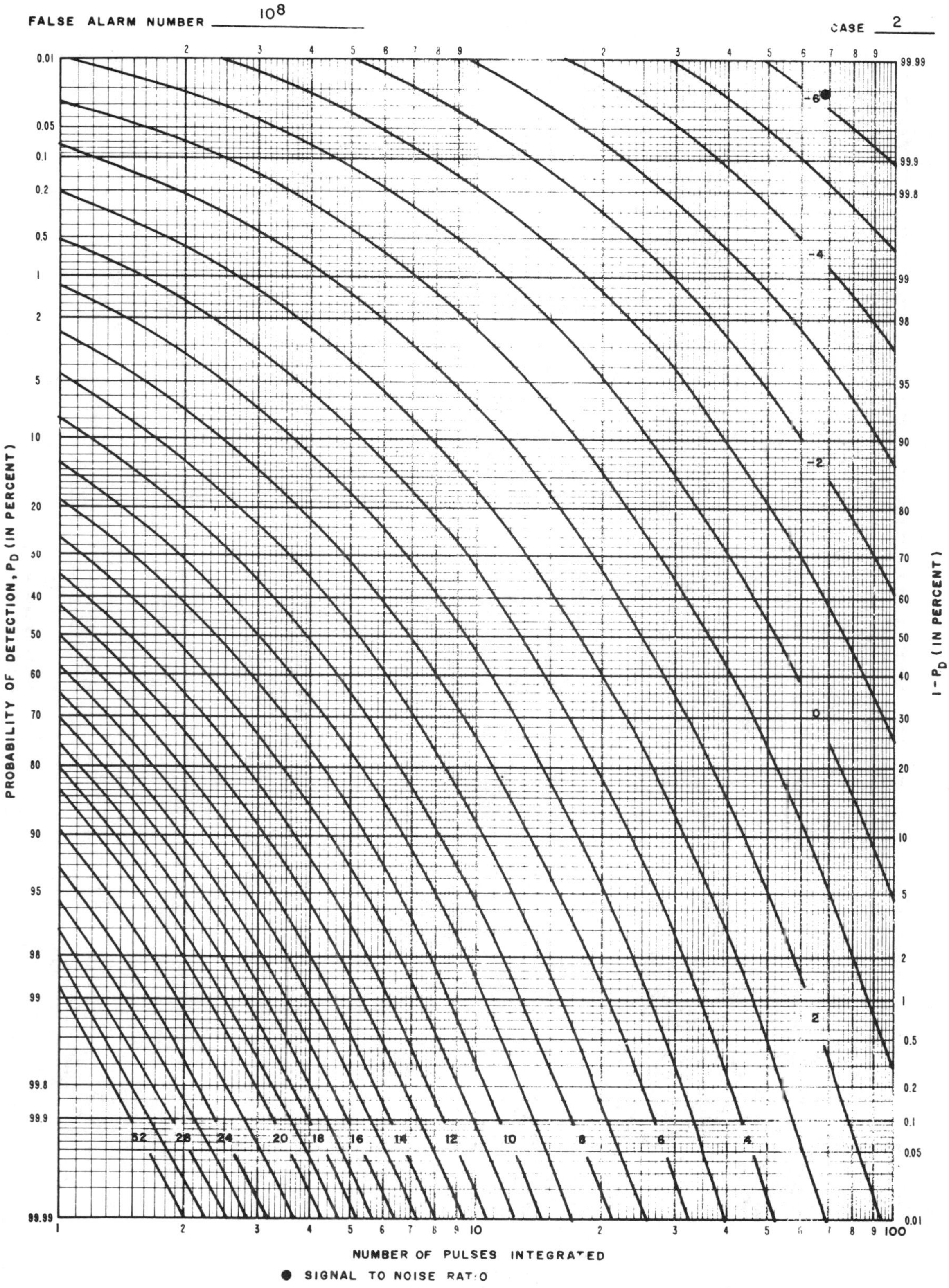
FALSE ALARM NUMBER 10^8
CASE 2
PROBABILITY OF DETECTION, P_D (IN PERCENT)
1 - P_D (IN PERCENT)
NUMBER OF PULSES INTEGRATED
● SIGNAL TO NOISE RATIO
32 28 24 20 18 16 14 12 10 8 6 4 2 0 -2 -4 -6

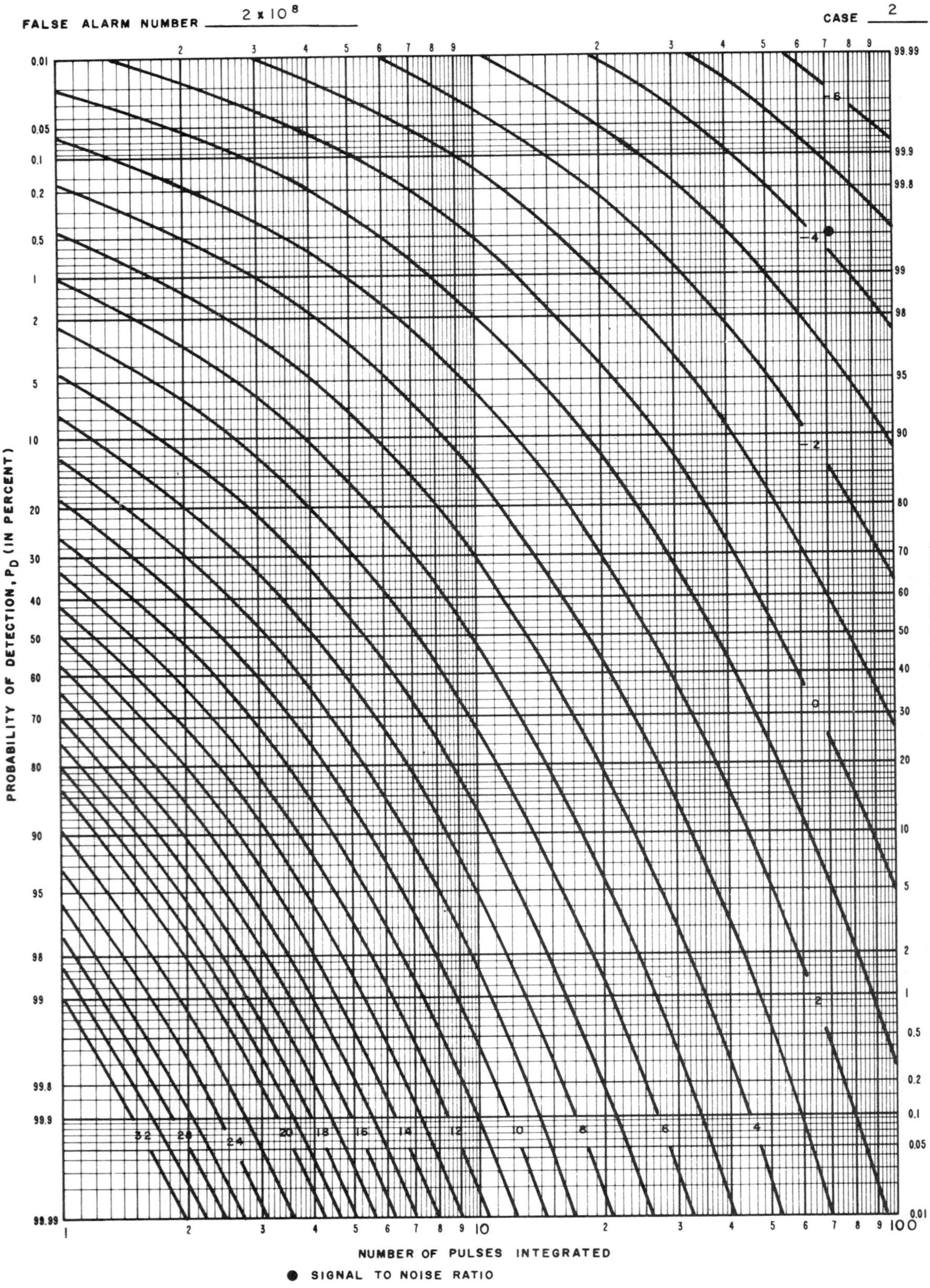
FALSE ALARM NUMBER 2 x 10^8
CASE 2
PROBABILITY OF DETECTION, P_D (IN PERCENT)
1 - P_D (IN PERCENT)
-6
-4
-2
0
2
32
28
24
20
18
16
14
12
10
8
6
4
NUMBER OF PULSES INTEGRATED
● SIGNAL TO NOISE RATIO

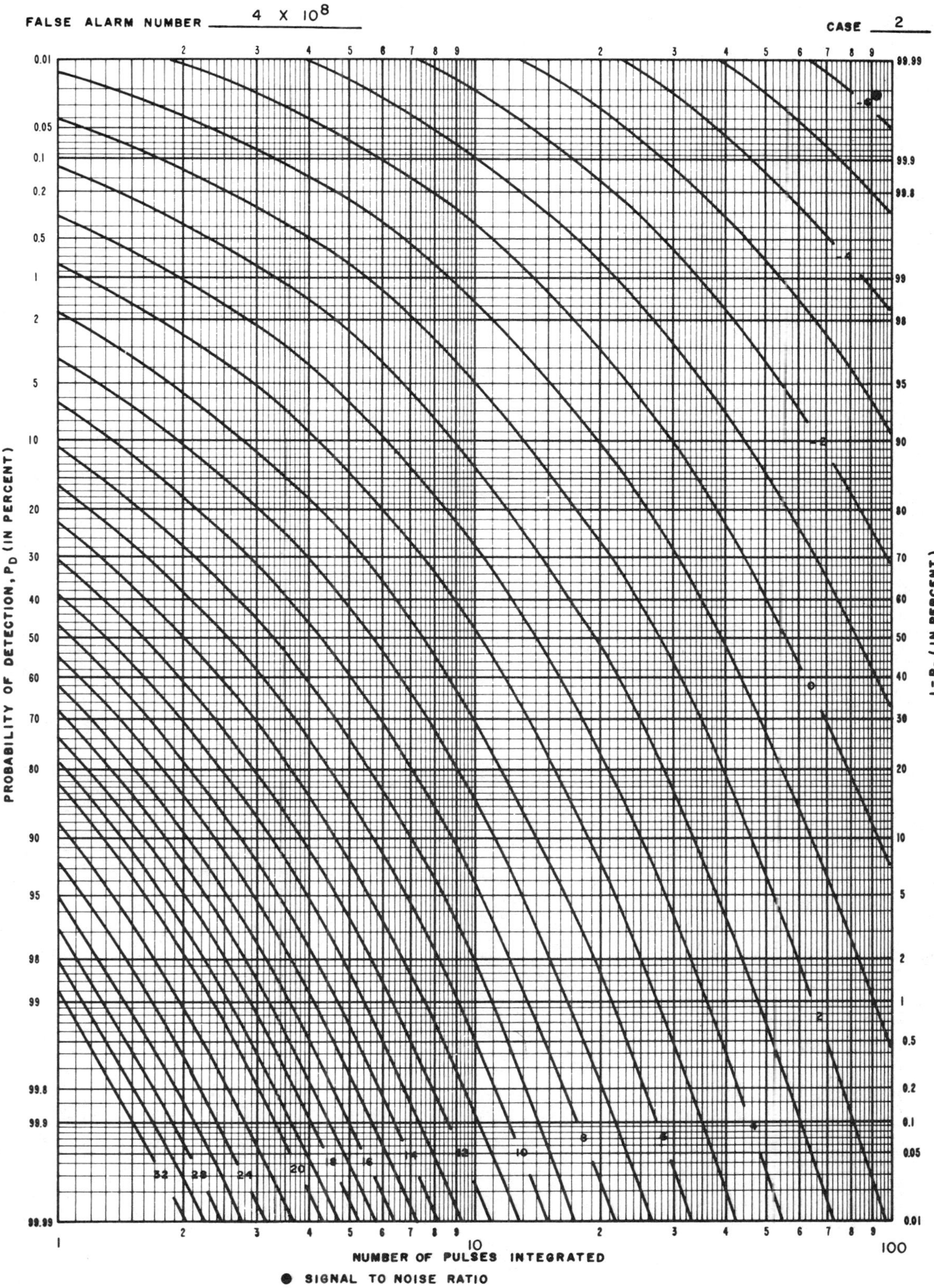
FALSE ALARM NUMBER 4 X 10^8
CASE 2
PROBABILITY OF DETECTION, P_D (IN PERCENT)
1 - P_D (IN PERCENT)
NUMBER OF PULSES INTEGRATED
● SIGNAL TO NOISE RATIO
1
10
100
32
28
24
20
18
16
14
12
10
8
6
4
2
0
-2
-4
-6

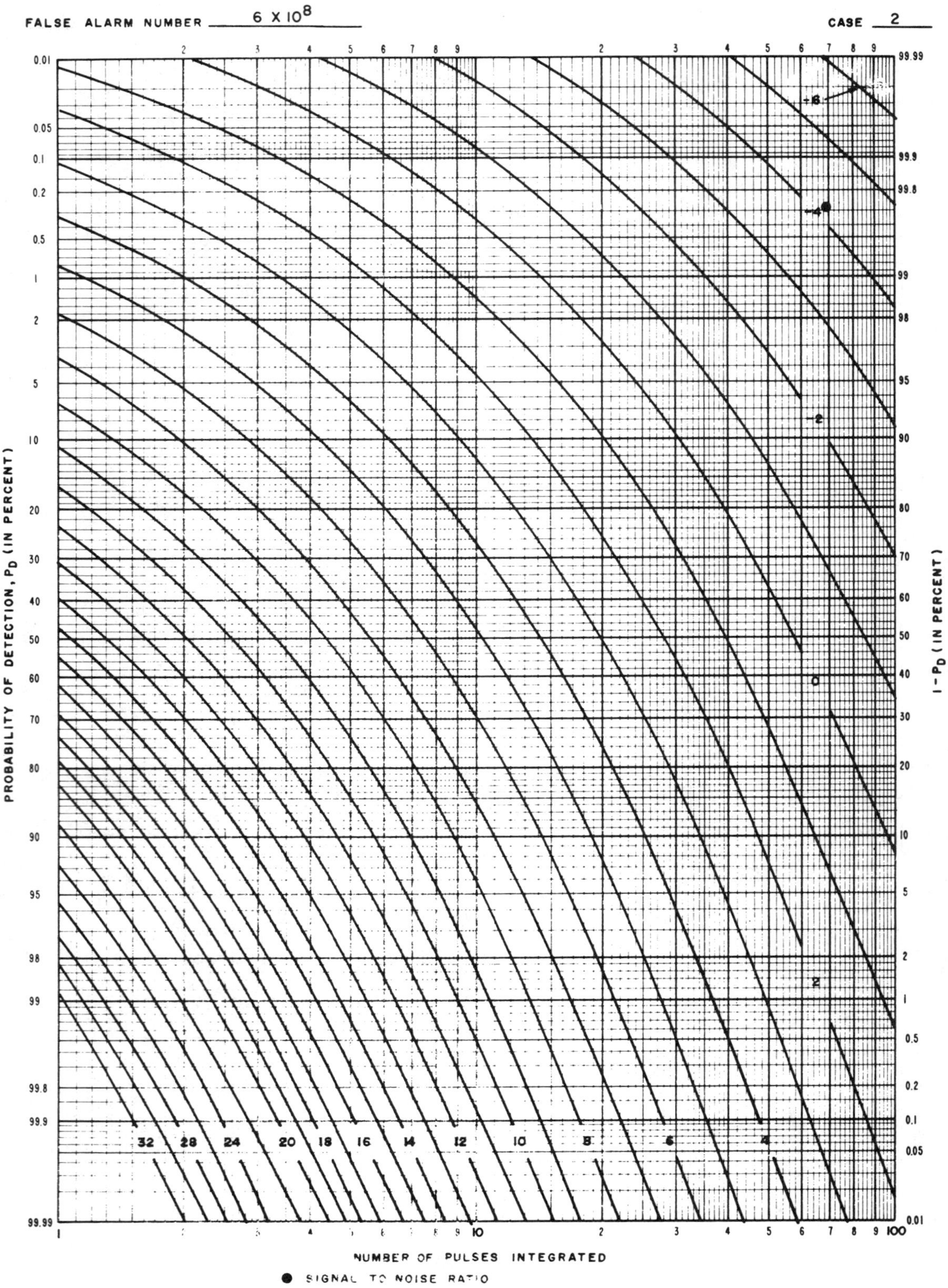
FALSE ALARM NUMBER 6 X 10^8
CASE 2
PROBABILITY OF DETECTION, P_D (IN PERCENT)
1 - P_D (IN PERCENT)
NUMBER OF PULSES INTEGRATED
● SIGNAL TO NOISE RATIO
32
28
24
20
18
16
14
12
10
8
6
4
2
0
-2
-4
-6

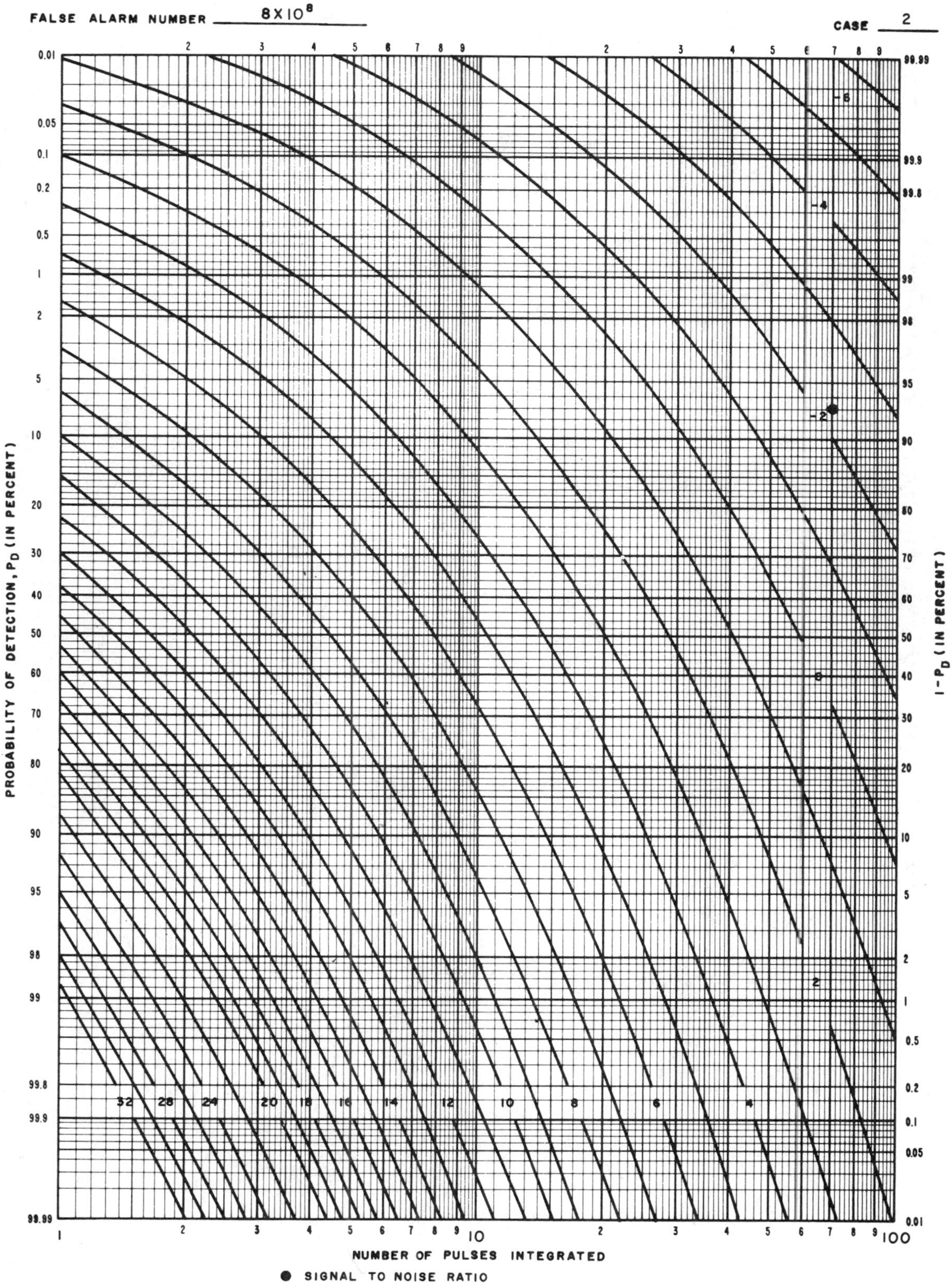
FALSE ALARM NUMBER 8X10^8
CASE 2
PROBABILITY OF DETECTION, P_D (IN PERCENT)
1-P_D (IN PERCENT)
NUMBER OF PULSES INTEGRATED
32
28
24
20
18
16
14
12
10
8
6
4
2
-2
-4
-6
● SIGNAL TO NOISE RATIO

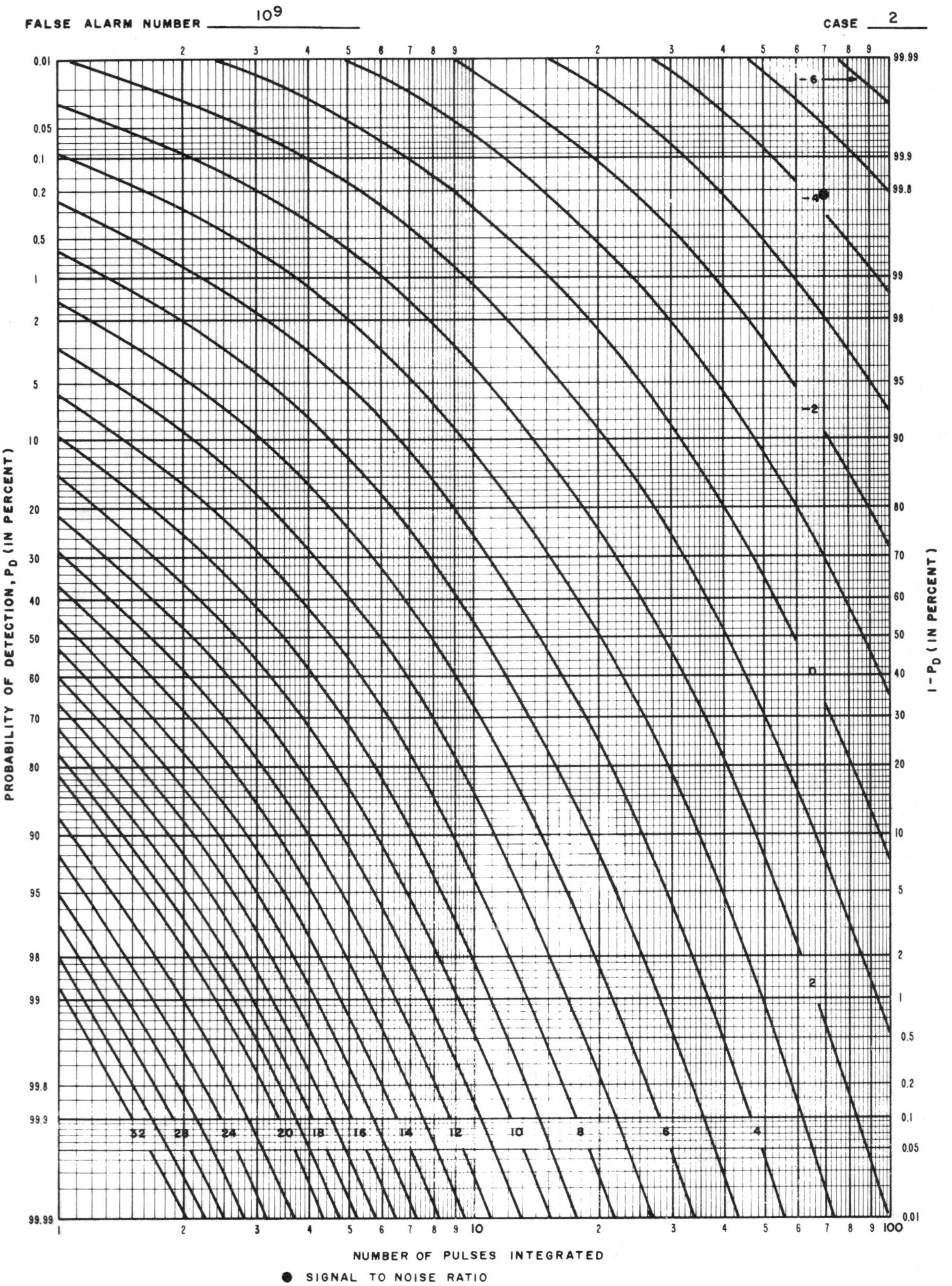
FALSE ALARM NUMBER 10^9
CASE 2
PROBABILITY OF DETECTION, P_D (IN PERCENT)
$1-P_D$ (IN PERCENT)
0.01
0.05
0.1
0.2
0.5
1
2
5
10
20
30
40
50
60
70
80
90
95
98
99
99.8
99.9
99.99
99.99
99.9
99.8
99
98
95
90
80
70
60
50
40
30
20
10
5
2
1
0.5
0.2
0.1
0.05
0.01
-6
-4
-2
0
2
32
28
24
20
18
16
14
12
10
8
6
4
1 2 3 4 5 6 7 8 9 10 2 3 4 5 6 7 8 9 100
NUMBER OF PULSES INTEGRATED
● SIGNAL TO NOISE RATIO

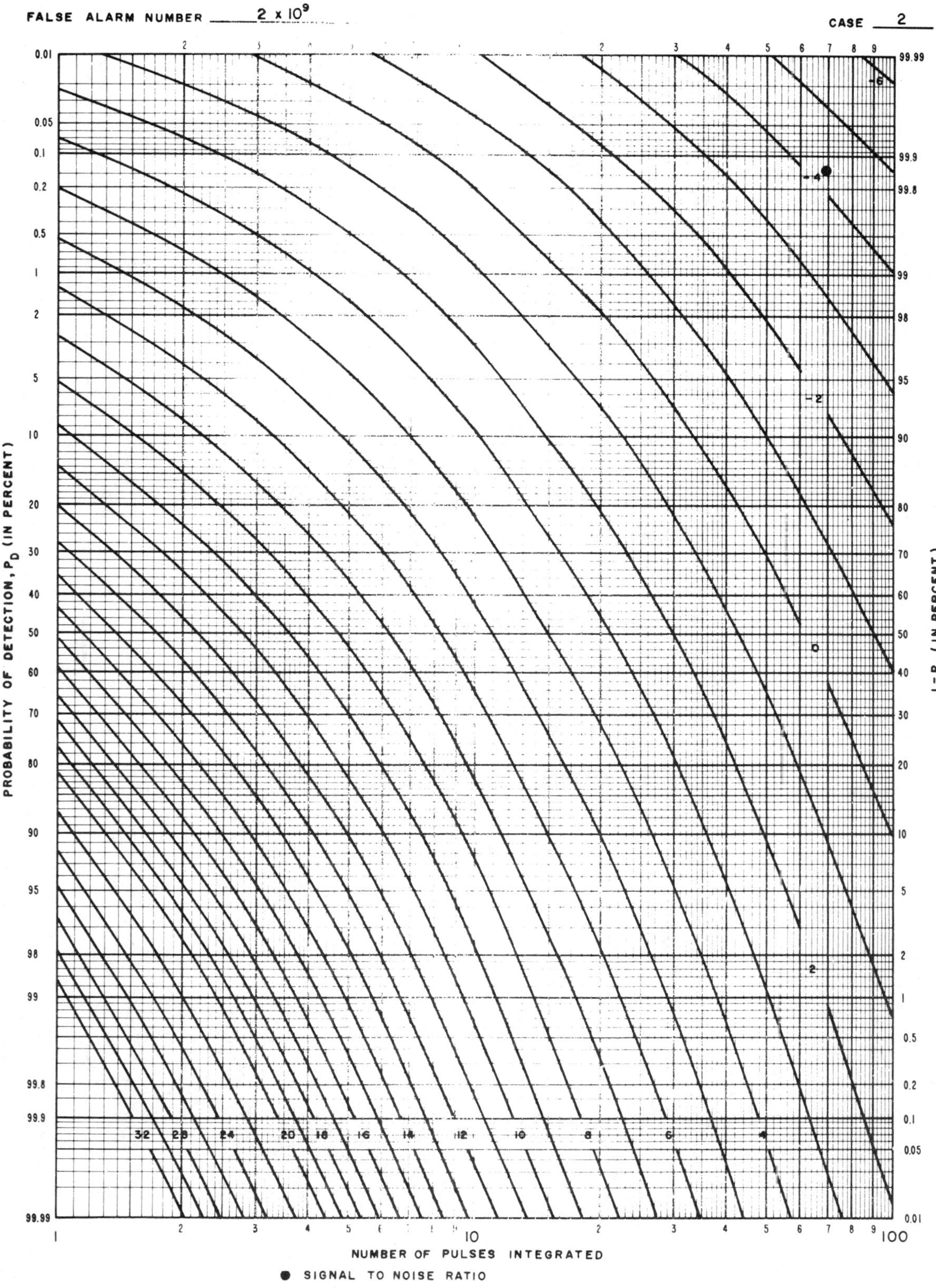
FALSE ALARM NUMBER 2 x 10^9
CASE 2
PROBABILITY OF DETECTION, P_D (IN PERCENT)
1 - P_D (IN PERCENT)
NUMBER OF PULSES INTEGRATED
● SIGNAL TO NOISE RATIO
32
28
24
20
18
16
14
12
10
8
6
4
2
0
-2
-4
-6

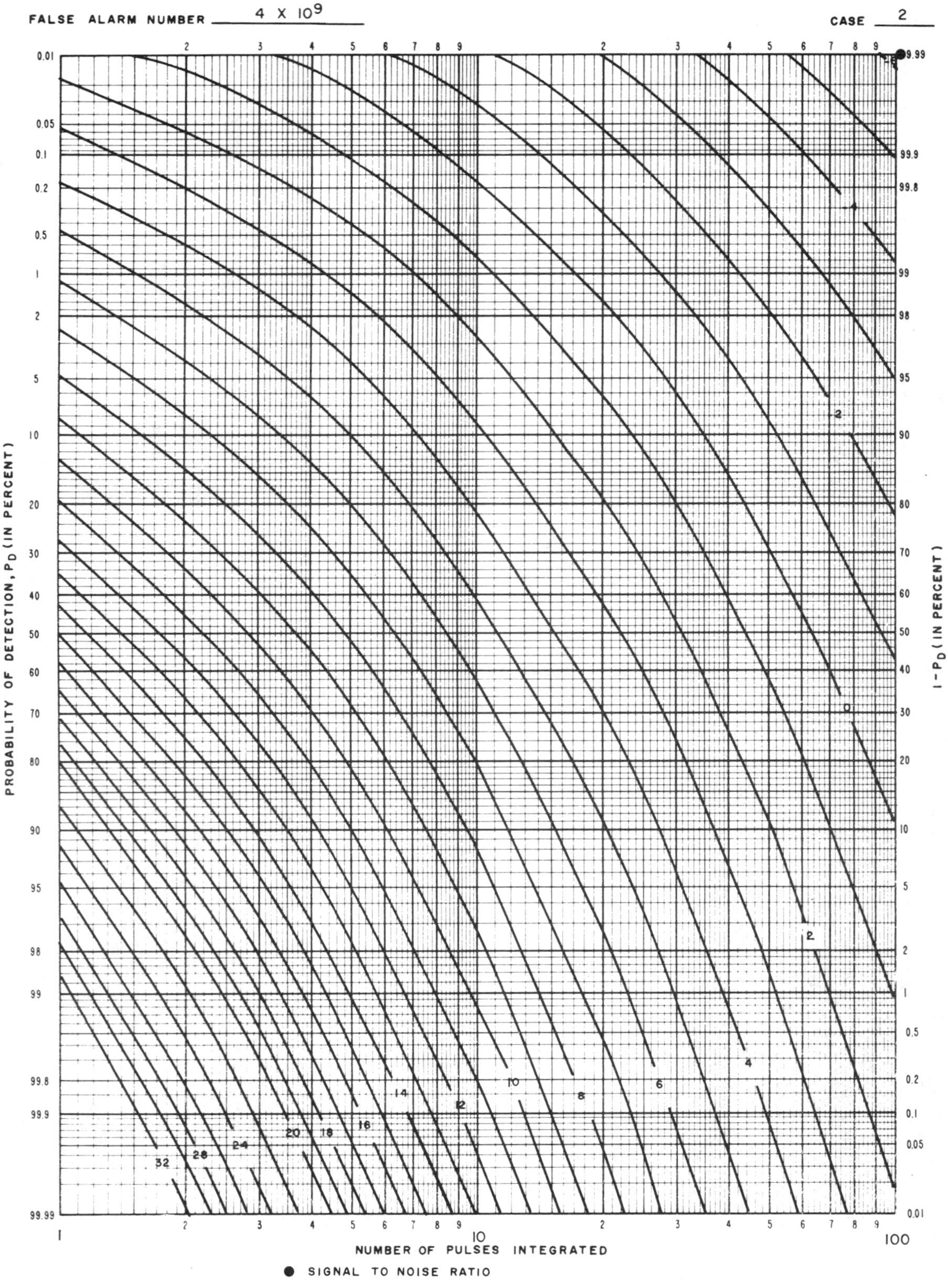
FALSE ALARM NUMBER 4 X 10^9
CASE 2
PROBABILITY OF DETECTION, P_D (IN PERCENT)
1-P_D (IN PERCENT)
NUMBER OF PULSES INTEGRATED
1
10
100
● SIGNAL TO NOISE RATIO
32
28
24
20
18
16
14
12
10
8
6
4
2
0
-2
-4

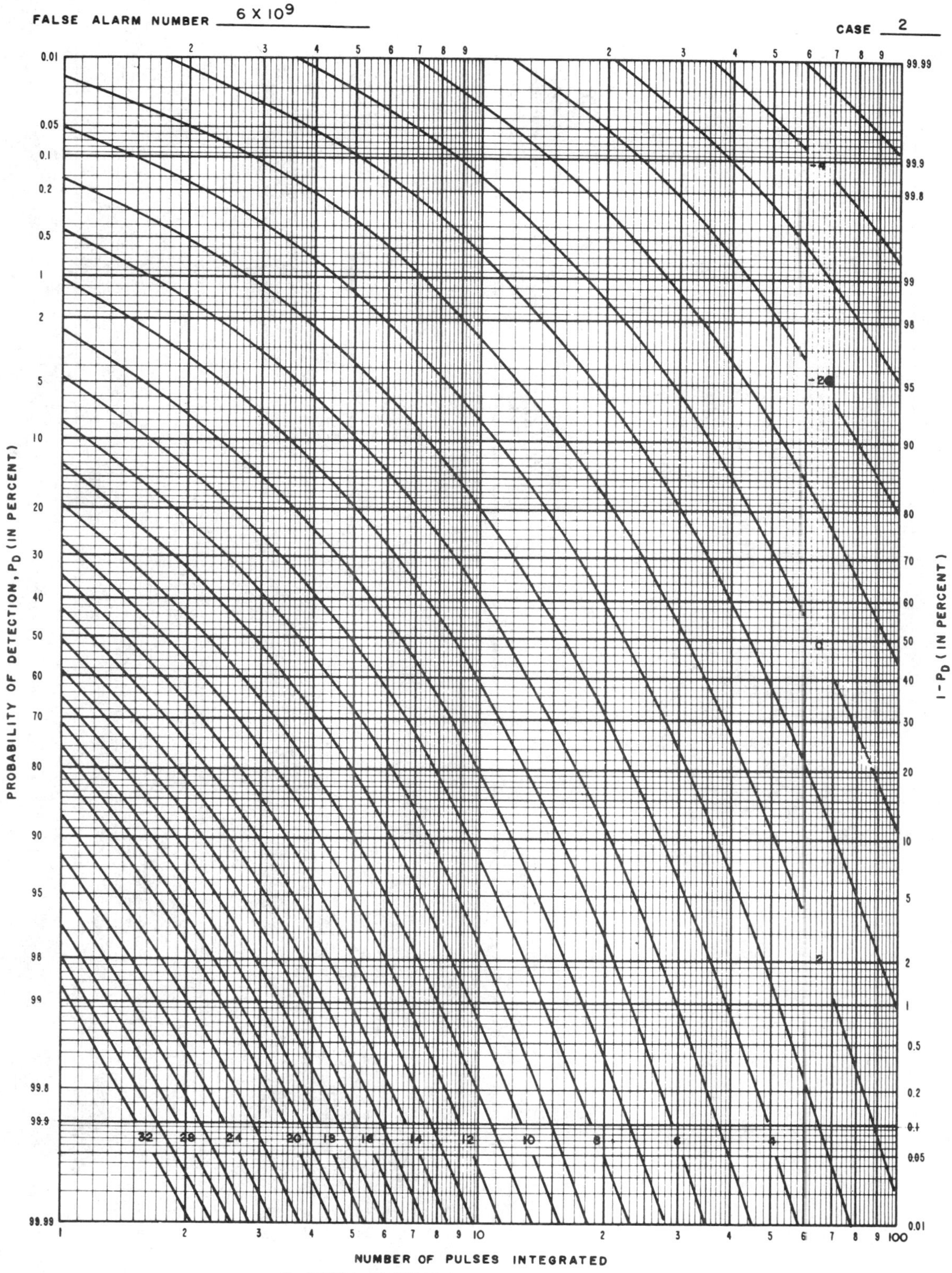
FALSE ALARM NUMBER 6 X 10^9
CASE 2
PROBABILITY OF DETECTION, P_D (IN PERCENT)
1 - P_D (IN PERCENT)
NUMBER OF PULSES INTEGRATED
● SIGNAL TO NOISE RATIO
32
28
24
20
18
16
14
12
10
8
6
4
2
0
-2
-4

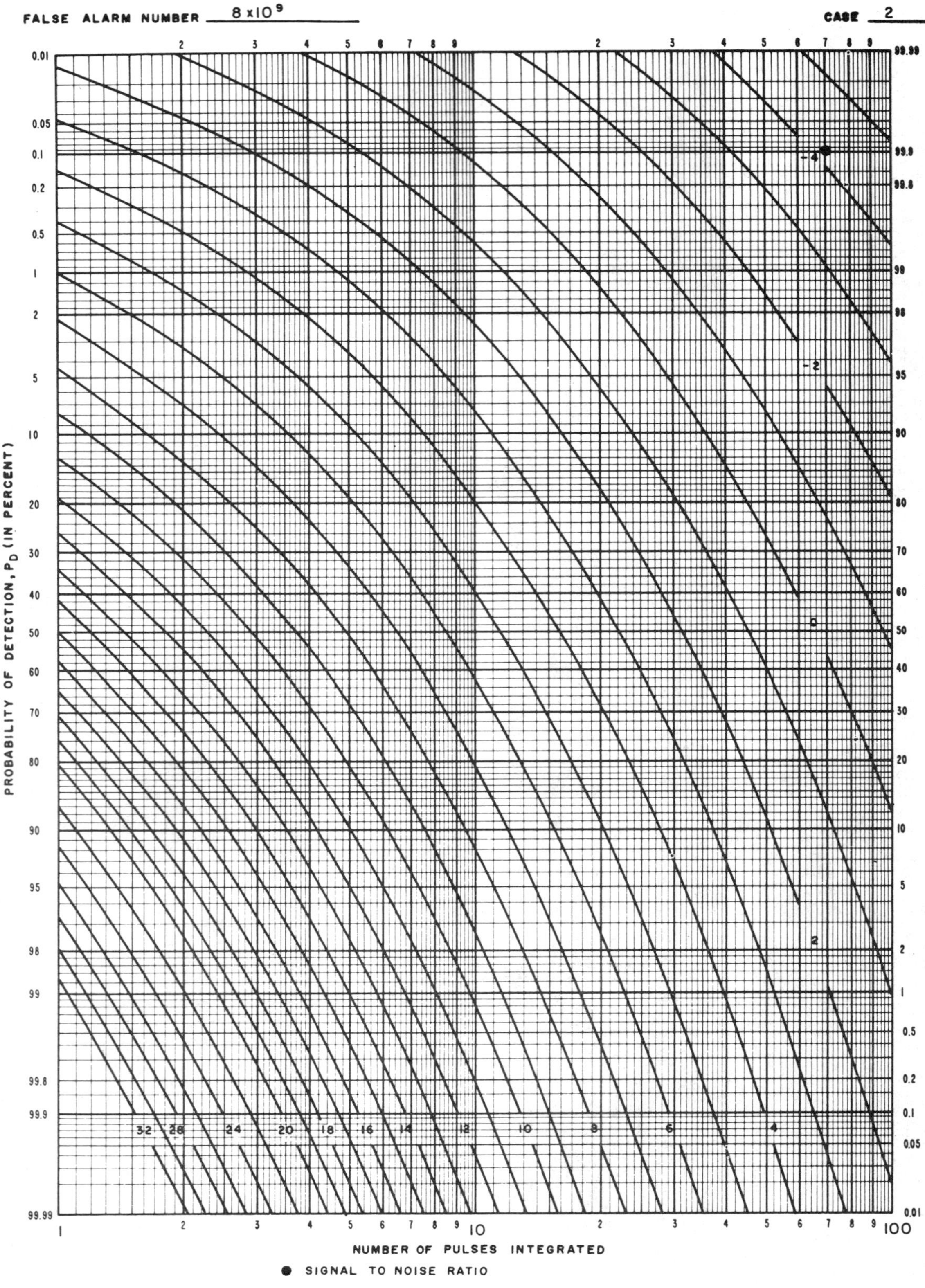

FALSE ALARM NUMBER 8 x 10⁹
CASE 2
PROBABILITY OF DETECTION, P_D (IN PERCENT)
1 - P_D (IN PERCENT)
0.01 0.05 0.1 0.2 0.5 1 2 5 10 20 30 40 50 60 70 80 90 95 98 99 99.8 99.9 99.99
99.99 99.9 99.8 99 98 95 90 80 70 60 50 40 30 20 10 5 2 1 0.5 0.2 0.1 0.05 0.01
-4 -2 0 2
32 28 24 20 18 16 14 12 10 8 6 4
1 2 3 4 5 6 7 8 9 10 2 3 4 5 6 7 8 9 100
NUMBER OF PULSES INTEGRATED
● SIGNAL TO NOISE RATIO

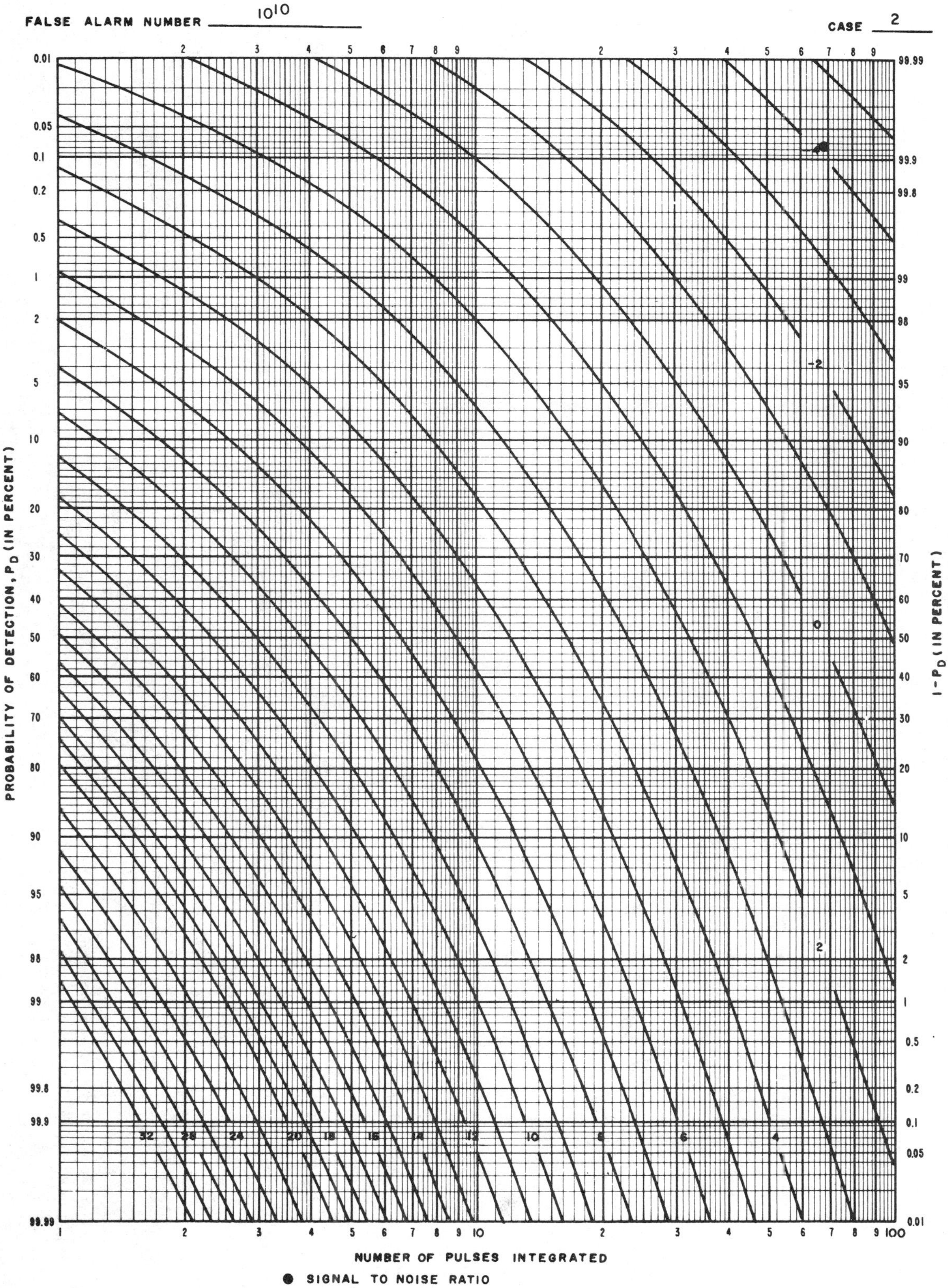
FALSE ALARM NUMBER 10^{10}
CASE 2
PROBABILITY OF DETECTION, P_D (IN PERCENT)
$1-P_D$ (IN PERCENT)
NUMBER OF PULSES INTEGRATED
● SIGNAL TO NOISE RATIO
32
28
24
20
18
16
14
12
10
8
6
4
2
0
-2
-4

FALSE ALARM NUMBER 10 CASE 3

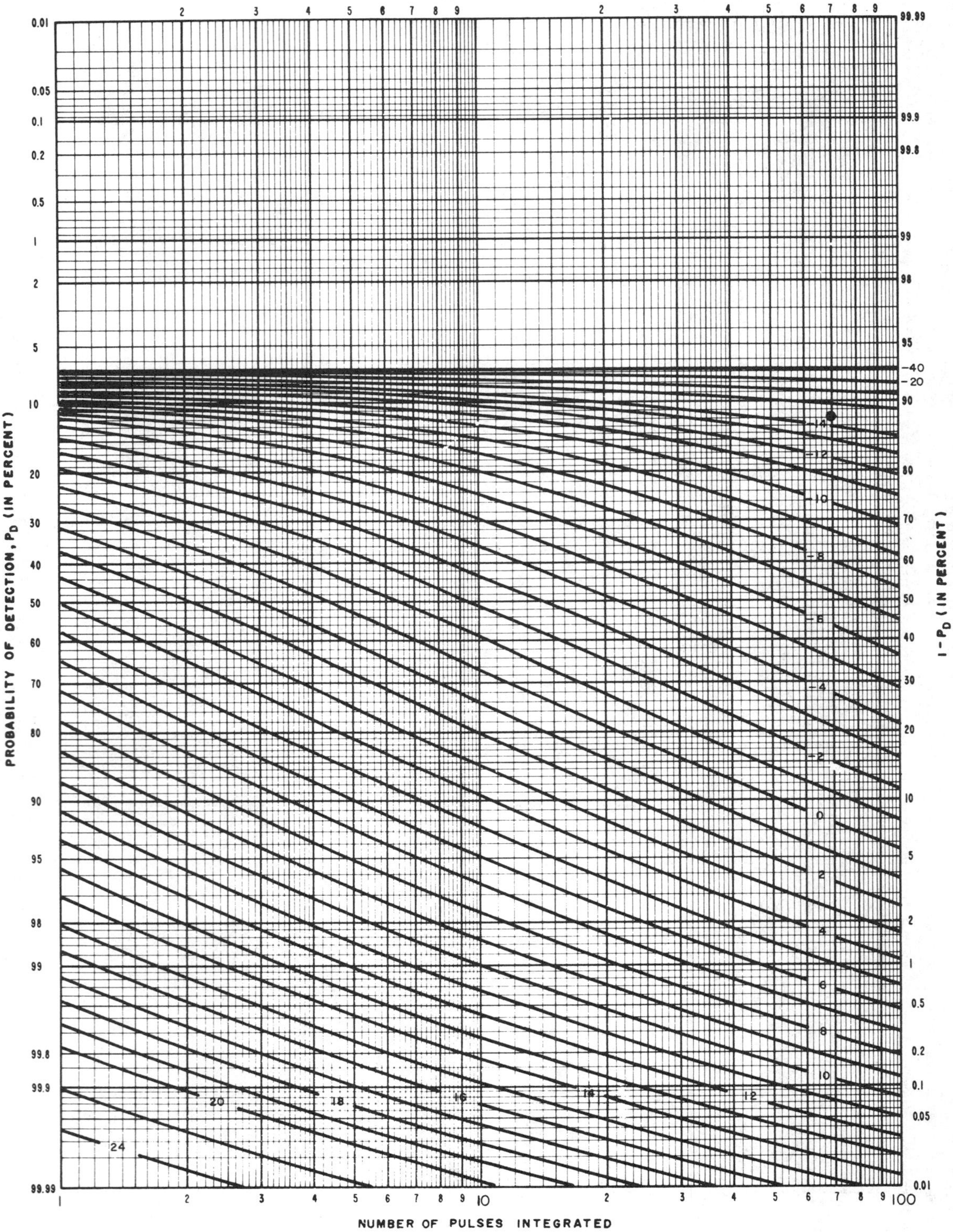

● SIGNAL TO NOISE RATIO

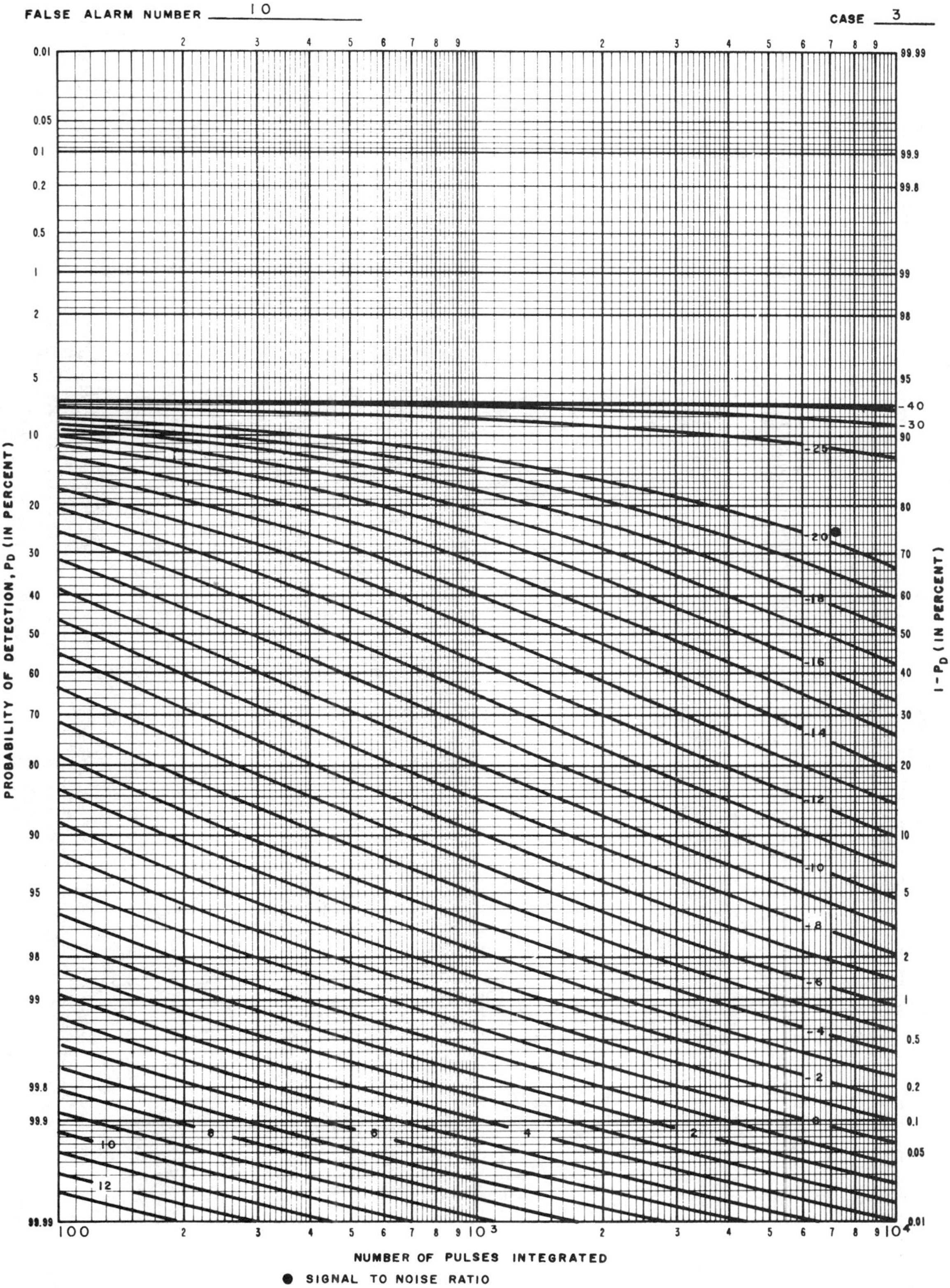
FALSE ALARM NUMBER 10
CASE 3
PROBABILITY OF DETECTION, P_D (IN PERCENT)
$1-P_D$ (IN PERCENT)
NUMBER OF PULSES INTEGRATED
● SIGNAL TO NOISE RATIO

FALSE ALARM NUMBER 20 CASE 3

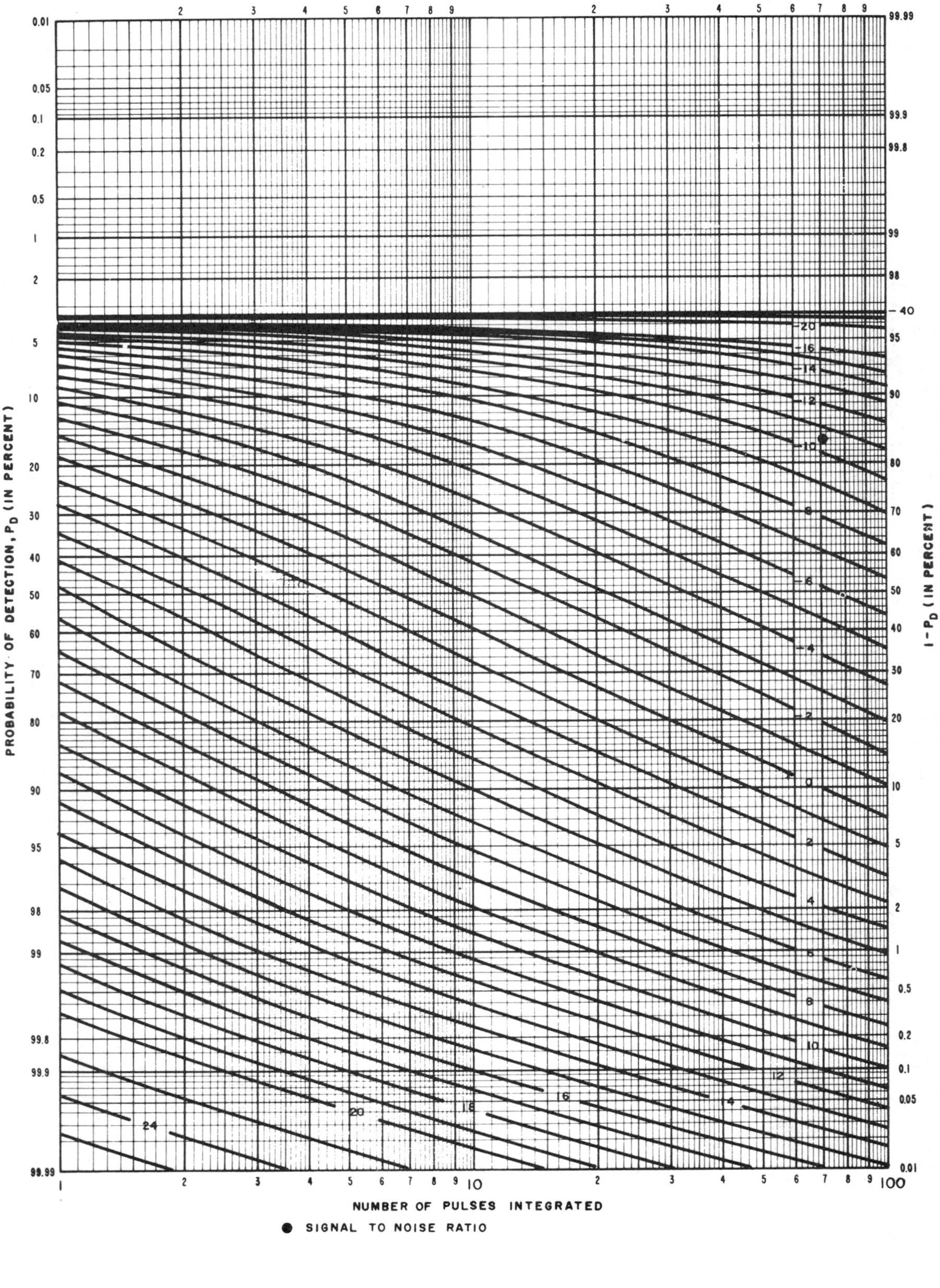

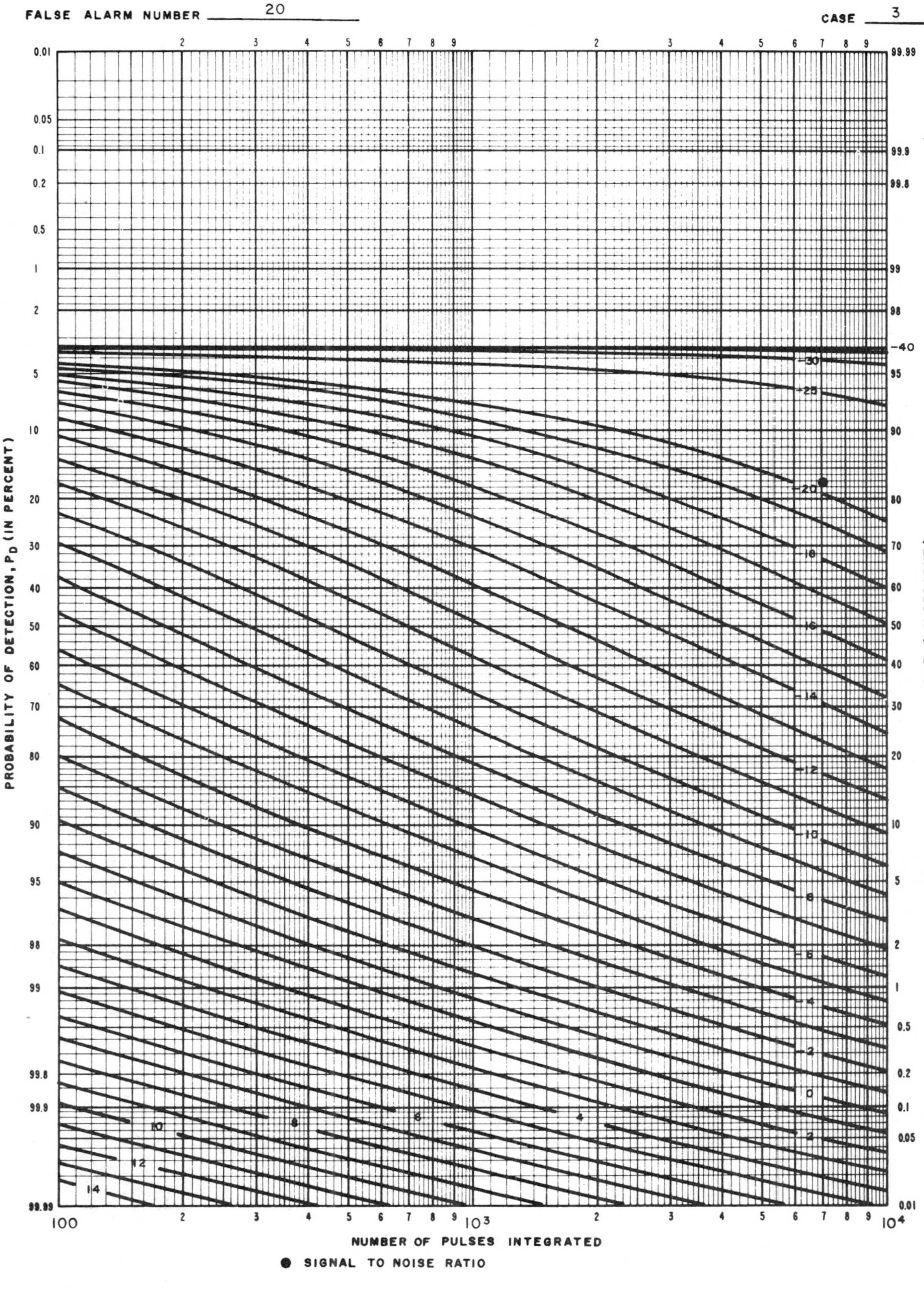
FALSE ALARM NUMBER 20
CASE 3
PROBABILITY OF DETECTION, P_D (IN PERCENT)
$1-P_D$ (IN PERCENT)
NUMBER OF PULSES INTEGRATED
100
10^3
10^4
-40
-30
-25
-20
18
16
14
12
10
8
6
4
2
0
-2
● SIGNAL TO NOISE RATIO

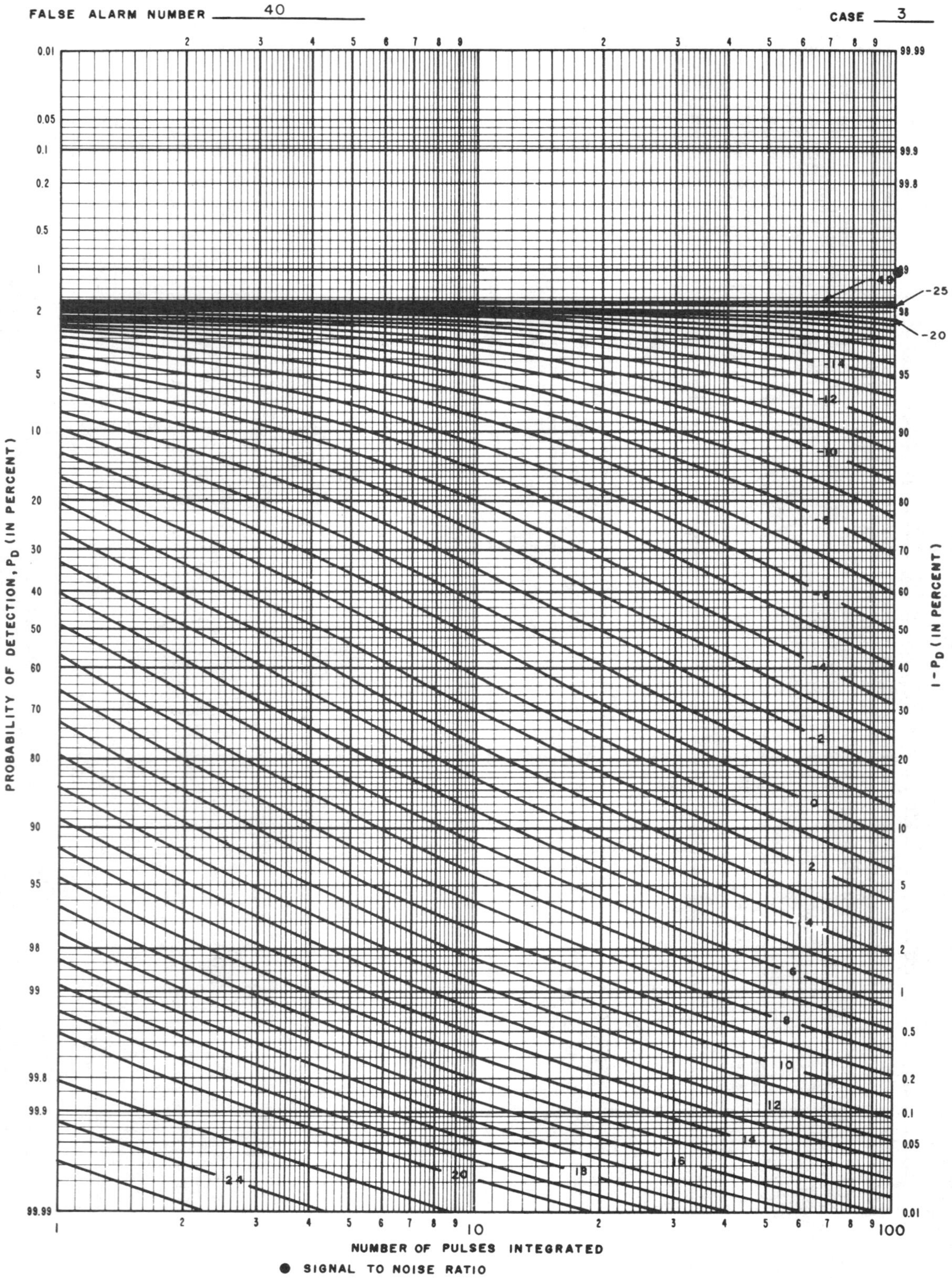
FALSE ALARM NUMBER 40
CASE 3
PROBABILITY OF DETECTION, P_D (IN PERCENT)
$1-P_D$ (IN PERCENT)
NUMBER OF PULSES INTEGRATED
● SIGNAL TO NOISE RATIO
-25
-20
-14
-12
0
2
6
8
10
12
14
16
18
20
24

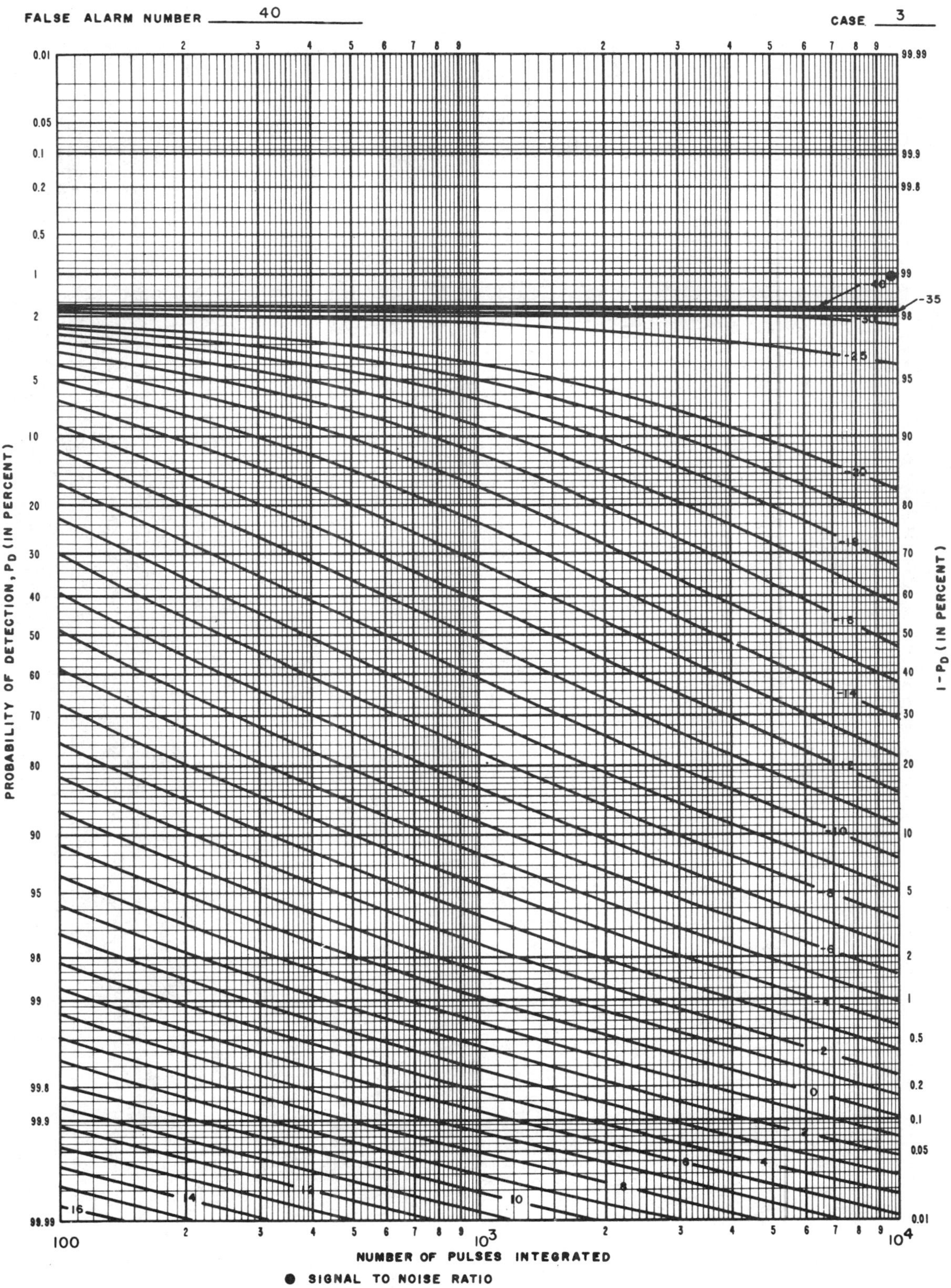
FALSE ALARM NUMBER 40
CASE 3
PROBABILITY OF DETECTION, P_D (IN PERCENT)
1 - P_D (IN PERCENT)
NUMBER OF PULSES INTEGRATED
● SIGNAL TO NOISE RATIO

FALSE ALARM NUMBER 60

CASE 3

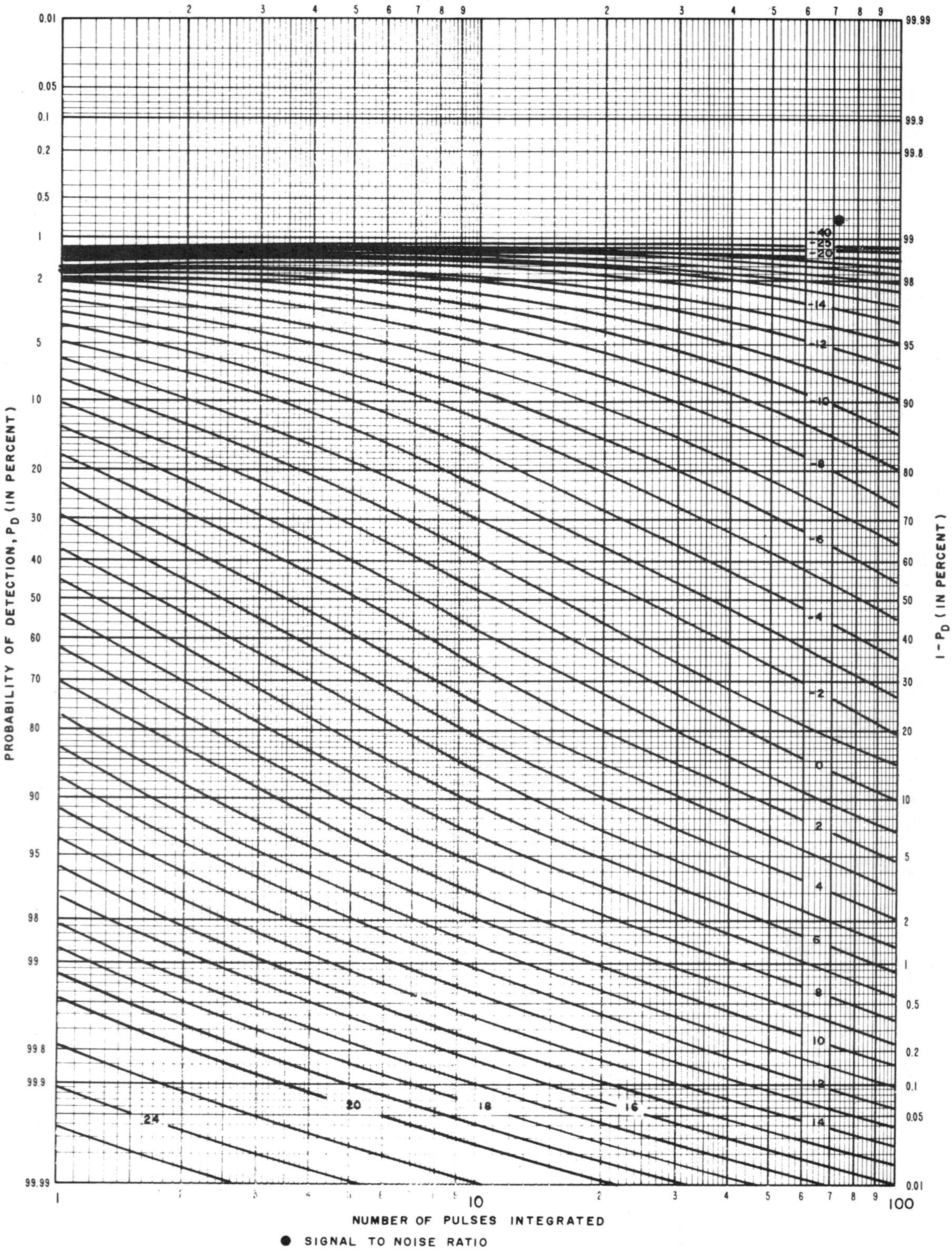

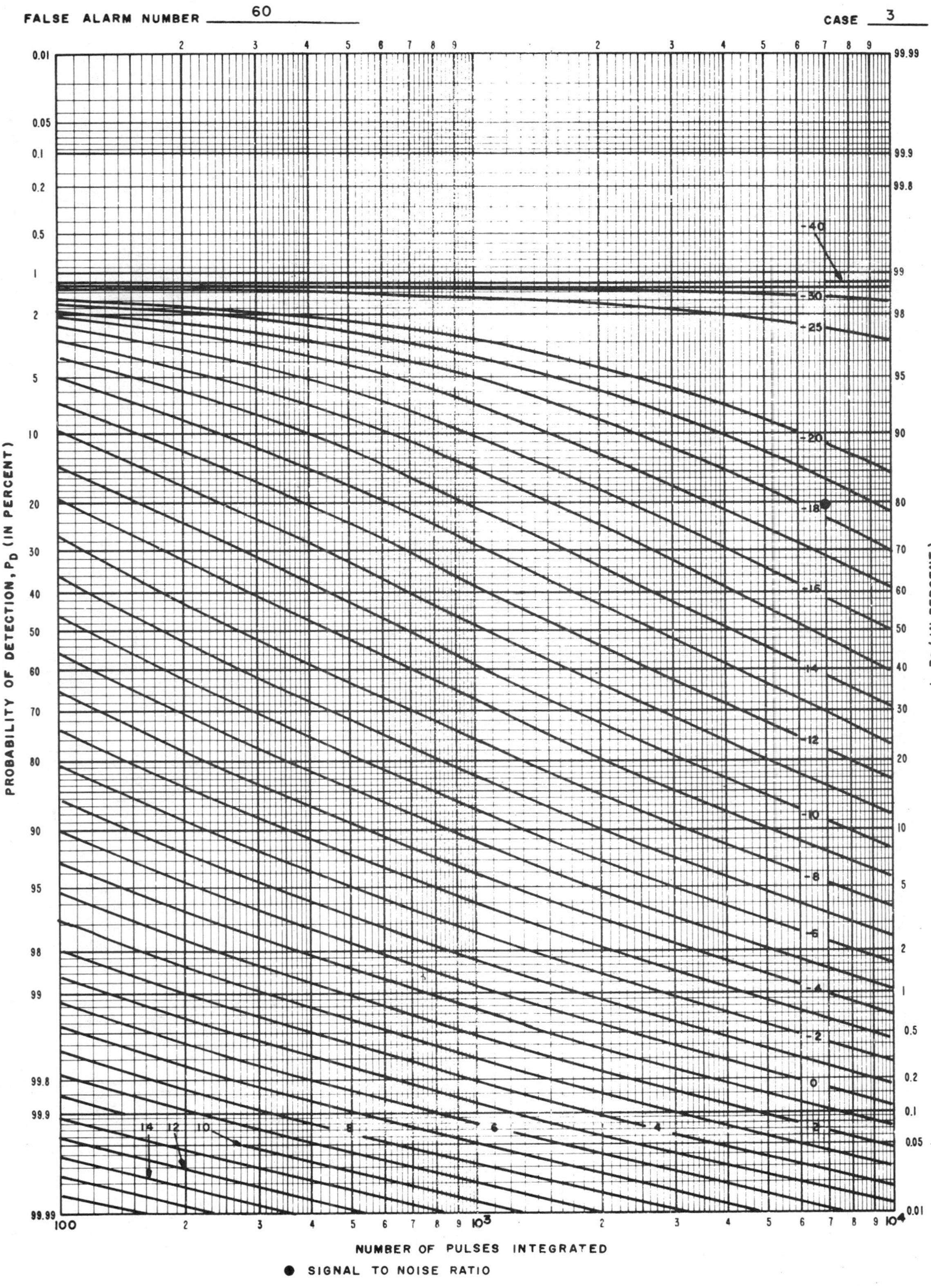
FALSE ALARM NUMBER 60
CASE 3
PROBABILITY OF DETECTION, P_D (IN PERCENT)
1 - P_D (IN PERCENT)
NUMBER OF PULSES INTEGRATED
● SIGNAL TO NOISE RATIO

FALSE ALARM NUMBER 80 CASE 3

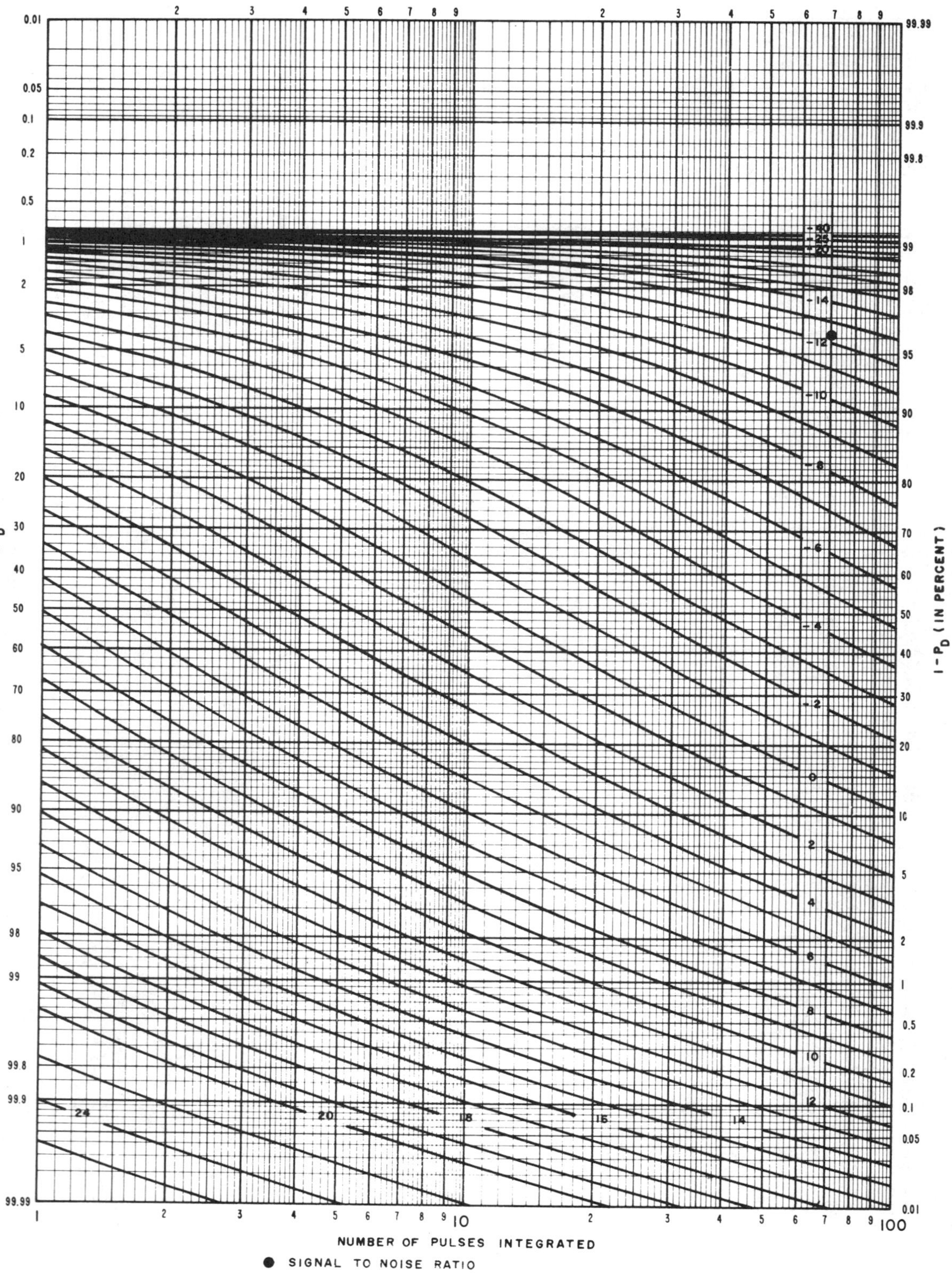

FALSE ALARM NUMBER 80 CASE 3

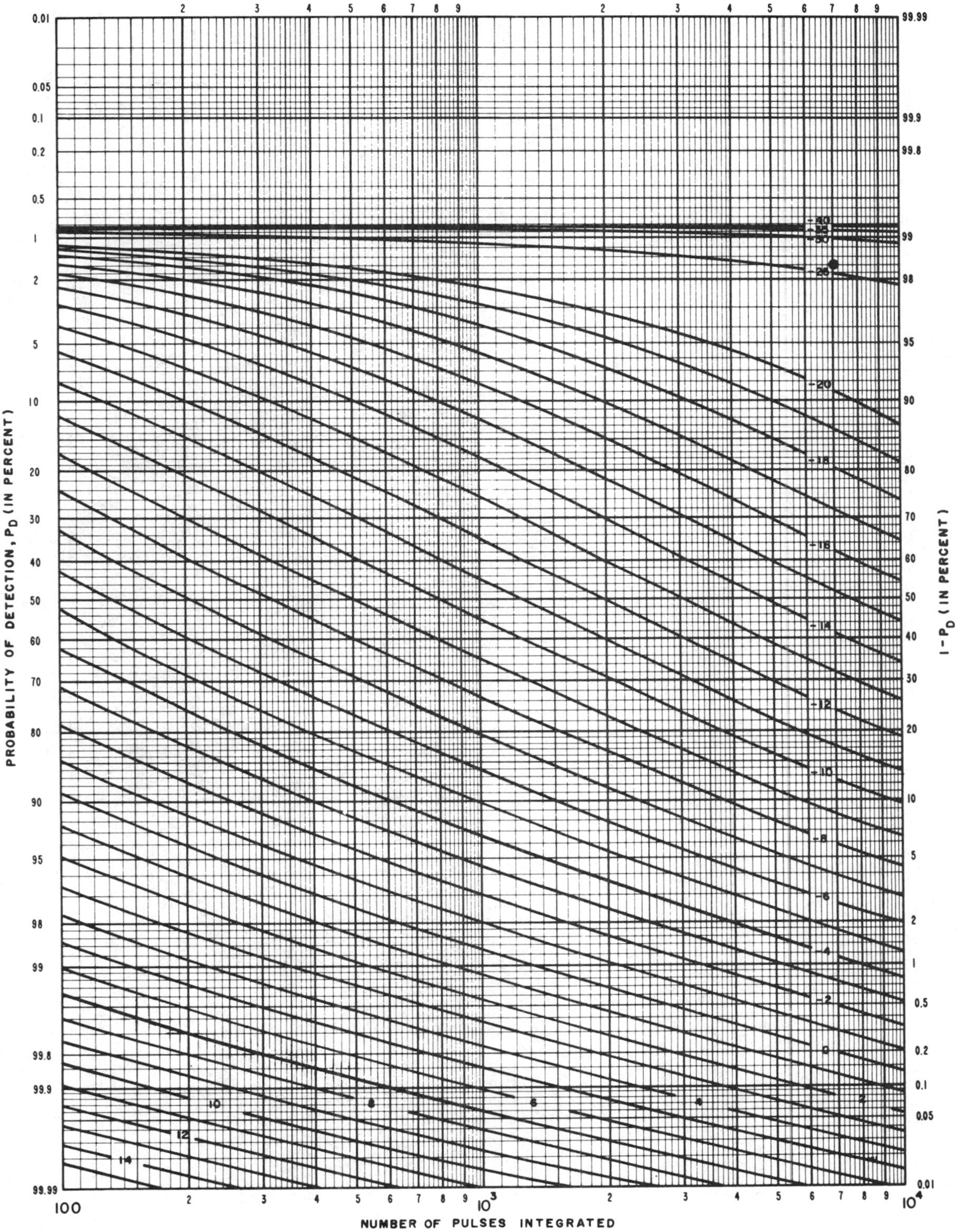

● SIGNAL TO NOISE RATIO

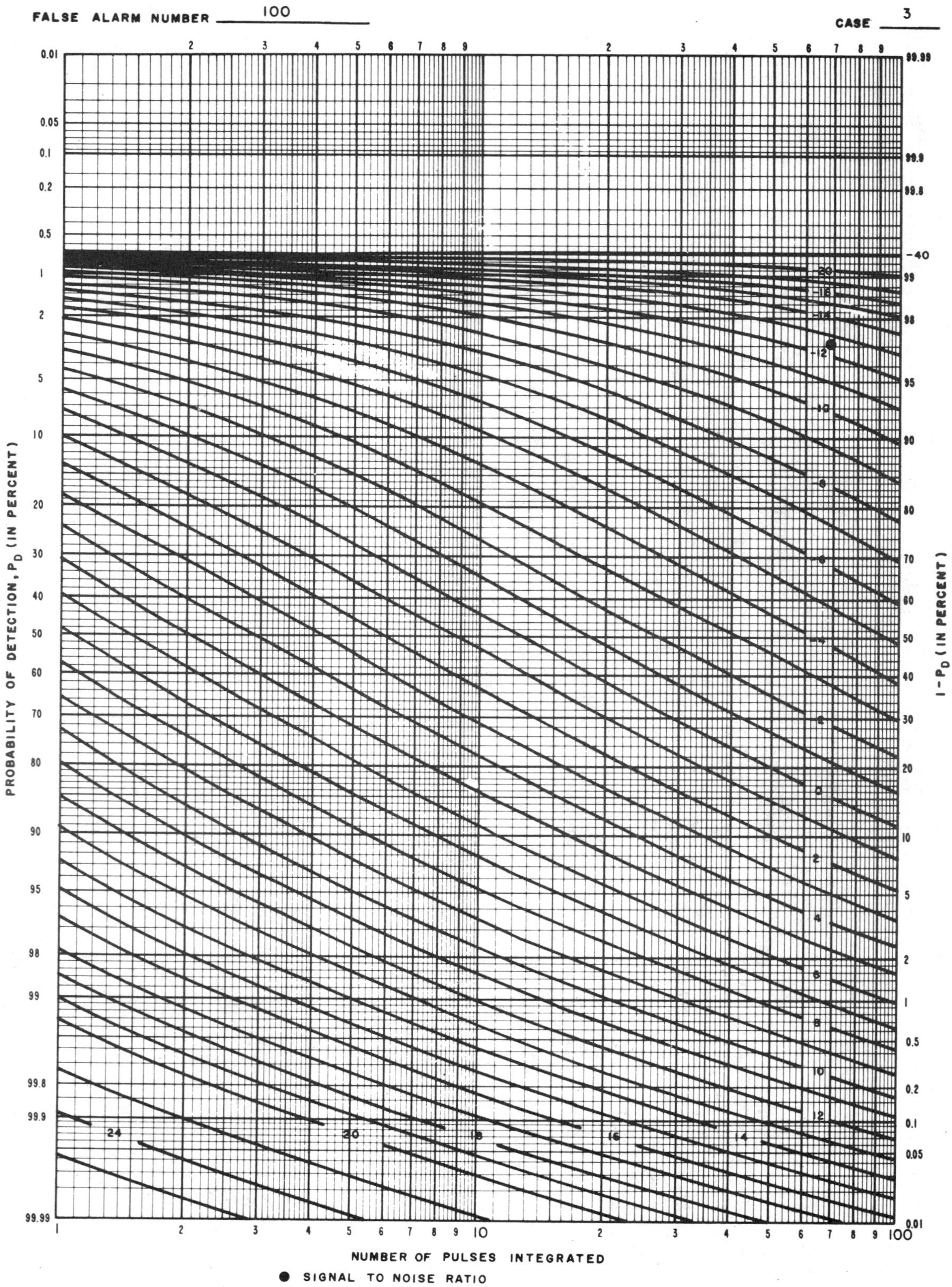
FALSE ALARM NUMBER 100
CASE 3
PROBABILITY OF DETECTION, P_D (IN PERCENT)
1 - P_D (IN PERCENT)
-40
-20
-18
-16
-12
-10
-8
-6
-4
-2
0
2
4
6
8
10
12
14
16
18
20
24
NUMBER OF PULSES INTEGRATED
● SIGNAL TO NOISE RATIO

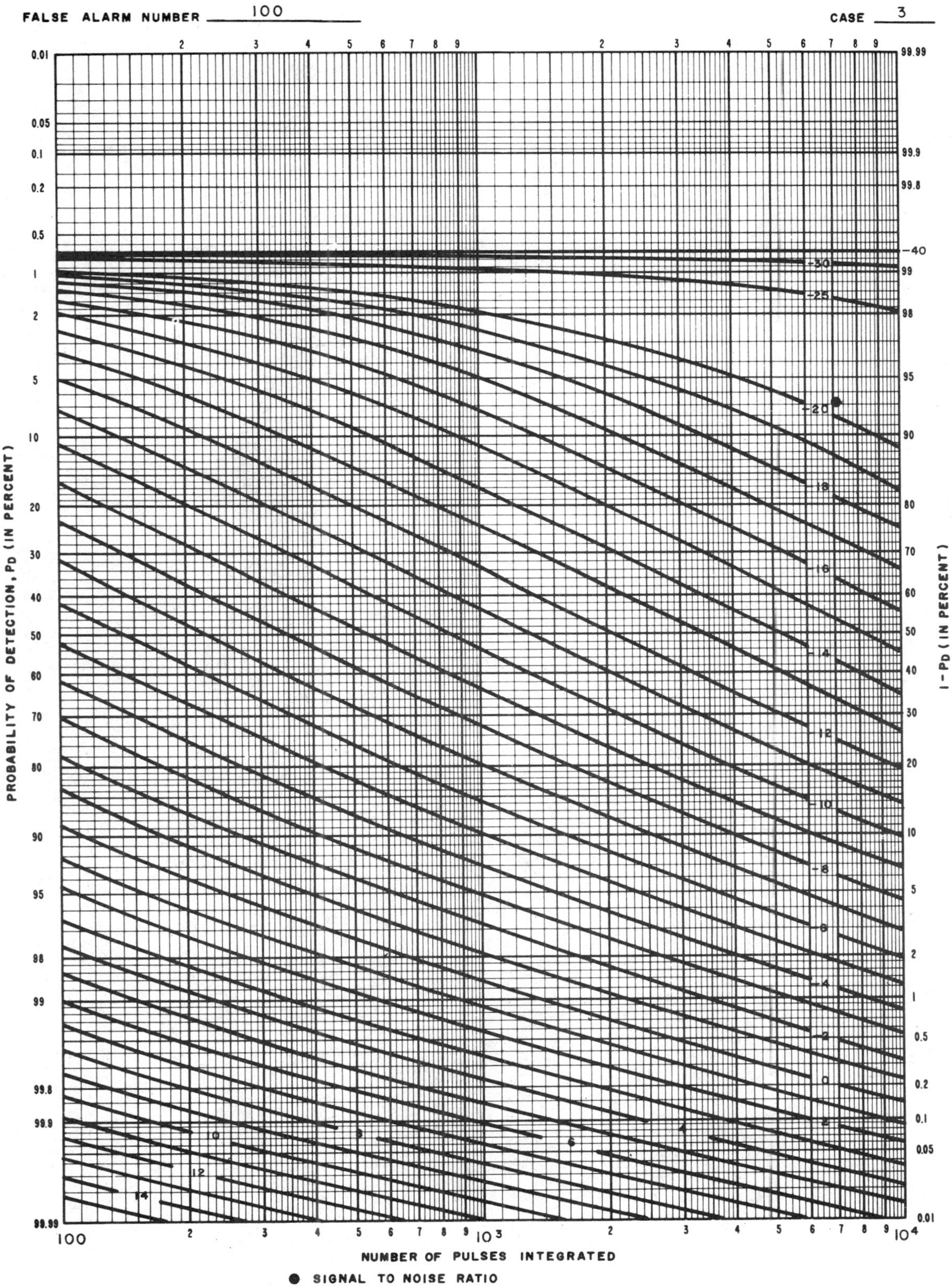
FALSE ALARM NUMBER 100
CASE 3
PROBABILITY OF DETECTION, P_D (IN PERCENT)
$1-P_D$ (IN PERCENT)
NUMBER OF PULSES INTEGRATED
● SIGNAL TO NOISE RATIO
100
10^3
10^4
0.01
0.05
0.1
0.2
0.5
1
2
5
10
20
30
40
50
60
70
80
90
95
98
99
99.8
99.9
99.99
-40
-30
-25
-20
-18
-16
-14
-12
-10
-8
-6
-4
-2
0
2
4
6
8
10
12
14

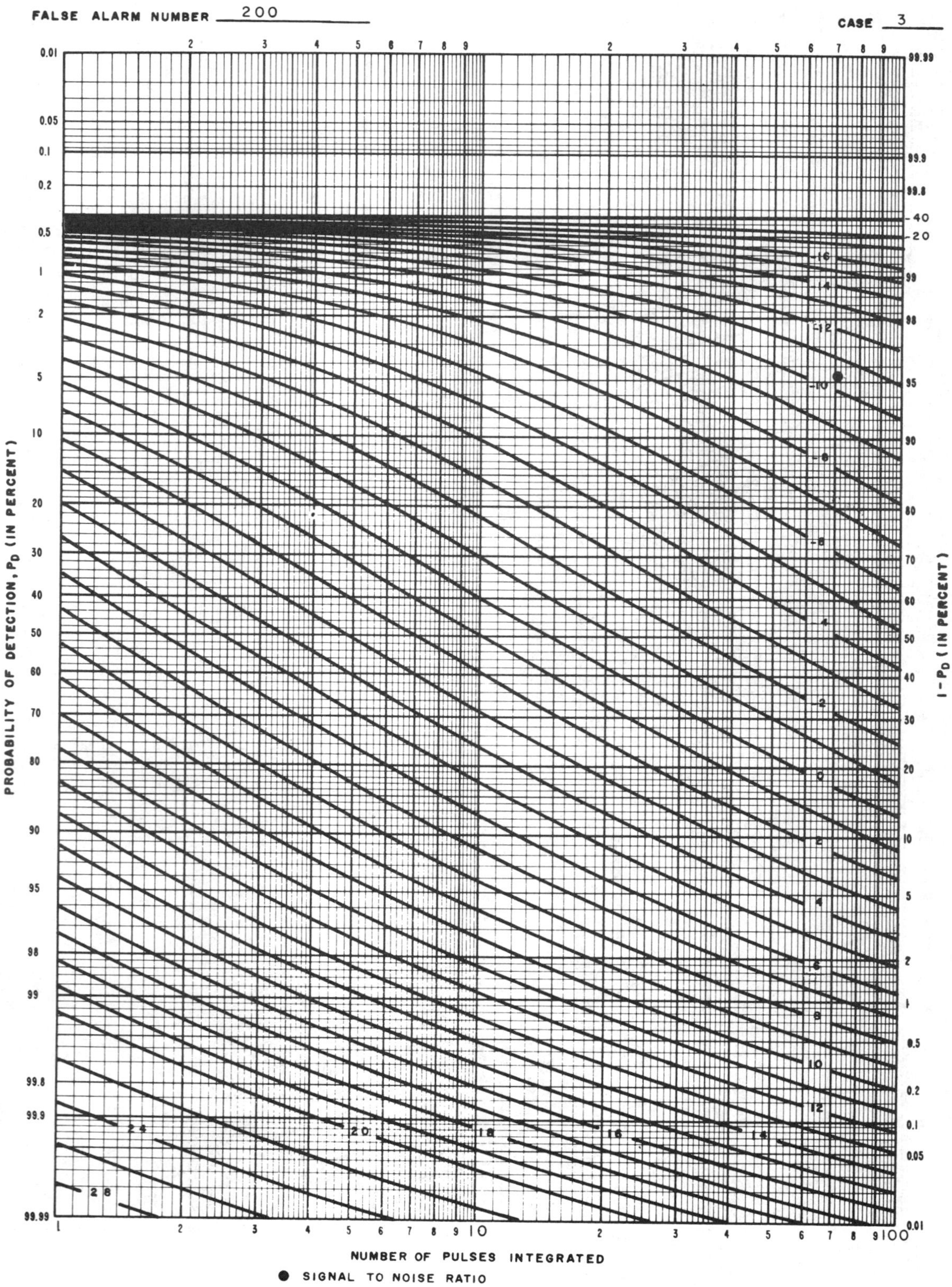
FALSE ALARM NUMBER 200
CASE 3
PROBABILITY OF DETECTION, P_D (IN PERCENT)
$1 - P_D$ (IN PERCENT)
NUMBER OF PULSES INTEGRATED
● SIGNAL TO NOISE RATIO

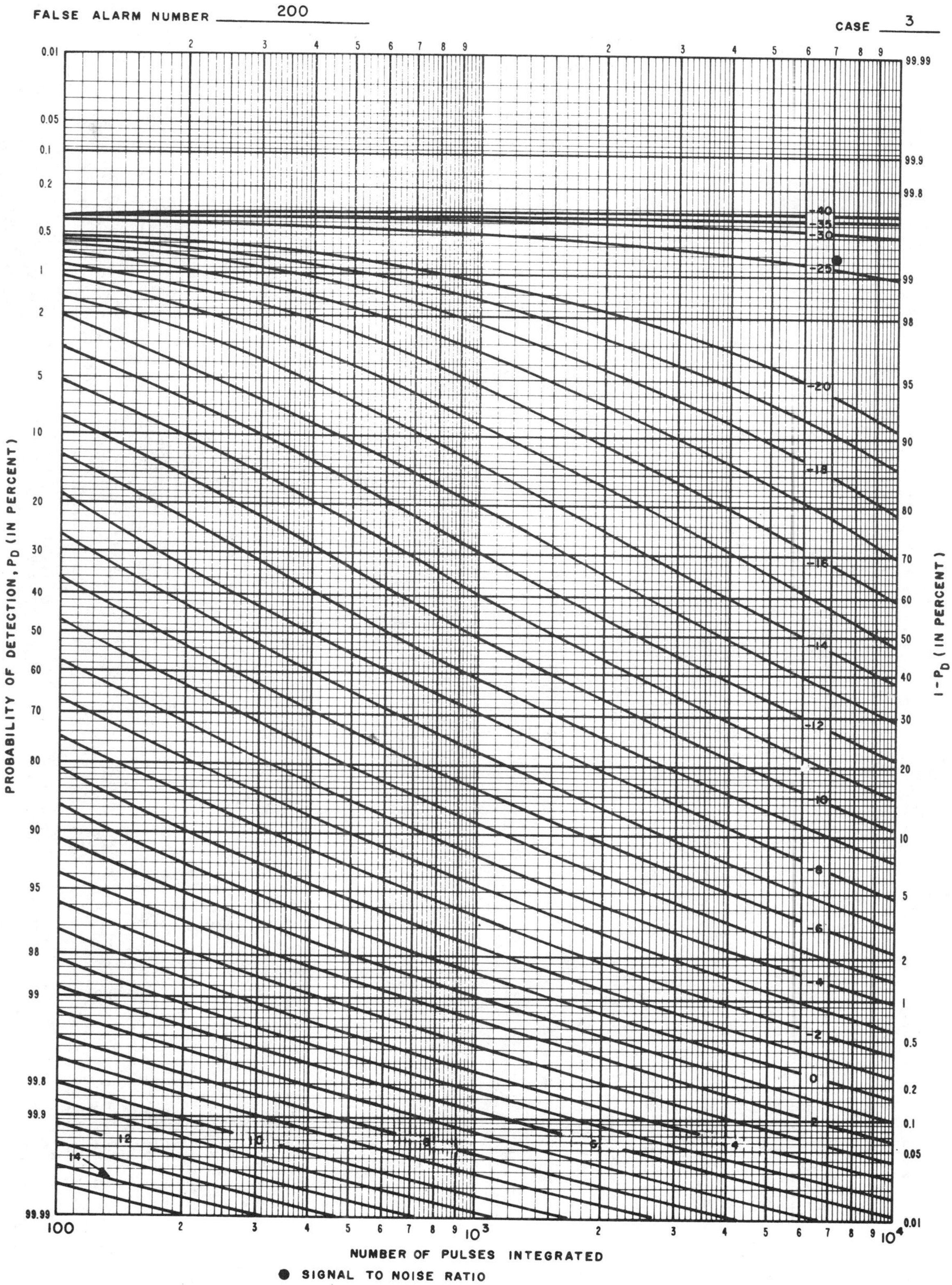
FALSE ALARM NUMBER 200
CASE 3
PROBABILITY OF DETECTION, P_D (IN PERCENT)
$1-P_D$ (IN PERCENT)
NUMBER OF PULSES INTEGRATED
● SIGNAL TO NOISE RATIO

FALSE ALARM NUMBER 400

CASE 3

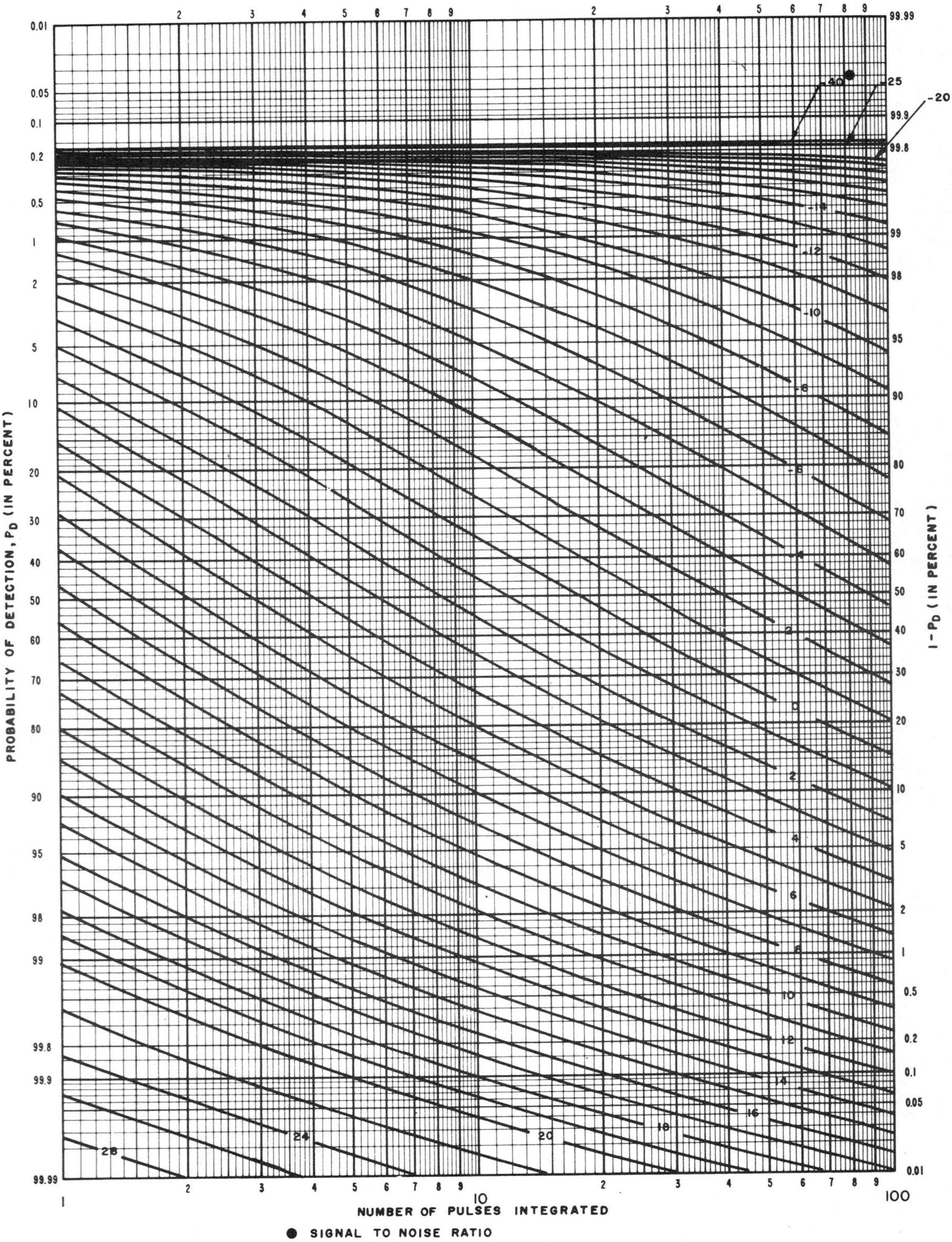

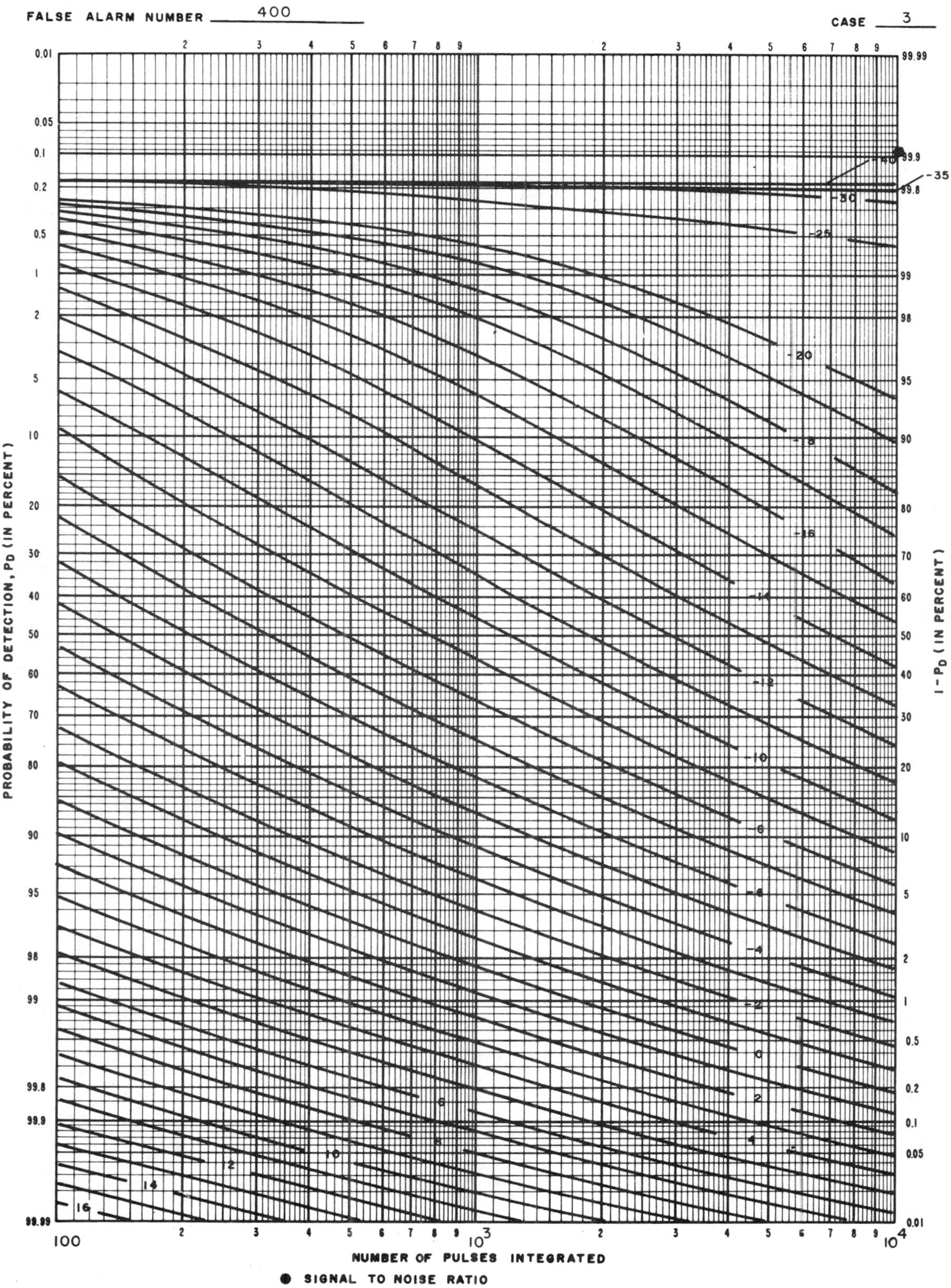
FALSE ALARM NUMBER 400
CASE 3
PROBABILITY OF DETECTION, P_D (IN PERCENT)
1 - P_D (IN PERCENT)
NUMBER OF PULSES INTEGRATED
● SIGNAL TO NOISE RATIO
100
10^3
10^4
-40
-35
-30
-25
-20
-18
-16
-14
-12
-10
-8
-6
-4
-2
0
2
4
6
8
10
12
14
16

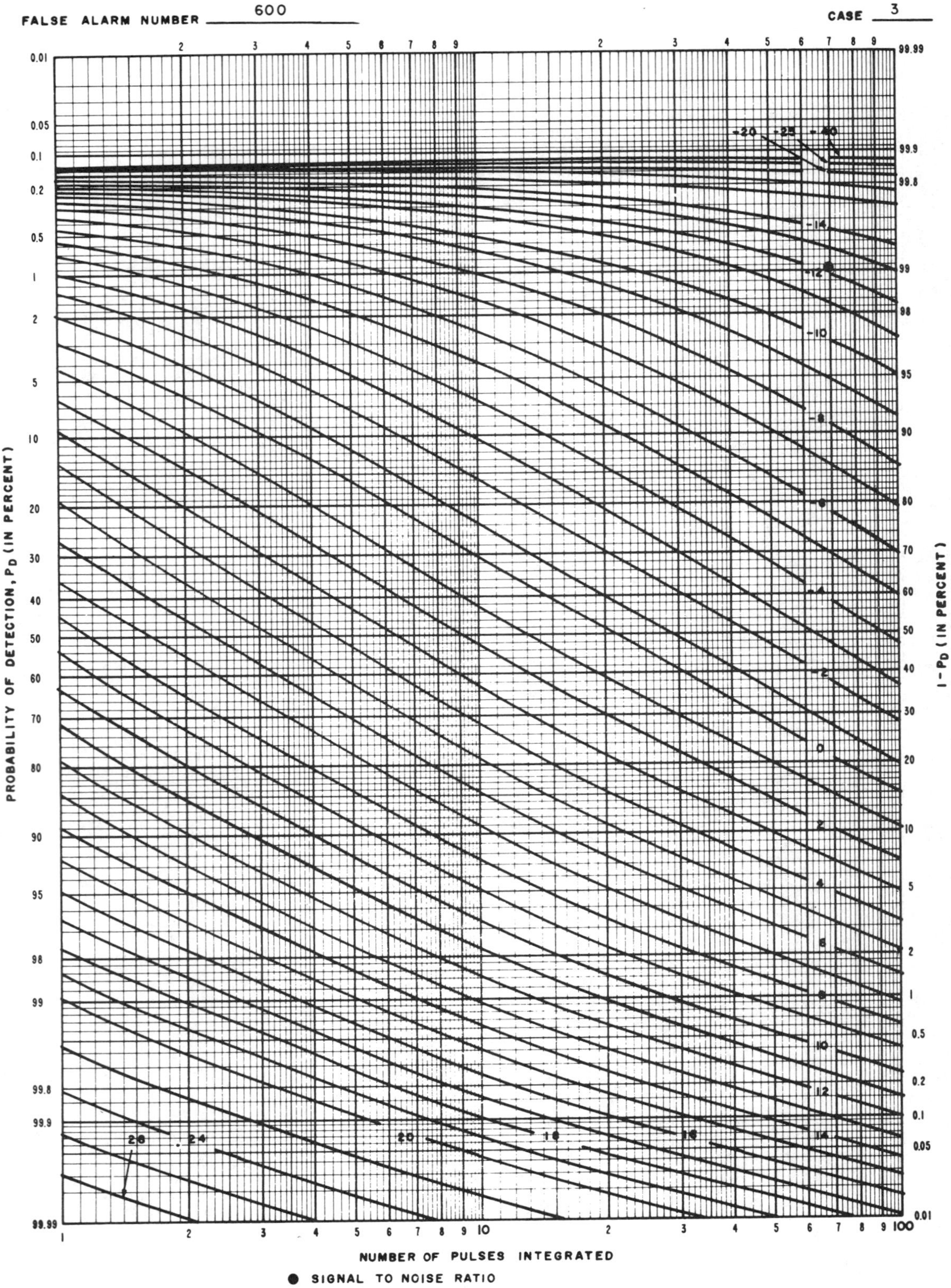
FALSE ALARM NUMBER 600
CASE 3
PROBABILITY OF DETECTION, P_D (IN PERCENT)
1 - P_D (IN PERCENT)
NUMBER OF PULSES INTEGRATED
● SIGNAL TO NOISE RATIO
-20
-25
-40
-14
-12
-10
-8
-6
-4
-2
0
2
4
6
8
10
12
14
16
18
20
24
26

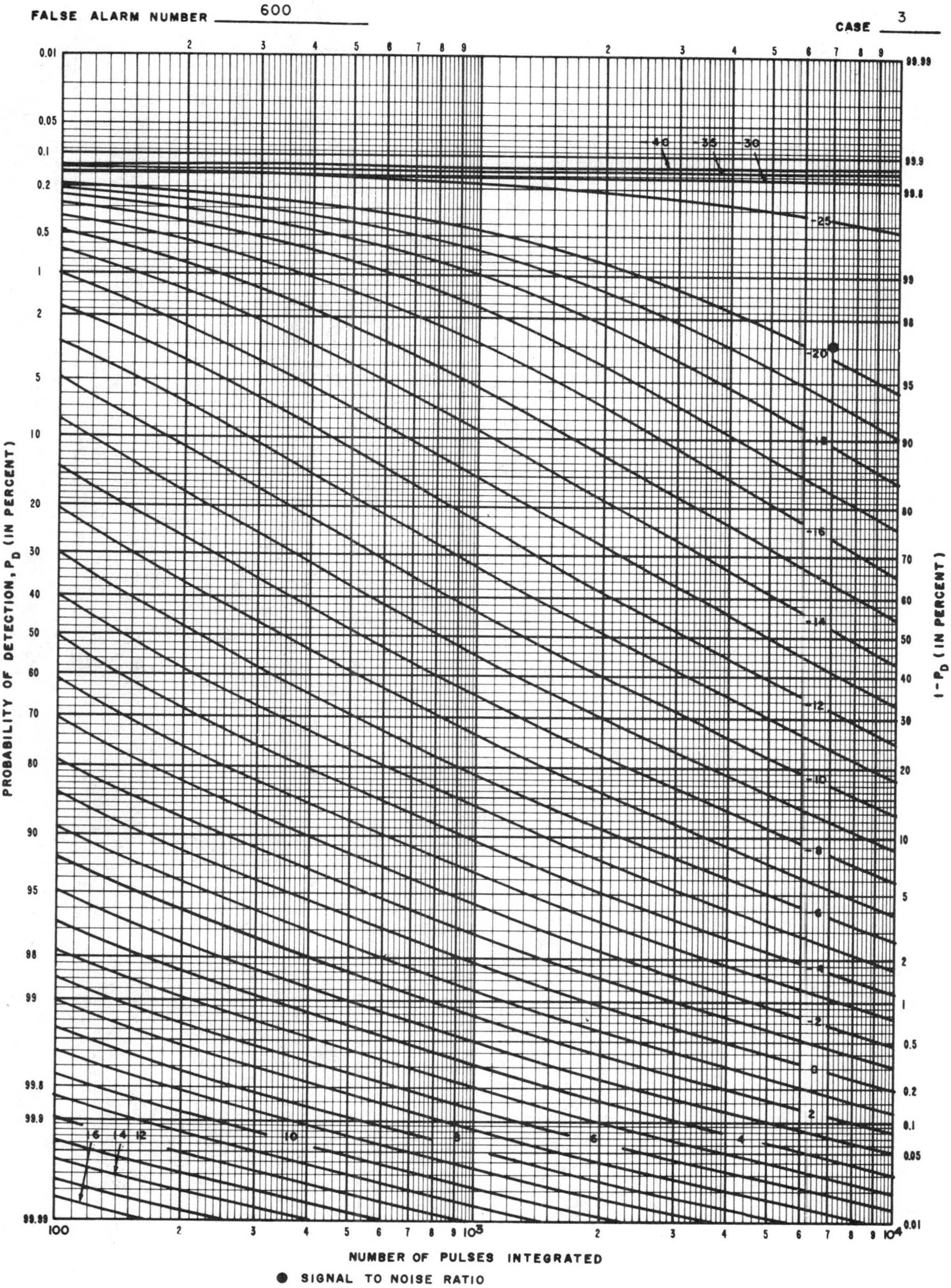
FALSE ALARM NUMBER 600
CASE 3
PROBABILITY OF DETECTION, P_D (IN PERCENT)
1 - P_D (IN PERCENT)
NUMBER OF PULSES INTEGRATED
● SIGNAL TO NOISE RATIO
-40
-35
-30
-25
-20
-14
-12
-10
-8
-6
-4
-2
0
2
4
6
8
10
12

FALSE ALARM NUMBER 800 CASE 3

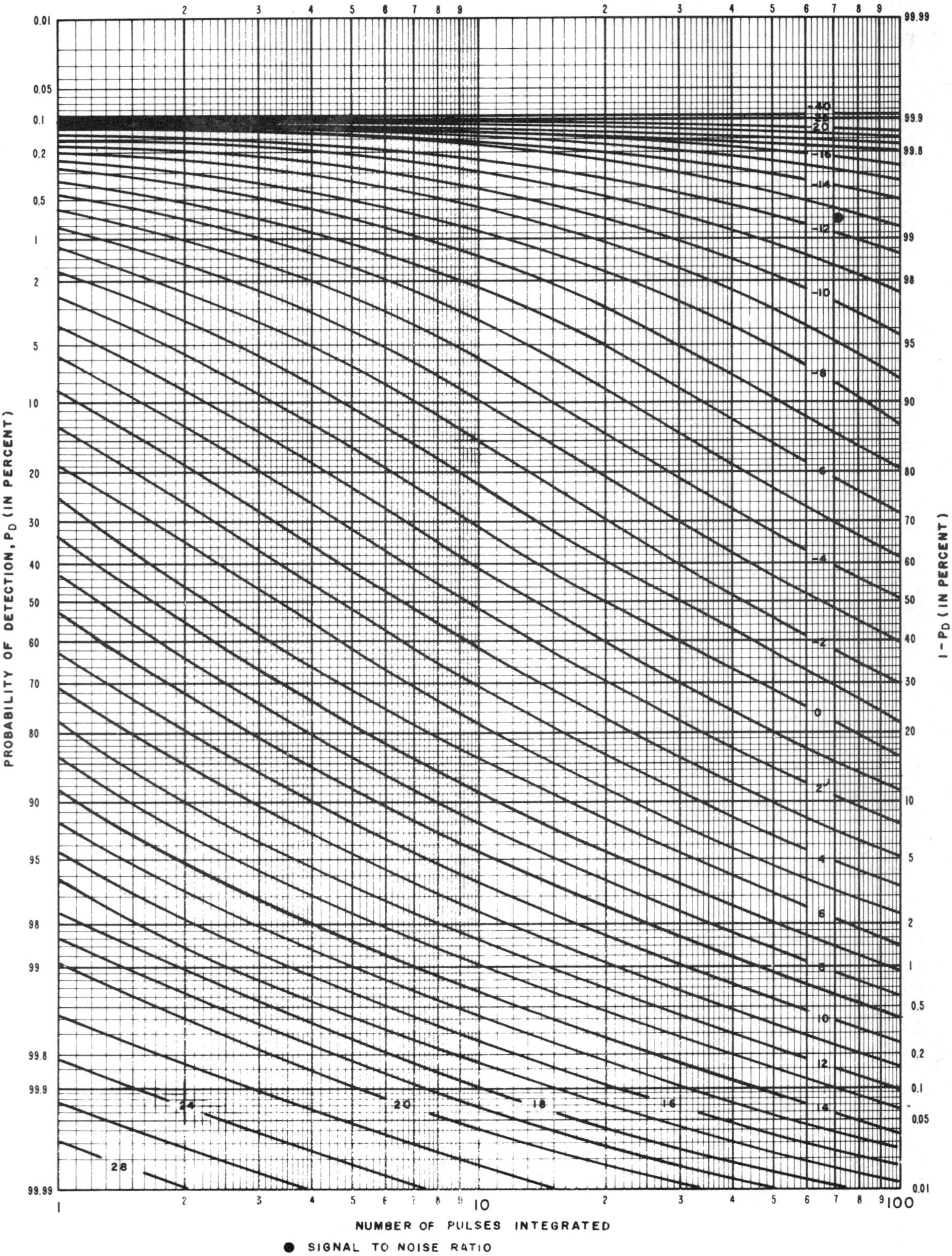

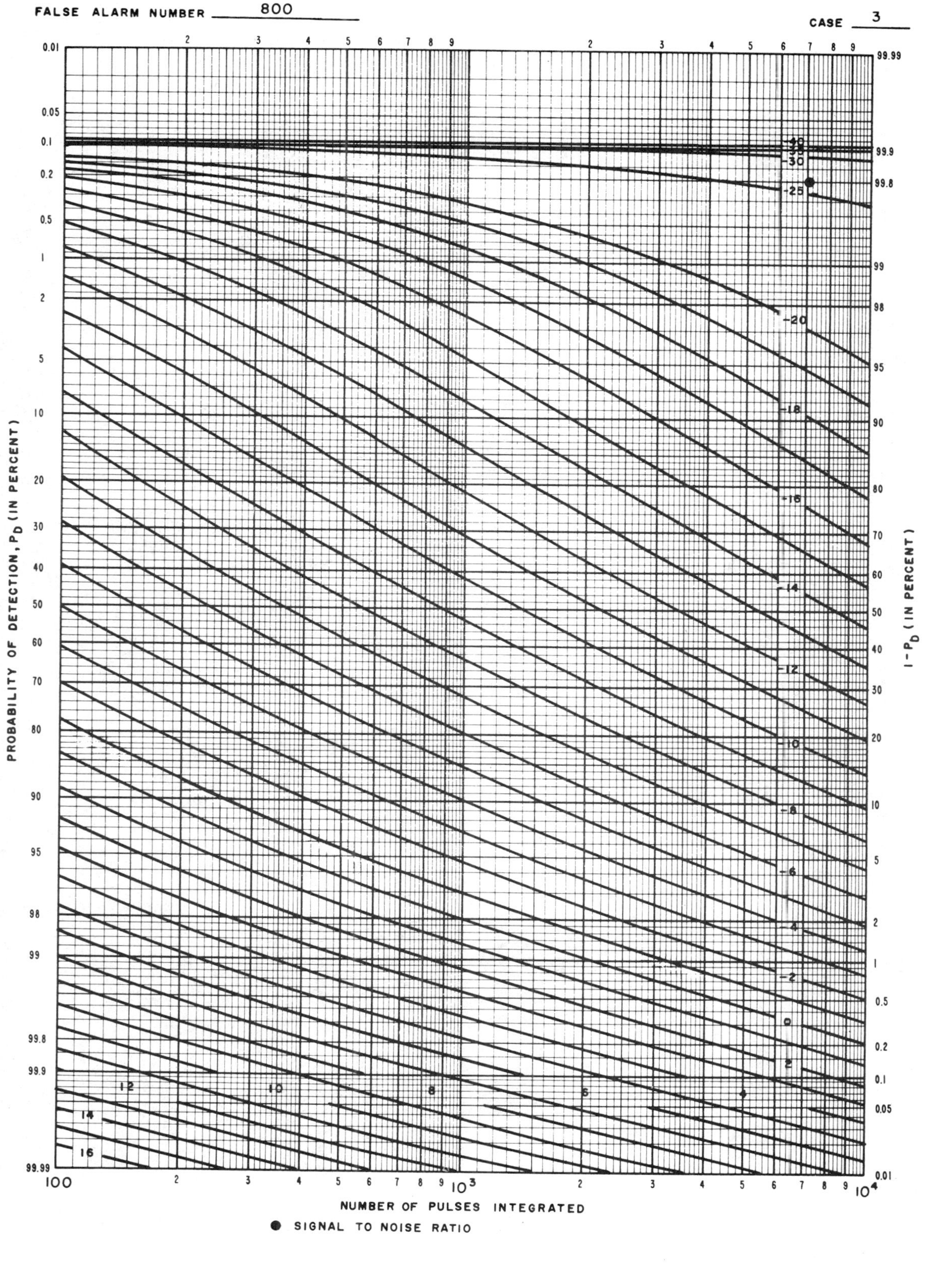
FALSE ALARM NUMBER 800
CASE 3
PROBABILITY OF DETECTION, P_D (IN PERCENT)
1 - P_D (IN PERCENT)
NUMBER OF PULSES INTEGRATED
SIGNAL TO NOISE RATIO
100
10^3
10^4
40
35
30
25
20
18
16
14
12
10
8
6
4
2
0
2
12
10
8
6
4
14
16

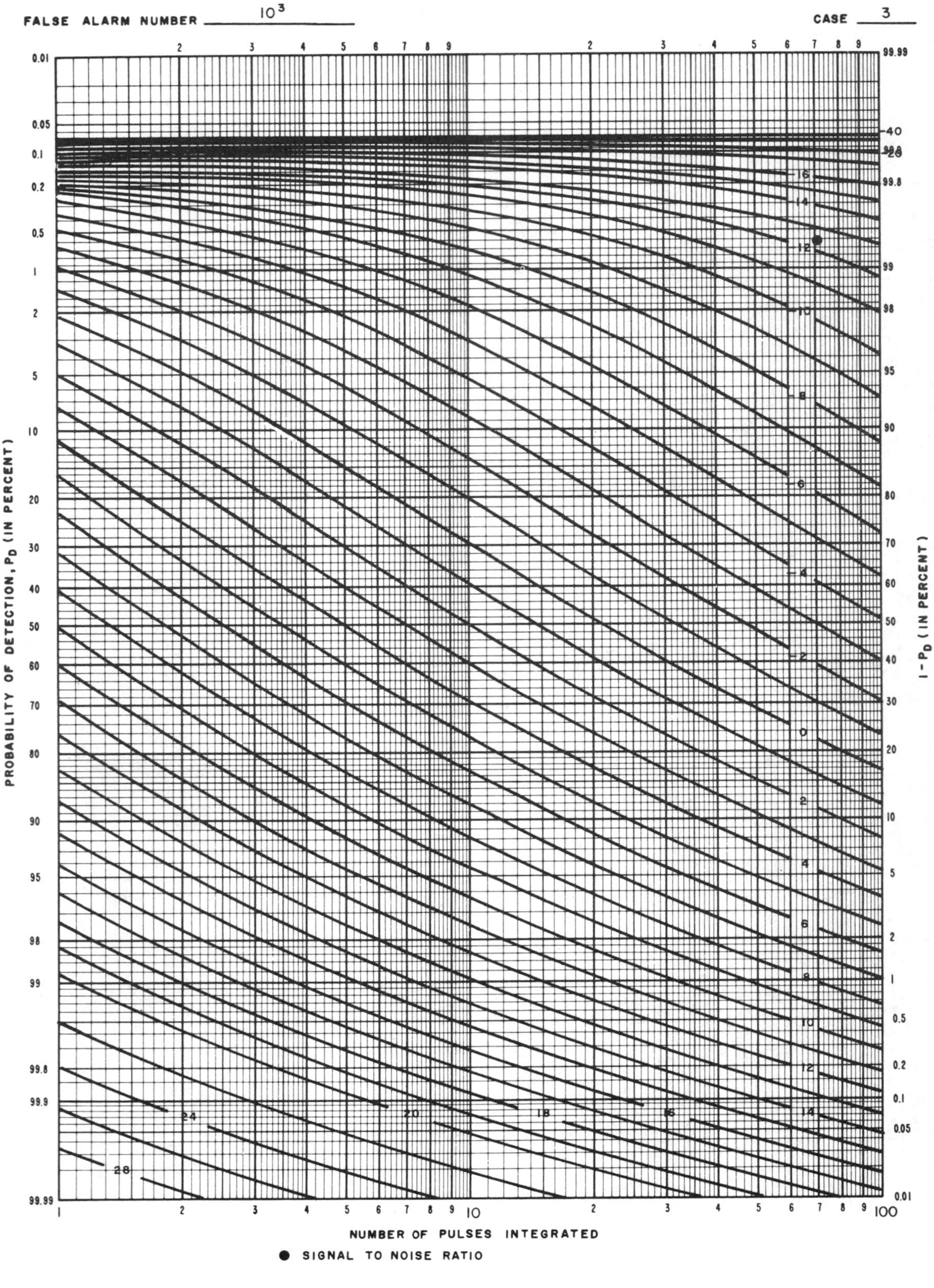
FALSE ALARM NUMBER 10^3
CASE 3
PROBABILITY OF DETECTION, P_D (IN PERCENT)
$1-P_D$ (IN PERCENT)
NUMBER OF PULSES INTEGRATED
● SIGNAL TO NOISE RATIO

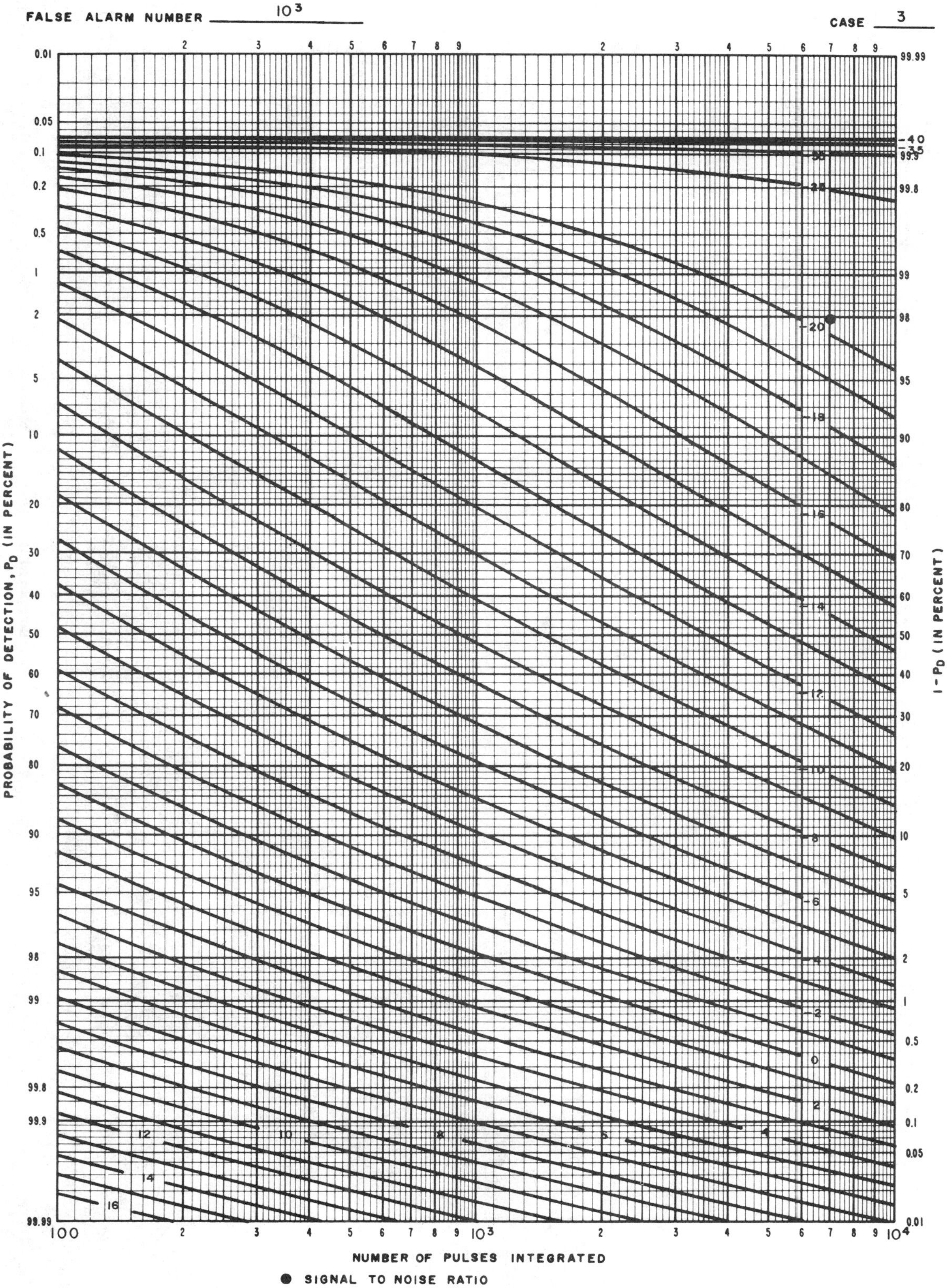
FALSE ALARM NUMBER 10^3
CASE 3
PROBABILITY OF DETECTION, P_D (IN PERCENT)
$1 - P_D$ (IN PERCENT)
NUMBER OF PULSES INTEGRATED
● SIGNAL TO NOISE RATIO

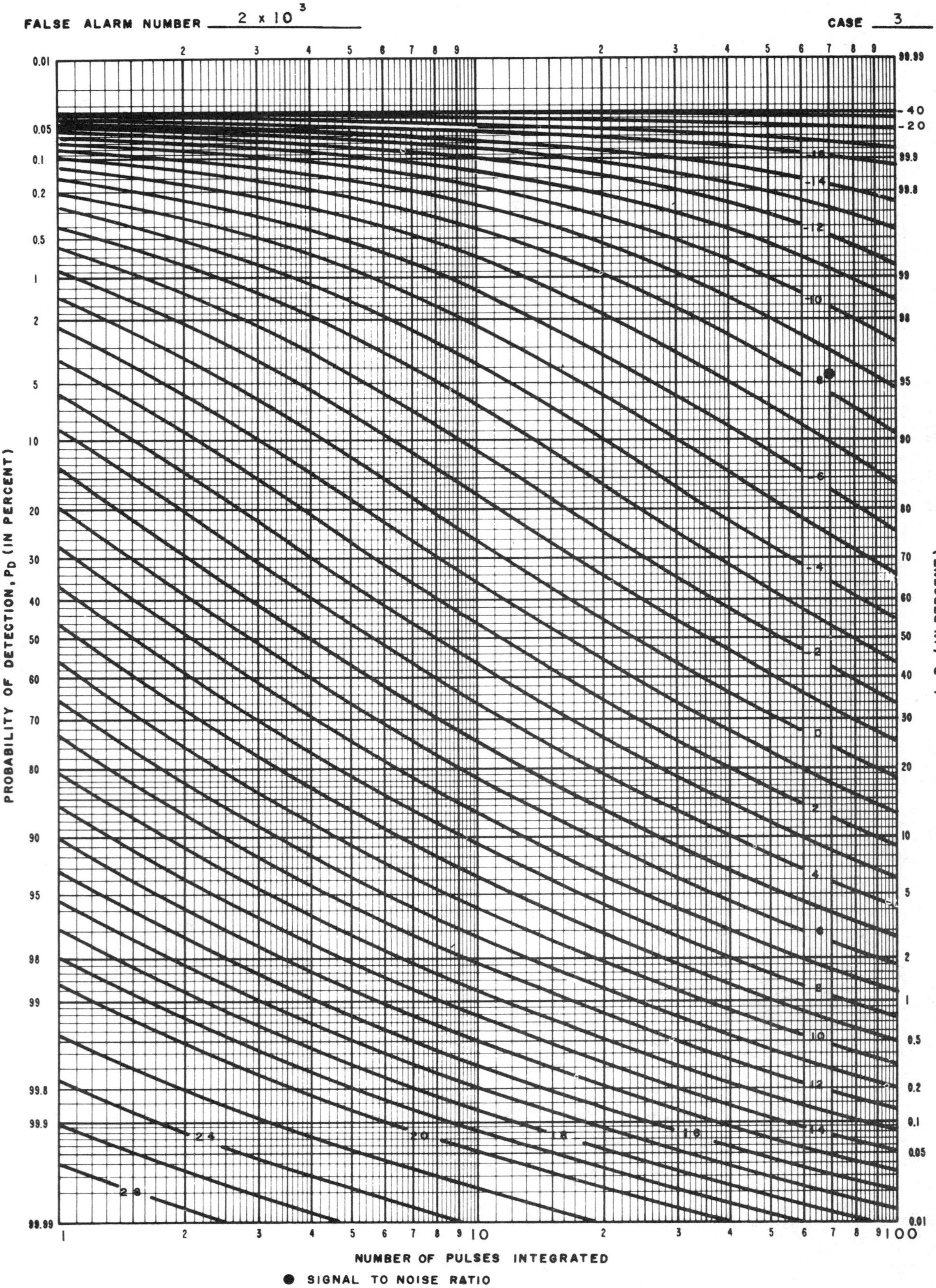
FALSE ALARM NUMBER 2 x 10^3
CASE 3
PROBABILITY OF DETECTION, P_D (IN PERCENT)
1 - P_D (IN PERCENT)
NUMBER OF PULSES INTEGRATED
● SIGNAL TO NOISE RATIO

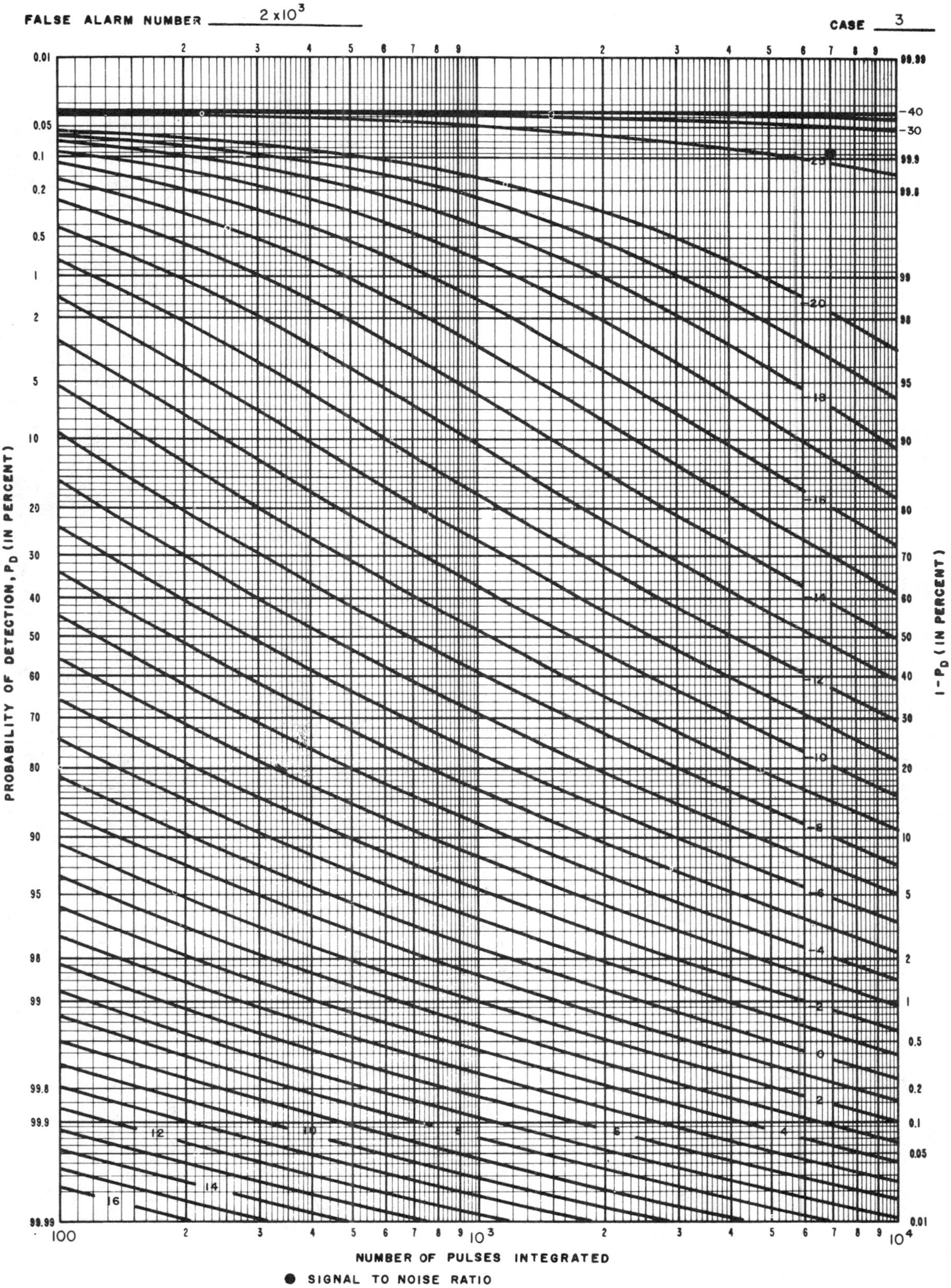
FALSE ALARM NUMBER 2×10^3
CASE 3
PROBABILITY OF DETECTION, P_D (IN PERCENT)
$1-P_D$ (IN PERCENT)
NUMBER OF PULSES INTEGRATED
● SIGNAL TO NOISE RATIO

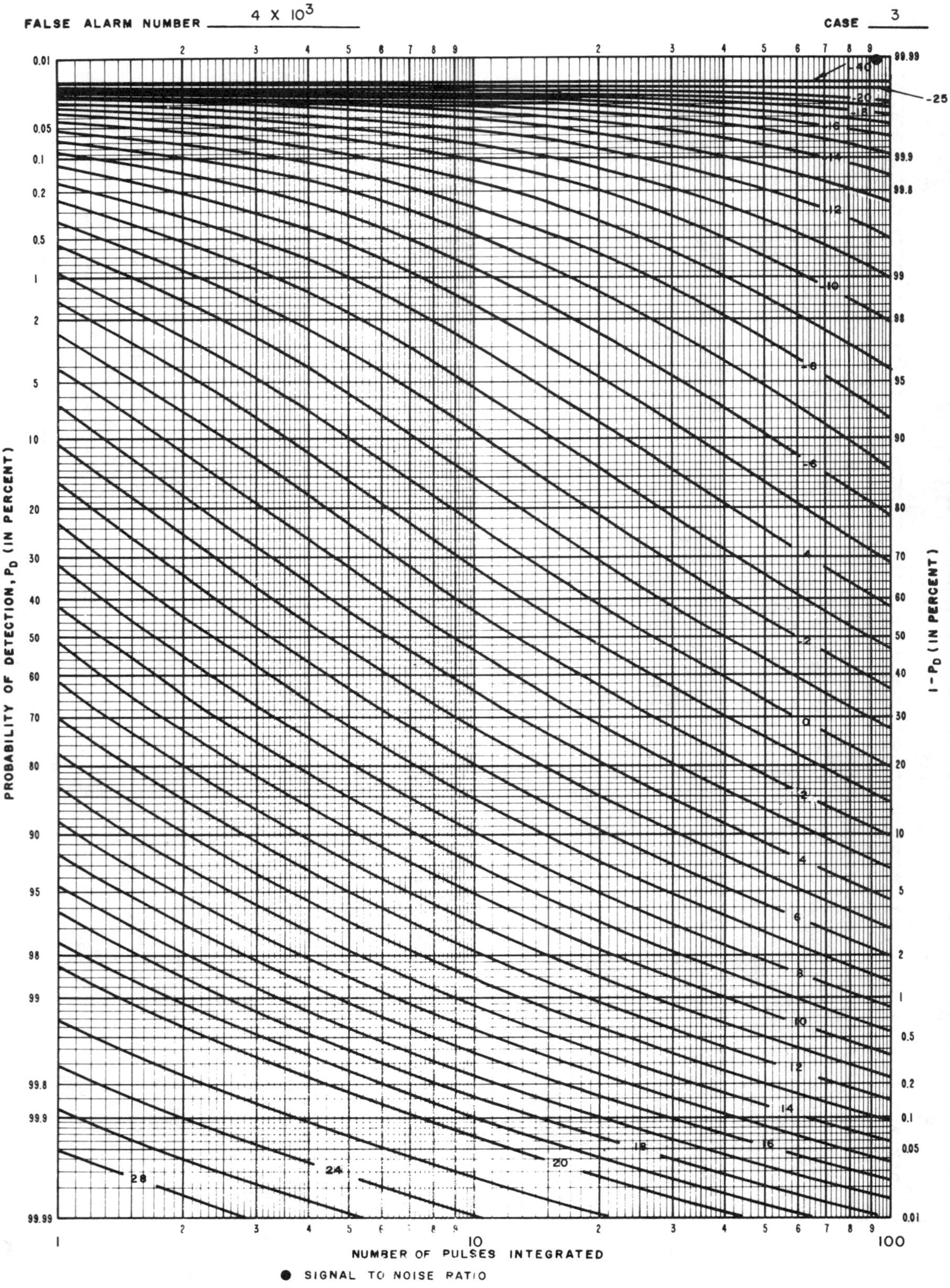
FALSE ALARM NUMBER 4 X 10^3
CASE 3
PROBABILITY OF DETECTION, P_D (IN PERCENT)
1-P_D (IN PERCENT)
NUMBER OF PULSES INTEGRATED
● SIGNAL TO NOISE RATIO
-25
28
24
20
18
16
14
12
10

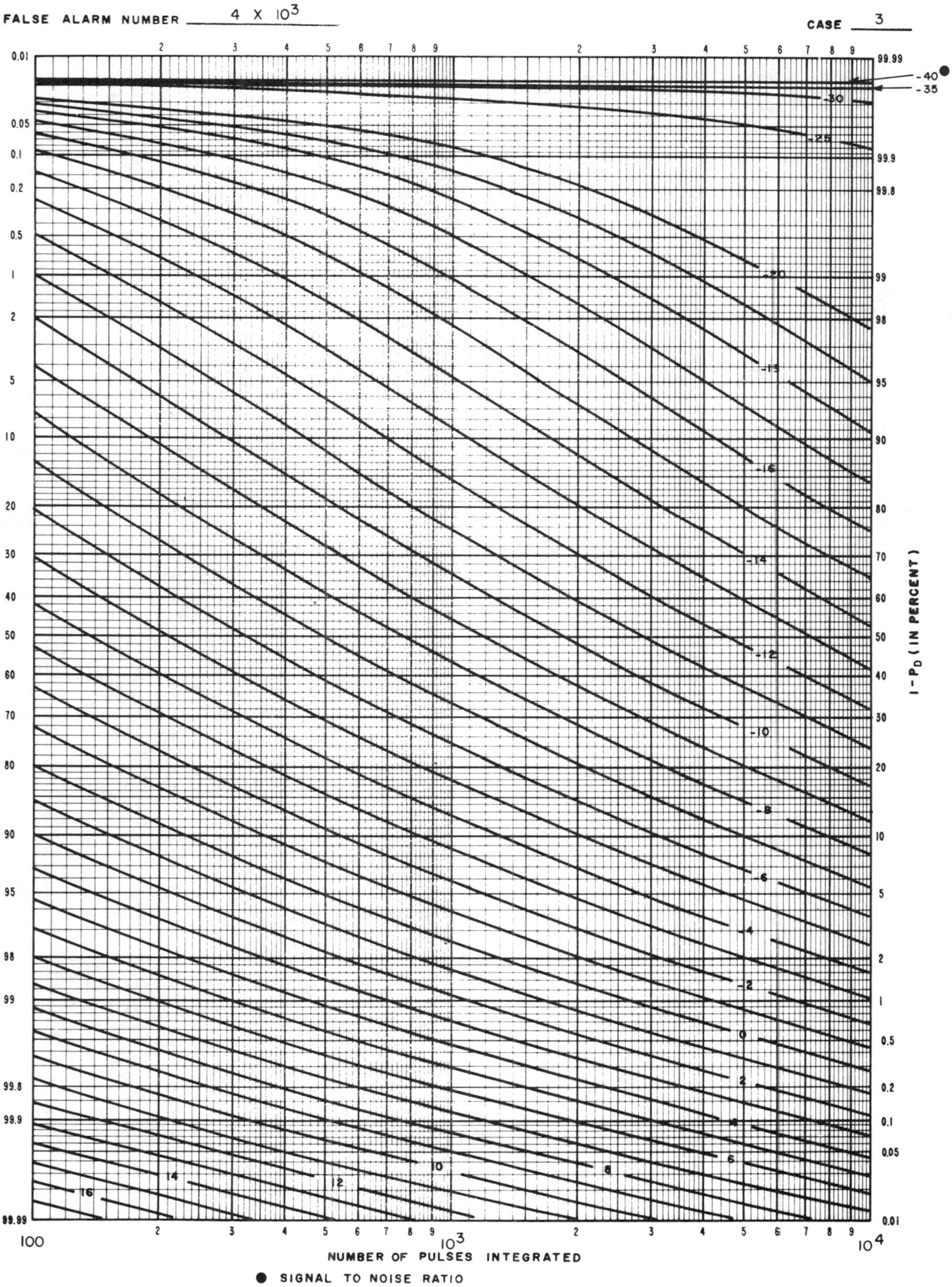
FALSE ALARM NUMBER 4 X 10^3
CASE 3
PROBABILITY OF DETECTION, P_D (IN PERCENT)
1 - P_D (IN PERCENT)
-40
-35
-30
-25
-20
-18
-16
-14
-12
-10
-8
-6
-4
-2
0
2
4
6
8
10
12
14
16
100
10^3
10^4
NUMBER OF PULSES INTEGRATED
● SIGNAL TO NOISE RATIO

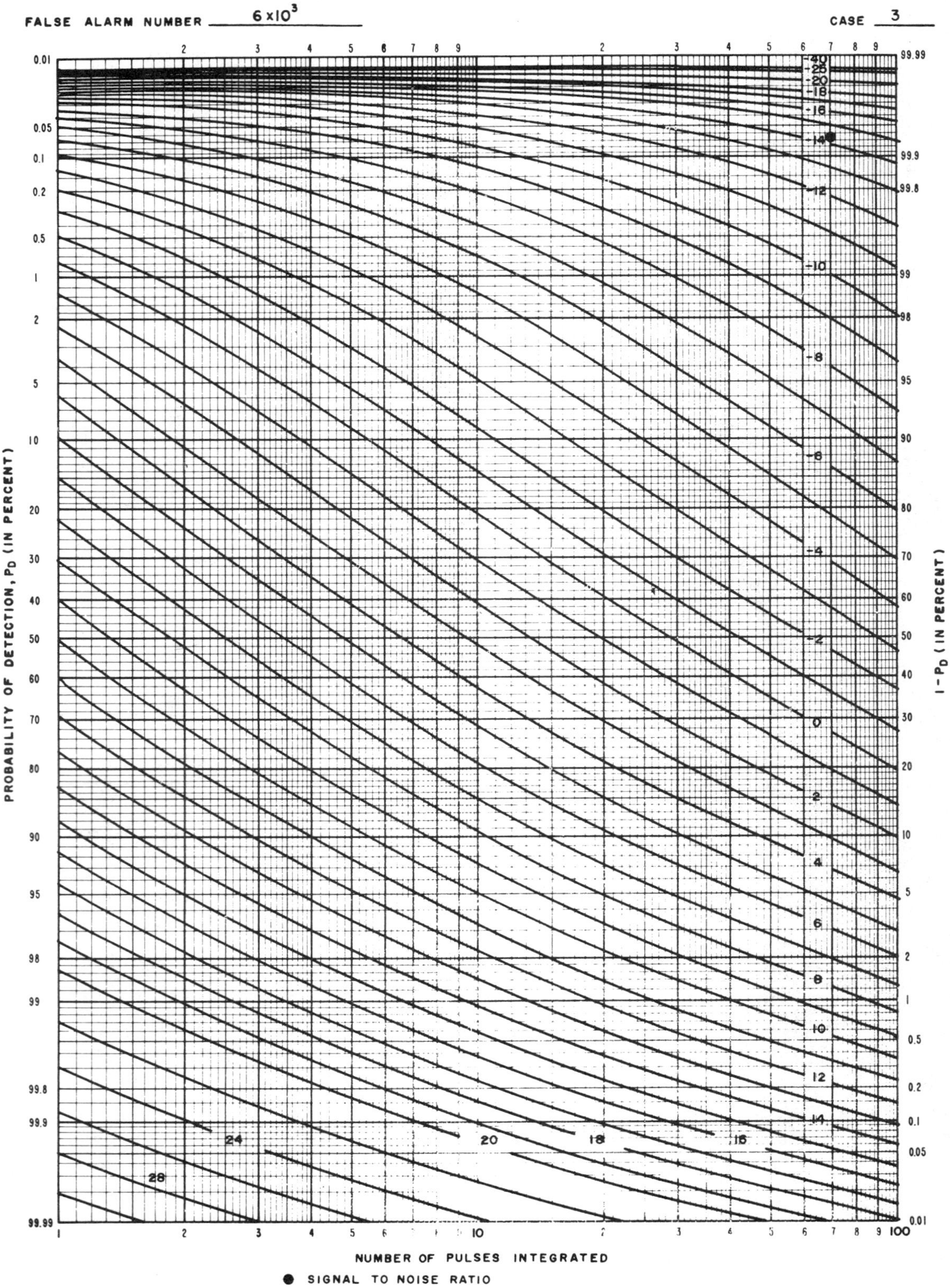
FALSE ALARM NUMBER 6 x10^3
CASE 3
PROBABILITY OF DETECTION, P_D (IN PERCENT)
1 - P_D (IN PERCENT)
NUMBER OF PULSES INTEGRATED
● SIGNAL TO NOISE RATIO
-40
-25
-20
-18
-16
-14
-12
-10
-8
-6
-4
-2
0
2
4
6
8
10
12
14
16
18
20
24
28

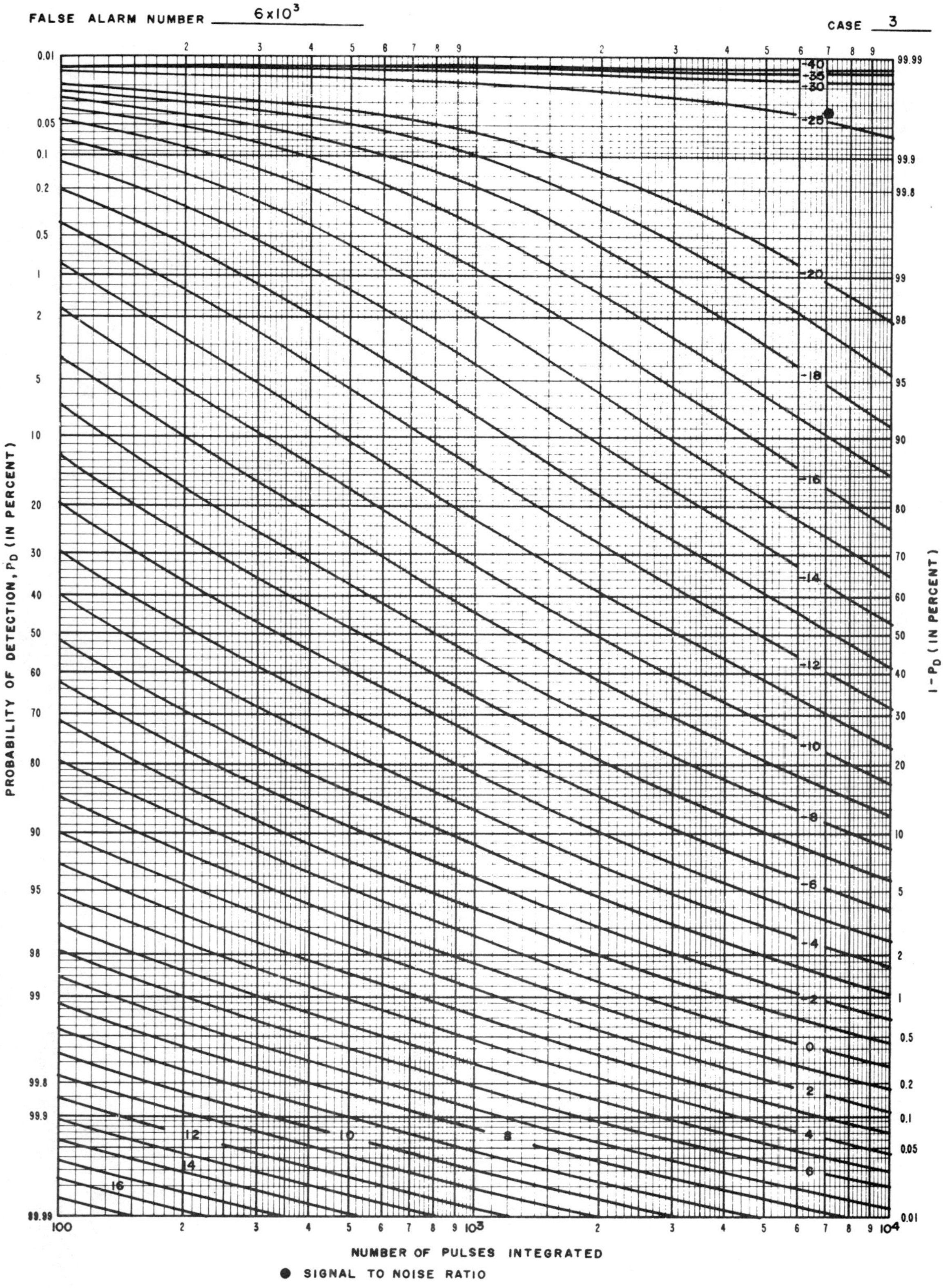
FALSE ALARM NUMBER 6x10³
CASE 3
PROBABILITY OF DETECTION, P_D (IN PERCENT)
1-P_D (IN PERCENT)
NUMBER OF PULSES INTEGRATED
● SIGNAL TO NOISE RATIO

FALSE ALARM NUMBER 8×10^3 CASE 3

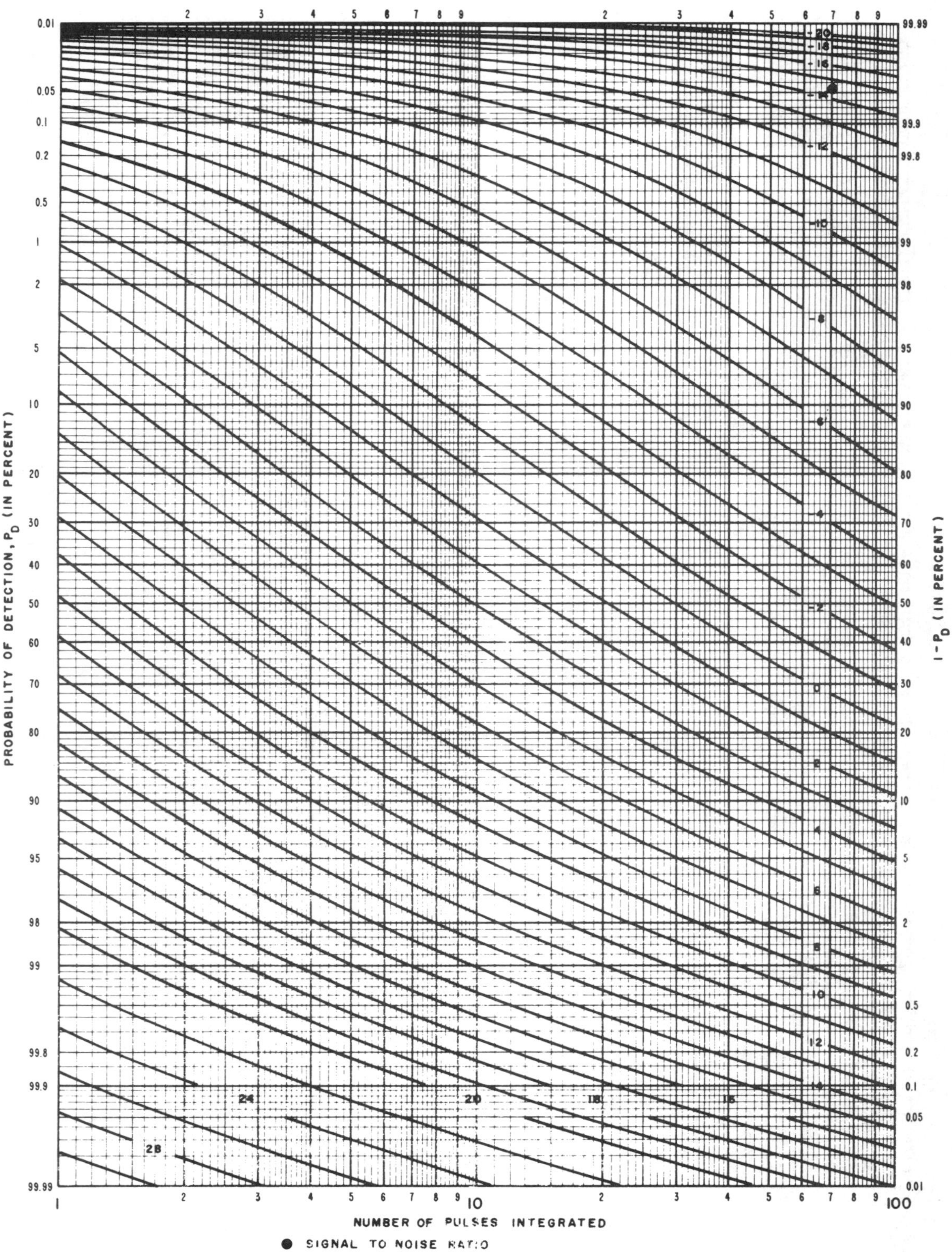

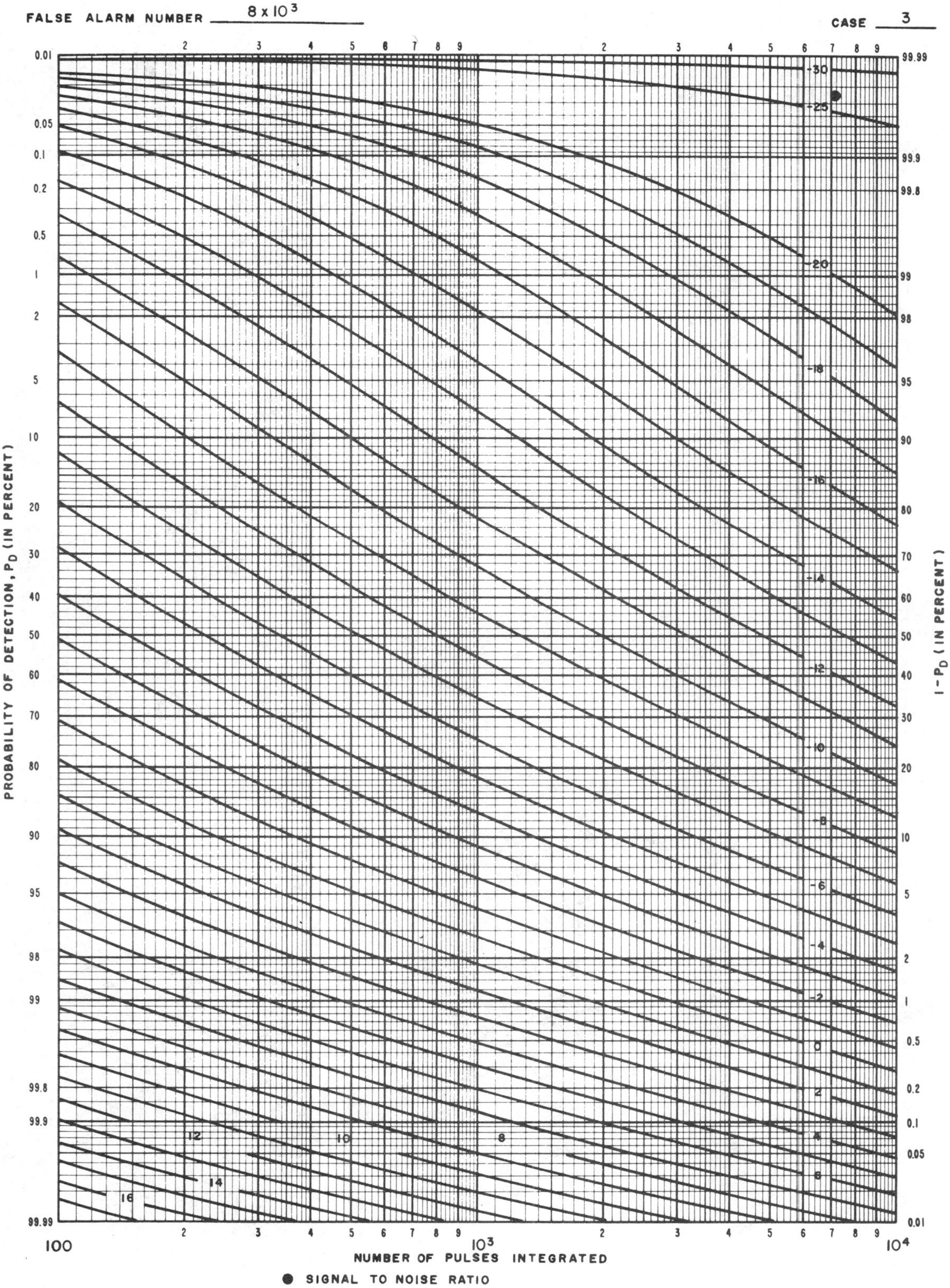
FALSE ALARM NUMBER 8 x 10³
CASE 3
PROBABILITY OF DETECTION, P_D (IN PERCENT)
1 - P_D (IN PERCENT)
NUMBER OF PULSES INTEGRATED
100
10³
10⁴
● SIGNAL TO NOISE RATIO
-30
-25
-20
-18
-16
-14
-12
-10
-8
-6
-4
-2
0
2
4
6
8
10
12
14
16

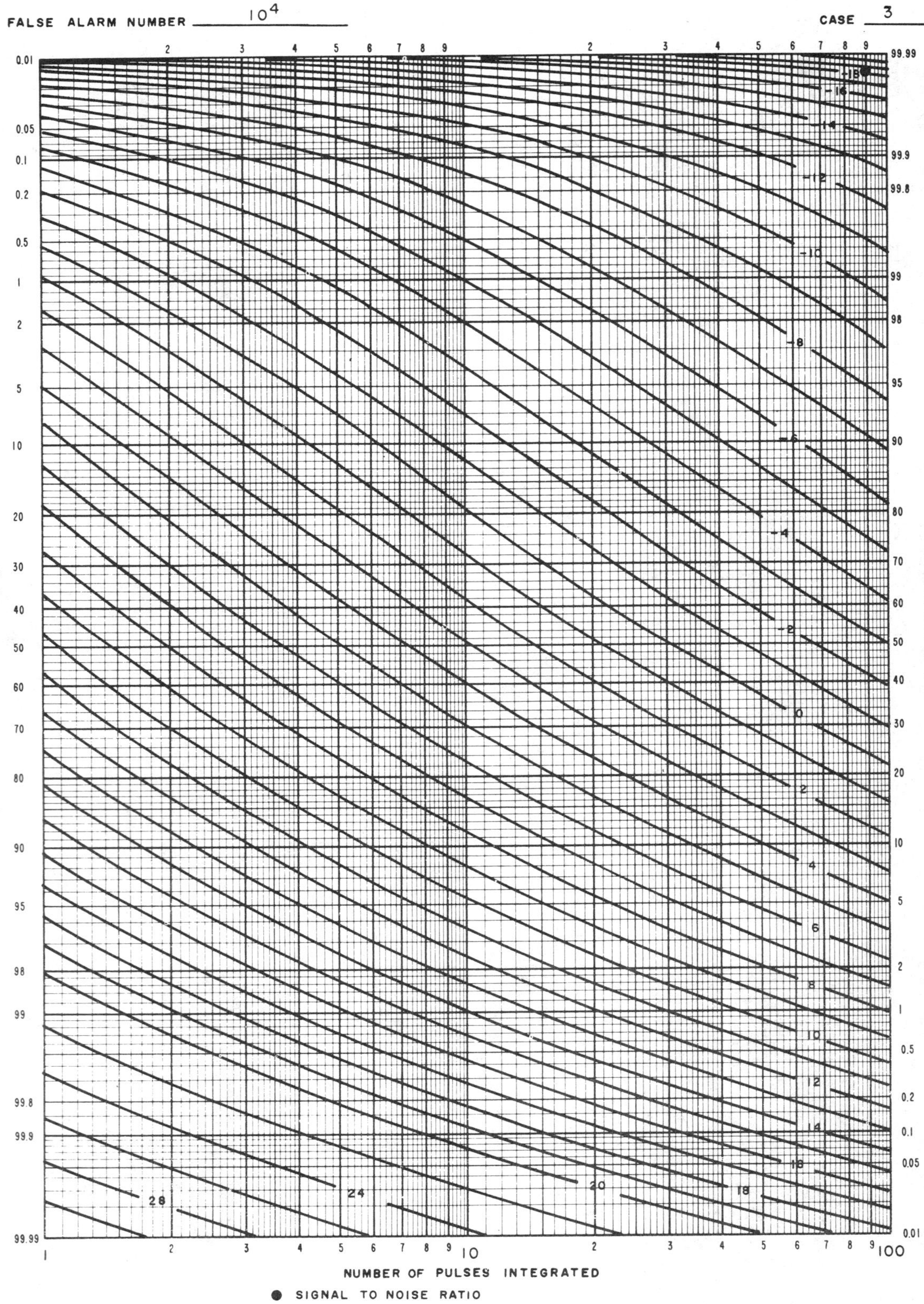
FALSE ALARM NUMBER 10^4
CASE 3
PROBABILITY OF DETECTION, P_D (IN PERCENT)
1 - P_D (IN PERCENT)
NUMBER OF PULSES INTEGRATED
SIGNAL TO NOISE RATIO
0.01
0.05
0.1
0.2
0.5
1
2
5
10
20
30
40
50
60
70
80
90
95
98
99
99.8
99.9
99.99
-18
-16
-14
-12
-10
-8
-6
-4
-2
0
2
4
6
8
10
12
14
16
18
20
24
28
1
10
100

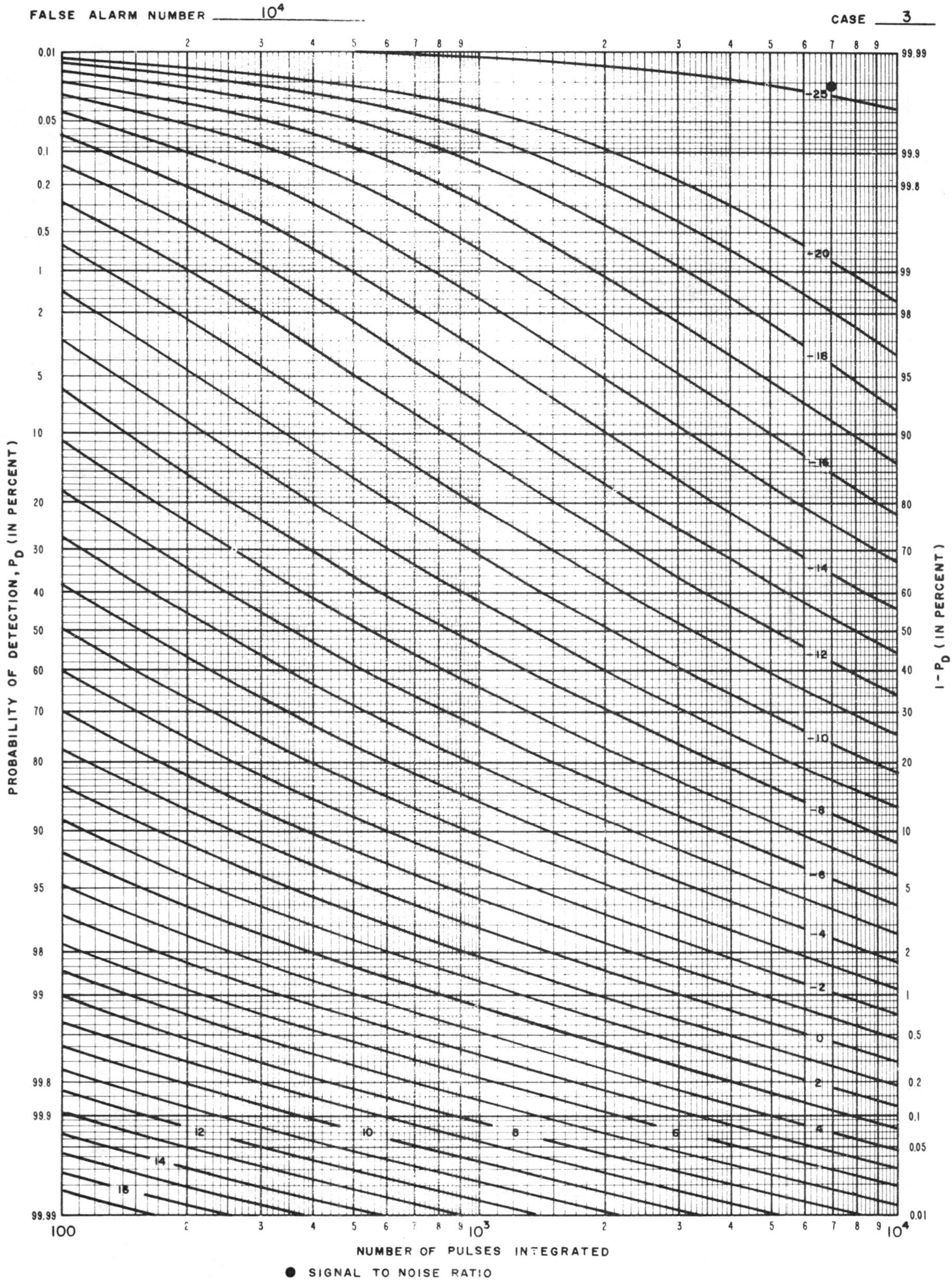
FALSE ALARM NUMBER 10^4
CASE 3
PROBABILITY OF DETECTION, P_D (IN PERCENT)
1 - P_D (IN PERCENT)
NUMBER OF PULSES INTEGRATED
● SIGNAL TO NOISE RATIO
-25
-20
-18
-16
-14
-12
-10
-8
-6
-4
-2
0
2
4
6
8
10
12
14
16

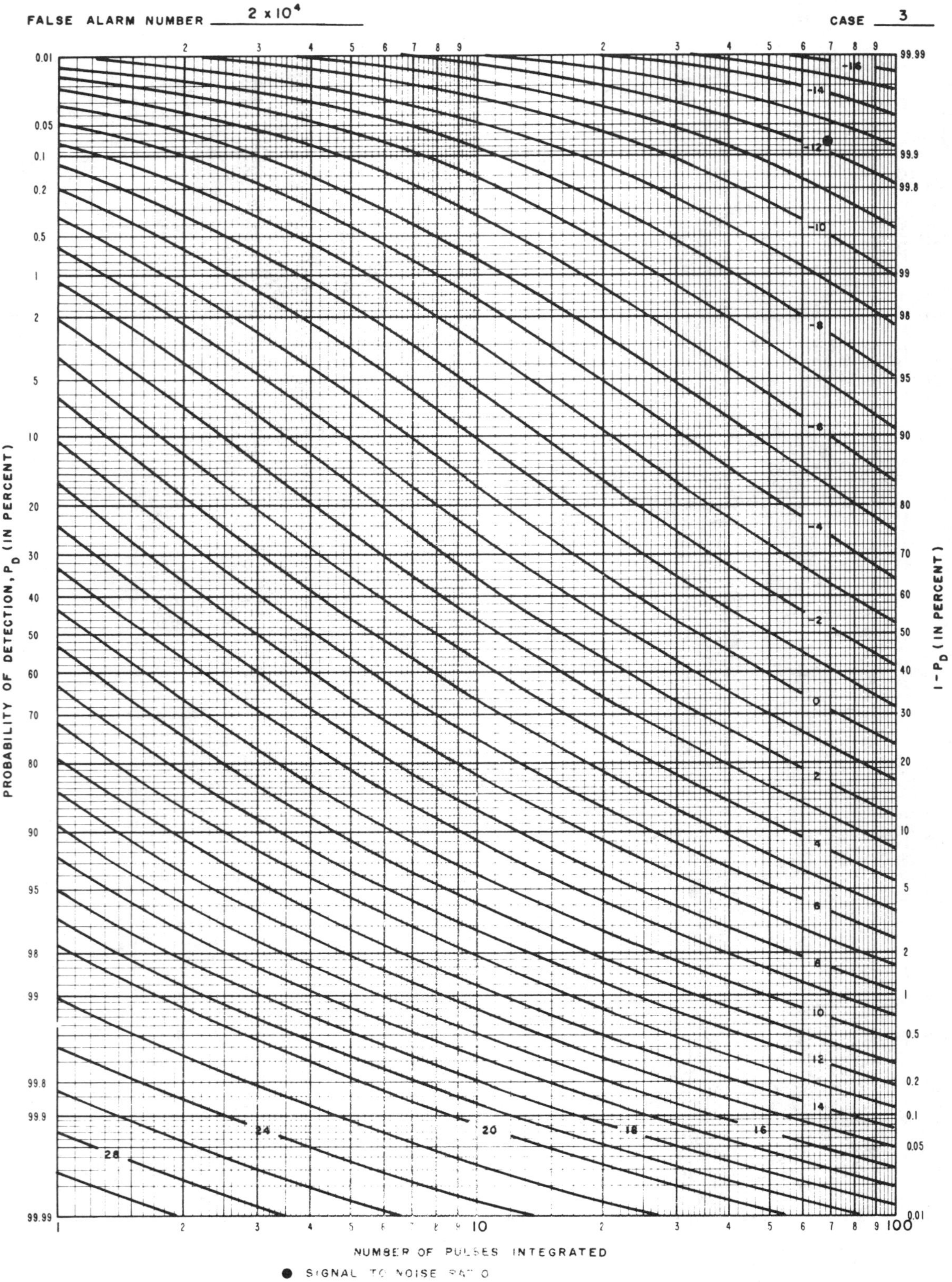
FALSE ALARM NUMBER 2 x 10⁴
CASE 3
PROBABILITY OF DETECTION, P_D (IN PERCENT)
1 - P_D (IN PERCENT)
NUMBER OF PULSES INTEGRATED
● SIGNAL TO NOISE RATIO
-16
-14
-12
-10
-8
-6
-4
-2
0
2
4
6
8
10
12
14
16
18
20
24
28

FALSE ALARM NUMBER 2 x 10^4

CASE 3

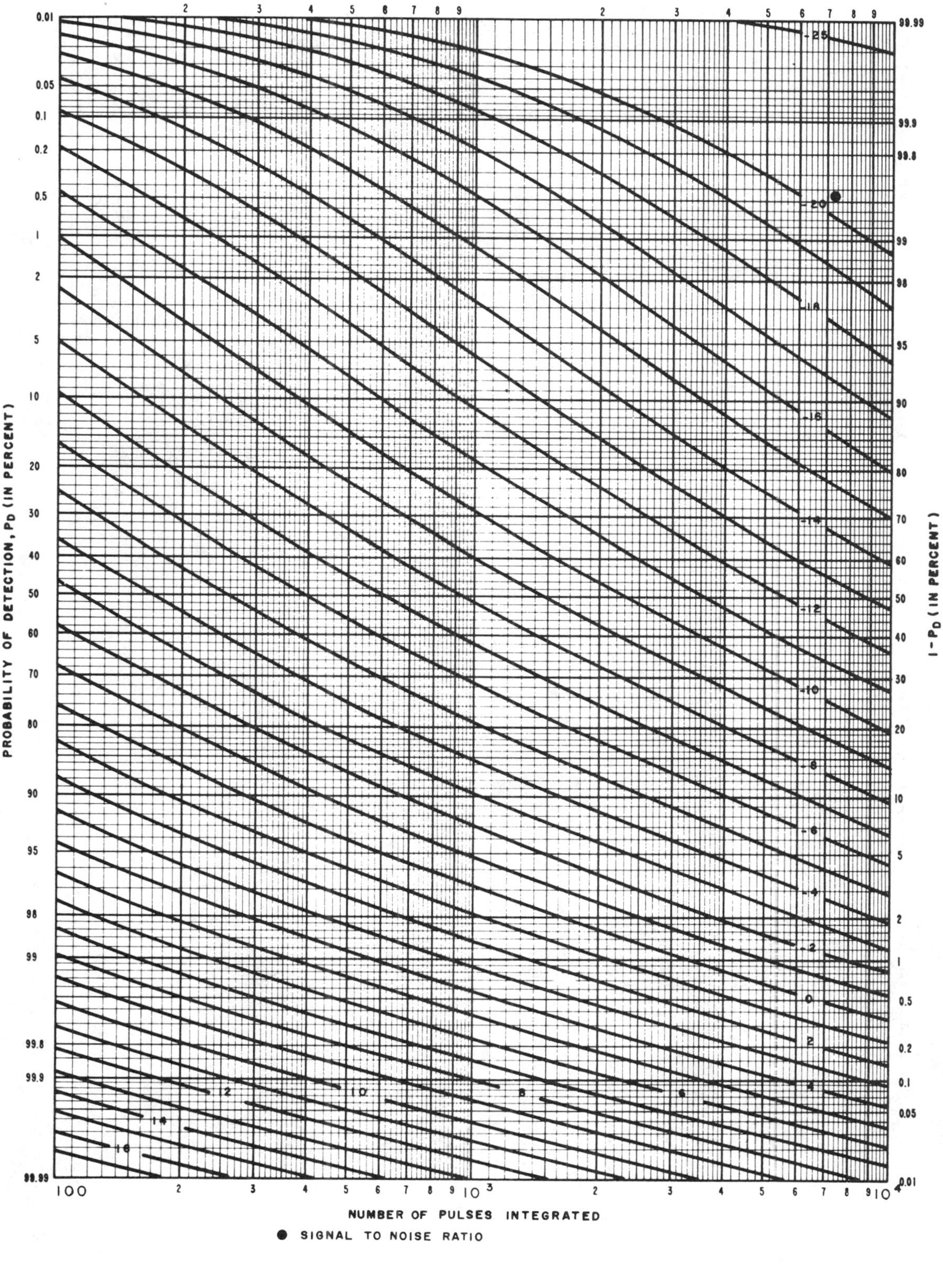

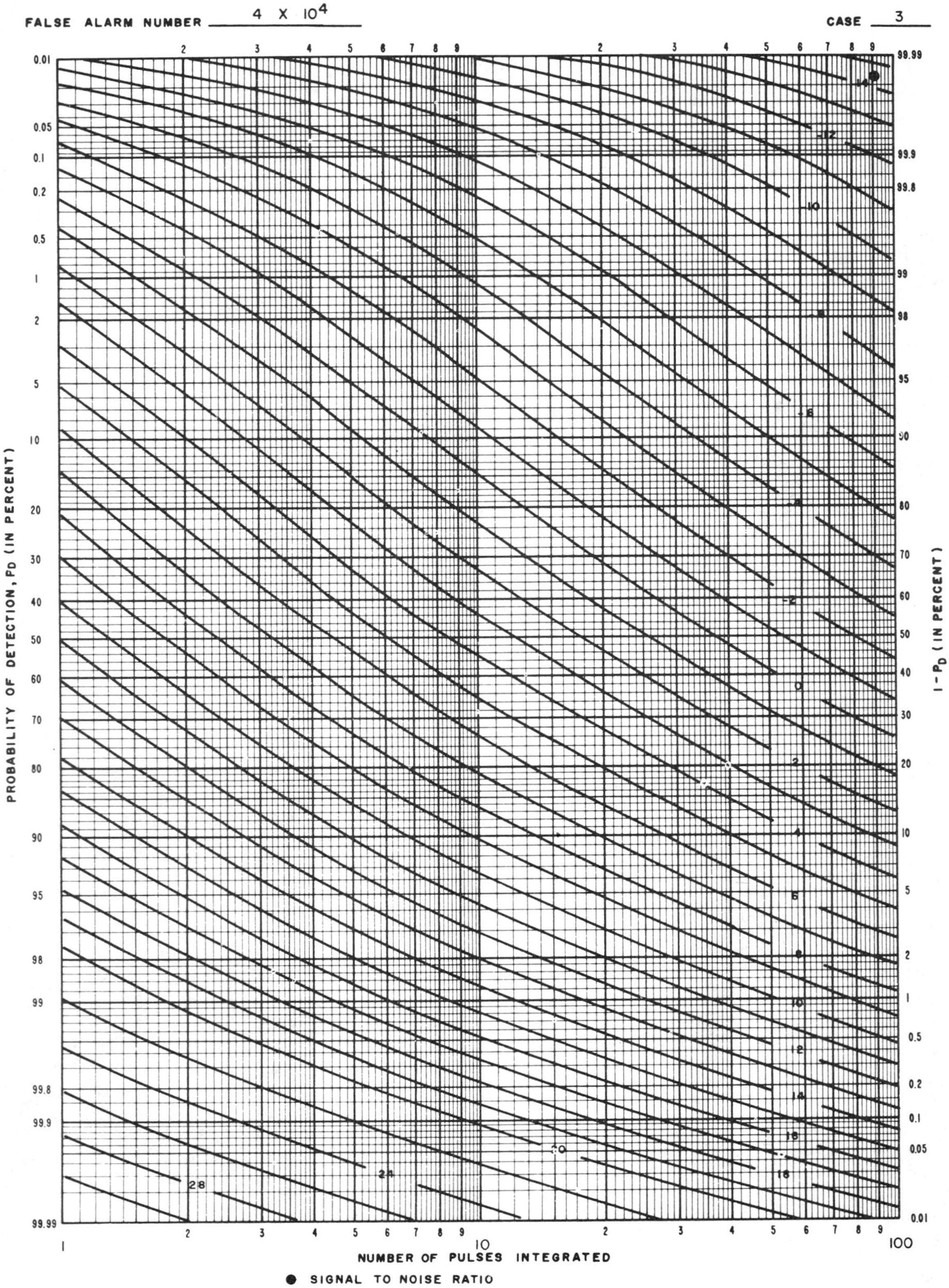
FALSE ALARM NUMBER 4 X 10^4
CASE 3
PROBABILITY OF DETECTION, P_D (IN PERCENT)
1 - P_D (IN PERCENT)
NUMBER OF PULSES INTEGRATED
● SIGNAL TO NOISE RATIO

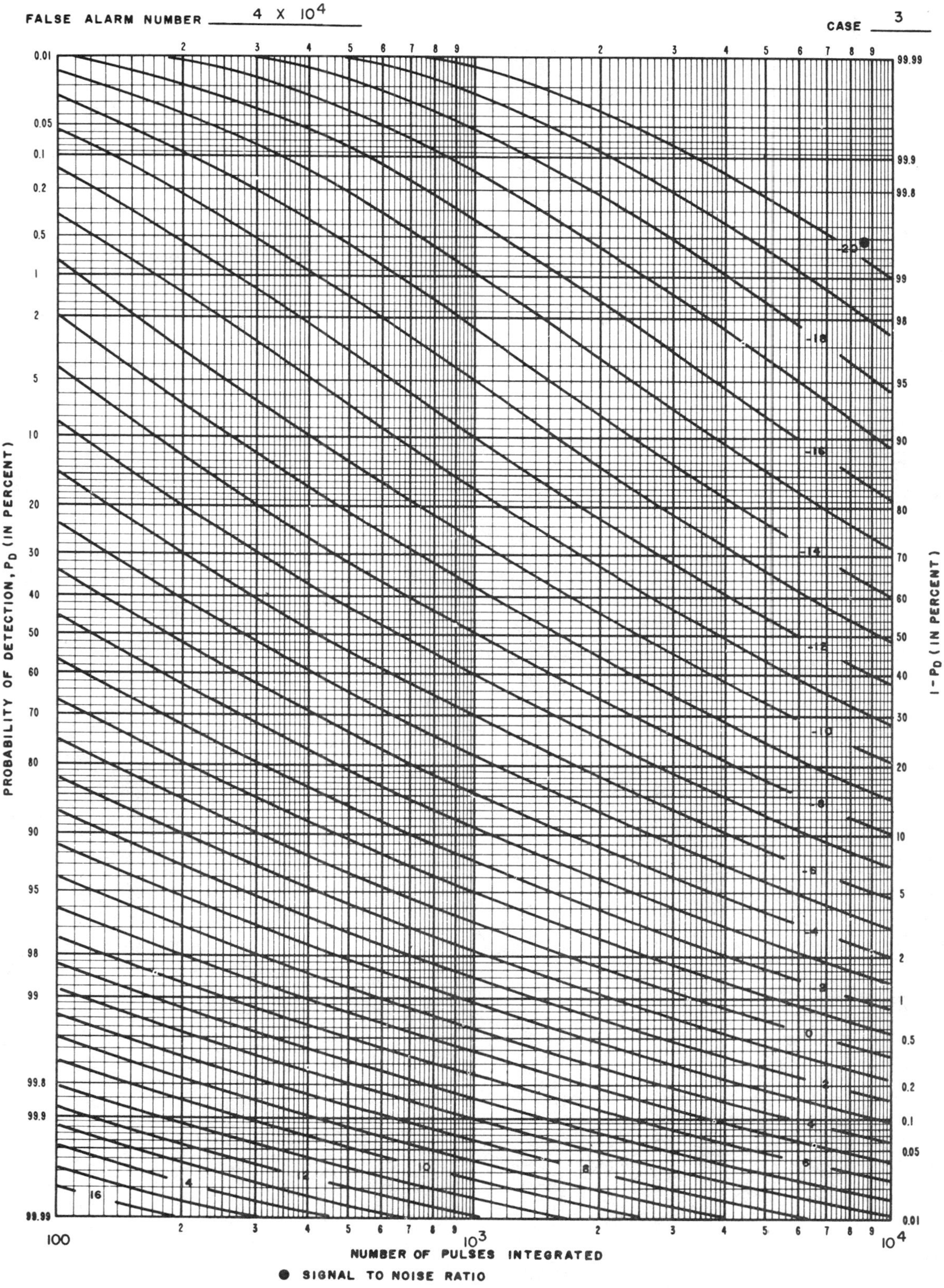
FALSE ALARM NUMBER 4 X 10^4
CASE 3
PROBABILITY OF DETECTION, P_D (IN PERCENT)
1 - P_D (IN PERCENT)
NUMBER OF PULSES INTEGRATED
100
10^3
10^4
● SIGNAL TO NOISE RATIO

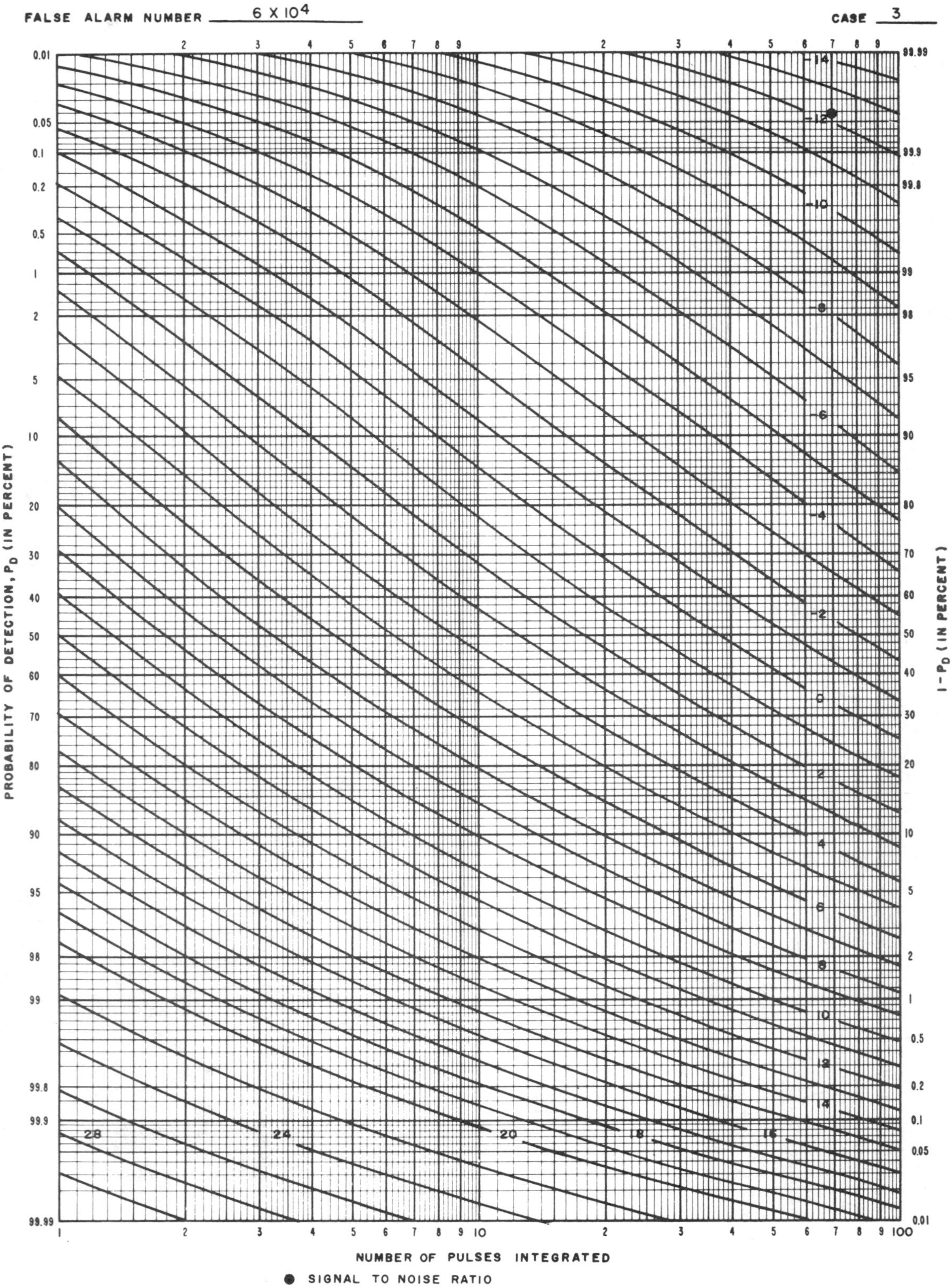
FALSE ALARM NUMBER 6 X 10^4
CASE 3
PROBABILITY OF DETECTION, P_D (IN PERCENT)
1 - P_D (IN PERCENT)
NUMBER OF PULSES INTEGRATED
SIGNAL TO NOISE RATIO
-14
-12
-10
-8
-6
-4
-2
0
2
4
6
8
10
12
14
16
18
20
24
28

FALSE ALARM NUMBER 6 X 10^4 CASE 3

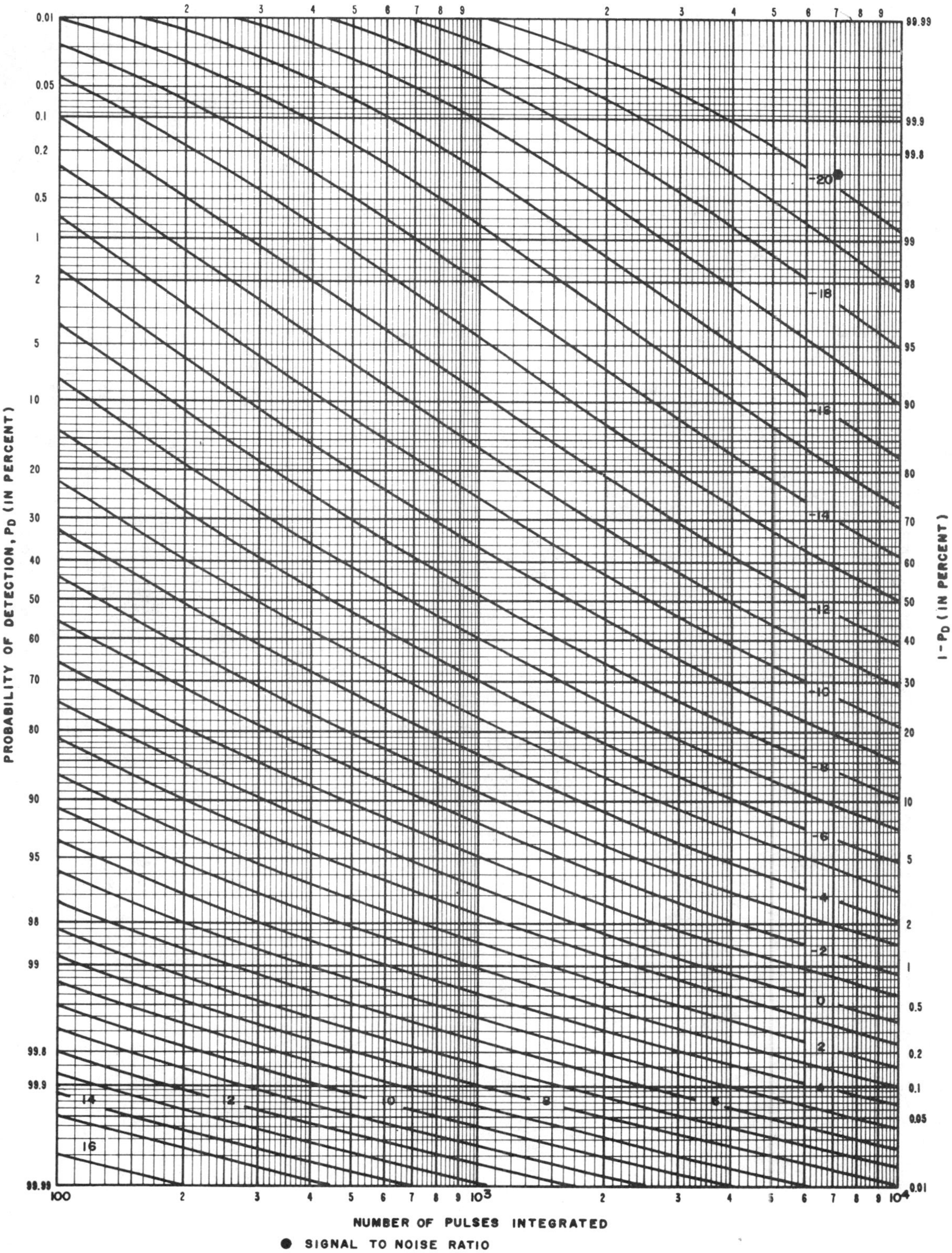

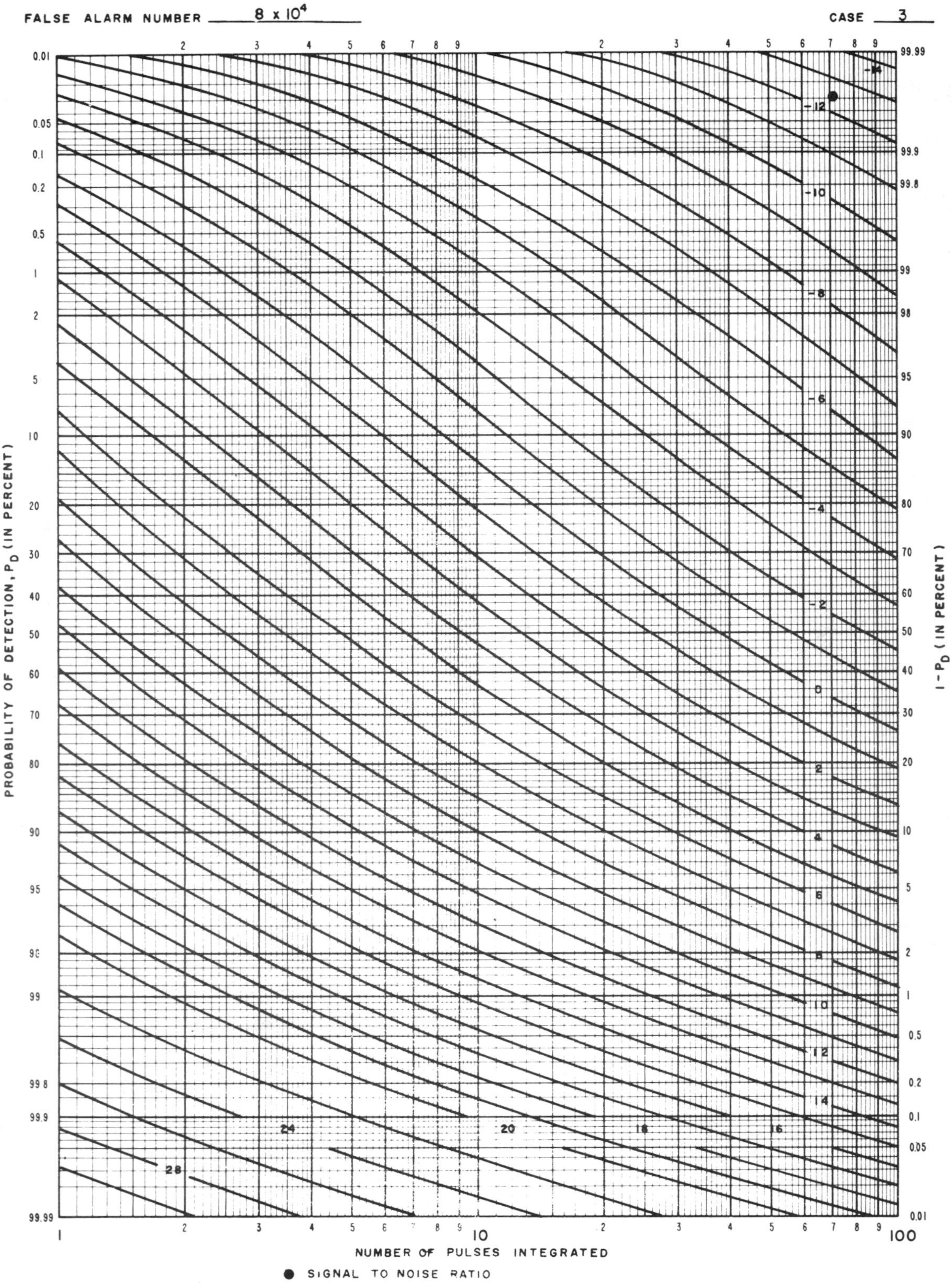
FALSE ALARM NUMBER 8 x 10^4
CASE 3
PROBABILITY OF DETECTION, P_D (IN PERCENT)
1 - P_D (IN PERCENT)
NUMBER OF PULSES INTEGRATED
● SIGNAL TO NOISE RATIO
-14
-12
-10
-8
-6
-4
-2
0
2
4
6
8
10
12
14
16
18
20
24
28

FALSE ALARM NUMBER 8 x 10^4 CASE 3

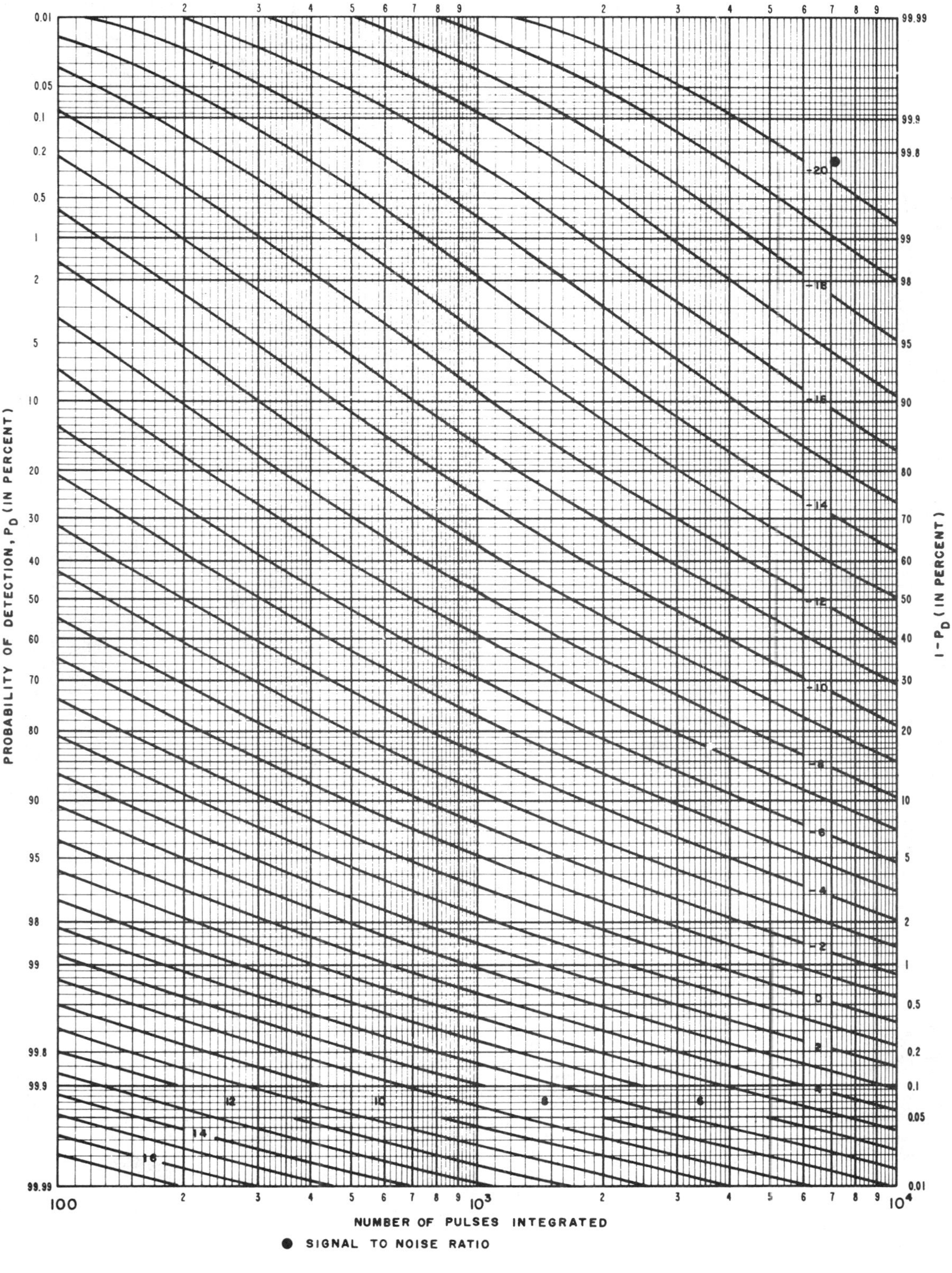

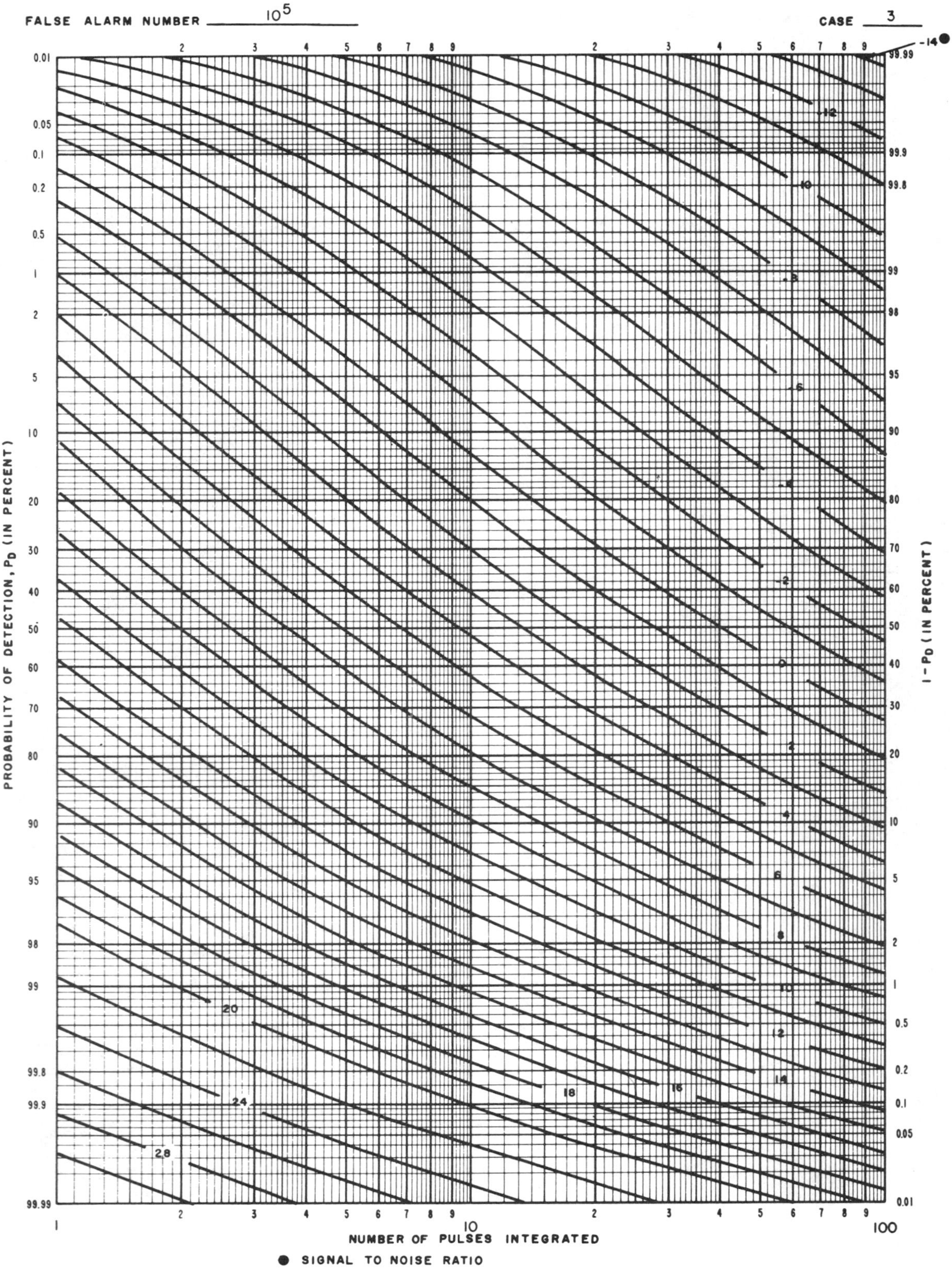
FALSE ALARM NUMBER 10^5
CASE 3
PROBABILITY OF DETECTION, P_D (IN PERCENT)
1 − P_D (IN PERCENT)
NUMBER OF PULSES INTEGRATED
● SIGNAL TO NOISE RATIO
-14
-12
-10
-8
-6
-4
-2
0
2
4
6
8
10
12
14
16
18
20
24
28

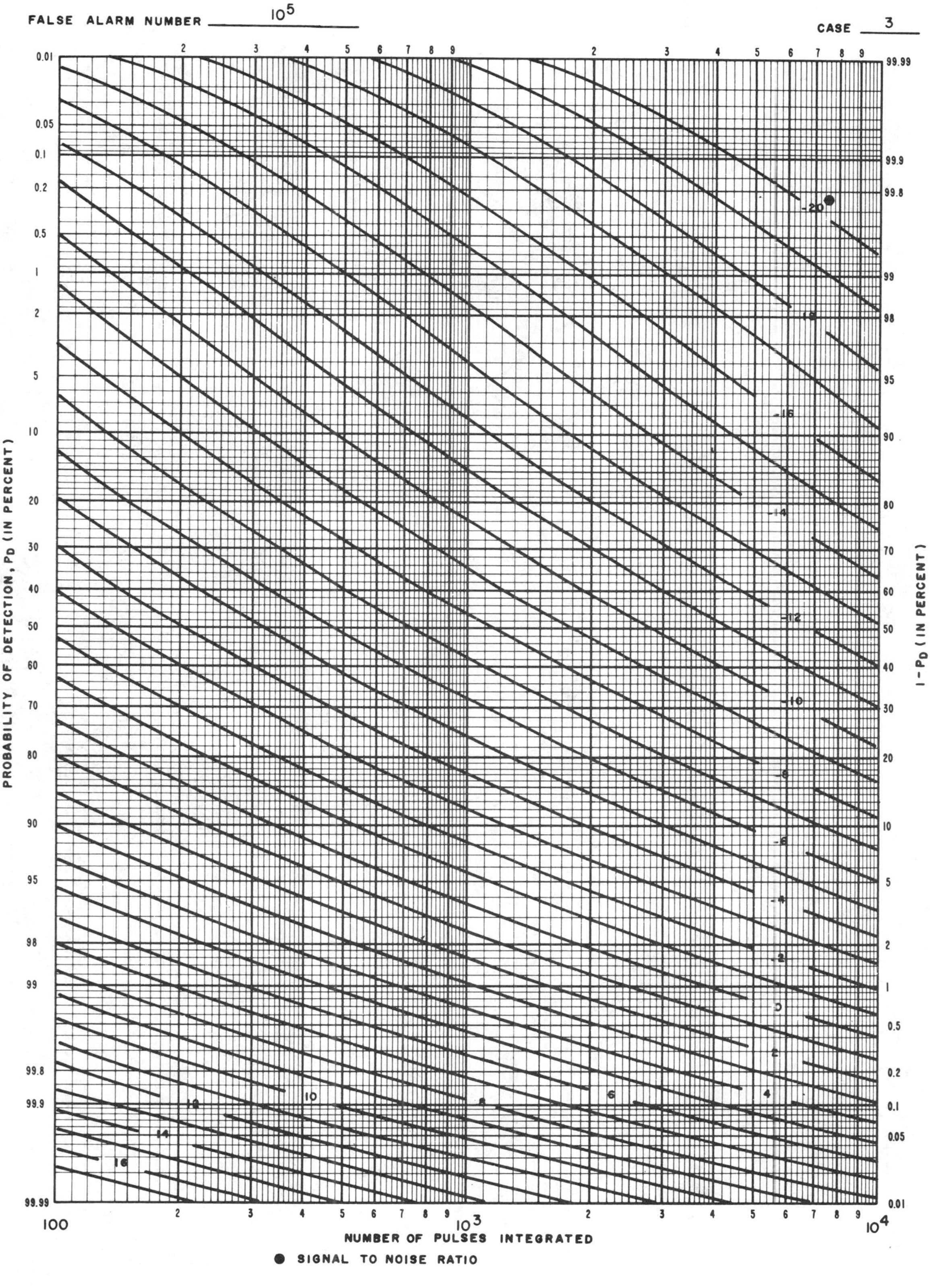
FALSE ALARM NUMBER 10^5
CASE 3
PROBABILITY OF DETECTION, P_D (IN PERCENT)
1 - P_D (IN PERCENT)
NUMBER OF PULSES INTEGRATED
● SIGNAL TO NOISE RATIO
100
10^3
10^4
-20
-18
-16
-14
-12
-10
-8
-6
-4
-2
0
2
4
6
8
10
12
14
16

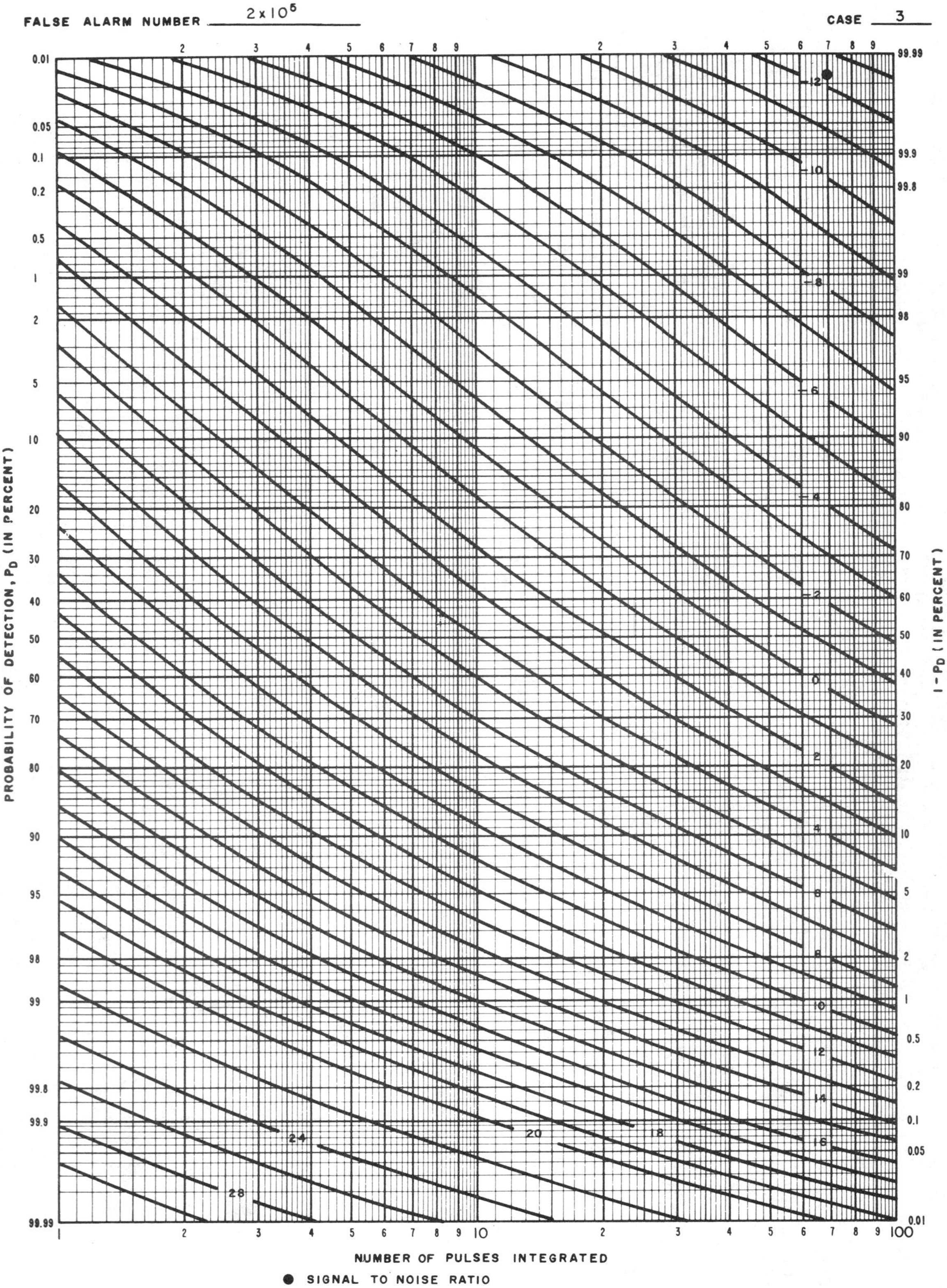
FALSE ALARM NUMBER 2x10^5
CASE 3
PROBABILITY OF DETECTION, P_D (IN PERCENT)
1 - P_D (IN PERCENT)
NUMBER OF PULSES INTEGRATED
● SIGNAL TO NOISE RATIO

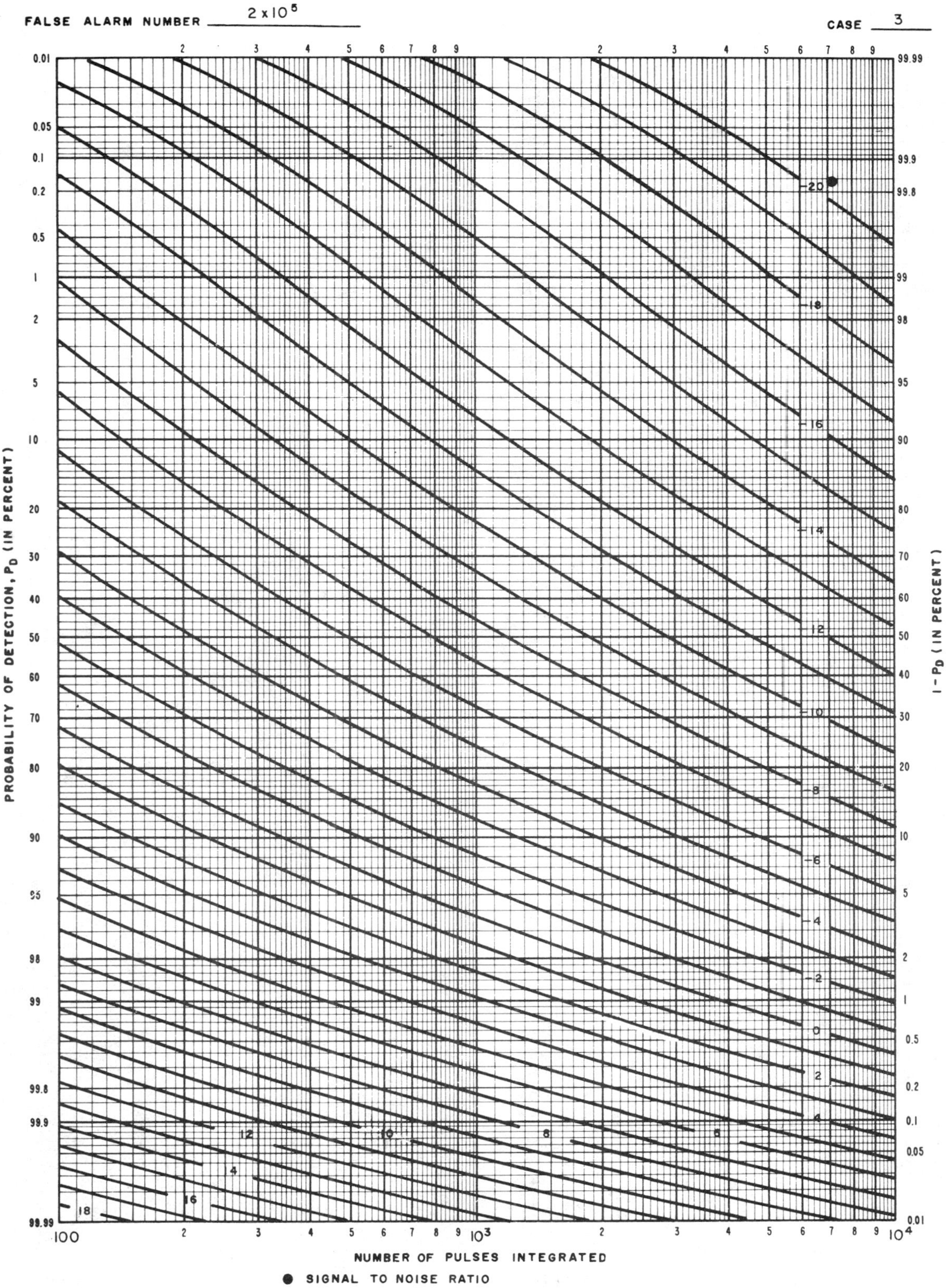
FALSE ALARM NUMBER 2 x 10^5
CASE 3
PROBABILITY OF DETECTION, P_D (IN PERCENT)
1 - P_D (IN PERCENT)
NUMBER OF PULSES INTEGRATED
● SIGNAL TO NOISE RATIO

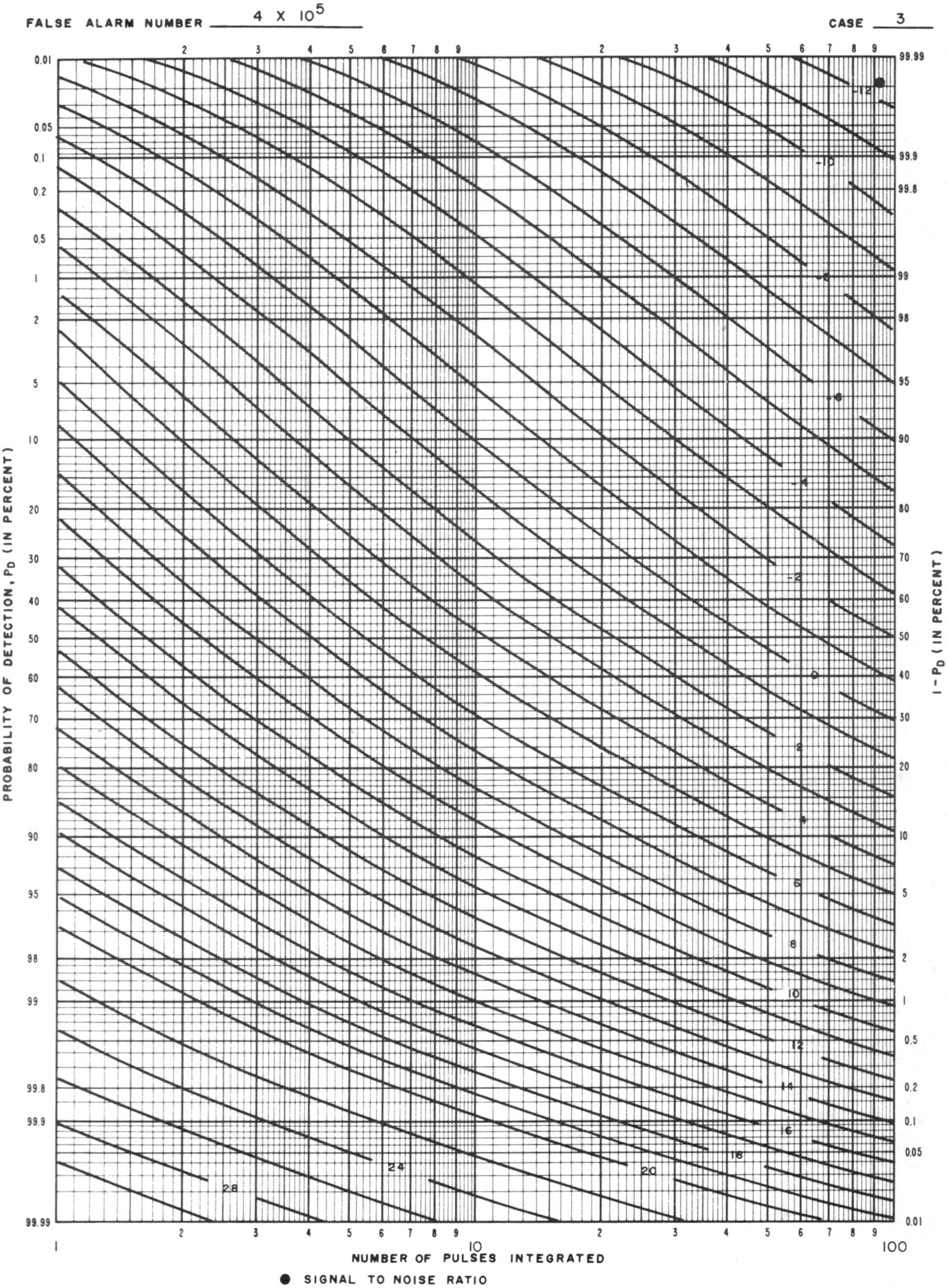
FALSE ALARM NUMBER 4 X 10^5
CASE 3
PROBABILITY OF DETECTION, P_D (IN PERCENT)
1 - P_D (IN PERCENT)
NUMBER OF PULSES INTEGRATED
● SIGNAL TO NOISE RATIO
-12
-10
-8
-6
-4
-2
0
2
4
6
8
10
12
14
16
18
20
24
28

FALSE ALARM NUMBER 4 X 10^5 CASE 3

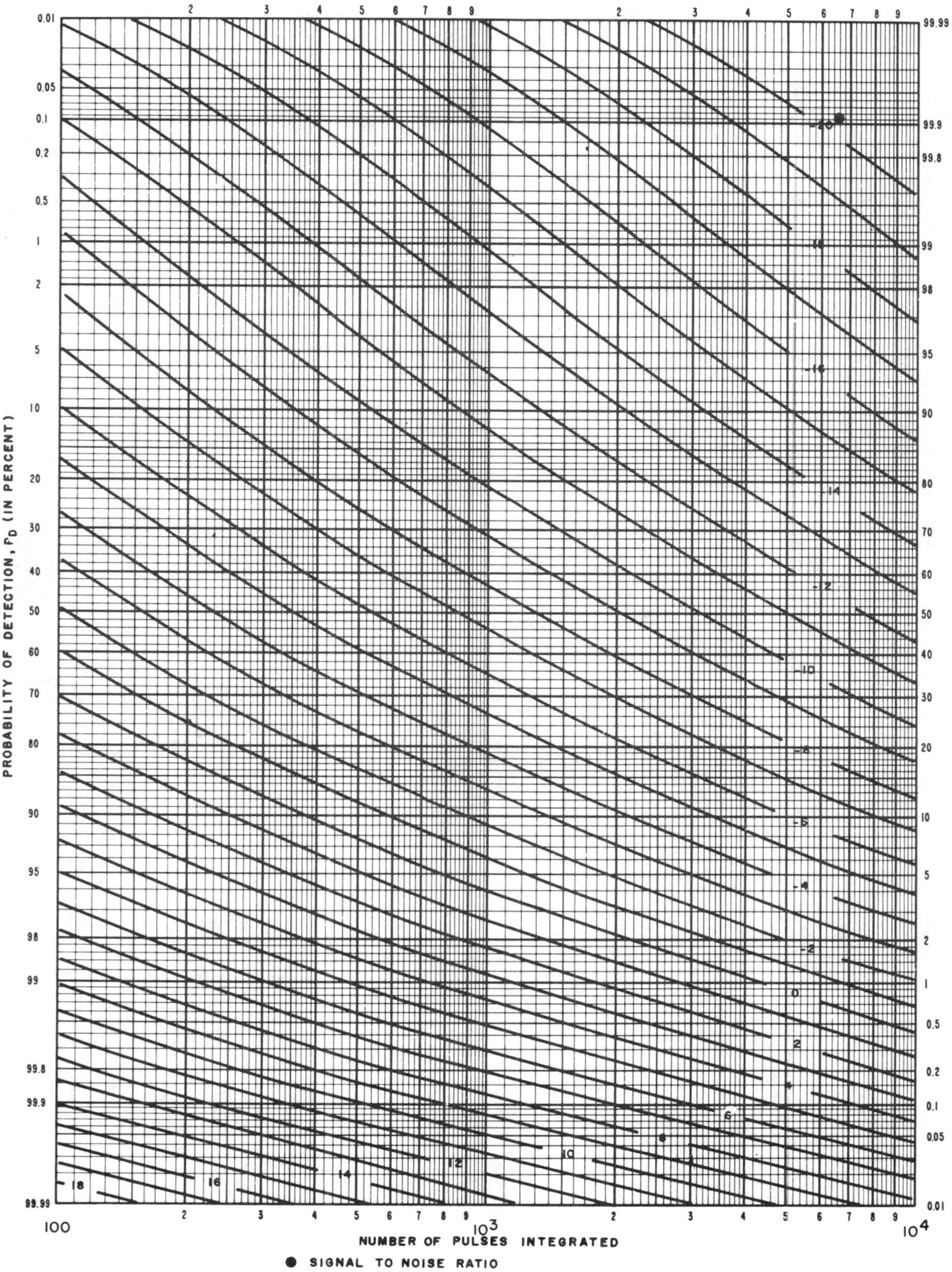

FALSE ALARM NUMBER 6×10^5 CASE 3

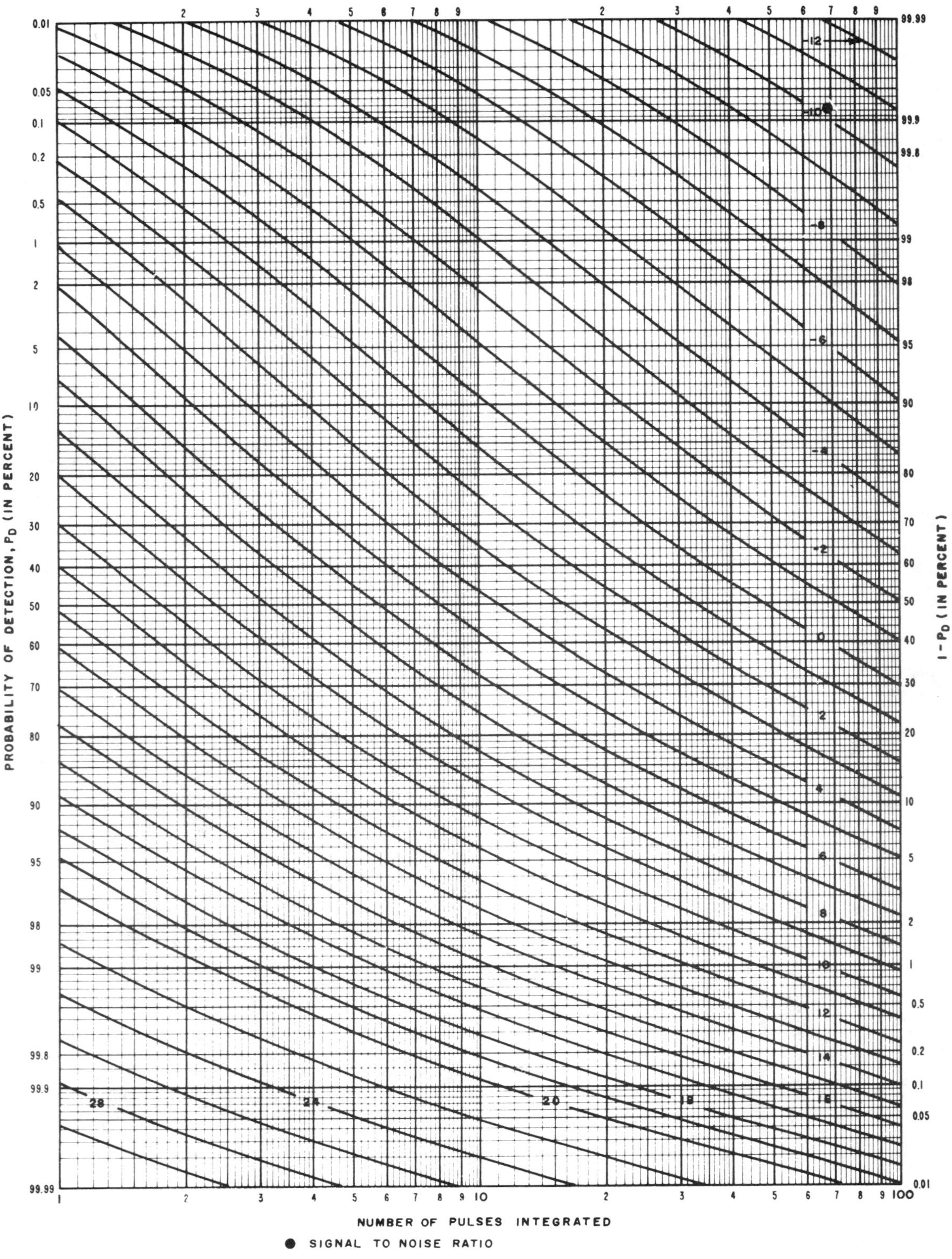

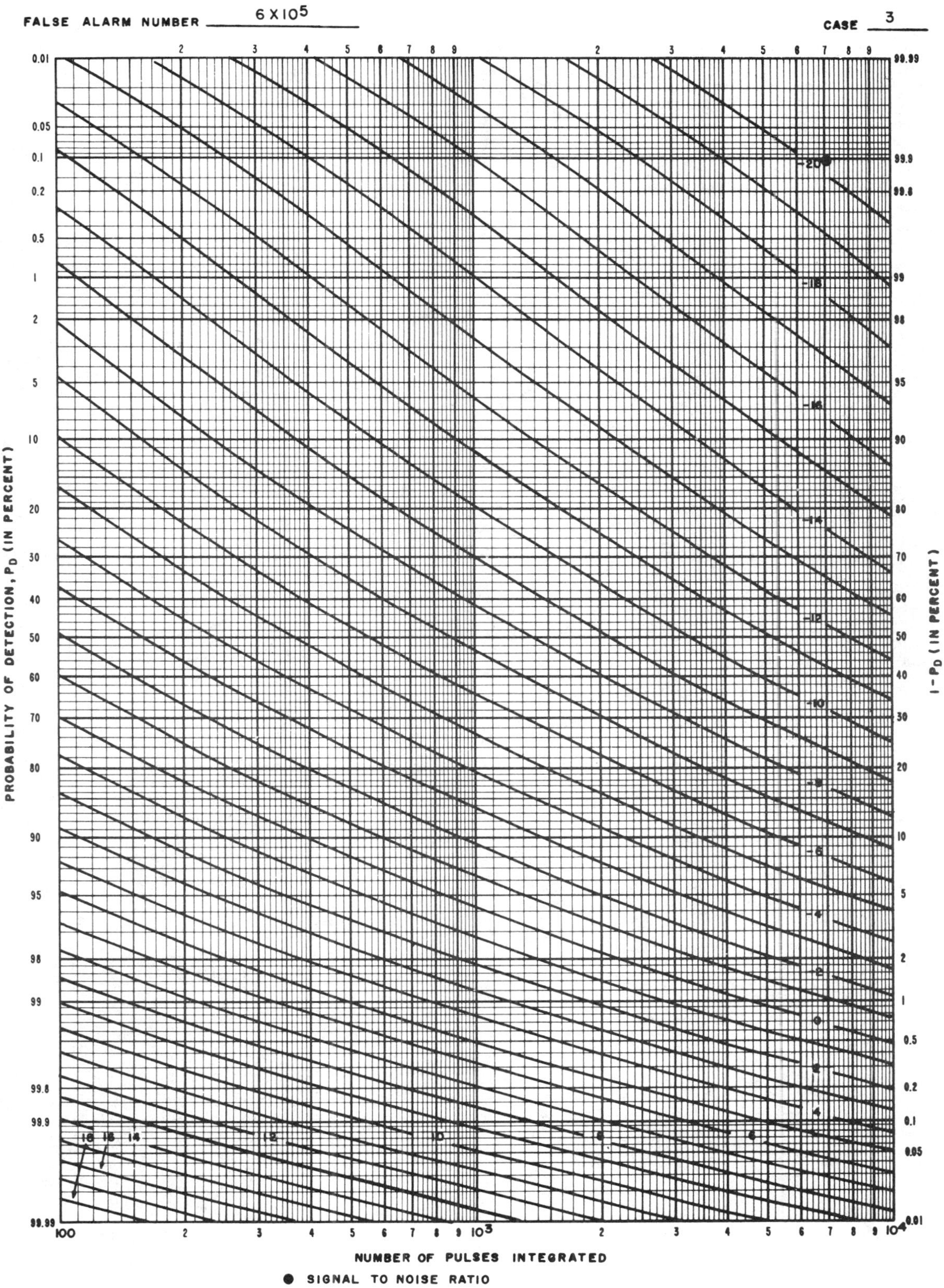
FALSE ALARM NUMBER 6 X 10^5
CASE 3
PROBABILITY OF DETECTION, P_D (IN PERCENT)
1 - P_D (IN PERCENT)
NUMBER OF PULSES INTEGRATED
● SIGNAL TO NOISE RATIO
100
10^3
10^4

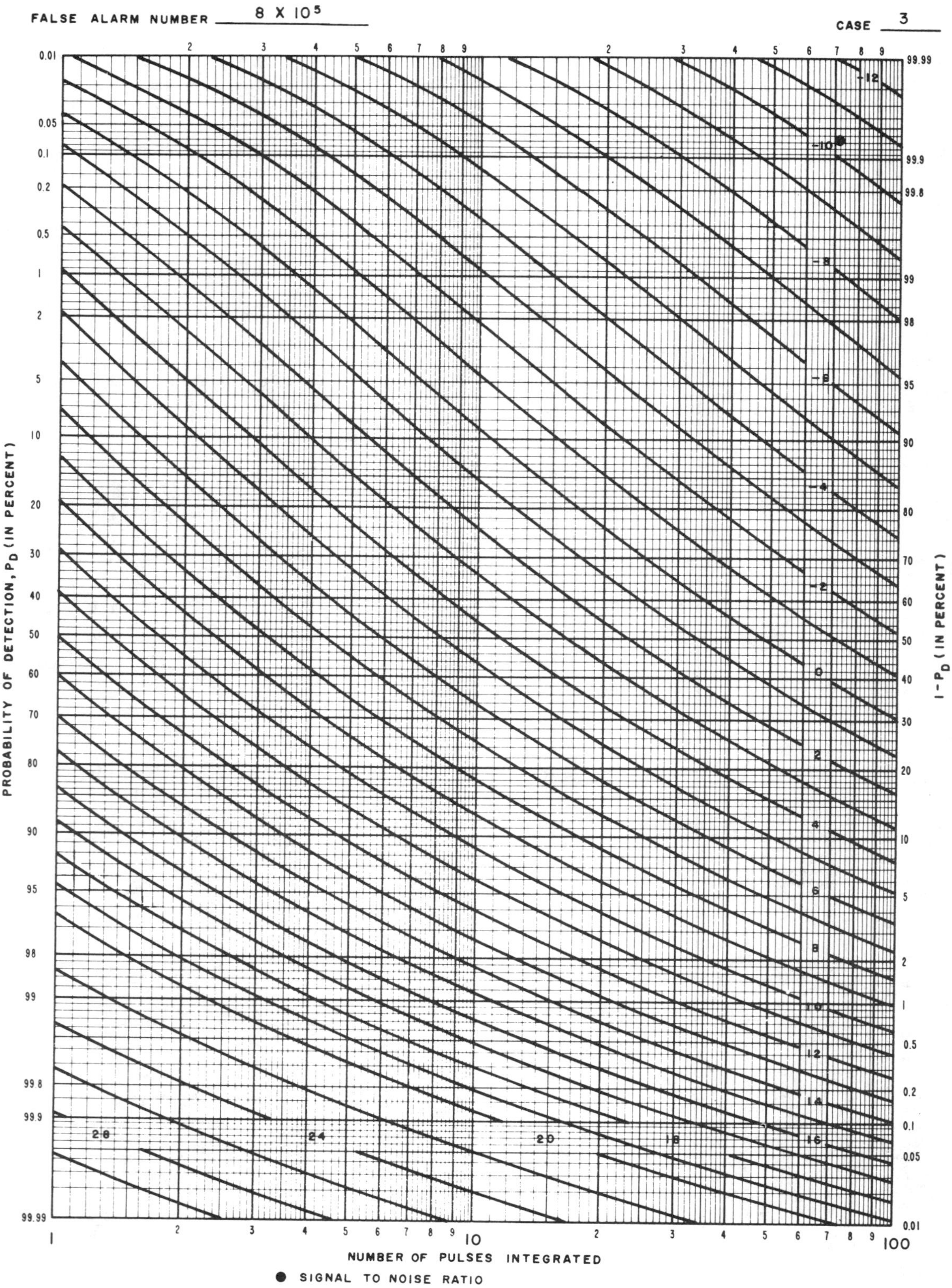
FALSE ALARM NUMBER 8 X 10^5
CASE 3
PROBABILITY OF DETECTION, P_D (IN PERCENT)
1-P_D (IN PERCENT)
-12
-10
-8
-6
-4
-2
0
2
4
6
8
10
12
14
16
18
20
24
28
NUMBER OF PULSES INTEGRATED
● SIGNAL TO NOISE RATIO

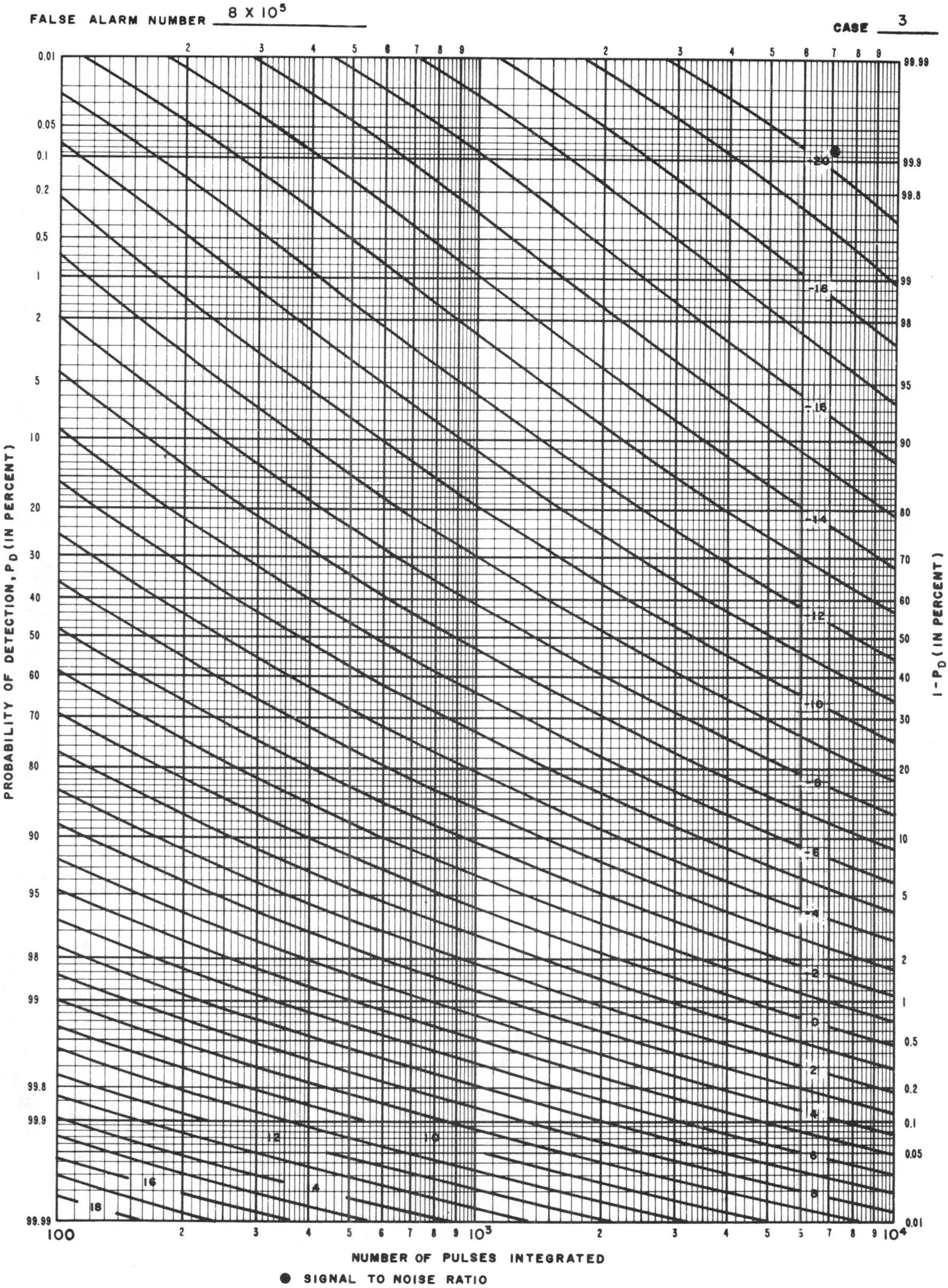
FALSE ALARM NUMBER 8 X 10^5
CASE 3
PROBABILITY OF DETECTION, P_D (IN PERCENT)
1 - P_D (IN PERCENT)
NUMBER OF PULSES INTEGRATED
● SIGNAL TO NOISE RATIO
-20
-18
-16
-14
-12
-10
-8
-6
-2
0
2
4
6
8
12
10
14
16
18

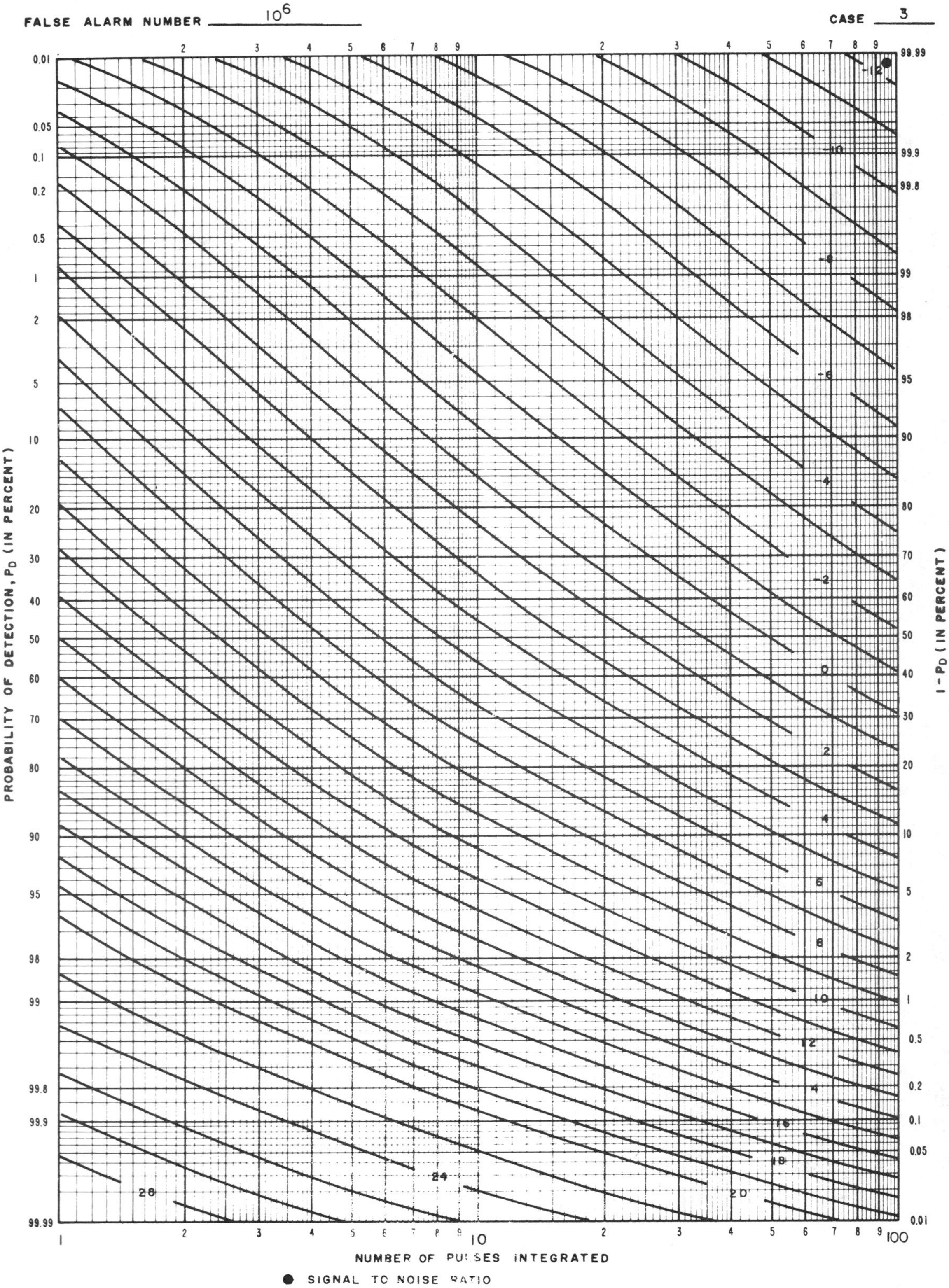

FALSE ALARM NUMBER 10^6
CASE 3
PROBABILITY OF DETECTION, P_D (IN PERCENT)
1 - P_D (IN PERCENT)
NUMBER OF PULSES INTEGRATED
● SIGNAL TO NOISE RATIO

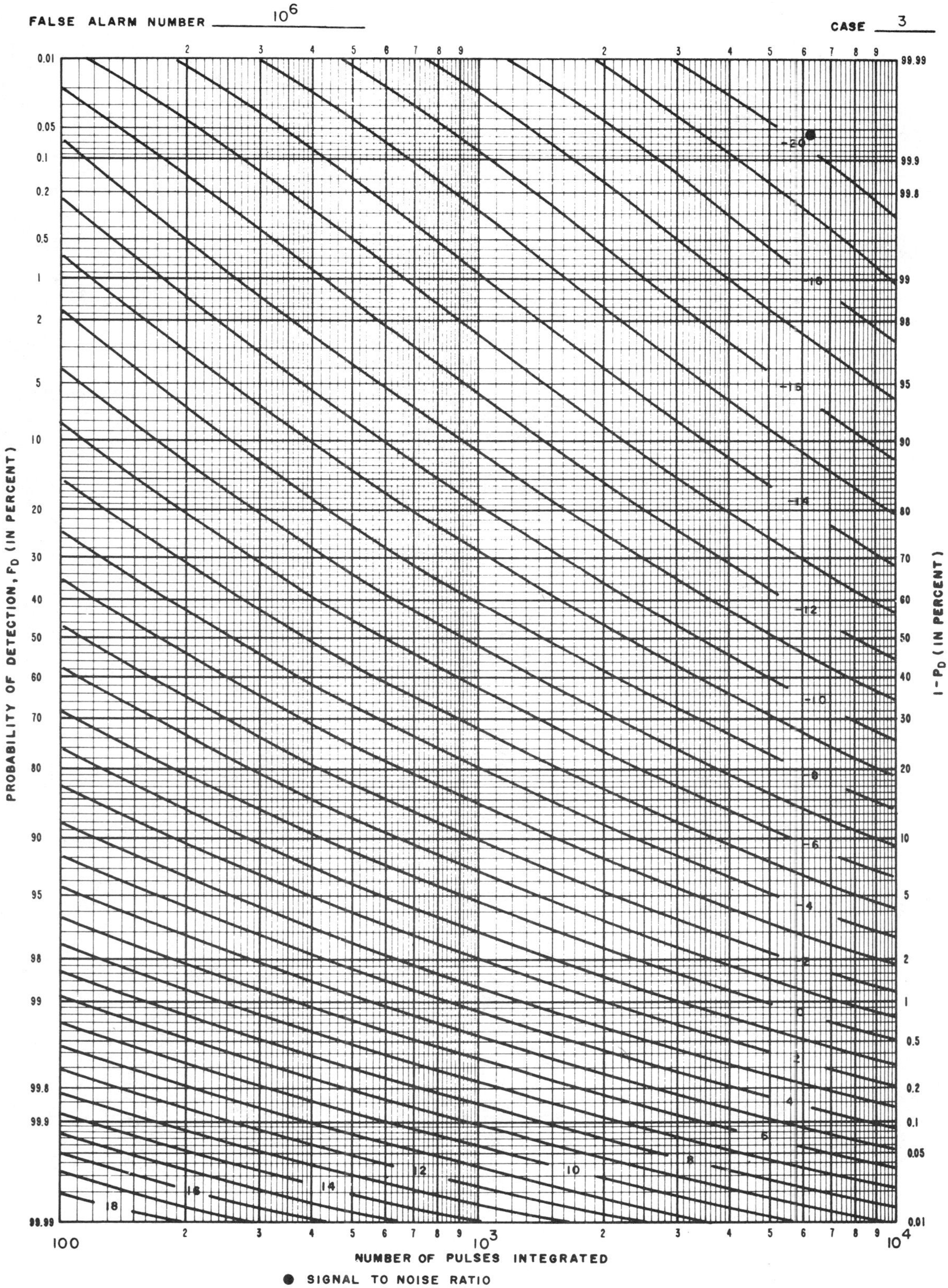
FALSE ALARM NUMBER 10^6
CASE 3
PROBABILITY OF DETECTION, P_D (IN PERCENT)
$1-P_D$ (IN PERCENT)
NUMBER OF PULSES INTEGRATED
● SIGNAL TO NOISE RATIO
100
10^3
10^4
-20
-18
-16
-14
-12
-10
-8
-6
-4
-2
0
2
4
6
8
10
12
14
16
18

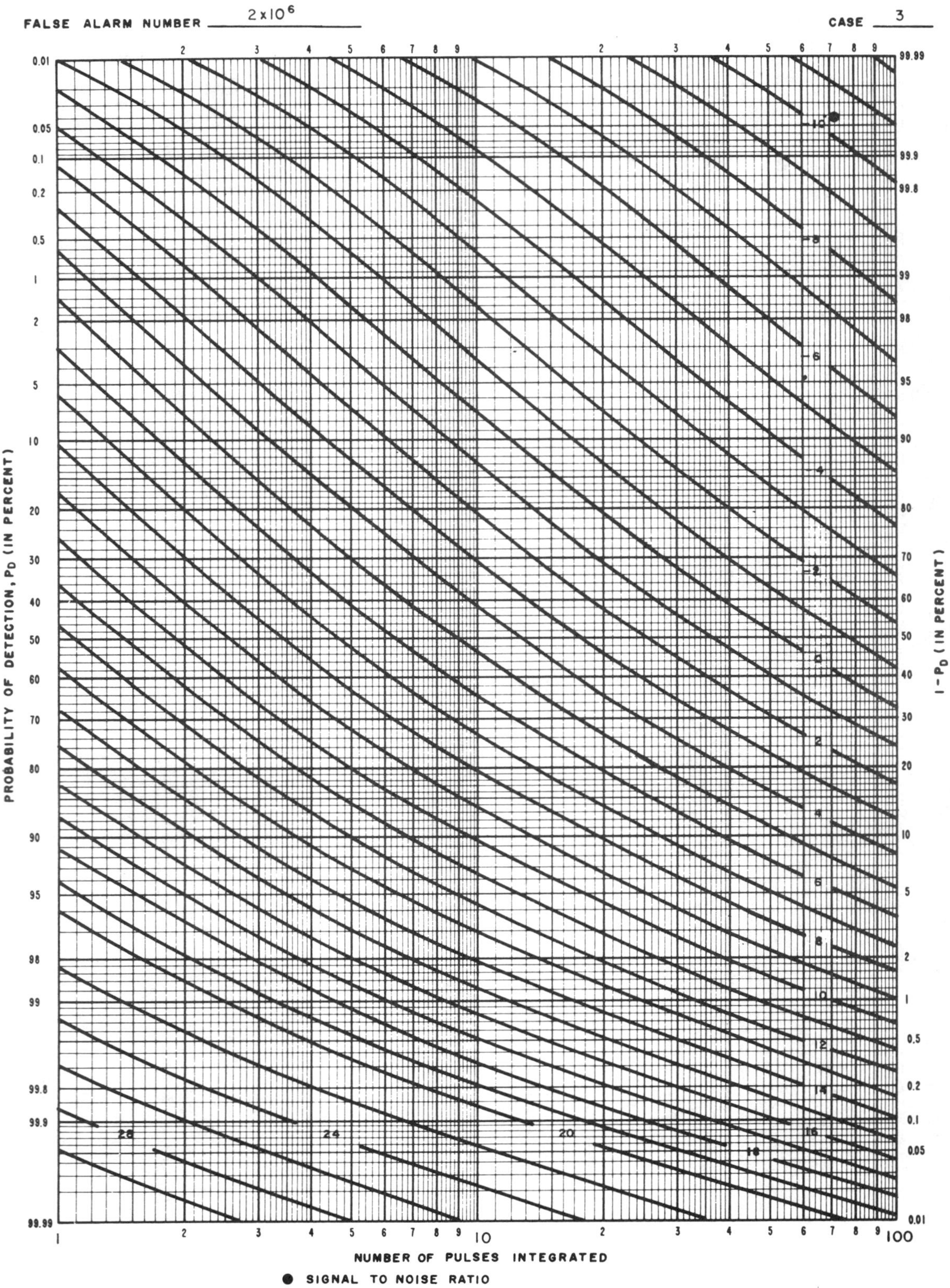
FALSE ALARM NUMBER 2x10^6
CASE 3
PROBABILITY OF DETECTION, P_D (IN PERCENT)
1 - P_D (IN PERCENT)
NUMBER OF PULSES INTEGRATED
● SIGNAL TO NOISE RATIO

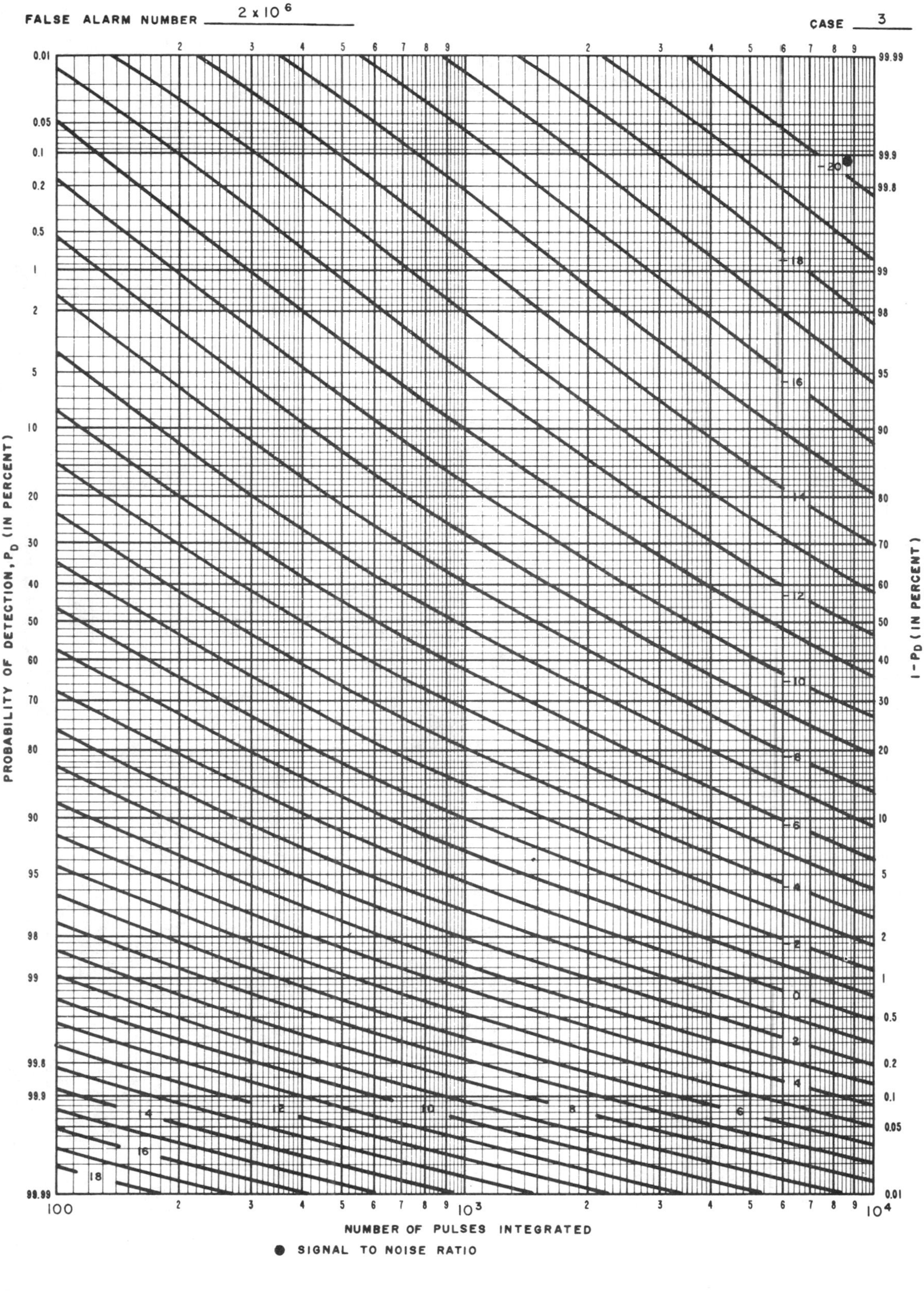
FALSE ALARM NUMBER 2 x 10^6
CASE 3
PROBABILITY OF DETECTION, P_D (IN PERCENT)
1 - P_D (IN PERCENT)
NUMBER OF PULSES INTEGRATED
● SIGNAL TO NOISE RATIO
100
10^3
10^4

FALSE ALARM NUMBER 4 X 10⁶ CASE 3

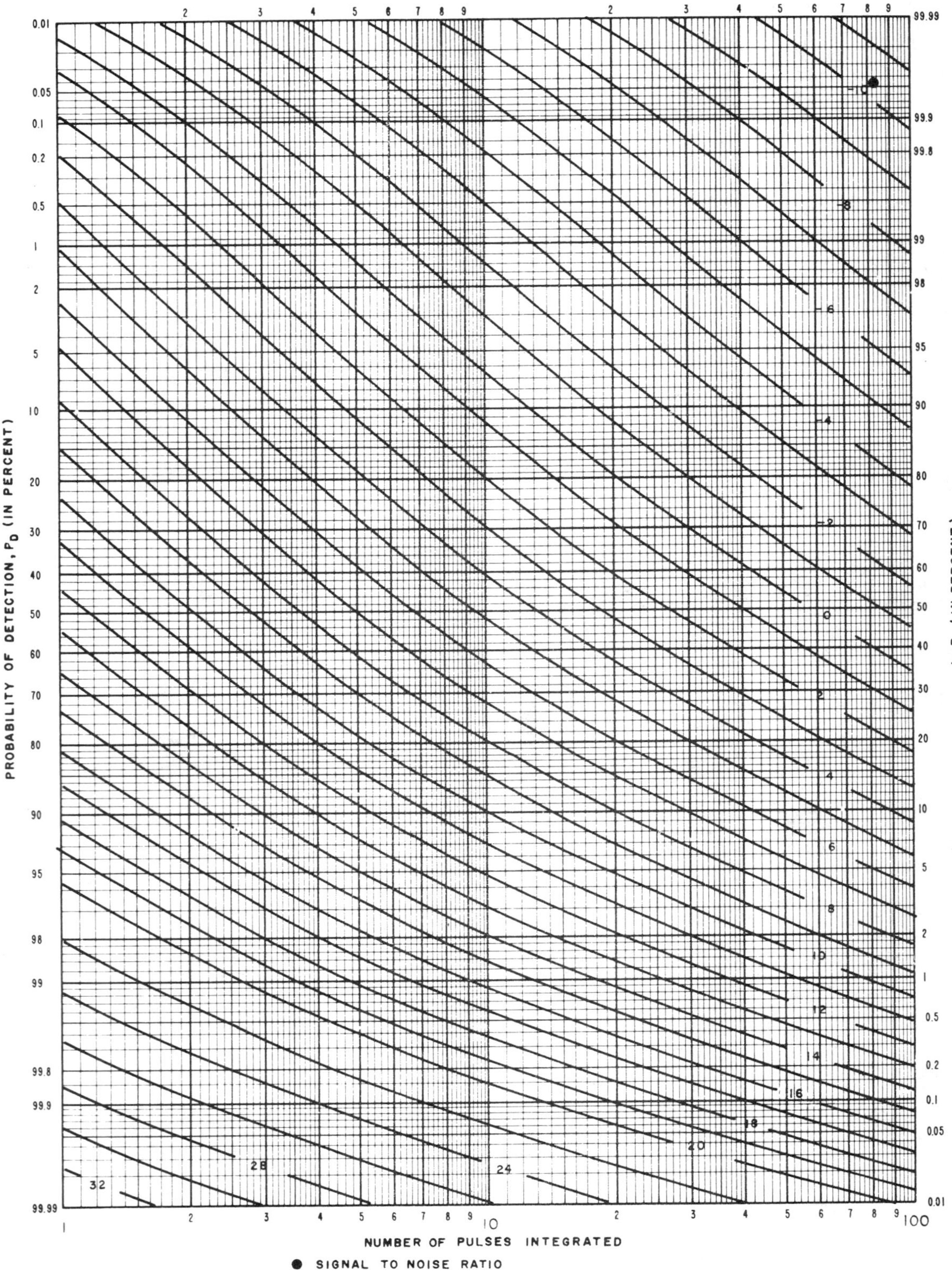

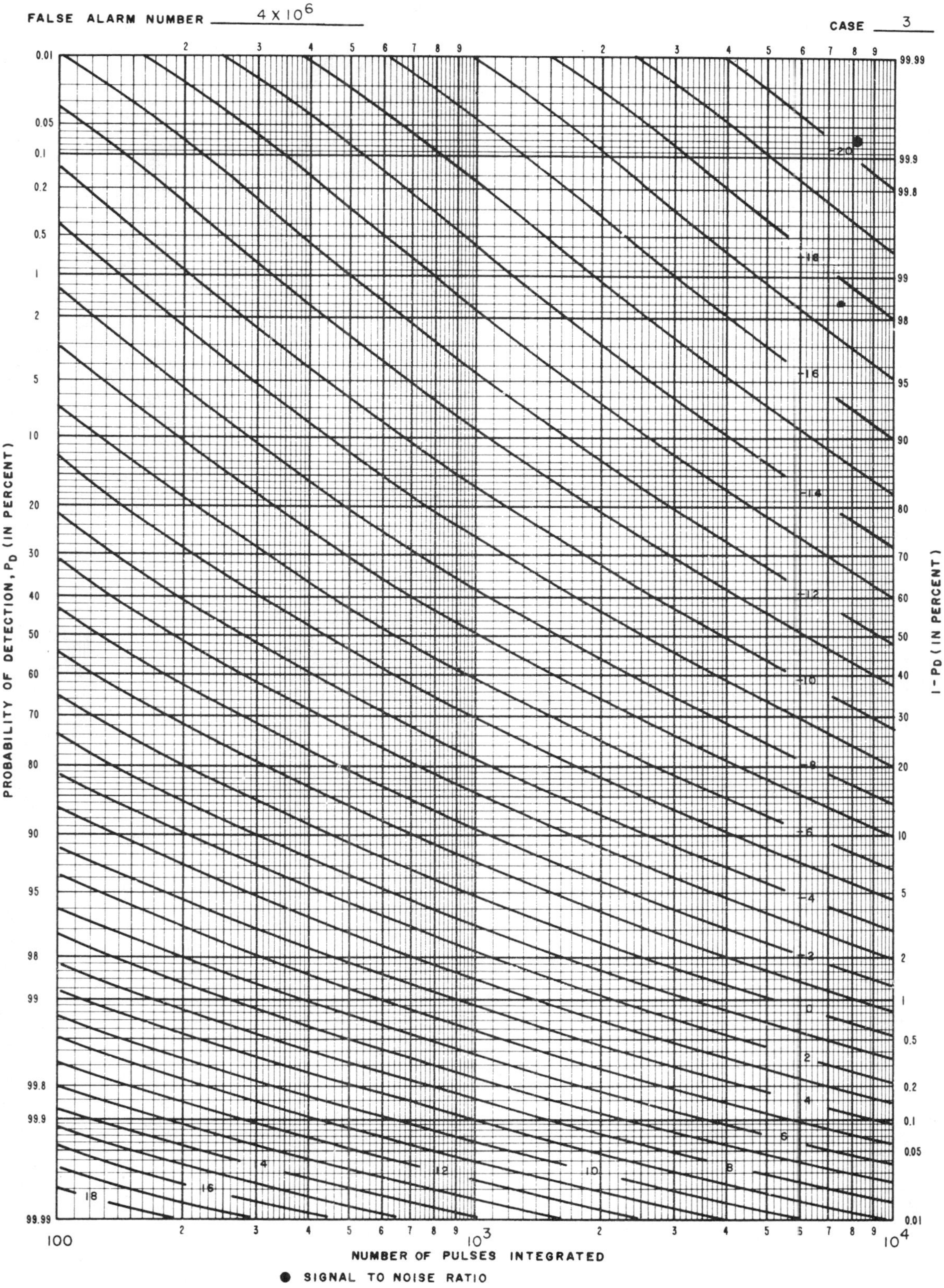
FALSE ALARM NUMBER 4 X 10^6
CASE 3
PROBABILITY OF DETECTION, P_D (IN PERCENT)
1 - P_D (IN PERCENT)
NUMBER OF PULSES INTEGRATED
● SIGNAL TO NOISE RATIO
100
10^3
10^4

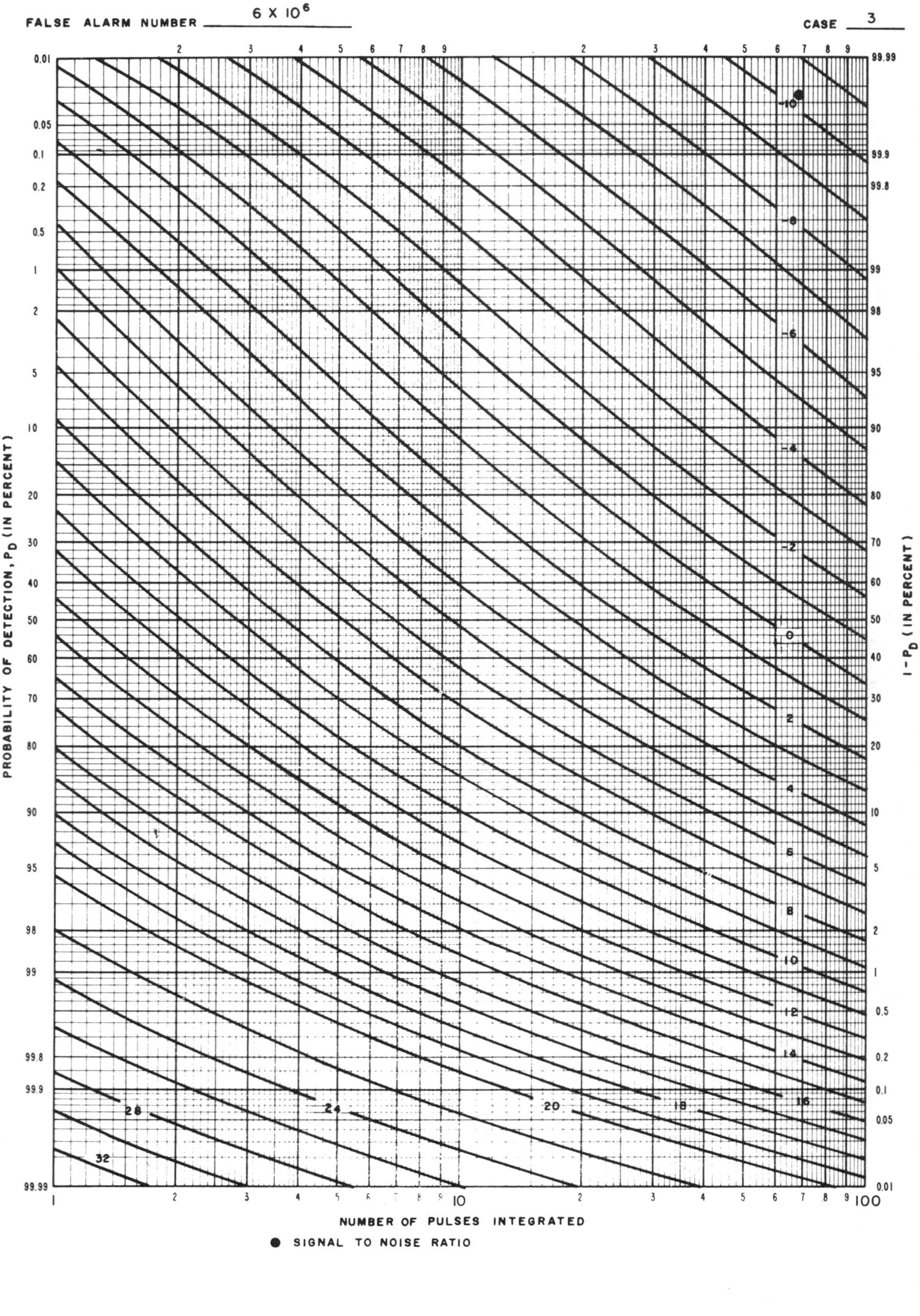
FALSE ALARM NUMBER 6 X 10^6
CASE 3
PROBABILITY OF DETECTION, P_D (IN PERCENT)
1 - P_D (IN PERCENT)
NUMBER OF PULSES INTEGRATED
● SIGNAL TO NOISE RATIO
-10
-8
-6
-4
-2
0
2
4
6
8
10
12
14
16
18
20
24
28
32

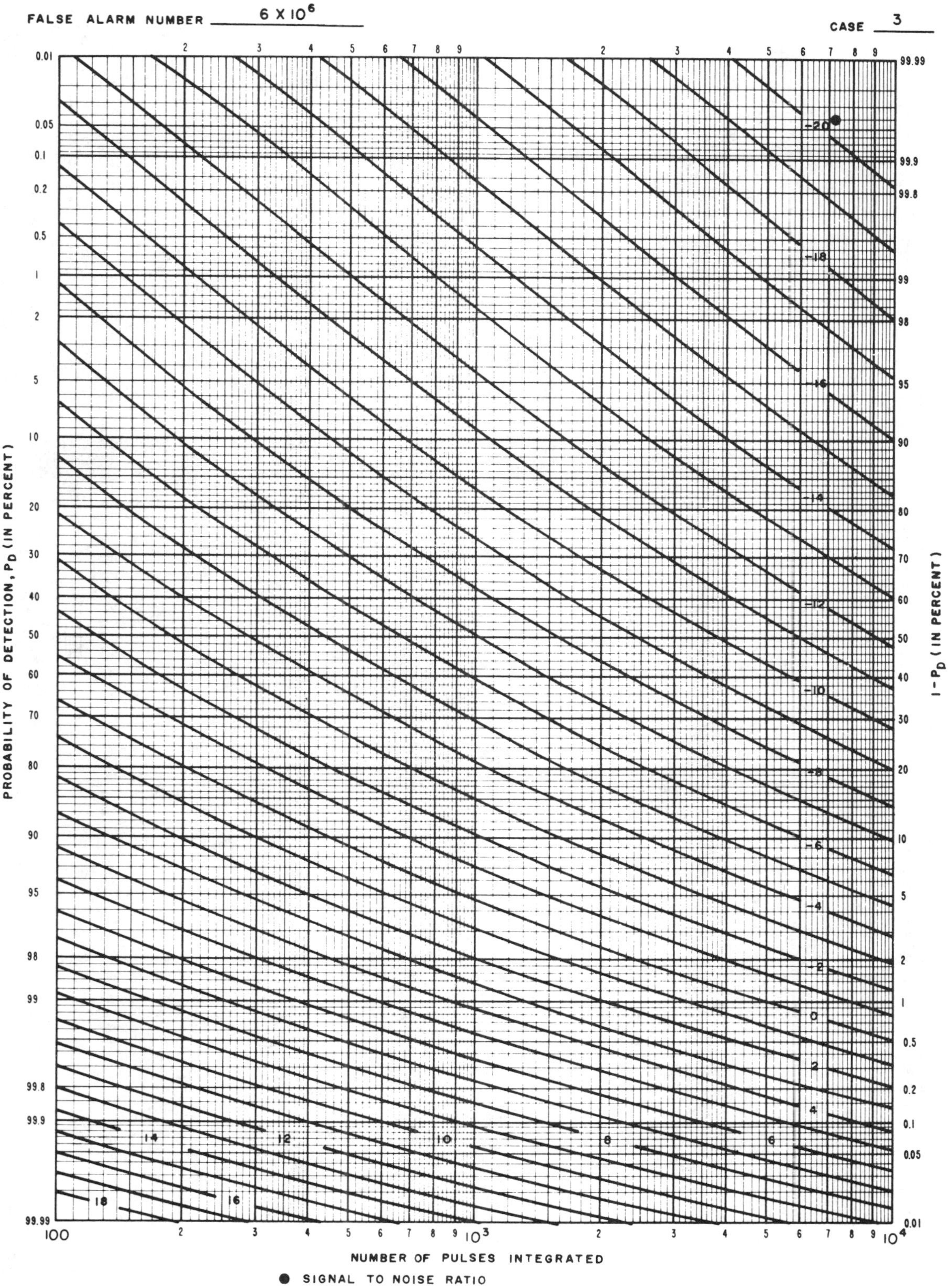
FALSE ALARM NUMBER 6 X 10^6
CASE 3
PROBABILITY OF DETECTION, P_D (IN PERCENT)
1 - P_D (IN PERCENT)
-20
-18
-16
-14
-12
-10
-8
-6
-4
-2
0
2
4
6
8
10
12
14
16
18
100
10^3
10^4
NUMBER OF PULSES INTEGRATED
● SIGNAL TO NOISE RATIO

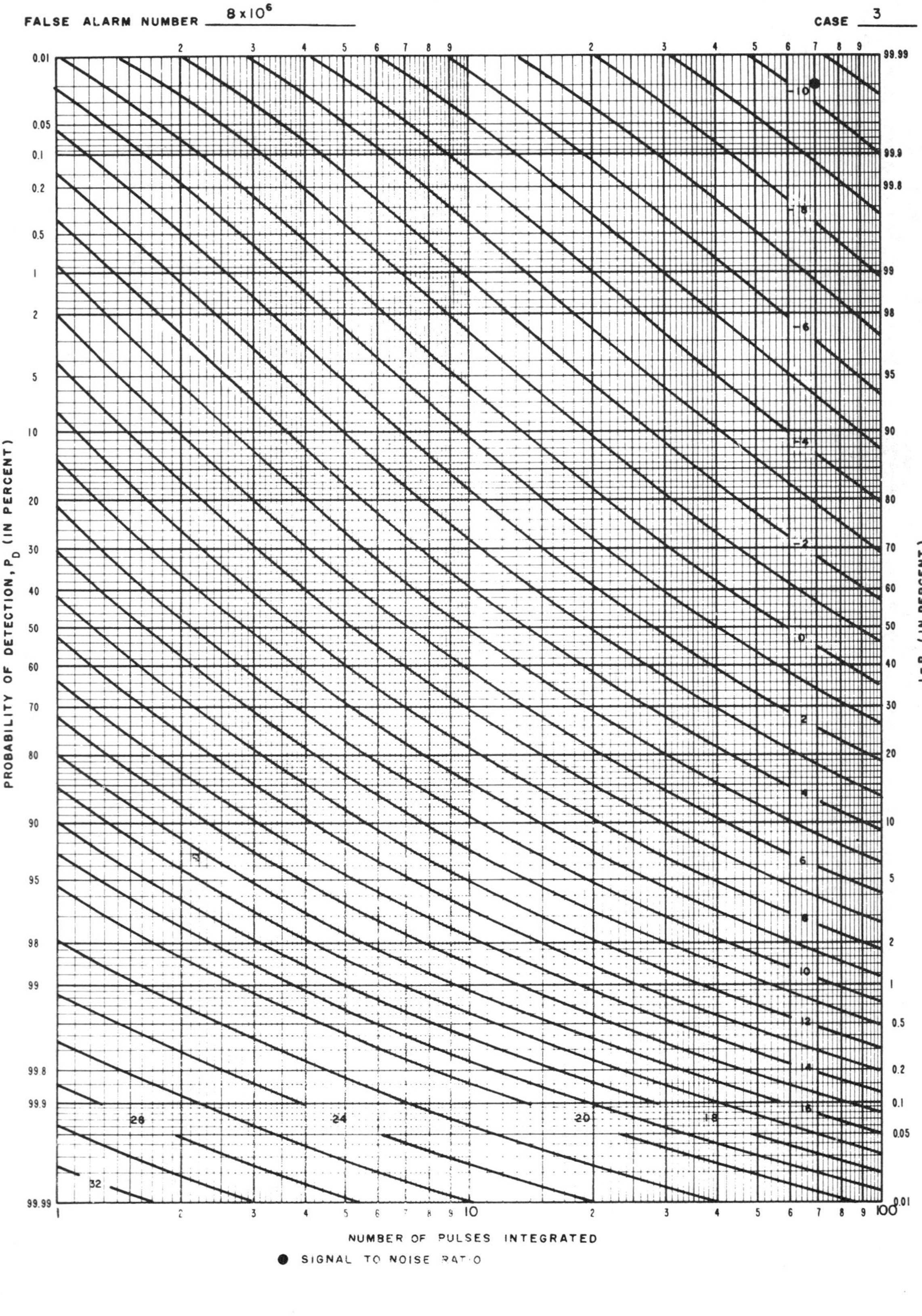
FALSE ALARM NUMBER 8 x 10^6
CASE 3
PROBABILITY OF DETECTION, P_D (IN PERCENT)
1 - P_D (IN PERCENT)
0.01
0.05
0.1
0.2
0.5
1
2
5
10
20
30
40
50
60
70
80
90
95
98
99
99.8
99.9
99.99
99.99
99.9
99.8
99
98
95
90
80
70
60
50
40
30
20
10
5
2
1
0.5
0.2
0.1
0.05
0.01
-10
-8
-6
-4
-2
0
2
4
6
8
10
12
14
16
18
20
24
28
32
1
2
3
4
5
6
7
8
9
10
100
NUMBER OF PULSES INTEGRATED
SIGNAL TO NOISE RATIO

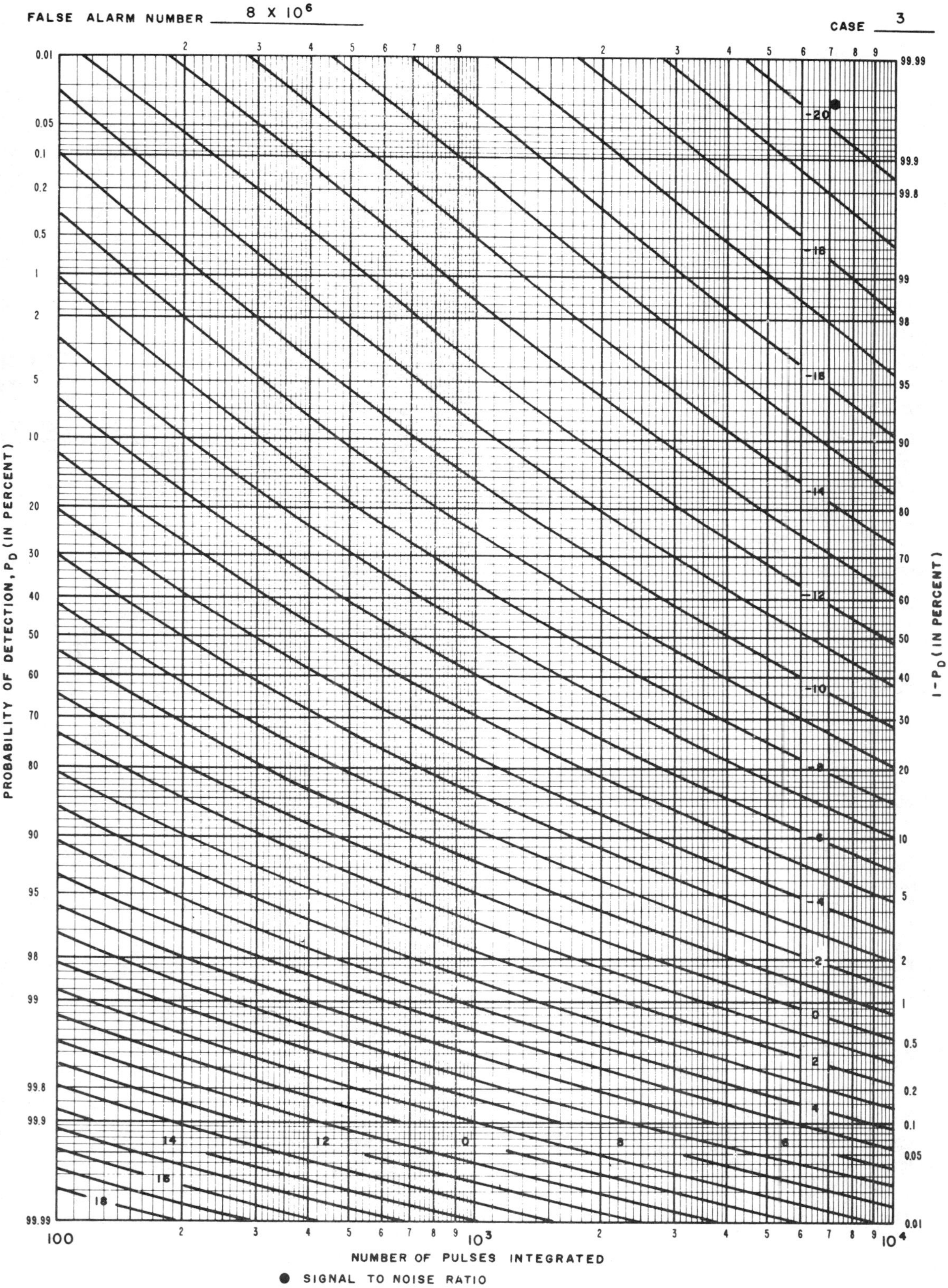
FALSE ALARM NUMBER 8 X 10^6
CASE 3
PROBABILITY OF DETECTION, P_D (IN PERCENT)
1-P_D (IN PERCENT)
NUMBER OF PULSES INTEGRATED
● SIGNAL TO NOISE RATIO
100
10^3
10^4
-20
-18
-16
-14
-12
-10
-8
-6
-4
-2
0
2
4
6
8
10
12
14
16
18

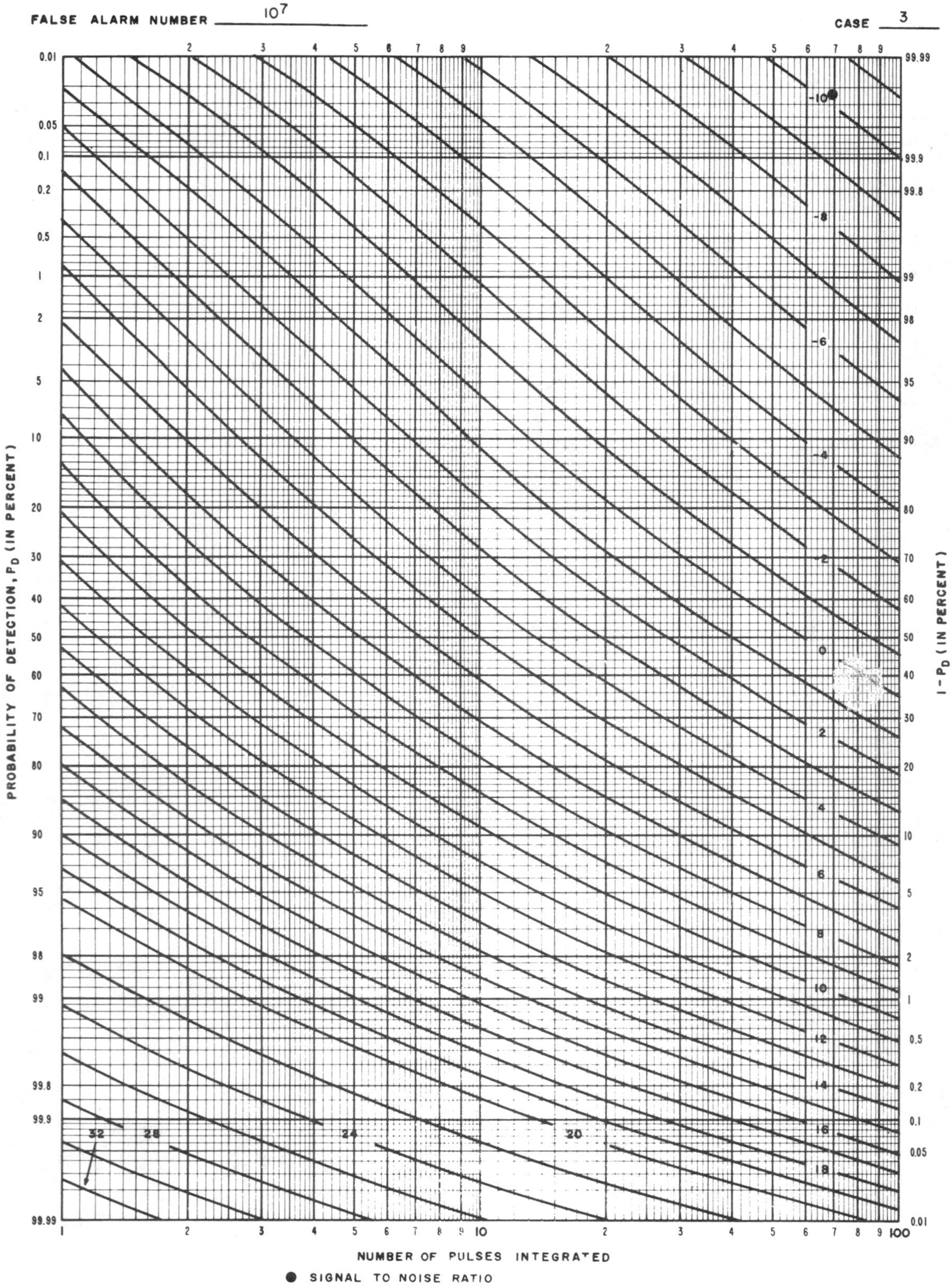
FALSE ALARM NUMBER 10^7
CASE 3
PROBABILITY OF DETECTION, P_D (IN PERCENT)
$1-P_D$ (IN PERCENT)
NUMBER OF PULSES INTEGRATED
● SIGNAL TO NOISE RATIO

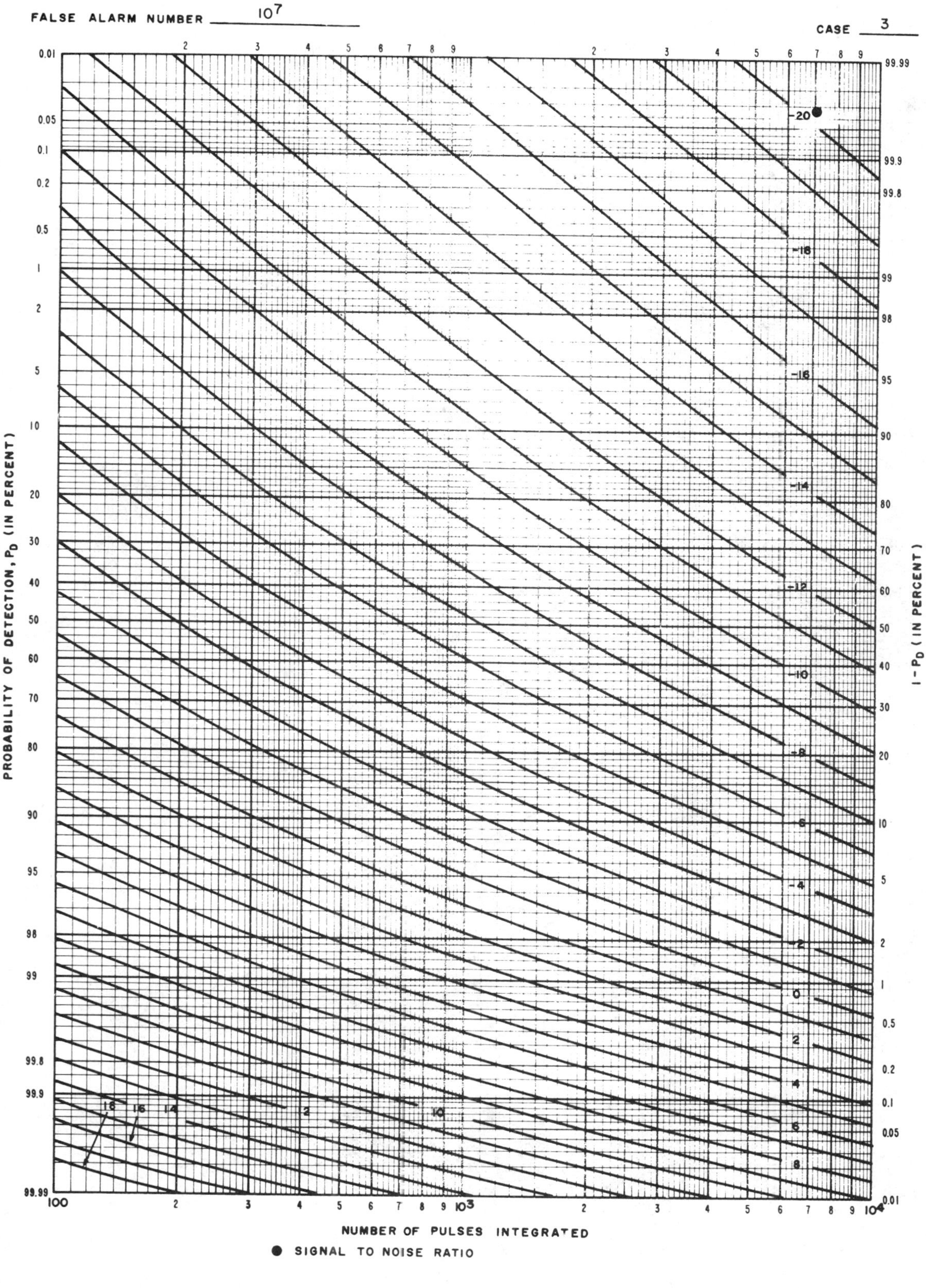
FALSE ALARM NUMBER 10^7
CASE 3
PROBABILITY OF DETECTION, P_D (IN PERCENT)
1 - P_D (IN PERCENT)
NUMBER OF PULSES INTEGRATED
● SIGNAL TO NOISE RATIO

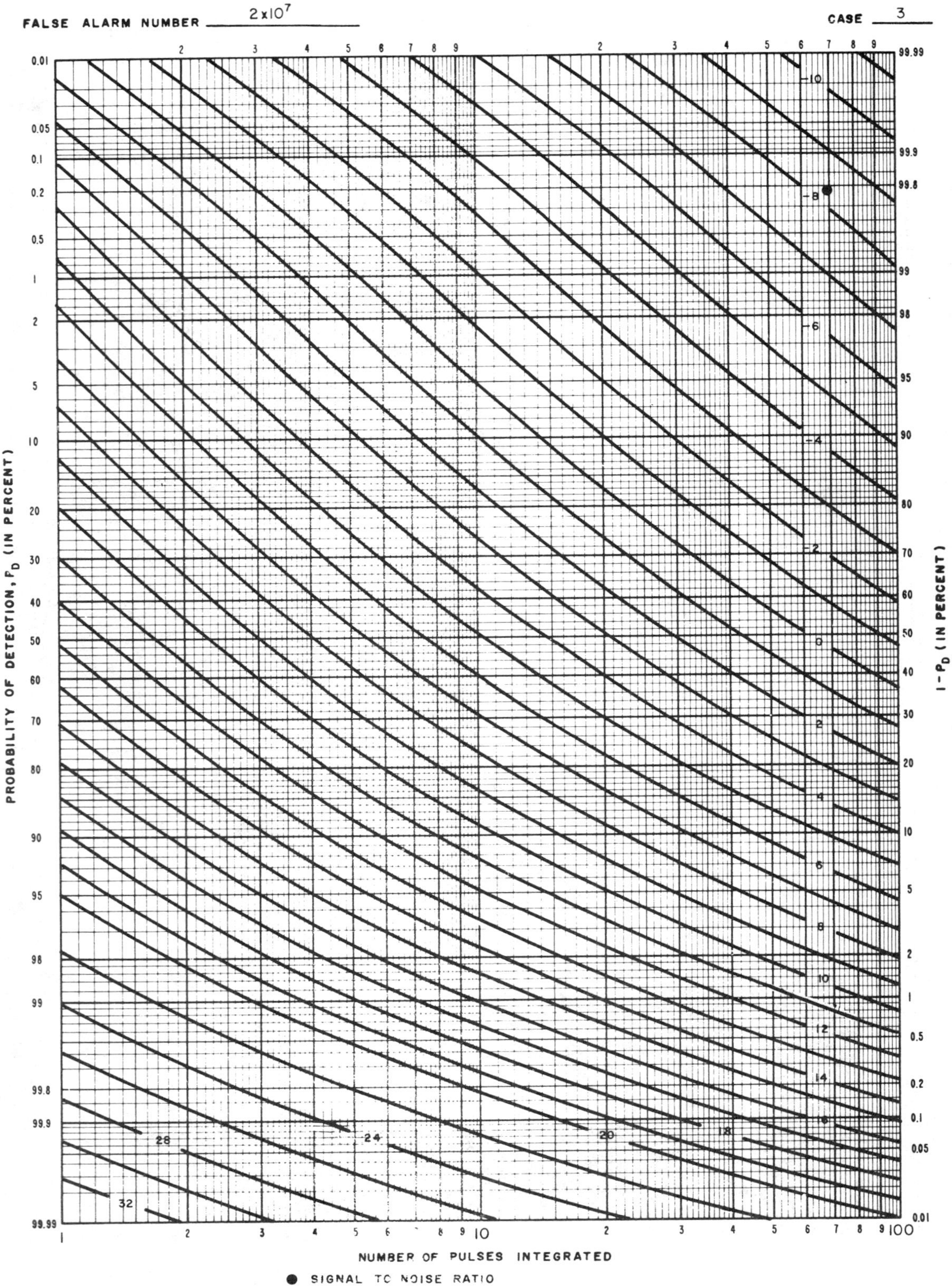
FALSE ALARM NUMBER 2x10^7
CASE 3
PROBABILITY OF DETECTION, P_D (IN PERCENT)
1-P_D (IN PERCENT)
NUMBER OF PULSES INTEGRATED
● SIGNAL TO NOISE RATIO

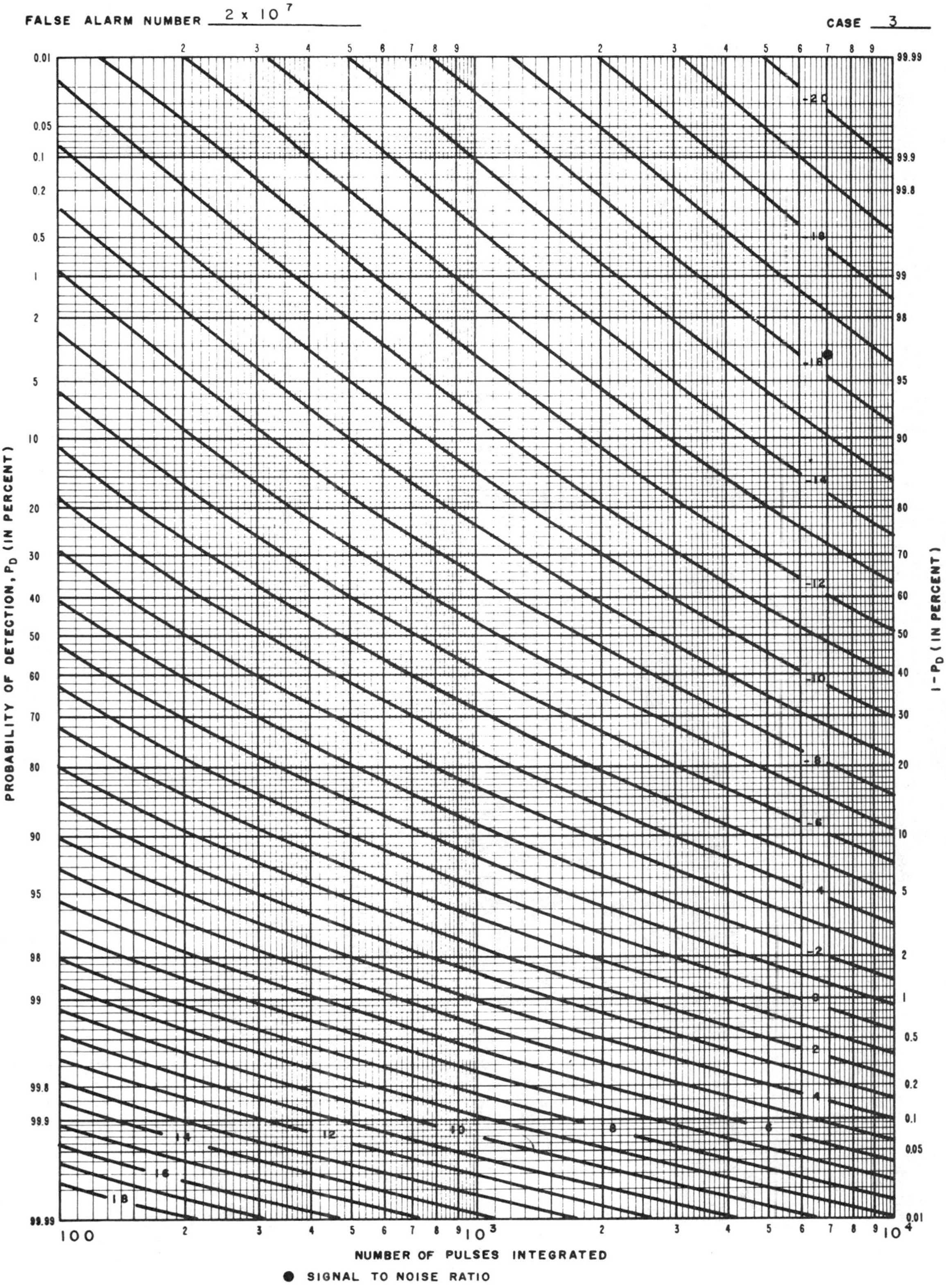
FALSE ALARM NUMBER 2 x 10^7
CASE 3
PROBABILITY OF DETECTION, P_D (IN PERCENT)
1 - P_D (IN PERCENT)
NUMBER OF PULSES INTEGRATED
100
10^3
10^4
● SIGNAL TO NOISE RATIO

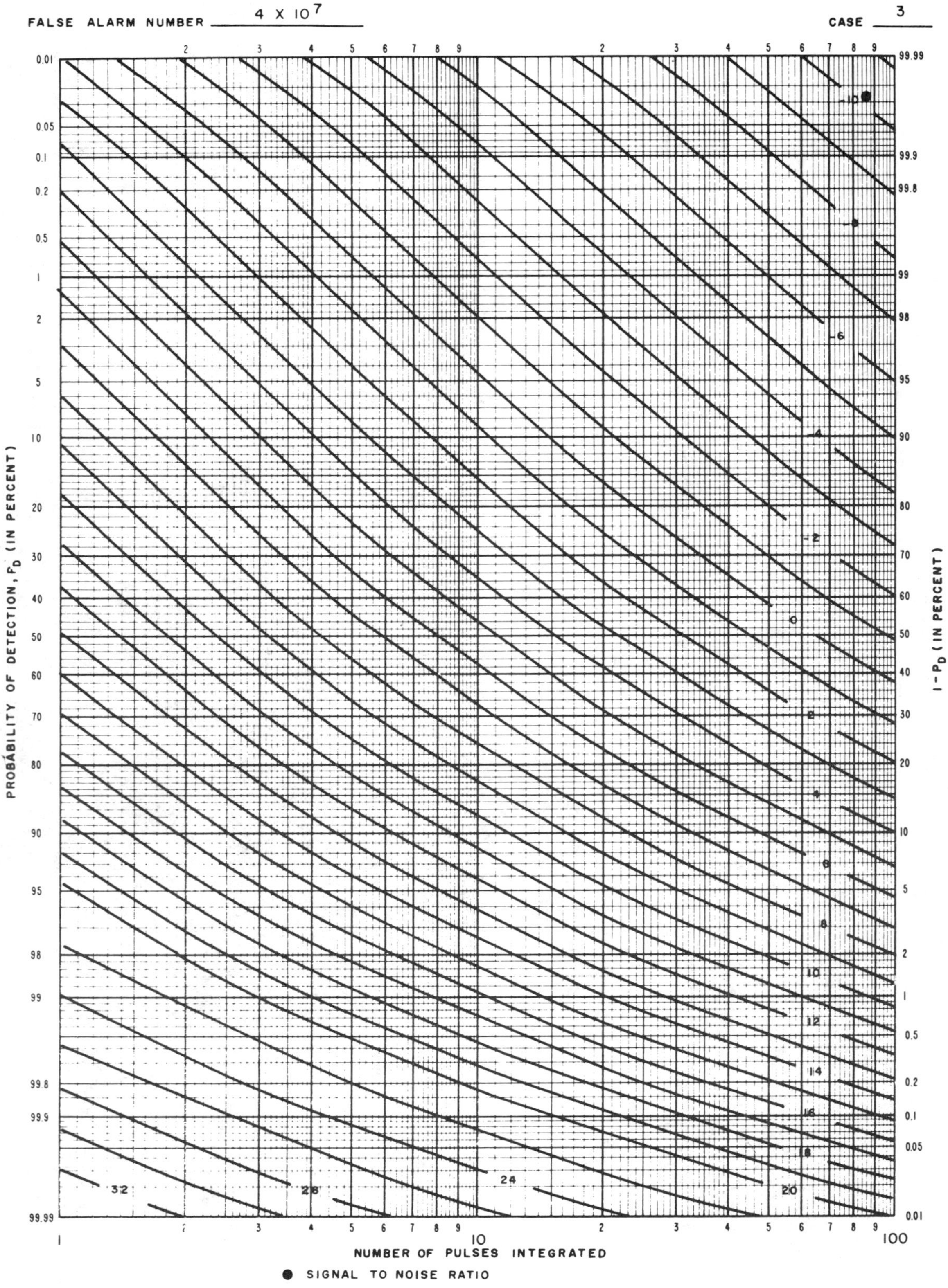
FALSE ALARM NUMBER 4 X 10^7
CASE 3
PROBABILITY OF DETECTION, P_D (IN PERCENT)
$1-P_D$ (IN PERCENT)
NUMBER OF PULSES INTEGRATED
● SIGNAL TO NOISE RATIO
-10
-8
-6
-4
-2
0
2
4
6
8
10
12
14
16
18
20
24
28
32

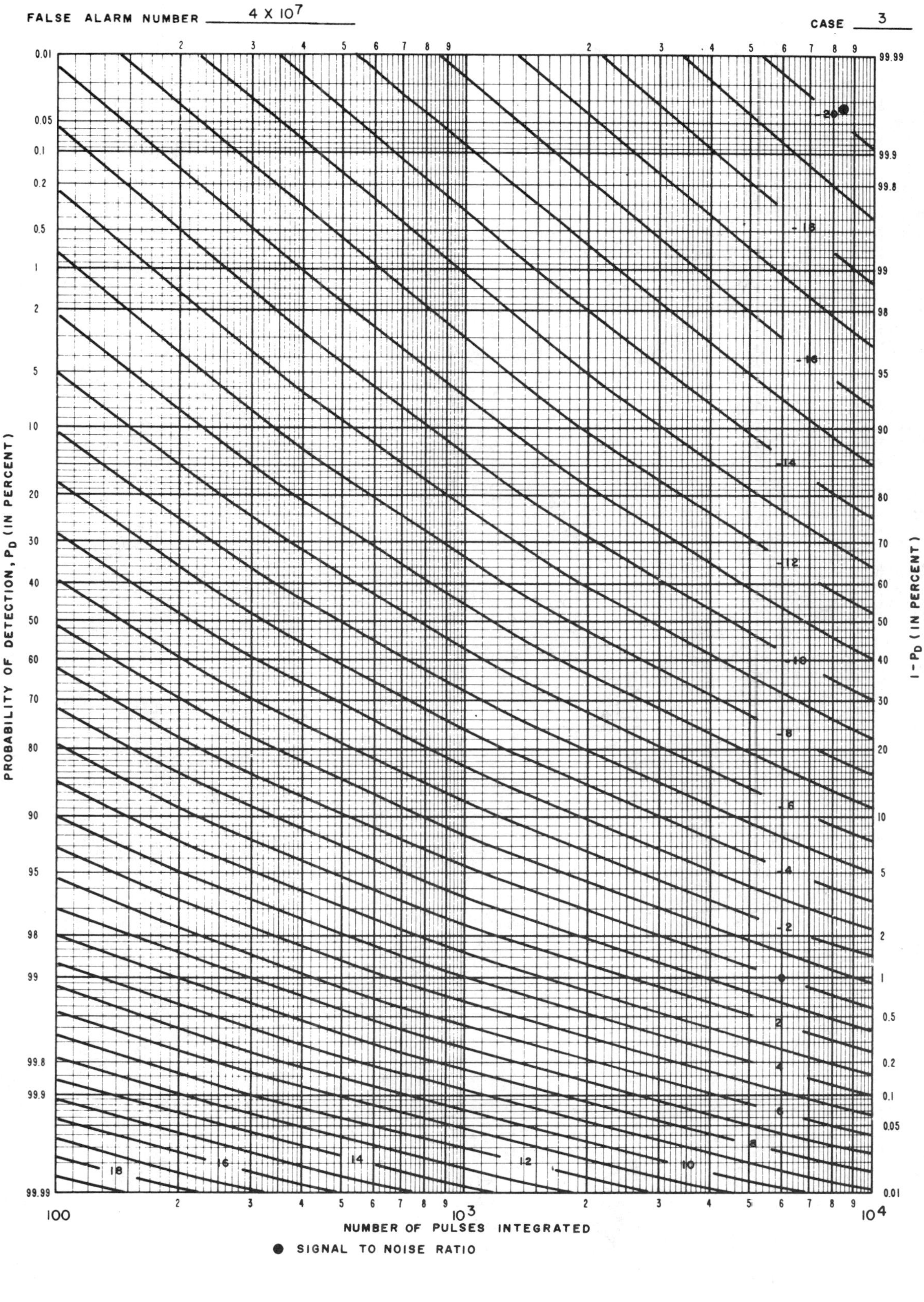
FALSE ALARM NUMBER 4 X 10^7
CASE 3
PROBABILITY OF DETECTION, P_D (IN PERCENT)
1 - P_D (IN PERCENT)
NUMBER OF PULSES INTEGRATED
● SIGNAL TO NOISE RATIO
100
10^3
10^4
-20
-18
-16
-14
-12
-10
-8
-6
-4
-2
0
2
4
6
8
10
12
14
16
18

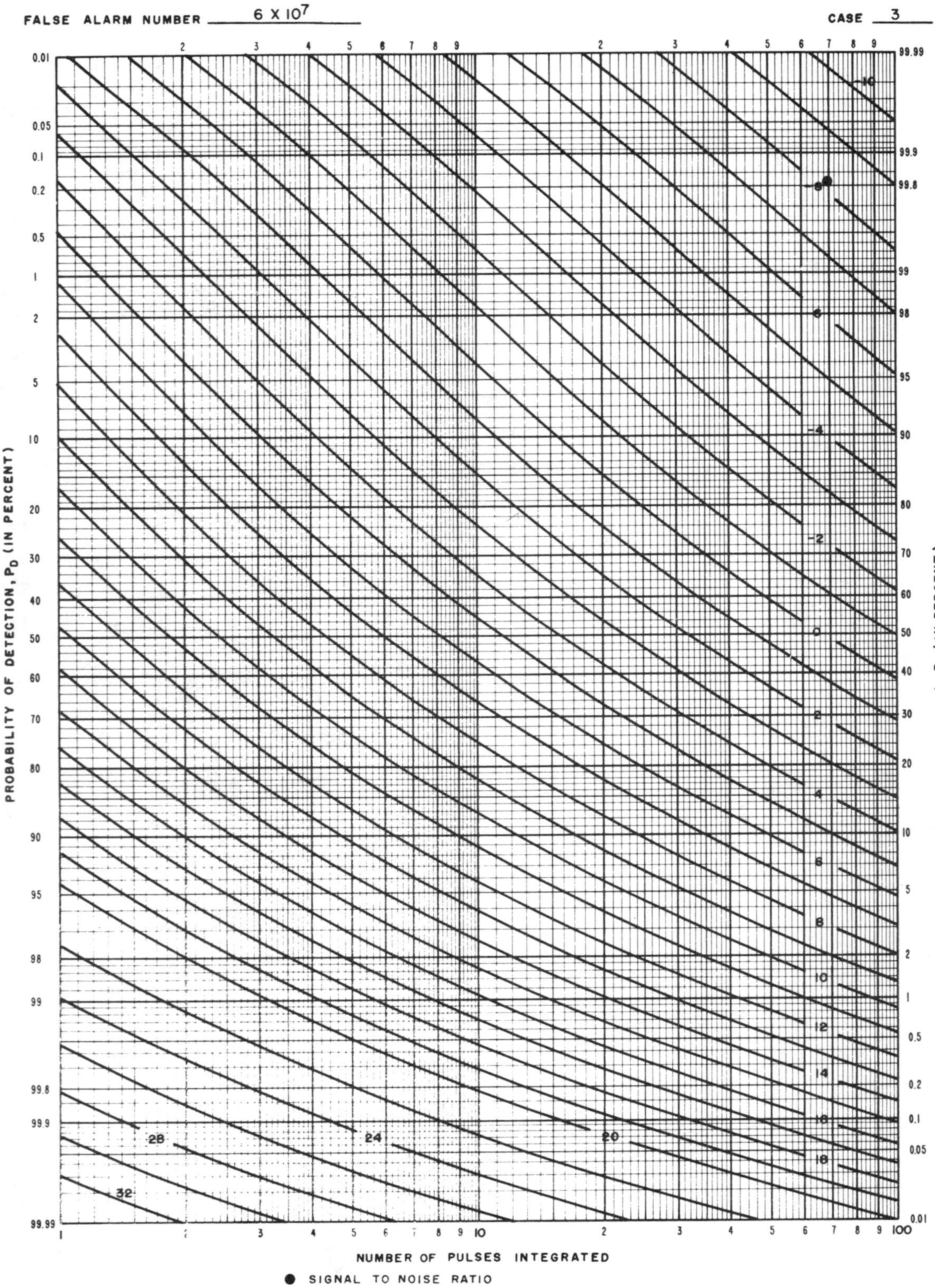
FALSE ALARM NUMBER 6 X 10^7
CASE 3
PROBABILITY OF DETECTION, P_D (IN PERCENT)
$1-P_D$ (IN PERCENT)
-10
-8
-6
-4
-2
0
2
4
6
8
10
12
14
16
18
20
24
28
32
NUMBER OF PULSES INTEGRATED
● SIGNAL TO NOISE RATIO

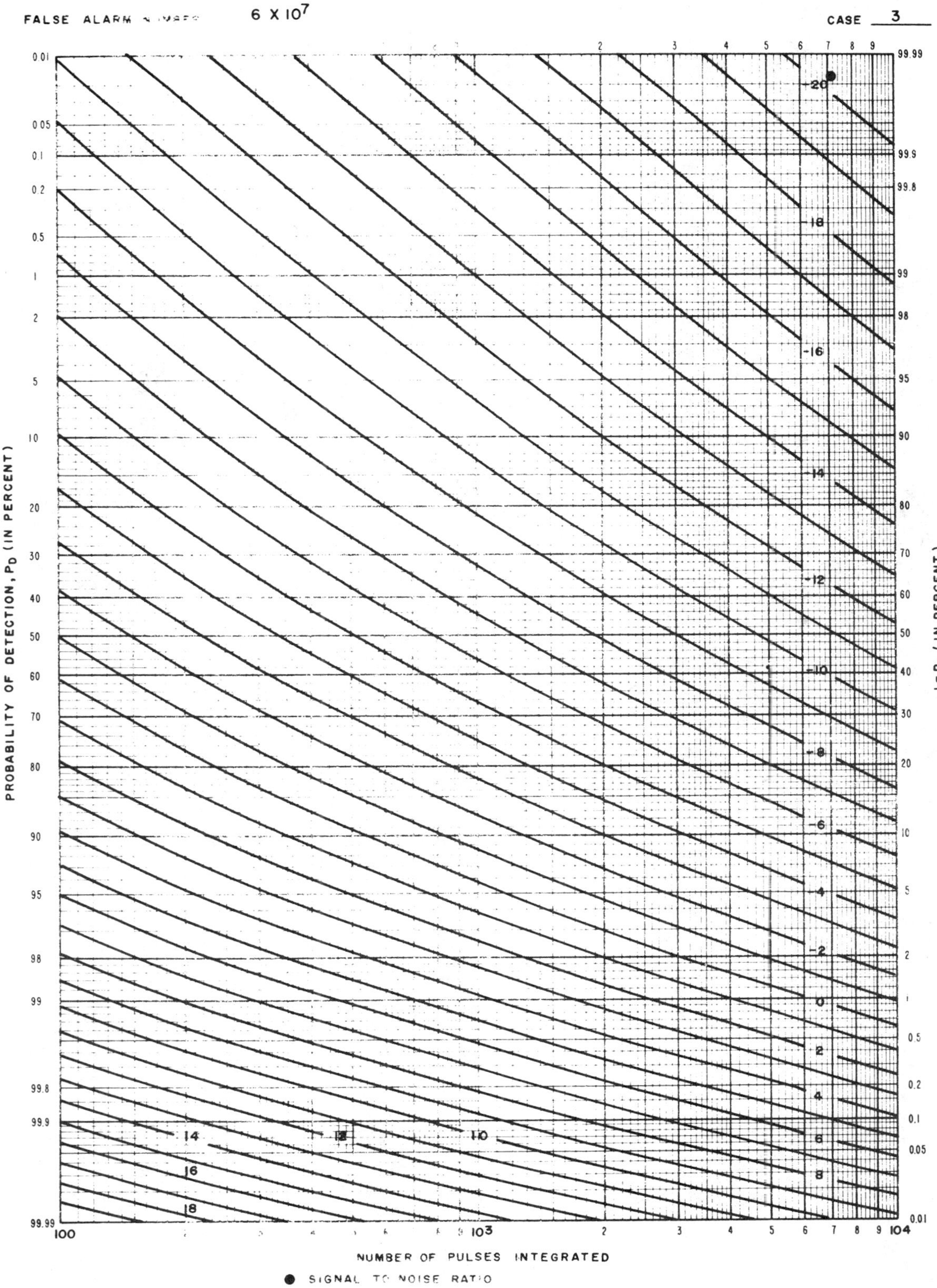
FALSE ALARM
6 X 10^7
CASE 3
PROBABILITY OF DETECTION, P_D (IN PERCENT)
1-P_D (IN PERCENT)
NUMBER OF PULSES INTEGRATED
SIGNAL TO NOISE RATIO

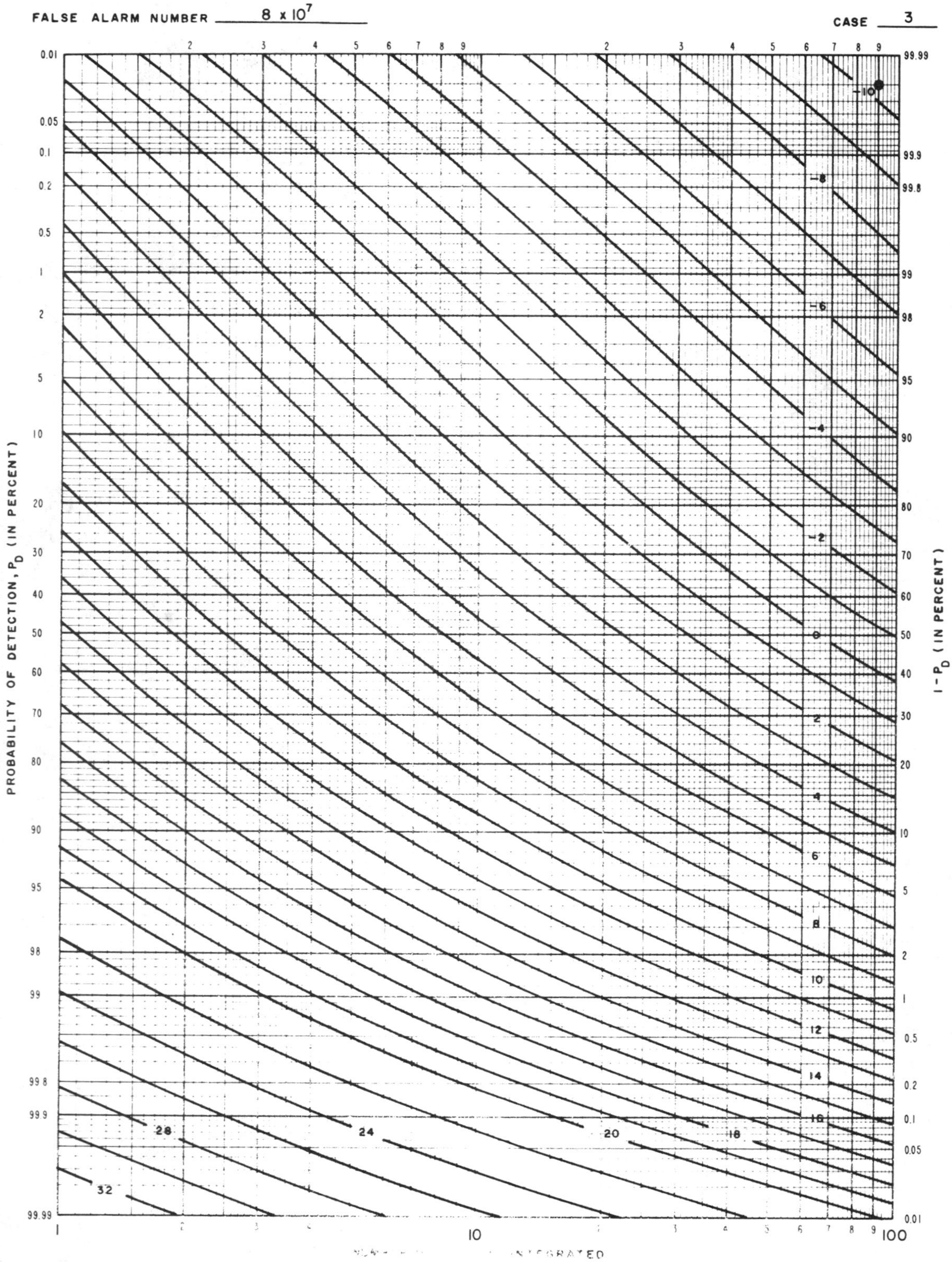
FALSE ALARM NUMBER 8 x 10^7
CASE 3
PROBABILITY OF DETECTION, P_D (IN PERCENT)
1 - P_D (IN PERCENT)
0.01
0.05
0.1
0.2
0.5
1
2
5
10
20
30
40
50
60
70
80
90
95
98
99
99.8
99.9
99.99
99.9
99.8
99
98
95
90
80
70
60
50
40
30
20
10
5
2
1
0.5
0.2
0.1
0.05
0.01
-10
-8
-6
-4
-2
0
2
4
6
8
10
12
14
16
18
20
24
28
32
1
10
100
INTEGRATED

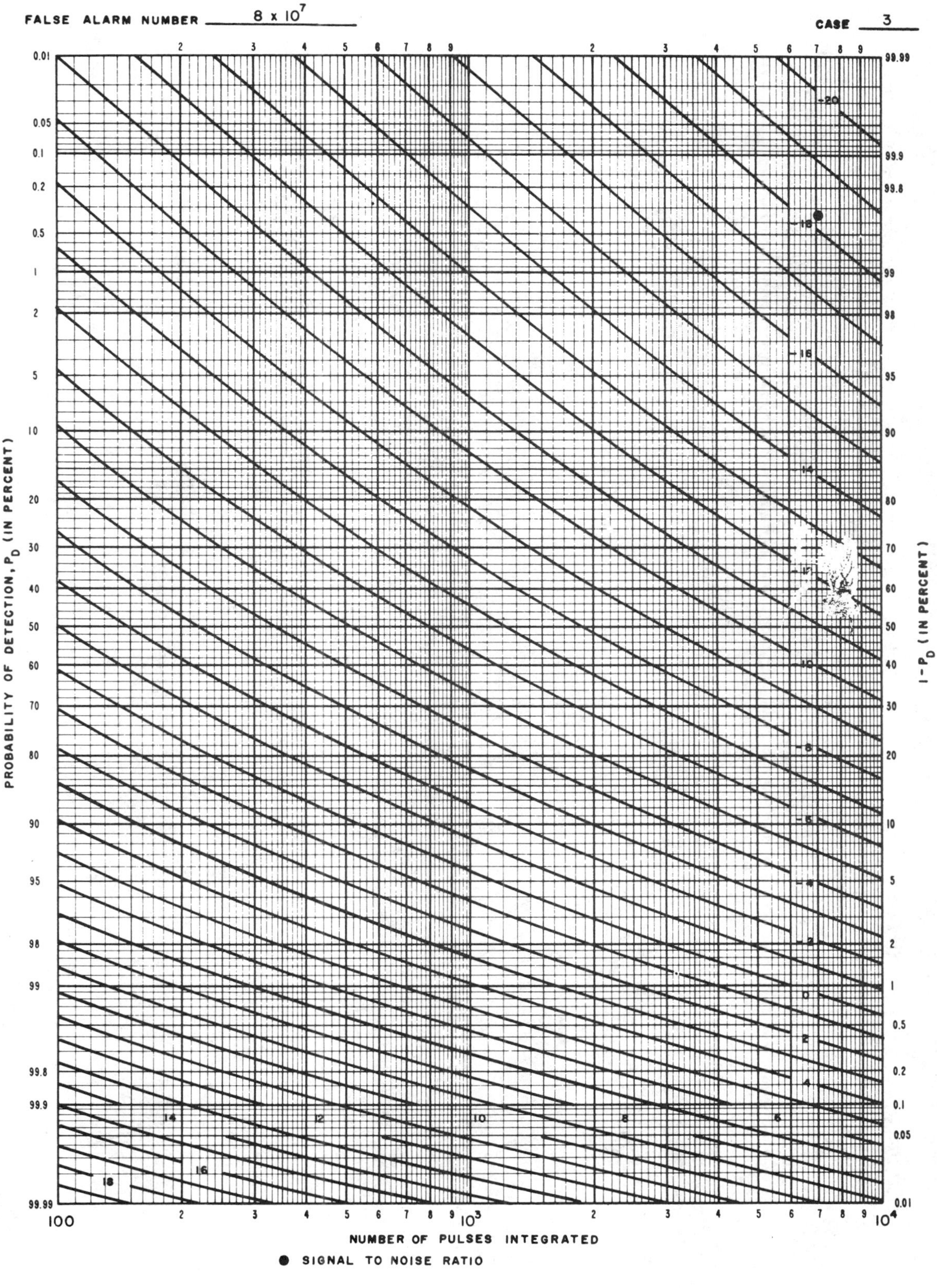
FALSE ALARM NUMBER 8 x 10^7
CASE 3
PROBABILITY OF DETECTION, P_D (IN PERCENT)
1 - P_D (IN PERCENT)
NUMBER OF PULSES INTEGRATED
● SIGNAL TO NOISE RATIO

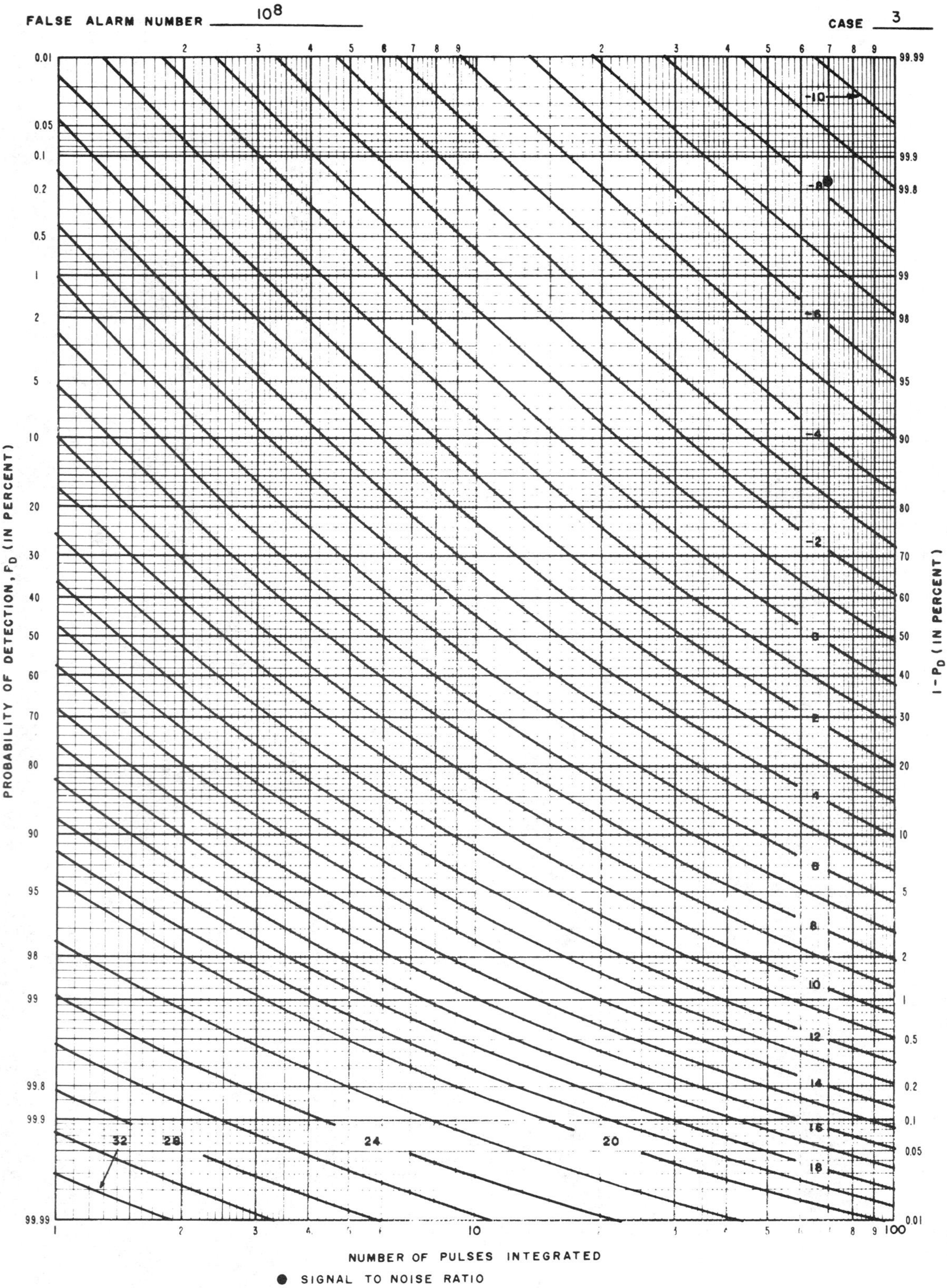
FALSE ALARM NUMBER 10^8
CASE 3
PROBABILITY OF DETECTION, P_D (IN PERCENT)
1 - P_D (IN PERCENT)
NUMBER OF PULSES INTEGRATED
● SIGNAL TO NOISE RATIO
-10
-8
-6
-4
-2
0
2
4
6
8
10
12
14
16
18
20
24
28
32

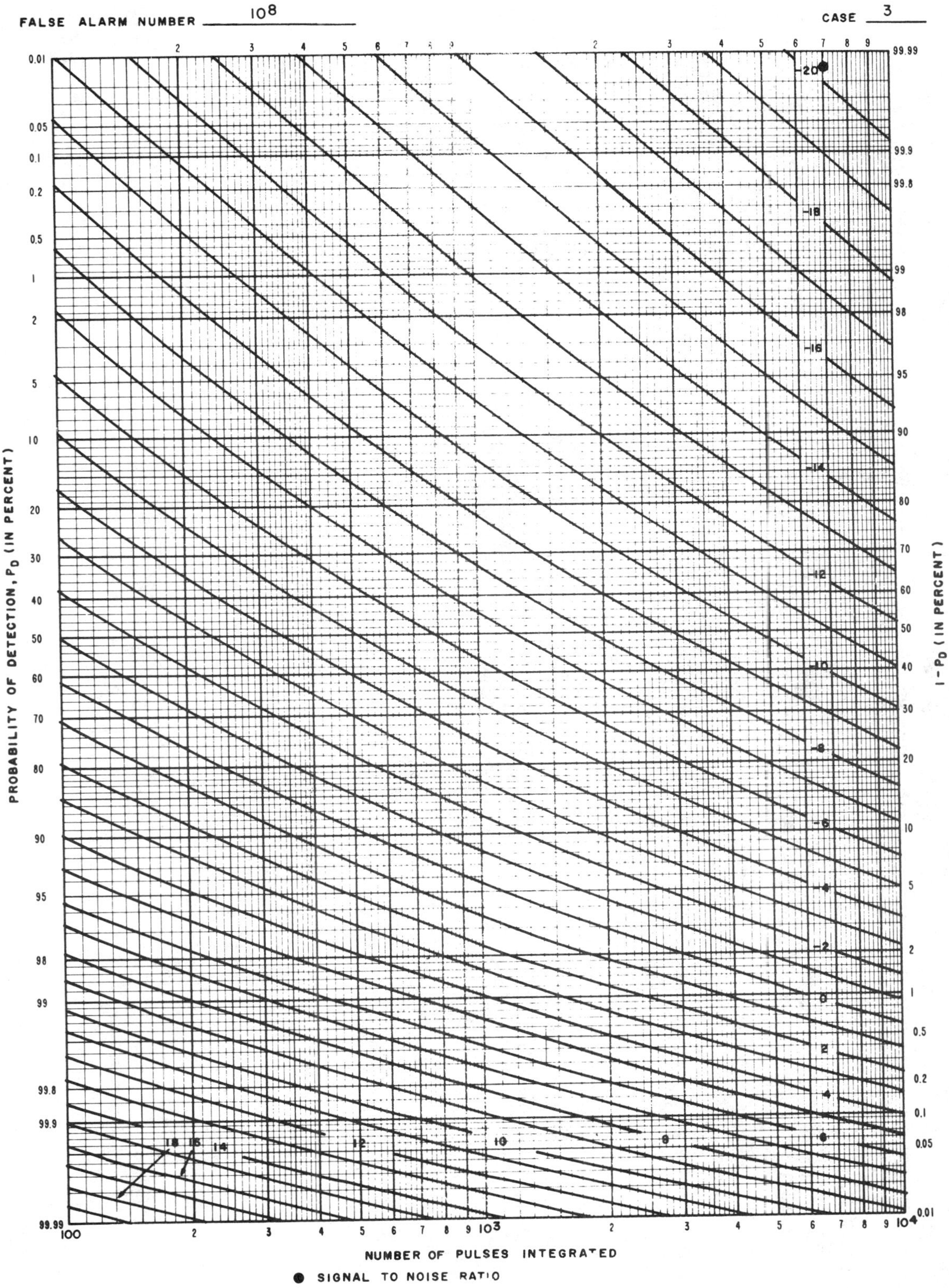
FALSE ALARM NUMBER 10^8
CASE 3
PROBABILITY OF DETECTION, P_D (IN PERCENT)
1 - P_D (IN PERCENT)
NUMBER OF PULSES INTEGRATED
SIGNAL TO NOISE RATIO

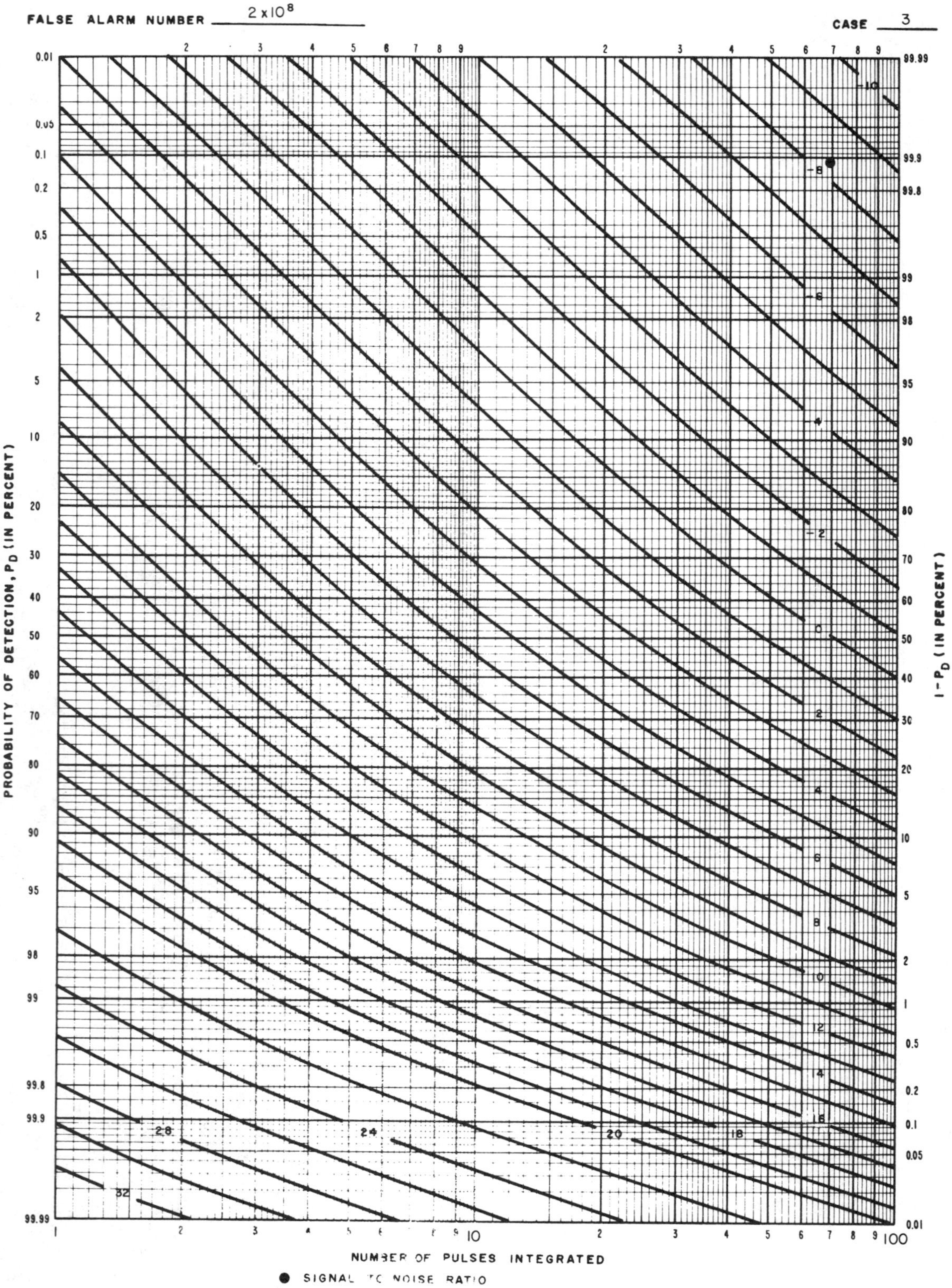
FALSE ALARM NUMBER 2 x 10^8
CASE 3
PROBABILITY OF DETECTION, P_D (IN PERCENT)
1 - P_D (IN PERCENT)
NUMBER OF PULSES INTEGRATED
● SIGNAL TO NOISE RATIO

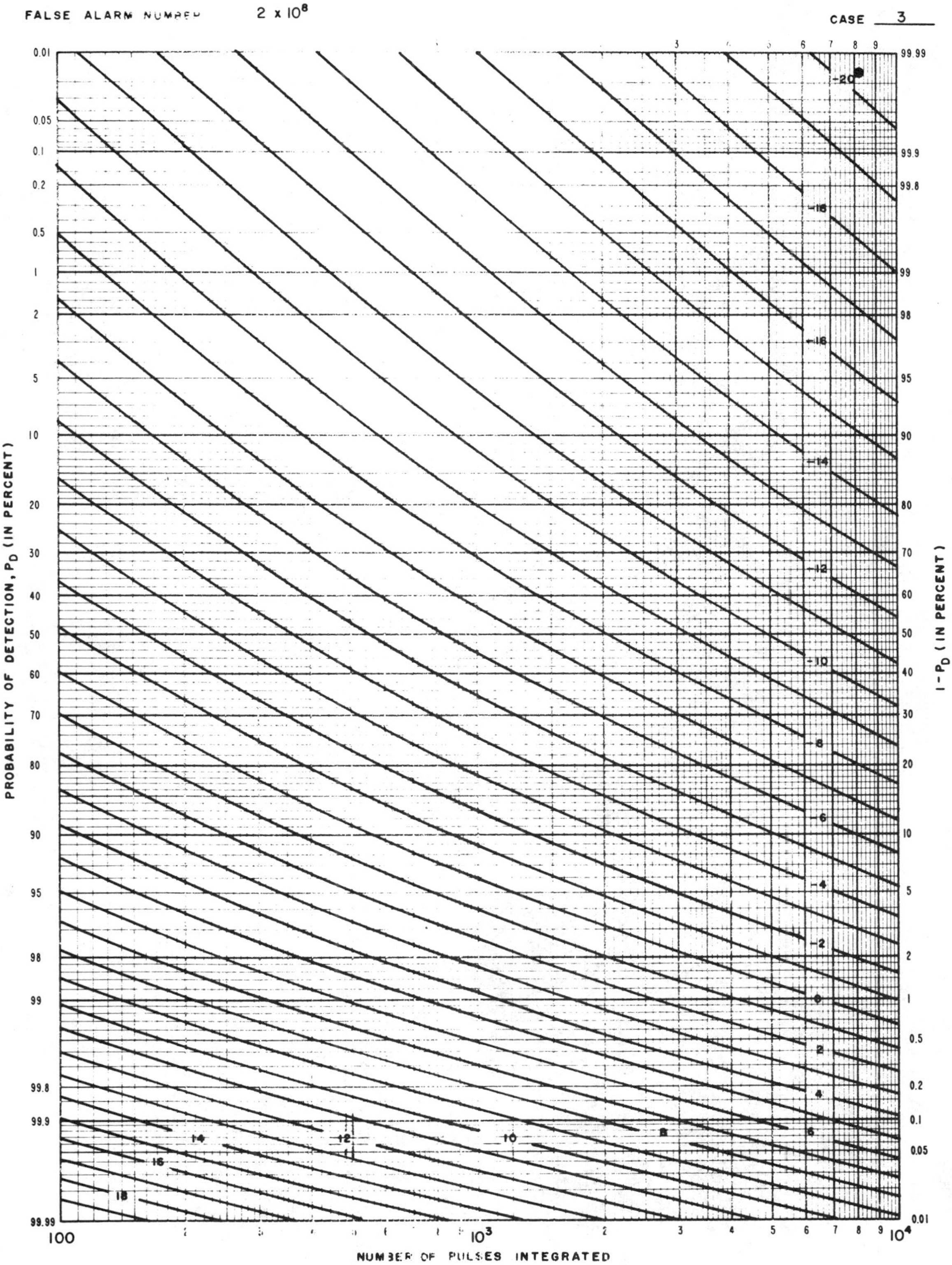

● SIGNAL TO NOISE RATIO

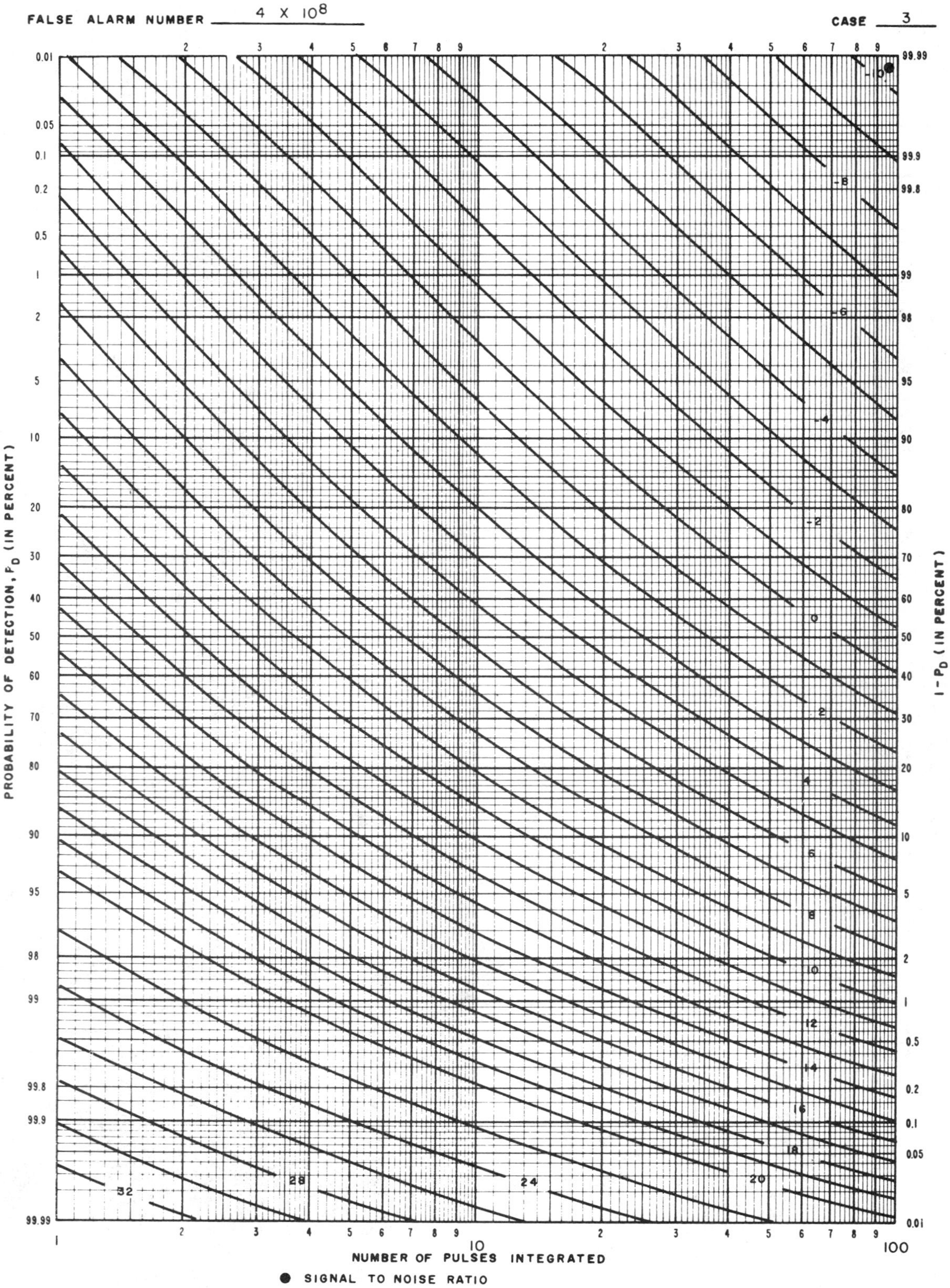
FALSE ALARM NUMBER 4 X 10^8
CASE 3
PROBABILITY OF DETECTION, P_D (IN PERCENT)
1 - P_D (IN PERCENT)
NUMBER OF PULSES INTEGRATED
● SIGNAL TO NOISE RATIO
-10
-8
-6
-4
-2
0
2
4
6
8
10
12
14
16
18
20
24
28
32
0.01
0.05
0.1
0.2
0.5
1
2
5
10
20
30
40
50
60
70
80
90
95
98
99
99.8
99.9
99.99
1
10
100

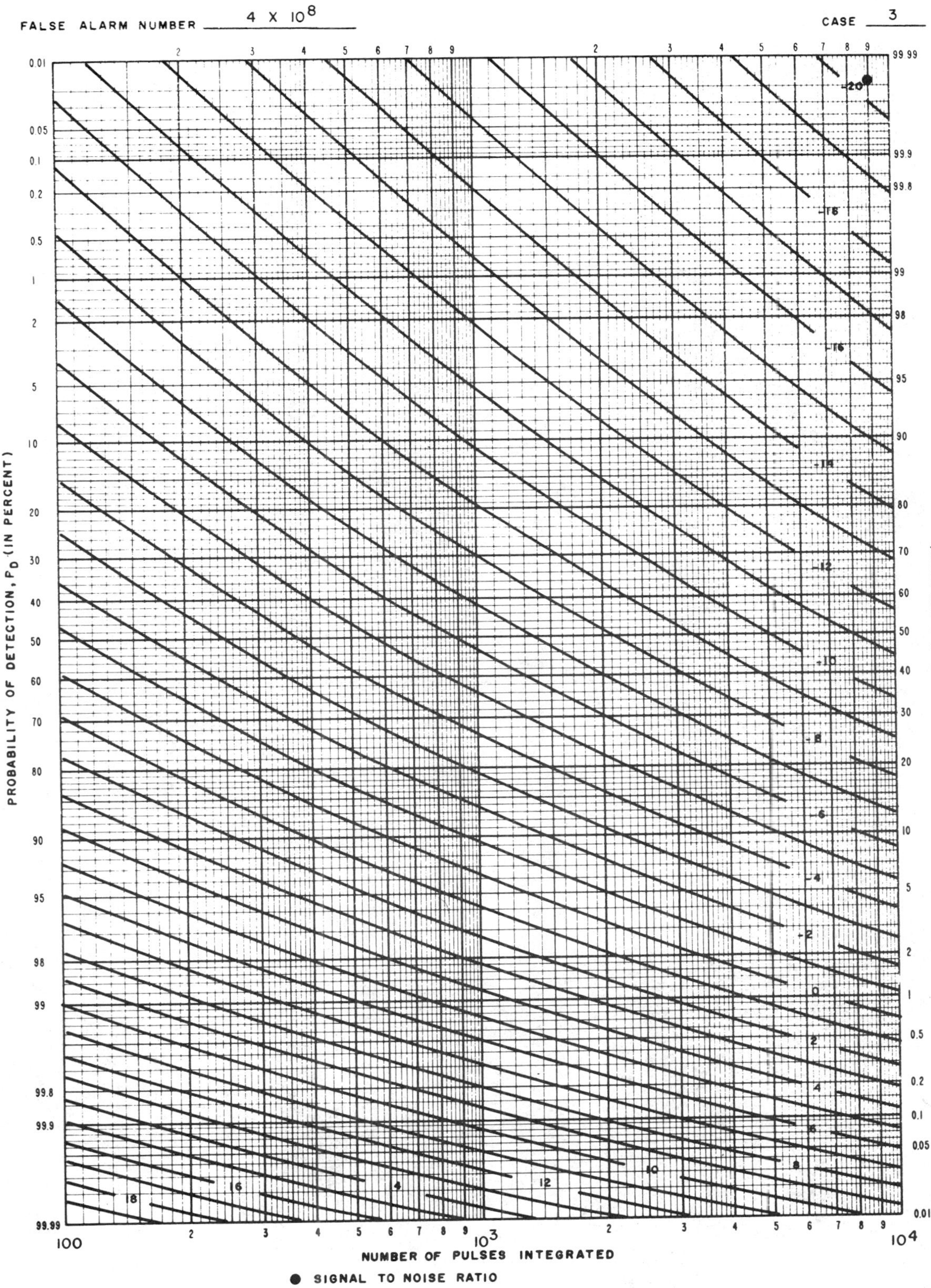
FALSE ALARM NUMBER 4 X 10^8
CASE 3
PROBABILITY OF DETECTION, P_D (IN PERCENT)
1-P_D (IN PERCENT)
NUMBER OF PULSES INTEGRATED
● SIGNAL TO NOISE RATIO
100
10^3
10^4

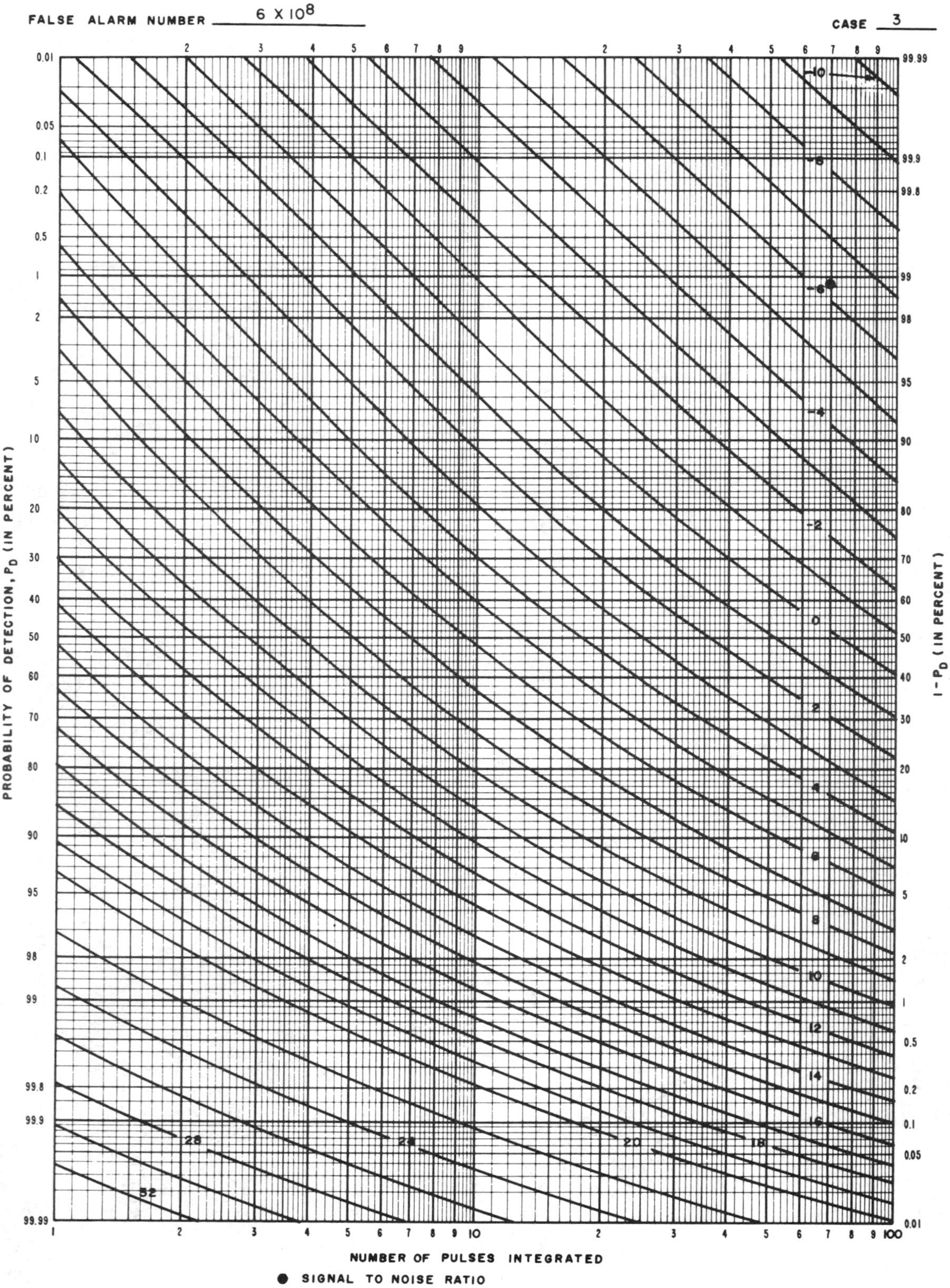
FALSE ALARM NUMBER 6 X 10^8
CASE 3
PROBABILITY OF DETECTION, P_D (IN PERCENT)
1 - P_D (IN PERCENT)
NUMBER OF PULSES INTEGRATED
● SIGNAL TO NOISE RATIO
-10
-8
-6
-4
-2
0
2
4
6
8
10
12
14
16
18
20
24
28
32

FALSE ALARM NUMBER 6×10^8 CASE 3

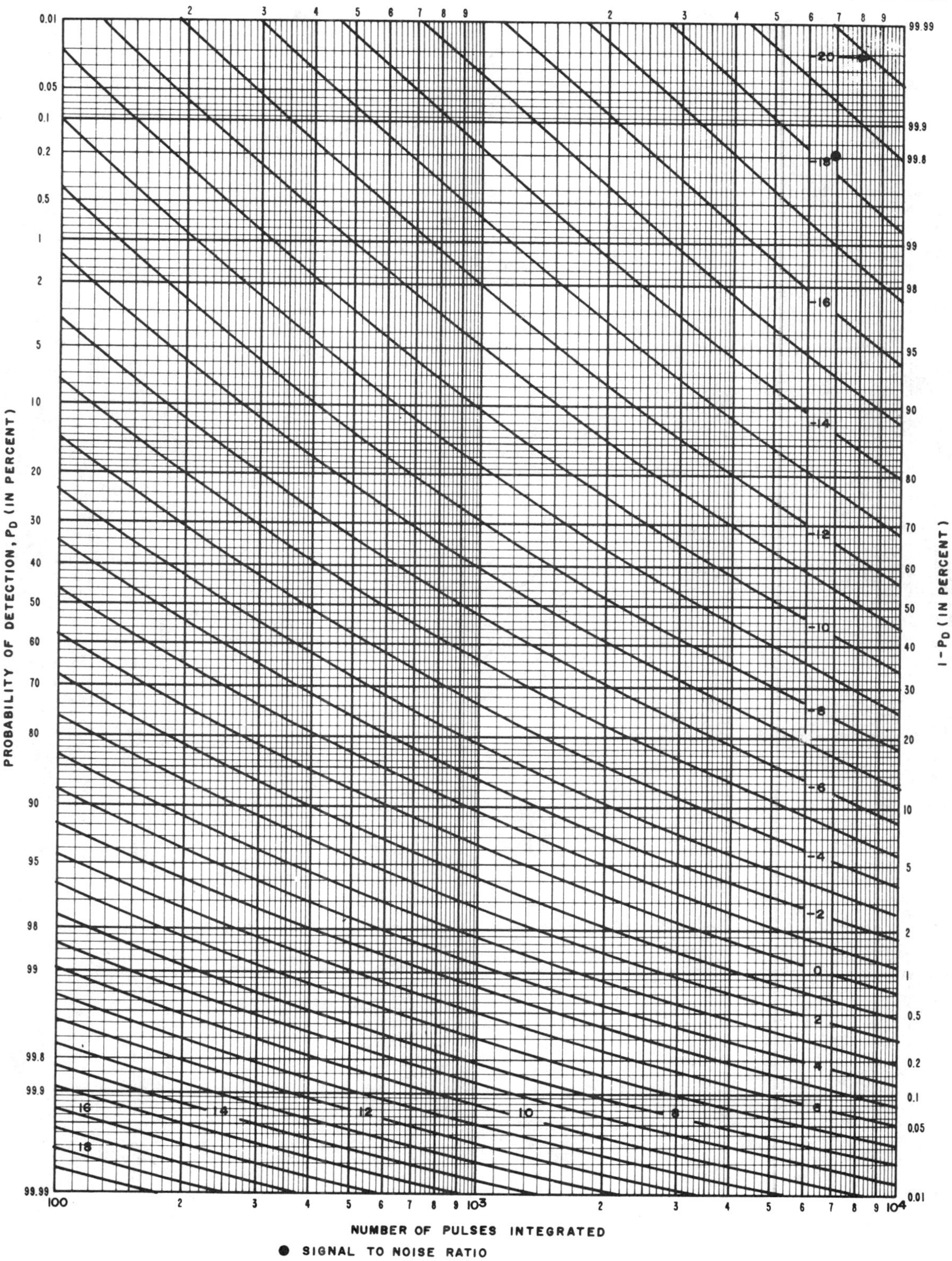

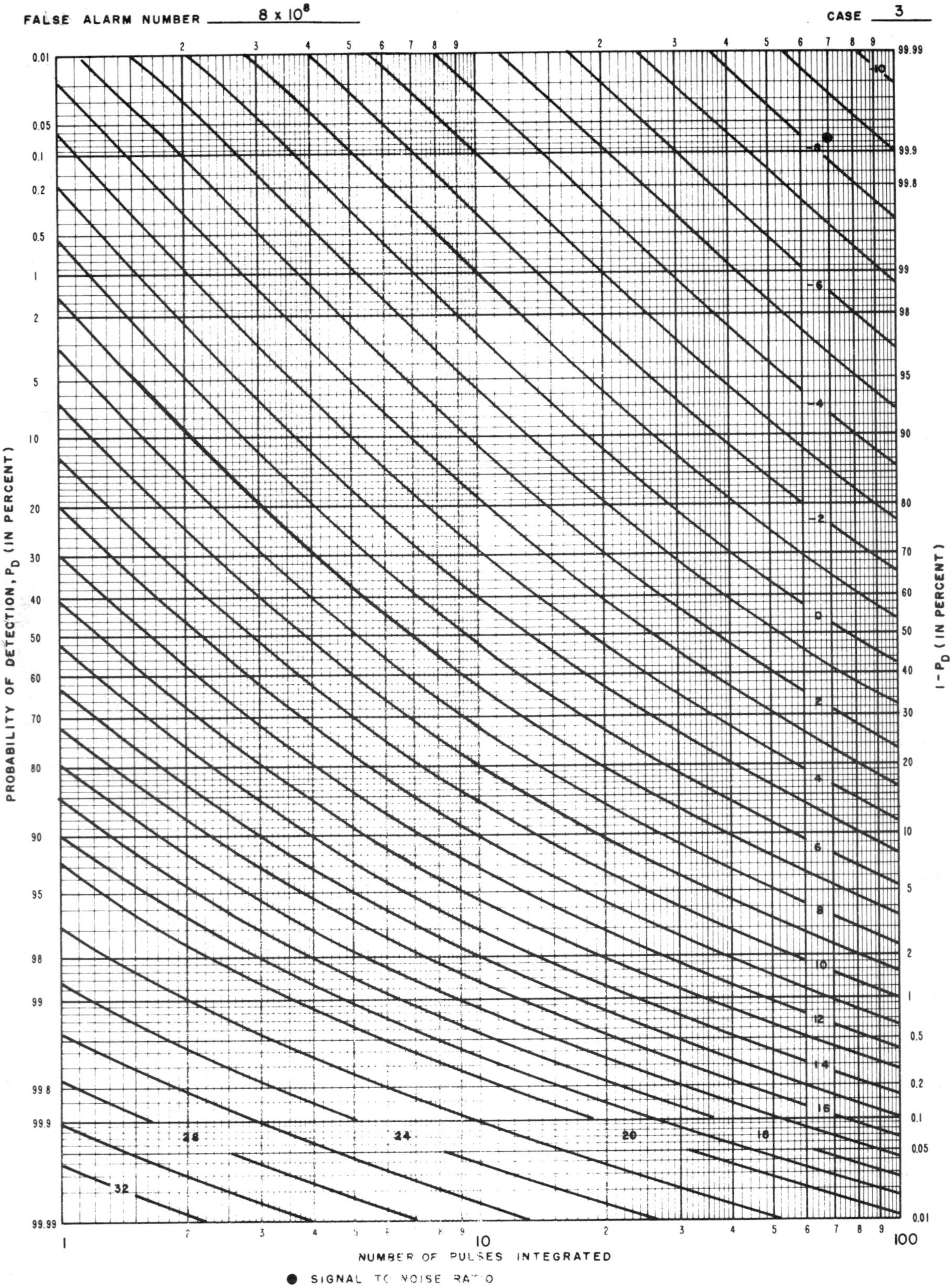
FALSE ALARM NUMBER 8 x 10^8
CASE 3
PROBABILITY OF DETECTION, P_D (IN PERCENT)
1 - P_D (IN PERCENT)
NUMBER OF PULSES INTEGRATED
● SIGNAL TO NOISE RATIO
32
28
24
20
18
16
14
12
10
8
6
4
2
0
-2
-4
-6
-8
-10

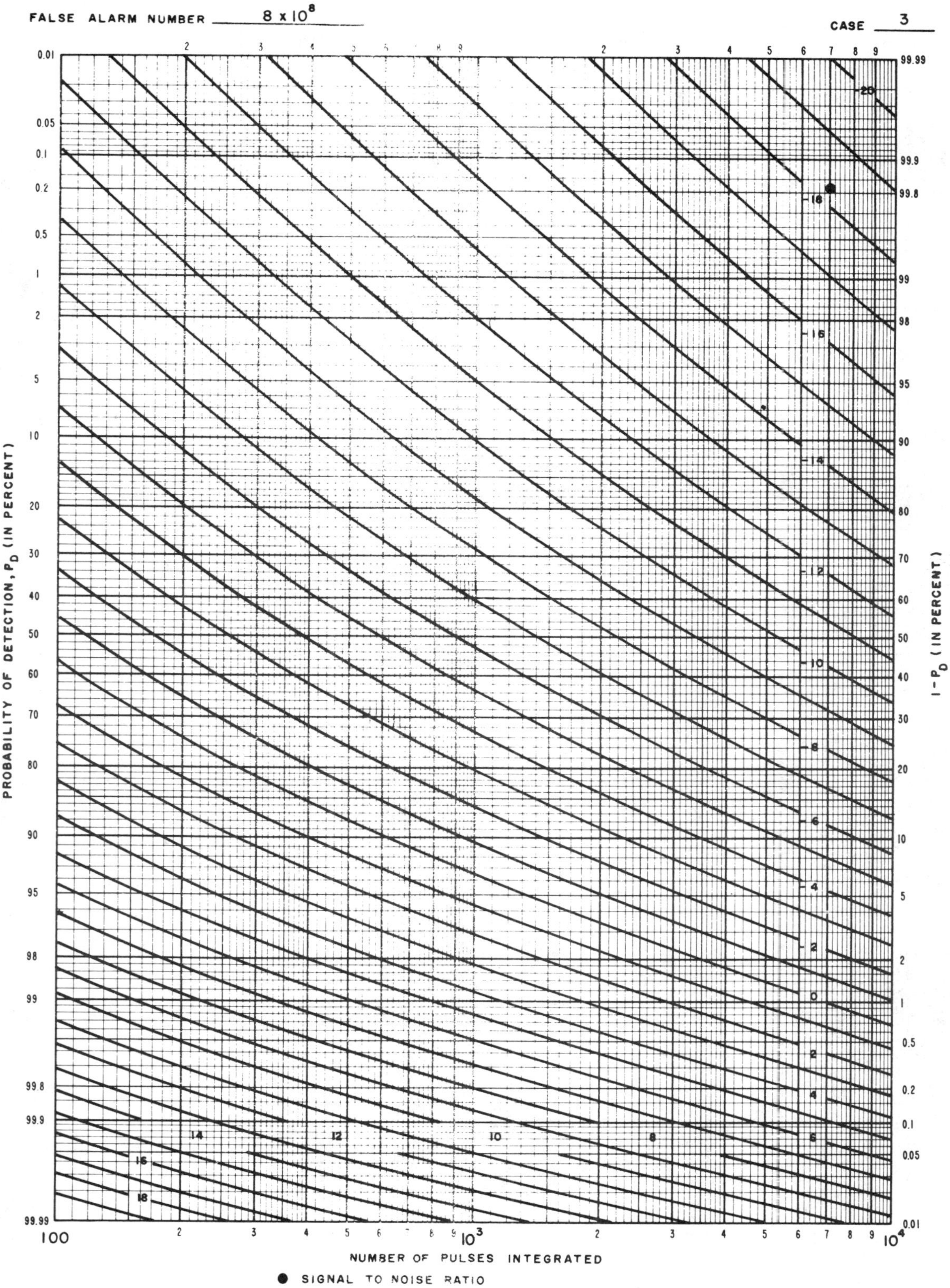
FALSE ALARM NUMBER 8 x 10^8
CASE 3
PROBABILITY OF DETECTION, P_D (IN PERCENT)
1 - P_D (IN PERCENT)
NUMBER OF PULSES INTEGRATED
● SIGNAL TO NOISE RATIO
100
10^3
10^4
20
18
16
14
12
10
8
6
4
2
0
-2
-4
-6

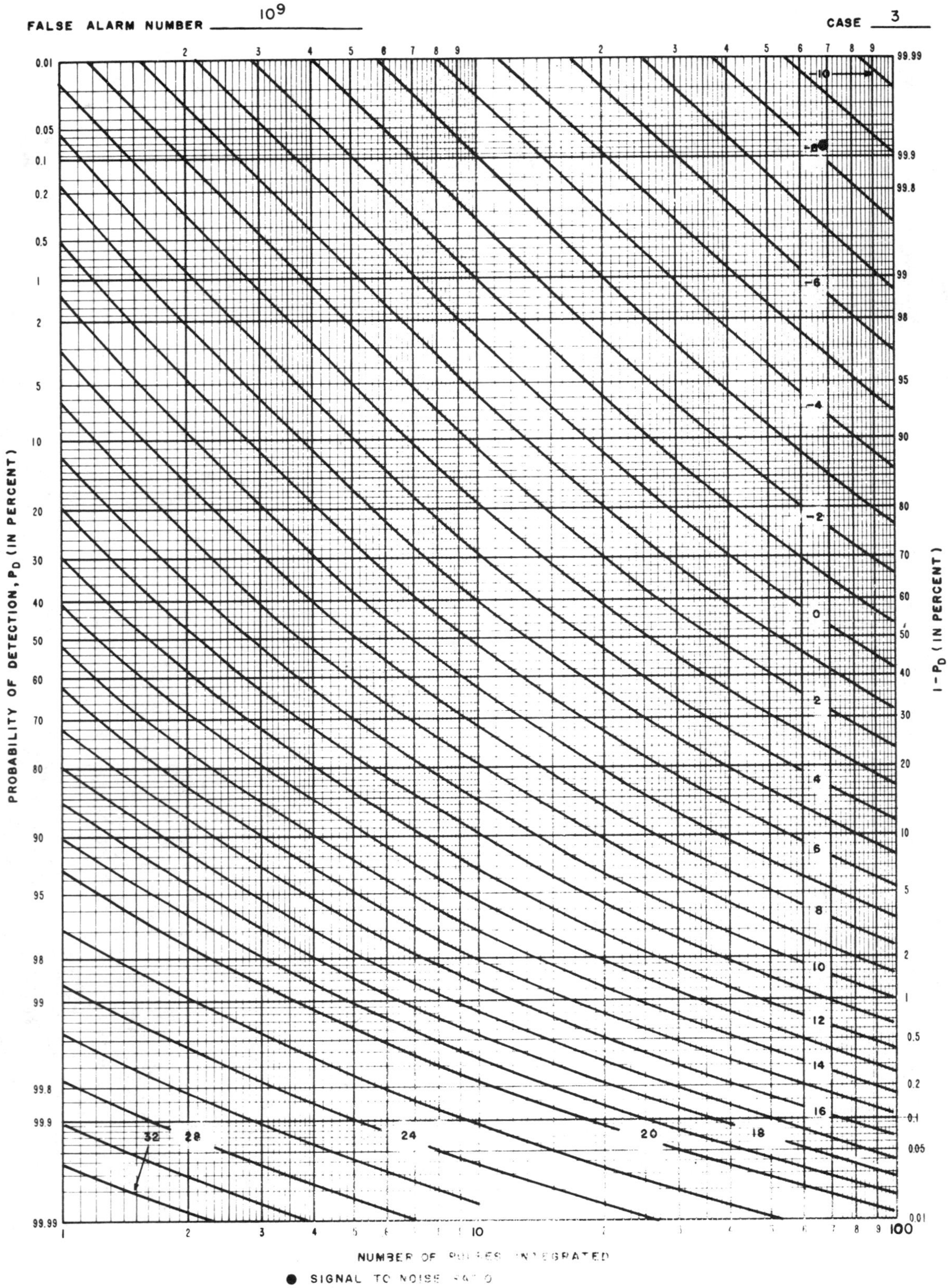
FALSE ALARM NUMBER 10^9
CASE 3
PROBABILITY OF DETECTION, P_D (IN PERCENT)
$1 - P_D$ (IN PERCENT)
-10
-6
-4
-2
0
2
4
6
8
10
12
14
16
18
20
24
28
32

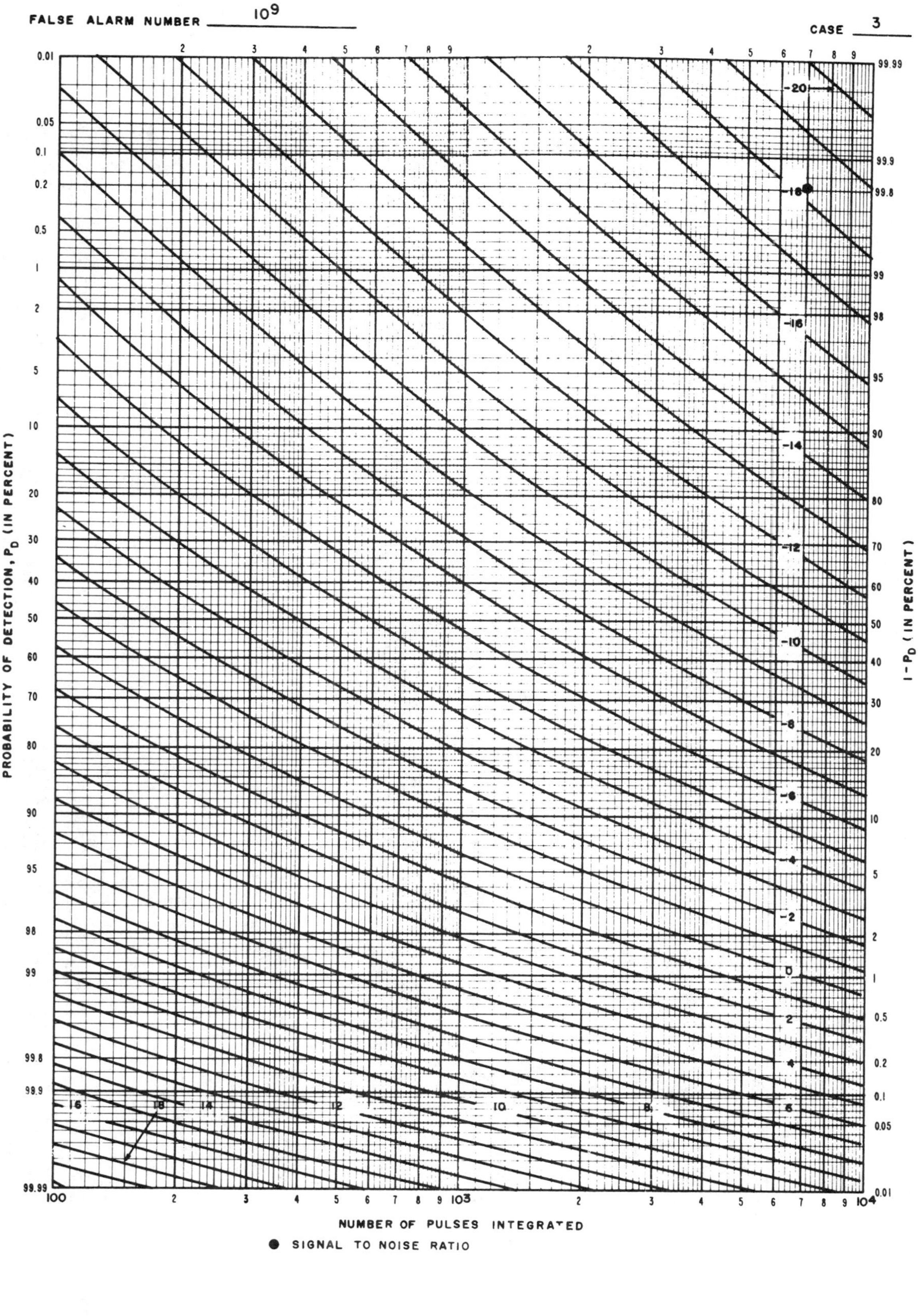
FALSE ALARM NUMBER 10^9
CASE 3
PROBABILITY OF DETECTION, P_D (IN PERCENT)
$1-P_D$ (IN PERCENT)
NUMBER OF PULSES INTEGRATED
● SIGNAL TO NOISE RATIO

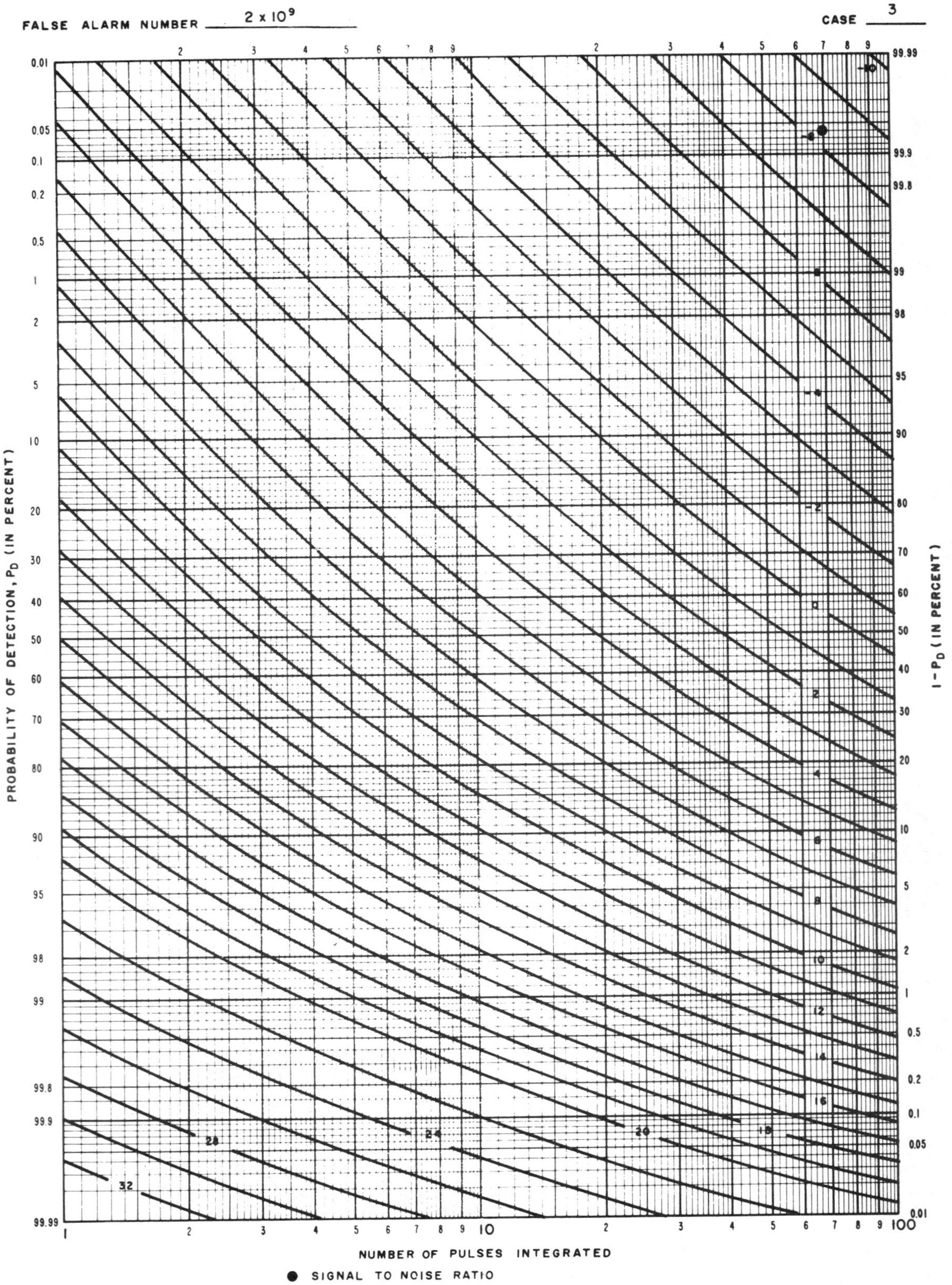
FALSE ALARM NUMBER 2 x 10^9
CASE 3
PROBABILITY OF DETECTION, P_D (IN PERCENT)
1 - P_D (IN PERCENT)
NUMBER OF PULSES INTEGRATED
● SIGNAL TO NOISE RATIO
-10
-8
-6
-4
-2
0
2
4
6
8
10
12
14
16
18
20
24
28
32

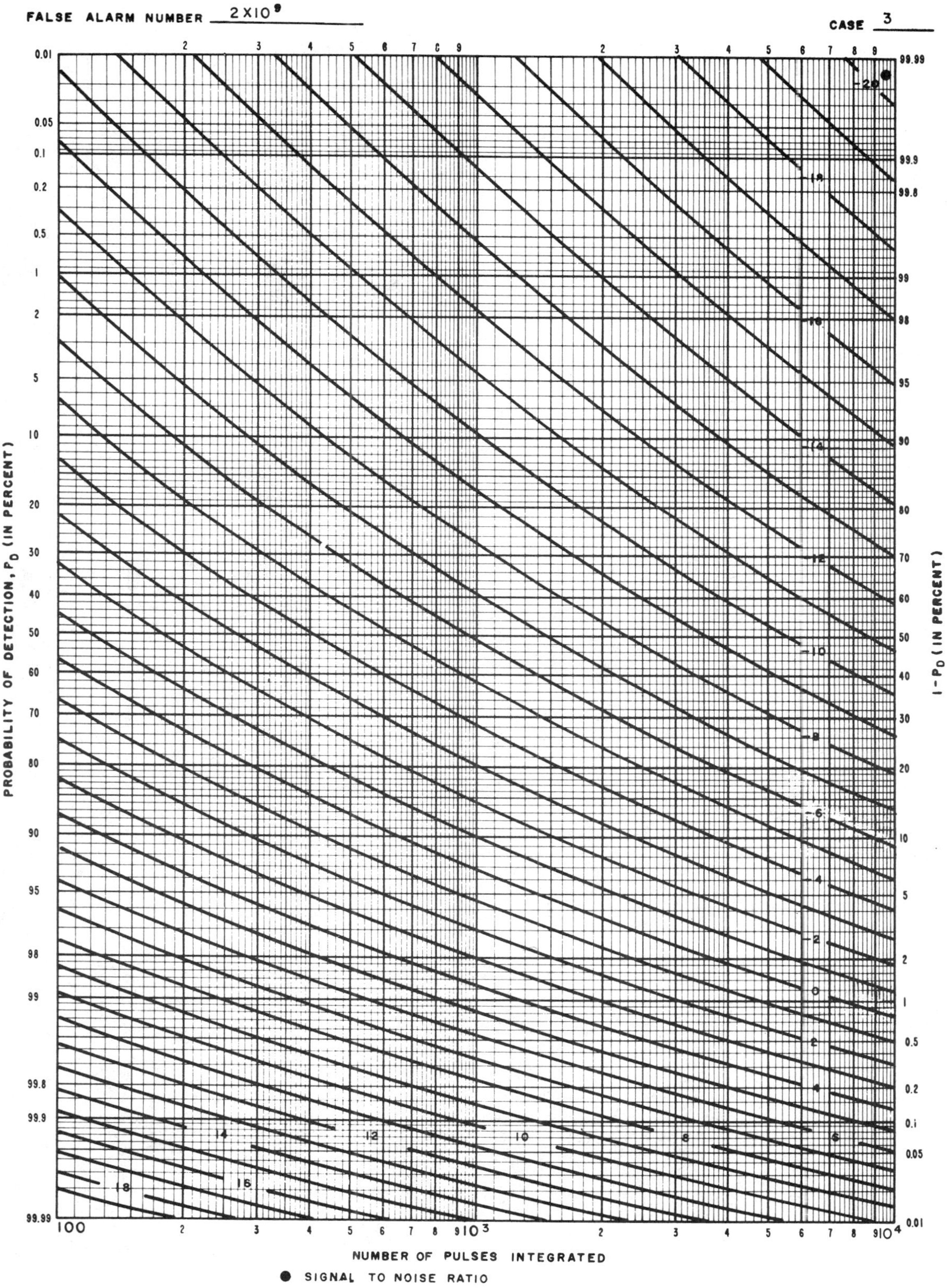
FALSE ALARM NUMBER 2X10⁹
CASE 3
PROBABILITY OF DETECTION, P_D (IN PERCENT)
1 - P_D (IN PERCENT)
NUMBER OF PULSES INTEGRATED
● SIGNAL TO NOISE RATIO
100
10^3
10^4
-20
18
16
14
12
10
8
6
4
2
0

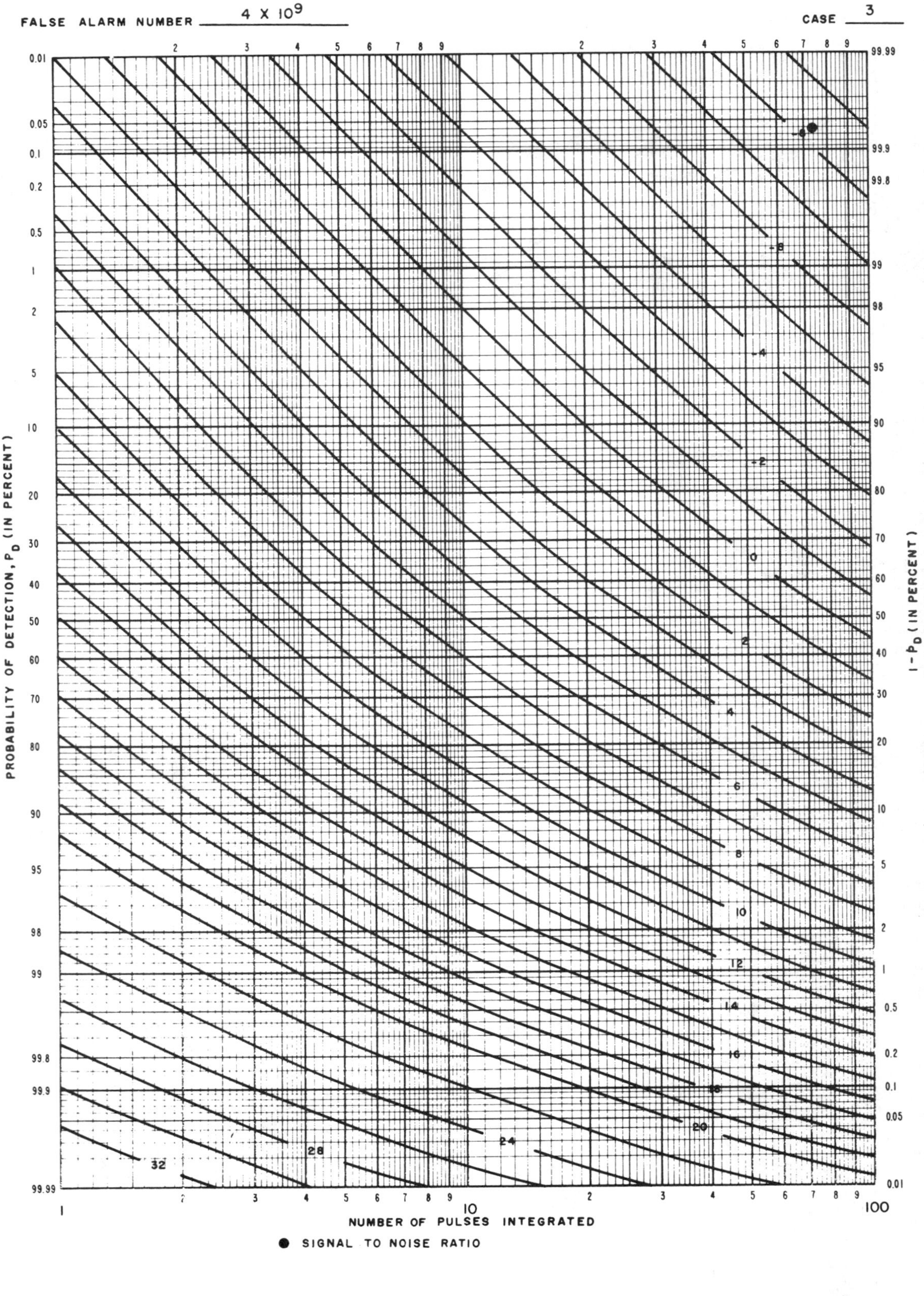
FALSE ALARM NUMBER 4 X 10^9
CASE 3
PROBABILITY OF DETECTION, P_D (IN PERCENT)
$1-P_D$ (IN PERCENT)
NUMBER OF PULSES INTEGRATED
● SIGNAL TO NOISE RATIO
-6
-4
-2
0
2
4
6
8
10
12
14
16
18
20
24
28
32

FALSE ALARM NUMBER 4×10^9 CASE 3

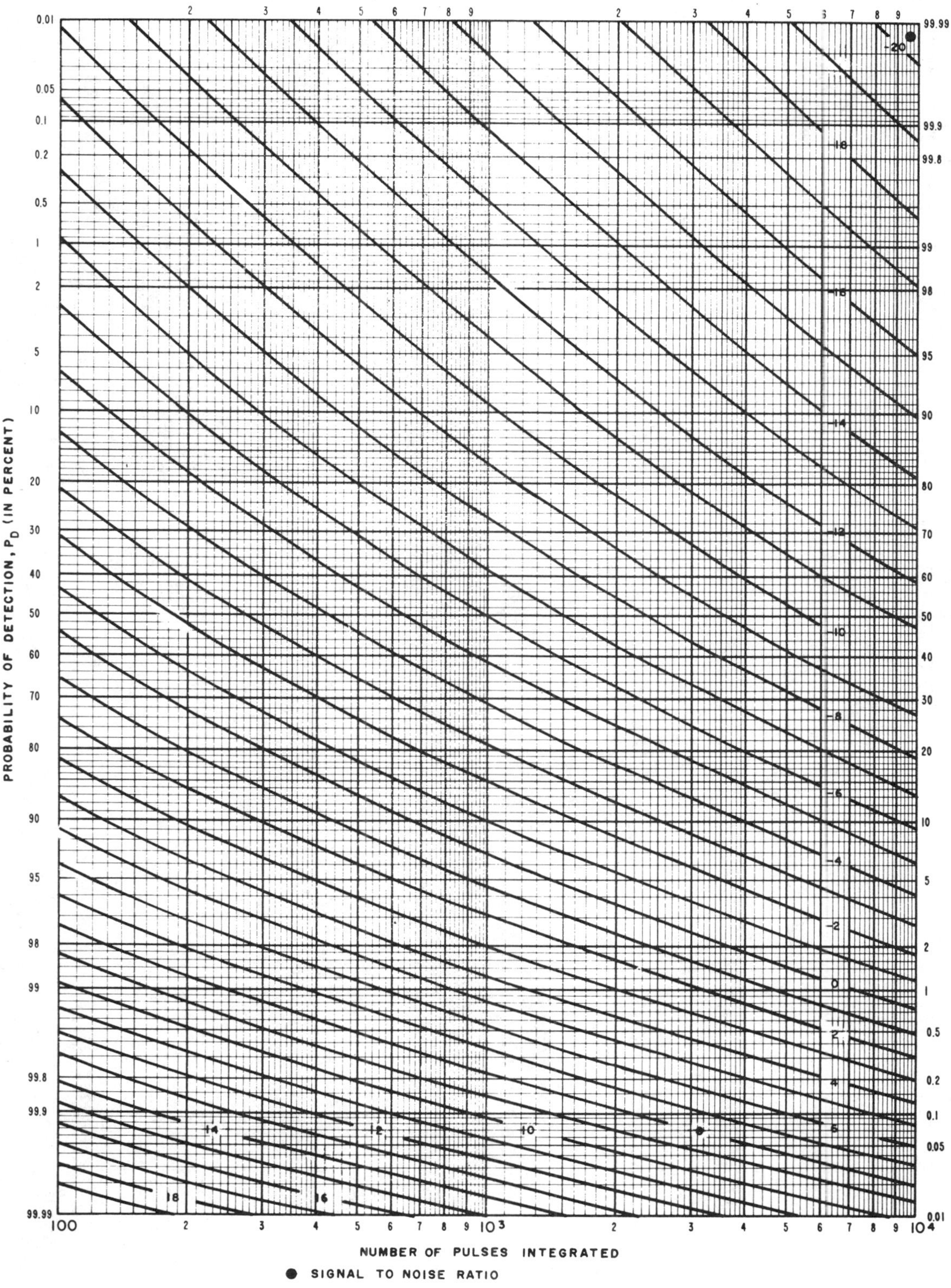

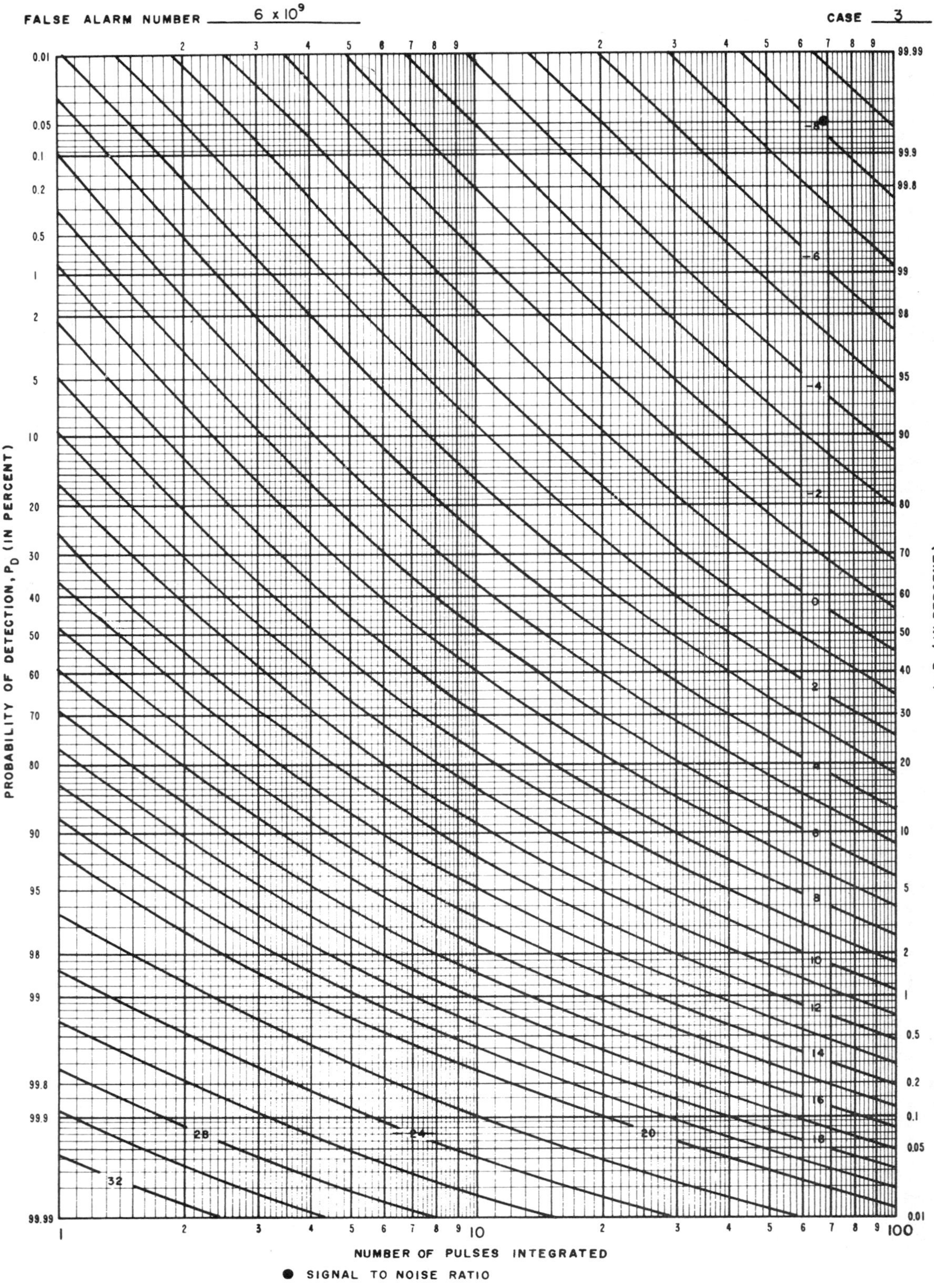
FALSE ALARM NUMBER 6 x 10^9
CASE 3
PROBABILITY OF DETECTION, P_D (IN PERCENT)
1 - P_D (IN PERCENT)
NUMBER OF PULSES INTEGRATED
● SIGNAL TO NOISE RATIO
-8
-6
-4
-2
0
2
4
6
8
10
12
14
16
18
20
24
28
32

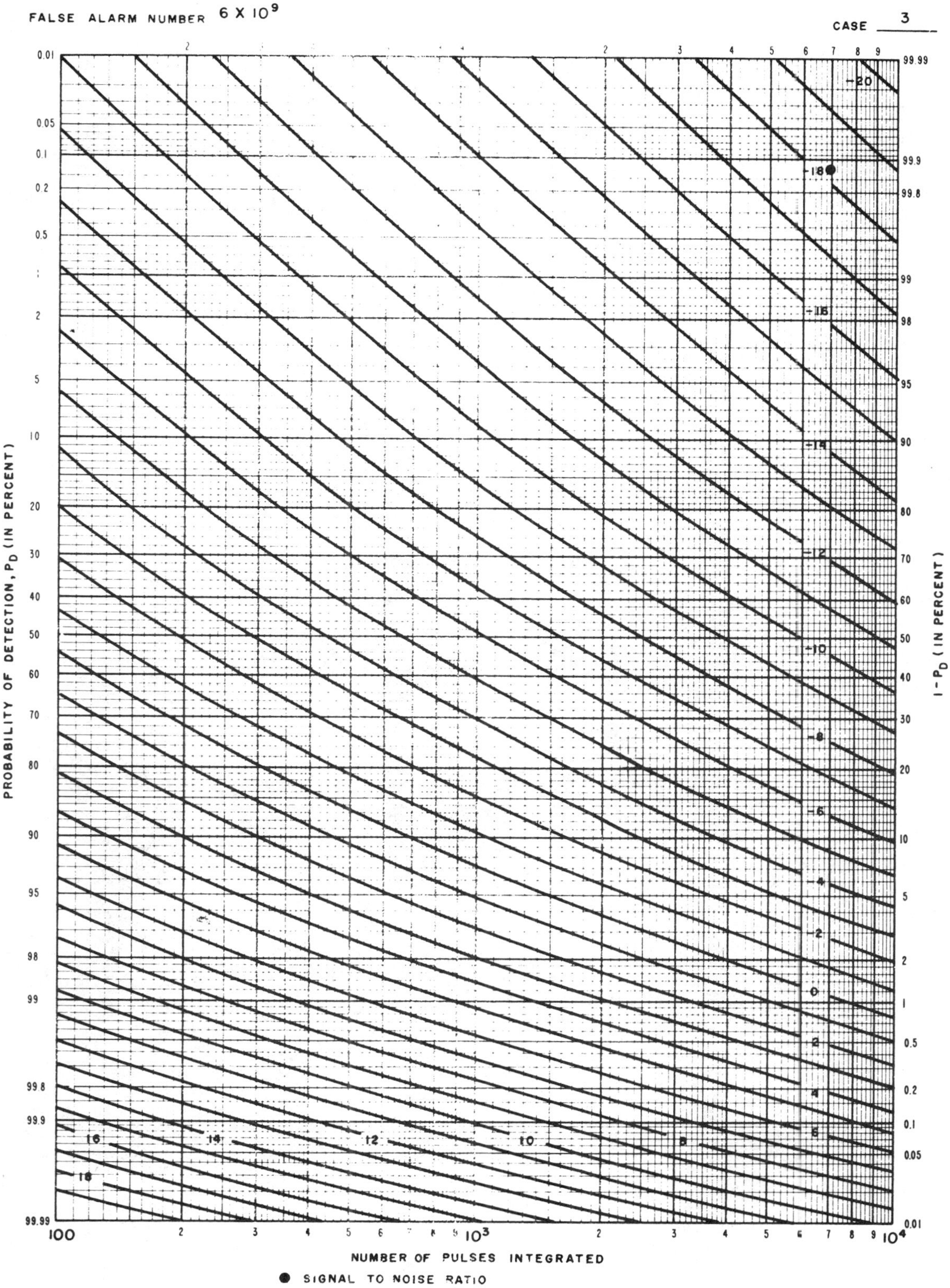

FALSE ALARM NUMBER 6 X 10^9
CASE 3
PROBABILITY OF DETECTION, P_D (IN PERCENT)
1 - P_D (IN PERCENT)
NUMBER OF PULSES INTEGRATED
SIGNAL TO NOISE RATIO
-20
-18
-16
-14
-12
-10
-8
-6
-4
-2
0
2
4
6
8
10
12
14
16
18

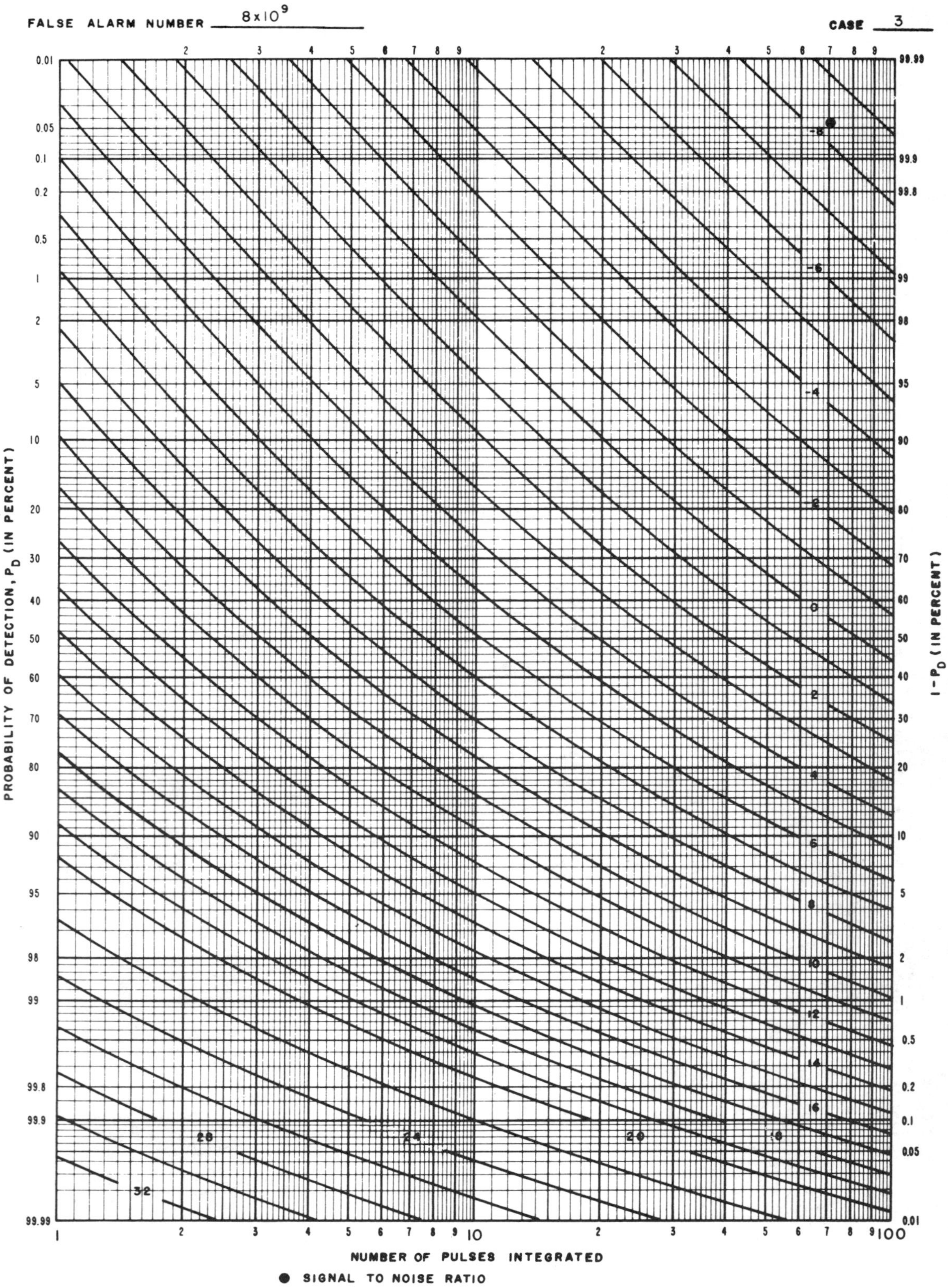
FALSE ALARM NUMBER 8 x 10^9
CASE 3
PROBABILITY OF DETECTION, P_D (IN PERCENT)
$1 - P_D$ (IN PERCENT)
NUMBER OF PULSES INTEGRATED
● SIGNAL TO NOISE RATIO

FALSE ALARM NUMBER 8×10^9

CASE 3

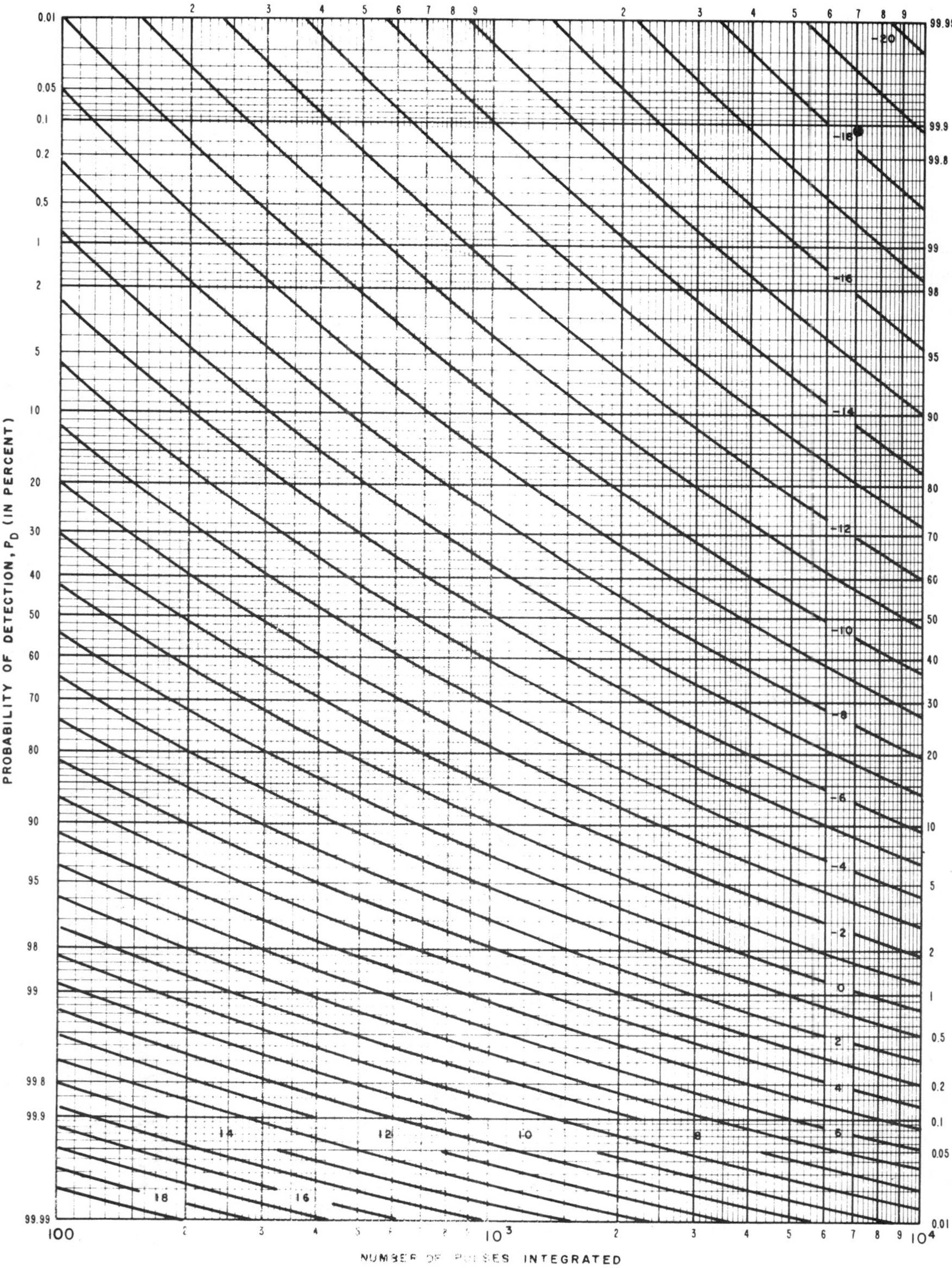

● SIGNAL TO NOISE RATIO

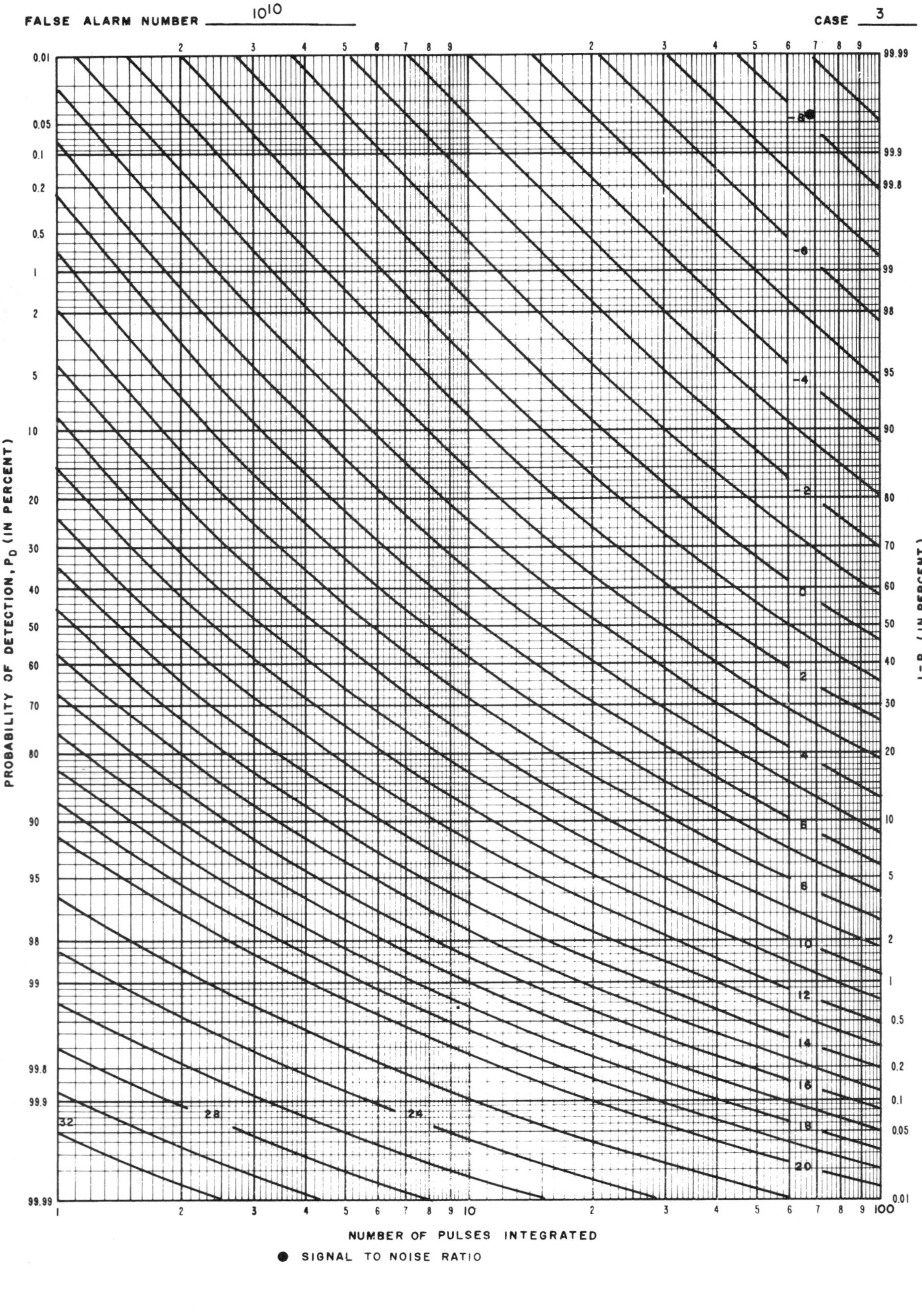
FALSE ALARM NUMBER 10^{10}
CASE 3
PROBABILITY OF DETECTION, P_D (IN PERCENT)
1 − P_D (IN PERCENT)
NUMBER OF PULSES INTEGRATED
● SIGNAL TO NOISE RATIO
-8
-6
-4
-2
0
2
4
6
8
10
12
14
16
18
20
24
28
32

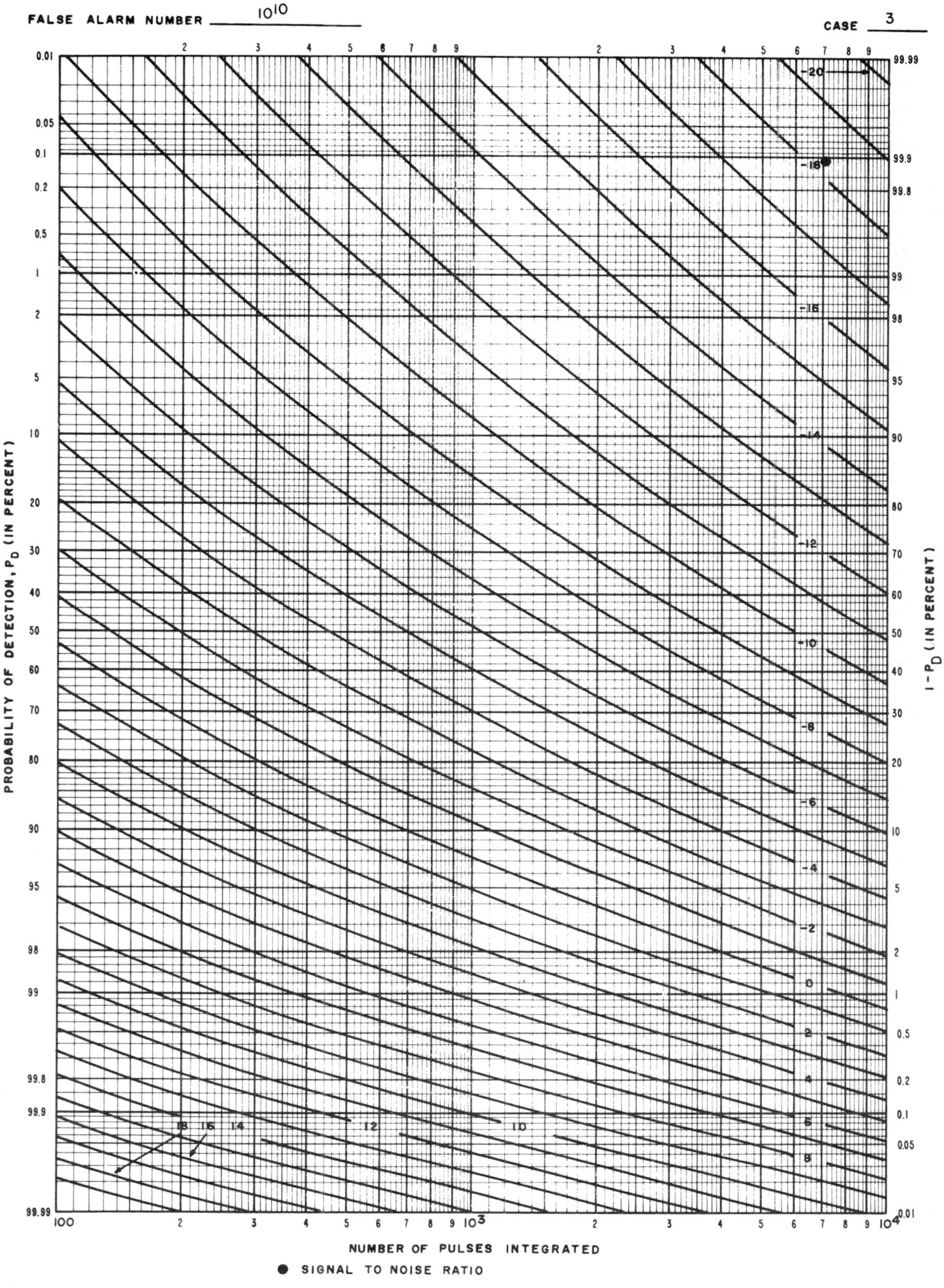
FALSE ALARM NUMBER 10^{10}
CASE 3
PROBABILITY OF DETECTION, P_D (IN PERCENT)
$1-P_D$ (IN PERCENT)
NUMBER OF PULSES INTEGRATED
● SIGNAL TO NOISE RATIO

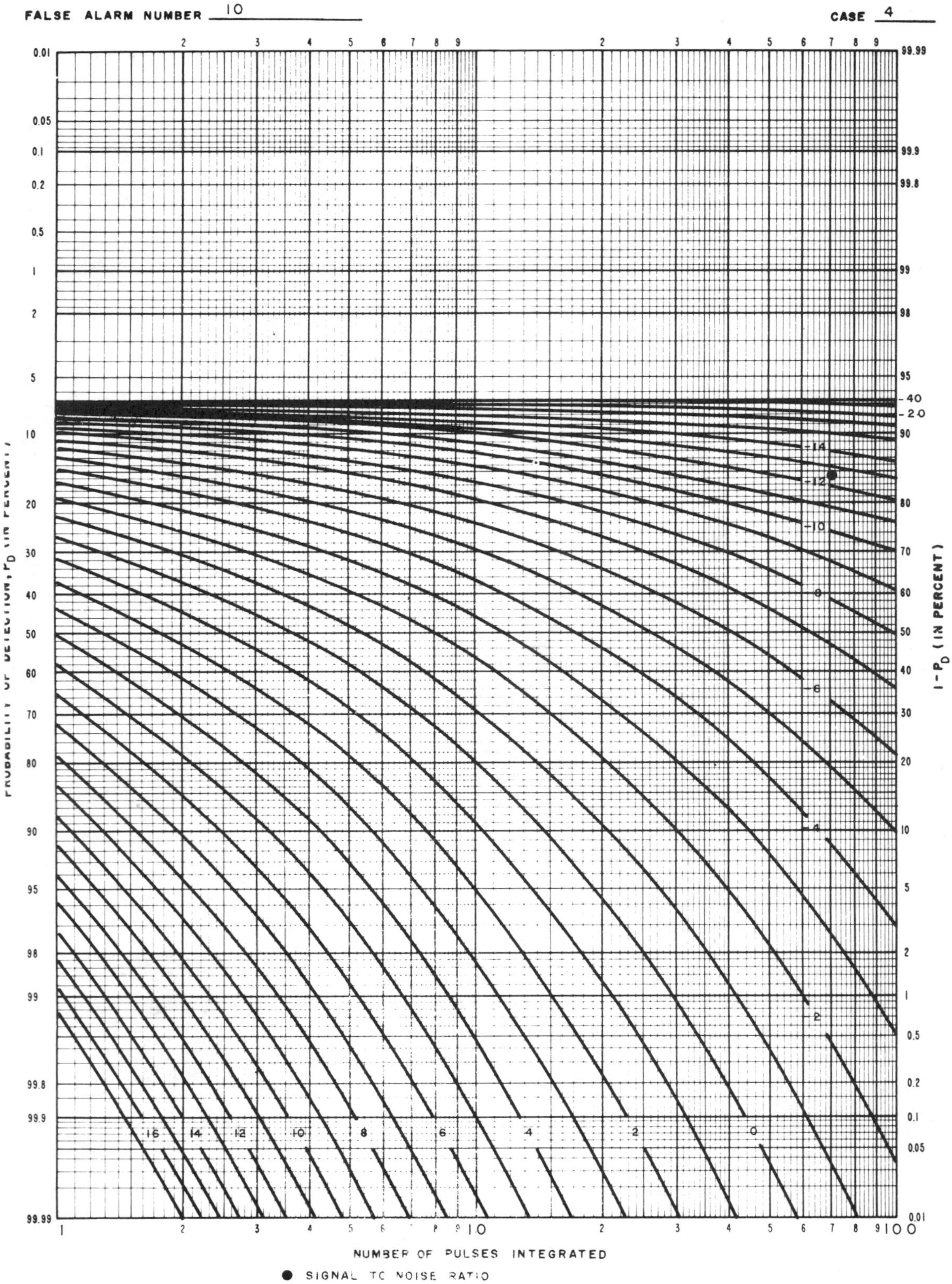
FALSE ALARM NUMBER 10
CASE 4
PROBABILITY OF DETECTION, P_D (IN PERCENT)
1 - P_D (IN PERCENT)
NUMBER OF PULSES INTEGRATED
● SIGNAL TO NOISE RATIO

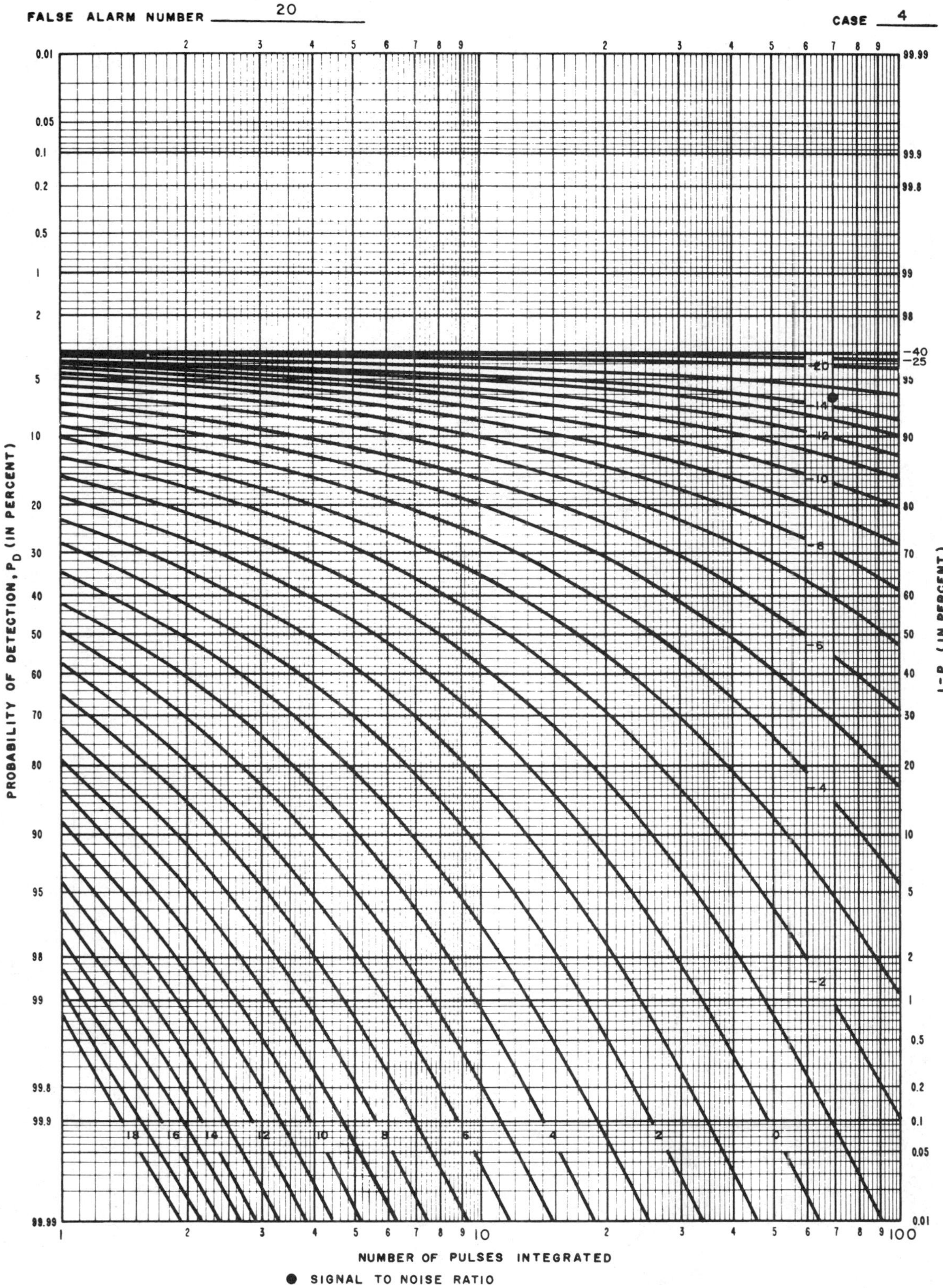
FALSE ALARM NUMBER 20
CASE 4
PROBABILITY OF DETECTION, P_D (IN PERCENT)
$1-P_D$ (IN PERCENT)
NUMBER OF PULSES INTEGRATED
● SIGNAL TO NOISE RATIO

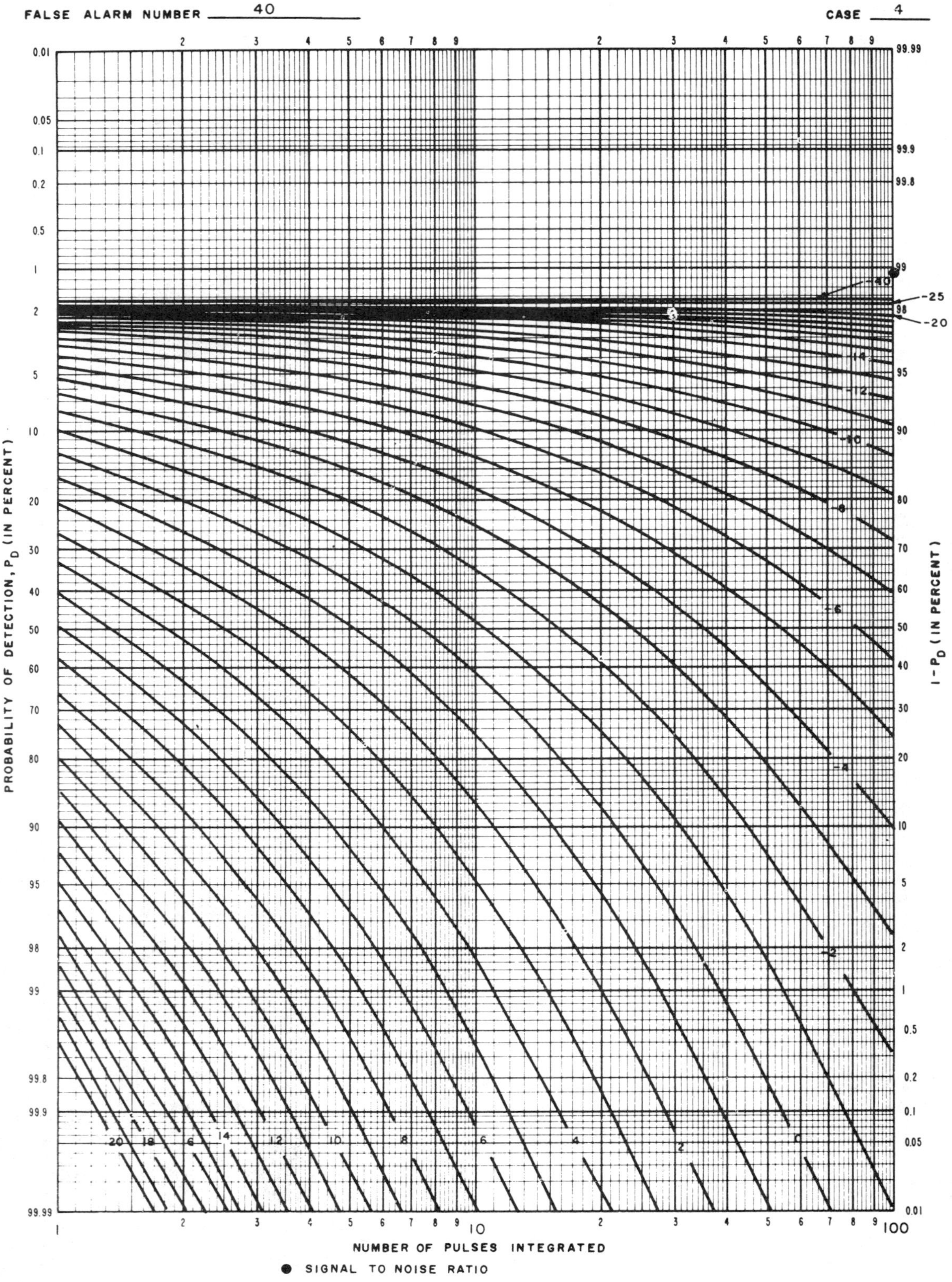
FALSE ALARM NUMBER 40
CASE 4
PROBABILITY OF DETECTION, P_D (IN PERCENT)
1 - P_D (IN PERCENT)
NUMBER OF PULSES INTEGRATED
● SIGNAL TO NOISE RATIO
-40
-25
-20
-14
-12
-10
-8
-6
-4
-2
0
2
4
6
8
10
12
14
16
18
20

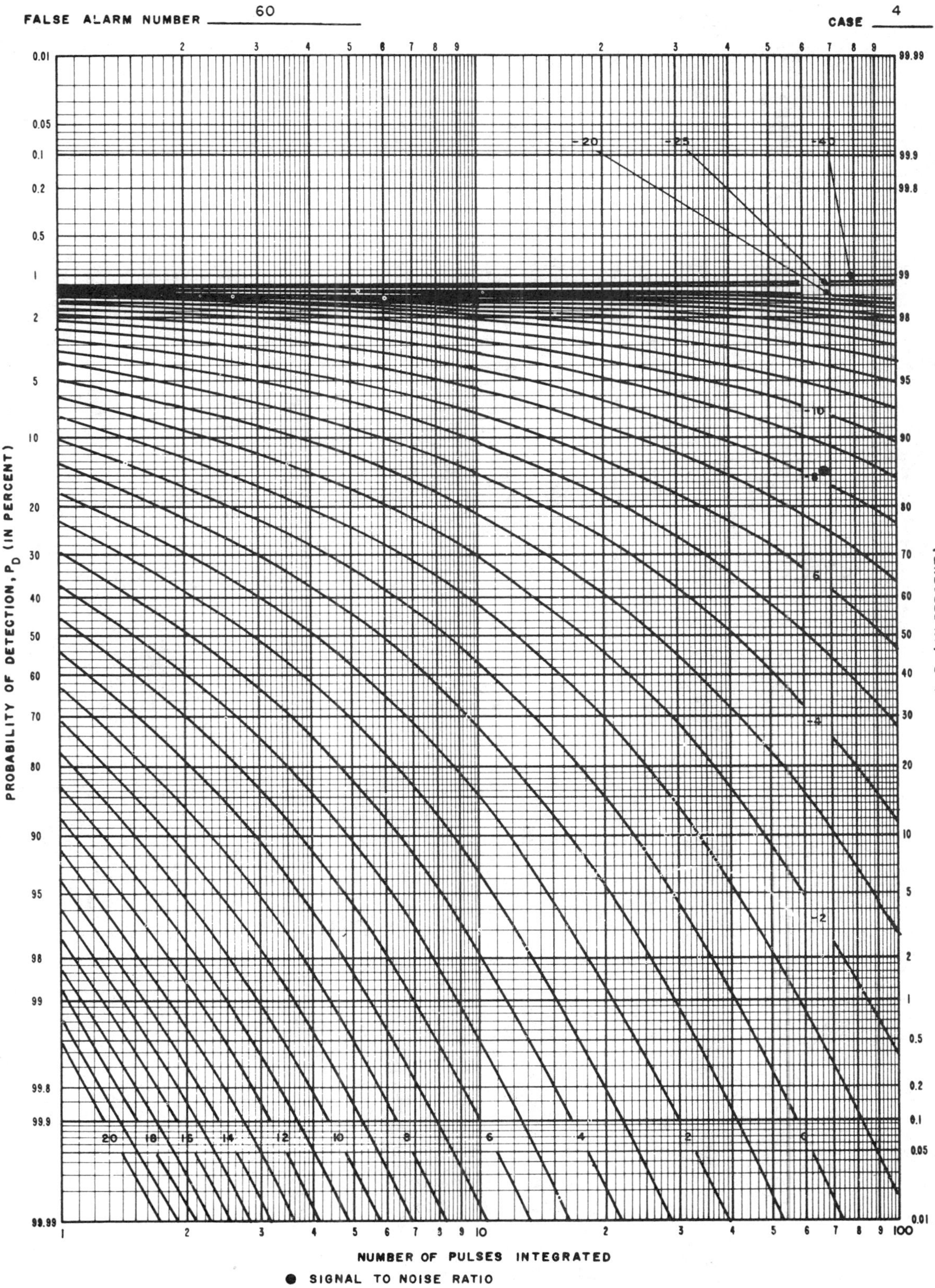
FALSE ALARM NUMBER 60
CASE 4
PROBABILITY OF DETECTION, P_D (IN PERCENT)
$1-P_D$ (IN PERCENT)
NUMBER OF PULSES INTEGRATED
● SIGNAL TO NOISE RATIO
-20
-25
-40
-10
-8
-6
-4
-2
20
18
16
14
12
10
8
6
4
2
0

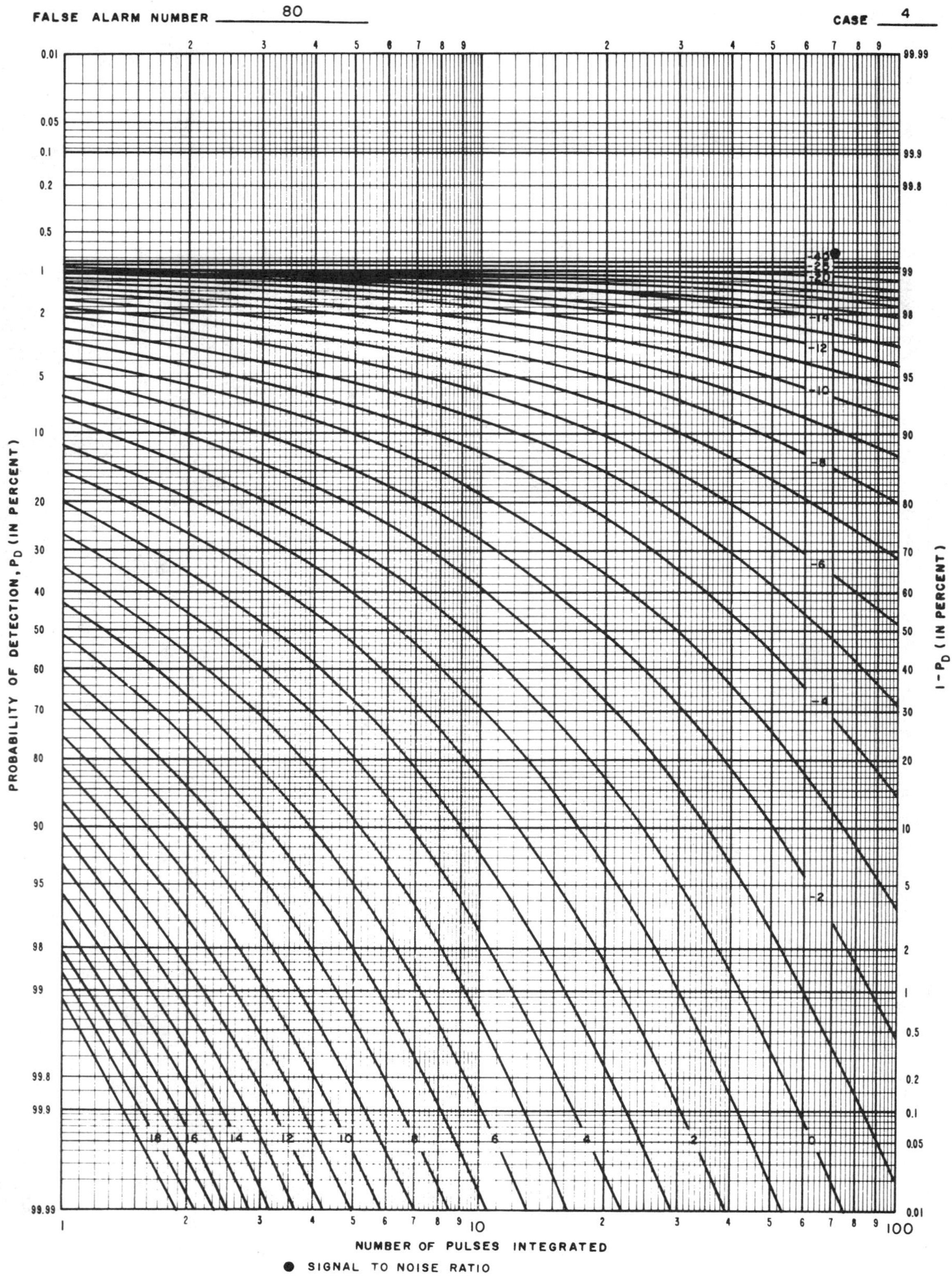
FALSE ALARM NUMBER 80
CASE 4
PROBABILITY OF DETECTION, P_D (IN PERCENT)
$1-P_D$ (IN PERCENT)
NUMBER OF PULSES INTEGRATED
● SIGNAL TO NOISE RATIO

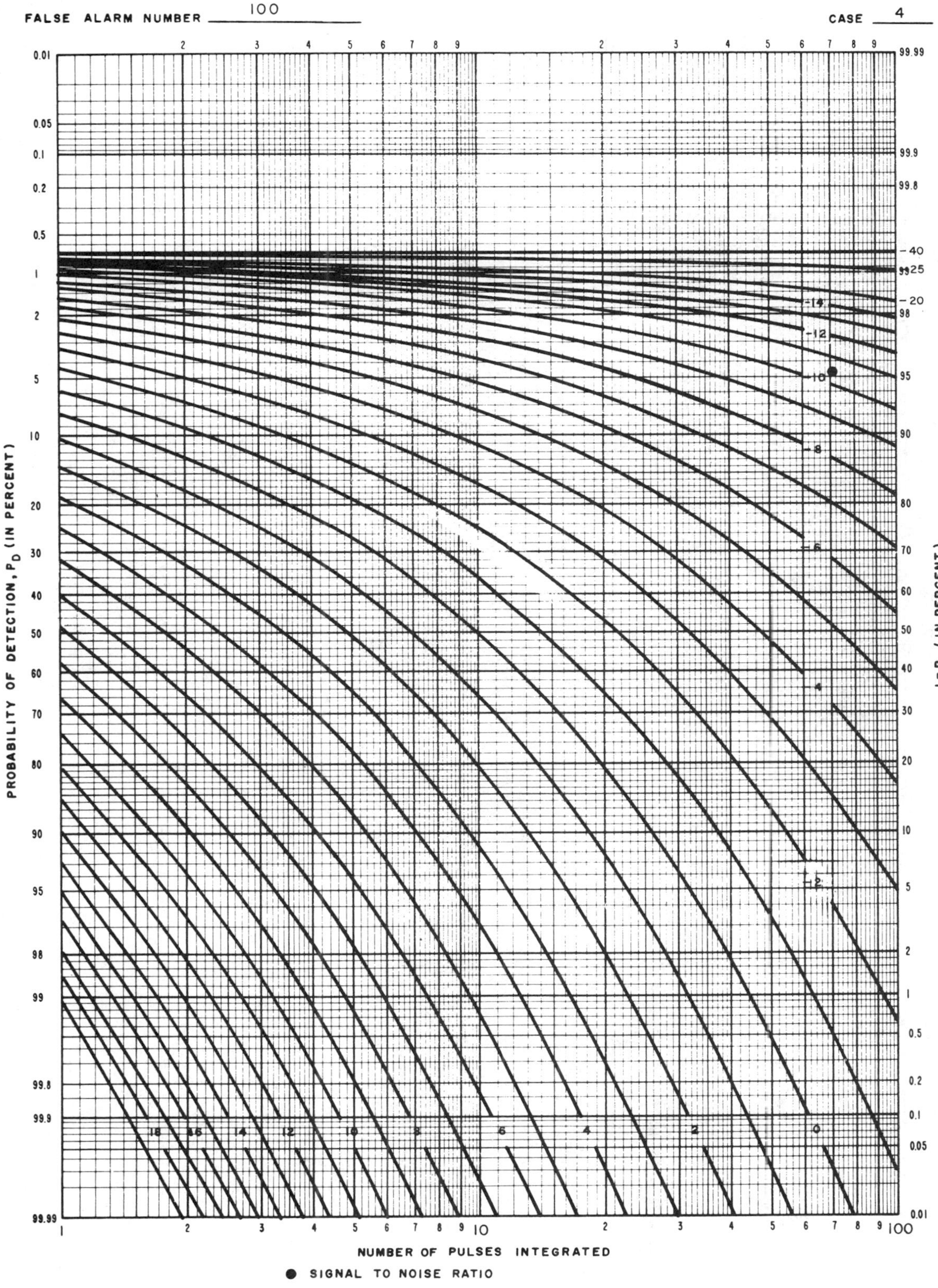
FALSE ALARM NUMBER 100
CASE 4
PROBABILITY OF DETECTION, P_D (IN PERCENT)
$1-P_D$ (IN PERCENT)
NUMBER OF PULSES INTEGRATED
● SIGNAL TO NOISE RATIO
-40
-25
-20
-14
-12
-10
-8
-6
-4
-2
0
2
4
6
8
10
12
14
16
18

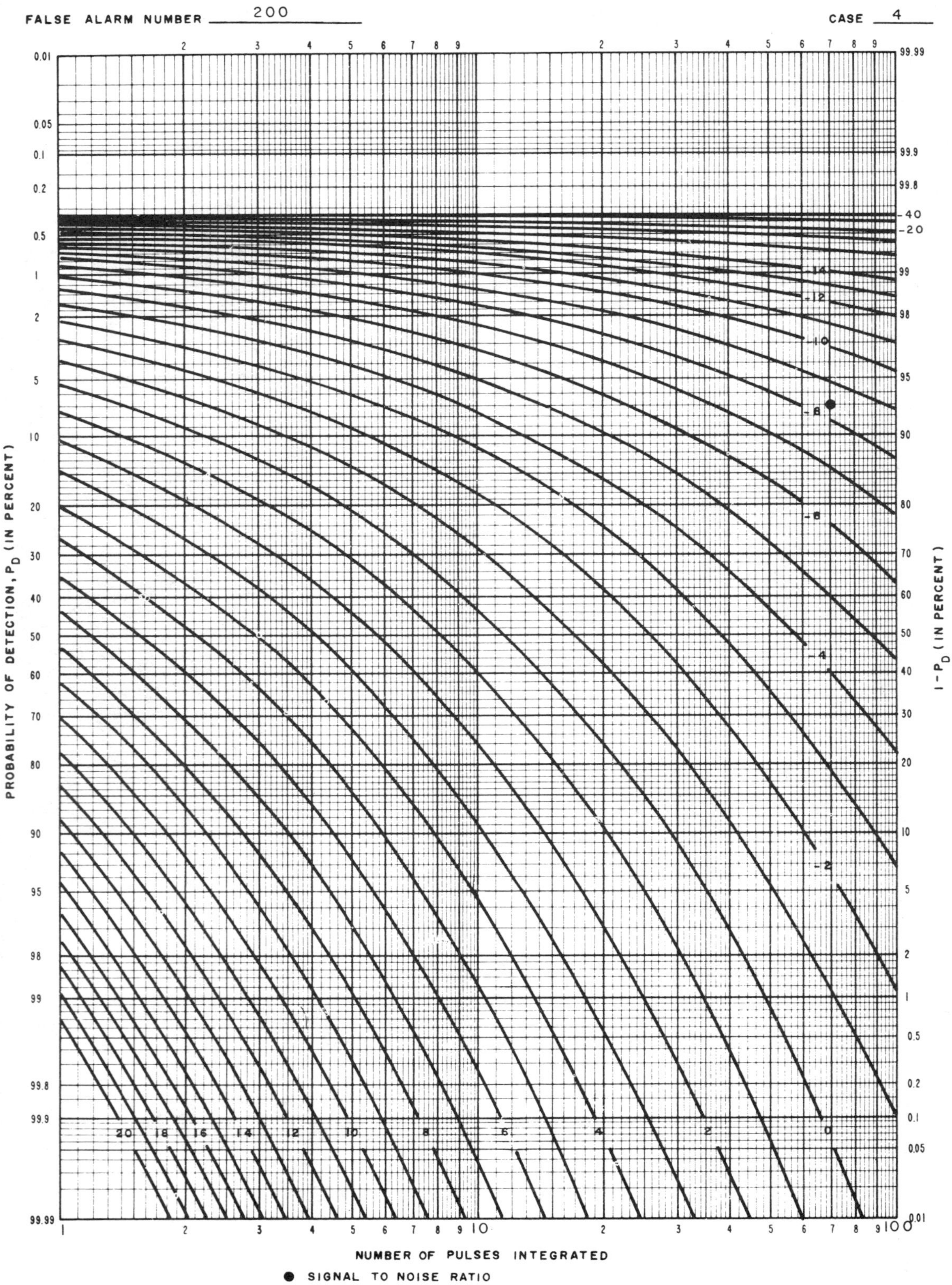
FALSE ALARM NUMBER 200
CASE 4
PROBABILITY OF DETECTION, P_D (IN PERCENT)
1 - P_D (IN PERCENT)
NUMBER OF PULSES INTEGRATED
● SIGNAL TO NOISE RATIO
-40
-20
-14
-12
-10
-8
-6
-4
-2
20 18 16 14 12 10 8 6 4 2 0

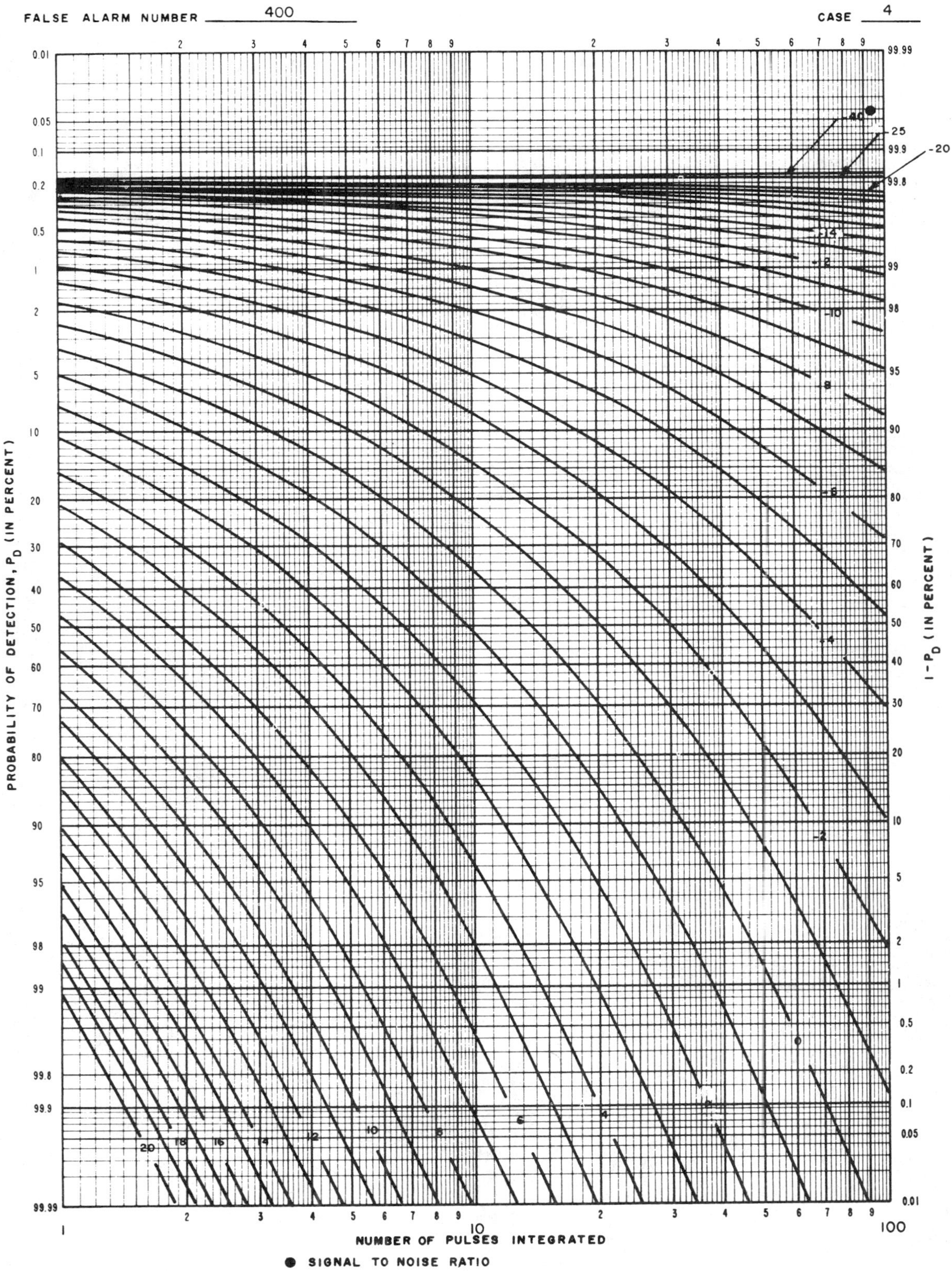
FALSE ALARM NUMBER 400
CASE 4
PROBABILITY OF DETECTION, P_D (IN PERCENT)
$1-P_D$ (IN PERCENT)
NUMBER OF PULSES INTEGRATED
● SIGNAL TO NOISE RATIO
40
-25
-20
-14
-12
-10
-8
-6
-4
-2
0
2
4
6
8
10
12
14
16
18
20

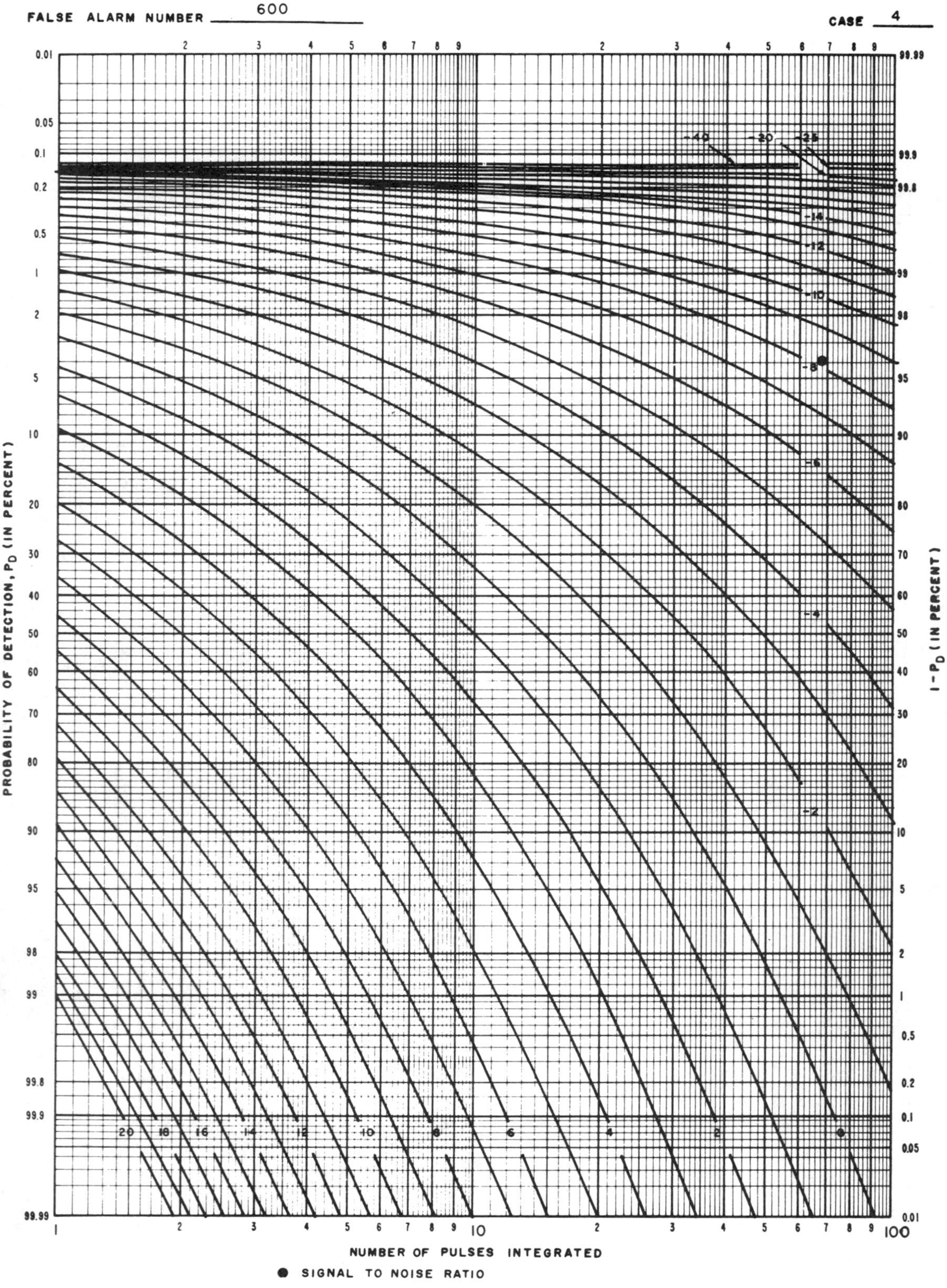
FALSE ALARM NUMBER 600
CASE 4
PROBABILITY OF DETECTION, P_D (IN PERCENT)
$1-P_D$ (IN PERCENT)
NUMBER OF PULSES INTEGRATED
● SIGNAL TO NOISE RATIO
20
18
16
14
12
10
8
6
4
2
0
-2
-4
-6
-8
-10
-12
-14
-20
-25
-40

FALSE ALARM NUMBER 800

CASE 4

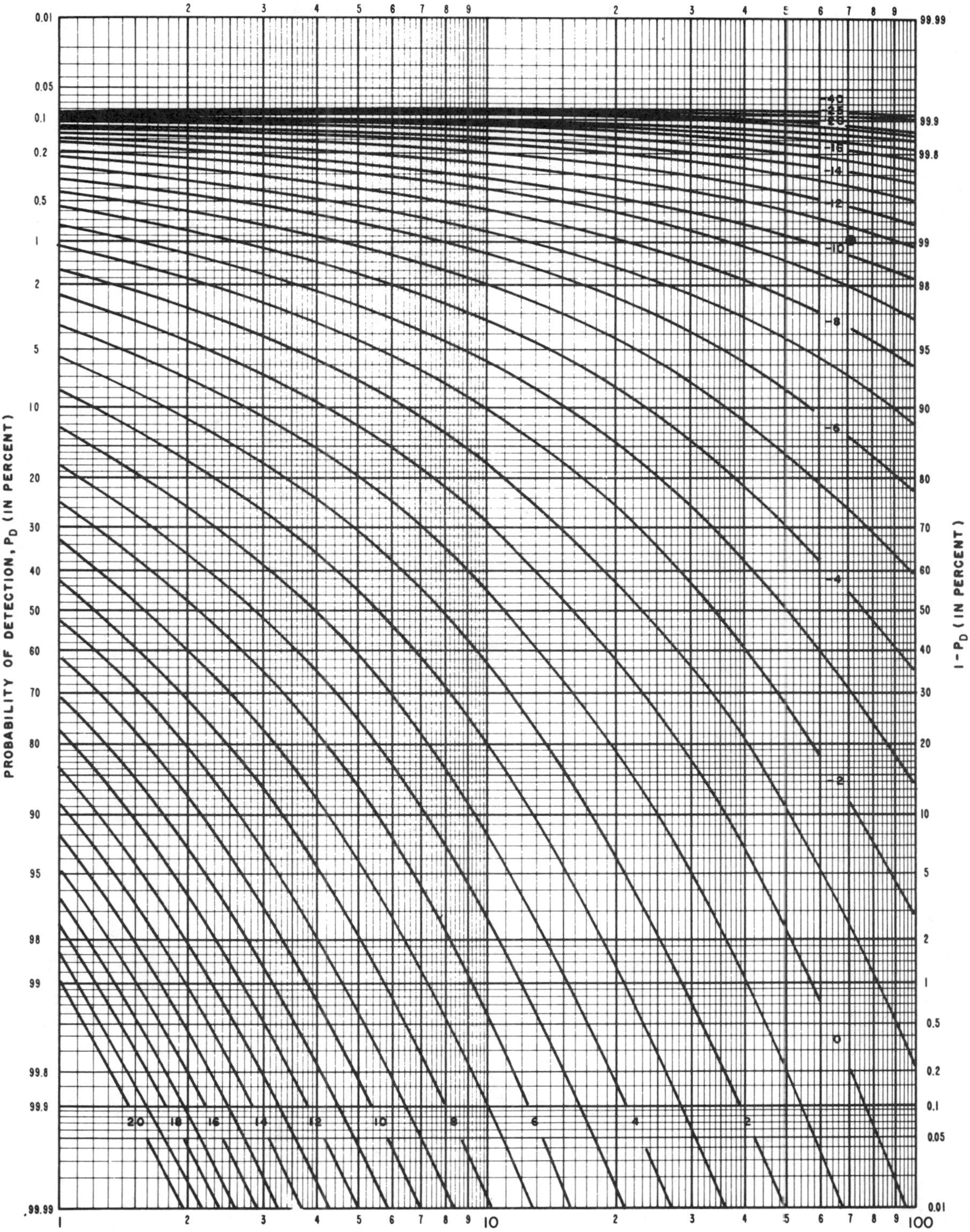

● SIGNAL TO NOISE RATIO

FIGURE

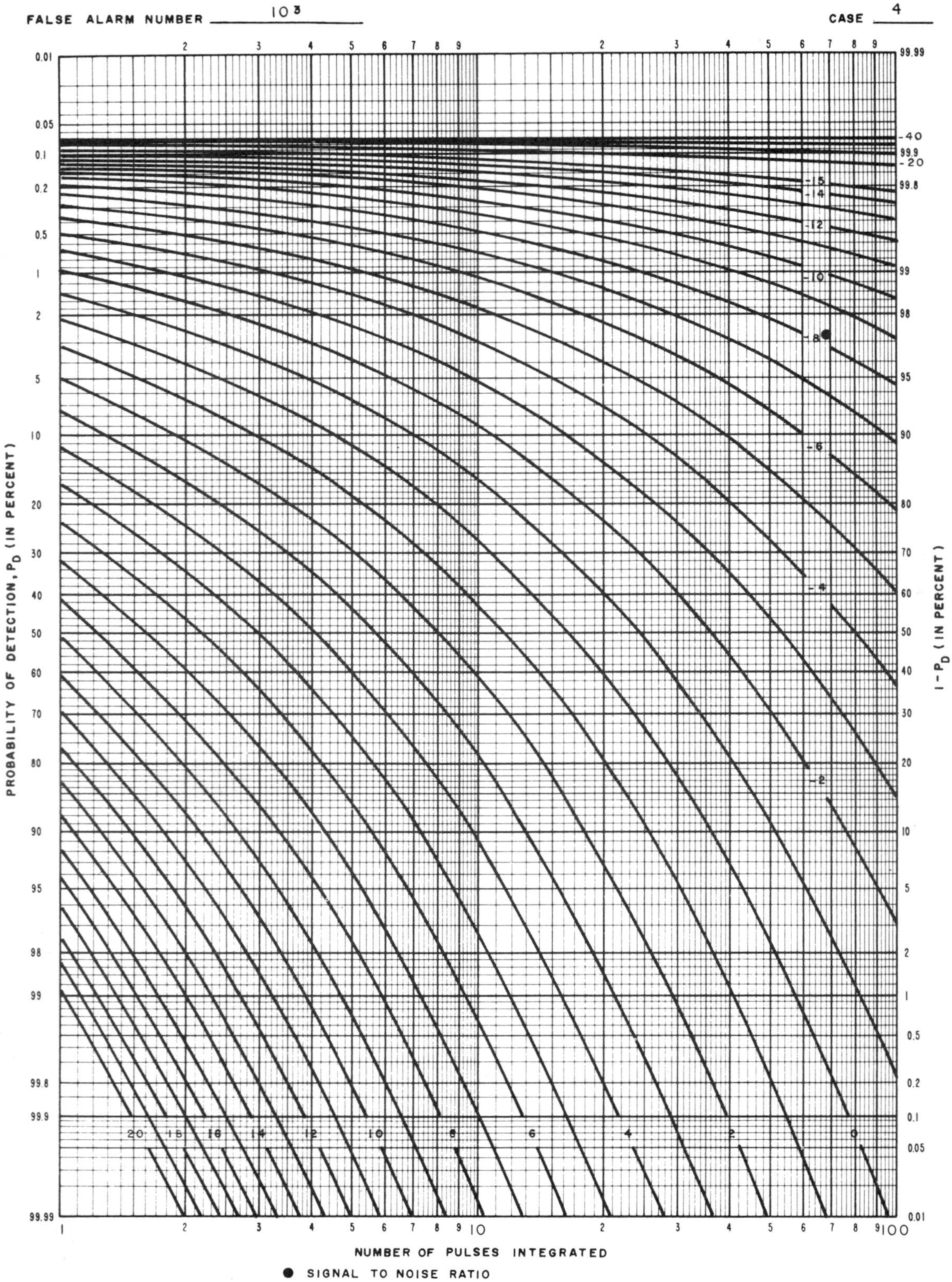
FALSE ALARM NUMBER 10³
CASE 4
PROBABILITY OF DETECTION, P_D (IN PERCENT)
1 - P_D (IN PERCENT)
NUMBER OF PULSES INTEGRATED
● SIGNAL TO NOISE RATIO
-40
-20
-15
-14
-12
-10
-8
-6
-4
-2
20
18
16
14
12
10
8
6
4
2
0

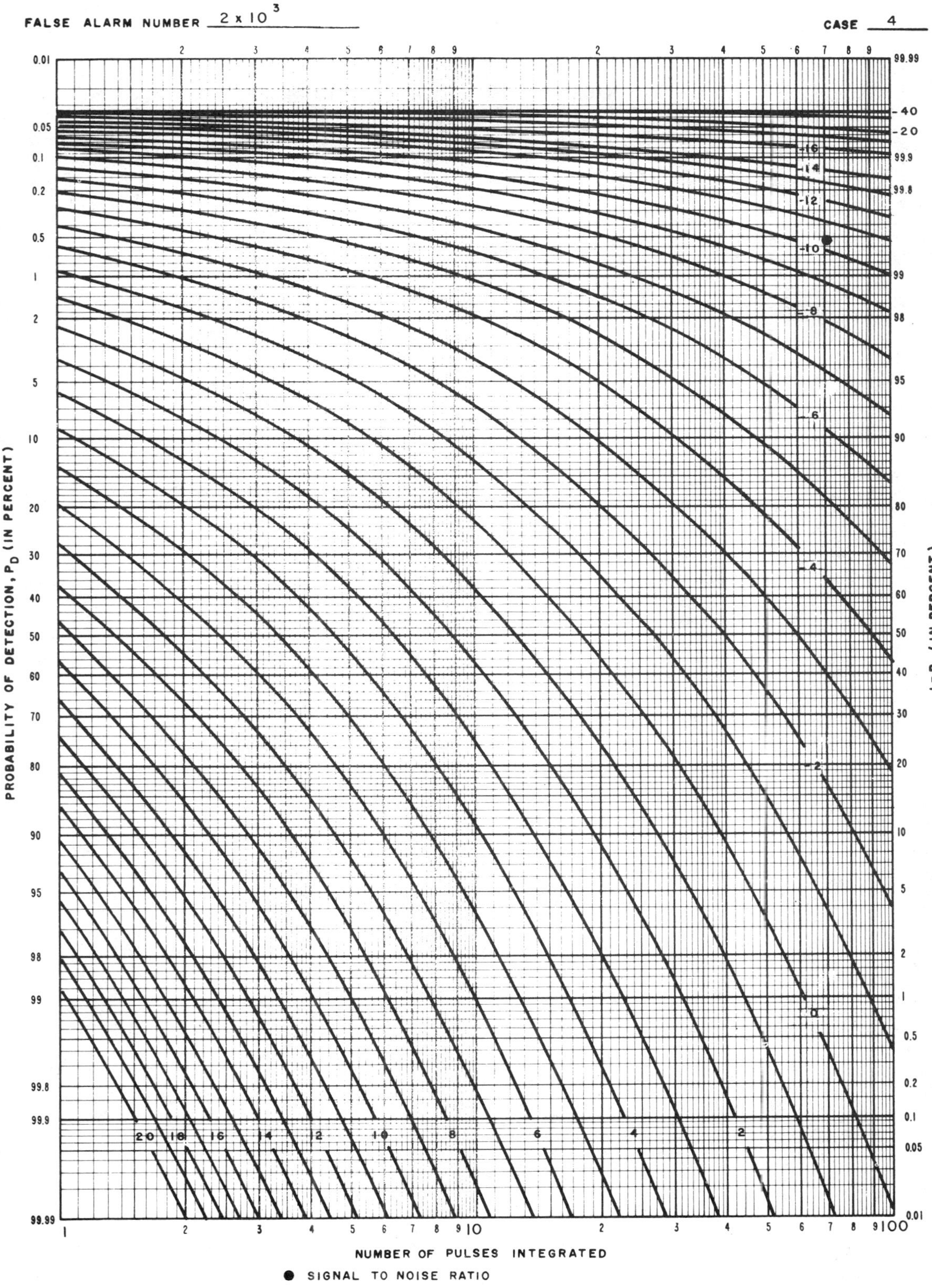
FALSE ALARM NUMBER 2 x 10^3
CASE 4
PROBABILITY OF DETECTION, P_D (IN PERCENT)
1 - P_D (IN PERCENT)
NUMBER OF PULSES INTEGRATED
● SIGNAL TO NOISE RATIO
-40
-20
-16
-14
-12
-10
-8
-6
-4
-2
0
20
18
16
14
12
10
8
6
4
2

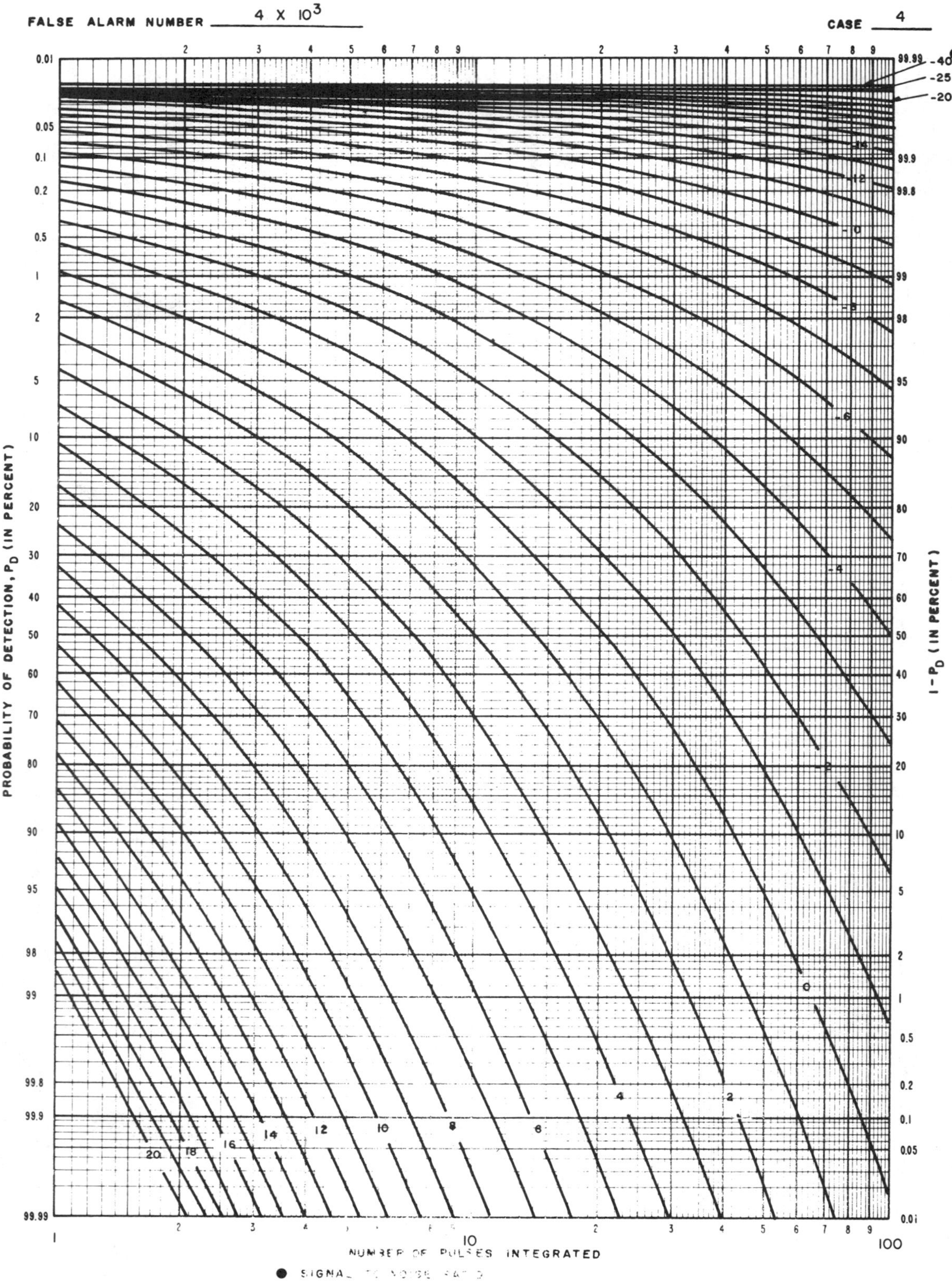
FALSE ALARM NUMBER 4 X 10^3
CASE 4
PROBABILITY OF DETECTION, P_D (IN PERCENT)
1 - P_D (IN PERCENT)
NUMBER OF PULSES INTEGRATED
● SIGNAL TO NOISE RATIO
-40
-25
-20
-16
-12
-10
-8
-6
-4
-2
0
2
4
6
8
10
12
14
16
18
20
1
10
100

FALSE ALARM NUMBER 6 x 10^3 CASE 4

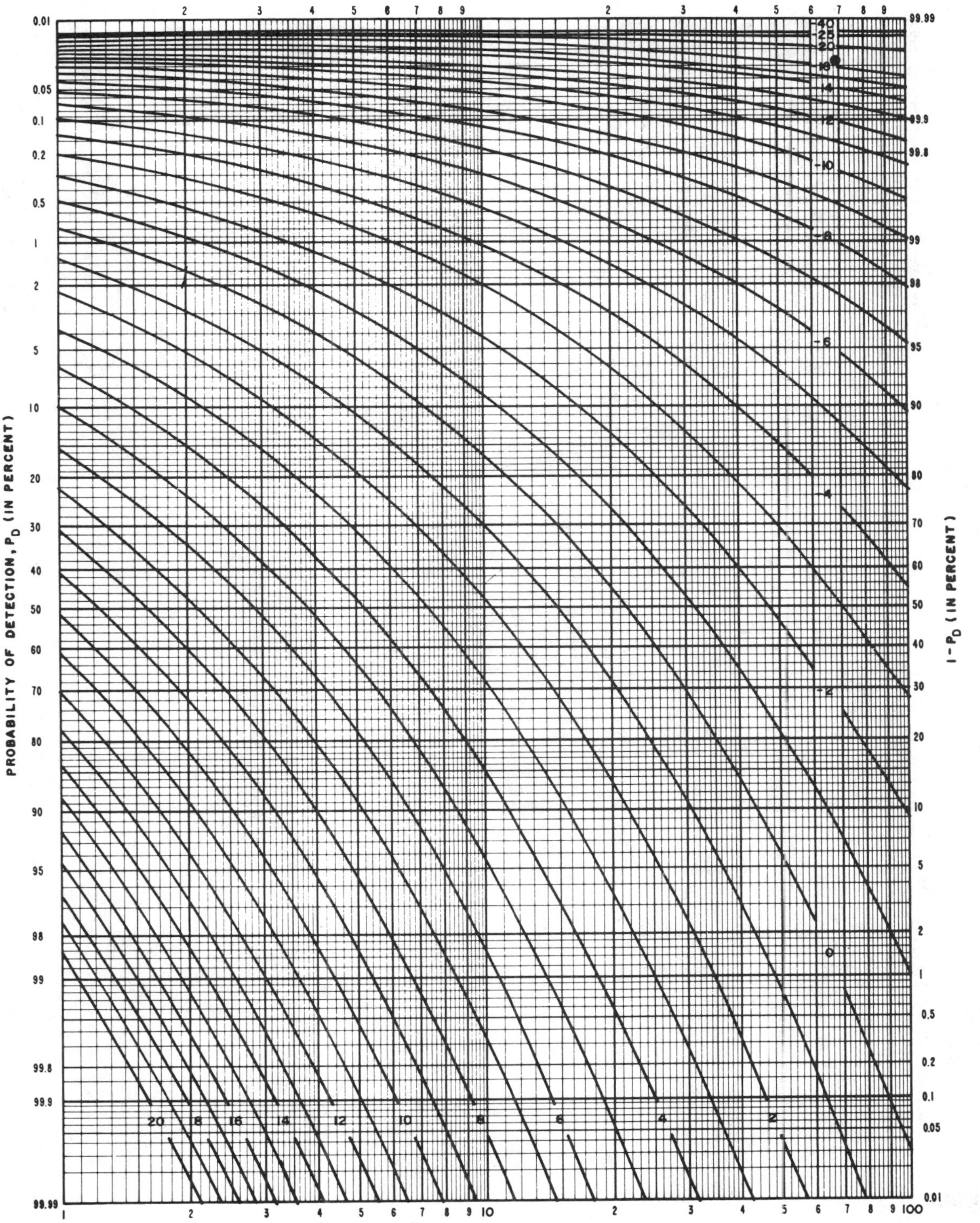

● SIGNAL TO NOISE RATIO

FALSE ALARM NUMBER 8×10^3 CASE 4

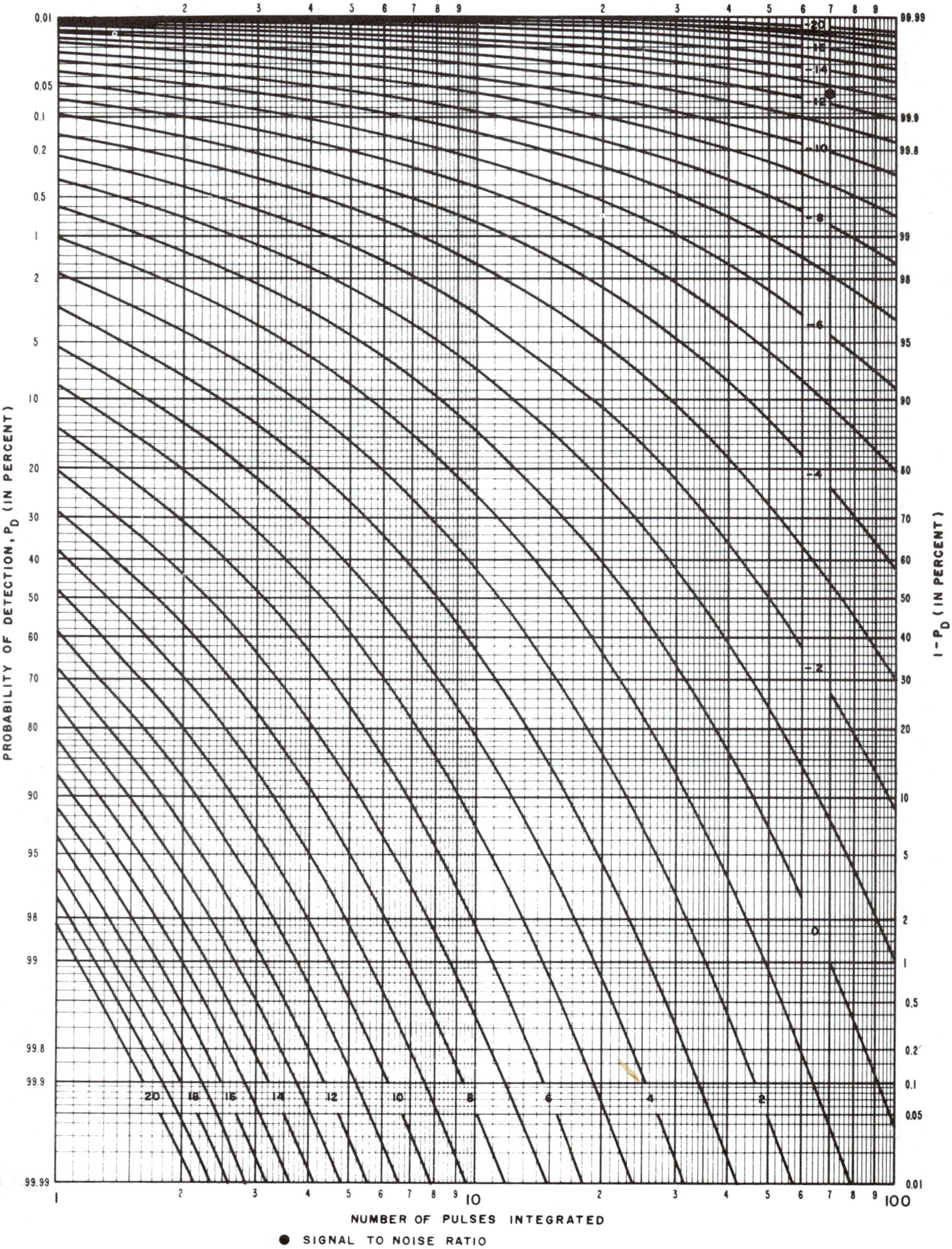

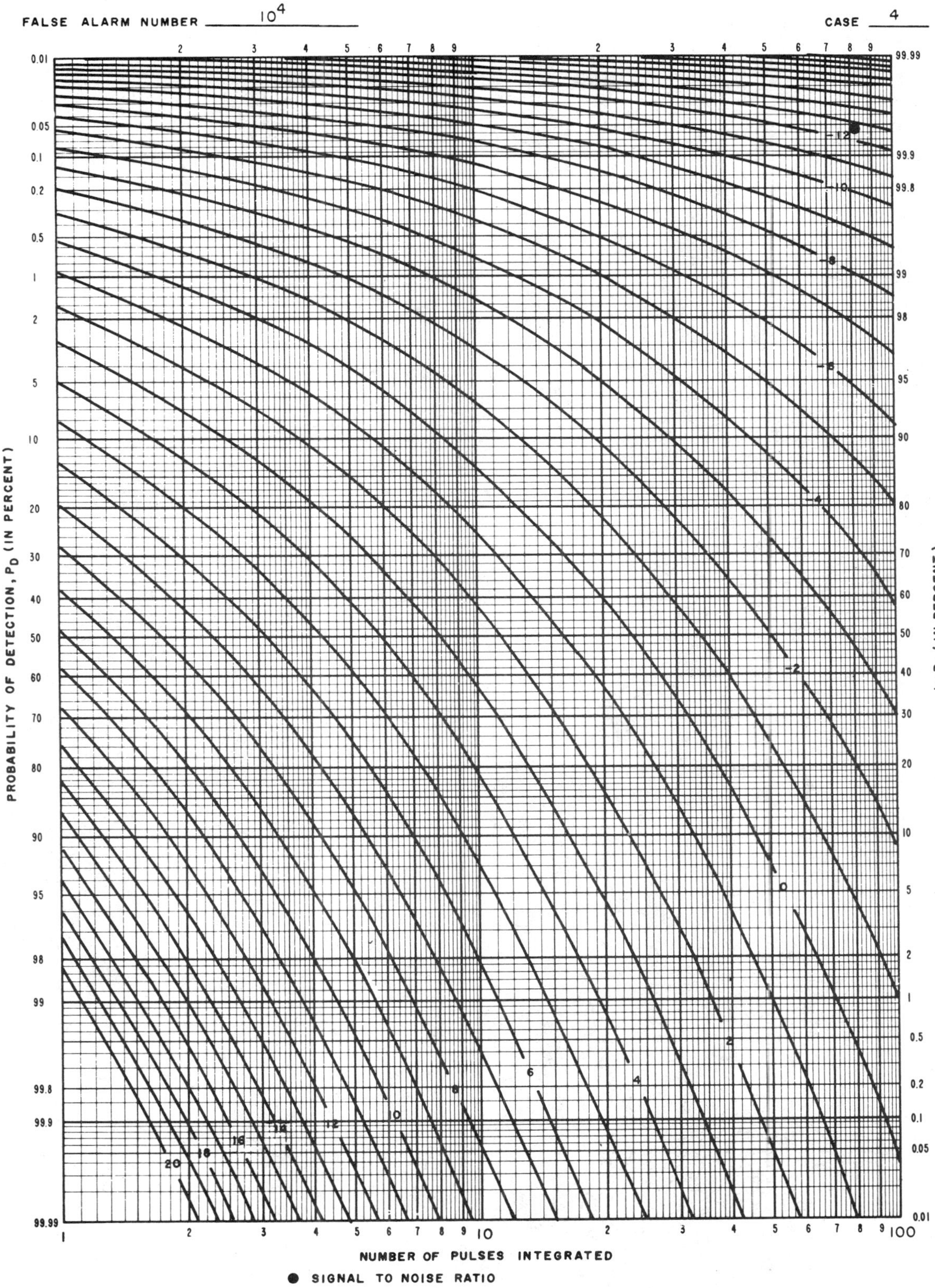
FALSE ALARM NUMBER 10^4
CASE 4
PROBABILITY OF DETECTION, P_D (IN PERCENT)
$1-P_D$ (IN PERCENT)
NUMBER OF PULSES INTEGRATED
● SIGNAL TO NOISE RATIO
-12
-10
-8
-6
-4
-2
0
2
4
6
8
10
12
14
16
18
20

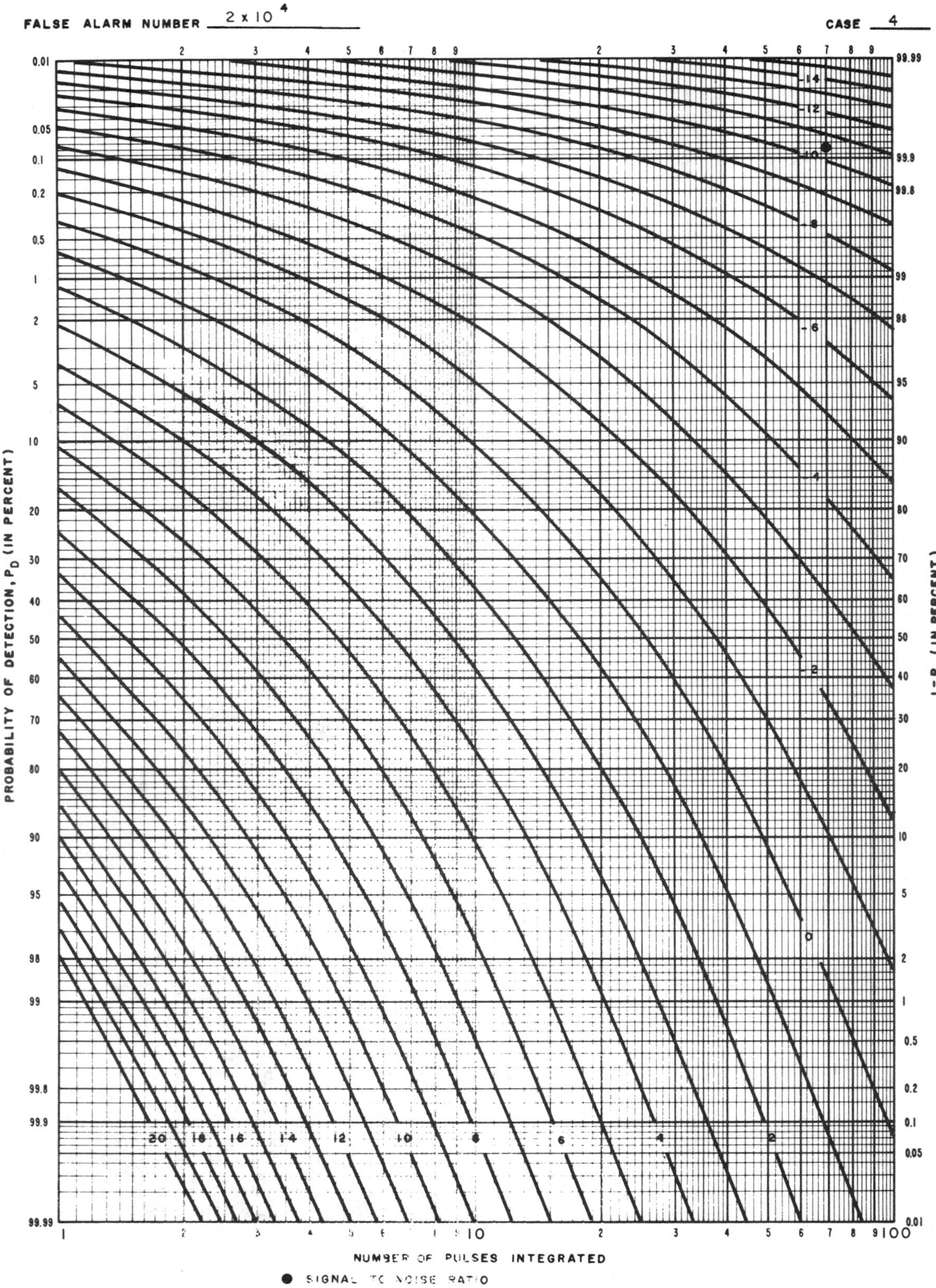
FALSE ALARM NUMBER 2 x 10^4
CASE 4
PROBABILITY OF DETECTION, P_D (IN PERCENT)
1 - P_D (IN PERCENT)
NUMBER OF PULSES INTEGRATED
● SIGNAL TO NOISE RATIO
20
18
16
14
12
10
8
6
4
2
0
-2
-4
-6
-8
-10
-12
-14

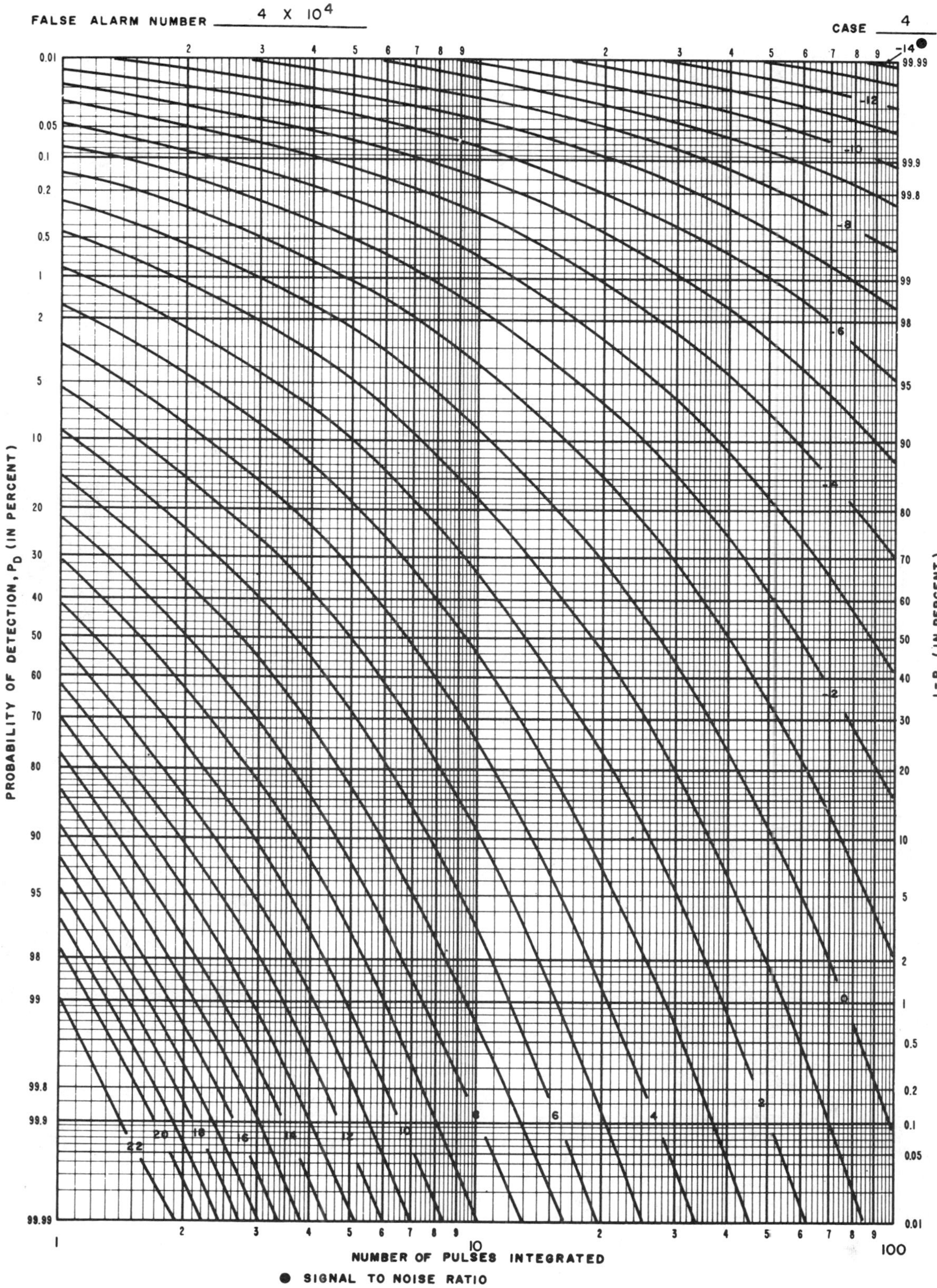
FALSE ALARM NUMBER 4 X 10^4
CASE 4
PROBABILITY OF DETECTION, P_D (IN PERCENT)
1 - P_D (IN PERCENT)
NUMBER OF PULSES INTEGRATED
● SIGNAL TO NOISE RATIO
22
20
18
16
14
12
10
8
6
4
2
0
-2
-4
-6
-8
-10
-12
-14

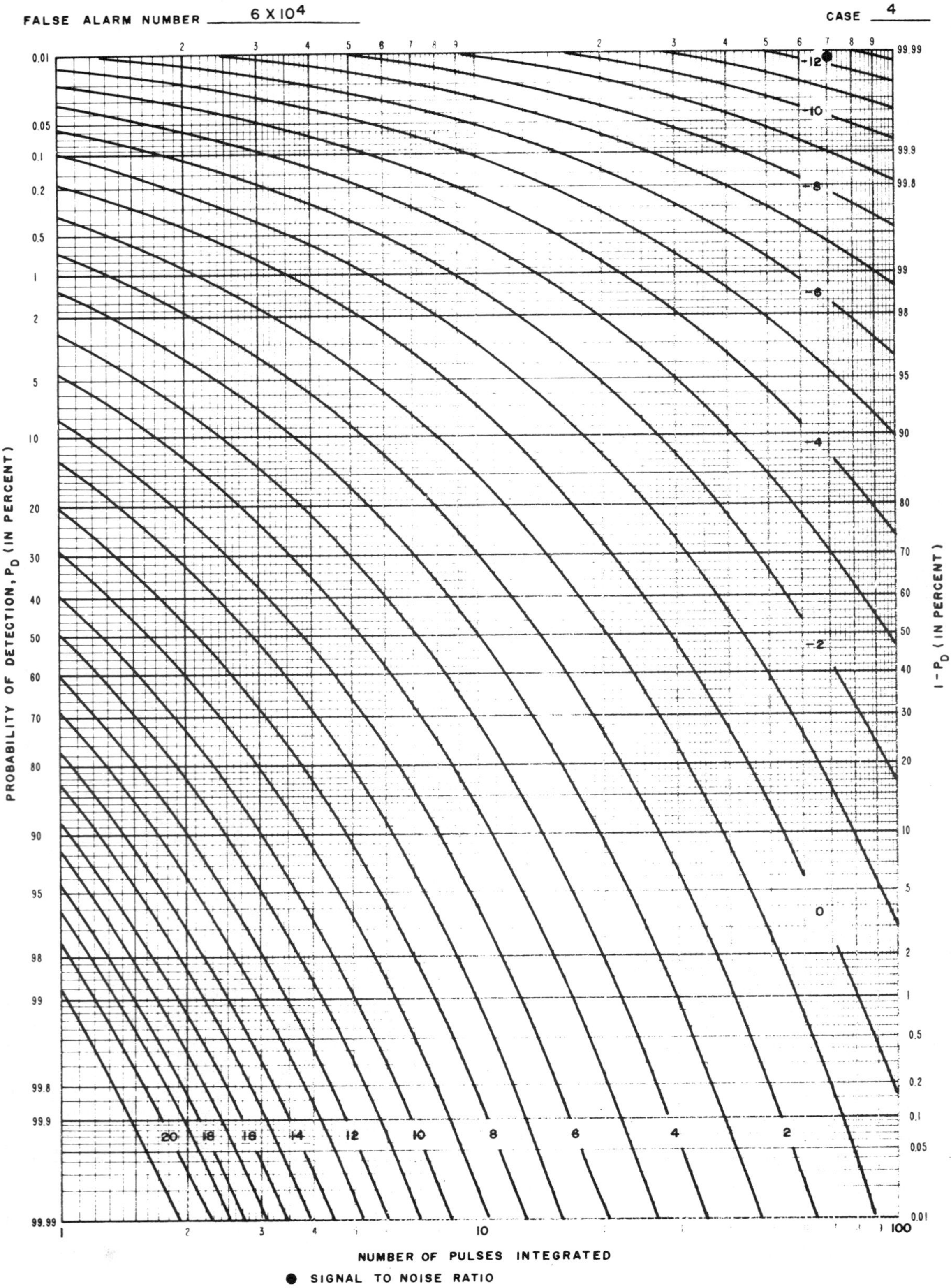
FALSE ALARM NUMBER 6 X 10^4
CASE 4
PROBABILITY OF DETECTION, P_D (IN PERCENT)
1 - P_D (IN PERCENT)
NUMBER OF PULSES INTEGRATED
● SIGNAL TO NOISE RATIO
-12
-10
-8
-6
-4
-2
0
2
4
6
8
10
12
14
16
18
20

FALSE ALARM NUMBER 8 x 10^4

CASE 4

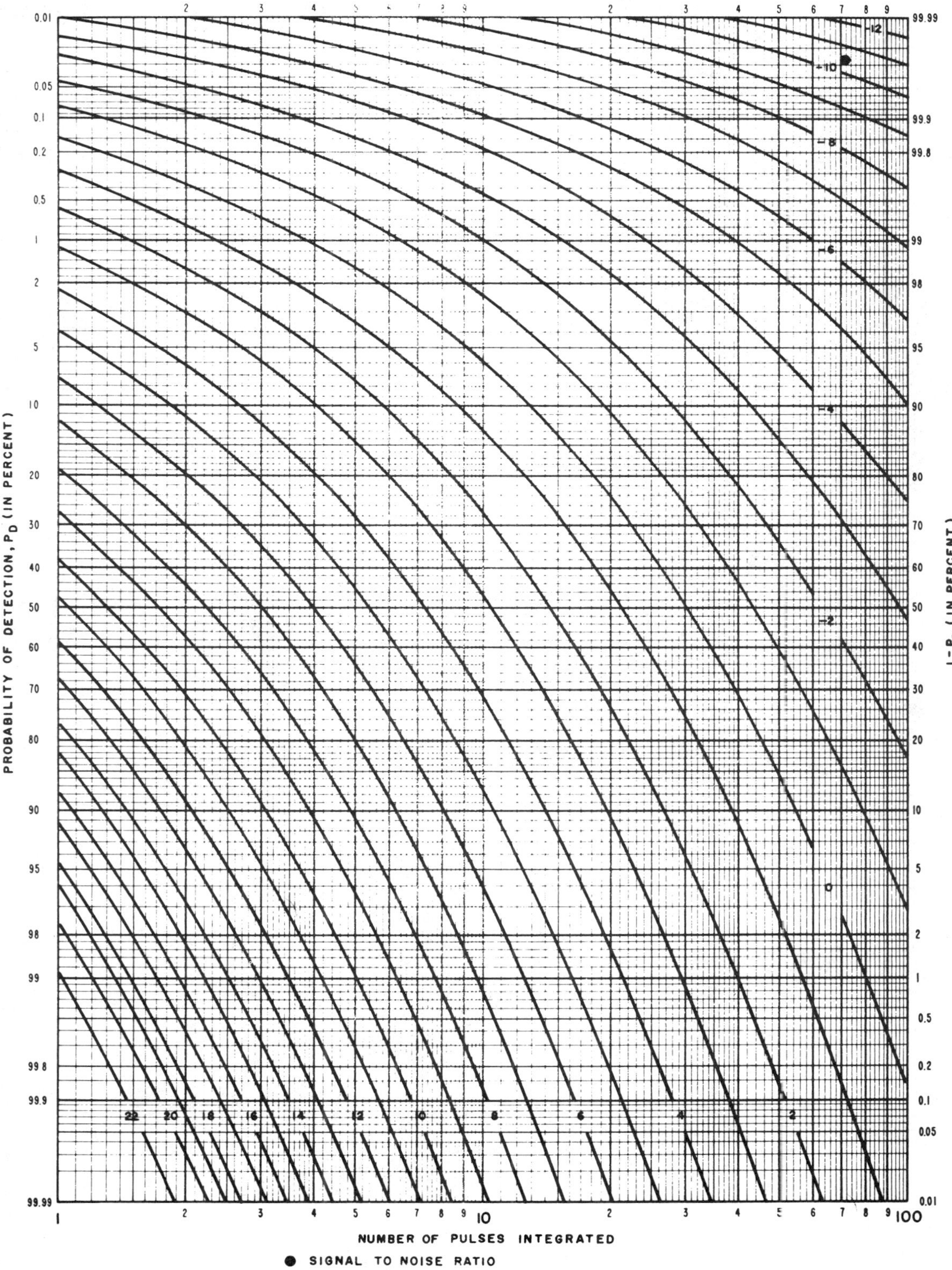

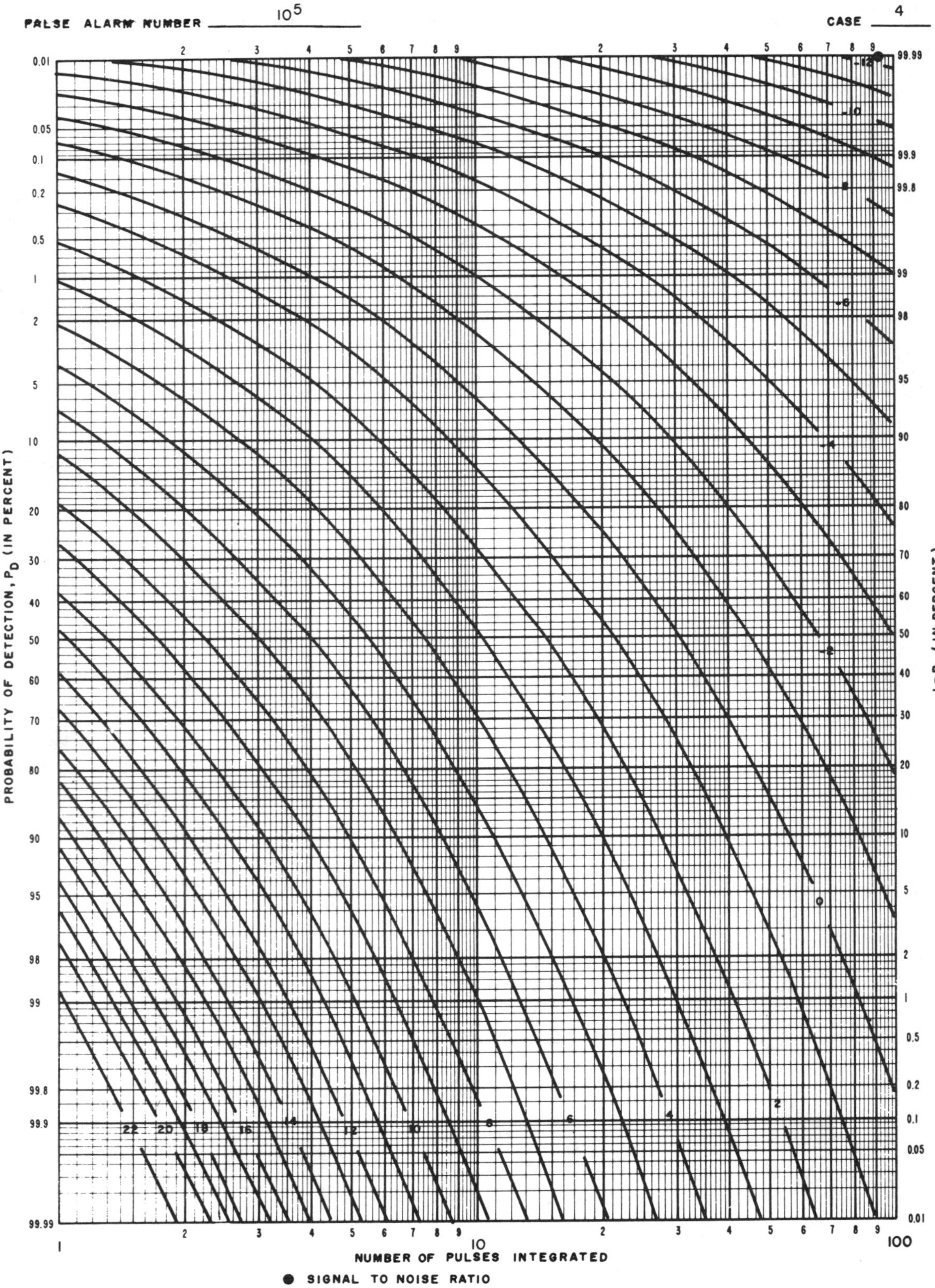
FALSE ALARM NUMBER 10^5
CASE 4
PROBABILITY OF DETECTION, P_D (IN PERCENT)
$1-P_D$ (IN PERCENT)
NUMBER OF PULSES INTEGRATED
● SIGNAL TO NOISE RATIO

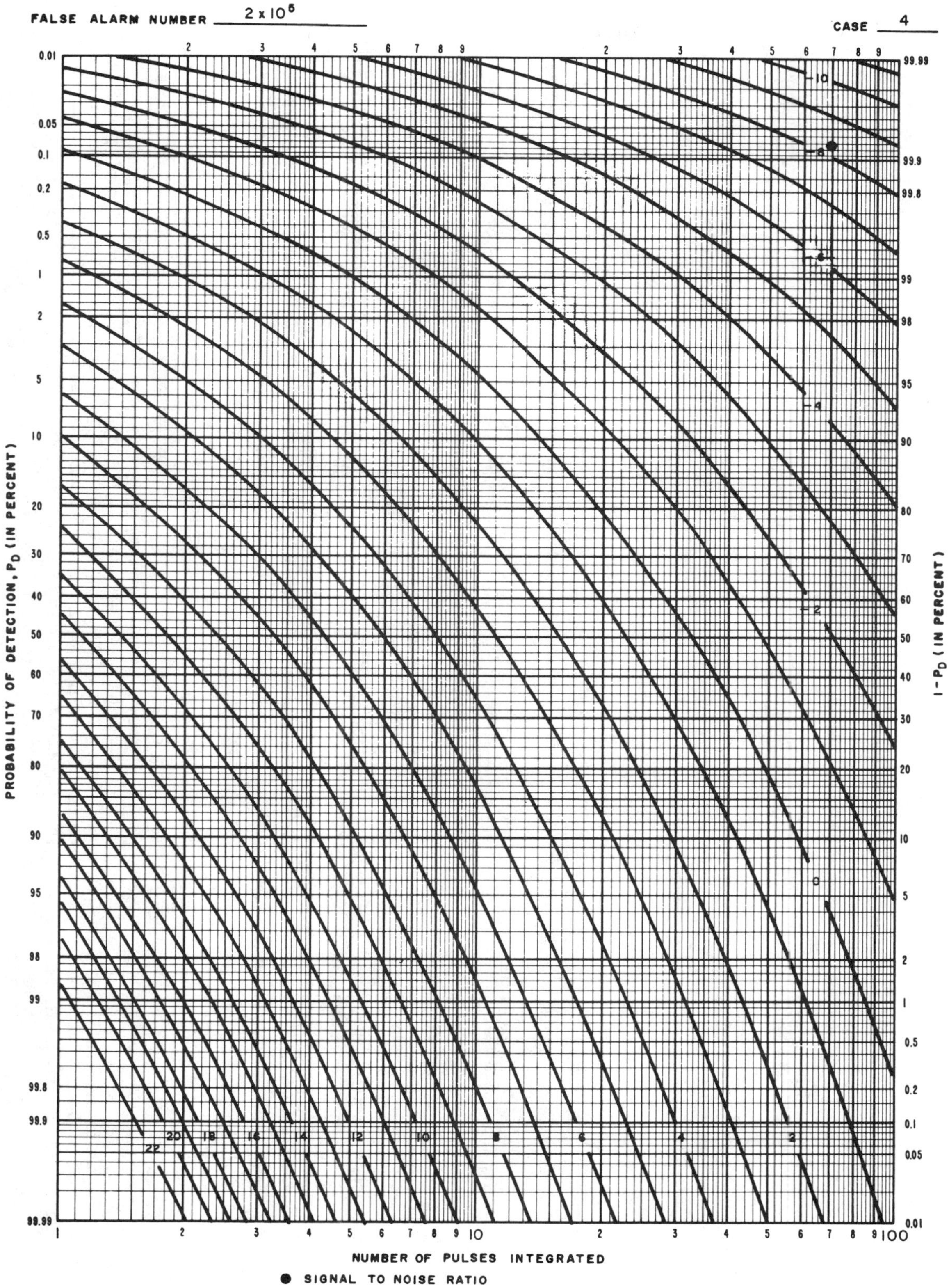
FALSE ALARM NUMBER 2 x 10^5
CASE 4
PROBABILITY OF DETECTION, P_D (IN PERCENT)
1 - P_D (IN PERCENT)
NUMBER OF PULSES INTEGRATED
● SIGNAL TO NOISE RATIO
22 20 18 16 14 12 10 8 6 4 2 0 -2 -4 -6 -8 -10

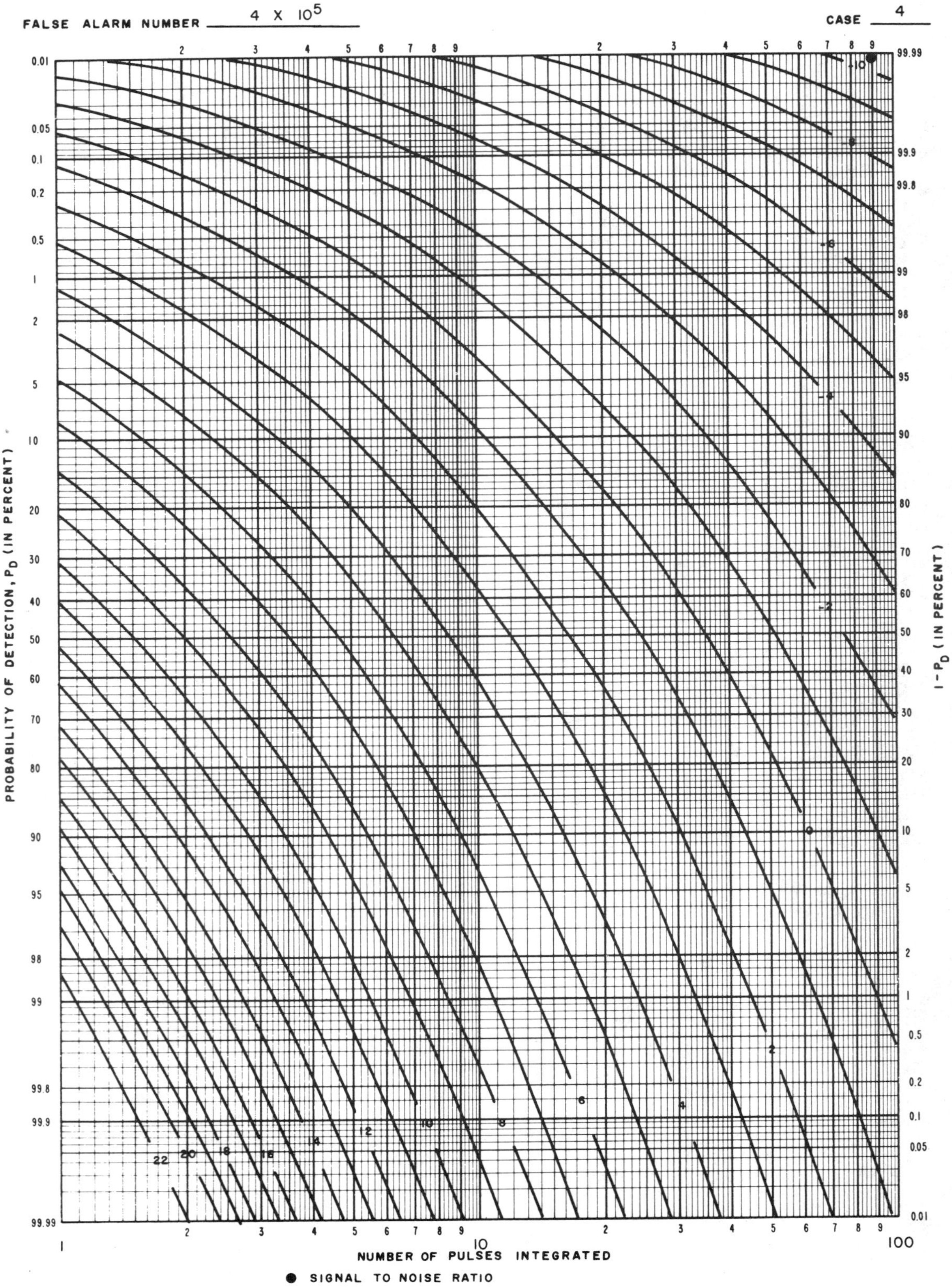
FALSE ALARM NUMBER 4 X 10^5
CASE 4
PROBABILITY OF DETECTION, P_D (IN PERCENT)
1 - P_D (IN PERCENT)
NUMBER OF PULSES INTEGRATED
● SIGNAL TO NOISE RATIO
-10
-8
-6
-4
-2
0
2
4
6
8
10
12
14
16
18
20
22

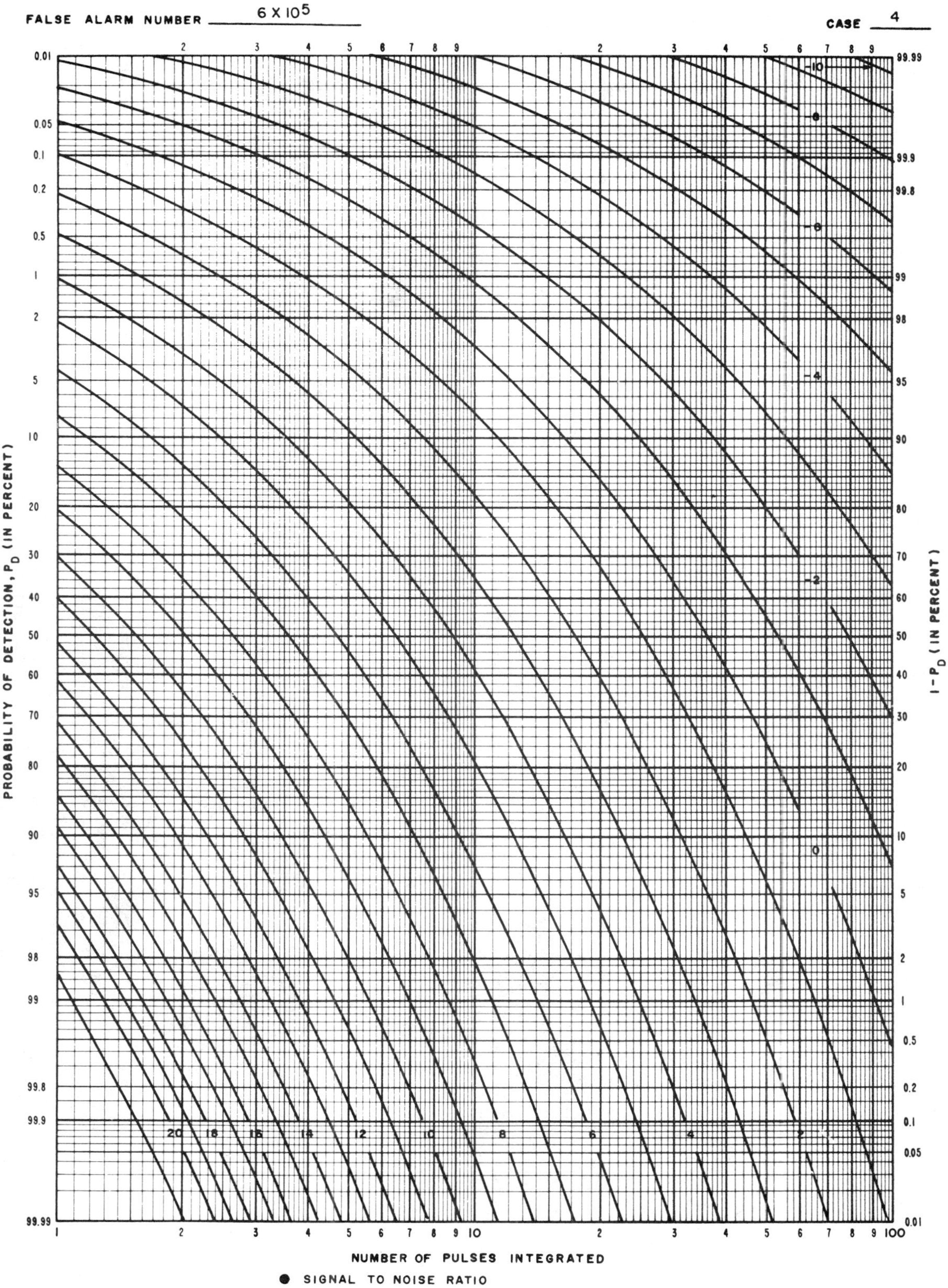
FALSE ALARM NUMBER 6 X 10^5
CASE 4
PROBABILITY OF DETECTION, P_D (IN PERCENT)
1 - P_D (IN PERCENT)
NUMBER OF PULSES INTEGRATED
● SIGNAL TO NOISE RATIO
20 18 16 14 12 10 8 6 4 2 0 -2 -4 -6 -8 -10

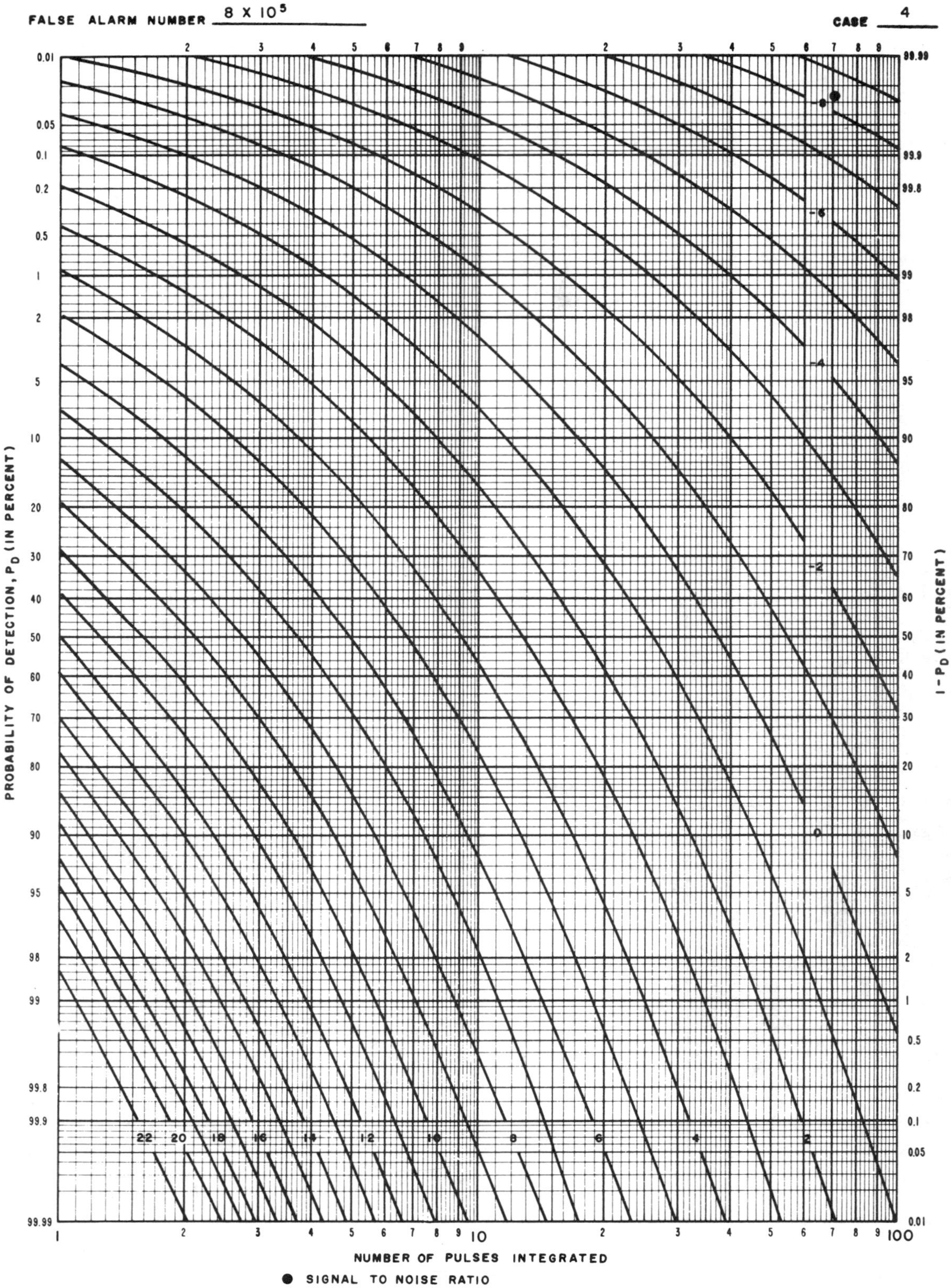
FALSE ALARM NUMBER 8 X 10^5
CASE 4
PROBABILITY OF DETECTION, P_D (IN PERCENT)
1 - P_D (IN PERCENT)
NUMBER OF PULSES INTEGRATED
● SIGNAL TO NOISE RATIO
-8
-6
-4
-2
0
22
20
18
16
14
12
10
8
6
4
2

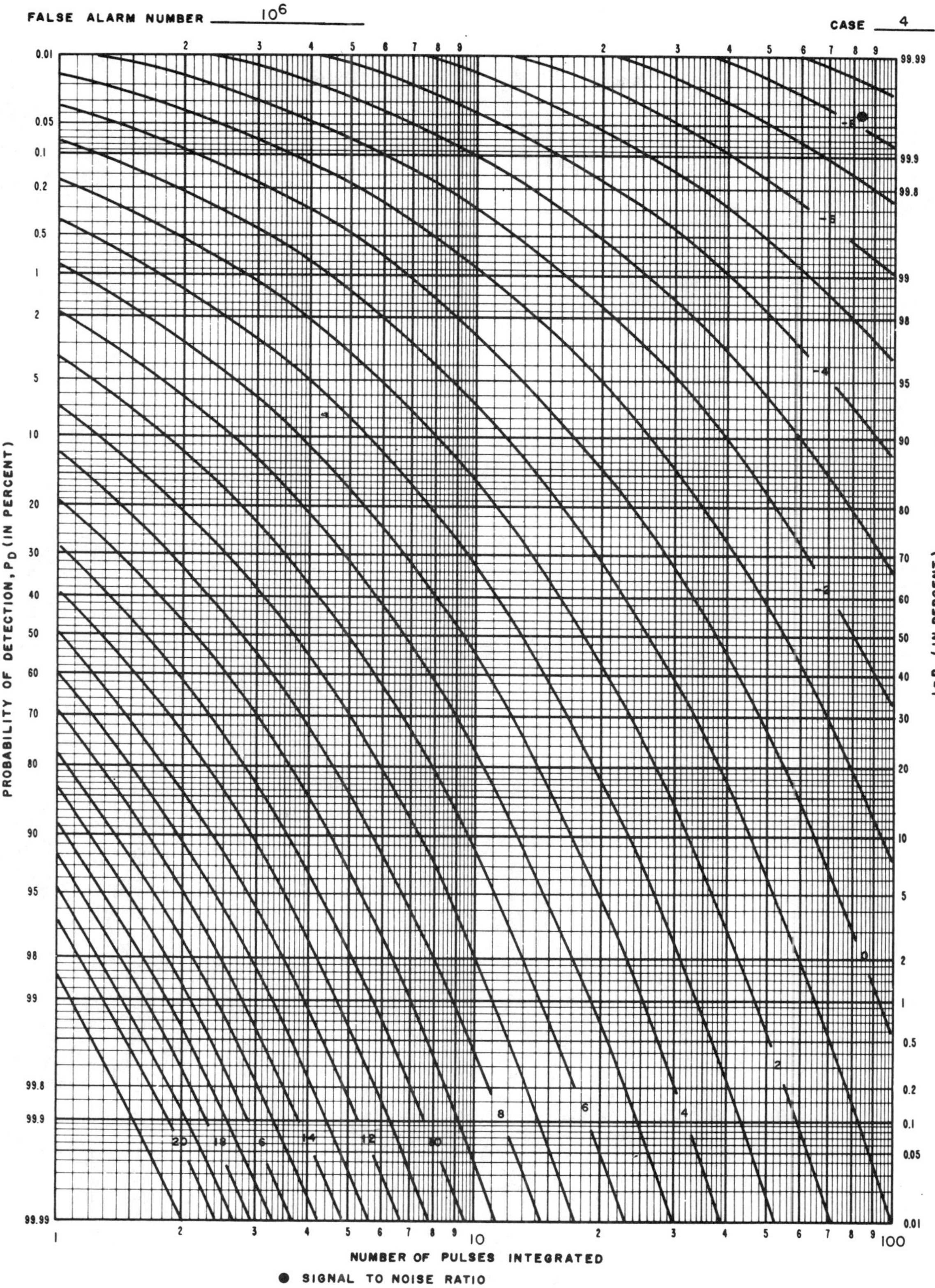
FALSE ALARM NUMBER 10^6
CASE 4
PROBABILITY OF DETECTION, P_D (IN PERCENT)
$1-P_D$ (IN PERCENT)
NUMBER OF PULSES INTEGRATED
● SIGNAL TO NOISE RATIO

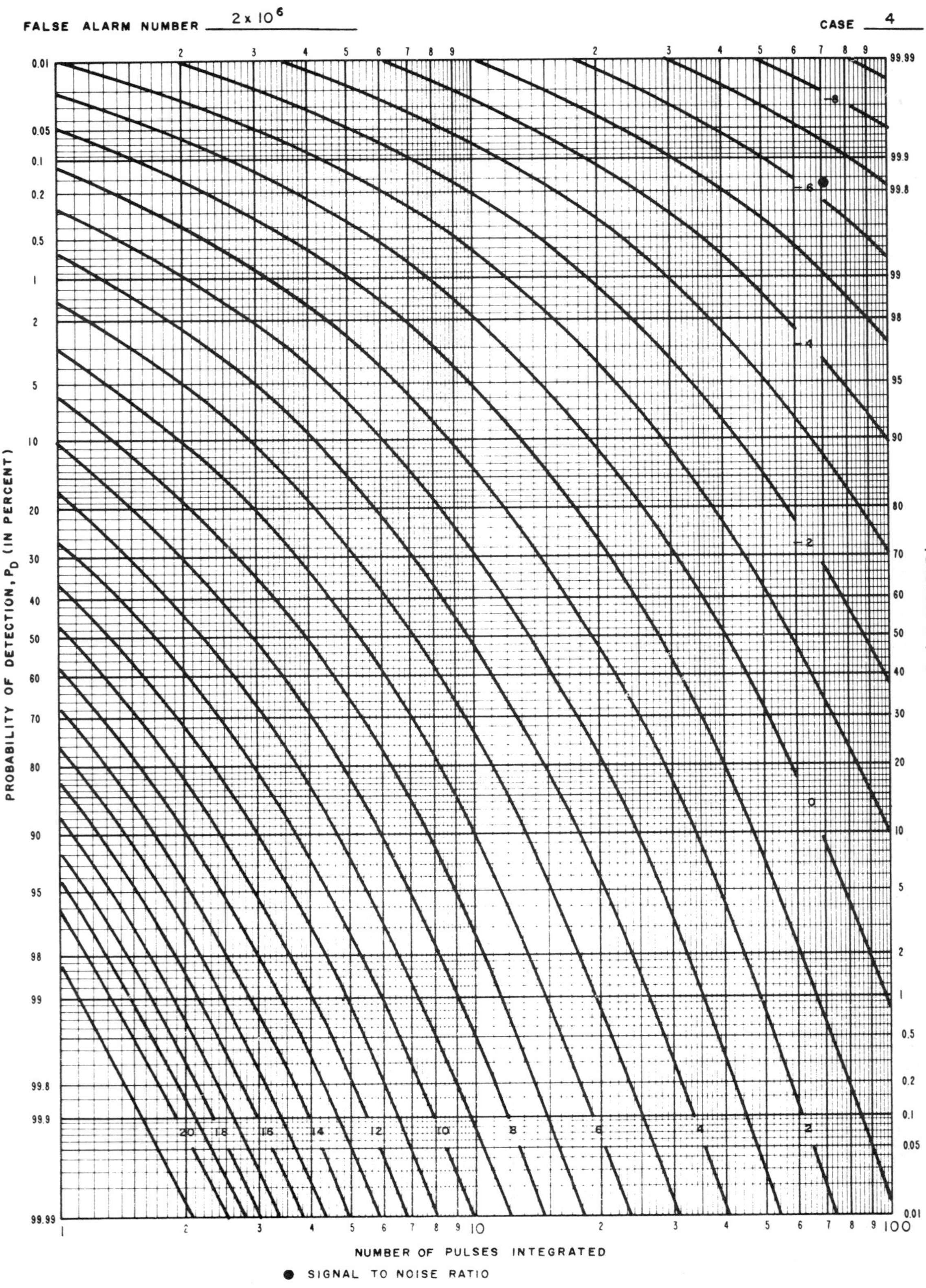
FALSE ALARM NUMBER 2 x 10^6
CASE 4
PROBABILITY OF DETECTION, P_D (IN PERCENT)
1 - P_D (IN PERCENT)
NUMBER OF PULSES INTEGRATED
● SIGNAL TO NOISE RATIO
20
18
16
14
12
10
8
6
4
2
0
-2
-4
-6
-8

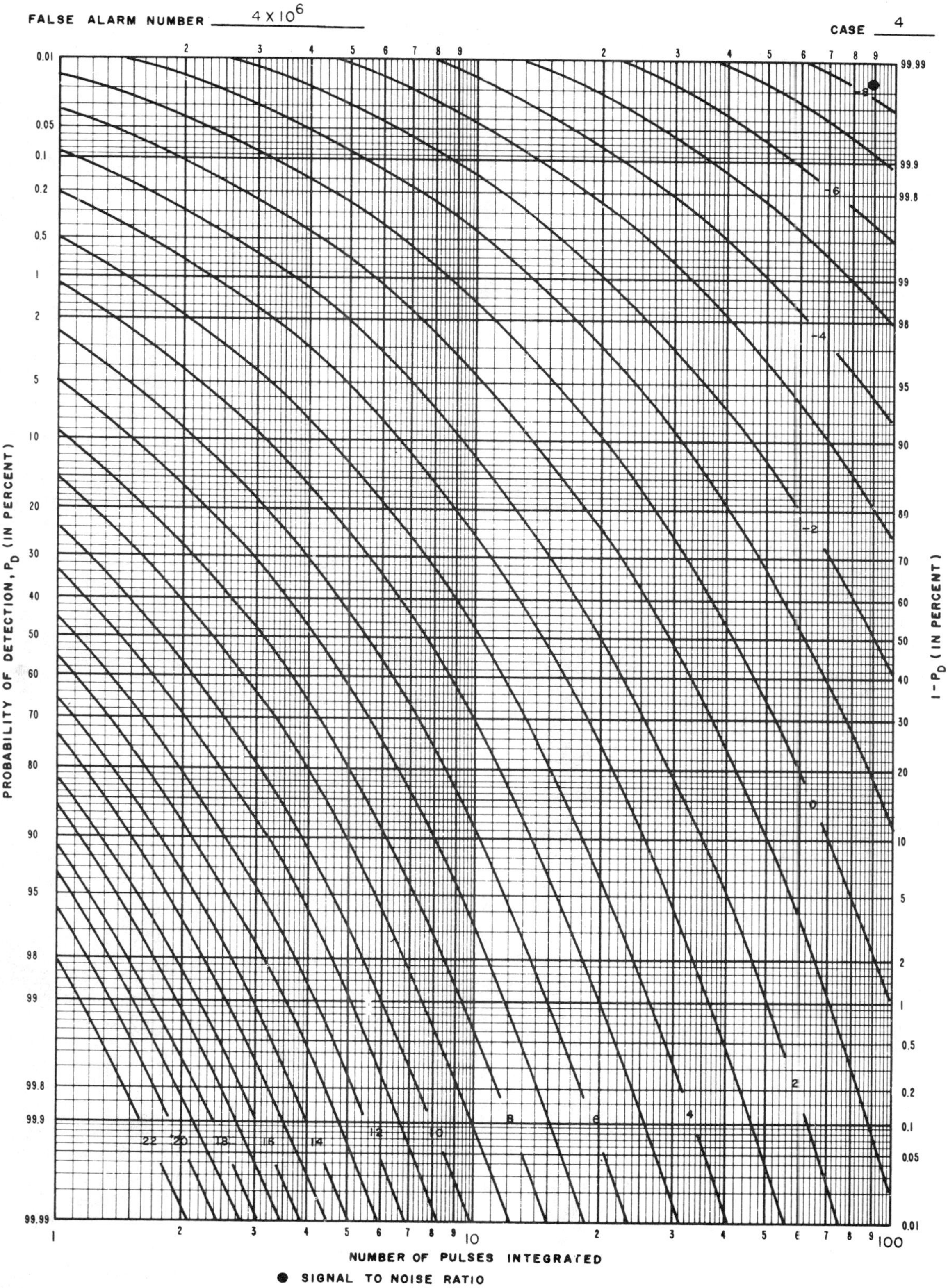
FALSE ALARM NUMBER 4 X 10^6
CASE 4
PROBABILITY OF DETECTION, P_D (IN PERCENT)
1 - P_D (IN PERCENT)
NUMBER OF PULSES INTEGRATED
● SIGNAL TO NOISE RATIO
-8
-6
-4
-2
0
2
4
6
8
10
12
14
16
18
20
22

FALSE ALARM NUMBER 6×10^6 CASE 4

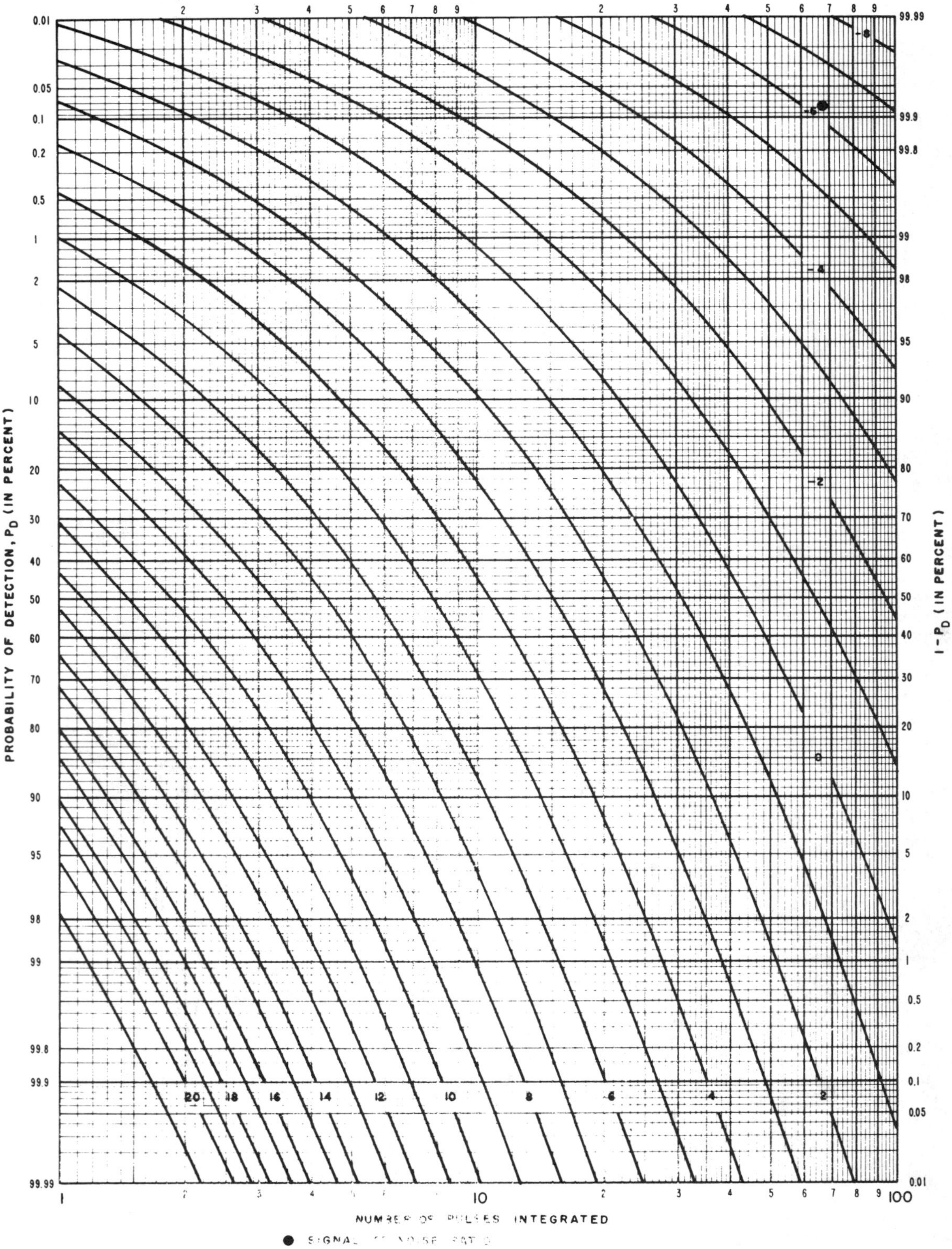

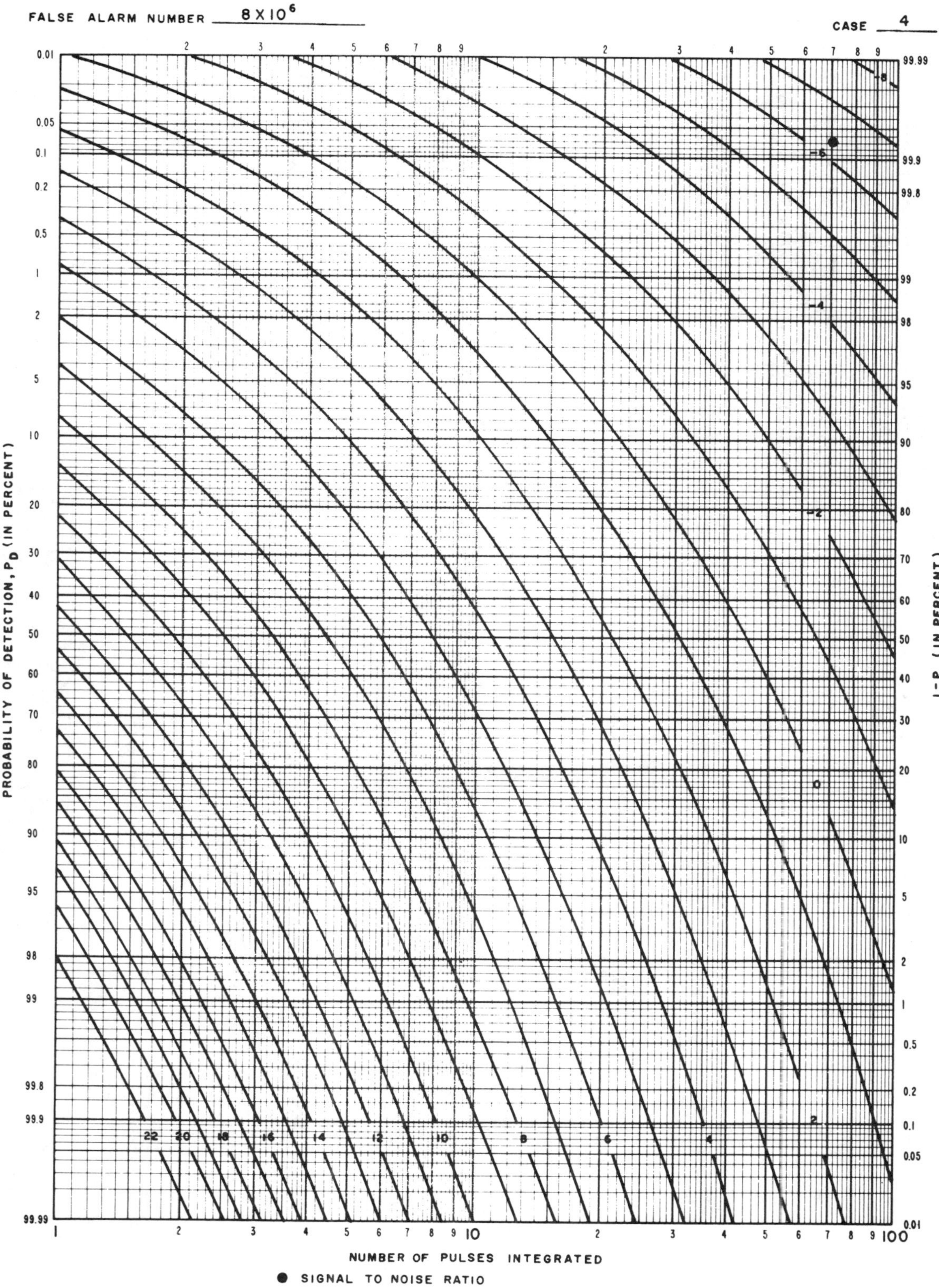
FALSE ALARM NUMBER 8 X 10^6
CASE 4
PROBABILITY OF DETECTION, P_D (IN PERCENT)
1 - P_D (IN PERCENT)
NUMBER OF PULSES INTEGRATED
22 20 18 16 14 12 10 8 6 4 2 0 -2 -4 -6 -8
● SIGNAL TO NOISE RATIO

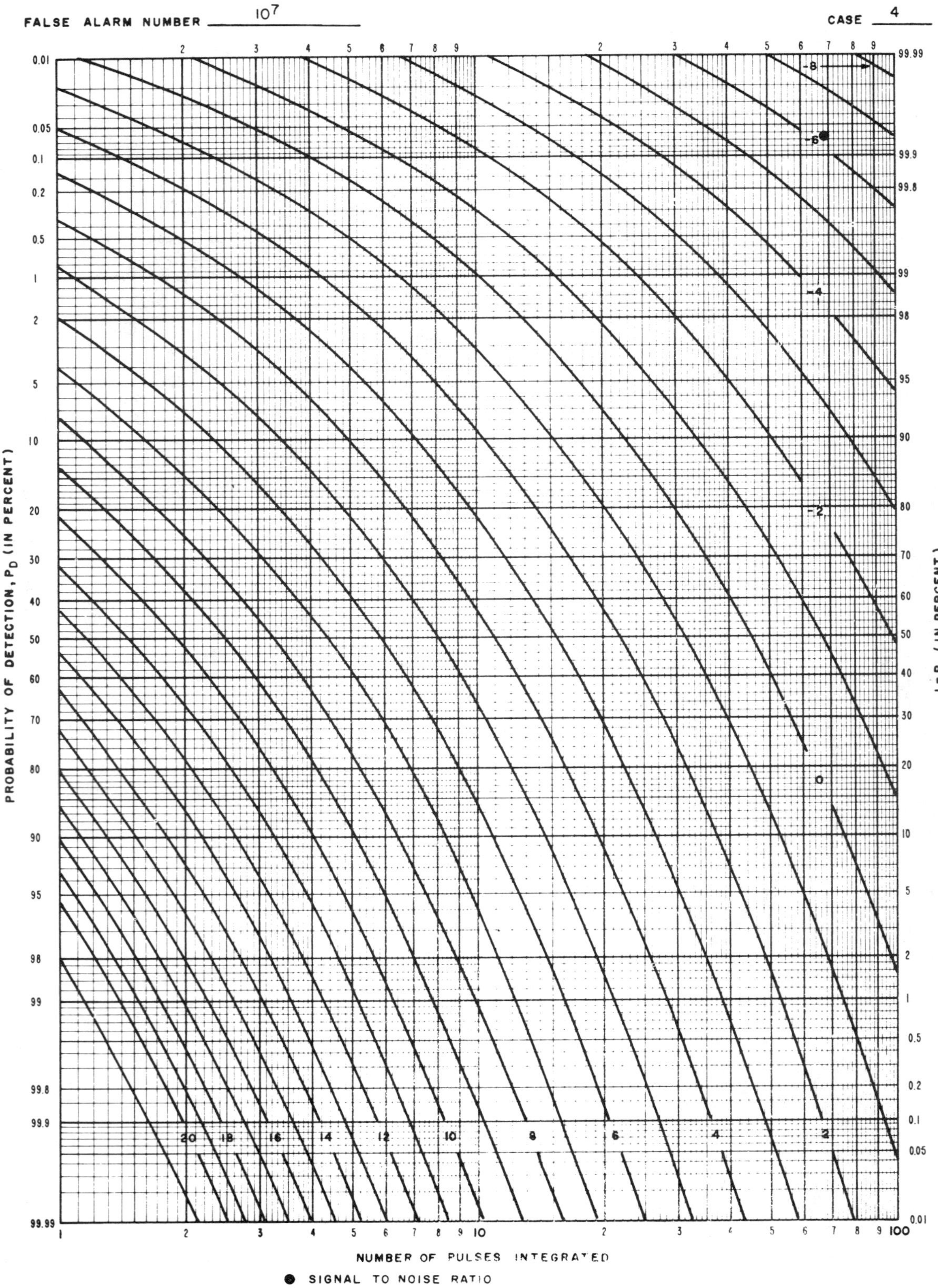
FALSE ALARM NUMBER 10^7
CASE 4
PROBABILITY OF DETECTION, P_D (IN PERCENT)
$1 - P_D$ (IN PERCENT)
NUMBER OF PULSES INTEGRATED
● SIGNAL TO NOISE RATIO
-8
-6
-4
-2
0
2
4
6
8
10
12
14
16
18
20

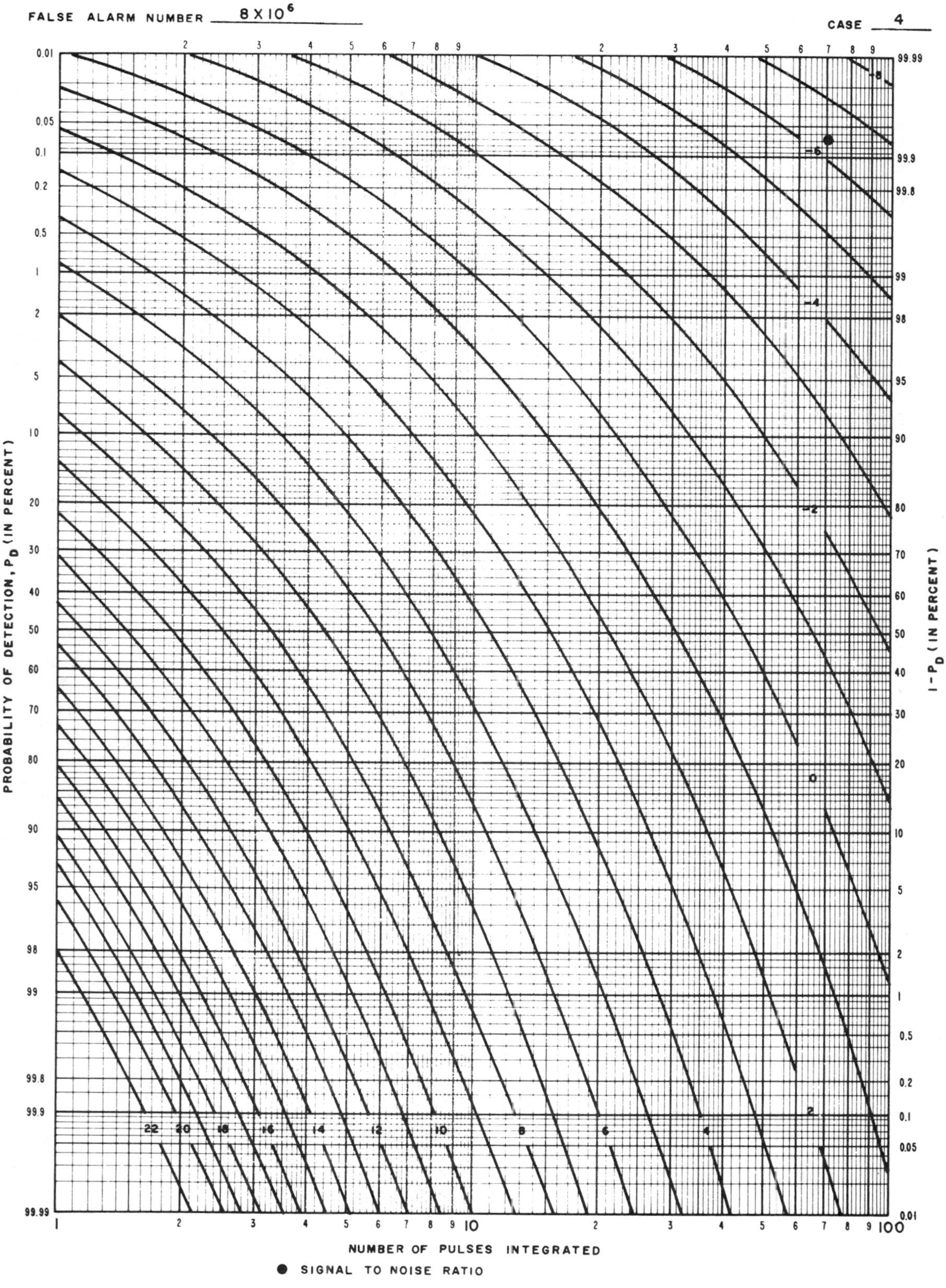
FALSE ALARM NUMBER 8 X 10^6
CASE 4
PROBABILITY OF DETECTION, P_D (IN PERCENT)
1 - P_D (IN PERCENT)
NUMBER OF PULSES INTEGRATED
● SIGNAL TO NOISE RATIO
22
20
18
16
14
12
10
8
6
4
2
0
-2
-4
-6
-8

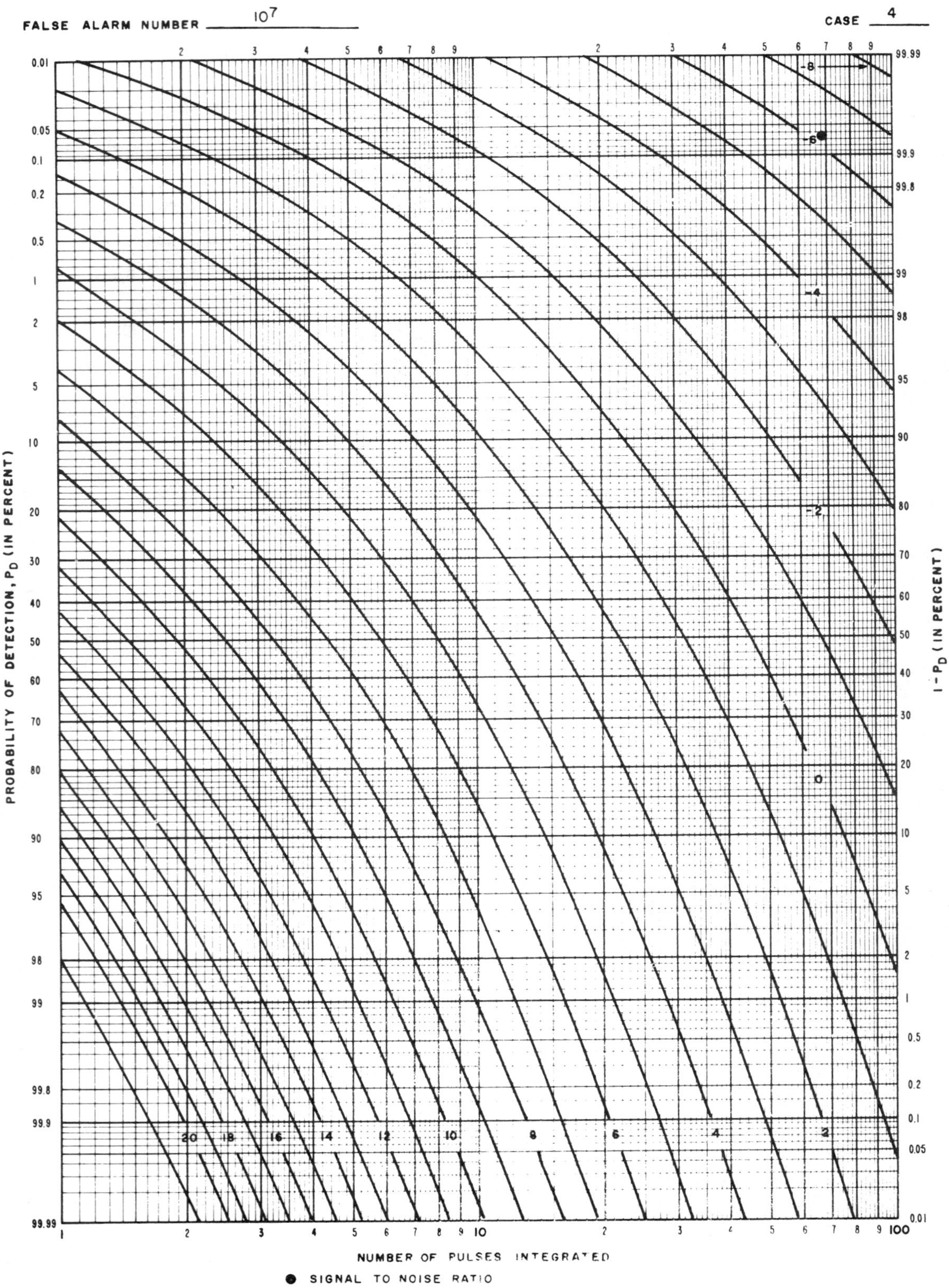
FALSE ALARM NUMBER 10^7
CASE 4
PROBABILITY OF DETECTION, P_D (IN PERCENT)
$1-P_D$ (IN PERCENT)
NUMBER OF PULSES INTEGRATED
● SIGNAL TO NOISE RATIO
-8
-6
-4
-2
0
2
4
6
8
10
12
14
16
18
20

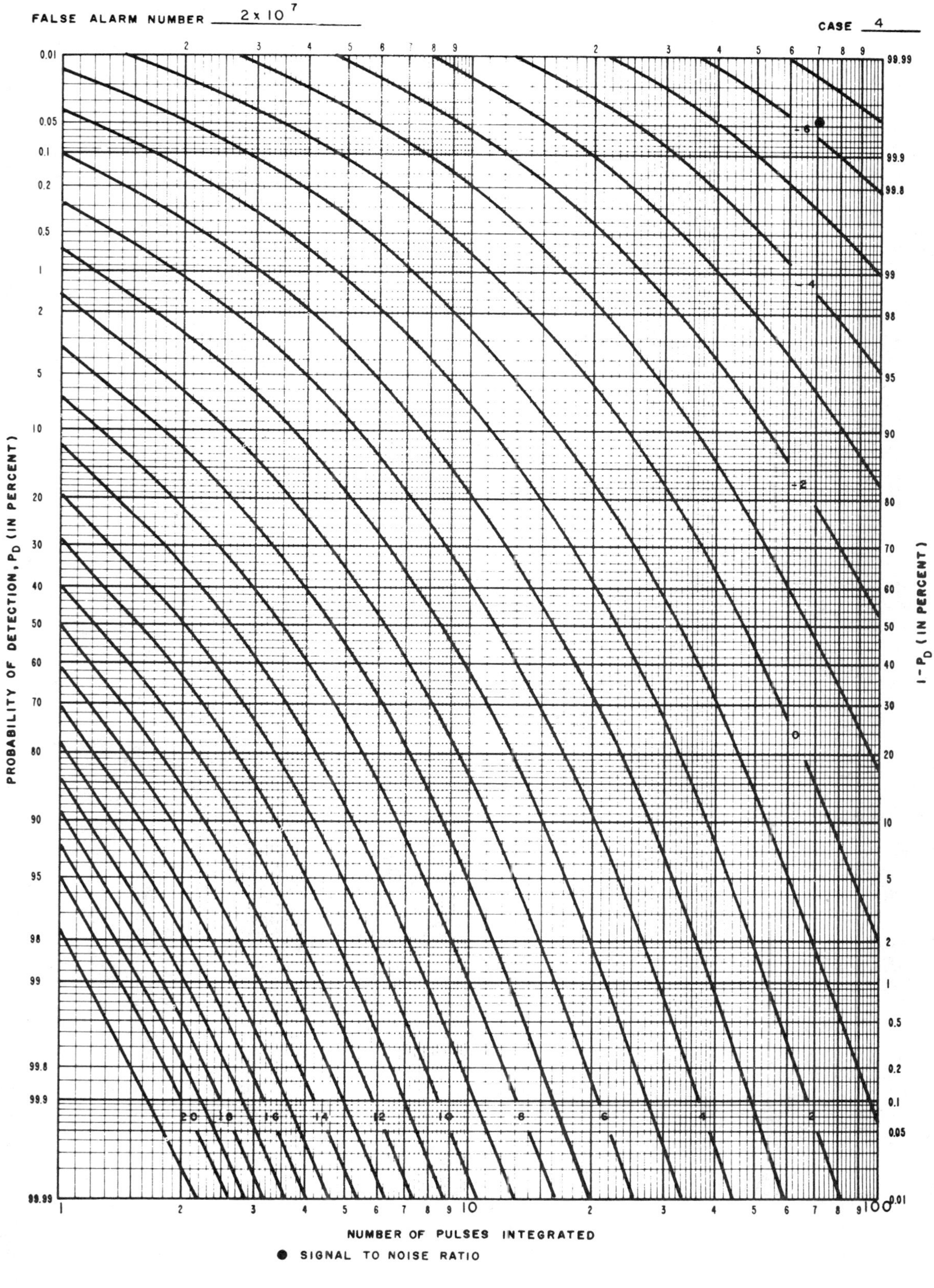
FALSE ALARM NUMBER 2 x 10^7
CASE 4
PROBABILITY OF DETECTION, P_D (IN PERCENT)
1 - P_D (IN PERCENT)
NUMBER OF PULSES INTEGRATED
● SIGNAL TO NOISE RATIO
20
18
16
14
12
10
8
6
4
2
0
-2
-4
-6

FALSE ALARM NUMBER 4×10^7 CASE 4

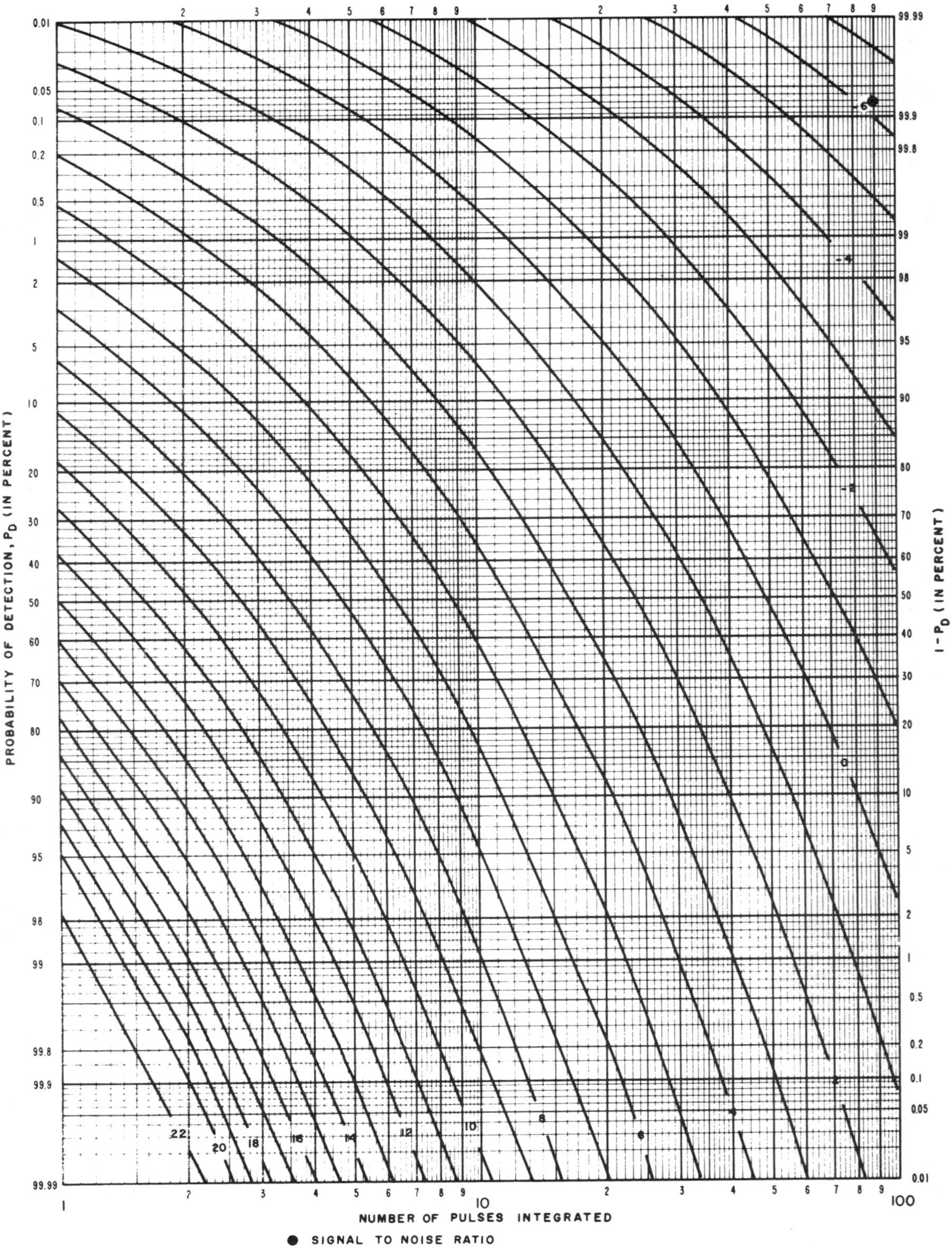

FALSE ALARM NUMBER 6 X 10^7 CASE 4

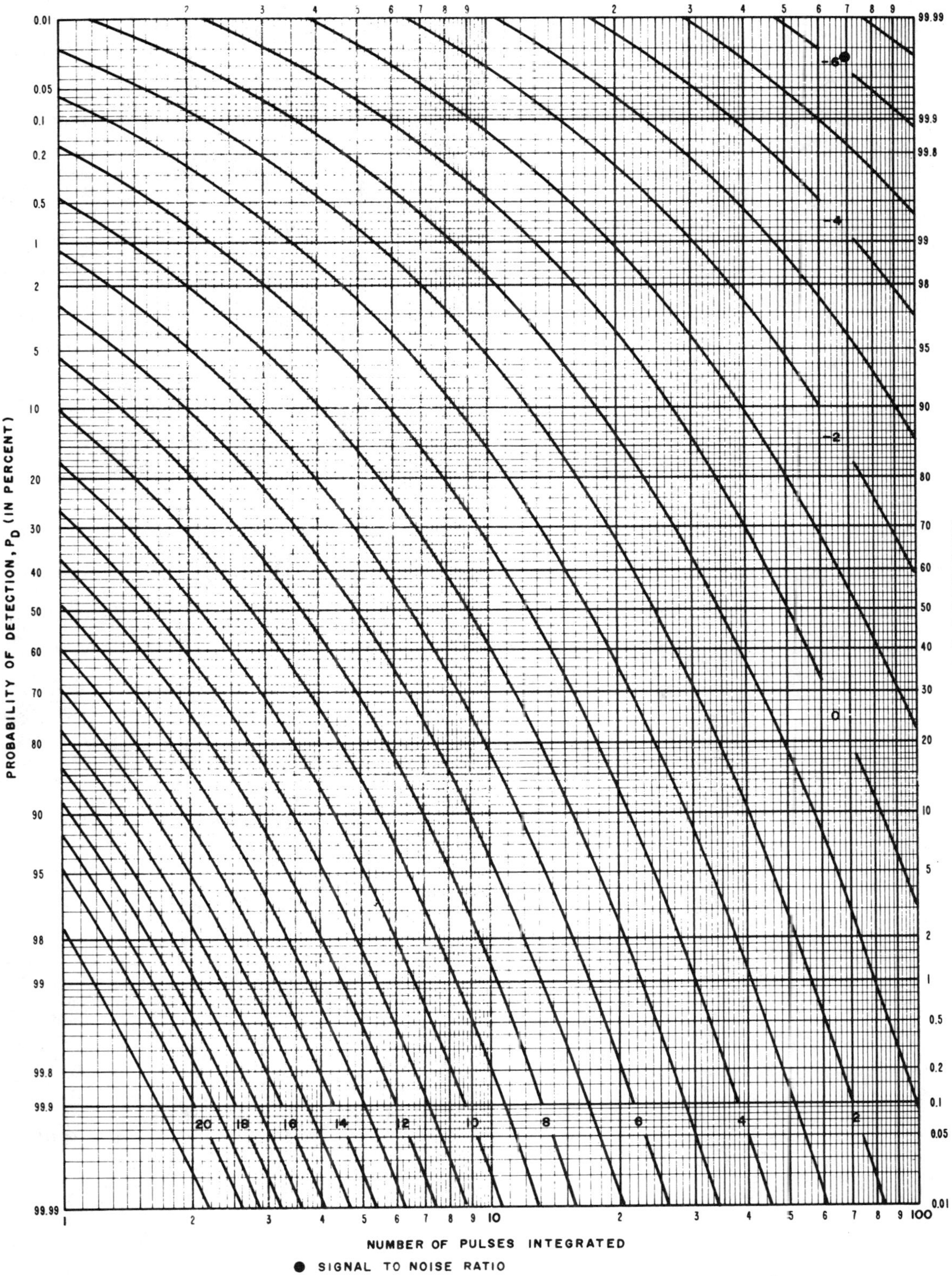

FALSE ALARM NUMBER 8×10^7 CASE 4

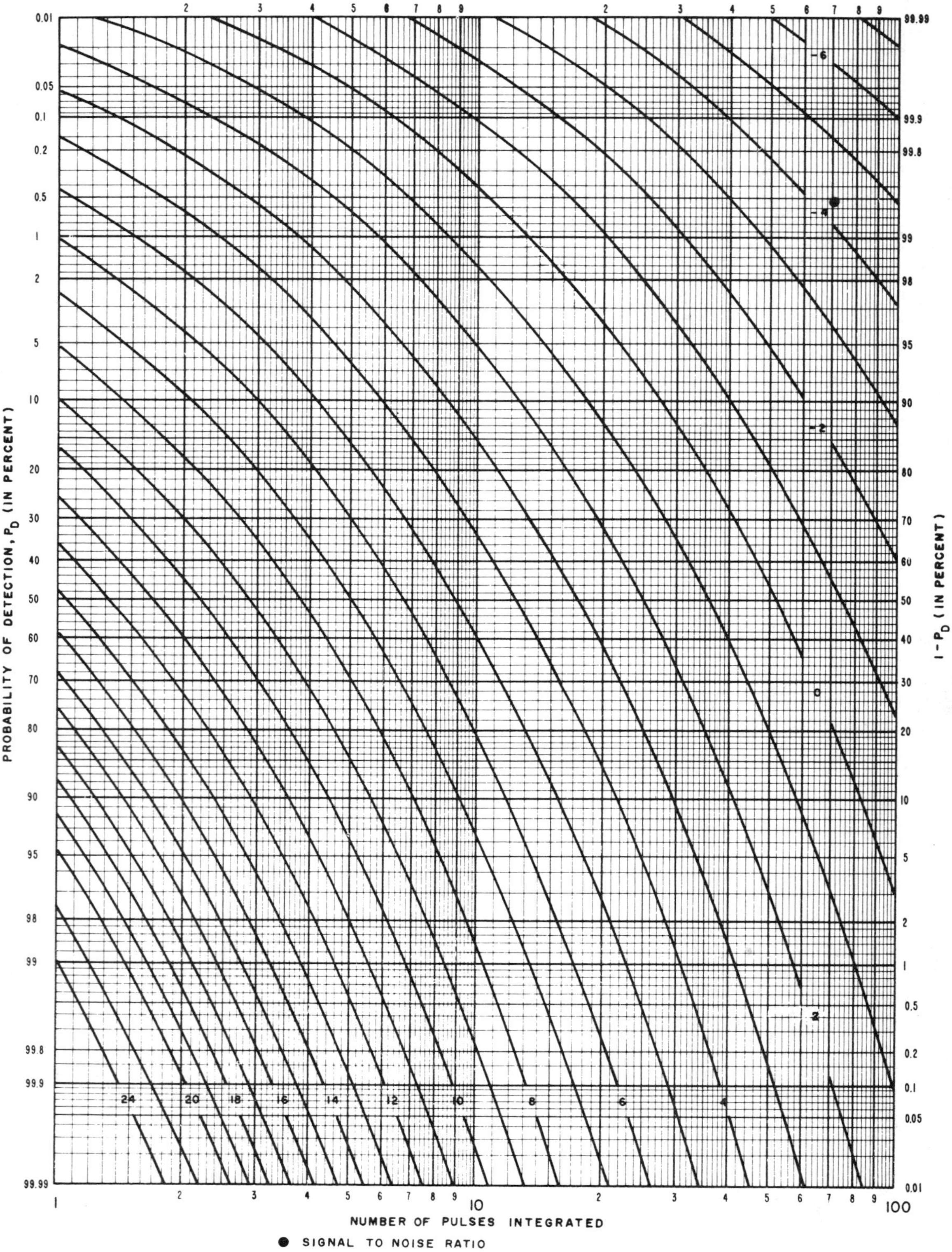

FALSE ALARM NUMBER 10^8 CASE 4

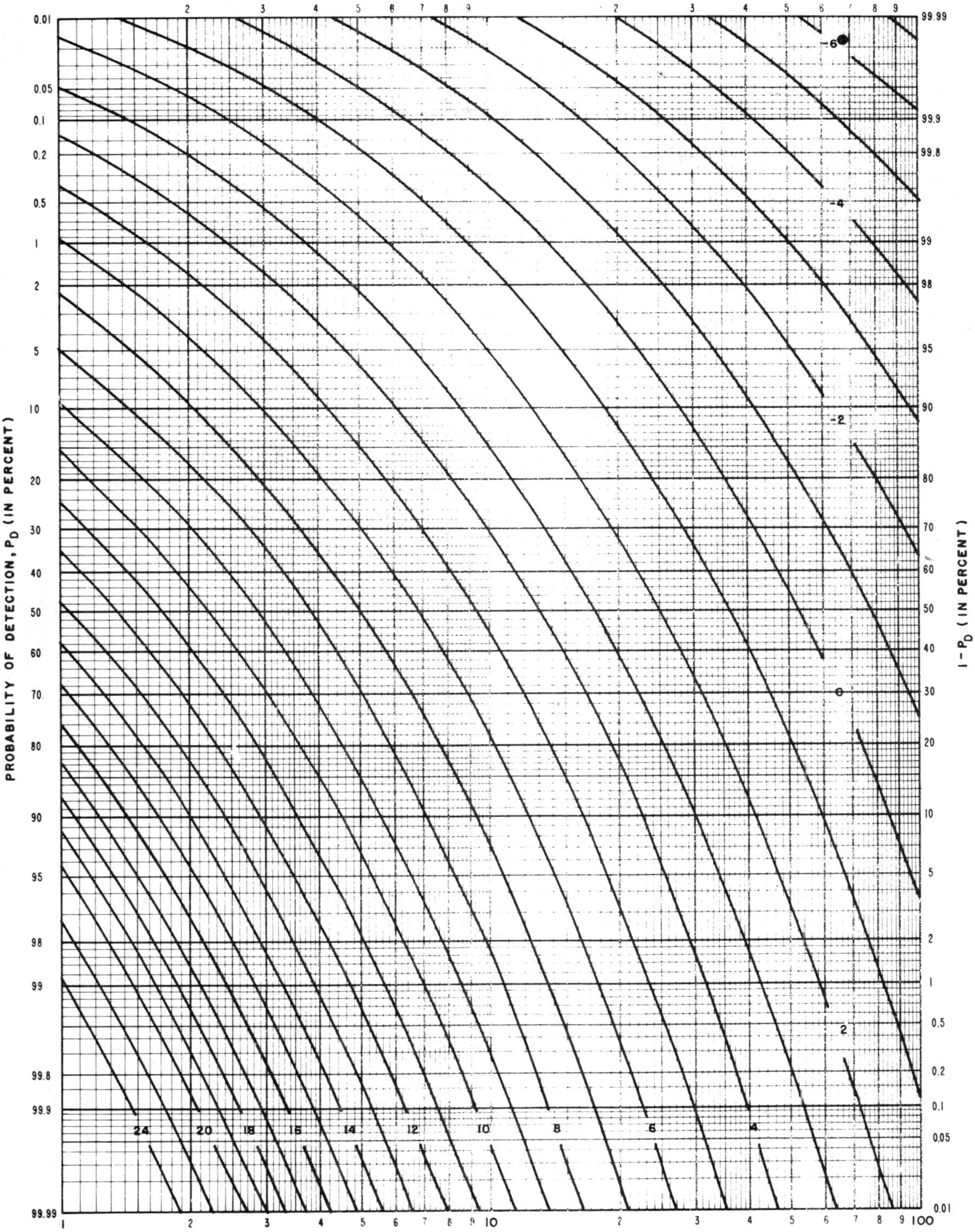

● SIGNAL TO NOISE RATIO

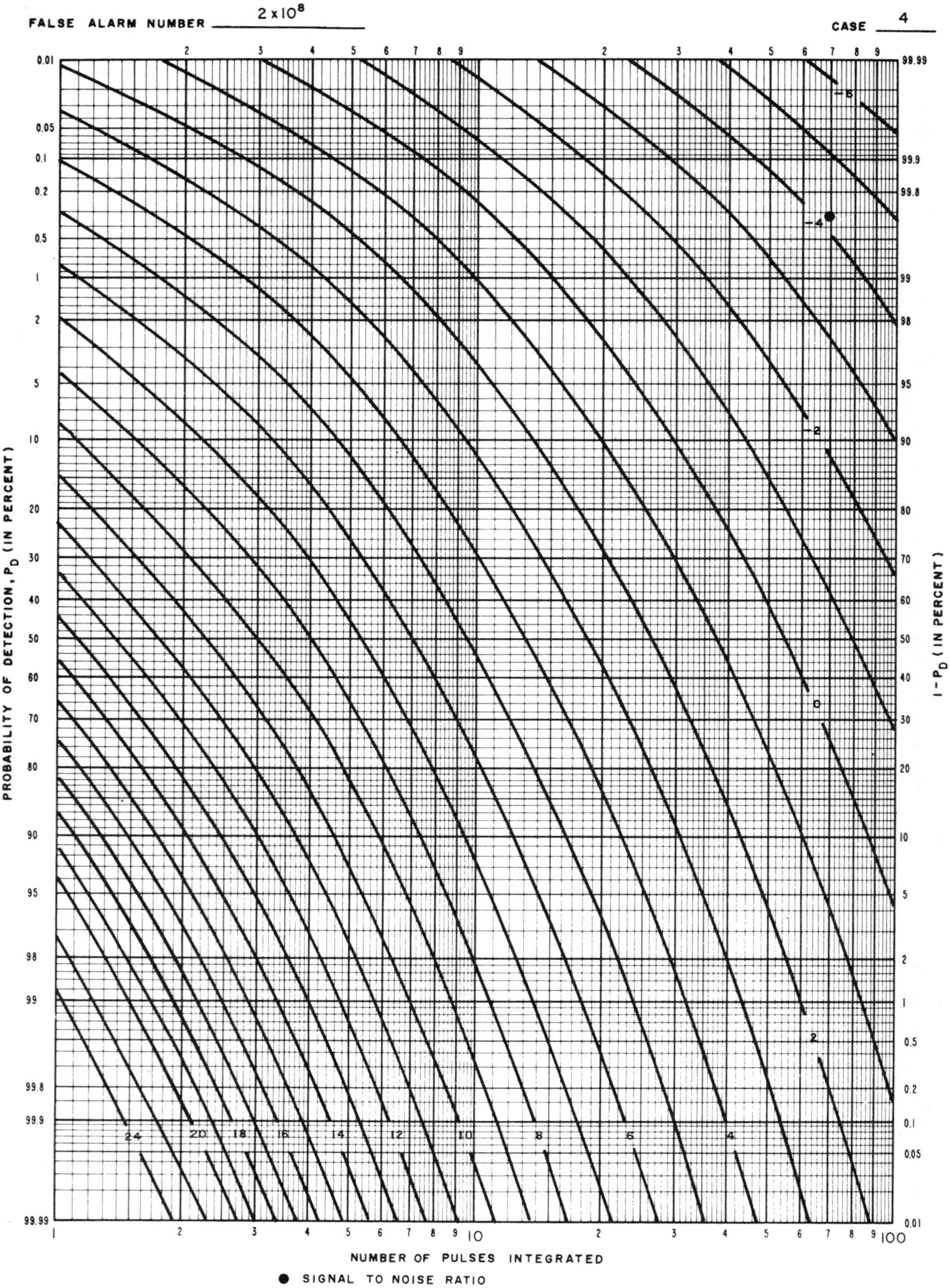
FALSE ALARM NUMBER 2×10^8
CASE 4
PROBABILITY OF DETECTION, P_D (IN PERCENT)
$1-P_D$ (IN PERCENT)
NUMBER OF PULSES INTEGRATED
● SIGNAL TO NOISE RATIO
-6
-4
-2
0
2
24
20
18
16
14
12
10
8
6
4

FALSE ALARM NUMBER 4 X 10^8 CASE 4

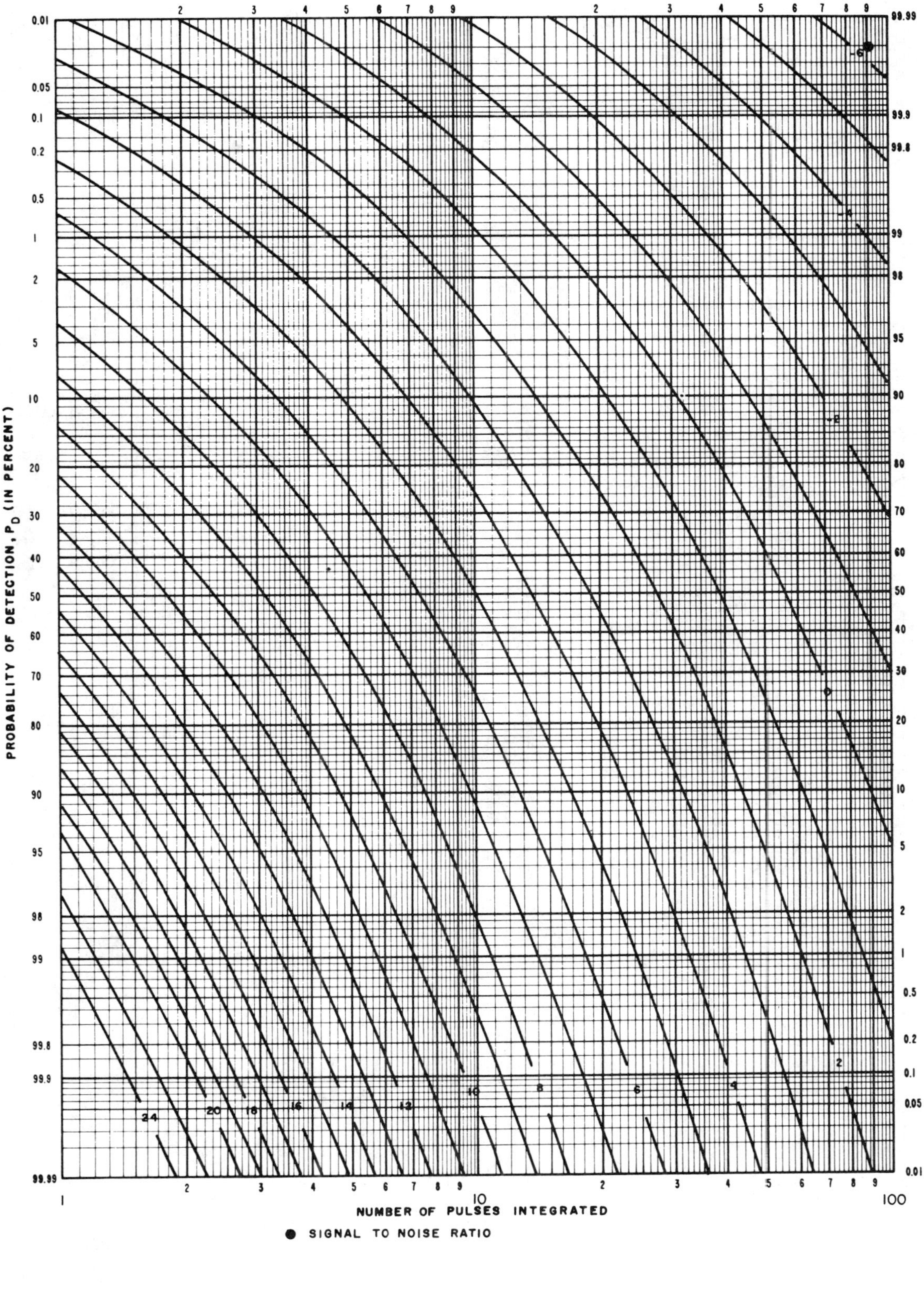

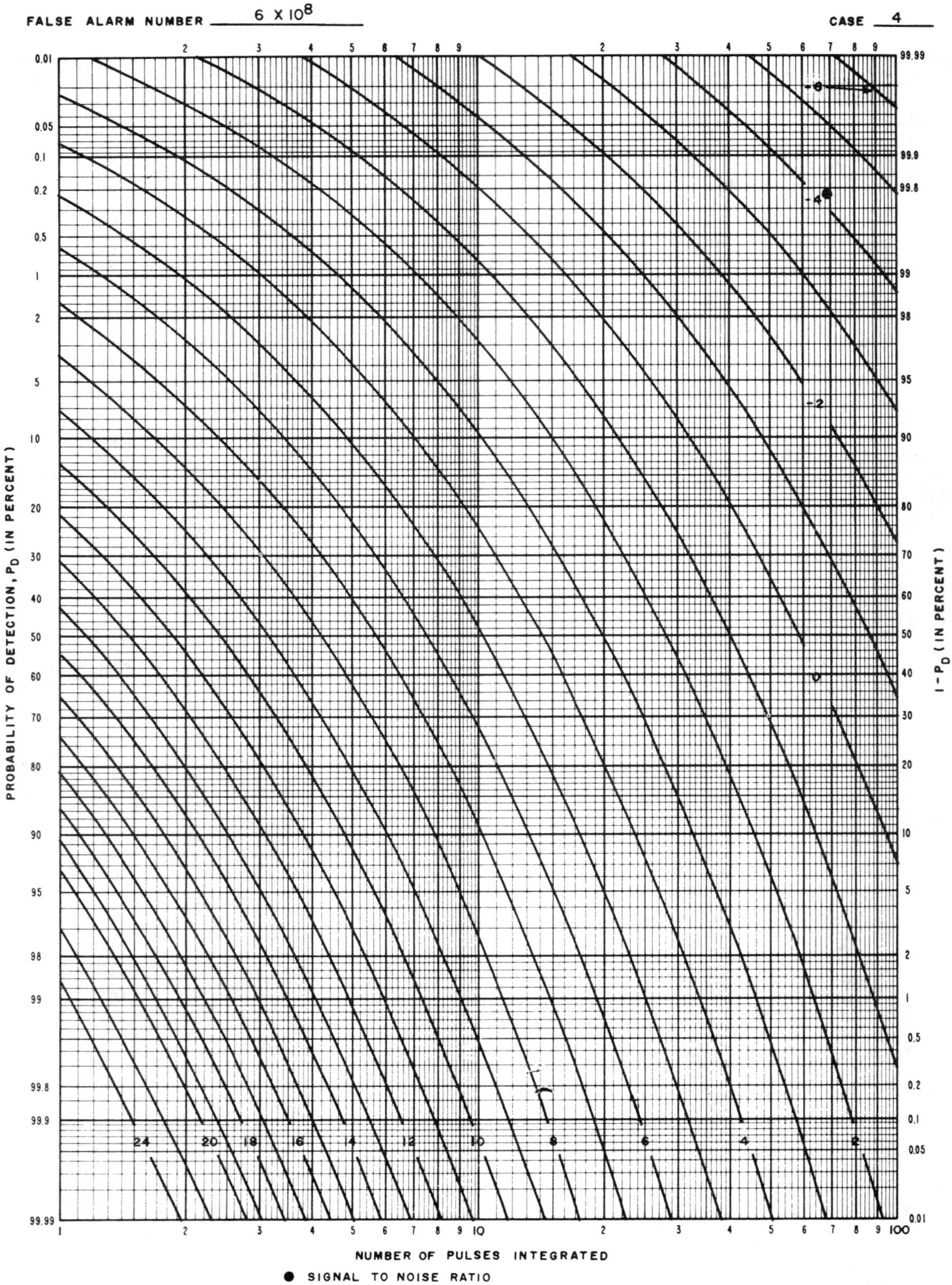
FALSE ALARM NUMBER 6 X 10^8
CASE 4
PROBABILITY OF DETECTION, P_D (IN PERCENT)
1 - P_D (IN PERCENT)
NUMBER OF PULSES INTEGRATED
● SIGNAL TO NOISE RATIO
-6
-4
-2
0
24
20
18
16
14
12
10
8
6
4
2

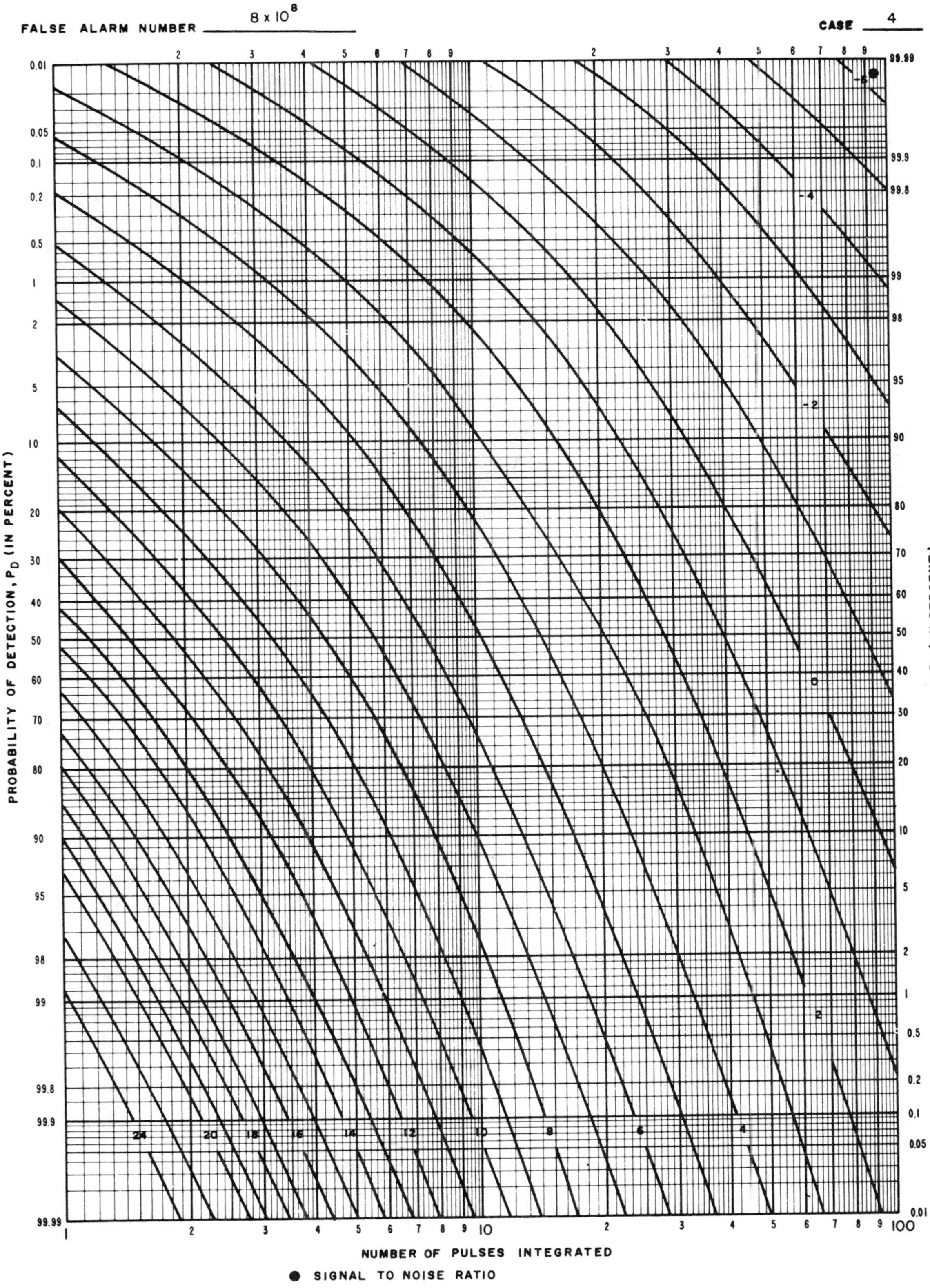
FALSE ALARM NUMBER 8 x 10[8]
CASE 4
PROBABILITY OF DETECTION, P_D (IN PERCENT)
1 - P_D (IN PERCENT)
NUMBER OF PULSES INTEGRATED
● SIGNAL TO NOISE RATIO
24
20
18
16
14
12
10
8
6
4
2
0
-2
-4
-6

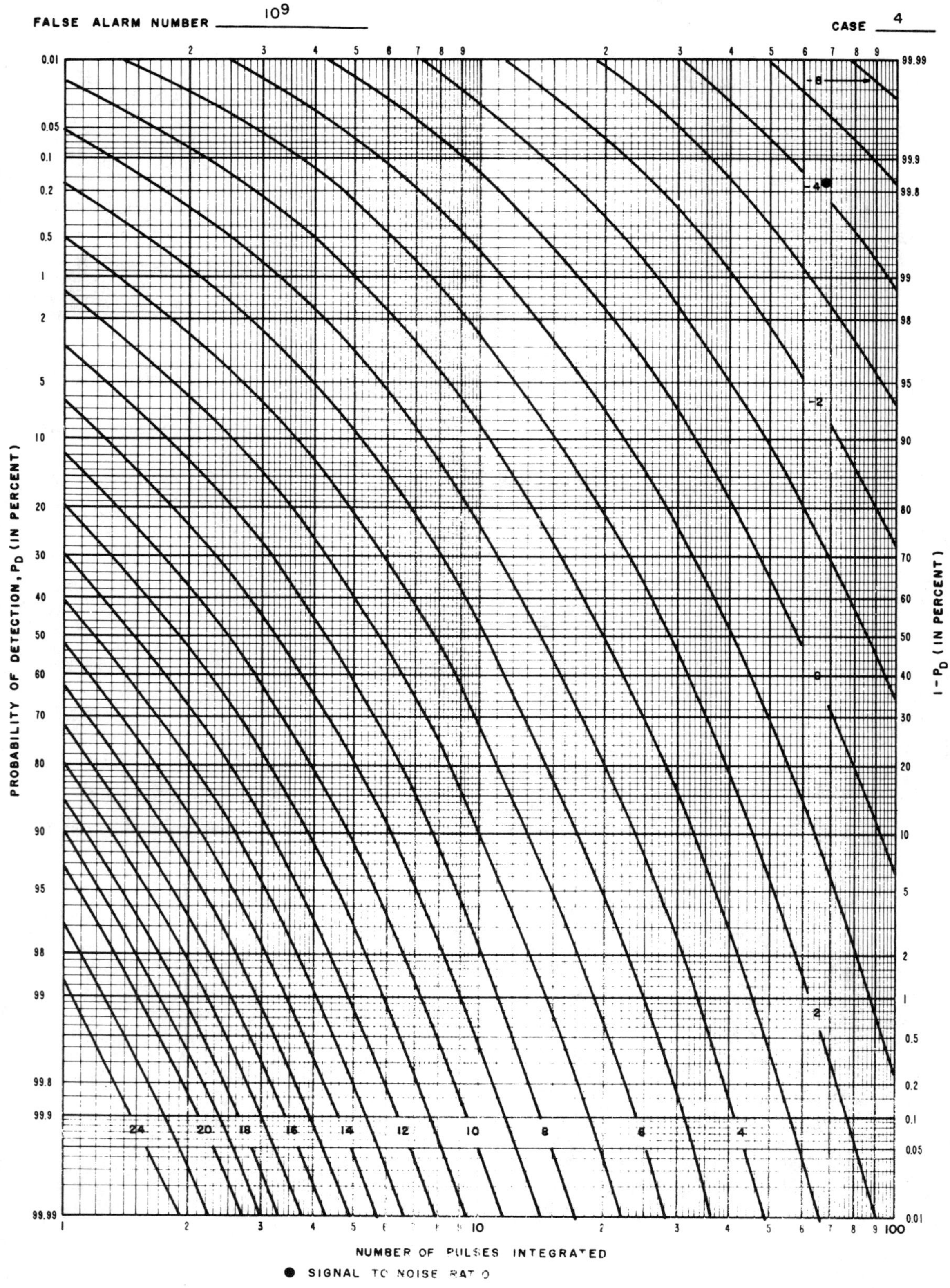
FALSE ALARM NUMBER 10^9
CASE 4
PROBABILITY OF DETECTION, P_D (IN PERCENT)
$1-P_D$ (IN PERCENT)
NUMBER OF PULSES INTEGRATED
● SIGNAL TO NOISE RATIO
24
20
18
16
14
12
10
8
6
4
2
0
-2
-4
-8

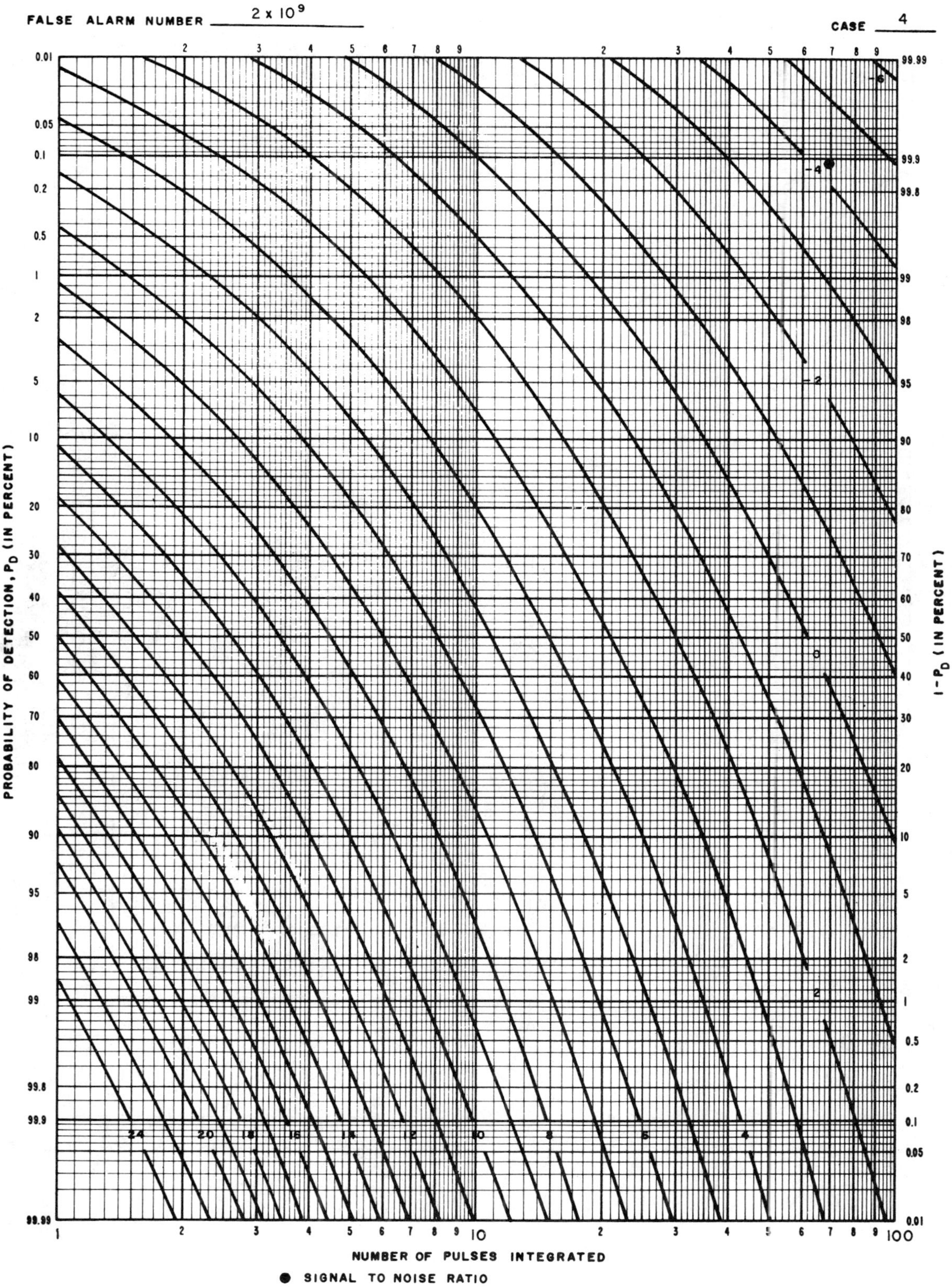
FALSE ALARM NUMBER 2 x 10^9
CASE 4
PROBABILITY OF DETECTION, P_D (IN PERCENT)
1 - P_D (IN PERCENT)
24
20
18
16
14
12
10
8
6
4
2
0
-2
-4
-6
NUMBER OF PULSES INTEGRATED
● SIGNAL TO NOISE RATIO

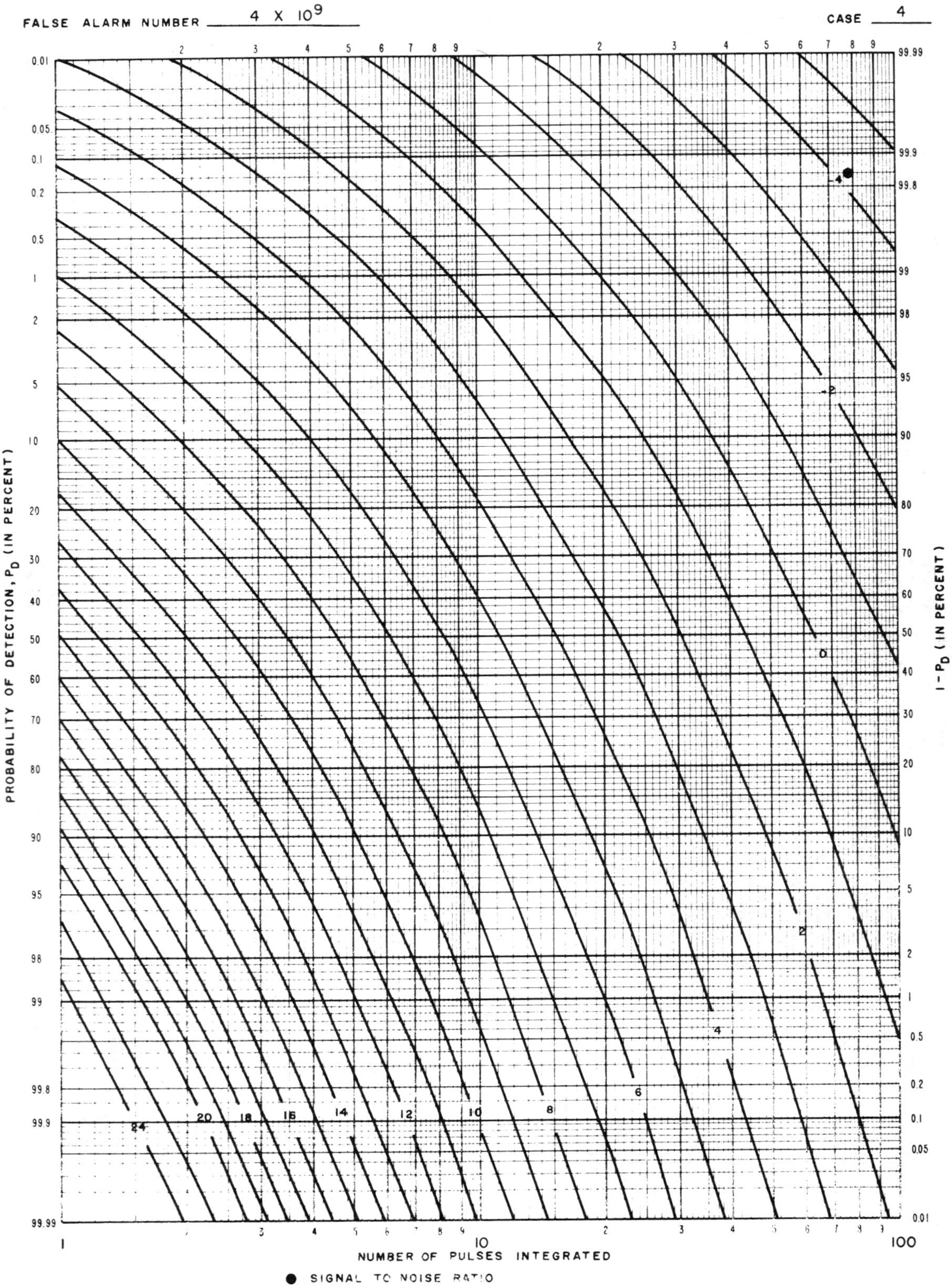
FALSE ALARM NUMBER 4 X 10^9
CASE 4
PROBABILITY OF DETECTION, P_D (IN PERCENT)
$1-P_D$ (IN PERCENT)
NUMBER OF PULSES INTEGRATED
-4
-2
0
2
4
6
8
10
12
14
16
18
20
24
● SIGNAL TO NOISE RATIO

FALSE ALARM NUMBER 6 X 10⁹ CASE 4

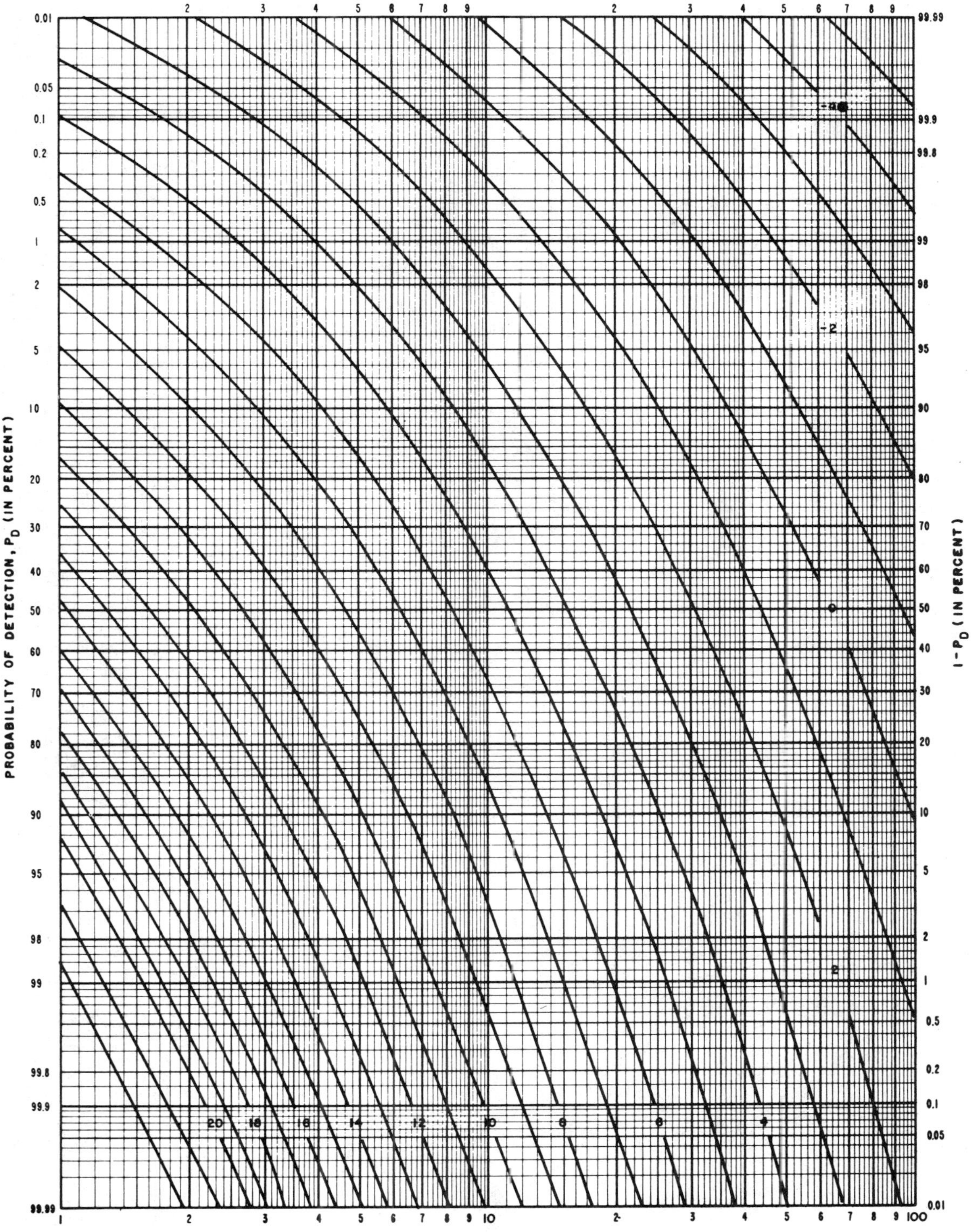

● SIGNAL TO NOISE RATIO

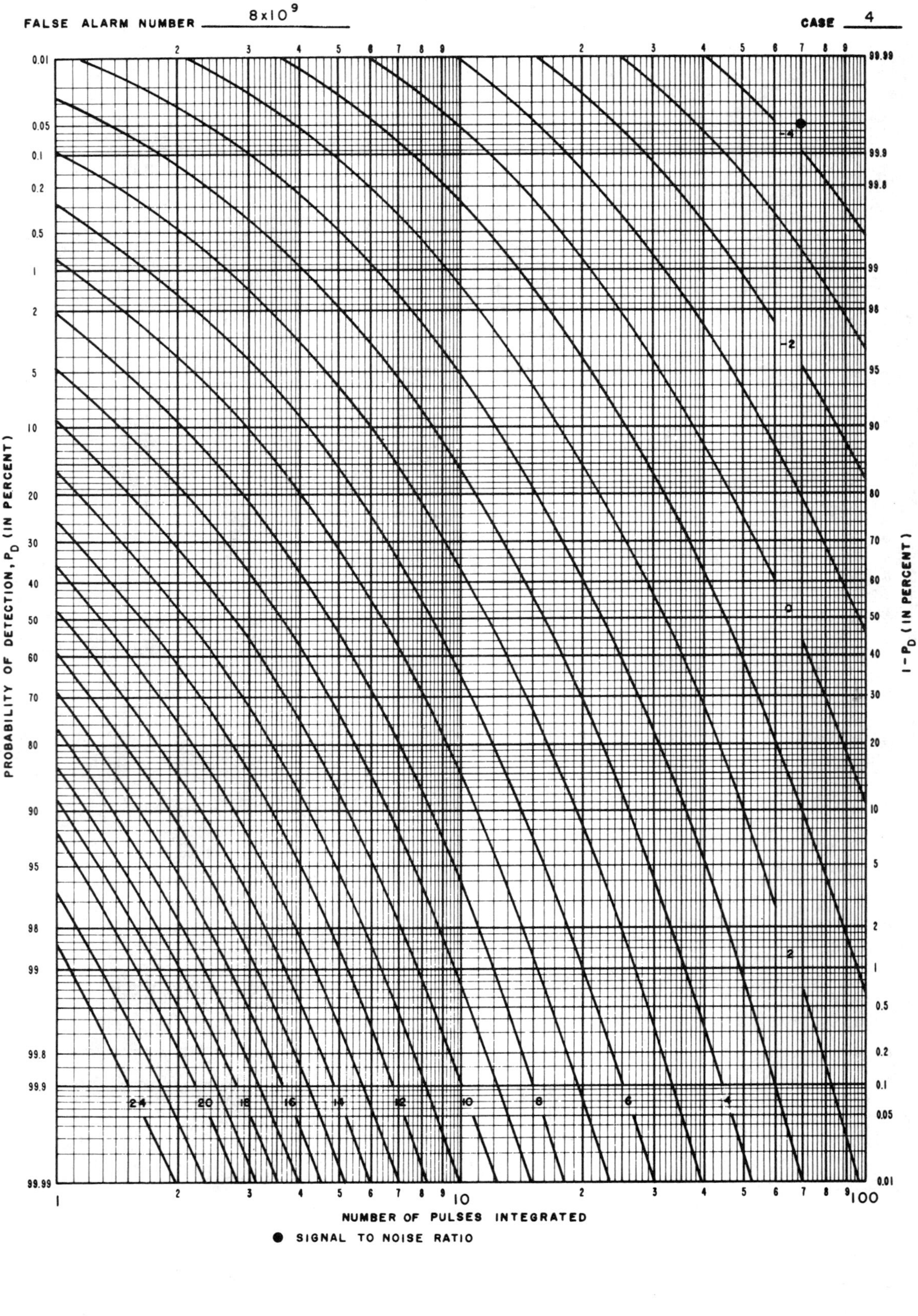
FALSE ALARM NUMBER 8x10^9
CASE 4
PROBABILITY OF DETECTION, P_D (IN PERCENT)
1-P_D (IN PERCENT)
NUMBER OF PULSES INTEGRATED
● SIGNAL TO NOISE RATIO
24
20
18
16
14
12
10
8
6
4
2
0
-2
-4

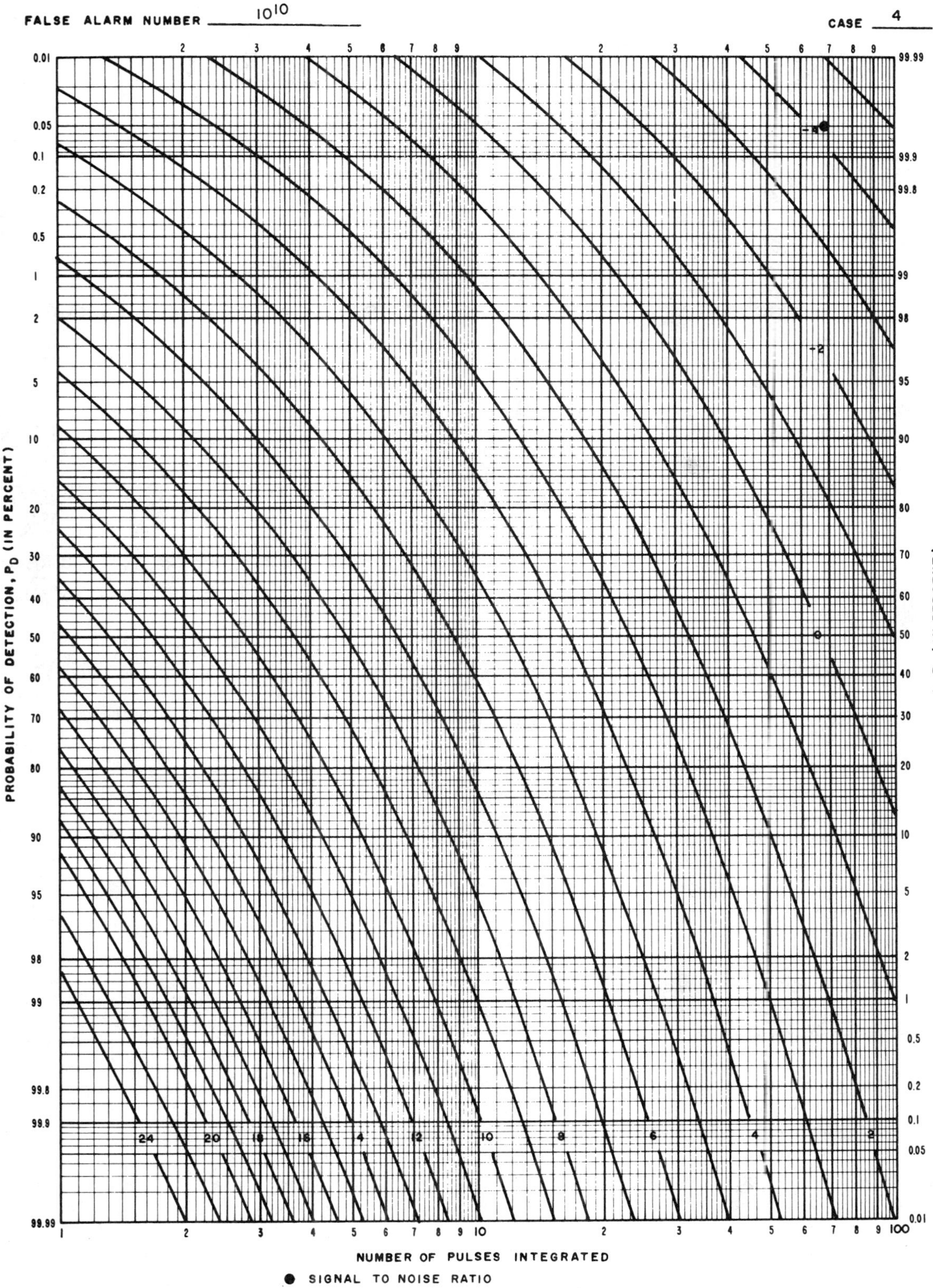
FALSE ALARM NUMBER 10^{10}
CASE 4
PROBABILITY OF DETECTION, P_D (IN PERCENT)
$1-P_D$ (IN PERCENT)
NUMBER OF PULSES INTEGRATED
● SIGNAL TO NOISE RATIO
24
20
18
16
14
12
10
8
6
4
2
0
-2
-4

APPENDIX

MEYER PLOTS: TECHNIQUES USED IN THEIR DEVELOPMENT

A-1 Introduction

The derivations for the probability of detection presented in Chapter 3 are based on the work of Marcum [1, 2] and Swerling [3]. From a standpoint of accuracy and efficiency, however, these expressions, as given by Marcum for the Case-0 nonfluctuating target model and by Swerling for the Case-3 and Case-4 chi-square distribution target models, are not in the best form for their numerical evaluation by digital computer. In Section A-3, the probability of detection for a nonfluctuating target model is rederived to produce a series which can be evaluated to any desired accuracy; in Sections A-6 and A-7, respectively, the probability-of-detection functions for the Case-3 completely correlated (scan-to-scan fluctuation) and the Case-4 completely uncorrelated (pulse-to-pulse fluctuation) chi-square distribution with four-degrees-of-freedom target models are similarly rederived in series form. The detailed derivations presented in this section were made by Belluck and Meyer [4], but are based on the original work of Fehlner [5]. They differ from the derivations originally presented in the referenced Sperry report in that they contain certain shortcuts suggested by J. Zellers, a co-worker of the authors. The series form solutions for the Case-1 completely correlated (scan-to-scan fluctuation) Rayleigh-distributed target model probability-of-detection function and the Case-2 completely uncorrelated (pulse-to-pulse fluctuation) Rayleigh-distributed target model probability-of-detection function are straightforward and presented, respectively, in Sections A-4 and A-5.

A review of the incomplete gamma function is presented in Section A-2, as it is the basis of the series form solutions that follow. The review is based on work published initially by Pearson [6].

A-2 The Incomplete Gamma Function

For P an integer, the incomplete gamma function is [6]†

$$f(a,P) = \int_0^a (e^{-u}u^P/P!)\,du \tag{A-1}$$

Alternately

$$f(a,P) = 1 - \int_a^\infty (e^{-u}u^P/P!)\,du \tag{A-2}$$

Equation (A-2) follows from Eq. (A-1) and the fact that $f(\infty, P) = 1$. To show that $f(\infty, P) = 1$, Eq. (A-1) is integrated by parts as follows:

$$f(a,P) = \frac{e^{-u}u^P}{P!}\bigg|_0^a + \int_0^a \frac{e^{-u}u^{(P-1)}}{(P-1)!}\,du \tag{A-3}$$

Repeated integration by parts yields the series

$$f(a,P) = \frac{-a^P e^{-a}}{P!} - \frac{a^{(P-1)}e^{-a}}{(P-1)!} - \cdots + \int_0^a e^{-u}\,du \tag{A-4}$$

and a final integration yields

$$f(a,P) = \frac{-a^P e^{-a}}{P!} - \frac{a^{(P-1)}e^{-a}}{(P-1)!} - \cdots + 1 - e^{-a} = 1 - \sum_{k=0}^{P} \frac{e^{-a}a^k}{k!} \tag{A-5}$$

from which it follows that $f(\infty, P) = 1$, since

$$\lim_{a\to\infty} \sum_{k=0}^{P} \frac{e^{-a}a^k}{k!} \to 0 \tag{A-6}$$

† This form of the incomplete gamma function, $f(a,P)$, is related to the form used in Chapters 2 and 3, $I(u,P)$, as follows: $f(a,P) = I[a/(P+1)^{1/2}, P]$; therefore, $a = u(P+1)^{1/2}$.

An alternate expression for $f(a,P)$ is

$$f(a,P) = \sum_{k=P+1}^{\infty} \frac{e^{-a}a^k}{k!} \quad \text{(A-7)}$$

If $a > P$, Eq. (A-5) converges more rapidly and is used for computation; if $a < P$, Eq. (A-7) is preferred.

A-3 The Case-0 Nonfluctuating Target Model

The characteristic function for Case 0 as given by Marcum and restated in Eq. (3-30) is

$$C_N(p) = e^{-Nx}e^{Nx/(p+1)}/(p+1)^N \quad \text{(A-8)}$$

where

x is the power signal-to-noise ratio (S/N),
N the number of variants integrated,
C the characteristic function, and
p the transformation variable.

The corresponding density function is obtained by contour integration; that is

$$f(t) = 1/2\pi j\int_{-j\infty}^{j\infty} F(p)\,e^{pt}\,dp \quad \text{(A-9)}$$

Replacing $F(p)$ with $C_N(p)$, we have

$$f(t) = \frac{1}{2\pi j}\int_{-j\infty}^{j\infty} \frac{e^{-Nx}e^{Nx/(p+1)}e^{pt}}{(p+1)^N}\,dp \quad \text{(A-10)}$$

Rearranging terms in Eq. (A-10) yields

$$f(t) = \frac{e^{-Nx}e^{-t}}{2\pi j}\int_{-j\infty}^{j\infty} \frac{e^{Nx/(p+1)}e^{t(p+1)}}{(p+1)^N}\,dp \quad \text{(A-11)}$$

The two exponentials within the integral of Eq. (A-11) can be represented, respectively, by the two infinite series

$$e^{Nx/(p+1)} = \sum_{k=0}^{\infty}\left(\frac{Nx}{p+1}\right)^k\left(\frac{1}{k!}\right) \quad \text{(A-12)}$$

and

$$e^{(p+1)t} = \sum_{L=0}^{\infty}\frac{(p+1)^L t^L}{L!} \quad \text{(A-13)}$$

Their product is

$$e^{Nx/(p+1)}[e^{(p+1)t}] = \sum_{k=0}^{\infty}\sum_{L=0}^{\infty}\frac{(Nx)^k t^L (p+1)^{(L-k)}}{k!L!} \quad \text{(A-14)}$$

The coefficients of the terms in Eq. (A-15) below, for which $L-k = N-1$, are the residues of the contour integral in Eq. (A-11)

$$f(t) = e^{-Nx}e^{-t}\sum_{k=0}^{\infty}\left(\frac{Nx^k}{k!}\right)\left[\frac{t^{(N-1+k)}}{(N-1+k)!}\right] \quad \text{(A-15)}$$

The probability distribution function is given by

$$P_N(Y_b) = 1 - \int_0^{Y_b} f(t)\,dt = \int_{Y_b}^{\infty} f(t)\,dt \quad \text{(A-16)}$$

and substituting Eq. (A-15) into Eq. (A-16) yields

$$P_N(x, Y_b) = \int_{Y_b}^{\infty} e^{-Nx}e^{-t}\sum_{k=0}^{\infty}\frac{(Nx)^k}{k!}\frac{t^{(N-1+k)}}{(N-1+k)!}\,dt \quad \text{(A-17)}$$

or

$$P_N(x, Y_b) = \sum_{k=0}^{\infty}\frac{(Nx)^k e^{-Nx}}{k!}\int_{Y_b}^{\infty}\frac{e^{-t}t^{(N-1+k)}}{(N-1+k)!}\,dt \quad \text{(A-18)}$$

The integral of Eq. (A-18) is the complement of the incomplete gamma function of Eq. (A-1) with

$$t = u \quad \text{(A-19)}$$

$$P = N - 1 + k \quad \text{(A-20)}$$

$$a = Y_b \quad \text{(A-21)}$$

$$dt = du \quad \text{(A-22)}$$

Therefore

$$1 - f[Y_b, (N-1+k)] = \int_{Y_b}^{\infty}\frac{e^{-t}t^{(N-1+k)}}{(N-1+k)!}\,dt \quad \text{(A-23)}$$

from which it follows that

$$\int_{Y_b}^{\infty}\frac{e^{-t}t^{(N-1+k)}}{(N-1+k)!}\,dt = 1 - \left(1 - \sum_{j=0}^{N-1+k}\frac{e^{-Y_b}Y_b^{\,j}}{j!}\right) \quad \text{(A-24)}$$

$$= \sum_{j=0}^{N-1+k}\frac{e^{-Y_b}Y_b^{\,j}}{j!} \quad \text{(A-25)}$$

or alternately

$$\int_{Y_b}^{\infty} \frac{e^t t^{(N-1+k)}}{(N-1+k)!} dt = 1 - \sum_{j=N+k}^{\infty} \frac{e^{-Y_b} Y_b^j}{j!} \tag{A-26}$$

Therefore, for $Y_b > N(x+1)$,

$$P_N(Y_b, x) = e^{-Nx} \sum_{k=0}^{\infty} \frac{(Nx)^k}{k!} \sum_{j=0}^{N-1+k} \frac{e^{-Y_b} Y_b^j}{j!} \tag{A-27}$$

and for $Y_b < N(x+1)$,

$$P_N(Y_b, x) = e^{-Nx} \sum_{k=0}^{\infty} \frac{(Nx)^k}{k!} \left(1 - \sum_{j=N+k}^{\infty} \frac{e^{-Y_b} Y_b^j}{j!}\right) \tag{A-28}$$

$$= 1 - e^{-Nx} \sum_{k=0}^{\infty} \frac{(Nx)^k}{k!} \sum_{j=N+k}^{\infty} \frac{e^{-Y_b} Y_b^j}{j!} \tag{A-29}$$

If $Y_b > N(x+1)$, Eq. (A-27) converges more rapidly and, therefore, is used; for $Y_b < N(x+1)$, Eq. (A-29) converges more rapidly and is preferred.

A-4 The Case-1 Target Model—Rayleigh-Distributed Signal with Scan-to-Scan Fluctuation

The probability density function for Case 1 as given by Swerling and restated in Eq. (3-54) can be rewritten as follows by use of Eq. (A-7):

$$f(v) = \left(1 + \frac{1}{N\bar{x}}\right)^{N-2} \left(\frac{1}{N\bar{x}}\right) \times \sum_{k=N-1}^{\infty} \frac{\exp\left[\frac{-v}{1+(1/N\bar{x})}\right] \left[\frac{v}{1+(1/N\bar{x})}\right]^k}{k!} \times \exp\left(\frac{-v}{1+N\bar{x}}\right) \tag{A-30}$$

by letting

$$a = \frac{v}{1+(1/N\bar{x})} \tag{A-31a}$$

and

$$P = N - 2 \tag{A-31b}$$

Similarly, using Eq. (A-7) to put the expression for probability of detection for Case 1 as given by Eq. (3-56) in series form yields

$$P_D = 1 - \sum_{k=N-1}^{\infty} \frac{\exp(-Y_b)}{k} (Y_b)^k + \left(1 + \frac{1}{N\bar{x}}\right)^{N-1} \sum_{k=N-1}^{\infty} \frac{\exp\left[\frac{-Y_b}{1+(1/N\bar{x})}\right]}{k!} \times \left[\frac{Y_b}{1+(1/N\bar{x})}\right]^k \exp\left[\frac{-Y_b}{1+N\bar{x}}\right] \tag{A-32}$$

by letting for the first summation

$$a = Y_b \tag{A-33a}$$

and

$$P = N - 2 \tag{A-33b}$$

and, similarly, for the second summation

$$a = \frac{Y_b}{1+(1/N\bar{x})} \tag{A-34a}$$

and

$$P = N - 2 \tag{A-34b}$$

A-5 The Case-2 Target Model—Rayleigh-Distributed Signal with Pulse-to-Pulse Fluctuation

The probability of detection for Case 2 as given by Swerling and restated in Eq. (3-60) can be rewritten as follows by use of Eq. (A-7):

$$P_D = 1 - \sum_{k=N}^{\infty} \frac{\exp\left(-\frac{Y_b}{1+\bar{x}}\right)}{k!} \times \left(\frac{Y_b}{1+\bar{x}}\right)^k \tag{A-35}$$

by letting

$$a = \frac{Y_b}{(1+\bar{x})} \tag{A-36a}$$

and

$$P = N - 1 \tag{A-36b}$$

A-6 The Case-3 Fluctuating Target Model—Four Degrees of Freedom Chi-Square Distributed Signal with Scan-to-Scan Fluctuation

For Case 3, the characteristic function as given by Swerling and restated in Eq. (3-63) is

$$\bar{C}_N(p) = \frac{1}{(1+p)^{N-2}\{1+p[1+(N\bar{x}/2)]\}^2} \tag{A-37}$$

If we let

$$c = 1/[1+(N\bar{x}/2)] \tag{A-38}$$

the corresponding density function may be obtained from

$$f(t) = \frac{1}{2\pi j}\int_{-j\infty}^{+j\infty} \frac{e^{pt}}{(1+p)^{N-2}[1+(1/c)\,p]^2}\,dp \tag{A-39}$$

$$= \frac{c^2}{2\pi j}\int_{-j\infty}^{+j\infty} \frac{e^{pt}}{(1+p)^{N-2}(c+p)^2}\,dp \tag{A-40}$$

For this contour integral the residue contribution association with the pole at $p = -c$ is

$$\left(\frac{c^2}{1!}\right)\frac{d}{dp}\left[\frac{e^{pt}}{(1+p)^{N-2}}\right]\Bigg|_{p=-c}$$

$$= \frac{c^2}{1!}\left\{\frac{e^{pt}[-(N-2)]}{(1+p)^{N-1}} + \frac{te^{pt}}{(1+p)^{N-2}}\right\}\Bigg|_{p=-c}$$

$$f(t)\Big|_{p=-c} = \frac{-c^2e^{-ct}(N-2)}{(1-c)^{N-1}} + \frac{c^2te^{-ct}}{(1-c)^{N-2}} \tag{A-41}$$

and the residue contribution due to the pole at $p = -1$ is

$$\frac{c^2}{(N-3)!}\frac{d^{N-3}}{dp^{N-3}}\left[\frac{e^{pt}}{(p+c)^2}\right]\Bigg|_{p=-1} \tag{A-42}$$

By letting $p+c = y$, this expression becomes

$$\frac{c^2e^{-ct}}{(N-3)!}\frac{d^{N-3}}{dy^{N-3}}\left(\frac{e^{yt}}{y^2}\right)\Bigg|_{y=c-1} \tag{A-43}$$

and by letting $x = yt$, the above expression becomes

$$\frac{c^2e^{-ct}}{(N-3)!}t^2t^{N-3}\frac{d^{N-3}}{dx^{N-3}}\left(\frac{e^x}{x^2}\right)\Bigg|_{x=(c-1)t} \tag{A-44}$$

According to Leibnitz's rule, the derivatives of e^xx^{-2}, concisely stated, are

$$\frac{d^j(e^xx^{-2})}{dx^j} = \sum_{k=0}^{j}\binom{j}{k}(-1)^k(k+1)!\,e^xx^{-(k+2)} \tag{A-45}$$

Hence, for $j = N-3$,

$$\frac{d^{N-3}}{dx^{N-3}}\left(\frac{e^x}{x^2}\right) = \sum_{k=0}^{N-3}\binom{N-3}{k}(-1)^k(k+1)! \times e^xx^{-(k+2)} \tag{A-46}$$

$$= \frac{e^x}{x^{N-1}}\sum_{k=0}^{N-3}(-1)^k(k+1) \times \frac{(N-3)!}{(N-3-k)!}x^{N-3-k} \tag{A-47}$$

and the residue at $x = (c-1)t$ is

$$f(t)\Big|_{x=(c-1)t} = \frac{c^2e^{-ct}}{(N-3)!}t^2t^{(N-3)}\frac{e^{(c-1)t}}{[(c-1)t]^2[(c-1)t]^{N-3}}\sum_{k=0}^{N-3}(-1)^k \times\frac{(k+1)(N-3)!}{(N-3-k)!}[(c-1)t]^{(N-3-k)} \tag{A-48}$$

$$= \frac{c^2e^{-t}}{(c-1)^{N-1}}\sum_{k=0}^{N-3}(-1)^k\frac{(k+1)\,t^{(N-3-k)}}{(N-3-k)!}(c-1)^{N-3-k} \tag{A-49}$$

$$= c^2e^{-t}\sum_{k=0}^{N-3}\frac{(k+1)\,t^{(N-3-k)}}{(N-3-k)!}\frac{1}{(-1)^k(c-1)^{k+2}} \tag{A-50}$$

$$= c^2e^{-t}\sum_{k=0}^{N-3}\frac{(k+1)\,t^{(N-3-k)}}{(N-3-k)!(1-c)^{k+2}} \tag{A-51}$$

Combining Eqs. (A-41) and (A-51) results in

$$f(t) = \frac{-(N-2)c^2e^{-ct}}{(1-c)^{N-1}} + \frac{c^2te^{-ct}}{(1-c)^{N-2}} + c^2e^{-t}\sum_{k=0}^{N-3}\frac{(k+1)t^{N-3-k}}{(N-3-k)!(1-c)^{k+2}} \quad \text{(A-52)}$$

The probability distribution function is

$$P_N(Y_b) = 1 - \int_0^{Y_b} f(t)\,dt = \int_{Y_b}^{\infty} f(t)\,dt \quad \text{(A-53)}$$

The integral of $f(t)$ can be represented as

$$\int f(t) = \int \mathrm{I}(t) + \int \mathrm{II}(t) + \int \mathrm{III}(t) \quad \text{(A-54)}$$

in which

$$\mathrm{I}(t) = \frac{-(N-2)c^2e^{-ct}}{(1-c)^{N-1}} \quad \text{(A-55)}$$

$$\mathrm{II}(t) = \frac{c^2te^{-ct}}{(1-c)^{N-2}} \quad \text{(A-56)}$$

$$\mathrm{III}(t) = c^2e^{-t}\sum_{k=0}^{N-3}\frac{(k+1)t^{(N-3-k)}}{(N-3-k)!(1-c)^{k+2}} \quad \text{(A-57)}$$

The integral of $\mathrm{I}(t)$ is evaluated as follows:

$$\int_{Y_b}^{\infty}\mathrm{I}(t)\,dt = \frac{-(N-2)c^2}{(1-c)^{N-1}}\int_{Y_b}^{\infty}e^{-ct}\,dt \quad \text{(A-58)}$$

$$= \frac{-(N-2)c^2}{(1-c)^{N-1}} \times \left.\frac{e^{-ct}}{-c}\right|_{Y_b}^{\infty}$$

$$= \frac{-(N-2)ce^{-Y_bc}}{(1-c)^{N-1}} \quad \text{(A-59)}$$

The integral of $\mathrm{II}(t)$ is evaluated as follows:

$$\int_{Y_b}^{\infty}\mathrm{II}(t) = \frac{c^2}{(1-c)^{N-2}}\int_{Y_b}^{\infty}te^{-ct}\,dt \quad \text{(A-60)}$$

$$= \frac{c^2}{(1-c)^{N-2}}\left[\left.\left(\frac{te^{-ct}}{-c}\right)\right|_{Y_b}^{\infty} + \int_{Y_b}^{\infty}\frac{e^{-ct}}{c}\,dt\right] \quad \text{(A-61)}$$

$$= \frac{e^{-Y_bc}}{(1-c)^{N-2}}(Y_bc+1) \quad \text{(A-62)}$$

The integral of $\mathrm{III}(t)$ is evaluated as follows:

$$\int_{Y_b}^{\infty}\mathrm{III}(t) = c^2\sum_{k=0}^{N-3}\frac{(k+1)}{(1-c)^{k+2}} \times \int_{Y_b}^{\infty}\frac{t^{N-3-k}e^{-t}}{(N-3-k)!}\,dt \quad \text{(A-63)}$$

This last integral is recognized as the incomplete gamma function [see Eq. (A-1)] with

$$a = Y_b \quad \text{(A-64a)}$$

$$u = t \quad \text{(A-64b)}$$

$$P = N-3-k \quad \text{(A-64c)}$$

and

$$[1 - f(Y_b, N-3-k)] = \int_{Y_b}^{\infty}\frac{t(N-3-k)e^{-t}}{(N-3-k)!}\,dt \quad \text{(A-64d)}$$

Since

$$f(Y_b, N-3-k) = 1 - \sum_{j=0}^{N-3-k}\frac{e^{-Y_b}Y_b^{\,j}}{j!} \quad \text{(A-65)}$$

it follows that

$$\int_{Y_b}^{\infty}\frac{e^{-t}t^{N-3-k}}{(N-3-k)!}\,dt = \sum_{j=0}^{N-3-k}\frac{e^{-Y_b}Y_b^{\,j}}{j!} \quad \text{(A-66)}$$

Substituting Eq. (A-66) into Eq. (A-57) results in

$$\int_{Y_b}^{\infty}\mathrm{III}(t)\,dt = c^2\sum_{k=0}^{N-3}\frac{k+1}{(1-c)^{k+2}}\sum_{j=0}^{N-3-k}\frac{e^{-Y_b}Y_b^{\,j}}{j!} \quad \text{(A-67)}$$

By letting $k+1 = L$, Eq. (A-67) becomes

$$\int_{Y_b}^{\infty}\mathrm{III}(t)\,dt = c^2\sum_{L=1}^{N-2}\frac{L}{(1-c)^{L+1}}\sum_{j=0}^{N-2-L}\frac{e^{-Y_b}Y_b^{\,j}}{j!} \quad \text{(A-68)}$$

The double summation of Eq. (A-68) is a convolution process for which the limits of the summations can be interchanged without altering the results; therefore

$$\int_{Y_b}^{\infty}\mathrm{III}(t)\,dt = \frac{c^2}{(1-c)}\sum_{j=0}^{N-2}\left\{\frac{e^{-Y_b}Y_b^{\,j}}{j!}\left[\sum_{L=0}^{N-2-j}L(1-c)^{-L}\right]\right\} \quad \text{(A-69)}$$

The inner series of Eq. (A-69) can be summed to give the following closed form expressions:

$$\sum_{L=0}^{N-2-j} L(1-c)^{-L} = (1-c)^{-1} + 2(1-c)^{-2} + \cdots + (N-2-j)(1-c)^{-(N-2-j)} \tag{A-70}$$

$$[(1-c)^{-1}-1]\sum_{L=0}^{N-2-j} L(1-c)^{-L} = \{(N-2-j)(1-c)^{-(N-1-j)} - (1-c)^{-1} - [(1-c)^{-2} + (1-c)^{-3} + \cdots + (1-c)^{-(N-2-j)}]\} \tag{A-71}$$

$$\sum_{L=0}^{N-2-j} L(1-c)^{-L} = \frac{(N-2-j)(1-c)^{-(N-1-j)} - (1-c)^{-1}}{[(1-c)^{-1}-1]} - \frac{[(1-c)^{-2} + (1-c)^{-3} + \cdots + (1-c)^{-(N-2-j)}]}{[(1-c)^{-1}-1]} \tag{A-72}$$

Now let

$$R = (1-c)^{-2} + (1-c)^{-3} + \cdots + (1-c)^{-(N-2-j)} \tag{A-73}$$

where R represents the numerator of the second term in Eq. (A-72) and

$$(1-c)^{-1}R = (1-c)^{-3} + \cdots + (1-c)^{-(N-1-j)} \tag{A-74}$$

$$R\left(\frac{c}{1-c}\right) = (1-c)^{-(N-1-j)} - (1-c)^{-2} \tag{A-75}$$

or

$$R = \frac{[(1-c)^{-(N-1-j)} - (1-c)^{-2}]}{\left(\frac{c}{1-c}\right)} \tag{A-76}$$

Combining Eqs. (A-72) and (A-76) results in

$$\sum_{L=0}^{N-2-j} L(1-c)^{-L} = \frac{(N-2-j)(1-c)^{-(N-1-j)} - (1-c)^{-1}}{\left(\frac{c}{1-c}\right)} - \frac{(1-c)^{-(N-1-j)} - (1-c)^{-2}}{\left(\frac{c}{1-c}\right)^2} \tag{A-77}$$

Therefore,

$$\sum_{L=0}^{N-2-j} L(1-c)^{-L} = \frac{(N-2-j)(1-c)^{-(N-2-j)} - 1}{c} - \frac{(1-c)^{-(N-3-j)} - 1}{c^2} \tag{A-78}$$

Substituting Eq. (A-78) into Eq. (A-69) results in

$$\int_{Y_b}^{\infty} \mathrm{III}(t)\,dt = \frac{c^2}{1-c}\sum_{j=0}^{N-2}\frac{e^{-Y_b}Y_b^{\,j}}{j!}\left[\frac{(N-2-j)(1-c)^{-(N-2-j)} - 1}{c} - \frac{(1-c)^{-(N-3-j)} - 1}{c^2}\right] \tag{A-79}$$

$$= \frac{1}{(1-c)^{N-2}}\sum_{j=0}^{N-2}\frac{e^{-Y_b}Y_b^{\,j}}{j!}\left[\frac{(N-2)c(1-c)^j}{(1-c)} - (1-c)^j - \frac{jc(1-c)^j}{(1-c)}\right] - \frac{c}{1-c}\sum_{j=0}^{N-2}\frac{e^{-Y_b}Y_b^{\,j}}{j!} + \frac{1}{1-c}\sum_{j=0}^{N-2}\frac{e^{-Y_b}Y_b^{\,j}}{j!} \tag{A-80}$$

$$= \sum_{j=0}^{N-2}\frac{e^{-Y_b}Y_b^{\,j}}{j!} + \frac{e^{-Y_b}}{(1-c)^{N-2}}\left[\frac{(N-2)c}{(1-c)} - 1\right]\sum_{j=0}^{N-2}\frac{e^{-Y_b}Y_b^{\,j}}{j!}(1-c)^j - \frac{e^{-Y_b}}{(1-c)^{(N-2)}}\sum_{j=0}^{N-2}\frac{jcY_b^{\,j}(1-c)^j}{(1-c)\,j!} \tag{A-81}$$

The last term of Eq. (A-81) can be reexpressed as follows:

$$\frac{e^{-Y_b c}}{(1-c)^{N-2}} \sum_{j=0}^{N-2} \frac{Y_b^j (1-c)^{j-1}}{j!} j = \frac{Y_b e^{-Y_b c}}{(1-c)^{N-2}} \sum_{j=1}^{N-2} \frac{Y_b^{j-1}(1-c)^{(j-1)}}{(j-1)!} \quad \text{(A-82)}$$

By letting $m = j - 1$, the right-hand side of Eq. (A-82) becomes

$$\frac{Y_b e^{-Y_b c}}{(1-c)^{N-2}} \sum_{m=0}^{N-3} \frac{Y_b^m (1-c)^m}{m!} = \frac{Y_b e^{-Y_b c}}{(1-c)^{N-2}} \sum_{j=0}^{N-2} \frac{Y_b^j (1-c)^j}{j!} - \frac{Y_b e^{-Y_b c}}{(1-c)^{N-2}} \left[\frac{Y_b^{N-2}(1-c)^{N-2}}{(N-2)!} \right] \quad \text{(A-83)}$$

Substituting Eq. (A-83) into Eq. (A-81) results in the following:

$$\int_{Y_b}^{\infty} \text{III}(t)\, dt = \sum_{j=0}^{N-2} \frac{e^{-Y_b} Y_b^j}{j!} + \frac{c Y_b^{(N-1)} e^{-Y_b}}{(N-2)!} - \frac{e^{Y_b}}{(1-c)^{N-2}} \left[1 - \frac{(N-2)c}{(1-c)} + cY_b \right] \sum_{j=0}^{N-2} \frac{Y_b^j (1-c)^j}{j!} \quad \text{(A-84)}$$

Equations (A-59), (A-63), and (A-84) can be introduced into Eq. (A-54), the expression for $f(t)$—that is, the probability density distribution. When this density is integrated in the expression for the probability of detection [Eq. (A-53)], the result is

$$P_N(c, Y_b) = \frac{Y_b^{N-1} e^{-Y_b c}}{(N-2)!} + \sum_{j=0}^{N-2} \frac{e^{-Y_b} Y_b^j}{j!} + \frac{e^{-cY_b}}{(1-c)^{N-2}} \left[1 - \frac{(N-2)c}{(1-c)} + cY_b \right] \left[1 - \sum_{j=0}^{N-2} \frac{e^{-Y_b(1-c)} Y_b^j (1-c)^j}{j!} \right] \quad \text{(A-85)}$$

A-7 The Case-4 Fluctuating Target Model—Four Degrees of Freedom Chi-Square Distributed Signal with Pulse-to-Pulse Fluctuation

For Case 4, the characteristic function as given by Swerling and restated in Eq. (3-71) is

$$\bar{C}_N(p) = \frac{(1+p)^N}{\{1 + p[1 + (\bar{x}/2)]\}^{2N}} \quad \text{(A-86)}$$

Let,

$$c = 1/[1 + (\bar{x}/2)] \quad \text{(A-87)}$$

Then

$$\bar{C}_N(p) = \frac{c^{2N}(1+p)^N}{(c+p)^{2N}} \quad \text{(A-88)}$$

The density function is obtained from

$$f(t) = \frac{1}{2\pi j} \int_{-j\infty}^{j\infty} \frac{c^{2N}(1+p)^N}{(c+p)^{2N}} e^{pt}\, dp \quad \text{(A-89)}$$

For this contour integral the residue contribution associated with the pole at $p = -c$ is

$$f(t)\Big|_{p=-c} = \frac{1}{(2N-1)!} [c^{2N} e^{pt} (1+p)^N] \Big|_{p=-c}^{(2N-1)} \quad \text{(A-90)}$$

$$= \frac{c^{2N}}{(2N-1)!} \times \frac{d^{2N-1}}{dp^{2N-1}} [e^{pt}(1+p)^N] \quad \text{(A-91)}$$

For $(1+p) = y$, $dy = dp$, $p = (y-1)$, Eq.(A-91) can be rewritten as follows:

$$f(t)\Big|_{y=1-c} = \frac{c^{2N}}{(2N-1)!} \times \frac{d^{2N-1}}{dy^{2N-1}} [e^{-t} e^{yt} y^N] \Big|_{y=1-c} \quad \text{(A-92)}$$

If $yt = x$, then

$$f(t)\Big|_{x=t(1-c)} = \frac{c^{2N}}{(2N-1)!} \times \frac{e^t t^{2N-1}}{t^N} \times \left[\frac{d^{2N-1}}{d_x^{2N-1}} e^x x^N \right] \Big|_{x=t(1-c)} \quad \text{(A-93)}$$

Simplifying Eq. (A-93) yields

$$f(t)\Big|_{x=t(1-c)} = \frac{c^{2N}}{(2N-1)!} e^t t^{N-1} \times \left(\frac{d^{2N-1}}{dx^{2N-1}} e^x x^N\right)\Big|_{x=(t)(1-c)} \tag{A-94}$$

According to Leibnitz rule, the derivatives of $e^x x^n$, concisely stated, are

$$\frac{d^k(e^x x^n)}{dx^k} = \sum_{j=0}^{k} \binom{k}{j} \frac{n!}{(n-j)!} x^{n-j} e^x \tag{A-95}$$

If $n \geqslant k$, then all terms in Eq. (A-95) are required. If $n < k$, then only terms up to the nth need be considered.

For the case under consideration (Case 4), $k = 2N-1$ and $n = N$. Therefore, the second condition, $n < k$, is applicable. Hence

$$\frac{d^{(2N-1)}(e^x x^N)}{dx^{(2N-1)}} = e^x \sum_{j=0}^{N} \binom{2N-1}{j} \frac{N!}{(N-j)!} x^{N-j} \tag{A-96}$$

Remembering that $x = t(1-c)$, Eq. (A-96) becomes

$$\frac{d^{(2N-1)}(e^x x^N)}{dx^{(2N-1)}} = e^{t(1-c)} \sum_{k=0}^{N} \left[\frac{(2N-1)!}{(2N-1-k)!}\right] \times \frac{N!\,t^{(N-k)}}{k!(N-k)!}(1-c)^{(N-k)} \tag{A-97}$$

Substituting Eq. (A-97) into Eq. (A-94), the density function becomes

$$f(t) = \frac{c^{2N}}{(2\mathrm{N}-1)!} e^{-t} t^{(N-1)} e^{t(1-c)} \times \sum_{k=0}^{N} \left[\frac{(2N-1)!}{(2N-1-k)!}\right] \frac{N!\,t^{N-k}(1-c)^{N-k}}{k!(N-k)!} \tag{A-98}$$

Equivalently,

$$f(t) = c^{2N} e^{-ct} \sum_{k=0}^{N} \frac{N!(1-c)^{(N-k)}}{k!(N-k)!} \frac{t^{(2N-1-k)}}{(2N-1-k)!} \tag{A-99}$$

As given earlier, the general expression for the cumulative probability of detection is

$$P_N(c, Y_b) = \int_{Y_b}^{\infty} f(t)\,dt = 1 - \int_0^{Y_b} f(t)\,dt \tag{A-100}$$

Substituting Eq. (A-99) into Eq. (A-100) results in

$$P_N(c, Y_b) = c^{2N} \sum_{k=0}^{N} \frac{N!(1-c)^{N-k}}{k!(N-k)!} \times \int_{Y_b}^{\infty} \frac{e^{-ct} t^{2N-1-k}}{(2N-1-k)} dt \tag{A-101}$$

$$= 1 - c^{2N} \sum_{k=0}^{N} \frac{N!(1-c)^{N-k}}{k!(N-k)!} \times \int_0^{Y_b} \frac{e^{-ct} t^{(2N-1-k)}}{(2N-1-k)!} dt \tag{A-102}$$

Letting I_1 equal the portion of Eq. (A-101) to be integrated gives

$$I_1 = \int_{Y_b}^{\infty} \frac{e^{-ct} t^{(2N-1-k)}}{(2N-1-k)!} dt \tag{A-103}$$

Letting $ct = x$ modifies this expression to

$$I_1 = \frac{1}{c^{(2N-k)}} \int_{cY_b}^{\infty} \frac{e^x x^{(2N-1-k)}}{(2N-1-k)!} dx \tag{A-104}$$

Equation (A-104) is in the form of the incomplete gamma function [see Eq. (A-2) and let $n = x$, $a = cY_b$, and $P = (2N-1-k)$]. Equation (A-104), therefore, can be written as

$$I_1 = \frac{1}{c^{(2N-k)}} \sum_{j=0}^{(2N-1-k)} \frac{e^{-cY_b}(cY_b)^j}{j!} \tag{A-105}$$

The probability of detection to be computed for when $Y_b > N(2-c)$ can be formulated by the substitution of Eq. (A-105) into Eq. (A-101):

$$P_N(c, Y_b) = c^{2N} \sum_{k=0}^{N} \frac{N!}{k!(N-k)!} [(1-c)^{N-k}] \times \frac{1}{c^{(2N-k)}} \sum_{j=0}^{2N-1-k} \frac{e^{-cY_b}(cY_b)^j}{j!} \tag{A-106}$$

$$= c^N \sum_{k=0}^{N} \frac{N!}{k!(N-k)!} \left(\frac{1-c}{c}\right)^{(N-k)} \times \sum_{j=0}^{2N-1-k} \frac{e^{-cY_b}(cY_b)^j}{j!} \tag{A-107}$$

To formulate the probability of detection to be computed for the case of $Y_b < N(2-c)$ Eq. (A-102) is selected. The integral in Eq. (A-102) is

$$I_2 = \int_0^{Y_b} \frac{e^{-ct} t^{2N-1-k}}{(2N-1-k)!} dt \tag{A-108}$$

Letting $ct = x$ modifies this expression to

$$I_2 = \frac{1}{c^{(2N-k)}} \int_0^{cY_b} \frac{(e^{-x}) x^{(2N-1-k)}}{(2N-1-k)!} dx \qquad \text{(A-109)}$$

Equation (A-109) is in the form of an incomplete gamma function [see Eq. (A-1) and let $u = x$, $a = cY_b$, and $P = (2N-1-k)$]. Equation (A-109), therefore, can be written as

$$I_2 = \frac{1}{c^{(2N-k)}} \sum_{j=2N-k}^{\infty} \frac{e^{-cY_b}(cY_b)^j}{j!} \qquad \text{(A-110)}$$

Substituting Eqs. (A-110) into Eq. (A-102) results in

$$P_N(c, Y_b) = 1 - c^N \sum_{k=0}^{N} \frac{N!}{k!(N-k)!} \left(\frac{1-c}{c}\right)^{N-k} \times \sum_{j=2N-k}^{\infty} \frac{e^{-cY_b}(cY_b)^j}{j!} \qquad \text{(A-111)}$$

A-8 Computation of Bias Level Values

The bias level (Y_b) values were determined for specific values of N and n' by means of a computer program that employs an iteration process and numerical integration to solve the fundamental relationship:

$$f(Y_b) = P_0^{1/n'} = \int_0^{Y_b} \frac{u^{N-1}e^{-u}}{(N-1)!} du = 1 - P_{fa} \qquad \text{(A-112)}$$

where

$$P_0 = P(0) = P(r = 0) = 0.5 \qquad \text{(A-113)}$$

Double precision computer arithmetic is required to evaluate this function in the region of interest, that is, from $f(Y_b) = 0.5^{-0.1}$ to $f(Y_b) = 0.5^{-0.0000000001}$. Fehlner's explanation of his program is as follows [5]:

This equation [Eq.(A-112)] was reduced to a differential equation of the form:

$$f'(Y) = \frac{Y^{N-1}e^{-Y}}{(N-1)!} \qquad \text{(A-114)}$$

[The starting point for the integration is]

$$f_0 = f(Y_0) = \int_0^{Y_0} \frac{e^{-u}u^{N-1}}{(N-1)!} du \qquad \text{(A-115)}$$

where Y_0 is an initial approximation to Y_b, read to about two significant digits from Marcum's curves.† This differential equation was integrated by Heun's method:

$$f_{j+1} = f_j + \frac{h}{2}[f'(Y_j) + f'(Y_{j+1})], \quad j = 0, 1, 2, \ldots \qquad \text{(A-118)}$$

where $Y_{j+1} = Y_j + h$. [The term] h was set arbitrarily to 0.01 when $f(Y_0) < f(Y_b)$ or -0.01 when $f(Y_0) > f(Y_b)$. The iteration was carried out up to $j = k$, such that $f(Y_k) \approx f(Y_b)$. Then $Y_b = Y_k$.

The integration was stopped when $f(Y_k)$ passed $f(Y_b)$ and then the final value of Y_k was determined by interpolation. The function $f(Y_0)$ was evaluated using the series representation of the incomplete gamma function described in Section A-2.

† Actually the initial approximations were made by use of the following. For $N \leqslant 12$,

$$Y_0 = N\left[1 + 2.2\left(\frac{\log_{10} n'}{N^{(2/3+0.015 \log_{10} n')}}\right)\right] \qquad \text{(A-116)}$$

and for $N > 12$,

$$Y_0 = N\left[1 + 1.3\left(\frac{\log_{10} n'}{N^{(0.5+0.011 \log_{10} n')}}\right)\right] \qquad \text{(A-117)}$$

References

1. J. I. Marcum, Studies of Target Detection by a Pulsed Radar. Rand Res. Memo, RM-754 (Dec. 1947); reprinted in *IRE Trans. Inform. Theory* **IT-6** (1960); also AD-101287.
2. J. I. Marcum, Studies of Target Detection by a Pulsed Radar—Mathematical Appendix. Rand Res. Memo RM-753 (1948); reprinted in *IRE Trans. Inform. Theory* **IT-6** (1960); also AD-101882.
3. P. Swerling, Probability of Detection for a Fluctuating Target. Rand Res. Memo RM-1217 (March 17, 1954); reprinted in *IRE Trans. Inform. Theory* **IT-6** (1960).

4. R. Belluck and D. P. Meyer, Method for Digital Machine Computation, Appendix A, in D. P. Meyer, Further Studies of Target Detection by a Pulsed Radar. Final Rep. PN-51601-21, Sperry Gyroscope Co., Sperry Rand Corp., Great Neck, N.Y. (July 1967).
5. L. F. Fehlner, Marcum's and Swerling's Data on Target Detection by a Pulsed Radar, Appendix A (R. G. Rolls and G. T. Trotter, Methods for Digital Machine Computation of Detection Probability), Rep. TG-451, Appl. Phys. Lab., The Johns Hopkins University, Silver Spring, Maryland; also AD-602-121 (July 1962).
6. K. Pearson, "Tables of Incomplete Gamma Function." Cambridge Univ. Press, London and New York, published by "Biometrika," 1946.

INDEX

A

Aitchison, J., 65, 82
Alperin, H. A., 18
Antenna
  angular beam ratio, 77
  azimuth half-power beam width, 75
  beam pattern 2, 75, 77
  effective aperture, 2, 10, 13, 15
    relative to squint loss, 81
  efficiency, 13
  elevation half-power beam width, 75
  gain, 2, 10, 11, 13, 76, 77
Anticharacteristic function, *see* Characteristic function, anti-
Atmospheric-absorption loss, 81

B

Back-scattering coefficient, 37
Bandwidth
  correction factor, 11
  effective, 39
  human information, 81
  I.F., 40
  video, effect of insufficient, 40
Barton, D. K., 3, 18, 32, 33, 37, 82
Beam shape, 13, 75, 76
  loss, *see* Loss, beam shape
Belluck, R., 478, 487
Berkowitz, R. S., 18
Bernfeld, M., 6, 12, 18
Bessel function, modified, *see* Function
Bias level, 19, 21-23, 26, 39, 58, 59
  determination of, 23, 486
  linear detector, 68
  plot of
    for linear-law detector, 70
    for square-law detector, 23
Bias setting, *see* Bias level
Binomial distribution, *see* Distribution
Blake, L. V., 14, 18, 76, 81, 82
Blip-scan ratio, 39
Block correlation, *see* Correlation, block
Boltzmann's constant, 4
Bronski, R. R., 18
Brown, J. A. C., 65, 82
Burington, R. S., 82

C

Campbell, G. A., 27, 59, 61-63, 66, 82
Capon, J., 107, 108
Characteristic function, 27, 59
  anti-, 59, 71
  log-normal, 68
  Marcum Case 0, 58, 479
    including collapsing loss, 30
  Rice target model, 67
  for single-hit, 63
  single-hit density of x, 42
  Swerling Cases, 61-63, 481, 484
    including collapsing loss, 30
Chi-square
  approximation for log-normal case, 90
  approximation for Rice Power case, 90
  approximation of other target models, 95
  distribution, *see* Distribution, chi square
  number of duo-degrees of freedom, 64
  probability density function, *see* Probability density function
  target model, *see* Target model
Clutter, 3
Coherent integration, *see* Integration
Collapsing loss, *see* Loss, collapsing
Collapsing ratio, 31-34
Continuous wave, *see* CW
Contour integral, 481
Cook, C. E., 6, 12, 18
Correlation
  block, 31, 84-85, 88-91
  effect on chi-square parameters, 66
  pulse-to-pulse, 86
  scan-to-scan, 86
Correlation time, 39
Cramer, H., 71, 82
Crispin, J. W., 18
Cross section, scattering, *see* Scattering cross section
CW radar
  analysis of, 74
  correlation interval, 39
  system application of pulse radar results, 74

D

Deflection criterion, *see* Detection
Delano, R. H., 61, 82
Detection, 1, 13, 19
  amplification limitations, 1
  curves for single-hit probability of, 43-57
  deflection criterion, 73
  double threshold, 99-107
  envelope, 39
  hybrid, 99-107
  median, 73-74, 99-107
  miss-, 21

probability of, 14, 15, 26, 28, 39, 41, 42, 58-63, 76, 83
graph of for median detector, 102
linear-law detector, 70
Marcum Case 0, computation of, 479-480
Meyer Plots, 83
Miss-, 83
single-hit, 38-57, 81
single-hit, *x* large, 104
Swerling Cases, computation of, 480-486
versus available time, 40-41
signal-plus-noise exceeding noise alone, 73
signal-plus-noise minus noise, 72
single-hit, 62
statistical decision theory error type, 21
trimmed-mean, 73-74
Detector, 40
best possible, 71-72
bias values, *see* Bias level
definition of square-law, 58
double threshold, 105, 106
envelope, 40
first threshold, 99
hybrid, 101, 103, 105
linear-law, 68, 72, 108
mean, 101, 103, 105
median, 73, 101, 103, 105
quadratic, *see* square law
second threshold, 99
square-law, 27, 58, 68, 69, 72, 99, 108
trimmed-mean, 73
Detector law, constant of, 23
Difranco, J. V., 35, 82
Distribution
binomial, 19
chi-square, 36, 62, 65, 96, 98
log-normal probability, 67
of envelope of signal-plus-noise, 71
Poisson, 20
Rayleigh, 60, 61, 478
Rice, 67
Doppler data, 25, 26
Double threshold detection, *see* Detection
Duwaldt, B. J., 82

E

Echoing area, effective, 37
Effective antenna aperture, *see* Antenna, effective aperture
Effective bandwidth, *see* Bandwidth, effective
Effective noise temperature, *see* Temperature
Envelope detection, *see* Detection

F

False alarm, 19, 21
probability of, 19, 20-22, 24, 25, 29, 40, 41, 68, 100, 108
False alarm number, 19-21, 23, 25, 28, 74
Fehlner's, 21, 22, 25, 27, 28, 31, 83, 108
Marcum's, 21, 28, 31, 68, 83, 108
False alarm time, 19, 21, 39, 71
Kaplan and McFall's definition, 19, 20
Marcum's definition, 20, 22, 40
Fehlner, L. F., 21, 25, 478, 486, 487
Fehlner's false alarm number, *see* False alarm number
Filter, frequency response of, 15
Filter bandwidth, comparison of types, 3
Fine, A. M., 8, 18
Foster, R. M., 27, 35, 59, 61-63, 66, 82
Fourier transform, 6, 7
Function
confluent hypergeometric, 63
Gaussian, 75
incomplete gamma, 22, 63, 478, 479, 482
modified bessel of the first kind, 39, 59, 66
Q, 41, 59

G

Gent, H., 37, 38, 82
George, S. F., 73, 82
Goldman, A. S., 82
Goodrich, R. F., 18

H

Hall, W. M., 68, 82
Hansen, V. G., 108
Heidbreder, G. R., 82, 95, 108
Hey, J. S., 37, 38, 82
Hoel, P. G., 65, 82
Hollis, R., 20, 35

I

Integration
coherent or predetection, 14, 25, 26
efficiency of, 14, 28, 29
general theory of, 25
noncoherent or postdetection, 14, 25, 26, 28
number of pulses, 14, 21-23, 25, 31, 78, 83
video, 107
Integrator
law of, 26
linear-law, 27, 58
linear video, 99
optimum, 27
recirculating, 107
weighting function, 26
I.F. amplifier, stagger tuned, 15
I.F. band-pass characteristics, 8

K

Kaplan, S. M., 18-21, 35
Kerr, D., 82

L

Laguerre polynomial, 64
Lawson, J. L., 3, 8, 18, 73, 82
Leibnitz's rule, 481, 485
Linear-law detector, *see* Detector
Log-normal probability density function, *see* Probability density function
Log-normal target model, *see* Target model
Loss, 5, 15, 16, 29, 75, 79-81
  antenna efficiency, 15
  atmospheric-absorption, 81
  beam shape, 15, 74-79
  collapsing, 29-34, 40
    approximation for cw radar system, 74
    with n constant, 34
    with n' constant, 34
  integration, 15, 28, 29
  limiting, 79-81
  operator, 79-81
  scanning, 15, 16, 74, 76
  target fluctuation, 15

M

Marcum, J. I., 2, 8, 11, 18-22, 25-28, 31, 35, 36, 39, 58-59, 68-74, 82, 478, 486
Marcum's false alarm number, *see* False alarm number
Matched-filter, 6, 8, 9, 13
  signal relationships of, 7
  transfer function, 6
Mayer, H. A., 108
May, Jr., D. C., 82
McFall, R. W., 18-21, 35
Mean, *see* Distribution
Mean detector, *see* Detector
Median detection, *see* Detection
Median detector, *see* Detector
Meyer Plots, 107-477
  application procedures, 84-85
  bistatic radar system, 13
  block correlation, 88-91
  chi-square target models, 88-90
  collapsing loss, 30, 31, 34
  computation of, 478-486
  description of, 83, 107-108
  double threshold detection, 99-107
  forward scattering cross section, 13
  hybrid detection, 99-107
  illustrative examples
    to determine P_D, 27, 28
    to optimize N relative to beam-shape loss, 76-79
  intergration loss, 15, 29
  interpolation, methods of, 86-88
  linear detector, 68
  log-normal target models, 90-95
  Marcum Case 0, 59
  median detection, 99-107
  other target models, 95-99
  recirculating integrator, 107
  Rice Target Model, 67, 90-91
  search radar, noncoherent integration, 14, 16
  Swerling Case 4, 63
  Swerling Cases 1 and 2, 62
  target fluctuation loss, 15
  use of to determine collapsing loss, 31
  Weinstock target models, 88-90
Meyer, D. P., 41, 43, 48, 53, 82, 478, 487
Middleton, D., 10, 18
Mitchell, R. L., 82, 95, 108
Moffett, A. L., 18
Moment matching, 97
Muchmore, R. B., 82

N

Nathanson, F. E., 18, 61, 82
Noise
  amplitude, instantaneous, 22
  correlation interval, 19
  cosmic, 3, 39
  distribution of the envelope of noise alone, 71
  effective temperature of a cascade network, 6
  Gaussian, 8, 108
  Johnson, 3
  power generated in receiver, 4
  shot effect, 3, 39
  standard deviation of, 25
  stationary statistics, 19
  thermal, 3, 39
Noise bandwidth, 3
Noise factor, 3-5
Noise figure, 3
Noise temperature, *see* Temperature
Noise voltage, instantaneous video, 39
Noncoherent integration, *see* Integration
North, D. O., 6, 18
North filter, 6

O

Operator coefficient, 81
Operator efficiency, 81

P

Pachares, J., 23, 35
Palmieri, C., 18
Paolillo, J., 18
Pearson, K., 22, 35, 61, 82, 108, 478, 487
Peters, Jr., L., 81
Poisson distribution, *see* Distribution
Power density, 2
Power transmitted, 2, 10
PRF, *see* Pulse repetition frequency
Probability density function, 59, 63, 487

based on empirical data, 95
chi-square, 62, 64, 65, 83, 100
of linear-law detector, 71
log-normal, 65
Marcum Case 0, 479
of noise alone, 21, 39
normal, 65
Rayleigh, 61, 83
Rice target model, 66
of signal-plus-noise, 39
Swerling Cases, 480-484
of target models, 84
Probability of detection, *see* Detection
Probability of false alarm, *see* False alarm, probability of
Pulse repetition frequency, 14, 21

Q

Q function, *see* Function

R

Radar
energy in transmitted pulse, 10
idealized characteristics, 13
optimum performance of, 41
optimum pulse-width bandwidth product, 15
power aperture product, 17
pulse-width bandwidth product, 26
search range equation, illustrative example, 16
transmitted pulse, 12
Radar cross-section, 37, 38
of spherical target, 38
Radar range equation, 1, 2, 6, 9-15, 75
Range bin, 19
Range delay time, 76
Range gate width, 29
Range resolution, 19
Range, transmission, 2
Rayleigh distribution, *see* Distribution
Rayleigh probability density function, *see* Probability density function
Rayleigh scattering, *see* Scattering, Rayleigh
Rayleigh target model, *see* Target model
RCS, *see* Radar cross section
Receiver, 4, 8, 9, 11
Receiver bandwidth, 8, 9, 29
Reflection characteristics, 38
Rice, S. O., 36, 82
Rice target model, *see* Target model
Ridenour, L. N., 9, 18
Rolls, R. G., 478, 487
Rubin, W. L., 35, 82

S

Scanning rate, 15
Scattering cross section, 10, 12, 37-38, 60-61
Schensted, C. E., 18
Schensted, I. V., 18
Scholefield, P. H. R., 82, 90, 108
Search pattern
total area to be searched, 14
uniform, 13
Search radar, performance of, 13
Search radar range equation, *see* Radar range equation
Series
edgeworth, 63
gram-charlier, 59, 68, 70
Siegel, K. M., 18
Siegman, A. E., 4, 18
Signal-to-noise ratio
average, 97
Marcum's, 11
relative to radar range equation, 6
Single-hit detection, *see* Detection
Single-hit detection curves, 41
Marcum Case 0, 43-47
Swerling Cases, 48-52
Single-hit probability of detection, *see* Detection
Skolnik, M. I., 8, 18, 33, 35, 81, 82
Slattery, T. B., 82
Smith, P. G., 37, 38, 82
Source resistor, 4
Sperber, A. O., 82
Square-law detector, *see* Detector
Stirling, J., 22, 35
Stirling's approximation, 22
Swerling, P., 30, 36, 41, 61, 62, 82, 90, 97, 108, 478, 480, 486

T

Target cross-section, 64
Target fluctuation, 84
Target model, *see also* Detection, probability of
aircraft, 60
best chi-square fit by curve matching, 97-99
chi-square, 66, 83, 84
arbitrary degrees of freedom, 88
general, 64-66
cylinder class, 64
historical background, 36-37
log-normal, 65, 67-68, 83, 84, 97
graphical approximation of, 93-95
Marcum Case 0, 41, 58-59, 61, 66, 83-85
Meyer Plots
chi-square, 88-90
log-normal, 90-95
other, 95-99
Rice Power, 90-91
Weinstock, 88-90
modern, 64-68
nonfluctuating, 83, 99
other, 84
physical interpretation of, 85
prolate spheroid, 64

Rayleigh, 61, 83, 84
Rice Power, 66, 67, 83, 84, 97
graphical approximation of, 91
limit of the dominant to Rayleigh component, 90
smooth bodies randomly oriented, 64
steady state, Marcum Case 0, 84
summary table of, 84-85
Swerling Cases, 27, 30, 36, 37, 41, 59-64, 66, 83-85, 103
Weinstock, 64, 83, 84
Temperature
effective antenna, 5
effective noise, 3-5
plumbing, 5
standard, 4
Thevenin equivalent resistor, 4
Threshold, *see* Bias level
Trimmed-mean detector, *see* Detector
Trotter, G. T., 478, 487
Trunk, G. V., 73, 82

U

Uhlenbeck, G. E., 3, 18, 73, 81

V

Van Vleck, J. H., 10, 18
Visibility factor, 10, 11, 74

W

Walker, J. F., 107, 108
Watson, G. N., 82
Weiner, F. C., 82
Weinstock target model, *see* Target model
Weinstock, W., 36, 64, 82
Woodward, P. M., 12, 18

ELECTRICAL SCIENCE

A Series of Monographs and Texts

Editors

Henry G. Booker
UNIVERSITY OF CALIFORNIA AT SAN DIEGO
LA JOLLA, CALIFORNIA

Nicholas DeClaris
UNIVERSITY OF MARYLAND
COLLEGE PARK, MARYLAND

Joseph E. Rowe. Nonlinear Electron-Wave Interaction Phenomena. 1965

Max J. O. Strutt. Semiconductor Devices: Volume I. Semiconductors and Semiconductor Diodes. 1966

Austin Blaquiere. Nonlinear System Analysis. 1966

Victor Rumsey. Frequency Independent Antennas. 1966

Charles K. Birdsall and William B. Bridges. Electron Dynamics of Diode Regions. 1966

A. D. Kuz'min and A. E. Salomonovich. Radioastronomical Methods of Antenna Measurements. 1966

Charles Cook and Marvin Bernfeld. Radar Signals: An Introduction to Theory and Application. 1967

J. W. Crispin, Jr., and K. M. Siegel (eds.). Methods of Radar Cross Section Analysis. 1968

Giuseppe Biorci (ed.). Network and Switching Theory. 1968

Ernest C. Okress (ed.). Microwave Power Engineering:
Volume 1. Generation, Transmission, Rectification. 1968
Volume 2. Applications. 1968

T. R. Bashkow (ed.). Engineering Applications of Digital Computers. 1968

Julius T. Tou (ed.). Applied Automata Theory. 1968

Robert Lyon-Caen. Diodes, Transistors, and Integrated Circuits for Switching Systems. 1969

M. Ronald Wohlers. Lumped and Distributed Passive Networks. 1969

Michel Cuenod and Allen E. Durling. A Discrete-Time Approach for System Analysis. 1969

K. Kurokawa. An Introduction to the Theory of Microwave Circuits. 1969

H. K. Messerle. Energy Conversion Statics. 1969

George Tyras. Radiation and Propagation of Electromagnetic Waves. 1969

Georges Metzger and Jean-Paul Vabre. Transmission Lines with Pulse Excitation. 1969

C. L. Sheng. Threshold Logic. 1969

Dale M. Grimes. Electromagnetism and Quantum Theory. 1969

Robert O. Harger. Synthetic Aperture Radar Systems: Theory and Design. 1970

M. A. Lampert and P. Mark. Current Injection in Solids. 1970

W. V. T. Rusch and P. D. Potter. Analysis of Reflector Antennas. 1970

Amar Mukhopadhyay. Recent Developments in Switching Theory. 1971

A. D. Whalen. Detection of Signals in Noise. 1971

J. E. Rubio. The Theory of Linear Systems. 1971

Keinosuke Fukunaga. Introduction To Statistical Pattern Recognition. 1972

Jacob Klapper and John T. Frankle. Phase-Locked and Frequency-Feedback Systems: Principles and Techniques. 1972

Kumpati S. Narendra and James H. Taylor. Frequency Domain Criteria for Absolute Stability. 1973

Daniel P. Meyer and Herbert A. Mayer. Radar Target Detection: Handbook of Theory and Practice. 1973

T. R. N. Rao. Error Coding for Arithmetic Processors. 1974

C. A. Desoer and M. Vidyasagar. Feedback Systems: Input-Output Properties. 1975